Precalculus Concepts

Preliminary Edition

Precalculus Concepts

Preliminary Edition

Warren W. Esty

Department of Mathematics
Montana State University

PRENTICE HALL, Upper Saddle River, New Jersey 07458

Editorial Director: **TIM BOZIK**
Editor-in-Chief: **JEROME GRANT**
Acquisitions Editor: **SALLY DENLOW**
Assistant Vice-President of
Production and Manufacturing: **DAVID W. RICCARDI**
Executive Managing Editor: **KATHLEEN SCHIAPARELLI**
Managing Editor: **LINDA BEHRENS**
Production Editor: **RICHARD DeLORENZO**
Marketing Manager: **JOLENE HOWARD**
Creative Director: **PAULA MAYLAHN**
Art Director: **JAYNE CONTE**
Cover Designer: **BRUCE KENSELAAR**
Manufacturing Manager: **TRUDY PISCIOTTI**
Prepress / Manufacturing buyer: **ALAN FISCHER**
Supplements Editor: **AUDRA WALSH**
Editorial Assistant: **JOANNE WENDELKEN**

Simon & Schuster / A Viacom Company
Upper Saddle River, NJ 07458

Printed in the United States of America

10 9 8 7 6 5 4 3 2 1

ISBN 0-13-261694-7

Prentice-Hall International (UK) Limited, London
Prentice-Hall of Australia Pty. Limited, Sydney
Prentice-Hall Canada Inc., Toronto
Prentice-Hall Hispanoamericana, S.A., Mexico
Prentice-Hall of India Private Limited, New Delhi
Prentice-Hall of Japan, Inc., Tokyo
Simon & Schuster Asia Pte. Ltd., Singapore
Editora Prentice-Hall do Brasil, Ltda., Rio de Janeiro

To Norah

Precalculus Concepts

by Warren W. Esty

Contents

Preface

The goal of this course is to develop your ability to read, write, think, and do mathematics and to give you command of the facts and methods of algebra and trigonometry. It emphasizes everything you need to understand to be prepared for calculus.

A unique feature of this course is its **emphasis on developing fluency in the abstract and symbolic language of algebra.** That "foreign" language is extremely important because it is the most effective and efficient language in which to learn, understand, recall, and think mathematical thoughts.

This text is also different because **it makes many mathematical connections between things you already know and things you are about to learn.** Rather than beginning each section with entirely new material, many sections begin with familiar material which is closely related to the new material.

This Class is Different. This class is different from any other math class you have taken. You will immediately notice that the reading is different and some of the homework is different. This is because the goals are different.

You are expected to

1) Learn to read symbolic mathematics fluently (and, in the process, learn how to learn math by reading it)
2) Learn to explain (symbolically, and in English) key general results from each section
3) Learn to illustrate (with illuminating pictures) key general results from each section
4) Remember (with the help of symbols, English explanations, and pictures) key general results from each section [What good does it do to have "taken" math if you don't remember it?]
5) Learn to work abstractly with symbols and functions (as comfortably as you now work with numbers)
6) Understand what you do
7) Become good at word problems
8) Learn the methods and facts of algebra so well that you have them at your command (even without recent review)
9) Learn the methods and facts of trigonometry so well that you have them at your command (even without recent review)
10) Learn how and when to use graphing technology

Do not be fooled. The apparent content (algebra and trig) may be familiar, but this course asks you to learn it, and mathematical language as well, in a new way that should raise you to a far higher level of mathematical ability.

Prerequisite. This is a "Precalculus" course which is intended to serve as a bridge from high-school Algebra II and trigonometry to college calculus. You should have taken a course equivalent to Algebra II, and possibly also some trigonometry. **A graphics calculator is required.**

Homework. The homework is different. In most math classes, this one included, you practice methods with problems requiring you to compute numbers. But this class also has other types of problems designed to promote *understanding and retention*. For example, problems may ask you to *illustrate* (with a memorable picture) an important fact, *explain* a method, or *state* a method in general using the language of mathematics.

Homework is categorized into "A" and "B" problems. "A" problems should be straightforward. If a problem merely requires you to "plug in" to a given formula or to solve a straightforward equation, the problem is an "A" problem. You should be able to do the "A" problems *as a bare minimum*.

Your goal should be to be able to answer the "B" problems,

which emphasize concepts, interpretations, explanations, visual images, notation, and applications of methods to more difficult problems.

Some problems ask about important concepts that are worth knowing by heart. These are starred (with an asterisk, *). You must learn the answers to these important problems. It would be possible to give exams based solely on starred problems.

To preserve essential mathematical connections some sections are long and worth two or more days of study. First learn the terms and basics so that you can answer the straightforward "A" problems. Then "put it all together" so you can see the "big picture" and answer the "B" problems.

Answers in the Text. Many homework problems do not have numerical answers, but many "A" problems do. When the answer is a number (as opposed to a formula, picture, or explanation), some problems have the solution given with two significant digits right alongside the problem. This is to give you immediate feedback about whether you used the proper method. However, when you hand in your homework, you must **report your answers rounded to three or more significant digits**. That way, if you also get the third significant digit right, we can be confident you used the proper method.

Connections and Symbolic Mathematics. New information is easily forgotten unless it is connected to familiar things. Therefore, familiar material is valuable not only as review, but also because the new higher-level material can and should be connected to it. Each section has material you already know, and a higher level of similar material that you do not know. There is a danger that you will see the review material and think that there is nothing new there. You would be mistaken. Most students have to work very hard to move up to the higher level of mathematical thought. The difference is visible in the "B" problems.

For example, this text promotes learning to read symbolic mathematics. You have seen a lot of mathematics written in symbols; symbols are familiar. But, can you learn math by reading it? Or do you learn math only by from your instructor? People who are "good at math" can learn it by reading it. Part of the goal of the course is to make you comfortable with reading math. Learning *how* to read is hard work, even if *what* you read is familiar.

Expect your instructor to lecture only on the higher-level new material.

Pace. The material has not been artificially subdivided into single-day chunks. On the contrary, material which is mathematically connected is grouped together and the connections are emphasized, not severed. The inherent unity of some substantial topics produces numerous sections that are two or three days long. Here is a pace I use in a four-credit one-semester course (with 50-minute classes) at an open-admission university. There is time for half-day weekly quizzes in this schedule, but exam days would be extra.

Section and numbers of days (excluding exams):

Chapter 1	Chapter 2	Chapter 3	Chapter 4	Chapter 5
1.1: 1-	2.1: 1	3.1: 2½	4.1: 2	5.1: 2
1.2: 1	2.2: 1	3.2: 2	4.2: 2	5.2: 2
1.3: 1½	2.3: 2	3.3: 1½	4.3: 1	5.3: 1+
1.4: 1	2.4: 2	3.4: 1½	4.4: 2	5.4: omitted
1.5: 1		3.5: 1	4.5: 1½	
1.6: 2		review ½	4.6: 2	
review 1			review ½	
------	------	-------	-------	-------
8½	6	9	11	5+

Chapter 6	Chapter 7
6.1: 1-	7.1: 1
6.2: 2	7.2: 3
6.3: 2	7.3: 1
6.4: 2	
-------	------
7	5

Chapters 1-7 constitute a one-semester four-credit course at Montana State University.

Sections 8.1 and 8.2 could be interchanged. Section 8.3 requires 8.2, but not 8.1. Chapters 8 and 9 could be interchanged, and Sections 9.1 and 9.2 could be inserted after Section 4.3 or even after Section 3.2. Section 9.3 requires trigonometry and Sections 8.2, 9.1, and 9.2, but not 8.1 or 8.3.

Supplements. Students may purchase a solutions manual with solutions (not just answers) to three-fourths of the problems (omitting problems with numbers divisible by four). In it solutions to numerical problems have numbers rounded to two significant digits. The steps and rounded numbers in the manual should make it easy for students to duplicate the correct work. Nevertheless, they cannot simply copy the manual because solutions with three or more significant digits are required. Therefore, they must at least retrace the proper steps and do the computations themselves.

Adopters may obtain a complete solutions manual and an instructor's manual with section-by-section comments and suggestions for instructors, as well as a potential syllabus including daily homework assignments. Sample exams and quizzes are also available.

Acknowledgements. Some good ideas take years to bear fruit. Anyone developing such an idea appreciates encouragement and moral support. I got a lot from my wife, Najaria, and my daughter, Norah. In addition, I appreciate the contributions and faith of Anne Teppo, Jean Schmittau, Ken Tiahrt, John Lund, Sally Denlow, and Rick DeLorenzo. Finally, I thank the numerous anonymous reviewers who loved the manuscript and made Prentice Hall happy to join me in this project.

Precalculus Concepts

Preliminary Edition

CHAPTER 1

Fundamental Concepts

Section 1.1. The Conceptual Level of Algebra

The purpose of this chapter is to give you the proper frame of mind for learning and doing mathematics at the level of algebra and calculus. The language of mathematics is especially designed to express the key ideas. The symbolic language of mathematics is a foreign language -- and studying it has difficulties and rewards similar to those of studying any foreign language. Chapter 1 emphasizes the symbolic language of mathematics (which I call Mathematics, spelled with a capital letter like other languages).

Suppose you want to communicate with a person who speaks only Japanese. You can get information from the Japanese-speaker by having someone translate for you. But that is obviously very awkward compared to just being able to speak and understand Japanese. And, even if you think that learning Japanese would be hard, remember that every 10-year-old in Japan can speak it. No one would say you have to be "bright" to be able to communicate in Japanese or any other language. You simply have to practice and be around those who use it.

You don't have to be bright to speak Mathematics either. But you have to try. You have to be around others who already speak or write it so you can learn from them. But let's distinguish between a teacher of Japanese and a translator of Japanese. A teacher helps you help yourself, and someday you will be able to get along without the teacher. In contrast, a translator just gives you today's information; tomorrow you will still need the translator.

I want you to become fluent. I want you to be able to get along without a translator. If you learn the language, mathematics will be easier to learn and use. Furthermore, you will enjoy math more and you will get more satisfaction and reward from it.

The language I am talking about is the symbolic language of algebra, which is basically the same as the language of trigonometry and calculus. Whether or not you are comfortable with algebra, it may be that this chapter can help you see it in a new light.

What is algebra *about*? Many students think that algebra is about numbers. This is entirely natural -- for years students are asked to use procedures to find numbers which answer problems. But algebra is not about numbers -- algebra is about *procedures* used to find numbers. Algebraic procedures require doing various mathematical operations (such as adding, multiplying, or taking the square root) in various orders. Actually, algebra is about *operations* and the *order* in which they should be used in various problems. Numbers are a primary concept of arithmetic; in algebra the concepts are one level up. *Operations* and *order* are essential concepts of algebra.

Interpretation of Expressions. Expressions are the nouns and pronouns of Mathematics. Numerical expressions include numbers such as "7", "$5(3^2)$", and "log 32", as well as expressions with variables such as "x", "$3y + 9$", and "$\sin x$". Expressions with variables can be understood at two distinct conceptual levels. One, the arithmetic level, regards an expression such as "$2(x + 5)$" as representing a *number*. You might evaluate it for a given value of x. Or, if it were in an equation such as "$2(x + 5) = 24$," you might try to solve for x, treating x as an unknown number.

The other, higher, conceptual level regards "$2(x + 5)$" as expressing a *sequence of operations*. It expresses a relationship in a functional manner. The following examples illustrate which levels of thought belong in the various parts of algebra.

Example 1: Notice the operations used to solve "$2(x + 5) = 24$."

		Operation
Typical solution:	$2(x + 5) = 24$ is equivalent to	[divide by 2]
	$x + 5 = 12$ is equivalent to	[subtract 5]
	$x = 7$.	

Your ability to solve this type of problem is dependent upon your ability to see the *sequence of operations* expressed in "$2(x + 5)$". The operations expressed are "Add 5" and then "Multiply by 2." This sequence of operations can be "undone" using "Divide by 2" and then "Subtract 5." The key to the solution is the *sequence of operations*, not the *number* 24. If the right-hand side were changed to 95.6, the sequence of operations in the solution process would be the same.

I like to say that

Algebra is not about numbers!

This sounds shocking, but it's basically true. Algebra is about operations (functions) and sequences of operations -- a conceptual step up from numbers. Algebra concerns numbers, but

Algebra is *about* operations and order.

Operations and Identities. There is an important distinction between equations and identities which cannot be understood at the conceptual level of arithmetic. Identities are about operations and order. Recall that an identity is an equation that is true for all values of the variable.

Example 2: The Distributive Property implies: $2(x + 5) = 2x + 2(5)$.

What does this equation (identity) say about the number "x"? Nothing! It's not about x.

The equation "$2(x + 5) = 24$" in Example 1 can be called a "conditional" equation because it is true only under certain conditions. It is true for certain values of x and false for others. On the other hand, the equation "$2(x + 5) = 2x + 2(5)$" in Example 2 is an identity which is true for *all* values of x. Both assert the equality of numbers, but at much different conceptual levels. For the equation "$2(x + 5) = 24$" to be true, the unknown "x" must be a particular number. So it says something about a number. On the other hand, the identity "$2(x + 5) = 2x + 2(5)$" is true for *all* numbers, so it does not say anything about "x", the number. Therefore it must say something on a different conceptual level, and of course, it does. It says something about the *functions* (operations) "Add 5" and "Multiply by 2," and the order in which they are executed.

Therefore, if you interpret the expression "$2(x + 5)$" as a number, you are missing the point of the identity "$2(x + 5) = 2x + 2(5)$." To understand it you must interpret the expressions as expressing operations in a certain order.

Identities expressed using the letter "x" are not about "x".

Textbooks are rarely explicit on the point, but all algebraic identities are actually about alternative sequences of operations.

Identities are about operations and order.

Example 3. $(x + 5)(x - 5) = x^2 - 25$.

This identity provides two alternative sequences in which to evaluate the expression. Given x, you may multiply two factors obtained by adding and subtracting 5 from x, or you may square x and then subtract 25. Which sequence of operations you prefer depends upon the context. Section 1.6, "Four Ways to Solve Equations," discusses various motivations for preferring one order to another.

Example 4. One "property" (identity) of powers is: $x^a x^b = x^{a+b}$, for $x > 0$.

The left side expresses two powers and then multiplication. The right side expresses addition and then one power. This property expresses two alternative operational sequences. Again, the one you encounter depends upon the context. If you prefer the other sequence, use it.

Formulas and Story Problems. Formulas express processes by exhibiting operations. Formulas use well-chosen letters to help indicate the real-world application of a given relationship. The formula for the circumference, C, of a circle with diameter d, $C = \pi d$, uses "C" and "d" to help you remember the application of the formula. The expression "πd" expresses the operation "Multiply by π." The formula for the area, A, of a circle with radius r, $A = \pi r^2$, uses "A" and "r" in a helpful manner and expresses "Square and then multiply by π." The next example shows that word (story) problems require this operational (functional) level of thought.

Example 5A: A field is in the shape of a square with a semicircular cap on one side (Figure 1). Its perimeter is 400 meters. How long is a side of the square?

By "perimeter" we mean the distance around three sides and the semicircle.

Preliminary thoughts: This problem asks you to find the side given the perimeter. It's the reverse of the easier problem: Find the perimeter given the side. Let's try one of that easier type first.

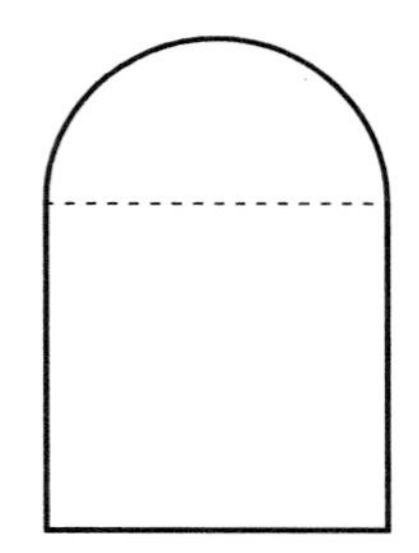

Figure 1. A field in the shape of a square with a semicircular cap.

Example 5B: A field is in the shape of a square with a semicircular cap on one side (Figure 1). The side is 100 meters. What is the perimeter of the field?

Solution to Example 5B: The three sides of the square contribute 300. The semicircle contributes half the circumference of a circle of diameter 100. So it contributes $100\pi/2$.

Total length: $300 + 100\pi/2 = 457.08$ (meters).

I hope this reminds you of arithmetic. Computing the perimeter does not require algebra. It is a direct calculation which requires use of formulas, and formulas are a basic way to express mathematical processes. Most students understand how to use formulas. This

Convention: In this text, numbers (such as 457.08) expressed with three or more significant digits may be approximations and not exact.

is an example where the circumference is of a semicircle, not a circle, but most students simply adapt the formula for a circle by dividing by two.

The same *operations* we used to calculate the perimeter from the side will work for *any* length side. Do you see the operations and their sequence in the above calculation?

Take the side and *multiply by three*. Separately, take the side and *multiply by* π and then *divide by two*. *Add* those two separate contributions.

This is where algebraic notation comes in. We can use *one* algebraic expression to represent *many* calculations (regardless of the unknown number x).

So, if the perimeter is 400 meters, how long is a side?

Solution Process: Call the length of a side "x". The operation "multiply by three" yields "$3x$." The sequence of operations *multiply by* π and then *divide by two* is represented by "$\pi x/2$." Adding, we obtain the formula:

$$P = 3x + \frac{\pi x}{2},$$

for the perimeter, P. It's not an important shape, so we do not memorize this formula, but it is just as valid as the formula for the perimeter of a square or the formula for the circumference of a circle. Note that the fact "Its perimeter is 400 meters," in the original problem had nothing to do with the very important step of creating the correct formula.

Now the word-problem sentence about "400 meters" translates to the equation

$$3x + \frac{\pi x}{2} = 400.$$

The problem asks us to solve this. We might wish to "simplify" the formula first. We will have to reorder the expression with x sooner or later. To simplify it you must continue to think about operations and order. The idea is to "isolate" x or "Consolidate like terms." The method is to rewrite the expression from the formula so it expresses a different sequence of operations. Recall that "$\pi x/2$" is equal to "$(\pi/2)x$". Now we can use the Distributive Property to factor out "x" and replace the expression on the left with an equivalent expression which expresses a different, more convenient, sequence of operations. The original formula expressed addition last. We prefer this formula which expresses multiplication last:

$$P = (3 + \frac{\pi}{2})x.$$

Therefore the sentence about "400 meters" can be rewritten:

$$(3 + \frac{\pi}{2})x = 400.$$

Now $3 + \pi/2$ is just a number, so we can find x by dividing by it.

$$x = \frac{400}{3 + \frac{\pi}{2}} = 87.51 \quad (\textit{meters}).$$

The solution *process* (but not the solution) would be identical if the perimeter were some other number besides 400 and the answer were some other value of x. The particular value of 400 has nothing to do with the formula and neither does the particular value of x which will eventually solve the equation. Similarly, in the solution process those numbers have nothing to do with the change of "$3x + \pi x/2$" to "$(3 + \pi/2)x$."

Apparently, *most of the problem is about operations and the order in which they occur, not numbers.* Finding the perimeter given the side (Example 5B) resembled arithmetic because it was direct. A problem is direct when the given words, symbols, or basic formulas express the operations you actually do to solve the problem. Direct problems are basically arithmetic computations. You use the operations, but your focus is on the numbers you compute. In contrast, finding the side given the perimeter (Example 5A) is much harder because it is "indirect." A problem is indirect when the given words, symbols, or basic formulas suggest operations that you are *not* supposed to actually *do*. Instead, you *represent* the operations in symbolic notation and then manipulate the operations (instead of just manipulating numbers). That is a characteristic of algebra. Finding the side given the perimeter is algebra.

> A problem is direct when the given words, symbols, or basic formulas express the operations you actually do to solve the problem. A problem is indirect when the given words, symbols, or basic formulas suggest operations that you are *not* supposed to actually *do*. Instead, you *represent* them in symbolic notation and then manipulate the operations.

Here is one final example to make the point that word-problems are mostly about operations, not numbers.

Example 6A: A field is (again) in the shape of a square with a semicircular cap on one side (Figure 1, again). Its area is 40,000 square meters. How long is a side of the square?

Preliminary thoughts: This problem asks you to find the side given the area. It's the reverse of the easier problem: Find the area given the side. Let's try one of those first.

Example 6B: A field is in the shape of a square with a semicircular cap on one side (Figure 1, again). Its side is 100 meters. What is its area?

Solution: The square contributes 100^2. The semicircle contributes half a circle. To find its area we need its radius, which is half the diameter, that is, half of 100, 50. Therefore, the semicircle contributes $\pi 50^2/2$.

Total area: $100^2 + \pi 50^2/2 = 13927$ (square meters).

Almost all my students can do that calculation correctly because they are comfortable with numbers. But, to do the original word problem they need to see the operations, not just the numbers.

Notice the *operations* used to calculate the number in Example 6B: *Square* the side. Take the side and *divide by 2* to obtain the radius and *square* the radius and *multiply by* π and then *divide by 2* because it's only half a circle. Finally, *add* the two contributions.

Operations and order are what algebraic notation was designed to express.

You must build a formula for the area that expresses those operations:

$$A = x^2 + \frac{\pi(\frac{x}{2})^2}{2}$$

Now the rest of the problem parallels Example 5A. The sentence expressing the area translates to the equation

$$x^2 + \frac{\pi(\frac{x}{2})^2}{2} = 40{,}000.$$

Because the left side is more complicated here, it takes more steps to rearrange it into an operational sequence we like. But the key idea is the same: **The sequence of operations must be changed.**

In a few steps the equation becomes

$$(1 + \frac{\pi}{8})x^2 = 40{,}000$$

which can be easily solved (problems B18-19) . "$1 + \pi/8$" is just a number (1.3927). The expression expresses squaring and then multiplication. "Undoing" the multiplication and then undoing the squaring, we find

$$x = \sqrt{\frac{40000}{1 + \frac{\pi}{8}}} = 169.47 \quad (\textit{meters}).$$

If the area were some other number besides 40,000 the solution *process* (but not the solution) would be identical. In the solution process the "40,000" and the actual value of "x" have nothing to do with developing the formula for area or its simplification from $x^2 + \frac{\pi(\frac{x}{2})^2}{2}$ *to* $x^2(1 + \frac{\pi}{8})$. Again, most of the problem is about operational sequences, not numbers. Too many students busy themselves looking for *the answer* (a number) when the key to solving the problem is to use algebraic symbolism to *represent* the operations (without actually doing those operations). The key is at the conceptual level of operations and order.

Conclusion. The symbolic language of algebra is a foreign language. It is designed to express thoughts about operations and order, which are concepts essential to algebraic methods (procedures). Identities and "properties" of functions are about alternative sequences of operations. Even when they are expressed in terms of x, they are, nevertheless, not about x. They are at a higher conceptual level -- the level of methods. One method may apply to many examples.

Direct story problems require computations that resemble arithmetic. Algebra is used in indirect problems where operations are *represented* symbolically, but not actually executed. Algebra can be *applied* to numbers, but it is at a higher conceptual level than numbers.

Therefore, it is correct to say, "Algebra is not about numbers," although, at first, that may appear to be nonsense. Algebra is about operations and order.

Terms: expression, identity, direct, indirect.

****

Exercises for Section 1.1, "The Conceptual Level of Algebra":

A1.* What are identities about?

A2. Obviously, the numbers which solve "$2(x + 5) = 24$" and "$2(x + 5) = 95.6$" are different. But something about solving them is the *same* for both. What?

A3. An expression such as "$2x + 5$" can be regarded at two conceptual levels. What are its two interpretations?

A4. An expression such as "$x^2 - 1$" can be regarded at two conceptual levels. What are its two interpretations?

A5. Give two distinct interpretations of the expression "$x/2 + 7$."

A6. What do solving "$3x - 5 = 80$" and "$3x - 5 = 47$" have in common?

A7. What do solving "$x/2 + 4 = 80$" and "$x/2 + 4 = 47$" have in common?

A8. State (with variables) the Distributive Property (multiplication distributes over addition).

^ ^ ^ ^ ^ ^ ^ ^ ^

B1.* a) Is algebra about numbers? b) What is algebra about?
c) What is algebraic notation with variables designed to express?

B2.* Expressions with variables can be interpreted at two conceptual levels. What are the two interpretations?

B3.* Algebra concerns indirect problems, as opposed to direct problems. What makes a problem "indirect"?

B4. A circular walk is inscribed in a square block. The area outside the circle and inside the square will be a garden (the shaded area in the figure). a) If the side of the square is 100 feet, what is the area of the garden?
b) Note the operations and the order in which you applied them in part (a). Create a formula for the area of the garden in terms of the side of the square, where the side of the square is "x".
c) If the area of the garden is 1400 square feet, how long is the side of the square?

B5. An animal pen has the pictured shape. a) If the length of the top side in the picture is 3.5 feet, what is the enclosed area of the pen?
b) Note the operations and the order in which you applied them in part (a). Create a formula for the area of the pen in terms of the length of the top side, "x".
c) If the area of the pen is 32 square feet, how long is the top side?

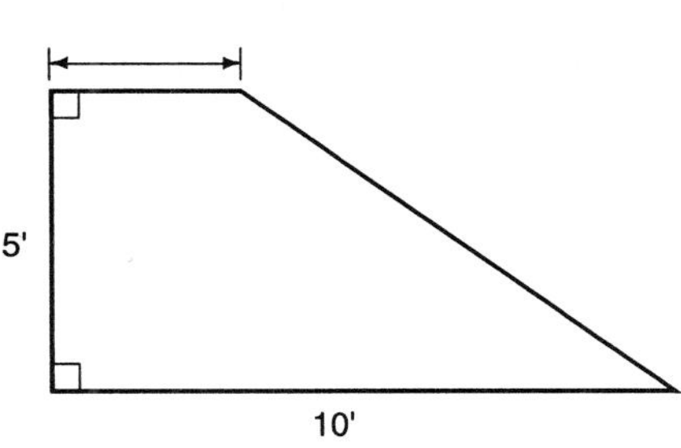

B6. One side of a rectangle is 3 inches longer than the other. The area of the rectangle is 50 square inches. How long are the sides?

B7. One side of a rectangle is twice as long as the other. The area of the rectangle is 60 square inches. How long are the sides?

B8. Sylvia runs the first part of her daily run at 10 miles per hour and then she slows down to 7 miles per hour. If she runs 6 miles and it takes 3/4 of an hour, how far did she run at 10 miles per hour?

B9. Jane runs the first three fourths of her daily 5-mile run at a certain speed and then finishes at half that speed. She runs it in 7/10 hour. How fast does she run the first part?

B10. a) The perimeter of a semicircular freestanding enclosure (see the picture) is made from p linear feet of fencing and gates. Express the relationship between its diameter and p.
b) Express the relationship between its area and p.

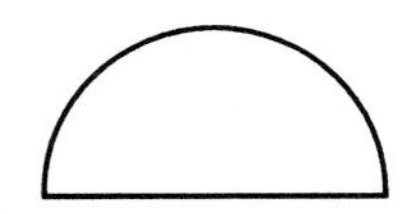

B11. If the perimeter of a freestanding semicircular enclosure is 350 feet, what is the length of its straight side (see the picture)?

B12. A farmer makes a rectangular dog kennel by using one side of the barn for a side and the other three sides from 20 feet of gate and fencing.
a) If the kennel projects out 5 feet from the side of the barn, what is its area?
b) Write an expression which expresses the operations employed to calculate the answer to part a).
c) How far out from the barn does the kennel project if its area is 48 square feet?

B13.* If you change a number in a word problem something important is, nevertheless, likely to remain the same. What?

B14.* Suppose you solve a particular type of word problem on a quiz, and on the exam there is an almost identical problem, only the numbers have been changed. From your point of view, what remains the same?

B15.* The Distributive Property states: $ab + ac = a(b + c)$. What is it about?

B16. Is the sentence "$x^2 + 2x + 1 = (x + 1)^2$" about x? Why or why not? What is it about?

B17. a) State another equation (your choice) using "x" which is not about x.
b) What is your equation about?

B18. State the general form of the identity which permits replacing $(x/2)^2$ with $x^2/4$.

B19. State the general form of the identity which permits replacing $(x^2/4)/2$ with $x^2/8$.

B20. How does the conceptual level of algebra differ from that of arithmetic?

Section 1.2. Order Matters!

If an algebraic expression (such as "$\pi + 3x$") indicates more than one operation (such as addition and multiplication), the order in which the operations are carried out usually makes a difference.

Order Matters!

On a calculator, the keystroke sequence required to evaluate an expression may differ substantially from the left-to-right order of the written expression. The keystrokes you use will depend upon the type of calculator you use, so you must learn the conventions of your calculator.

Consider an expression such as "$2 + 3x$." There are two different orders which we must distinguish:

1) the left-to-right *order in which it is written*, and
2) the *order in which the operations are to be executed.*

Algebraic notation is a language and the rules and conventions of interpreting it must be learned. For example, although the first operation you *see* when reading the expression "$2 + 3x$" left-to-right is addition, the first operation you *do* is multiplication, according to Convention 2, below. And, order matters:

If $x = 5$, $2 + (3x) = 17$.
If $x = 5$, $(2 + 3)x = 25$.

The two different orders yield two different values.

Order matters!

Parentheses are used to indicate that the enclosed operation is to be carried out before other operations in the same expression.

Example 1: To evaluate the expression "$3(x + 5)$," the value of "x" must be given. Then, addition is executed first, even though the multiplication is encountered first when reading left-to-right. This is a result of Convention 1 about parentheses. Read aloud, this is, "three times the quantity x plus five."

<u>Convention 1</u>: Operations enclosed in parentheses are executed first.

Example 2: In the expression "$\pi + 5x$" multiplication is executed first.

<u>Convention 2</u>: Multiplication and division are performed before addition or subtraction, unless parentheses indicate otherwise.

Example 3: The expression "$5 + x/4$" indicates that x is divided by 4 first and then 5 is added.

Note that division comes first, even when the addition is on the left and appears first in the natural left-to-right order.

There are many different models of calculators. You will have to learn how the keystrokes of your calculator correspond to written mathematics.

<u>Calculators</u>. For this course you should have a "graphics" calculator. This section has exercises which will help you determine how to use your calculator to evaluate expressions. We will emphasize the conventions of *written* mathematics. You, yourself, must determine how to make *your* calculator do the calculations in any given written expression.

Calculator Exercise 1: Determine your calculator's conventions by reproducing these results. Let $x = 3.7$. Then

$2(x + 1) = 9.4$ [not 8.4, which is $2x + 1$.]
$5 + \pi x = 16.62$. [Find and use the "π" key.]
$(2 + \pi)x = 19.02$ [not 13.62, which is $2 + \pi x$].

Convention of This Text: Numbers expressed with three or more significant digits may be approximations and are not necessarily exact.

Now you can see why we do not read "$2x + 1$" aloud as "two times x plus one." The insertion of the word "times" merely confuses matters. The listener may hear "two times (x plus one)." Actually, it is hard to hear parentheses, but pauses can indicate them. We reserve the English "two times [pause] x plus one" for "$2(x + 1)$." Even better would be, "Two times the quantity x plus one."

<u>Convention 3</u>: Squares, reciprocals, and other powers (that is, functions with exponents) are to be evaluated before multiplication or division, unless parentheses indicate otherwise.

Example 4: "$3x^2$" indicates that x is squared and then the square is multiplied by 3, even though the multiplication is on the left. "$(3x)^2$" indicates that x is multiplied by 3 and then the product is squared.

$5x^{-1} = 5(x^{-1})$, not $(5x)^{-1}$.

In the expression, "$5+4(x-5)^2$", the subtraction is executed first. Then the difference is squared (by Convention 3). Then the square is multiplied by 4, and finally 5 is added (by Convention 2). In this example the order of operations is certainly not the written left-to-right order!

If x is negative, x^2 is positive. The notation can be tricky. For example, -3^2 is negative, because the negative sign is treated as multiplication by -1, and Convention 3 says powers are executed first. So, $-3^2 = -(3^2)$, which is negative. If you want $(-3)^2$, use parentheses [or omit the negative sign altogether, since $(-c)^2 = c^2$ for all c].

Calculator Exercise 2: Check the use of your calculator by reproducing these results.

Let $x = 4.1$. Then $3x^2 = 50.43$ and $(3x)^2 = 151.29$.
Let $x = -5.6$. Then $3x^2 = 94.08$ [not -94.08].
Let $x = 1.7$. Then $(x + 4)^2 = 32.49$ [not 17.7].
Let $x = 2$. Then $(5x)^{-1} = .1$ [not 2.5].

These conventions are called the algebraic conventions. To remember the order in which operations are executed according to these algebraic conventions, some students use the acronym: PEMA. It abbreviates the operations in order: Parentheses, Exponents (such as squaring), Multiplication, Addition.

Some students like the phrase, "Please Excuse My Dear Aunt Sally": Parentheses, Exponents, Multiplication, Division, Addition, Subtraction.

Special Functions. Functions such as logarithmic functions, exponentials, and trigonometric functions are special functions. Graphics calculators have keys for these functions. Unfortunately, the order conventions for special functions are not as clear as they are for arithmetic functions. The appropriate calculator keystroke sequence may depend upon the brand of calculator. However, all agree that special functions are executed before addition or subtraction, unless parentheses indicate otherwise.

Example 5. $\log x + 2 = (\log x) + 2$, not $\log(x + 2)$.
$\sin x - 5 = (\sin x) - 5$, not $\sin(x - 5)$.

Calculator Exercise 3: Reproduce these results.

$\log(\pi + 2) = .71$.
$\log \pi + 2 = 2.497$.

Mixing other operations with special functions is not so easy.

Example 6. In the written language, $\sin 2x = \sin(2x)$, which is not $(\sin 2)x$. The side-by-side appearance of "$2x$" gives the multiplication priority. To avoid parentheses, if we wish to express $(\sin 2)x$ we could write "$x \sin 2$", in which there is no danger of misinterpretation.

$\log 5x = \log(5x)$, not $(\log 5)x$, which we would write "$x \log 5$".

Try to avoid handwriting "$\sin x/2$" with a slanting division bar. Some calculators treat this as "$(\sin x)/2$," although you may intend "$\sin(x/2)$." They are different. The precise position of the fraction bar can indicate what is meant. $\sin \frac{x}{2}$ indicates division is first; $\frac{\sin x}{2}$ indicates that sine is applied first.

Calculator Exercise 4: Check your calculator's conventions by reproducing these results. Let $x = 6.7$. $\log 2x = \log(2x) = 1.127$.

$(\log 2)x = 2.017$.

$$\frac{\log 297}{2} - 1.236. \qquad \log(\frac{297}{2}) - 2.17.$$

<u>Convention 4</u>. Functions with special names (which have function keys, such as trig functions, logs, and exponentials) are executed before addition and subtraction. Various ways may be used to indicate when functions with special names are executed before powers, multiplication, and division.

Example 7: $(\sin x)^2 \neq \sin(x^2)$, in general. So which order is intended when parentheses are omitted?

$\sin x^2 = \sin(x^2)$. The power right next to the x indicates that the power comes first. If you wish to handwrite $(\sin x)^2$ without parentheses, you may write "$\sin^2 x$", but this is understood as an abbreviation and is not acceptable for calculators.

Calculator Exercise 5: Check your calculator's conventions by reproducing these results. Let $x = 6.7$.

$\log x^2 = 1.652$. [$\log x^2 = \log(x^2)$, but, by convention, parentheses may be omitted.]

$(\log x)^2 = .68$.

<u>Other Notation for Grouping</u>. There are two common written grouping symbols that do not have keys on many calculators: the long fraction bar and the extended overhead line of the square root symbol. Calculators require parentheses (or a functional approach) to indicate the grouping that these symbols can indicate without parentheses. Superscripts (exponents) are a third, less common, way to indicate grouping without parentheses. You must learn how to tell your calculator what you mean.

<u>Fraction Bars</u>. The fraction bar is a grouping symbol.

$$\frac{7 + 10}{5} = \frac{(7 + 10)}{5} = (7 + 10)/5 = 3.4, \quad not \quad 7 + 10/5 = 9.$$

$$\frac{a + b}{c} = \frac{(a + b)}{c} = (a + b)/c, \quad not \quad a + \frac{b}{c}$$

$$\frac{a}{b + c} = \frac{a}{(b + c)} = a/(b + c), \quad not \quad a/b + c$$

Note that each error could result from poor pronunciation. The pronunciation "*a* plus *b* over *c*" is ambiguous and might be used for either "$(a + b)/c$" or "$a + (b/c)$." Similarly, "*a* over *b* plus *c*," is also ambiguous and might serve for either "$a/(b + c)$" or "$(a/b) + c$." To avoid this ambiguity, I would pronounce $\frac{a + b}{c}$ as "*a* plus *b*, the quantity, over *c*," or "*a* plus *b* [said quickly, followed by a long pause] over *c*." Then $a + \frac{b}{c}$ is different: "*a* plus the quantity *b* over *c*," or "*a* plus [pause, and then more quickly] *b* over *c*."

Calculator Exercise 6. Check your use of your calculator by reproducing these results.

$$\frac{4+\pi}{7} = 1.02, \quad \text{[not 4.45, which is } 4 + \pi/7\text{]}.$$

$$\frac{19}{2 + \pi} = 3.695, \quad \text{[not 12.64, which is } 19/2 + \pi\text{]}.$$

<u>Square Roots</u>. The extended square root symbol is a grouping symbol which many calculators cannot use.

$$\sqrt{a + b} = \sqrt{(a + b)} = \surd\,(a + b), \text{ not } \surd a + b.$$

Grouping symbols are hard to pronounce. The first expression is "The square root of the quantity *a* plus *b*," where the phrase "the quantity" indicates the grouping. The last is "The square root of *a* [pause], plus *b*." Note that, aloud, only the phrase "the quantity" distinguishes these two different expressions.

Calculator Exercise 7. Check the use of your calculator by reproducing these results.

$$\sqrt{\pi + 9} = 3.48 \quad [\textit{not } 10.77, \textit{ which is } \sqrt{\pi} + 9\,].$$

Let $b = -3.7$ and $d = 20$. $\sqrt{b^2 + d} = 5.80.$

Superscripts. Superscripts are generally written in a smaller size. The intended order can be discerned by the size and position.

$$2^{m+n} = 2^{(m+n)}, \textit{ not } 2^m + n\,.$$

$$5^{\frac{m}{n}} = 5^{(\frac{m}{n})}, \textit{ not } \frac{5^m}{n}\,.$$

Calculator Exercise 8: Check the use of your calculator by reproducing these results.

$2^5 = 32.$
$8^{2/3} = 4$ [not 21.333, which is $8^2/3$].
$2^{1.2+\pi} = 20.27$ [not 5.439, which is $2^{1.2} + \pi$].

Parentheses For the Quadratic Theorem. The quadratic theorem is often written with an extended fraction bar and an extended square root bar as grouping symbols. To inform your calculator of the intended order of operations, you will have to use extra pairs of parentheses.

The Quadratic Theorem (1.2.1): $ax^2 + bx + c = 0$ is equivalent to

$$x = \frac{-b \pm \sqrt{b^2 - 4ac}}{2a}\,.$$

"Negative b plus or minus the square root of (the quantity) b squared minus four a c, all over two a."

The entire top is grouped (by the long fraction bar) before the division and so is the "$2a$" in the denominator. Also, the interior of the square root is grouped, by the long square root bar.

This is tricky to enter into your calculator correctly because you need extra grouping symbols. Given numerical values for a, b, and c, most students get it wrong the first time. Try these examples with your calculator and check your answers. I'm warning you, this is more complicated than it first appears.

Example 8: Solve $3.67x^2 + 20x + 5 = 0$. $a = 3.67$, $b = 20$, and $c = 5$.

There are two answers, one with a plus sign and one with a minus sign. We will do only the case with the plus sign. Use the formula, but remember the grouping. Here is the formula with the numbers plugged in as you might write it using * for multiplication:

> A program in your calculator is not a substitute for knowing how to enter complex expressions correctly. Use these Quadratic Theorem examples for practice.

$$x = \frac{-20 + \sqrt{20^2 - 4*3.67*5}}{2*3.67}$$

. But your calculator may need this entered as if the whole formula were on a single line:

(-*b* + √(*b*² - 4**a***c*))/(2**a*) .
Extra grouping symbols: (( )) ( ) .

Note the extra grouping symbols in the one-line version that are not necessary in the usual written over-and-under version.

Calculator Exercise 9. Check the use of your calculator by reproducing these results. Try to enter the whole formula at once, rather than breaking it up into separate calculations. Do the "plus" term and use $a = 3.67$, $b = 20$, and $c = 5$.

$$\frac{-b + \sqrt{b^2 - 4ac}}{2a} = -.2627.$$

The solution to the equation in Example 8 is $x = -.2627$ or $x = -5.1869$.

The next one is a bit trickier.

Calculator Exercise 10. Let $a = 4.69$, $b = -5.3$, and $c = -7$. Evaluate the "plus" term of the Quadratic Formula (1.2.1).

$$\frac{-b + \sqrt{b^2 - 4ac}}{2a} = 1.91.$$

In the quadratic formula be careful with the final "divided by two times *a*." The grouping matters. Suppose the numerator were 12 and *a* were 3. The expression "12 ÷ (2 * 3)" yields 2, but "12 ÷ 2 * 3" yields 18. The former sequence is what we want.

There is still one more thing that can go wrong when you evaluate the quadratic formula. It can happen when b is negative, for example when b = -13. b^2 is positive, but -13^2 is not b^2. -13^2 is negative. Don't forget the convention which says that squaring is executed before multiplication (A negative sign is interpreted as multiplication by negative one). Of course, since $(-13)^2 = 13^2$, the shortcut of just using 13^2 instead of $(-13)^2$ saves keystrokes and avoids the problem.

Calculator Exercise 11: Try evaluating the quadratic formula for the solutions to: $5.67x^2 - \pi x - 40 = 0$. Beware of negative signs.

The solutions are x = 2.9475 and x = -2.3934.

Your calculator may have the Quadratic Formula programmed in it, but you still need to know how to enter expressions so that your calculator will evaluate them correctly. Try the next example.

Calculator Exercise 12: Let c = -9.1, d = 4.7, and k = 3.6. Evaluate

$$\frac{\sqrt{c^2 - 2k}}{3d}.$$

The value of the expression is .617.

****

Guide to Pronunciation of Mathematical Expressions

$5x$	five x [avoid "five times x"]
$x + 2$	x plus two
$x - 2$	x minus two
-5	negative five [or] minus five
$5 - (-2)$	five minus negative two [or] five minus minus two
$3x + 2$	three x plus two [avoid "three times x plus two]
$3(x + 2)$	three times the quantity x plus two [or] three times (pause) x plus two

[In spoken Mathematics, parentheses are hard to say aloud. "The quantity" is a phrase used to alert the listener to parentheses. Also, pauses can be used to try to indicate parentheses.]

$x/2$	x over two [or] x divided by two
$3x^2$	three x squared
$(3x)^2$	three x, the quantity, squared [or] the quantity three x (pause) squared
$\sqrt{2}$	the square root of two [some say, "root 2"]
$\sqrt{x+2}$	the square root of the quantity x plus two

[Sometimes the awkward phrase, "the quantity," is omitted, in which case this and the next expression are not easily distinguished aloud.]

$\sqrt{x} + 2$	the square root of x (pause) plus two
$\|x\|$	the absolute value of x [some say "absolute x"]
$\|x + 2\|$	the absolute value of the quantity x plus two
$\|x\| + 2$	the absolute value of x (pause) plus two
x^3	x cubed [or] x to the third [or] x to the third power
x^p	x to the p*th* power [or] x to the p
ax^2	a x squared
$=$	equals [or] is equal to [not "equals to"]
$<$	is less than [avoid "is smaller than"]
$\leq$	is less than or equal to
$>$	is greater than [avoid "is bigger than"]
$\geq$	is greater than or equal to
$-2 < x < 2$	negative two is less than x is less than two [or] minus two is less than x is less than two [or] negative two is less than x and x is less than two

****

The next table gives some expressions and some mistakes about order commonly made by students.

expression	correct alternative	common mistake	comment
$a(b + c)$	$ab + ac$	$ab + c$	Distributive Property
$(a + b)^2$	$a^2 + 2ab + b^2$	$a^2 + b^2$	squaring does not distribute over addition (use FOIL).
$\sqrt{a^2 + b^2}$	none	$a + b$	taking the square root does not distribute over addition
$a + (b + c)$	$a + b + c$	none	associative property
$a - (b - c)$	$a - b + c$	$a - b - c$	subtraction is not associative
$a/(b/c)$	ac/b	ab/c	division is not associative
$(a/b)/c$	$a/(bc)$	ac/b	division is not associative
$\frac{a}{b} + \frac{a}{c}$	$\frac{ac + ab}{bc}$	$\frac{a}{b + c}$	adding numerators over a common denominator is ok; adding denominators is not.
$(ab)^2$	a^2b^2	ab^2	powers "distribute" over multiplication
$\sqrt{ab}$	$\sqrt{a}\ \sqrt{b}$	$\sqrt{a}\ b$	roots are powers which "distribute" over multiplication
$(a/b)^2$	a^2/b^2	a/b^2 or a^2/b	powers "distribute" over division
$\sqrt{\frac{a}{b}}$	$\frac{\sqrt{a}}{\sqrt{b}}$	$\frac{\sqrt{a}}{b}$	powers "distribute" over division
$x^m x^n$	x^{m+n}	x^{mn}	multiplication and division do not "distribute" over powers
x^m/x^n	x^{m-n}	$x^{m/n}$	
$a(b/c)$	$(ab)/c$	none	in this case, order does not matter
e^{2x}	$(e^x)^2$	$2e^x$	order matters!
$\log(2x)$	$\log 2 + \log x$	$2 \log x$	order matters!
$\sin(2x)$	$2(\sin x)(\cos x)$	$2 \sin x$	order matters!

Note that the original expression and the common mistake for it are, unfortunately, often pronounced similarly in the cases of a - (b - c), $(ab)^2$, $(a/b)^2$, and $\sqrt{\dfrac{a}{b}}$. Correct pronunciation, however, would make the proper grouping clear and help avoid mistakes.

Conclusion. **Order matters!** Now you have seen the use and importance of parentheses and conventions in arithmetic, algebra, and calculators. All scientific calculators use the algebraic conventions, but different types may differ in the keystroke sequences required to evaluate functions. The main point of this section is that

Order matters!

Term: algebraic conventions.

****

Exercises for Section 1.2, "Order Matters!":

Note: When the problem is to evaluate an expression using your calculator, give three or more significant digits. An approximate value with two significant digits is given in brackets. This will help you notice when your keystroke sequence is incorrect.

A1.* What is the main lesson of this section?

A2.* True or False? [No reason required.]
a) $(a + b) + c = a + (b + c)$. b) $(a - b) - c = a - (b - c)$. c) $(ab)c = a(bc)$
d) $(a/b)/c = a/(b/c)$. e) $a + b = b + a$ f) $a - b = b - a$.
g) $ab = ba$. h) $a/b = b/a$.

A3. a) How do you (personally) remember the algebraic conventions?
b) Give an acronym which can help you remember the algebraic conventions.

A4.* In algebra we avoid the symbol "×". a) Why?
b) How can we write "6 × 5" without using "×", still leaving it in factored form?

A5.* True or false? [No reason required.]
a) $(x + 4)^2 = x^2 + 16$. b) $(x - 3)^2 = x^2 - 9$. c) $(x + 5)(x - 5) = x^2 - 25$.
d) $1/2 + 1/5 = 1/7$. e) $(1/x)^2 = 1/(x^2)$. f) $-x^2 = x^2$.

g) $\sqrt{1/x}$ - $1/\sqrt{x}$. h) $\sqrt{x^2 + 25}$ - $x + 5$.

^ ^ ^ ^ ^ In all the following calculator exercises evaluate the expressions to <u>three or more significant digits</u>. To help you check your keystroke sequence, a two-significant-digit answer is given in brackets.

A6. a) 3.4567^4 [140] b) $4^{3.4567}$. [120]

A7. a) $\dfrac{3+\pi}{5}$ [1.2] b) $\dfrac{3}{4+\pi}$ [.42]

A8. a) $\sqrt{15+\pi}$ [4.3] b) $1.2^{1+\pi}$ [2.1]

A9. a) $4^{2/3}$ [2.5] b) $2^3/5$ [1.6]

A10. a) $3^{2/5}$. [1.6] b) $\dfrac{17}{2+\pi}$ [3.3]

A11. a) $\sqrt{51+6\pi}$ [8.4] b) $\sqrt{30}+\pi$ [8.6]

A12. a) $\sin 2 + \pi$ (use radians) [4.1] b) $\sin(2+\pi)$. [-.91]

A13. a) $\log 2 + \pi$ [3.4] b) $\log(2+\pi)$ [.71]

A14. a) $10^{2+\pi}$ [140,000] b) $10^2+\pi$ [100]

A15. Write out, in English, the proper pronunciation of the expressions in A7.

A16. Write out, in English, the proper pronunciation of the expressions in A11.

A17. Write out, in English, the proper pronunciation of the expressions in A12.

A18. Let b=2, c=3, d=4. Evaluate

a) bc/d b) b+c/d c) d/b+d d) b/c+d e) bc^2

A19. a) Is bc/d open to misinterpretation? b) Is $b + c - d$?

A20. The expression "$3(b + c)$" indicates that the addition is executed first. Find an equivalent (equal) expression without parentheses in which multiplication is executed first and addition last.

A21. The expression "$(b + c)/3$" indicates that the addition is executed first. Find an equivalent (equal) expression without parentheses in which division is executed first and addition last.

A22. Give an equivalent expression without parentheses: $100 - (20 - \pi)$.

A23. Give an equivalent expression without parentheses: $100/(20/\pi)$.

A24. How can you distinguish "2^{3+4}" from "$2^3 + 4$"?

A25. Give the sequence of operations in the expression:

a) $3 + 5x$ b) $5(x + y)$

A26. Give the sequence of operations in the expression:

a) $3^{2/7}$ b) $\sqrt{x^2+9}$

A27. Give a sequence of keystrokes for evaluating "5(4 + π)" on your calculator *without* using parentheses keys. You should get 35.708.

A28. Evaluate $\frac{3 + \pi}{\pi + 6}$. [.67]   A29. Evaluate $\sqrt{10 - 2\pi}$. [1.9]

A30. Let b=4, c=5, and d=2. Create a simple expression using all three variables (once each) and some combination of the operations of arithmetic that produce the following value:
a) 4.5 b) 10 c) 2.5 d) .4 e) 7 f) 2.8

^ ^ ^ ^ Evaluate the quadratic formula with
A31. a = 3.7, b = -1.2, and c = -7.8. [1.6, -1.3].
A32. a = 14.4, b = 19, and c = -95. [2.0, -3.3].

^ ^ ^ ^ ^ ^ ^ ^ ^

B1.* Name three ways to group expressions that are available in written mathematics that are not available on your calculator.

^ ^ ^ ^ ^ In all the following calculator exercises evaluate the expressions to three or more significant digits. To help you check your keystroke sequence, a two-significant-digit answer is given in brackets.

B2. Let b = 23.4, c = -4.5, and d = 6. Evaluate $\frac{-d + \sqrt{c^2 + bd}}{5bc}$. [-.013]

B3. Let a = -4.1, b = 2.34, and c = -5.6. Evaluate $\frac{-b - \sqrt{a^2 + c}}{7b}$. [-.35]

B4. Let a = 1.4 and b = 7.8. Evaluate $\frac{3^{a/b} - 3}{b - a}$. [-.28]

B5. Let b = 2.3, c = 4.1 and k = -6. Evaluate $\frac{\sqrt{c^2 - k}}{5b}$. [.42]

B6. Let c = 3.1, d = -5.6, and k = 2.3. Evaluate $\frac{k + \sqrt{d^2 + c}}{cd}$. [-.47]

B7.* True or false? [No reason required.]
a) $(x + a)^2 = x^2 + a^2$. b) $(x - a)^2 = x^2 - a^2$. c) $(x + a)(x - a) = x^2 - a^2$.
d) $x/a + x/b = x/(a + b)$ e) $(a/x)^2 = a^2/x^2$ f) $-b^2 = b^2$

g) $\sqrt{x^2 + a^2} = x + a$. h) $\sqrt{\frac{x^2}{a^2}} = \frac{x}{a}$, if x and a are positive.

B8.* Determine if these assertions are identities. If not, change the right hand assertion to the correct expression. Note if there is no common alternative equivalent expression that can be used on the right side. Assume $a \geq 0$ and $b \geq 0$.

a) $(a + b)^2 = a^2 + b^2$. b) $(ab)^2 = a^2b^2$. c) $\frac{a}{b} + \frac{c}{b} = \frac{a + c}{b}$.

d) $\frac{a}{b} + \frac{a}{c} = \frac{a}{b + c}$

B9.* Determine if these assertions are identities. If not, change the right hand assertion to the correct expression. Note if there is no common alternative equivalent expression that can be used on the right side. Assume $a \geq 0$ and $b \geq 0$.

a) $\sqrt{a^2 + b^2} = a + b$. b) $\sqrt{ab} = \sqrt{a}\sqrt{b}$

c) $a - (b - c) = a - b - c$ d) $(a/b)(c/b) = ac/b$.

^ ^ ^ ^ Use the quadratic formula and your calculator to evaluate the "plus" solution to:

B10. $3.45x^2 + 2x - 15 = 0$. [1.8] B11. $6.54x^2 - 4x - 12.3 = 0$. [1.7]

B12. $-2.3x^2 + 6x - 1.5 = 0$. [.28] B13. $1.23x^2 + 3x = 2 - 5x$. [.24]

B14. Inspect the table with expressions, correct alternatives, and common mistakes. Which of the listed mistakes might be promoted by poor pronunciation which omits expressing the proper grouping?

B15. What order mistake might a student make with these expressions?
a) $3x^2$ b) $15/(x/3)$

B16. What simplification mistake might a student make with these expressions?

a) $\frac{(3x)(3y)}{3}$ b) $\sqrt{9 + x^2}$

B17. Identify two expressions which are not equivalent, but which might be pronounced similarly.

B18. Express without parentheses: a) $b - (c - d)$. b) $b/(c/d)$.

B19. Some calculators use "reverse Polish notation." What is the keystroke sequence to add 3 plus 5? To multiply 4×6? To take 3^5? log 12? $e^{2.2}$?

B20. Write an expression without parentheses in which multiplication is encountered first but its execution is delayed until two other operations have been executed.

B21. Learn how to program your calculator. Program your calculator with the Quadratic Formula. a) What are the solutions if $a = 2.3$, $b = 4.56$, and $c = -7.8$?
b) Solve for b if $3^2 = 7^2 + b^2 - 2b(7)\cos 25°$.

B22. "$3x$" means 3 times x. Explain what "37", "3(7)", "3½", "3(½)", and "$f(7)$" mean. Compare and contrast the interpretation of putting two symbols side-by-side in the six cases.

Section 1.3. Abstraction and Written Mathematics

This section is about expressing mathematical processes in symbolic mathematics. I expect you already know how to *use* the algebraic processes discussed in this section. Now learn how to *read* and *write* them.

This text is entitled *Precalculus Concepts*. The concepts are mostly concepts from algebra. By definition, a concept is an abstract idea that summarizes the essence of *many* similar examples by regarding the similarity to be *one* mental object. The opposite of *abstract* is *concrete*. Particular instances of an abstract concept may be "concrete" examples. Here are some non-mathematical abstract concepts.

Example 1: Children have an idea of what behavior is "fair" and what is "unfair" from many occasions in which they share toys, candy, and responsibility, even though "fair" is an abstract concept. Few concepts receive more attention than "love," but "love" does not exist as an object by itself, it must be abstracted from examples of real or fictional behavior. Turning to physics for another instance, "gravity" is an abstract concept which serves as an explanation for many phenomena, including why apples fall and why the Australians don't drop off the earth even though they are on the bottom! None of these abstractions are concrete, but they are all quite real.

The point is that you are already quite familiar with abstraction, which is **the process of taking many examples with something in common and separating out what they have in common and regarding that as a new conceptual object**. Certainly written mathematics emphasizes abstraction.

Example 2: When you learned to multiply you noticed that $4 \times 3 = 3 \times 4$, $5 \times 8 = 8 \times 5$, $3 \times 7 = 7 \times 3$, and so on. We can write the *one* pattern which gives the essence of these many similar facts: $ab = ba$, for all a and b. This is called the commutative property of multiplication.

Example 3: $3^2 5^2 = 15^2$. $2^2 3^2 = 6^2$. $1.5^2 6^2 = 9^2$. Do you see the one pattern?

These facts are about numbers, but their generalization is a fact about squaring and multiplication: $a^2 b^2 = (ab)^2$, for all a and b.

Recognizing and being able to use a pattern is one stage of understanding. Then, to study the pattern it helps to be able to *express the pattern in abstract mathematical symbolism*. A person who sees one thing (one pattern) where others see many things (many different examples) is more capable of getting to the essential point and dealing with it directly.

Although algebra relies heavily upon abstraction, most algebra courses do not isolate and study it. This course does.

Example 4: You began to reason abstractly in mathematics as a child. As a toddler you learned to say the sounds "1, 2, 3, 4, 5" and you knew that *five* was a sound that pleased adults. With a lot of guidance and effort, you learned to count using those sounds. Then the number *five* became a way to describe a group of objects, for example, "five blocks." Your concept of *five* changed.

In first grade children are asked to add, and many still need to *count* physical objects to be able to *add*. When asked to add 3 and 2 they may need actual objects to find the sum. They may get 3 blocks and 2 blocks and then count the total. Furthermore, they may not be satisfied that the answer with 3 blocks and 2 blocks will necessarily be the same as when they try 3 marbles and 2 marbles! When they finally realize it doesn't matter whether they count blocks, marbles, or fingers, they have begun to abstract (that is, to separate out and consider apart) a new concept of *number*. *Five* becomes a mental object. Previously, the word "five" in "five blocks" described a group -- it was an adjective. We want children to be able to see "five" as a noun, that is, as an object in its own right.

Obviously, you can't see, touch, hear, smell, or taste the object, five. It is a mental object. It is a new reality at a higher conceptual level.

To become good at math, children must learn to add abstract numbers. Nevertheless, we have all seen young children who cling to the crutch of counting on their fingers to add. They resist changing their concept of *number* from adjective to noun.

Is a young child wrong to think of *five* as an adjective? Of course not. But, new concepts allow flexibility. The number *five* may be a noun (a mental object) in addition to retaining its important role as an adjective. This conceptual advance is essential to mathematics.

There is a parallel with advancing to a new concept of algebra. Many students "know" that algebra is about numbers, the same way that children "know" that *five* is for counting. They are right, but in both cases there is a higher conceptual level that is useful and effective.

In this course you will learn that the key concepts of algebra are operations and order, not numbers. You must grasp this higher conceptual level. If you resist moving to the new level (because you already "know" that algebra is about numbers), you will remain handicapped in math like the child who adds on his fingers because he or she "knows" numbers are for counting.

Advanced Concepts. The language of mathematics is specially designed to work with the concepts of algebra. Operations and order become "real" objects when you work with them enough in the proper language.

Example 5: You are familiar with algebraic expressions such as "$2(x + 5)$". Figure 1 illustrates a distinction between the numbers and the operations it expresses.

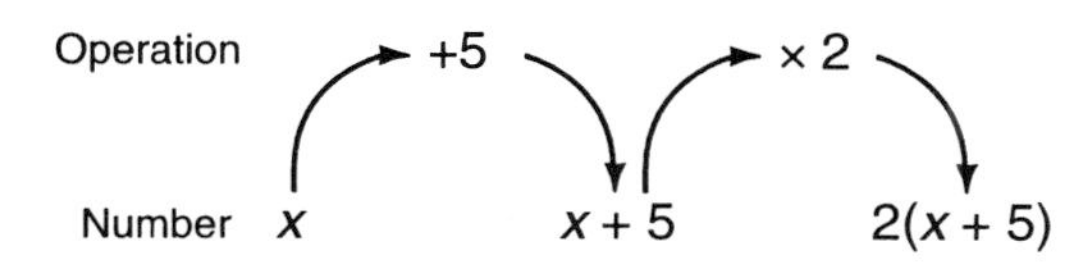

Figure 1. Whatever x is, add 5 to it first and then multiply the result by 2, to obtain $2(x + 5)$.

The algebraic expression "$2(x + 5)$" represents a number (which depends upon the number represented by "x"). At a higher level it also represents the operations "Add 5" and "Multiply by 2" in that order. The conceptual level of *operations* and *order* is extremely important for algebra. The subject which abstracts these ideas is *functions*. We will emphasize functions in this course.

The Relationship of Commands and Statements. In algebra numerous problems can be done by the same method. We can abstract the one method from the many examples of its use. Then we can study the method as a conceptual object. In written mathematics, the *method* will be expressed as a *fact*. We distinguish three primary ways of expressing methods as facts 1) **formulas**, 2) **identities**, and 3) **theorems that relate two equations**.

1. Methods as Formulas. Formulas express operations and methods. Most students find formulas easy to use. Formulas are facts that can tell you what to do.

Example 6: Find the circumference of a circle with diameter 15.

Solution: The circumference is $15\pi = 47.1$.

A formula expresses the method. The formula is "$C = \pi d$." That is a fact.

The meaning of "$C = \pi d$" is not self-contained. You must know the context (circles) and the meaning of the letters "C" and "d". The left side of the equation, C, expresses a problem-pattern ("What is the circumference?") and the right, πd, gives the solution-pattern ("Multiply the diameter by π").

2. Methods as Identities. Identities are a second way to express methods as facts. By definition, an equation with a variable is an identity when it is true for all values of the variable. Here is an example that describes how do a process from arithmetic.

Example 7: Find $5 - (-2)$.

Solution: $5 - (-2) = 5 + 2 = 7$.

An identity expresses the method. The problem requires subtracting a negative number. How do you subtract a negative number from a positive number?

You can express the method in two ways:

a) Express the method in English (as a command).
b) Express the method in mathematical notation (as a statement of fact).

Answer: a) "Drop the negative sign and add."

[This thought can be expressed with different words.]

b) "$b - (-c) = b + c$."

[This thought can be expressed with different letters.]

The sentence "$b - (-c) = b + c$" in part b) is a fact (an identity, which is a type of theorem). It may not look like a command, but there is a plus sign on the right side and that does tell you to add. The *left* side of the identity expresses a problem-pattern and the *right* side tells you the solution-pattern, which tells you what to *do*. "$b - (-c)$" is the pattern behind many problems: $12 -(-6)$, $16.5 -(-3.43)$, etc. If you recognize our way of expressing the pattern, then the identity does tell you what to do.

The equation "$b - (-c) = b + c$" is an identity because the two expressions "$b - (-c)$" and "$b + c$" are not only equal, they are equivalent. Two *expressions* are equivalent when they *always* express the same numerical value, although usually with a different order of operations.

The Use of "=". In algebra, the symbol "=" is used two different ways. It may denote *equality* in order to express facts about numbers, such as "$3 + 4 = 7$" or "$x + 5 = 9$" (which tells you $x = 4$). It may also denote *equivalence* in order to express facts about operations and order, such as "$b - (-c) = b + c$." Learn to grasp the difference.

Sometimes the symbol "=" is misused. In arithmetic, the fill-in-the-blank problem: "56 + 32 = ____" asks you to do the addition indicated on the left. This may teach you the incorrect lesson that "=" is the command, "Do it!" Many students erroneously use "=" to mean "I'm about to do the next step." For example, they may solve the equation "$2x = 30$" with the work "$2x = 30 = x = 15$," which technically asserts the contradiction that x is equal to both 30 and 15. The student *did* the right thing and *wrote* the wrong thing. The sentences "$2x = 30$" and "$x = 15$" *are connected*, but the connective is not "equals" (it is "is equivalent to" which can be denoted by "if and only if", "iff," or "$\Leftrightarrow$", but not by "="). In fill-in-the-blank problems you may learn to *do* what is on the *left* side of an equal sign. But identities are stated with a different intention. They often ask you to *recognize a pattern* on the left of the equal sign and *do* what is on the right!

Example 8: Explain how to divide a fraction by a fraction.

If you know already how to multiply fractions, the process can be expressed in words with the command, "Invert and multiply." For beginners, that command may be far more meaningful than the equivalent abstract formulation in the next theorem. But, as your ability with the language of mathematics improves, the theorem will become a very concise and precise way of expressing how to divide fractions.

Theorem (identity): If b, c, and d are not zero,

$$\frac{(\frac{a}{b})}{(\frac{c}{d})} = (\frac{a}{b})(\frac{d}{c}) = \frac{ad}{bc}.$$

This identity expresses how to divide numbers expressed as quotients or fractions. The *pattern* on the left tells you that. The theorem tells you that, instead of doing the three operations of division expressed on the left, you may do, instead, the two multiplications ("ad" and "bc") and one division (the quotient of those two) expressed on the right.

Instances: $\frac{(\frac{3}{4})}{(\frac{5}{8})} = \frac{3 \times 8}{4 \times 5}.$ $\frac{(\frac{2}{3})}{(\frac{1}{2})} = \frac{2 \times 2}{3 \times 1}.$

This identity and Example 7 are primarily about operations and order.

The purpose of identities
is to provide alternative sequences of operations
for evaluating expressions.

One side of the identity expresses a problem-pattern and the other expresses a solution-pattern. For example, when you study logarithms you will see the "property" (identity): $\log(ab) = \log a + \log b$. This property relates multiplying and then taking the log (on the left) to taking logs and then adding (on the right). The property gives an alternative sequence of operations.

Identities are about operations and order.

Identities express equivalence of expressions.

3. Methods as Theorems That Relate Two Equations. A third way to express a method as a fact is to state a theorem that relates two *equations* (as contrasted with the two *expressions* in a formula or identity). The first example below is a theorem that asserts two equations are equivalent. Two equations with the same unknown are said to be equivalent if they have the same solution.

Most equation-solving steps are thought of as commands: "Add 12 to both sides," or "Square both sides," or "Put all the terms on the left side." Doing and explaining math with commands is natural and fairly successful. It was the primary approach to mathematics for thousands of years and it still is for most students.

But *modern mathematics is not written as commands*. It is written as statements of facts. Teachers can convert it into commands for you, but you should learn to read it yourself.

Example 9: Express as a fact the process of the first step in solving these equations: "$x + 23 = 44$," "$3x + 172 = 586$," and "$x^2 + 40 = 60$."

In each case the first command is "Subtract!" (some number). The number to subtract was whatever number was added on the left. Here is a relevant theorem.

Theorem 1.3.1: For any x, a, and b, $x + a = b$ is equivalent to $x = b - a$.

This theorem relates two equations, "$x + a = b$" and "$x = b - a$". The first gives the problem-pattern: "Unknown plus something equals something else." The second gives the solution-pattern: "Unknown equals something else minus something."

	problem-pattern	solution-pattern
Theorem:	$x + a = b$	$x = b - a$, which tells you to subtract.
Problem 1:	$x + 23 = 44$	$x = 44 - 23$.
Problem 2:	$3x + 172 = 586$	$3x = 586 - 172$ ["$3x$" in the place of "x"].
Problem 3:	$x^2 + 40 = 60$	$x^2 = 60 - 40$ ["x^2" in the place of "x"].

You do not need to know what the letters mean; it is the places of the letters in the pattern that matter. This theorem could also be stated with different letters in the pattern: "For all b, c, and d, $b + c = d$ is equivalent to $b = d - c$." Sometimes we abbreviate theorems by omitting the "For all ..." part: "$x + a = b$ is equivalent to $x = b - a$."

Different theorems that relate two equations would be used to express the processes in remaining steps. For example, to solve "$x^2 = 20$" you would use:

Theorem 1.3.2: $x^2 = c$ is equivalent to $x = \pm\sqrt{c}$.

This says we may replace "$x^2 = 20$" with its solution, "$x = \pm\sqrt{20}$."

Learn how to read patterns expressed in the language of mathematics.

A common abstract problem-pattern is exhibited in one half of a theorem and a useful replacement solution-pattern is exhibited in the other half.

The abstract pattern is the key.

Math literacy largely depends upon pattern recognition.
Facts, written in symbolic patterns, can tell you what to do.

Distinguishing the Three. What is the difference between formulas, identities, and theorems that relate two equations? Formulas and identities are both individual equations, so they have that in common. But they are much different because identities are self-contained and formulas are not. To use a formula you must know the context and what the letters mean. When do you use the formula "$A = 4\pi r^2$"? The formula itself does not tell you (It is the formula for the surface area, A, of a sphere, when r is its radius).

Identities give their own context. You do not need to know what the letters mean to use an identity. The expression on the left side of "$a - (-b) = a + b$" gives the problem-pattern context, and the letters hold places in the pattern but have no meaning.

Theorems that relate two equations also give their own context. One of the two equations expresses the context. For example:

Theorem 1.3.3 (The Quadratic Theorem): The solutions to

$$ax^2 + bx + c = 0 \text{ are given by } \quad x = \frac{-b \pm \sqrt{b^2 - 4ac}}{2a}.$$

The solution-pattern by itself is known as the "Quadratic Formula," which would be useless if you didn't know the quadratic pattern to which it applies.

These three types of facts are profound because they abstract one method from many problems. They finish off entire subjects and permit us to move on to something new and different.

Maybe you don't want to learn to read and write math. Maybe you are like the child who continues to count on his fingers rather than learn to add. But, when you truly grasp how one fact can finish off a whole category of problems you may begin to see the awesome power of the symbolic language of mathematics. That language is worth learning to read and write.

Conclusion. Many students study algebra with a mind-set that is wrong for calculus. The view that algebra is primarily about *numbers* (or letters) must be replaced by an understanding that it is really about *operations* and *order*. Given an algebra problem, the solution *process* is more abstract than the solution itself. Problems using the same process fit a pattern. Abstraction separates out the process as an object of study in its own right.

Algebraic notation is well-designed to express mathematical methods as facts. Three primary ways are: 1) **formulas** (individual equations where the letters have meaning that must be understood from outside the equation), 2) **identities** (individual equations where the meaning is self-contained and the letters are placeholders that do not have meaning), or 3) **theorems that relate two equations.**

Learn to read symbolic mathematics. Learn to fully grasp algebra by reviewing it with an eye toward operations, order, and methods -- concepts above the conceptual level of numbers.

Terms: abstract (verb and adjective), abstraction (the process), concrete, concept, command, statement of fact, identity, equivalent (expressions, sentences).

****

Exercises for Section 1.3, "Abstraction":

A1. In previous math courses did you (personally) learn primarily by *reading* your text or by *doing* what your teacher showed you how to do? [Or some other way? There is no "correct" answer to this question.]

A2.* True or False: Methods can be expressed as facts.

A3.* This section lists three categories of ways in which methods can be expressed as facts. What are the three?

A4.* Identities express equivalence of ________________.

A5.* What word expresses the opposite of the adjective *abstract*?

A6. People sometimes resist thinking at a higher level of abstraction. Which example in the text illustrates that resistance in the context of low-level mathematics?

A7. Write out, in English, how you would say these aloud:
a) $a/b = c$. b) $a/(b/c) = ac/b$. c) $a(b + c) = ab + ac$.

A8. a) What will your calculator do if you use the subtraction key twice in the keystroke sequence for subtracting -2 from 5? [i.e. 5 - - 2]
b) What sequence of keystrokes should you use?

A9. Simplify: a) $a/(b/c)$ b) $(a/b)/c$.

A10. $1 \times 5 = 5$. $1 \times 97 = 97$. State the corresponding abstract fact about multiplication by 1.

A11. $17/1 = 17$. $324/1 = 324$. State the corresponding abstract fact about division by 1.

A12. $99 + 0 = 99$. $47 + 0 = 47$. State the corresponding abstract fact about addition of zero.

A13. Give one simple mathematical abstract concept and the context from which it is abstracted. [Use one noted in this section. See also Problem B42.]

^ ^ ^ ^ ^ ^ ^ ^

B1. a)* Define "abstract" (the adjective) or "abstraction" (the process) [You may look them up.] b) Name one abstract concept which is not mathematical.
c) Would you say that the concept you named in part b) was "real" (that is, a part of reality)?
d) Can you see, touch, hear, smell, or taste it?

B2.* a) Which types of things can be connected by the symbol "="? b) What are its two uses in algebra? [Answer with more that just the word "equals".] c) What is the meaning sometimes erroneously given it by students who work only with commands?

B3.* Name three ways in which methods can be expressed as facts and give one example of each.

B4.* a) What are identities about? b) What is the purpose of identities?

B5.* Students who think that algebra is about numbers must change their mind set and understand that algebra is really about ____________________.

^ ^ ^ ^ State the *algebraic formulation* of the procedure used to evaluate these expressions. [That is, state an identity with the appropriate problem-pattern and solution-pattern. The solution pattern should express only operations with positive integers. In the solution pattern, subtraction, if used, should express subtraction of smaller positive integers from larger positive integers. See Examples 7 and 8.]

B6. $19 - 40$.	B7. $(3/4)/5$.	B8. $(-3)7$.
B9. $4/(-6)$	B10. $12 - (-8)$.	B11. $-6 - 5$.
B12. $(-5)(-3)$.	B13. $2/(3/4)$.	B14. $1/3 + 2/5$.
B15. $-34 + 12$.	B16. $(5/6)/(7/8)$.	B17. $(-3)/(-6)$.

^ ^ ^ ^ State the *algebraic formulation* of the procedure used to evaluate these expressions (in terms of simpler procedures, as in Examples 7 and 8).

B18. $(4/5)/7$.	B19. $12 - 43$.	B20. $3/(-5)$.
B21. $23 - (-4)$.	B22. $5(-8)$.	B23. $-4 - 9$.
B24. $(2/3)/4$.	B25. $1/2 + 3/5$.	B26. $3/(5/7)$.
B27. $(5/7)/(3/5)$.	B28. $-54 + 19$.	B29. $(-52)/(-13)$.

B30. Give the symbolic fact which expresses how to subtract greater positive numbers from lesser positive numbers.

B31.* a) In the discussion of abstract numbers (Example 4), numbers treated as adjectives are abstracted to numbers treated as ____________.
b) In the discussion of methods and theorems, methods treated as commands are abstracted to methods treated as ____________________.

B32. How do we add a negative number to a negative number? a) Express your answer in English. b) Express your answer in mathematical notation.

B33. How do we multiply fractions? a) Express your answer in English.
b) Express your answer in mathematical notation.

B34. How do we divide a fraction by a number? a) Express your answer in English.
b) Express your answer in mathematical notation.

B35. a) What would you *do* to solve the equation "$x - 34 = 457$"? [Write your answer as a command.] b) State an abstract *fact* which corresponds to your method.

B36. a) What would you *do* to solve the equation "$x - 92 = 398$"? [Write your answer as a command.] b) State an abstract *fact* which corresponds to your method.

B37. a) What would you *do* first (one step) to solve the equation "$3(x - \pi) = 18$"? [Write your answer as a command.] b) State an abstract *fact* which corresponds to your method.
c) Division by zero is undefined. Does your abstract statement acknowledge that?

B38. a) What is the relationship between commands and statements of fact in written mathematics?
b) Give an example which illustrates you understand that distinction.

B39. What is the general purpose of abstraction?

B40. Some theorems assert the equivalence of two sentences. How are theorems stated in order to help solve problems?

B41. Give an example of a method in mathematics which is not in any of the three major categories discussed in this section.

B42. Short essay: Negative numbers are abstract. You cannot have a pile of -5 blocks. In what context do negative numbers become "real"?

B43. a) What would you *do* to solve the equation "$42x = 3218$"? [Write your answer as a command.]
b) State an abstract *fact* (with variables) which corresponds to your method.

B44. What does the *fact* "$|x| < c$ if and only if $-c < x < c$" tell you to *do* to begin to solve the inequality $|2x - 3| < 12$?

B45. What does the *fact*: if $b \neq 0$ and $d \neq 0$, then $(a/b)+(c/d) = (ad + bc)/(bd)$ say you may *do* to add fractions? Explain in complete English sentences.

Section 1.4. Variables, Patterns, and Theorems

This section is about how to read mathematics to learn mathematics. Most students learn mathematics by imitating their instructor. Many do not, or cannot, learn by reading their math texts. The reason is often lack of understanding of the profound, but simple, concept of a placeholder. This section explains and illustrates the use of placeholders, which are also called "dummy variables." This section also explains why the most important operation in an expression is the last operation.

Placeholders (Dummy Variables). "Placeholder" and "dummy variable" are synonyms. Dummy variables are variables (letters) used in sentences about operations and order to hold places where *any* number or expression can fit. There is no parallel in English to the mathematical concept of dummy variables. Unlike English, algebraic notation is well-designed to express thoughts about operations and order.

If the sentence is about a *particular* number, then the variable is not a "dummy" variable.

Example 1: "$x + 3 = 12$" is an equation about a number represented by the letter "x". In the equation, $x = 9$. In this example "x" is not a dummy variable; it is an "unknown" -- a name for a particular number.

Example 2: "$3(x + 1) = 3x + 3$" is a sentence which is not about the number represented by "x". It is true for all x, so it tells nothing about which number is represented by "x". This equation is about multiplication and addition. In it, "x" is a dummy variable which holds a place on each side of the equation. Any number or expression representing a number could replace "x" and the sentence would still be true. Notice the pattern and the placeholder:

$$3(x + 1) = 3x + 3$$
$$3(z + 1) = 3z + 3$$
$$3(17 + 1) = 3(17) + 3$$
$$3(y^2 + 1) = 3y^2 + 3$$
$$3(\blacksquare + 1) = 3\blacksquare + 3.$$

In the last equation any numerical expression at all can replace the black box, as long as it replaces the black box *every place* it appears. The black box is a placeholder. In the first line with "x", any numerical expression at all can replace the "x", as long as it replaces "x" *every place* it appears.

A dummy variable can be replaced by any expression,
(including a different letter)
as long as each occurrence of that dummy variable
is replaced by the same expression.

Dummy variables are used to represent patterns of operations and order. In Example 2, "$3(x + 1)$" is an example of the abstract pattern of operations "$a(b + c)$," which has the possible replacement pattern, "$ab + ac$," according to the next property.

(1.4.1) The Distributive Property: For all a, b, and c, $a(b + c) = ab + ac$.

A key to understanding the written form of any such theorem or property is that **any letter in it can represent any expression, no matter how simple or complex, and no matter what the letters are.**

Example 3: Write as a sum: $x(x + 7)$.

We know this is $x^2 + 7x$, which is a sum since addition is last.

The problem fits the pattern, $a(b + c)$, on the left side of the Distributive Property, and the right side gives the alternative pattern, $ab + ac$, of the solution. To "multiply out" an expression we read the property left-to-right.

Example 4: Factor "$(x^2 - 2)\cos x + (x^2 - 2)(1 - 3 \cos x)$."

To factor an expression means to write it as an equivalent product. The expression is long and messy, but, as it stands, the expression is regarded as a *sum*, because the last operation is addition. It fits the pattern of the right side, "$ab + ac$," of the Distributive Property. The key is that the factor "$x^2 - 2$" appears in both terms of the sum, just as "a" appears in both terms of the abstract version, "$ab + ac$."

$$(x^2 - 2)\cos x + (x^2 - 2)(1 - 3 \cos x)$$
$$= (x^2 - 2)[\cos x + 1 - 3 \cos x)$$
$$= (x^2 - 2)(1 - 2 \cos x)$$

which is a product, as desired. We read the property right-to-left for factoring.

Equality and Equivalence of Expressions. The type of equality in the Distributive Property is different from the type of equality in the sentence "$x + 3 = 12$." Only a certain value of "x" makes "$x + 3 = 12$" true (x must be 9). However, in the Distributive Property the two sides are *always* equal. If the two sides of an equality are always equal, the equality is called an identity, and the expressions on the two sides are said to be equivalent expressions.

You are responsible for recognizing the difference between equations about operations which are always true and equations about numbers which are only true for certain values of the variable. Equations about operations are expressed with dummy variables.

Example 5: Is the variable a dummy variable? (Unknowns are not dummy variables.)

a) $2x + x = 3x$.
b) $2x + x = 5x - 6$.
c) $x^2x = 8$.
d) $x^2x = x^3$.

Answers: In part (a), "$2x + x = 3x$," the letter is a dummy variable because the equation is always true. The same meaning could be expressed using different letters: "$2a + a = 3a$." "$2b + b = 3b$." These letters are dummy variables and these sentences are not about the numbers represented.

In part (b), "$2x + x = 5x - 6$," the letter is an unknown, not a dummy variable, because the equation is not always true. It expresses something about the number represented by "x": $x = 3$.

In part (c), "$x^2x = 8$," "x" is an unknown. We can solve it to discover $x = 2$.

In part (d), "$x^2x = x^3$," the variable is a dummy variable. Every value of x makes it true, so it is not about "x". The equation is about powers, multiplication, and order. Such facts about operations and order can be expressed with different letters: "$a^2a = a^3$." "$c^2c = c^3$."

Example 6: Which letters are used as placeholders (dummy variables)? (Unknowns are not placeholders, rather, they are names for particular numbers.)

sentence	answer
a) $x + 5 = 12$.	unknown
b) $x + a = b$ is equivalent to $x = b - a$.	all are placeholders
c) $1x = x$.	placeholder
d) $0x = 0$.	placeholder

One way to tell if a variable is a placeholder is to decide what the sentence is about. If it is about a particular number, the variable is not a placeholder. For example, Part (a), "$x + 5 = 12$," is about the number x: $x = 7$. However, if the sentence is a generalization about operations and order, then the variable is a dummy variable -- a placeholder. Part (b) is about the relationship between addition and subtraction. Part (c) is about

To say two equations are *equivalent* means they are true for the same values of the variables. Therefore they have the same solutions. Another convenient way to assert equivalence of equations is to use the phrase "if and only if," which is commonly abbreviated "iff" where the 2 f's are the key. "Iff" is short, but it is still pronounced "if and only if."

multiplication by 1.

The role of dummy variables is to hold the places of numbers in two equivalent patterns, as in Parts (b) and (c). Therefore, placeholders usually appear twice or more, but not always, as Part (d) shows. Any number could be put in for "x" and "$0x = 0$" would still be true. The same meaning is expressed by "$0b = 0$."

The term unknown is used to specify the variable for which you are supposed to solve.

Example 7: Consider the sentence, "$ax + b = c$." Which letter is the unknown?

According to traditional usage, "x" is the unknown. The sentence has four letters, and, technically, you could solve for any of them. But we usually use "x" to indicate which one is intended to be the unknown. The values of the other variables are not given, but treated as known in theory for purposes of solving for x. In this sentence "a", "b", and "c" are parameters. They play two different roles at two levels of abstraction in the same sentence. On the one hand, they can be regarded as known and fixed. On the other hand, they can vary to represent different problems, so they create a family of similar sentences. Variables which are treated as fixed in one context and variable in another are called parameters.

Unknowns are letters used as
***names* for particular numbers in sentences *about numbers*.**

Dummy variables are used in generalizations as
***placeholders* for general numbers (expressions)**
in sentences *about operations and order*.

One-Operation Patterns. Theorems are one way to express methods of mathematics. Many theorems employ dummy variables to express a pattern which involves only one operation.

Example 8: Here is a sample theorem about the relationship of exponentiation and taking the common logarithm:

Theorem 1.4.2: $10^a = b$ is equivalent to $a = \log b$.

The pattern "10^a" expresses only one operation, exponentiation. The pattern "$\log b$" expresses only one operation, taking logarithms. The letters "a" and "b" are dummy variables. You may replace "a" by "x" or any other letter as long as you do so in both places. Similarly with "b".

The two halves handle two types of problems. If a problem has the pattern "$10^a = b$," and the unknown is in the place of "a", you may replace the equation with the equivalent equation with pattern "$a = \log b$." On the other hand, if the problem has pattern "$\log b = a$," with the unknown in the location of "b", you may replace the equation with the equivalent equation with pattern "$b = 10^a$."

You are *not* expected to have already mastered logarithms and exponents. But you are expected to learn to read Mathematics. Theorems are sentences in Mathematics. This one tells you something about the relationship of logarithms to exponentiation. In English you can read and understand sentences about subjects you have never studied before. This section and this example are intended to help you learn to read Mathematics -- even new mathematics you have never seen before.

Example 8, left-to-right: Solve "$10^x = 70$."

This fits the problem-pattern, "$10^a = b$," on the left of Theorem 1.4.2. It tells us to replace "$10^x = 70$" with the equivalent equation "$x = \log 70$" ["x" is "a" and "70" is "b".]

Example 8, right-to-left: Solve "$\log x = 2.13$."

This fits the pattern on the right, "$a = \log b$." The theorem tells us to replace "$\log x = 2.13$" with the equivalent equation "$x = 10^{2.13}$" ["x" is "b" and "2.13" is "a".]

Calculator Exercise 1: Learn the keystroke sequences for evaluating "$\log x$" and "10^x." Try to reproduce these results:

$\log 70 = 1.845.$
$10^{2.13} = 134.9$
$10^{-1.2} = .063.$

The logarithm of a negative number is not a real number, so many calculators return an error message if you try to take "log(-2)."

Many texts would write our sample theorem as two theorems, one to emphasize "a" as the unknown, and another to emphasize "b" as the unknown.

<u>Theorem 1.4.2B</u>: $10^x = y$ iff $x = \log y$.

<u>Theorem 1.4.2C</u>: $\log x = y$ iff $x = 10^y$.

Versions B and C have the advantage that they emphasize the use of the result by using the traditional "x" for the unknown. However, together they say nothing that was not already said in the first version. In all three versions the letters are simply dummy variables. The variables can represent any number or expression, so "a" can be "x" and "b" can be "y" (Version B) or "a"

can by "y" and "b" be "x" (Version C). Furthermore, "a" and "b" can represent more complex expressions.

Example 8, continued: Solve $10^{x^2} - 4$.

$$10^{x^2} - 4 \quad \textit{iff} \quad x^2 - \log 4 \ .$$

The theorem takes us this far. Now we must use some other method to handle the squaring (problem B39).

This example makes the point that

Theorems are often about one operation at a time,
and are applicable when that operation is *last*.

To read expressions in mathematics, you must be able to recognize which operation is last. A one-operation pattern of an expression is an abstract symbolic expression which exhibits only the last operation.

Example 9: Identify the one-operation pattern:

expression	pattern
a) $\log(x - 7)$	$\log x$ or $\log a$
b) $\lvert x^2 \rvert$	$\lvert x \rvert$
c) $x(x + 7)$	ab
d) $x^2 + 7$	$x + a$ or $a + b$

Of course, since patterns are expressed using dummy variables, your answers may employ different letters than the answers above. What matters is that only the last operation is exhibited. The point is that any theorem expressing a one-operation pattern (such as Theorem 1.4.2 above) applies to more complex expressions with several operations, as long as the one-operation pattern is right.

Example 10: Solve "$\log(x - 3) = 1.4$."

Because the one-operation pattern is "$\log x$," Theorem 1.4.2C applies.

$\log(x - 3) = 1.4$ iff $x - 3 = 10^{1.4}$ (= 25.12).

So, $x = 28.12$.

Example 11: Solve "$3(10^{x-2}) = 60$."

On the left, multiplication is last. The pattern of that expression is *not* "10^a". The one-operation pattern is "ab" or "ax" because, most importantly, the expression is a product. Therefore, to solve "$3(10^{x-2}) = 60$," divide first.

$3(10^{x-2}) = 60$ iff $10^{x-2} = 20$.

Now, this new equation does have the one-operation pattern "$10^a = b$," so now Theorem 1.4.2 applies.

$10^{x-2} = 20$ iff $x - 2 = \log 20$
iff $x = \log 20 + 2 = 3.301$.
So, $x = 3.301$ is the solution.

Example 12: State the theorem which expresses how to do the first step in problems like Example 11: "$3(10^{x-2}) = 60$."

For the first step, only the multiplication is relevant. The one-operation pattern is a product. Here is how the relevant theorem might be stated:

Theorem 1.4.3: If $a \neq 0$, $ab = c$ iff $b = c/a$.

This could be rewritten with other letters, depending on which place holds the unknown. One version emphasizes that sentences with multiplication can be replaced by sentences with division:

Theorem 1.4.3B: If $a \neq 0$, $ax = c$ iff $x = c/a$.

This version is written to apply to problems like "$3(10^{x-2}) = 60$" where an unknown expression is multiplied by a number. The next version emphasizes problems where the unknown is divided by a number.

Theorem 1.4.3C: If $a \neq 0$, $x/a = b$ iff $x = ab$.

Note that the first version says everything versions B and C say. Versions B and C use "x" to emphasize the location of the unknown.

Example 13: Solve "$10^{x/3} = 100$."

Solution process: Recognize the one-operation pattern. Exponentiation is last. Therefore the pattern is "$10^a = b$" or "$10^x = y$", the pattern of Theorem 1.4.2 and 1.4.2B.

$10^{x/3} = 100$ iff $x/3 = \log 100$.
iff $x = 3 \log 100 = 6$.
The solution is $x = 6$. The process in the last step is expressed by version C.

Often, theorems are stated to address one operation and one step.

Example 14: State the theorem that expresses the process used in the first step of the following solution of "$(x - 3)(x - 4) = 0$."

Solution: $(x - 3)(x - 4) = 0$ is equivalent to

$$x - 3 = 0 \text{ or } x - 4 = 0.$$

These are easily solved. The solution is

$$x = 3 \text{ or } x = 4.$$

The one-operation pattern of the problem was "$ab = 0$" which represents a product equal to zero. The relevant theorem uses that problem-pattern:

<u>(1.4.4) The Zero Product Rule</u>: For all a and b, $ab = 0$ iff $a = 0$ or $b = 0$.

This theorem gives only the first step of the solution process. The process for solving "$x - 3 = 0$" and "$x - 4 = 0$" was expressed in Theorem 1.3.1 on subtraction and addition.

There is no requirement that dummy variables represent short expressions.

Example 15: Solve $(10^{x/3} - 100)[3(10^{x-2}) - 60] = 0$.

The expression on the left is long, but that is not important for the first step. The last operation is multiplication and the right side is zero. So the problem-pattern is simply "$ab = 0$." The Zero Product Rule applies:

$$(10^{x/3} - 100)[3(10^{x-2}) - 60] = 0$$
$$\text{iff } 10^{x/3} - 100 = 0 \text{ or } 3(10^{x-2}) - 60 = 0.$$

Consider the first new problem: $10^{x/3} - 100 = 0$. This is equivalent to

$10^{x/3} = 100$, which was solved in Example 13.

Similarly, the second problem arising from the Zero Product Rule,

$$3(10^{x-2}) - 60 = 0$$

is equivalent to

$3(10^{x-2}) = 60$, which was solved in Example 11.

The solution to Example 15 is, therefore, $x = 3.301$ or $x = 6$.

<u>**Identities**</u>. Identities are one of the three ways mentioned in Section 1.3 to express methods of mathematics. Identities are equations that state that two expressions are equivalent. The two equivalent expressions always express the same number, but with a different order of operations.

The purpose of identities
is to permit replacement of expressions
with equivalent expressions
which express a different order of operations.

Example 16: Solve $10^x 10^{2x} = 50$.

None of the one-operation solution patterns we have given permits us to replace this equation with a simpler equation. This is a common difficulty with equations. This is where identities play an important role.

Theorem 1.4.5 (an identity): For all a and b, $10^a 10^b = 10^{a+b}$.

Since "a" and "b" are dummy variables, they can stand for "x" and "$2x$" in the expression in Example 16. Therefore,

$$10^x 10^{2x} = 50 \text{ is equivalent to } 10^{3x} = 50.$$

By Theorem 1.4.2 on exponentiation,

$$3x = \log 50$$
$$x = (\log 50)/\ 3 = .566.$$

When a problem does not fit a pattern of a known theorem, identities are used to change the pattern of operations to a new pattern which does fit some theorem.

Conclusion. Dummy variables are used in theorems and identities as placeholders for *general* numbers in sentences about operations and order. Some important patterns (such as the Distributive Property) relate two or more operations, but many emphasize only one operation. If a pattern apparently expresses one operation, it also applies to more complicated expressions in which that operation is last.

Unknowns are letters used as names for *particular* numbers in sentences about numbers. Variables which are treated as particular in one context and general in another are called parameters.

Terms: dummy variable, placeholder, unknown, parameter, equivalent expressions, equivalent equations, identity, generalization, one-operation pattern, Zero Product Rule.

****

Exercises for Section 1.4, "Variables, Patterns, and Theorems":

A1.* a) Which type of variable is used to state theorems?
b) Which type of variable is used to state identities?
c) Is there any other type of variable?

A2.* When "x" is used in a sentence about an unknown number, what type or variable is "x"?

A3.* If "x" is used in a sentence about an operation, what type of variable is "x"?

A4.* A theorem about one operation applies to more complex expressions with several operations. When?

A5.* a) What types of thoughts are expressed by sentences with unknowns?
b) What types of thoughts are expressed by sentences with dummy variables?

A6.* True or false: "x" is always a placeholder.

A7.* What does the command, "Factor (some expression)" mean?

^ ^ ^ ^ Express the one-operation pattern:

A8. a) $x(x - 5)$ b) $x^2 - 5x$ c) $(x - 2)^2$. d) $x^2 - 2$

A9. a) $2 \sin x$ b) $\sin(2x)$ c) $|x - 2|$ d) $|x| - 2$

A10. a) $\log(3x + 1)$ b) $\log(3x) + 1$ c) $3 \log(x + 1)$ d) $3(\log x + 1)$

A11. a) $(x^2 - 2)/(x + 1)$ b) $(x^2 - 2)/x + 1$ c) $x^2 - 2/(x + 1)$

^ ^ ^ ^ Is "x" a placeholder in the given sentence?

A12. a) $3x = 60$ b) $5x = 2x + 3x$ c) $x(x + 1) = 2x + 2$.

A13. a) $4(x + 2x^2) = 4x + 8x^2$ b) $4(x + 2x^2) = 16$ c) $x + 9 = 42$.

^ ^ ^ ^ Use Theorem 1.4.2 to solve these equations. [You can use the given solution with two significant digits in brackets to check your method. Report your solution with at least three significant digits.]

A14. $10^x = 20$ [1.3].

A15. $10^x = 500$ [2.7].

A16. $10^{2x} = 3000$ [1.7].

A17. $10^{x-5} = .01$ [3.0].

A18. $\log x = 2.5$ [320].

A19. $\log x = -1.8$ [.016].

A20. $\log(4x) = 3$ [250].

A21. $\log(x - 2) = 1.3$ [22].

^ ^ ^ ^ Rewrite these expressions using the identity "$10^a 10^b = 10^{a+b}$."

A22. $10^3(10^{1.2})$.

A23. $10^4(10^{2.7})$.

A24. $10^x(10^2)$.

A25. $10^x(10^5)$.

A26. $10^x(10^{3x})$

A27. $10^x(10^{-1})$.

^ ^ ^ ^ ^ ^ ^ ^ ^

B1.* Medium Length Essay: Distinguish between dummy variables and unknowns. Which is a "placeholder," and what does "placeholder" mean?

B2.* Short essay: What is a "one-operation pattern"? Why are one-operation patterns important?

B3.* How can a pattern with only one operation apply to an expression which expresses several operations?

B4.* What are algebraic generalizations with dummy variables about?

B5.* What is the purpose of identities?

B6.* How important is the choice of letter when the variable is a "dummy" variable?

B7.* Explain how the type of variable in an equation relates to what the equation is about.

B8.* a) Which symbol is usually used to assert two expressions are equivalent?
b) Which symbol or phrase can be used to assert two sentences are equivalent?
c) (Short essay) Distinguish "equivalent" from "equal" expressions. Also, distinguish the types of variables they use.

^ ^ ^ ^ Here is a theorem for positive integers represented by "n":
"$1 + 2 + 3 + \ldots + n = n(n + 1)/2$."
Read it and use it to rewrite the value of the given expression.

B9. a) $1 + 2 + 3 + \ldots + 100$. b) $1 + 2 + 3 + \ldots + k$.
B10. a) $1 + 2 + 3 + \ldots + 250$. b) $1 + 2 + 3 + \ldots + j$.
B11. $1 + 2 + 3 + \ldots + 2n$.
B12. $1 + 2 + 3 + \ldots + n + n{+}1 + \ldots + n{+}5$.
B13. $101 + 102 + 103 + \ldots + 500$
B14. $2 + 4 + 6 + \ldots + 2n$.

^ ^ ^ ^ Here is a property of powers: $\dfrac{a^p}{a^r} = a^{p-r}$.

Read it and use it to rewrite the given expression.

B15. a) $\dfrac{x^3}{x^2}$ b) $\dfrac{x^a}{x^b}$ B16. a) $\dfrac{a^5}{a^2}$ c) $\dfrac{a^b}{a^c}$

B17. $\dfrac{x^2}{x^{1/3}}$ B18. $\dfrac{x^3}{x^{1/2}}$ B19. $\dfrac{x^{1/2}}{x^{1/3}}$ B20. $\dfrac{x^{2/3}}{x^{1/2}}$

B21. $\dfrac{x^2}{x^{-1}}$ B22. $\dfrac{x^4}{x^{-3}}$

B23. Here is a sample theorem about the natural logarithm function:
"$e^a = b$ iff $a = \ln b$." ["e" is the particular number 2.718... and not a dummy variable.]
a) Rewrite it to emphasize problems of the form "$e^x = c$."
b) Rewrite it to emphasize problems of the form "$\ln x = c$."

^ ^ ^ ^ Read the theorem in B23 and use it to solve the following equations. [You can use the given solution with two significant digits in brackets to check your method. Report your solution with at least three significant digits.]

B24. $e^x = 5$ [1.6]. B25. $e^x = 100$ [4.6].
B26. $\ln x = 7$ [1100]. B27. $\ln x = -3$ [.050].
B28. $e^{x-4} = 150$ [9.0]. B29. $e^{3x} = .34$ [-.36].
B30. $\ln(2x) = 5$ [74]. B31. $\ln(x + 4) = 1.1$ [-1.0].

^ ^ ^ ^ Here is a sample identity: "$\sin(2\theta) = 2(\sin\theta)(\cos\theta)$."
Read it and use it to rewrite the given expression.

B32. a) $\sin(2a)$. b) $\sin(40°)$.
B33. a) $\sin(2x)$. c) $\sin(100°)$.
B34. $\sin(2(x + h))$. B35. $\sin(2(\theta + 1))$.
B36. $\sin(4\theta)$. B37. $\sin(4x)$.

B38. Here is a sample theorem: $|x| < c$ iff $-c < x < c$.
a) Which letter is intended to represent the unknown?
b) Use the theorem to rewrite "$|2x - 7| < 12$.
c) Restate the theorem with some other letters.

B39. a) State the theorem used to solve problems like "x^2 = log 4." b) Solve it.

B40. In which of the following sentences is the variable x a placeholder (dummy variable)?
a) $x(x + 1) = 3x$.
b) $x(x + 1) = x^2 + x$.
c) $x + 0 = x$.
d) $x + 0 = 0x$.

B41. In which of the following sentences is the variable x a placeholder (dummy variable)?
a) $x^2 \geq 0$.
b) $x \geq 0$
c) $(x + a)(x - a) = 0$.
d) $(x + a)(x - a) = x^2 - a^2$.

B42. In which of the following sentences is the variable x a placeholder (dummy variable)?
a) $x - a = b$ iff $x = b + a$.
b) Let $f(x) = x^2$.
c) $3x - 15 = 50$.
d) $x - a = -(a - x)$.

B43. a) Invent your own sentence which uses "x" as an unknown.
b) Invent a second, similar, sentence which uses "x" as a dummy variable.

B44. a) In the equation "$mx + b = c$," which letter is the unknown?
b) How can you tell? c) In the equation "$a/b = c/d$" which letter is the unknown? Discuss your answer.

B45. a) In the equation "$ax^3 + bx = d$" which letter is the unknown?
b) How can you tell? c) In the equation "$ab + cd = 1$," which letter is the unknown? Discuss your answer.

B46. Authors tend to avoid using certain letters to express patterns in theorems about operations on the real numbers. Which ones?

B47. Define "unknown" (as a type of variable).

B48. True or false? Theorems usually use *capital* letters to represent numbers.

B49. Solve: $10^x 10^{x/2} = 40$.

B50. Solve: $10^x 10^{x-1} = 5000$.

B50. Here is a sample theorem: "If $n \neq 0$ and $f(x) = x^n$, then $f'(x) = nx^{n-1}$."
Use it to find $f'(x)$ if $f(x)$ is
a) x^3
b) x^{-1}
c) x^p.

B51. Use the identity in problems B32-36 and the fact that $\sin^{-1} c$ is a solution to "$\sin x = c$" find one solution to: $(\sin x)(\cos x) = .34$.

B52. Use the identity in problems B32-36 and the fact that $\sin^{-1} c$ is a solution to "$\sin x = c$" to find one solution to: $3(\sin x)(\cos x) = .82$.

Section 1.5. Expressions, Equations, and Graphs

In order to communicate well in Mathematics we need to agree on the usage of some terms. This section compares and contrasts the important terms *expression* and *equation*. We directly "evaluate" expressions; we "solve" equations. These two processes are "inverses" of one another (and we will explain what that means). Their connection can be illustrated with graphs.

Expressions. Expressions are the nouns and pronouns of Mathematics. Expressions include numbers. Expressions may be written with variables, which are letters used to represent numbers.

Example 1: "5", "72 - 12", and "log 2.3" are expressions. They are nouns. They express particular numbers without using variables.

"$x + 5$", "ab", and "log x" are expressions with variables. They are pronouns.

Expressions with a variable can be interpreted two ways at different levels of abstraction. At the lower level, we think of expressions as representing a number which you could *evaluate* (determine the value of) if you knew the value of the variable.

Example 2: Consider the expression "$3x + 5$." Without knowing the value of x, it cannot be evaluated. However, given $x = 1$, its value is 8. Given $x = \pi$, its value is $3\pi + 5$ (which cannot be simplified without using decimals). The *single* expression expresses a single evaluation process, but it creates *many* number pairs.

At the higher level of abstraction, an expression with a variable expresses the *one* thing that the *many* number pairs have in common: a sequence of operations (an evaluation *process*). This interpretation is fundamental to the major activities of algebra such as solving equations and doing word problems. The algebraic subject which concentrates attention on the operations is *functions*, which is why there will shortly be a section on functions (Section 2.1).

Sentences. An equation is a sentence in Mathematics in which the verb *equals* connects two expressions.

Example 3: "$x + 7 = 2x$" is a sentence and an equation.
"$x + 7$" is an expression.
"$2 \sin x$" is an expression.
"$2 \sin x = \cos x$" is an equation -- a sentence.

It is very important to distinguish between expressions and equations. Expressions tell you what to do to evaluate themselves, but equations do not tell you what to do to solve themselves.

Example 4: The *expression* "$3x + 1/x$" tells you how to find its value when you are given x. It says to "Multiply the number by 3 and add to that the reciprocal of the number." If $x = 4$, its value is $3(4)+ 1/4 = 12.25$. That computation is direct because the process for doing it is given in the problem itself.

In contrast, the *equation* "$3x + 1/x = 5$" does not tell you how to find x. The process given by the expression "$3x + 1/x$" is *not* the solution process. Relating equations to solution processes is part of the purpose of algebra (Problem B43). When the process you need to do is not indicated by the problem, we call the problem indirect. Solving is more difficult than evaluating.

In this text we use the words *evaluate* and *solve* with mathematical meanings. We "evaluate" expressions; we "solve" equations for unknowns.
If the process for obtaining a desired number is suggested by symbols, words, or a basic formula, we call that a direct process. Evaluating an expression is a direct process. If the process for obtaining a desired number is not given, the problem is indirect. Algebra provides indirect methods for solving equations.

Expressions and equations are also different because processes which apply to equations may not apply to expressions.

Example 5: In the context of solving an equation it is acceptable to apply the process, "Divide by 2." The equation "$10 = 2\pi r$" is equivalent to the equation "$5 = \pi r$." But it is not acceptable to simplify expressions by applying the process, "Divide by 2." The expression for the circumference of a circle, "$2\pi r$," is not equivalent to the expression "πr."

Example 6: To solve the *equation* "$x/\sqrt{x} = 7$" you may "Square it" to obtain "$x^2/x = 49$," which is then easy to solve: $x = 49$. However, to simplify the *expression* "$x/\sqrt{x}$" it is not legal to "Square it." "$x/\sqrt{x}$" is *not* equivalent to "x^2/x" (which is equivalent to "x"). ("$x/\sqrt{x}$" is equivalent to "$\sqrt{x}$".)

When two different expressions always represent the same number, regardless of the value of the variable, they are said to be equivalent. When two distinct expressions are equivalent their numerical values are the same but the order of operations is not.

Example 7: Explain how to evaluate "32 - 37."

To do it most people actually evaluate "37 - 32" and then change the sign. This process can be expressed abstractly with an identity, using letters to replace the particular numbers: "$a - b = -(b - a)$." The problem-pattern "$a - b$" is the pattern in "32 - 37." The solution-pattern is useful because it expresses subtraction of the smaller number, which we learn how to do before learning how to subtract the larger number. The numbers expressed in the two patterns are the same, but the order of operations is not.

Example 8: Explain how to multiply fractions.

" $(\frac{a}{b})(\frac{c}{d}) = \frac{ac}{bd}$. " The left expresses two divisions and then a multiplication. The right expresses two multiplications and then a division. For any values of a, b, c, and d the values of the two expressions are the same, but the order of operations is not.

The fact that two expressions can be different and yet always express the same numbers proves that **expressions do not merely express numbers**. They also express operations and order. We emphasize this higher conceptual level.

The Terminology of Graphs. The setup for a "rectangular coordinate graph" (or, simply, "graph") consists of two perpendicular number lines with their zeros at the same place, the origin. Each number line is called an axis and exhibits a scale, which is an indication of which locations correspond to which numbers. There is no requirement that the horizontal and vertical scales be the same. The combination of the axes and their scales is called the axis system (Figures 1A and 1B). The two axes naturally divide the plane into four regions called quadrants, which are sometimes designated by Roman numerals as in Figure 1A.

How are points located in the two-dimensional plane? Think about what you would do, using blank paper, to sketch a graph and mark the location of the point (3, 5) on it. First you would draw perpendicular number lines and then you would mark points on the axes (Figure 2). Then you would draw the vertical line $x = 3$ and the horizontal line $y = 5$. The unique point where those two lines meet is (3, 5) (Figure 2).

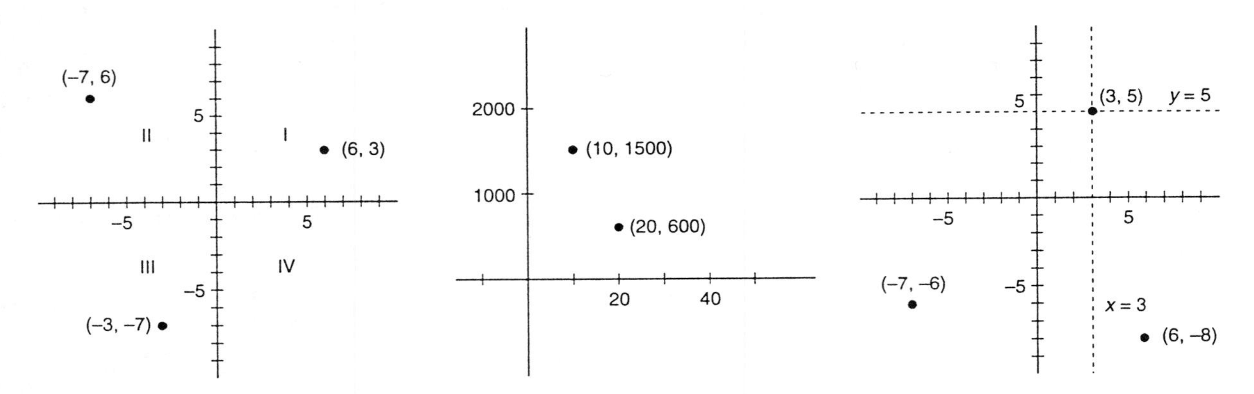

Figure 1A: An axis system. **Figure 1B**: Another axis system, with a different scale. **Figure 2**: The locations of (3, 5) and some other points (ordered pairs).

In one dimension, the equation "$x = 3$" describes a point on a number line. In two dimensions, with two variables, we interpret the equation "$x = 3$" as "$x = 3$ and $y =$ anything." In two dimensions the equation "$x = 3$" is the equation of a line.

The notation "(x, y)" serves for both points and ordered pairs. The number in the position of "x" is called the first-coordinate (or "x-coordinate") and the number in the position of y is called the second-coordinate (or "y" coordinate).

---

vertical line: $x = c$, for some c.
horizontal line: $y = c$, for some c.
x-axis: $y = 0$.
y-axis: $x = 0$.

We often do not bother to distinguish between a line and its equation. On the one hand, the x-axis is a line. On the other hand, the x-axis is $y = 0$. (At least it has equation "$y = 0$".) Perhaps we should say "The line *with equation* such-and such," rather than "The line such-and-such." But we presume that you, the reader, can tell what we mean.

Expressions and Graphs. Let "$f(x)$" denote an expression with one variable, "x", such as "x^2" or "$2x + 1$" (This notation anticipates functional notation which will be emphasized in Section 2.1). The *one* sequence of operations makes *many* number pairs -- ordered pairs.

Definition 1.5.1 (Graph): Technically, the graph of the expression $f(x)$ is the set of all points $(x, f(x))$ for all values of x. More commonly, the term graph is used to mean a *picture* of the set of all points $(x, f(x))$. In practice, because the size of a picture is limited, a graph actually consists of a picture of all the pairs that fit in the window (or viewing rectangle), which is the region shown.

Example 9: To graph the expression "$2x + 1$" mark all the pairs of the form $(x, 2x + 1)$ that fit in the window. Mark (0,1), (1,3), (2,5), and all the other pairs where x assumes all other real values including negative values, fractional values, and irrational values (Figure 3).

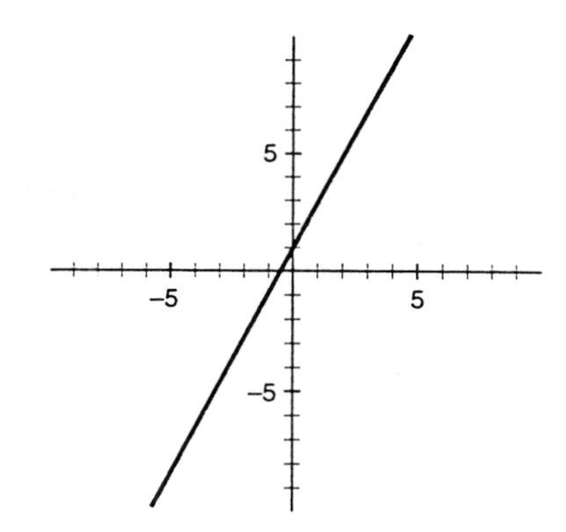

Figure 3. A graph of the expression $2x + 1$. The window is $-10 \leq x \leq 10$ and $-10 \leq y \leq 10$.

Graphs of expressions show (x, y) pairs. Even if no algebraic representation is given, an expression can be directly evaluated from its graph by reading the y-values of points with x-values of interest.

Example 10: Figure 4 gives the graph of an expression $f(x)$, without giving its algebraic representation. Tick marks are one unit apart. Find $f(2)$. Find $f(-3)$. Find $f(5) - f(0)$.

To find $f(2)$ look for a point on the graph above or below 2 on the x-axis. Read the y-value of the point by looking across to the scale on the y-axis. $f(2) = 1$. $f(-3) = 0$. $f(5) - f(0) = 2 - (-2) = 4$.

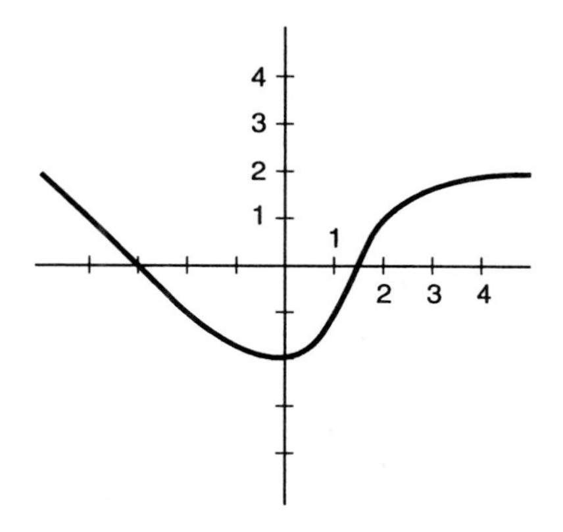

Figure 4: A graph of $f(x)$ with no algebraic representation given.

<u>Graphing with Calculators</u>. Graphing calculators can produce graphs in a few seconds that would have taken minutes or hours to produce with the technology available only a few years ago. Calculators are changing the face of mathematics.

Calculator Exercise 1: Learn how to graph expressions with your calculator. In particular, learn how to

1) enter expressions to graph,
2) select the desired window,
3) draw the graph, and
4) read the coordinates of points on the graph.

We will use the terms in the table below.

Term	Meaning
standard window	We will use $-10 \le x \le 10$ and $-10 \le y \le 10$. Other windows may work equally well.
zoom	Change the window, sometimes by magnifying the graph about a particular point
trace	Identify points on a graph and display their x- and y-values

Now, graph "$y = x^2$."

You will see a graph like Figure 5 if the window is our "standard" rectangle. If not, the window is simple to change with a few keystrokes.

Now use the *trace* feature to display the (x, y) pair at the cursor. You can move the cursor to the left or right with the arrow keys to exhibit different (x, y) pairs.

Most calculators have an "x" key and require you to use "x" for the variable. If you want to graph the expression "$16t^2$" you may rewrite it as "$16x^2$" and realize that the x-axis serves in place of the t-axis.

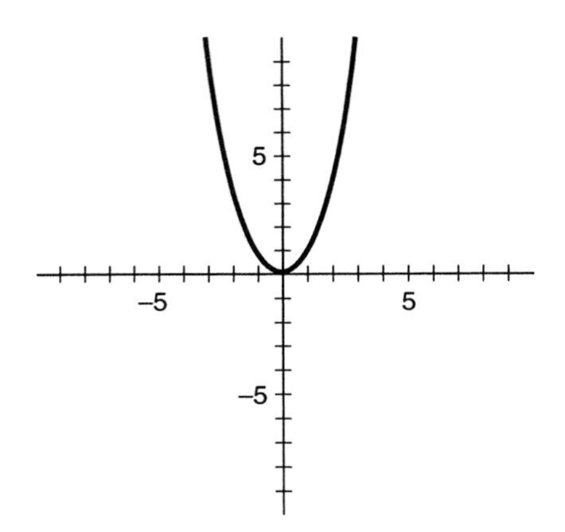

Figure 5: $y = x^2$.
[-10, 10] by [-10, 10].

> We will employ interval notation to express the x- and y-intervals used. "$a \le x \le b$ and $c \le y \le d$" will be written "$[a, b]$ by $[c, d]$."

Selecting a Window. Most graphs do not illustrate the whole function. For example, when you graph the expression x^2 (Figure 5), the points (7, 49) and (12, 144) will not appear in the standard window. But the standard window is often fine for creating a "representative" graph. A representative graph is one that indicates the general behavior of the expression, even for points not in the window. A representative graph might not include all the detail we want, but, by definition, **a "representative" graph is not misleading.**

For example, the graph of x^2 in Figure 5 clearly exhibits a steep increase in y-values as the x-values increase from 0 toward 3. This trend gives the right impression. Although we cannot use the standard scale to see the point associated with $x = 10$, we can be confident that it is very high above the top of the window.

When graphs do not exhibit the whole function we must decide what limited region to display by selecting an interval of x-values and an interval of y-values. If we do not make a choice, the calculator will make one for us, and its choice may not exhibit where the "action" is.

Calculator Exercise 2: Graph $10x^2 + 100$.

The picture on the standard window is blank. Try it and see for yourself. We need a different window to see any action. How can we select a "good" window?

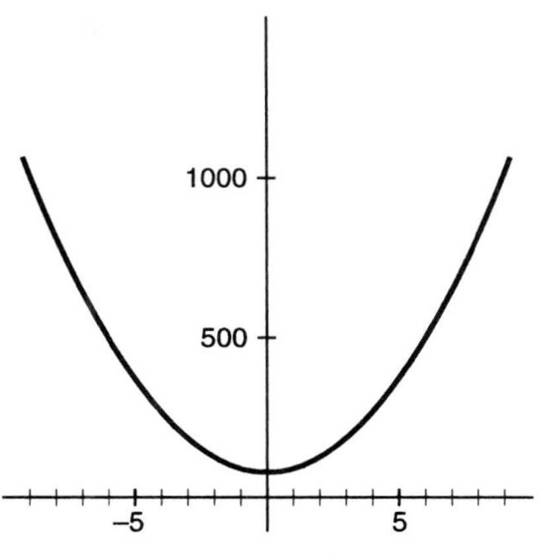

Figure 6: $10x^2 + 100$. [-10,10] by [0, 1500]

There is not just one "good" window. Many different choices are fine. The point is to first find one that is reasonable, and later modify it to an even better one if you want.

First select an interval of x-values -- domain values. If you don't like your first choice, it can be changed later. For example, begin with the "standard" $-10 \leq x \leq 10$. The important analysis comes now. Which y-values should be displayed?

Determine roughly which y-values correspond to the given interval of x-values. For example, in the middle of the picture $x = 0$. There (evaluating the expression $10x^2 + 100$), $y = 100$. So we want our y-interval to bracket 100.

Now consider other possible x-values, say, at the edges of the picture. When $x = 10$, $y = 1100$. Also, when $x = -10$, $y = 1100$. So we want the y-interval to contain 1100.

To select a window, determine the x-values of greatest interest. If you don't know which ones are of interest, begin with [-10, 10] or [0, 10]. Then evaluate the expression to find the y-values corresponding to some convenient x-values in the interval. The *trace* feature can be used to evaluate relevant y-values. Then use a y-interval that does an adequate job of displaying those y-values. You can always change the window easily if your first choice is not satisfactory.

So a picture that brackets 100 and 1100 will show some action. Overall, we might use, say, $0 \leq y \leq 1500$, and put tick marks every 500 units (Figure 6). On some calculators you can use the *trace* feature to do these calculations. Just select an x-interval and try to graph the expression. Even if nothing appears on your screen, the *trace* feature will display the y-values corresponding to the x-values. Note those y-values and then adjust the window to display them.

Calculator Exercise 3: Graph $200\sqrt{x}$.

The picture in the standard window is virtually blank. Try it and see for yourself. Which windows will provide illuminating pictures?

Ask yourself, "Which x-values are possible? Which y-values can occur?"

Deal with x-values first. Negative x-values are not possible, so we might as well omit them from our window. All positive values of x are possible, but we cannot graph them all. So pick a convenient maximum x-value, say $x = 10$.

Now comes the analysis. Which y-values are possible? At the left edge, when $x = 0$, $y = 0$. At the right edge, when $x = 10$, $y = 200\sqrt{x} = 200\sqrt{10} = 632.5$. So pick a y-interval which includes y-values from 0 to 632.5. We usually like a nice round numbers, so Figure 7 uses 0 to 1000, with tick marks every 100 units.

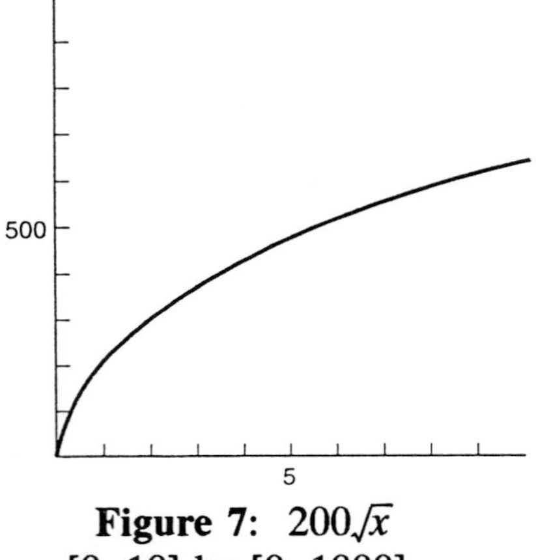

Figure 7: $200\sqrt{x}$
[0, 10] by [0, 1000].

Limitations of Calculator Graphs. Graph any simple graph and look closely at the picture. Graphics calculators are wonderful, but they are not perfect. They have some important limitations. By studying even a simple graph such as the graphs of "x^2" we can learn a lot about how they work. Use your calculator to graph x^2 (Figure 8). Figures 5 and 8 graph the same expression with the same window. Why does Figure 8 appear ragged?

A graphics calculator does not display all the points in the window. The calculator has a screen which is made up of discrete dots. Each dot, called a pixel, can be illuminated or not. Pixels have a small, but non-zero, width. If you look closely at your calculator screen you can see the pixels arranged in vertical columns and horizontal rows. Each column corresponds to an exact horizontal position -- an exact x-value. My calculator has, for example, 96 columns of pixels. To graph an expression the calculator selects x-values evenly spaced across the window and evaluates the expression at each of the 96 values. Points with other x-values in between those 96 x-values are omitted. Furthermore, my calculator has only 56 rows of pixels. There are only 56 possible heights. That means, for each x, whatever the value of y, the calculator must select from among only 57 possibilities. It can illuminate any of the 56 pixels above the x-value, or it can decline to illuminate any pixel in that column.

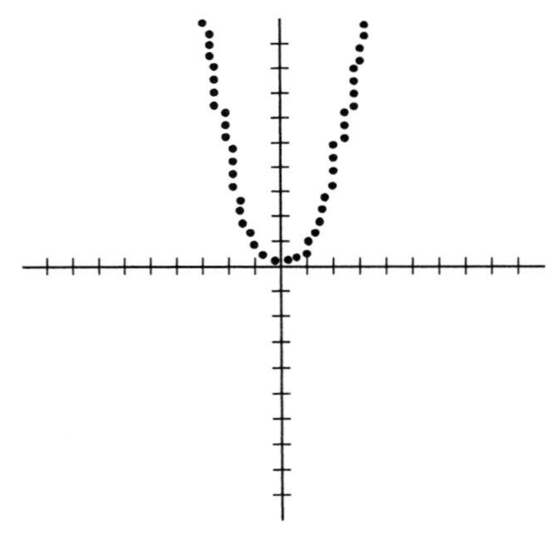

Figure 8: A screen-like picture of the graph of x^2, illustrating the discrete pixels.

Calculator Exercise 4: Graph x^2 on the standard scale. Put the cursor on the graph using the *trace* key. Move the cursor to the left and right slowly and note how the x-value changes. Each column of pixels corresponds to a particular x-value. Neighboring columns correspond to x-values some fixed distance apart (about 0.2 units on my calculator). Other x-values between those x-values do not yield points.

The picture is composed of discrete dots, which explains why curves which should be smooth may appear ragged on your screen. Consider two neighboring columns of pixels. Suppose the graph is steep and the left column illuminates the pixel in row 21 and the right column illuminates the pixel in row 25 (Figure 9, left). Should your calculator leave a gap between those two dots? Or should it "fill in" between them to visually connect them? Remember that the true graph would have points for the many x-values between the two x-values labeling the two columns. Most of the time the true graph would have y-values between the two given y-values and the graph would be connected (Figure 9, right).

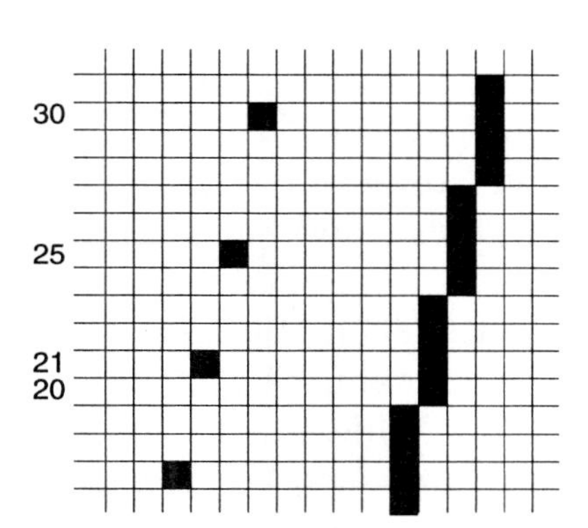

Figure 9: Pixels in neighboring columns not connected (left), and connected (right).

Most of the time it is right to "fill in" between neighboring columns of pixels, which is why graphics calculators usually do fill in between dots. Nevertheless, sometimes dots are connected which should not be connected. A part of a calculator's graph that appears to be incorrect because of the way the dots are, or are not, connected is said to be an <u>artifact</u> of the calculator's programming.

Calculator Exercise 5: Try these graphs to see if your calculator produces artifacts.

Graph $y = 1/(x - 1.5)$. Your picture should not show a vertical line through $x = 1.5$. If there is a vertical line there, it is an artifact.

Graph $y = \sqrt{83 - x^2}$. The picture should show the top half of a circle centered at the origin. However, many calculators show a gap between the curve and the x-axis. The gap should not be there. It is a result of the limited number of columns of pixels used to evaluate the expression. The x-value where $y = 0$ was not used, so the calculator did not connect the dots to the x-axis (Problem B40).

<u>Graphing Context</u>: Words are used in context. When we talk about graphs, are we talking about graphing expressions, equations, or functions? Actually, the term "graph" can be correctly used in all three contexts, but the next

discussion will show why I prefer to talk about graphing *expressions* first.

Graphing Expressions to Solve Equations. Terms require a context. Equations may be "solved"; expressions may be "evaluated." To solve the equation "$f(x) = c$," where c is some constant, means to find the values of x which make the equation *true*, which in turn means to find the values of x for which the expression "$f(x)$" has the value c. One elementary approach to solving is to *evaluate the expression* $f(x)$ for various values of x to try to *solve the equation* "$f(x) = c$." Note the different contexts for the terms "evaluate" and "solve." Graphs can be used to visually evaluate expressions at numerous x-values simultaneously.

Example 11: Suppose you ask a fourth grade child to answer the question, "I'm thinking of a number. Multiply it by three. Then add four. The result is 16. What was the number?"

Prior to studying algebra, many children simply do the process described when the number is 1, 2, 3, ... until they find that the result is 16 when the number is 4. They do the process and compare the result to the desired value, 16.

In algebraic terms, the process corresponds to an expression. The process, "Multiply by three. Then add four," can be represented by "$3x + 4$." Algebraically, we can think of evaluating this expression for various values of x and comparing the result to the desired value, 16.

Evaluating the expression to see when the equation is true can be called "guess and check" or "trial and error." I call it "evaluate-and-compare." You *evaluate* the expression and *compare* it to the desired value to see if they are equal.

Example 12: Solve $x^3 + x = 5$.

The expression on the left is a cubic (third power) polynomial. There is a cubic formula, somewhat like the quadratic formula, but it is so complex that hardly anyone knows it or uses it. So how does a mathematician solve that problem?

Mathematicians are likely to use the same technique young children naturally use: evaluate-and-compare. Of course, they don't do it all by hand; they use a calculator or computer to evaluate the expression "$x^3 + x$" for various values of x. When they find an x for which the value of the expression is close to 5, they may quit, figuring close is good enough. How many decimal places of accuracy do you want, anyway?

Calculator Exercise 6: Graph "$x^3 + x$" and use the *trace* feature to find an x-value where the y-value is slightly less than 5, and a neighboring x-value where the y-value is slightly greater than 5 (Figure 10). The solution you seek is somewhere between those two x-values.

A graphics calculator does not evaluate the expression for all values of x in the x-interval, so there is a good chance that the x-value which solves the equation will not even be tried. However, if the solution is between neighboring x-values that differ by a very small amount, you have determined the solution to within that amount. However, on the standard scale the solution is not closely determined because the neighboring x-values may differ by a substantial amount (about 0.2 on my calculator).

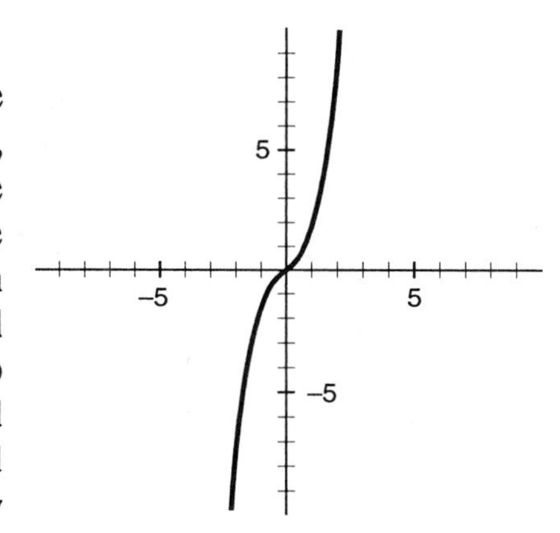

Figure 10: A representative graph of $x^3 + x$. [-10,10] by [-10,10].

Of course, we could change the scale -- magnify the window about the key point so that neighboring columns of x-values have x-values which are closer together so the solution can be bracketed more closely. Figure 11A illustrates the original standard picture, a point with y-value close to 5, and the region which we will magnify. Figure 11B then illustrates the picture magnified by four using the *zoom* feature.

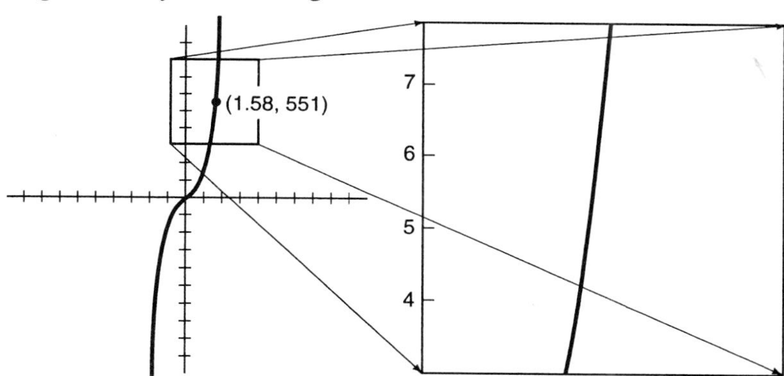

Figure 11A. A standard graph $x^3 + x$ and a region to be magnified by four.

Figure 11B. The region magnified by four.

Calculator Exercise 6, continued. Learn to "zoom in" on a point of a graph. Learn to do it at least two ways. One is to use the *zoom* feature. Another is to use the *window* feature to manually select a window.

In the magnified picture (Figure 11B) the x-interval of neighboring pixels which brackets the solution will be only one fourth as wide. Using this procedure repeatedly, we can bracket the solution with narrower and narrower intervals and obtain the solution to any desired degree of accuracy. The solution is, to two decimal places, $x = 1.52$.

Example 12, revisited: Again, solve "$x^3 + x = 5$."

To solve an equation it is common to "Put everything on the left." The equation is equivalent to

$$x^3 + x - 5 = 0.$$

Now, we can graph the expression "$x^3 + x - 5$" and solve the equation by finding the x-value where the expression has value zero, using the *evaluate-and-compare* method. This is the approach of many graphics calculators with a "solver." After you tell the calculator approximately where to look for a zero of an expression (by visually inspecting the graph), it will evaluate the expression at numerous x-values and "zero in" on the x-value that yields a zero. Again, $x = 1.52$.

> Some calculators have a *solve* key for solving equations. The *solve* key uses an automatic evaluate-and-compare procedure to find a solution. The method is very similar in spirit to this example. Be sure you understand the evaluate-and-compare process, even if your calculator has a *solve* key and you do not have to manually zoom in yourself.

Here is an example where the *evaluate-and-compare* method is necessary. There is no algebraic method for solving this equation.

Calculator Exercise 7: Solve "$\ln x + x = 4$" graphically. Obtain at least two decimal places of accuracy.

The only approach that works is "evaluate-and-compare." It is common to move the 4 to the left and solve the equivalent equation "$\ln x + x - 4 = 0$" instead. Obtain a representative graph of "$\ln x + x - 4$" (Figure 12) and look for height 0. Some calculators have a *solve* feature that does this automatically after an x-interval containing the zero has been found visually. Or, you may trace along the graph until you discover x-values which bracket the solution. Then zoom in until the x-interval bracketing the solution is sufficiently narrow that you are satisfied with the accuracy of your solution.

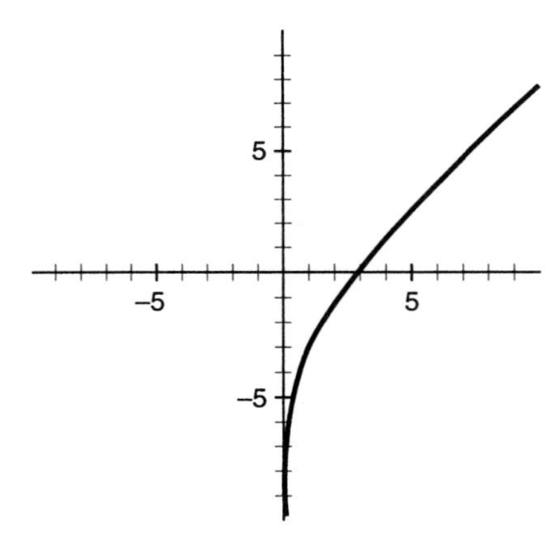

Figure 12: $\ln x + x - 4$. [-10,10] by [-10,10].

The solution is, to two decimal places, $x = 2.93$.

Calculator Exercise 8: Solve "10 ln x - x = 0" by the "evaluate-and-compare" method using a graph.

Figure 13 gives a graph of the expression "10 ln x - x" on the standard scale. From the graph, you expect a solution in the neighborhood of $x = 1$. Zoom in and you can find it: $x = 1.12$, to two decimal places.

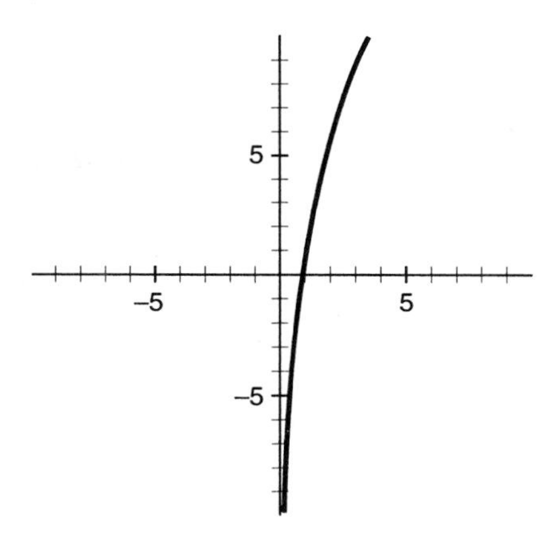

Figure 13: 10 ln x - x. [-10, 10] by [-10, 10].

This answer is right, as far as it goes, but it does not go far enough. This is why we introduced the concept of a "representative" graph. A representative graph is one which indicates all the important features of the expression and which is not misleading. The graph in Figure 13 is not representative. Zoom out to the interval $0 < x \le 40$ (Figure 14) and you will see that the strong upward trend toward the right of Figure 13 is misleading. The curve turns around and comes back down. The second solution is $x = 35.77$.

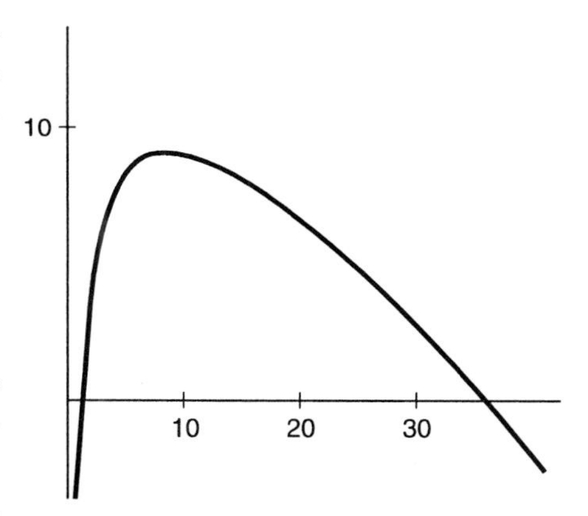

Figure 14: 10 ln x - x. [0, 40] by [-5, 15].

<u>Maximizing or Minimizing an Expression</u>. A major type of calculus problem is to maximize or minimize the value of an expression. Graphically, it is very easy to find the x-value which yields the maximum of an expression. Just graph the expression and look for the highest point on the graph.

Calculator Exercise 9: Use your calculator to find the maximum value of $x^2(5 - 2x)$ for $x > 0$, and find the corresponding x-value.

Calculus spends a great deal of time on developing a method of doing this problem. The calculus approach is very powerful, but unnecessarily complicated for this particular problem. The approximate graphical approach is far easier.

Graph the expression and identify the maximum y-value and corresponding x-value (Figure 15). Zoom in until you are satisfied with the accuracy of your approximate solution.

The maximum value for $x > 0$ appears to be about $y = 4.63$ when $x = 1.67$. More decimal places could be determined by zooming in further. The exact answer, $y = 125/27$ (about 4.6296) at $x = 5/3$ (about 1.6667), could be obtained using calculus.

Calculators that have a "solver" may have a "maximizer" as well. If yours does and you avoid this manual process, be sure you at least understand how the process depends upon *evaluating* the expression and *comparing* the value to nearby values.

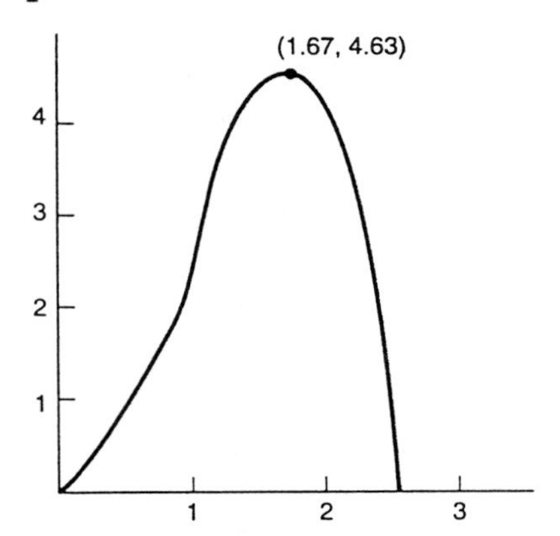

Figure 15: $x^2(5 - 2x)$. [0,3] by [0,5]. The highest point is identified.

Graphing Terminology. What is the proper context for the term "graph"? Do we graph expressions, equations, or functions?

The term "graph" is usually used in the context of equations and functions, not expressions. Nevertheless, the idea of graphing expressions is really what we want. Expressions correspond to functions, so the functional context is very appropriate and will be emphasized in Section 2.2. But the idea of graphing an "equation" can be misleading. We do not graph the same sort of equation that we solve.

We might "solve" the equation with one variable, "$2x + 1 = 7$." The solution is a single number, $x = 3$, and there is no point in graphing that on a 2-dimensional picture. If we want to graph anything in this context, it is the *expression* "$2x + 1$."

Many graphics calculators only accept equations of a single form: "$y = ...$" You fill in the blank with an *expression* with *one* variable, x. Problems that require you to maximize something (as in Calculator Exercise 9), ask you to maximize an expression, not an equation. Again, it is appropriate to think of graphing an expression rather than graphing an equation.

Nevertheless, in the two-variable context it is conceptually correct to speak of "graphing an equation" such as "$y = 2x + 1$." Just keep in mind that this is not the sort of equation we "solve."

Sentences such as "$y = 2x + 1$" or "$y = f(x)$" are equations which define the use of the symbol "y". These kinds of equations are simply true *by definition*. They merely assert that "y" will be used a certain way, at least for a while until the next problem. With this notation, the graph of the expression "$f(x)$" is the same as the graph of the equation "$y = f(x)$."

There are cases where the appropriate context for graphing *is* equations with two variables. These cases are when we do not have a functional form for the relationship between x and y, but they are related by an equation.

Example 13: "$x^2 + y^2 = 1$" is the equation of the important "unit circle" of radius 1 centered at the origin. Certain (x, y) pairs make this sentence true, most do not. The graph of the equation (this time *equation* is the proper context for the term *graph*) is a picture of the set of points which make the equation true (Figure 16).

Graphing calculators do not generally accept equations like "$x^2 + y^2 = 1$" to graph. They accept expressions. To graph this on a calculator we would need to break it up into separate expressions, each of which defines y values with a functional expression in x.

Solving this for y, we obtain:

$$y = \sqrt{1 - x^2} \quad or \quad y = -\sqrt{1 - x^2} \,.$$

These two graphs plotted together on one axis system will make the picture.

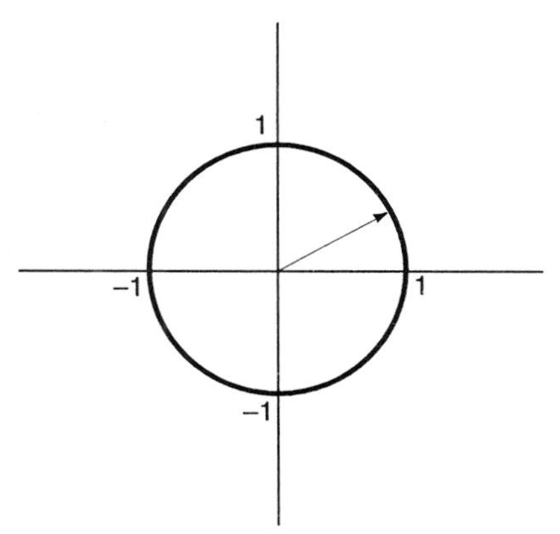

Figure 16:
Graph of the unit circle,
$x^2 + y^2 = 1$.
[-2, 2] by [-2, 2].

Graphs Without Expressions. The principle of the method of solving an equation with a graph still holds even when there is no algebraic expression to accompany the picture. Graphs can be read two ways. The direct way is for evaluation; the inverse way is for solving.

Example 14: Figure 17 is a representative graph of an arbitrary function, f, without an accompanying algebraic expression.

To evaluate $f(x)$ for a particular x, find the particular x (horizontal coordinate) and look above it for the vertical coordinate (height) on the graph. To find $f(5)$, look above 5 on the x-axis to the graph and note the y-coordinate, 3. $f(5) = 3$. Similarly, $f(2) = -3$, approximately. The scale does not permit the images to be determined with a great deal of accuracy, and we cannot zoom in.

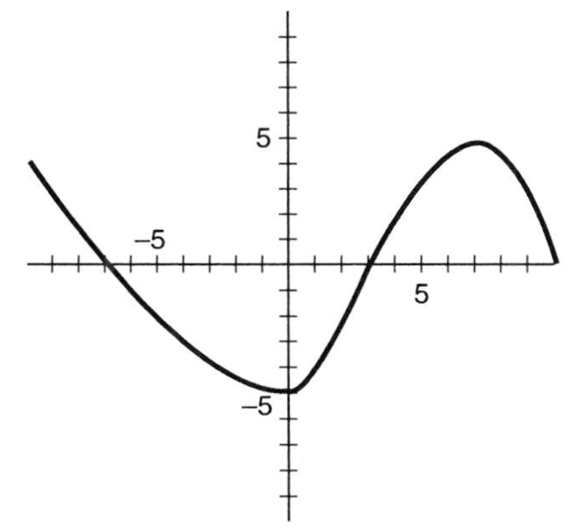

Figure 17: A graph without a given algebraic representation.

In the other, inverse, direction, we can read the graph to solve equations. To solve $f(x) = 3$, note *all* x-values for which a point on the graph has y-value 3. We already noted one such x-value, 5. There are two more solutions, as can be seen looking across to the graph at height $y = 3$. The solution is $x = -9$ or $x = 5$ or $x = 9$.

To solve the inequality "$f(x) < 0$" note the x-values of the points on the graph where the y-value is less than zero. That is, note the x-values of all points below the x-axis. (The x-axis has equation $y = 0$.) The solution is approximately: $-7 < x < 3.2$ or $x > 10$.

Conclusion. Distinguish between the terms *expression* and *equation*. To solve an equation we may graph an expression. Evaluating and solving are inverse activities. To "evaluate" an expression, $f(x)$, is to begin with x and find the value, y. To "solve" the equation $f(x) = y$ is to begin with y and find x. Evaluating is a direct process. Algebraic methods of solving equations are indirect processes. However, the guess-and-check ("trial-and-error" or "evaluate-and-compare") method of solving, which is not really algebraic, is direct.

Calculators facilitate both evaluating and solving. Graphics calculators graph expressions or functions and display x- and y-coordinate pairs. They can be said to graph equations in the context of an equation with two variables. By "zooming in" (which is effectively what a "solver" does), equations can be solved graphically using the method I call "evaluate-and-compare."

Terms: expression, variable, evaluate, equation, equivalent expressions, graph, window, standard window, representative graph, origin, axis, scale, quadrant, ordered pair, point, coordinate, trace, pixel, artifact, solve, evaluate-and-compare method.

****

Exercises for Section 1.5, "Expressions, Equations, and Graphs":

A1.* a) Thinking of Mathematics as a language, what parts of speech are expressions? b) Can an expression be true? c) Can an equation be true?

A2.* If two expressions are always equal, they are said to be __________.

A3.* What types of sentences express the equivalence of expressions?

A4.* When two expressions are equivalent, something they express is the same and something is not. What?

A5.* a) Define "window." b) Define "pixel."

A6.* Give the equation of the a) x-axis b) y-axis.

A7.* Give the equation of the
a) vertical line through (a, b). b) horizontal line through (a, b).

A8. Give the equation of the
a) vertical line through $(2, \pi)$. b) horizontal line through $(-1, -4)$.

A9.* Define the adjective "representative" in the context of graphs.

^ ^ ^ ^ Which number quadrant is
A10.* a) the upper left? b) the lower left?
A11.* a) the upper right? b) the lower right?

^ ^ ^ ^ If you sketch a graph with both on it, which is located first?
A12. The line $x = a$ or the point (a, b)?
A13. The line $y = b$ or the point (a, b)?

A14. Suppose the instructions on a problem are to "Simplify the expression $x/\sqrt{x}$" and a student squares to get $x^2/x = x$. "x" is simpler but it is not right. What went wrong?

A15. Which are expressions?
a) $3x + 2$ b) $3x + 2 = x^2$ c) $x^2 + 1$ d) $x^2 + 1 = 3x$ e) $mx + b$.

^ ^ ^ ^ ^ ^ ^ ^ ^

B1.* Medium length essay: What is the relationship between *solving* and *evaluating* (when the terms are used algebraically, as in this text)? Be sure to name the types of things do you "evaluate" and the types of things you "solve."

B2.* Given a picture of a representative graph of $f(x)$, explain how to (approximately)
a) Evaluate $f(a)$. b) Solve $f(x) = c$.
In your answer, be sure to distinguish points on graphs from their coordinates.

B3.* a) To solve "$f(x) = 0$" graphically, which expression or equation would we graph?
b) There is an important difference between the types of equations we "solve" and the types we "graph." What is it?

B4. Figure 18 gives a representative graph of $f(x)$ without giving its algebraic expression.
a) Find y for $x = 2$. b) Find x such that $y = 3$.
c) Use the "$f(\ )$" notation and the terms "solve" and "evaluate" to rephrase the questions in parts (a) and (b).

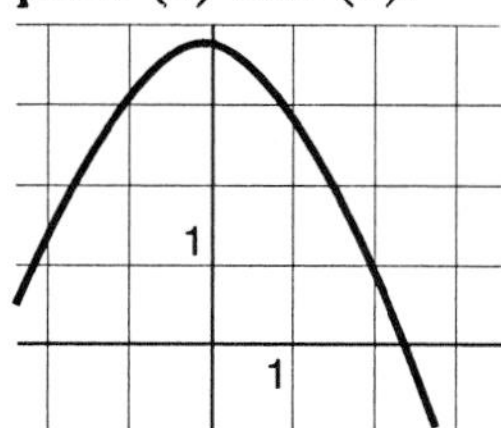

Figure 18.

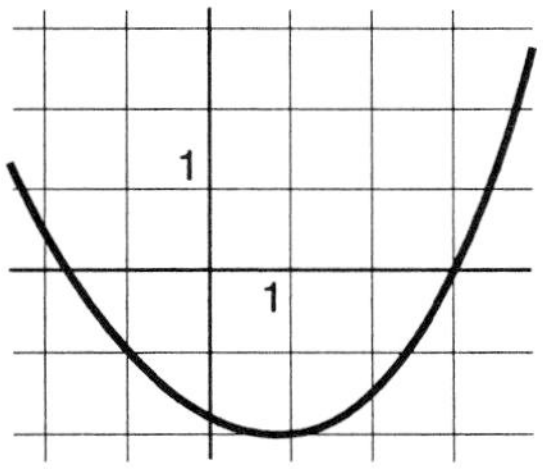

Figure 19.

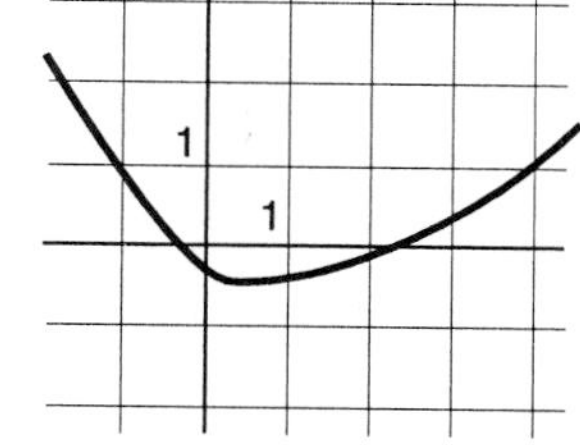

Figure 20.

B5. Figure 19 gives a representative graph of $f(x)$ without giving its algebraic expression.
a) Find y for $x = 1$. b) Find x such that $y = -1$.
c) Use the "$f(\ )$" notation and the terms "solve" and "evaluate" to rephrase the questions in parts (a) and (b).

B6. Figure 20 gives a representative graph of $f(x)$ without giving its algebraic expression.
a) Find y for $x = 3$. b) Find x such that $y = 1$.
c) Use the "$f(\ )$" notation and the terms "solve" and "evaluate" to rephrase the questions in parts (a) and (b).

^ ^ ^ ^ State the *algebraic formulation* of the procedure used to evaluate these expressions (in terms of simpler procedures, as in Examples 7 and 8).

B7. (3/7)/5.
B8. 9 - 15.
B9. 7/(-2).
B10. 19 - (-6).
B11. 4(-7).
B12. -4 - 12.
B13. (3/5)/9.
B14. 3/7 + 2/15.
B15. 2/(3/5).
B16. (2/7)/(5/3).
B17. -34 + 29.
B18. (-48)/(-12).
B19.* In the problems B7-18, what type of variable was used to express the answers?

^ ^ ^ ^ **Note:** This text occasionally supplies solutions with two significant digits in brackets (problems B20-29). These are given so you can immediately determine whether you used the correct solution process. For all such homework problems, prove that you know the correct process by giving a more accurate answer rounded to three or more significant digits.

B20. $x^3 - 5x - 6 = 0$ [2.7].
B21. $8\sqrt{x} + x^2 = 12$ [1.5].
B22. $x + \sin x = 1$ (use radians) [.51]
B23. $x + \ln x = 2$ [1.6].
B24. $x + e^x = 2$ [.44].
B25. $x + \tan x = 5$ (use radians, $0 < x < \pi/2$) [1.3].
B26. $x^3 - 3x < 1$ [$x < -1.5$ or ....].
B27. $\sin x > x^2$ (use radians) [$0 < x$ and ....]

^ ^ ^ ^ Find the x-value that yields the maximum y-value.

B28. $y = x^2 + 3x - x^4$ [1.1].
B29. $y = 6x + \sqrt{x} - x^2$ [3.1].

^ ^ ^ ^ Sometimes the "standard" scale is not the one we want. a) Sketch, in a window you draw, a "representative" graph of the equation. b) Give the x and y-intervals you used. c) find the minimum value of y for $x > 0$ in your window. [Great accuracy is not required, just get it close enough to show you know how to do it.]

B30. $y = (x - 20)(x - 50)$.
B31. $y = (2x - 10)(x - 100)$
B32. $y = x^3 - 20x^2 + 150$.
B33. $y = x^3 - 50x + 300$.

^ ^ ^ ^ Sometimes the "standard" scale is not the one we want. a) Sketch, in a window you draw, a "representative" graph for solving the equation. b) Give the x and y-intervals you used. c) Find the solution. [Great accuracy is not required, just get it close enough to show you know how to do it.]

B34. $10\sqrt{x + 7} - x - 20 - 0$.
B35. $x^3 + 26x^2 - 132x - 360 = 0$.

B36. Figure 21 locates (a, b) and (c, d) on an axis system. Locate (b, a) and (d, c).

B37. Figure 21 locates (a, b) and (c, d) on an axis system. Locate $(-a, b)$ and $(c, -d)$.

B38. Figure 22 gives (a, b) and (c, d). Find the coordinates of the point P.

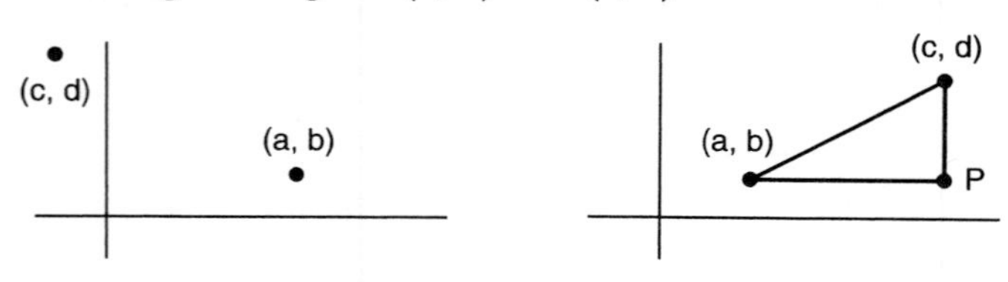

Figure 21. **Figure 22.**

B39.* What is an "artifact" (in the context of a graph produced by a graphics calculator)?

B40. a) Look at the graph of $1/(x - 1.5)$ on your graphics calculator. Does it produce an artifact? What is the artifact, if any? [Mention the model of calculator you use.]

b) Look at the graph of $\sqrt{83 - x^2}$ on your graphics calculator. Does it produce an artifact? What is the artifact, if any?

B41. Find an expression and a window such that your graphing calculator produces an artifact different from the ones given in the previous problem. Mention the model of calculator you use.

B42. (Short essay) Graphics calculators are wonderful, but they are not perfect. Discuss two things that are not perfect about calculator graphs.

B43. Solve "$3x + 1/x = 5$" using algebra and not a calculator-aided method.

B44.* What are the two ways to interpret expressions?

B45.* What is symbolic algebraic notation designed to express?

B46.* a) What are sentences with dummy variables about?
b) What are sentences with unknowns about?

B47.* Medium length essay: Distinguish clearly between the terms *expression* and *equation*. Why is it important to distinguish between them?

B48.* Short essay: Explain why algebra is not about numbers.

B49.* Define the (rectangular coordinate) "graph" of an expression.

^ ^ ^ ^ When you sketch a graph with these features, what order would they come in?
B50.* a) the point (a, b) b) the line $x = a$. c) the x-axis.
B51. a) the point (a, b) b) the line $y = b$. c) the y-axis.

B52.* What is the difference between graphing *equations* and graphing *expressions*?

B53. a) Solve $8\sqrt{x} - x = 12$ using a graphics calculator. Give at least one decimal place of accuracy. b) What mistake would be easy to make in part (a)?

B54. a) Solve $100x + 50\sqrt{x} - x^2 = 500$ using a graphics calculator.
b) What mistake would be easy to make in part (a)?

B55. Here is a graph that may produce an <u>artifact</u>. Let $f(x) = (1 - \cos(x^4))/x^8$.
The domain does not include $x = 0$. a) Graph it to approximate the value for x near, but not equal to, zero. Zoom in at most once to find an approximation. b) Now change to the scale [-.1, .1] by [-.1, .6]. Does your graph produce nearly a horizontal line at $y = 1/2$? It should, but some calculators show the height dropping to zero for $|x| < .01$. This is a mistake -- an artifact.

Section 1.6. Four Ways to Solve Equations

Over your mathematical career you have solved many equations. To solve an equation is to find an equivalent equation which exhibits the solution. In Section 1.4 we discussed two ways to exchange one equation for another, equivalent, equation. One was to employ an identity to change one side (using equivalence of expressions); the other way was to change both sides simultaneously (using a theorem which asserts the equivalence of two equations).

Identities and theorems express processes you *may* do to an equation. But what *should* you do? What guides your choice of what to do? The purpose of this section is to show you that, when all is said and done, there are only four basic ways to solve an equation. Therefore, all the manipulation we do to an equation should aim for one of the four basic formats that allow us to finish off easily.

Here are the four primary methods of solving equations:

1) Do the inverse operations in the reverse order,
2) Use the Zero Product Rule,
3) Use the Quadratic Theorem, and
4) Evaluate-and-Compare
   (Also known as "trial and error" or "guess and check."
   This includes graphical methods).

Algebraic expressions indicate operations in a particular sequence or order.

To solve an equation its component expressions must fit,
or be rearranged to fit,
certain patterns which express operations in convenient orders.

Look for the importance of order in the following discussions of the four equation-solving methods.

In this text, to solve an equation "algebraically" means to use one of the first three traditional algebraic methods. These three methods are said to be indirect because you do not actually do the operations represented in the problem. Algebraic methods are capable of yielding exact expressions for solutions, but we expect you to report the solutions as decimal numbers rounded off to three or more significant digits.

The "evaluate-and-compare" method is direct. It is a "numerical" method that relies on actually doing the operations represented in the problem, as opposed to traditional algebraic procedures.

The "Inverse-Reverse" Method. The "inverse-reverse" method is also known as "doing and undoing." The name "inverse-reverse" is short for "inverse operations in the reverse order."

Example 1: Solve $3(x + 4) = 27$.

First, look at the expression on the left and identify the operations and the order in which they are to be executed on x. There is "Add 4" and then "Multiply by 3," in that order. Thus, to solve it we may *divide* by 3, and then *subtract* 4. $27/3 - 4 = 5$. The solution is $x = 5$.

The process can be illustrated by a "function loop" in which the top shows the operations applied to "x" to obtain "$3(x + 4)$" and the bottom shows the functions which "undo" those operations.

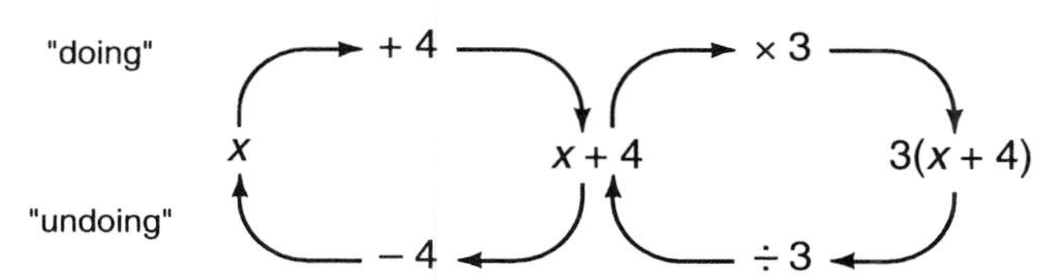

Figure 1: "Doing and Undoing."

Of course, these two steps ("Divide by 3" and "Subtract 4") are justified by theorems.

<u>Theorem 1.6.1</u>. If $a \neq 0$, $ax = b$ iff $x = b/a$.

This theorem applies to the "Divide by 3" step which undoes the "Multiply by 3" operation in the expression "$3(x - 4)$." It expresses the fact that multiplication and division are inverse operations of one another.

<u>Theorem 1.6.2A</u>. $x + b = c$ iff $x = c - b$.
<u>1.6.2B</u>. $x - b = c$ iff $x = c + b$.

Part A applies to the "Subtract 4" step which undoes the "Add 4" operation in "$3(x + 4)$." It expresses the fact that subtraction is the inverse of addition.

Theorem 1.6.1 about multiplication and division was used *first* because the expression "$3(x + 4)$" expresses multiplication *last*. That is the "reverse" order part of the "inverse-reverse" name. The *multiplication* in the expression caused us to *divide* in the solution process. Multiplication and division are *inverse* operations, which illustrates the "inverse" part of the "inverse-reverse" method name.

This inverse-reverse approach is very common. Often you would see it written out at greater length with each step creating a new equation:

	$3(x + 4) = 27$	
iff	$x + 4 = 9$	by Theorem 1.6.1
iff	$x = 5$	by Theorem 1.6.2A.

Any time you solve an equation by obtaining a sequence of equations leading toward the solution, the successive equations you obtain *are connected* by some logical connective (preferably "if and only if," symbolized by "iff" or "$\Leftrightarrow$", and which means "is equivalent to"). Also, if your steps are correct, *each step is justified* by some general result. Instead of just *doing* the procedure, look for the connections and justifications so you can *state* and *understand* the procedure.

The inverse-reverse method requires the equation to be in a specific format.

Inverse-Reverse Format Requirement: The equation must fit, or be altered to fit, the form

"$f(x) = c$," where

1) c is a constant, and
2) $f(x)$ is expressed with only one appearance of the unknown, "x".

When the *operations* and the *order* in which they are to be executed have been identified, to solve the equation we need only to do the inverse operations in the reverse order.

Some equations are not yet ready for use of the inverse-reverse method, but can be easily converted to inverse-reverse form.

Example 2: Solve $5x + 7x = 48$.

Obviously, we "consolidate like terms" and rewrite this as "$12x = 48$," which exhibits the operation "multiply by 12." Now it *is* in inverse-reverse form so we divide both sides by 12 to obtain the solution: $x = 4$.

Of course, you already know how to solve simple equations like Example 2. Its point is to illustrate the inverse-reverse format requirement. The original equation has a constant on the right like we want, but *two* appearances of "x" on the left. If you were to evaluate the left side *in the given order* you would have to use the value of "x" twice. The order of operations in the original equation is not correct for the inverse-reverse method. We need to apply an identity to express the left side in a different order so "x" appears only once.

This solution process illustrates the purpose and value of identities. Identities are used to replace the sequence of operations in an expression with a more convenient alternative sequence. This example used the simple Distributive Property.

Theorem 1.6.3 (Distributive Property): For all a, b, and c,

$$ac + bc = (a + b)c.$$

Note that the Distributive Property can take an expression where addition is last (as on the left) and turn it into an expression where multiplication is last (as on the right).

The next theorem formally states a use of identities.

Theorem 1.6.4 (Substitution): If "$f(x) = g(x)$" is an identity, then
"$f(x) = h(x)$" is equivalent to
"$g(x) = h(x)$."

Perhaps this is too obvious to need stating, but it does point out why identities are important. In Example 2 we traded "$5x + 7x$" for "$12x$" to obtain multiplication last. In the pattern of this theorem, "$5x + 7x$" is "$f(x)$", "$g(x)$" is "$12x$", and $h(x)$ is simply "48".

Many equations are not immediately ready for the inverse-reverse method, but the inverse-reverse format serves as a goal for the initial steps. Once that goal is reached, the solution is easy to obtain.

Example 3: Solve $x \tan 60° = 3 + x$.

For problems like this you learned to "Put all the terms with unknowns on one side." The inverse-reverse format requires this, and more. Note there is no variable in "tan 60°," it is just a number which can be treated like any other number.

$x \tan 60° = 3 + x$ iff $x \tan 60° - x = 3$ [subtracting x from both sides]
iff $x(\tan 60° - 1) = 3$ [Distributive Property]
iff $x = 3/(\tan 60° - 1) = 4.098$ [inverse-reverse]

Not only do we put all the unknowns on one side, we also consolidate like terms so that there is "only one appearance of 'x'."

Example 4: Solve $x^2 = 45$.

The only trick to this is that the inverse-operation relationship is not a function. The square-root function on our calculators gives the square root of 45 which is positive. We must remember that the negative of the square root is another solution.

$\sqrt{45} = 6.708$. The solution is $x = 6.708$ or $x = -6.708$.

Theorem 1.6.5 (Inverse Square): $x^2 = c$ iff $x = \sqrt{c}$ or $x = -\sqrt{c}$.

Recall that the square root symbol refers only to the non-negative solution of $x^2 = c$, so $\sqrt{c}$ is never negative. The solution to $x^2 = 25$ is not just $\sqrt{25}$. It is $\pm\sqrt{25}$ (that is, 5 or -5).

In summary, one method of equation-solving which works when the unknown appears only once is to

Do the inverse operations in the reverse order.

The Zero Product Rule Method. The Zero Product Rule method consists of replacing one equation with two related, but simpler, equations according to the Zero Product Rule.

Theorem 1.6.6. The Zero Product Rule. $ab = 0$ iff $a = 0$ or $b = 0$.

To use this rule, the equation must fit, or be altered to fit, a certain format. The name of this rule describes the two keys.

Zero Product Rule Format Requirement: One side of the equation is *zero* and the other is a *product* (that is, multiplication is last).

Example 5: Solve $(x - 5)(x^2 - 16) = 0$.

Because one side is zero and the other is a product, the Zero Product Rule applies. This equation is equivalent to

$$x - 5 = 0 \text{ or } x^2 - 16 = 0.$$

Now the original equation has been converted into two, simpler, equations. Using Theorem 1.6.2, these are equivalent to

$$x = 5 \text{ or } x^2 = 16.$$

The second can be solved using Theorem 1.6.5. These are equivalent to

$$x = 5 \text{ or } x = 4 \text{ or } x = -4,$$

which exhibit the solution.

One key to the Zero Product Rule is that it requires a product of two factors. Therefore, factoring may be necessary.

Example 6: Solve $x^2 + 5x - 6 = 0$.

Here is an identity relevant to this example: $x^2 + 5x - 6 = (x + 6)(x - 1)$, for all x. Both sides of the identity express addition and multiplication, but in different orders. For the Zero Product Rule we want multiplication *last*.

The purpose of algebraic identities is to give alternative orders in which the operations in expressions can be executed.

Therefore (by Substitution, Theorem 1.6.4) the original equation is equivalent to

$$(x + 6)(x - 1) = 0.$$

Now it fits the Zero Product Rule format. So

$$x + 6 = 0 \text{ or } x - 1 = 0.$$

The solution is

$$x = -6 \text{ or } x = 1.$$

This solution also could have been obtained using the Quadratic Theorem, which we will discuss later in this section.

The Zero Product Rule is the primary motivation for factoring.

It is why you spent so much time on factoring in school.

The Zero Product Rule converts a longer equation into two shorter equations. You still have to be able to solve the shorter equations.

Example 7: Solve $(x - 5)(\ln x + 2) = 0$.

"ln" is the symbol for the "natural logarithm" function which is very important in calculus.

To solve "$(x - 5)(\ln x + 2) = 0$" note that one side is 0 and the other is a product. The product is *last*. Therefore, the Zero Product Rule can be used. Solving this equation is equivalent to solving

$$x - 5 = 0 \text{ or } \ln x + 2 = 0.$$

The Zero Product Rule does not necessarily immediately solve equations -- it merely converts them into two shorter equations. Then you need to have methods for solving those. The solution to "$x - 5 = 0$" is obviously "$x = 5$." The solution to the second equation, "$\ln x + 2 = 0$" is less obvious. By the inverse-reverse method we subtract 2 to find the equivalent equation, "$\ln x = -2$," but now we need a fact about the natural logarithm function. We need its inverse. To solve this equation algebraically you need to know the inverse of the function expressed.

The next result would appear in any list of the basic properties of the "natural logarithm" function. A similar fact about logarithms and exponentials with base 10 already appeared as Theorem 1.4.2C in the discussion of "one-operation patterns."

Theorem 1.6.7 (The inverse of *ln*): $\ln x = c$ iff $x = e^c$.

$\ln x = -2$ iff $x = e^{-2} = 0.135$. Therefore, the solution to the original problem in Example 7 is

$$x = 5 \text{ or } x = 0.135.$$

Calculator Exercise 1: Reproduce these results:
$e^{-2} = .135$. $e^3 = 20.09$. $\ln 10 = 2.30$.

You will study the properties of logarithms and exponentials later. At this stage, this fact about logarithms is only intended to illustrate the idea of an inverse function. Right now your responsibility is merely to note that

To solve an equation you may need to know the inverses of the component functions expressed.

Quotients can be regarded as products because $a/b = a(1/b)$. Note that $1/b$ is never zero and division by zero is undefined.

Theorem 1.6.8 (Zero Quotient Rule):
$a/b = 0$ if and only if $a = 0$ and b is not zero.

Example 8: Solve $\dfrac{2}{x} + \dfrac{3}{x-5} = 0$.

One side is zero. Now write the other side as a quotient. To add fractions, use a common denominator.

$$\frac{2}{x} + \frac{3}{x-5} = \frac{2(x-5)+3x}{x(x-5)} = \frac{5x-10}{x(x-5)}$$

This is equal to zero if and only if $5x - 10 = 0$ (and $x \neq 0$ and $x \neq 5$). Therefore the solution is $x = 2$.

To use the Zero Product Rule one side must be zero. Equally important is that, on the other side, the product must be executed *last*.

Example 9: Solve $(x + 1)(x + 5) + 3 = 0$.

This equation is not in the proper form yet. The left side expresses addition last. We could use Theorem 1.6.2 to find the equivalent equation

$$(x + 1)(x + 5) = -3,$$

but that does not help. Now the right side is not zero. **In equations, expressions in factored form are useful if and only if the other side is zero.**

An approach that works is to multiply out the left side, consolidate all like terms on the left, and factor to express the product *last* as desired.

$$(x + 1)(x + 5) + 3 = x^2 + 6x + 5 + 3 = x^2 + 6x + 8 = (x + 4)(x + 2).$$

These equalities express identities. Thus, by Substitution (Theorem 1.6.4) the original equation "$(x + 1)(x + 5) + 3 = 0$" is equivalent to the new equation "$(x + 4)(x + 2) = 0$," which is (finally) in the proper form to apply the Zero Product Rule. The next equivalent equation is

$x + 4 = 0$ or $x + 2 = 0$. Thus the solution is $x = -4$ or $x = -2$.

Of course, at the stage when we knew the original left side was equivalent to $x^2 + 6x + 8$ we could have used the Quadratic Theorem, Theorem 1.6.9.

The Quadratic Theorem. The third primary way to solve an equation is to use the famous "Quadratic Theorem." Equations which are in the form

$$ax^2 + bx + c = 0,$$

or which can be rearranged into this form, where a, b, and c are constants and a is not zero, are called "quadratic" equations. All quadratic equations have either zero, one, or two real-valued solutions. They can be found easily using the famous "Quadratic Theorem."

Theorem 1.6.9 (The Quadratic Theorem): The solutions to

$$ax^2 + bx + c = 0$$

are given by

$$x = \frac{-b \pm \sqrt{b^2 - 4ac}}{2a},$$

where the symbol "±" (read "plus or minus") indicates two distinct solutions when the square root is positive.

Quadratic Theorem Format Requirement: The equation must be in the form

"$ax^2 + bx + c = 0$,"

where "a", "b", and "c" are parameters, a is not zero, and "x" is the unknown.

Note that **theorems express their own "format requirements."** One half of this theorem expresses a format in abstract algebraic notation using dummy variables. It is the form or pattern that matters, not the letters used to express the pattern. The other half expresses the alternative form into which the original may be changed.

Example 10: Solve $7x^2 - 5x = 3$.

The first step is to convert the given equation into the standard quadratic form with all terms on the left. It is equivalent to

$$7x^2 - 5x - 3 = 0.$$

This does not factor easily. $a = 7$, $b = -5$, and $c = -3$. Now we can go straight to the solution:

$$x = \frac{-(-5) \pm \sqrt{(-5)^2 - 4(7)(-3)}}{2(7)} = \frac{5 \pm \sqrt{109}}{14}$$

so

$$x = 1.103 \text{ or } x = -0.389.$$

To solve a quadratic equation it is *sometimes* possible to use the Zero Product Rule, but it is *always* possible to use the Quadratic Theorem.

Example 9, revisited: Recall that in Example 8 we factored to replace the equation "$x^2 + 5x - 6 = 0$" with "$(x + 6)(x - 1) = 0$", which has solution $x = -6$ or $x = 1$. Of course, if we used the Quadratic Theorem instead, we would obtain the same answer. Here $a = 1$, $b = 5$, and $c = -6$. Therefore

$$x = \frac{-5 \pm \sqrt{5^2 - 4(1)(-6)}}{2(1)} = \frac{-5 \pm \sqrt{49}}{2} = \frac{-5 \pm 7}{2} = -6 \ \textit{or}\ 1,$$

as before.

Example 11: "$(x + 2)(x - 5) = x^2 - 3x - 10$" is an identity stating the equivalence of two expressions. Which expression is preferable?

Equivalent expressions express different sequences of operations. Which expression is preferable depends upon which order of operations is preferable, and that depends upon the context. Watch.

Context 1: Solve $x^2 - 3x - 10 = 0$. In this context we might prefer the other expression (in factored form) because of the Zero Product Rule format requirement that one side be a product and the other zero. The equation "$(x + 2)(x - 5) = 0$" is easy to solve.

Context 2: Solve $(x + 2)(x - 5) = 4$. In Context 1 we liked the product; in this context we don't. Here the right side is not zero, so the product on the left is of no use for the Zero Product Rule. We will have to subtract 4 from both sides to obtain the zero we want, and then the left will not be a product. So we "multiply out" the left and consolidate like terms. Then we deal with the resulting quadratic however we can. The Quadratic Theorem will certainly work, and, if it factors, we could use the Zero Product Rule.

$(x + 2)(x - 5) = 4$ iff $x^2 - 3x - 10 = 4$ iff $x^2 - 3x - 14 = 0$.

This does not factor easily, so use the Quadratic Theorem.

Reconsider the original question: Which expression is preferable?

Without a particular context there is no answer. **The format requirements of the three traditional methods of solving equations tell you which expressions are preferable.**

Evaluate-and-Compare. I call the fourth basic method of equation-solving, "evaluate-and-compare." Other common names for it are "guess-and-check" and "trial-and-error."

Example 12: Solve $2(x + 5) = 18$.

Children who barely know the notation can solve this by trial and error. They try $x = 1$, then $x = 2$, etc. until they discover the solution ($x = 4$). The expression "$2(x + 5)$" tells you what to do to evaluate it. It expresses "Add 5 and then multiply by 2." It is not surprising if children want to solve this by trying what it says to do!

Fortunately, in modern classrooms this urge to try can be accommodated with calculators. Computer-aided solutions to equations are usually refinements of this *evaluate-and-compare* method. Graphics calculators facilitate this method by providing a very efficient way of trying many numbers and displaying the results visually.

The fact that children use *evaluate-and-compare* tells us that it is the most natural method for solving equations. *Evaluate-and-compare* will work for any equation with one unknown; it does not have any format requirements. (The other three methods all have format requirements.) Therefore the method you were probably discouraged from using in algebra is actually the one that is conceptually simplest.

Graphical Solutions. *Evaluate-and-compare* is the basic method behind solving solutions graphically. Calculators can rapidly evaluate an expression for many values of x and the display the results with a graph.

Calculator Exercise 2: Solve $xe^x = 7$ graphically.

For graphical solutions it is common to "Put everything on the left." Here this step yields the equivalent equation "$xe^x - 7 = 0$."

The previous three methods of indirect solution will not work. But a graphics calculator can rapidly *evaluate* the expression $xe^x - 7$ for many x's (corresponding to vertical columns of pixels) and display the images (Figure 2). *Compare* them to 0. Zooming in on the x-value which yields the image (y-value) 0 is easy.

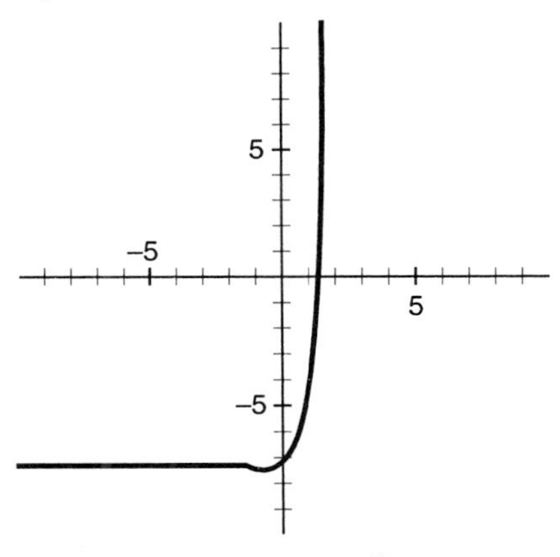

Figure 2. $xe^x - 7$. [-10,10] by [-10,10].

From the picture it looks as if the x-value which yields height 0 is somewhere between 1 and 2. But this graph will not yield the answer with much accuracy.

To obtain greater accuracy we could enlarge the picture to see finer detail. Methods for doing this were discussed in the previous section. A three-significant-digit approximation is $x = 1.52$.

Guess-and-check, also known as trial-and-error, is an excellent approach which can work even without a graphics calculator.

Calculator Exercise 3: Solve $xe^x = 7$ using guess-and-check (not graphically).

To solve this equation start with any initial number, say $x = 1$, evaluate xe^x, and check to see how close it is to 7. $1e^1 = 2.718$, too low. Now I could try $x = 2$ and see what its image is. $2e^2 = 14.778$, too high. Now try a number in between, say 1.5. $1.5e^{1.5} = 6.722$, closer, but too low. So try a slightly higher x, perhaps 1.6. $1.6e^{1.6} = 7.925$, too high. Repeat this process until you have, say, three significant digits.

You get the point. This is basically what a calculator "solver" does, only many times faster. Many calculators with a *solve* key ask you to "Put everything on one side" first. Then they know to aim for the value zero. Again, the three-significant-digit approximate solution is 1.52.

Evaluate-and-Compare Format Requirements: None. (However, it is easier to use if one side is zero or a constant.)

The number of real-world problems which naturally fit into this fourth category is far greater than you may have been led to believe in algebra and trigonometry. Examples include:

$$x^3 + x = 1, \quad x + \sin x = 3, \text{ and } \quad x + \ln x = 3.$$

These three are merely representative of vast categories of equations which are not usually solved by indirect algebraic methods. This evaluate-and-compare method is a part of the important subject area of mathematics called "numerical analysis."

<u>Comparing the Four Methods</u>. The four ways cover all types of precalculus equations. When you realize that there are only four essentially different routes to solving equations, you will find it easier to decide what to do to solve a new equation.

Evaluate-and-compare is conceptually the simplest method. All you need to know is the order of operations in expressions. It is a direct "numerical" method because you actually do the operations represented by the notation.

The other three methods (inverse-reverse, Zero Product Rule, and the Quadratic Theorem) are traditional algebraic methods that are indirect because you do *not* do the operations represented in the problem. They are capable of yielding exact symbolic expressions for solutions. Nevertheless, we usually give solutions in decimal form rounded off to a few significant digits.

> **When we ask you to "solve algebraically," we mean that you should use one of the three traditional, algebraic, indirect methods. We do not mean that you must report the exact symbolic solution which the method is capable of yielding. Just give the solution in decimal form rounded off to three or more significant digits.**

In many algebra classes the *evaluate-and-compare* method is regarded as not really algebraic and therefore missing the point, so, in those classes students need to choose between the three indirect methods. *Evaluate-and-compare* is a so-called "numerical" method which can handle many problems the other methods cannot, but which may yield approximate rather than exact solutions. For practical purposes, an approximate solution with a sufficient number of significant digits is all you could want. Therefore, the evaluate-and-compare method should not be considered inferior to the other methods, although it receives far less attention in school.

The "four ways" are intended to help determine how to solve an equation. If you don't want to use evaluate-and-compare, pick one of the three algebraic methods and, if necessary, manipulate the original equation so it fits the format requirement of that method.

Example 13: Solve $x \ln 25 - 7 = 2x \ln 4 + 10$.

You may subtract $2x \ln 4$ from both sides to obtain

$$x \ln 25 - 2x \ln 4 - 7 = 10.$$

Consolidate like terms.

$$x(\ln 25 - 2\ln 4) = 17.$$

This is now ready for the inverse-reverse method.

$$\text{Solution: } x = 17/(\ln 25 - 2\ln 4) = 38.09.$$

<u>**Format**</u>. If the format of the original equation is not right, as initially in Example 13, we must first take steps to rearrange it into an equivalent equation which does satisfy one of the format requirements. Equivalence is important in this context. Theorems, including identities, justify steps which are *not* intended to change the *unknown numbers*. But they do change the format of the equation and the order in which the operations are executed. *Order* is a key concept of this entire chapter.

Example 14: Let $x > 0$. Solve $x^2(x^3) = 243$.

If we stick with the given order, only the evaluate-and-compare method would solve this equation. But an identity ("property" of powers) can convert it into an equation with only one appearance of x.

Theorem 1.6.10: For $x > 0$, and any p and r, $x^p x^r = x^{p+r}$.

Therefore, the original equation is equivalent to "$x^5 = 243$." This equation has only one appearance of "x". Thus the *inverse-reverse* method will work, but to finish the problem we also need to know the inverse of the 5th power function. It is the 5th root function -- the one-fifth power function. Theorems about power functions and root functions include the following property.

Theorem 1.6.11: If p is an odd integer, $x^p = c$ iff $x = c^{1/p}$.

Returning to the equation "$x^5 = 243$," we see it is equivalent to "$x = 243^{1/5}$." Evaluating this, we find the solution, $x = 3$.

It is important to note that to reorder the operations may even require different operations. For instance, in Example 14 multiplication in one order (x^2x^3) is converted to addition in another ($x^{2+3} = x^5$). Then the inverse-reverse method is applicable.

Calculator Exercise 4: Find the keystroke sequences for evaluating powers and roots. Reproduce these results:

$5^3 = 125.$ $4^{-2} = .0625.$

The cube root of a number is also called the one-third power, as is justified by Theorem 1.6.11 above. When taking roots, you may treat 1/3 as 3^{-1} and $1/n$ as n^{-1}. Reproduce these results:

$243^{1/5} = 3$ [not 48.6. Caution: Distinguish between $243^{1/5}$ and $243^1/5$.]

$100^{5^{-1}}$ - 2.51.

Example 15: Solve algebraically: $(x + 4)(x - 1)^2 + (x - 1)(x + 4)^2 = 0$.

The intent of the instruction with the word "algebraically" is to have you use one of the three traditional methods. Which method will you use? Look at the arrangement of the variables and the order of operations.

Concisely stated, the three acceptable "algebraic" formats are: a quadratic, only one appearance of "x", or a zero product. Are we close to any of these?

It's not a quadratic (note the power of x, which is 3, not 2). I see four x's and I do not see how to rewrite this with only one. So, by elimination, I will try to write the expression as a product. Now it is a sum.

The Distributive Property can turn a sum into a product. Let's "factor out" (but not "cancel") common terms.

$$(x + 4)(x - 1)^2 + (x - 1)(x + 4)^2$$
$$= (x + 4)(x - 1)[(x - 1) + (x + 4)] \quad \text{[by the Distributive Property]}$$
$$= (x + 4)(x - 1)[2x + 3]$$

A product!

Now, $(x + 4)(x - 1)[2x + 3] = 0$ if and only if

$$x + 4 = 0 \text{ or } x - 1 = 0 \text{ or } 2x + 3 = 0.$$

Therefore the solution is: $x = -4$ or $x = 1$ or $x = -3/2$.

By the way, this type of problem arises frequently in calculus.

Simplifying. Now the purpose of "simplifying" can be made clear. There are some types of expressions we especially like, and some types we don't like. We like expressions that fit the format requirements. If an expression does not express one of the operational sequences we like, we can try to rewrite it. To do so we may employ identities, properties of functions, and other theorems. To simplify an expression means to rewrite it as an equivalent expression in a more convenient form. Precisely what is "convenient" depends upon what form you want (Problem A18).

"Nice". Sometimes equations have expressions I like, and sometimes they have expressions I don't like -- and I say so to my students. But many students don't think I should ever call an expression or equation "nice." What could be "nice" about an equation? They must think that math is a cut and dried subject with no room for subjective terms. But practice definitely shows that some equations are "nicer" than others. Some are not very friendly until you change them.

In Example 14 I did not like the expression "x^2x^3." It's not a nice expression. I like "x^5," an equivalent expression, much better, because it expresses a more convenient order of operations.

In Example 15, the original equation was

$$(x + 4)(x - 1)^2 + (x - 1)(x + 4)^2 = 0.$$

Looking at it again, there are parts I like and parts I don't like. The zero on the right is fine, but the sum on the left is not. A product on the left would be nicer. I responded to that feeling and solved the equation by rewriting the expression on the left as a product.

Example 16: Solve the equation "$(x - 3)(x - 5) = 10$."

The left side is factored. It expresses a product. Do you like that?

You spent a great deal of time in school learning to factor, so your first reaction might be to like this factored form. But *why* do we like factored form?

We like factored form because of the Zero Product Rule, which deals only with *zero* products. If you have a product that is not zero, like this one, factored form is useless. Not only do I not like this factored form, I am going to have to get rid of it. To solve this equation algebraically, I'm going to have to multiply out ("unfactor") the left, subtract 10 from both sides to obtain a zero on the right, and use the "$ax^2 + bx + c = 0$" form in the Quadratic Theorem (Problem A17).

The purpose of identities is to offer you options to rewrite expressions in different forms, and many theorems give you options to rewrite equations in different forms. You must learn which forms you like and dislike -- which are nice and which are not -- so you can decide when to exercise your options.

Conclusion. To solve an equation algebraically its component expressions must fit, or be rearranged to fit, certain patterns which express convenient orders. In the context of equation-solving, the purpose of identities and "properties" of functions is to permit reordering the operations so that the format requirements of equation-solving methods can be satisfied. This four-way classification scheme not only helps explain traditional equation-solving methods, it awards the newer high-tech methods their proper place.

Terms: direct, indirect, solve algebraically, inverse-reverse, Distributive Property, Zero Product Rule, Quadratic Theorem, evaluate-and-compare, format requirement, simplify.

****

Exercises for Section 1.6, "Four ways to solve an equation":

A1.* a) What are identities about? b) What is their use?

A2.* a) The name "inverse-reverse" is an abbreviation of what?
b) What is the format requirement for the *inverse-reverse* method?

A3.* a) State (with variables) the Zero Product Rule.
b) Give, in English, the format requirement for the Zero Product Rule method.

A4.* What is the format requirement for the *evaluate-and-compare* method?

A5.* Which property is commonly used to turn a sum or difference into a product?

A6. The format requirements of which methods are already met by the following expressions? a) $(x - 5)(\ln x + 7) = 5$. b) $xe^x = 0$.

A7. The format requirements of which methods are already met by the following expressions? a) $3(5 - \ln x^2) = 0$. b) $x(\sin x + 1) = 7$.

A8. a) Why can't you use the inverse-reverse method on "$5.4x + 3.2x = 76$" as it stands? b) Solve it algebraically.

A9. a) Why can't you use the Zero Product Rule on "$(x - 5)(x + 3) + 1 = 0$" as it stands? b) Solve it algebraically.

A10. a) Why can't you use the Zero Product Rule on "$(x - 5)(x + 3) = 1$" as it stands? b) Solve it algebraically.

A11. a) Why can't you use the Zero Product Rule on "$(x - 5)(x + 3) = 4x$" as it stands? b) Solve it algebraically.

A12. State an identity which allows us to regard "$5 + x$" as beginning with x (instead of beginning with 5).

A13. State an identity which allows us to regard "$5x$" as beginning with x (instead of beginning with 5).

A14. State the Distributive Property using "x", "y", and "z".

A15. State the Zero Product Rule (Theorem 1.6.6) using some other letters.

A16. State the Zero Quotient Rule (Theorem 1.6.8) using some other letters.

A17. Solve $(x - 3)(x - 5) = 10$.

A18. Suppose a student begins to solve the equation in Example 15 by multiplying out the terms. a) Are any of the four methods easier to use after the terms have been multiplied out? b) What does this teach you about multiplying out long expressions?

^ ^ ^ ^ ^ ^ ^ ^

B1.* Name the four ways to solve an equation and their corresponding format requirements.

B2.* Suppose you are to solve an equation which does not fit any of the three indirect-method formats and you are forbidden to use *evaluate-and-compare*. What can you do?

B3.* Medium length essay: Algebra is about order and operations. How is that important in this section?

B4.* Compare the following theorem, letter by letter, to part A of Theorem 1.6.2 and then to part B: $a + b = c$ iff $a = c - b$. In each case, what is the essential difference? Why was Theorem 1.6.2 stated the way it was?

B5.* What is the purpose of simplifying an expression?

B6.* State the version of the Distributive Property which applies to
a) multiplication and subtraction b) division and addition
c) division and subtraction

B7. State the Zero Quotient Rule (1.6.8) using different letters.

B8.* When do we like factored form?

B9. a)* Theorems express their own "format requirements." How?
b) Use the Zero Product Rule to illustrate your answer to part (a).

B10. Theorem 1.6.1 is stated with letters chosen to emphasize its role in solving problems in which the unknown is multiplied by a number.
a) Restate it, using its dummy variables differently, to emphasize problems in which the unknown is divided by a number.
b) Restate it in a neutral manner which does not use "x" to suggest which position holds the unknown.

B11. Theorem 1.6.7 is stated with letters chosen to emphasize its role in solving problems in which the natural logarithm function is applied to the unknown.
a) Restate it, using its dummy variables differently, to emphasize problems in which the exponential function is applied to the unknown.
b) Restate it in a neutral manner which does not use "x" to suggest which position holds the unknown.

B12. Solve algebraically: $(x - 5)(x^2 - 9) + 2x(x^2 - 9) = 0$.

B13. Solve algebraically: $4x(x^2 - 5) + (x^2 - 5)^2 = 0$.

B14. Solve algebraically: $(x - 3)^2(x + 5) - 3(x - 3)(x + 5)^2 = 0$.

B15. Solve algebraically: $4x^3(1 - x)^3 - 3(1 - x)^2x^4 = 0$.

B16. Solve: $10 \ln x = x$ [Hint: Did you find a representative graph?].

Note: This text occasionally supplies solutions with two significant digits in brackets (problems B17-20). These are given so you can immediately determine whether you used the correct solution process. On the homework, prove that you know the correct process by giving a more accurate answer rounded to three or more significant digits.

B17. Solve: $x + 2 = e^x$ [1.1 and a second solution].

B18. Put your calculator in "radian mode" so that sin 1 = 0.841471 (not 0.0157073, that would be in degrees). Solve $x + \sin x = 1$ to three significant digits on your calculator. [.51]

B19. Put your calculator in "radian mode" so that sin 1 = 0.841471 (not 0.0157073, that would be in degrees). Solve $x + \cos x = 1.3$ to three significant digits. [.37]

B20. Solve $x + \ln x = 3$ to three significant digits on your calculator. [2.2]

B21. State an identity which allows us to regard "$5 - x$" as beginning with x (instead of beginning with 5).

B22. State an identity which allows us to regard "$7/x$" as beginning with x (instead of beginning with 7).

B23. State an identity which allows us to regard "$120/x$" as beginning with x (instead of beginning with 120).

B24. State an identity which allows us to regard "99 - x" as beginning with x (instead of beginning with 99).

B25. Define "simplify" in the context of expressions.

^ ^ ^ ^There is a big difference in how the two equations usually would be solved, in spite of superficial similarities. Identify the different approaches (do not solve them).

B26. "$x^2 + 7x = 0$" and "$x^2 + 7x - 2 = 0$."

B27. "$x^2 - x = 7$" and "$x^3 - x = 7$."

B28. "$x^2 + \sqrt{x} - 3 = 0$" and "$x^2 + x - 3 = 0$."

B29. "$x(x + 3) = 0$" and "$x(x + 3) = 1$."

B30. Graph this by algebraically solving for y and then using a graphics calculator:
$3x + xy + 7 = 5y$.

B31. Graph this by algebraically solving for y and then using a graphics calculator:
$x^2 + 2x - 3xy + y^2 - 17 = 0$.

CHAPTER 2

Functions and Graphs

Section 2.1. Functions

Algebra is about operations and order. Chapter 1 emphasized that fact. Algebraic expressions such as "$2x + 9$" incorporate a sequence of operations. When you see such an expression, you should distinguish three (yes, 3) things: the number you begin with, x, the number you end with, $2x + 9$, and the sequence of operations for getting from the one to the other, "Multiply by 2 and then add 9." Such sequences are rules. To enable us to study this type of rule, as opposed to numbers, we give this type of rule a name, *function*. Terms are important because by use of a name we can bring an object (even an abstract object) to mind. The objective of this section is to raise your conception of functions up from the conceptual level of numbers to the level at which functions are objects in their own right.

In the previous section, "Four Ways to Solve Equations," we saw that to solve an equation it is often necessary to identify the operations and their order in an expression. To solve "$2x + 9 = 42$" we first note the operational sequence "Multiply by two and then add nine." Then we "undo" that sequence with the "inverse-reverse" method. In the terminology of this section, we identify the *function* "Multiply by two and then add nine." Whether or not you use the term *function*, to do algebra you must have that concept. By giving it a name we can bring this concept to mind and focus attention on it. *Function* is a fundamental concept which is more abstract than the concept of *number*. Be careful to make the distinction between *function* and *number*.

Example 1: An expression such as "$7(x + 3)$" expresses a rule for obtaining its calculated value (the "image" or "output") given an original number ("x", the "argument" or "input").

When the argument is 2, the image is $7(2 + 3) = 35$.

When the argument is 8, the image is $7(8 + 3) = 77$.

The expression creates *many* pairs of numbers by *one* rule. Separating out the rule for consideration as an object in its own right is an example of mathematical abstraction (Figure 1).

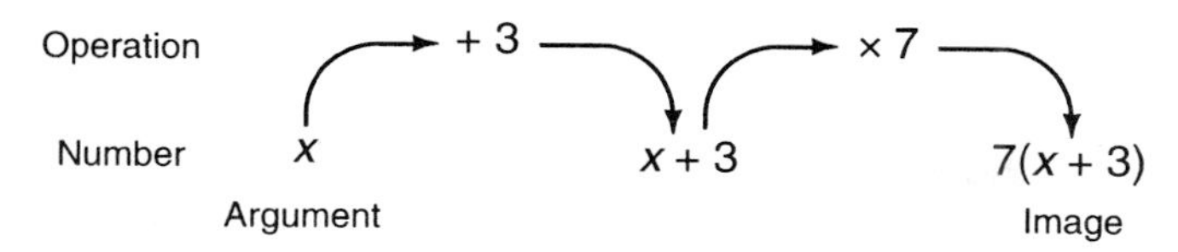

Figure 1: Whatever number x is, add 3 to it and then multiply the sum by 7, to obtain $7(x + 3)$.

"$7(x + 3)$" is the mathematical notation for the *rule* "Add 3 and then multiply by 7."

Example 2: Give a mathematical expression expressing the rule: "Subtract 5 and then square."

The answer in terms of "x" is "$(x - 5)^2$."

Now we are ready for a formal definition of *function*. Actually, there are several correct ways to define *function*, depending upon the mathematical sophistication of the audience. First I will give a correct and illuminating version which defines a function to be a rule.

Definition 2.1.1 (Function). A function is a rule that associates exactly one number with each number in a specified set.

The "specified set" is called the domain of the function and will often consist of all real numbers. In the usual "$y = f(x)$" notation, the domain is the set of all x values under consideration. A function is a rule that tells you how to obtain the image ($f(x)$ or y) associated with any argument (x) in the domain. Many rules are simple, such as "Multiply by ten." Many functions are far more complex. But simplicity or complexity is not the key idea in the definition of "function." The key idea is that a function is an abstract object in its own right. Functions apply to numbers and generate numbers, but they are not numbers themselves.

An important part of the definition of *function* is that each initial number (each argument) yields exactly *one* second number (one image). The function *determines* the image. A function never yields two images that might require a choice between them. If we want to discuss rules where there may be two or more images for a single argument, we can. They are called "relations" and will be discussed in Section 2.4. We study functions first because algebraic expressions work like functions. For a given value of "x", an expression with variable "x" takes on only *one* value (one image).

Because functions in algebra and calculus usually have numbers as arguments and images, I defined the term "function" to refer to rules for which both the arguments and images are numbers (rather than something else, like people). Technically, the term *function* could be defined to permit other types of arguments or images, but we will only use the usual numerical type of functions.

Often we denote the rule by "f ", the argument by "x", and the image by "$f(x)$" (read aloud, "f of x").

Terminology. If a function, f, takes a number x in the domain and yields the number $f(x)$, then the number x is called the argument and $f(x)$ is called the image or the value of f at x. Using computer terminology, the argument is the "input" and the image is the "output."

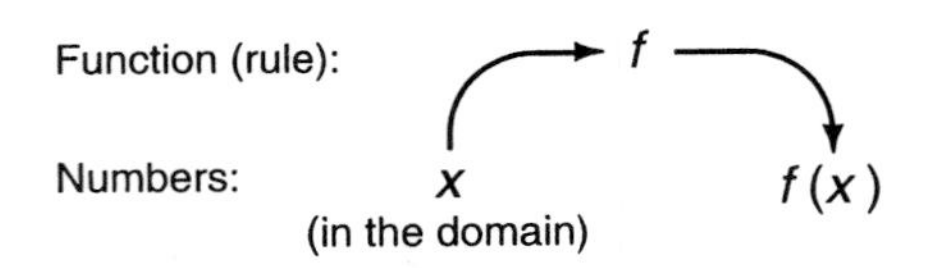

Figure 2: argument-function-image

The argument must be in the domain for the expression "$f(x)$" to make sense. The point of mentioning the domain is that we may wish to restrict the possible arguments. For instance, in the rule for calculating the area of a square (the image) from the length of a side (the argument), we would probably wish to forbid sides of negative lengths. Then the domain of the function would not include negative numbers, even though it is possible to take the square of a negative number.

Order is important in the concept of a function. Think of the argument as first, then the function as operating (that is, doing something to the argument), and then of the second number (the image) appearing (Figure 2). Note that the function is not a number.

The notation for some functions is, unfortunately, written right-to-left. For example, the symbol for the square root function, $\surd$, is usually written to the left of the argument: $\surd x$. Similarly "ln" is written to the left of "x" in "ln x." These notations are misleading, but too entrenched to ever change. Conceptually, you should still think of the x as first, the operation (take the square root, or take the natural logarithm) as next, and the image as last.

Formulas. You are familiar with functions already, but you may know them as "formulas."

Example 3: The formula for converting temperatures expressed in Celsius to temperatures expressed in Fahrenheit is $F = (9/5)C + 32$, where C represents the Celsius temperature and F represents the Fahrenheit temperature. For example, the formula tells us that 20 degrees Celsius is 68 degrees Fahrenheit.

We could rewrite the formula using functional notation: $F(C) = (9/5)C + 32$. For this function the letter "C" is the variable used as the argument and $F(C)$ is the image. The rule is, "Multiply by 9/5 and then add 32."

Example 4: The formula for the area, A, of a circle of radius r is given by $A = \pi r^2$. The area is a function of the radius. We might write "$A(r) = \pi r^2$" to emphasize the dependence of the area on the radius. The argument is "r" and the image is "$A(r)$". The domain is the set of all $r > 0$ [or $r \geq 0$].

To find the area of a circle given the radius, you "Square and multiply by π." The formula expresses the rule which is the function.

Functional versus Formula Notation. The formula notation "$A = \pi r^2$" corresponds to the functional notation "$A(r) = \pi r^2$." There are advantages to functional notation and other advantages to formula notation. Formula notation is shorter and therefore more convenient. But with formula notation when r changes, A changes. When $r = 2$, $A = 4\pi$. When $r = 3$, $A = 9\pi$. But both give the area of circles. What do those two calculations have in common? The rule. They have the process "Square and then multiply by π" in common.

Functional notation gives the rule a name of its own.

Using the notation, "$A(r) = \pi r^2$," the rule is "A". Area (the image) is "$A(r)$." Functional notation is abstract and emphasizes the essence of the calculation. Nevertheless, formula notation is so entrenched in mathematics that we will use it repeatedly.

In algebra it is common to avoid the advanced "$f(x)$" notation in favor of "y" notation. For example, the function "Multiply by two" is often expressed by the equation "$y = 2x$." When the purpose of an equation with two variables is to determine y functionally in terms of x, "x" is called the independent variable (the argument of the function) and "y" is called the dependent variable (the image). We will use both types of notation, but we emphasize the "$f(x)$" notation because it is essential for calculus. The equation "$y = 2x$" creates many ordered pairs. What do they have in common? The rule, "Multiply by 2." Writing it using "$f(x) = 2x$" emphasizes the rule, f.

Functions of Two Variables. In each of the first four examples the image depends upon the argument which is expressed with *one* variable. Some formulas relate three or more numbers to each other and determine functions of two or more variables.

Example 5: Consider the formula "distance equals rate times time." The formula, $d = rt$, determines the distance, d, traveled by an object moving at a constant rate, r, for a given amount of time, t. In this case we could think of the distance as a function of two arguments, rate and time. We might even write "$d(r, t) = rt$" to emphasize the dependence of the distance on both the rate and the time. Note that if one of the arguments is assigned a fixed value, then d becomes a function of the other argument. For example, if $r = 55$ (miles per hour), then $d = 55t$, which expresses a function of one variable.

The theory of functions with two or more arguments is similar to the theory of functions with only one argument; we will concentrate on functions with one argument.

"f" versus "$f(x)$". When you see the expression "$f(x)$" you should see three (yes, 3) things. The argument, x, the image, $f(x)$, and the function, f. If you call the function "$f(x)$" you are making a conceptual mistake that may come back to haunt you.

Example 6: Give a descriptive name, as a command, for the function defined by the statement, "Let $f(x) = 4x + 3$."

Answer: Multiply by 4 and then add 3.

Do *not* describe the function as "Multiply x by four and then add three." The reference to "x" should be omitted. The function has nothing to do with "x". Only the *notation* has to do with x. The function will take *anything* and "Multiply by four and add three." It is only a feature of our notation that we tend to use "x" to represent *anything* when writing statements. In our notation, "x" is merely a *placeholder* -- a dummy variable. "x" could be replaced by "z" as long as it is replaced in both places: "Let $f(z) = 4z + 3$" defines exactly the same f as "Let $f(x) = 4x + 3$." The letter is a dummy variable since it does not matter which letter is used (as long as certain conventions are not violated).

Note that functions are rules (processes), not numbers. **As a command we may think of a function as an evaluation process.**

Example 7: Suppose a mathematician wished to define a function which would square any given number and then subtract 7 from the square. How would that function be defined in mathematical notation?

A mathematician would probably define the function by stating "Let $f(x) = x^2 - 7$."

"Let" is a word commonly used in Mathematics to introduce a definition. "x" is the favorite letter used to represent any real number. Since it is merely a dummy variable, it may represent *any* real number, including 3, 10.2, -5, or even a number given the name c, y, $2x$, or $x - 5$.

Example 7, continued: Let $f(x) = x^2 - 7$. Determine $f(y)$.

This can fool beginners. They sometimes ask, "How can I tell? You only told me f of x." But, the point is, the function is "Square and then subtract 7" and has nothing to do with the name of the variable. "x" is merely a letter used to represent *any* real number. Remember, "x" is only a dummy variable -- a placeholder.

$f(x) = x^2 - 7$.
$f(3) = 3^2 - 7 \quad [= 2]$.
$f(y) = y^2 - 7$.
$f(2x) = (2x)^2 - 7$.
$f(x + h) = (x + h)^2 - 7$.
$f(x/2) = (x/2)^2 - 7$.

Note that $(f(x))/2$ is not the same as $f(x/2)$. Order matters! Also, $(2x)^2 - 7$ is not the same as $2x^2 - 7$. The parentheses are used to indicate the order of the operations. The parentheses in "$f(2x)$" are used to indicate that the given number "x" is doubled *first* and then is treated as a *new given* number. It is then $2x$ which is squared, not x, and then 7 is subtracted.

Placeholders. The fact that "x" need not be "x" is remarkable. Math has placeholders (dummy variables). You can often say the same thing two different ways by using different letters. For example, "Let $f(x) = x^2 - 7$" defines *exactly* the same f as "Let $f(z) = z^2 - 7$." Both apply to all real numbers. Whether you represent the real number by "x" or "z" or some other symbol is not important.

The notation for functions commonly uses a dummy variable, "x". It could use a black box as a placeholder instead.

Example 8: Define f by $f(\blacksquare) = \blacksquare(\blacksquare + 3)$.

Find $f(5)$, $f(17)$, $f(x)$, and $f(\sqrt{x})$.

$f(\blacksquare) = \blacksquare(\blacksquare + 3)$.
$f(5) = 5(5 + 3)$
$f(x) = x(x + 3)$.
$f(\sqrt{x}) = \sqrt{x}(\sqrt{x} + 3)$.
$f(x + h) = (x + h)(x + h + 3)$

This illustrates that "x" is merely a placeholder in the definition of a function.

Examples 6 through 8 are intended to emphasize that the function is the *rule* and the notation in which it is expressed is not an essential part of the function.

Functional Notation and Order: Although Mathematics is usually read from left to right, like English, the interpretation of the notation for functions is not.

Functional notation is often understood right to left.

The argument (whatever is in the parentheses) comes first and *then* the operation indicated by f is applied to it. Thus "$f(x^2)$" (read "f of x squared") first squares x (on the right) and then applies f to the square.

If we want to apply f first and square last, we want $[f(x)]^2$, which may be written "$f^2(x)$." "$f(x^2)$" is quite different from "$[f(x)]^2$." For example, if f is defined by $f(x) = x - 8$, then $f(x^2) = x^2 - 8$, but $[f(x)]^2 = (x - 8)^2 = x^2 - 16x + 64$, which is not the same.

In a very real sense the notation for functions is backwards. When read right-to-left, sentences like "$y = f(x)$" display the image y first, then the function f, and then the argument x. Completely backwards! However, when you get used to it, that is no problem. But it may be a big problem at first. Be sure to learn to think in the proper "argument - function - image" sequence.

"f " and "$f(x)$". In this section we take pains to distinguish between "f " and "$f(x)$." Most texts do not. Why is the difference worth emphasizing?

First of all, f and $f(x)$ are conceptually quite different. f is a function and $f(x)$ is not. Algebra is conducted at the functional level of thought. Whether or not you use the term *function*, you cannot do algebra without recognizing functions.

Example 9: Suppose the problem is to "Solve $6x + 4 = 87.5$." What will you *do*?

This is a very simple problem, but it makes a point. Compare that problem to "Solve $6x + 4 = 123.4$. What is the difference? Here are the two solution processes, side by side:

	$6x + 4 = 87.5$		$6x + 4 = 123.4$
iff	$6x = 87.5 - 4 = 83.5$	iff	$6x = 123.4 - 4 = 119.4$
iff	$x = 83.5/6 = 13.917$	iff	$x = 119.4/6 = 19.9$

To do these problems you must recognize the "Add 4" function, which was applied last, and undo it (by subtraction). Then, you must recognize the "Multiply by 6" function and undo it (by division). At each step you must think about the functions, not just the numbers. The same *process* applies to both problems, regardless of whether the number is 87.5 or 123.4. The solution *process* is identical, although the solution is different. **The process is determined by the function, not the numbers.** Note that abstraction is at work here. One process applies to many examples with an essential similarity.

Algebra is about operations and order, not about numbers. This is a very strong reason for abstracting operations and order into a single concept. **The one term we have that concentrates attention on operations and order is *function*.**

There is no law that says functions have to be simple. You can define a function to be as complicated as you wish; the only qualification is that each argument in the domain must have exactly one image.

Example 10: Let $h(x) = 1$ if x is rational, and $h(x) = 0$ if x is irrational.

With this definition of h, $h(2/5) = 1$, $h(1.4) = 1$, and $h(\sqrt{2}) = 0$.

It is important to grasp that a particular function is *one* thing. Even this function h gives only *one* rule. You may, at first, think that it gives two rules, one for rational numbers, and another for irrational numbers, and in some sense that is so, but, from the point of view of mathematics, these are merely two parts of the *one* rule, h.

Many important functions have definitions in which the domain is split into two or more parts. For example, the absolute value function is defined by "$|x| = x$ if $x \geq 0$ and $|x| = -x$ if $x < 0$," in which the numbers less than zero are treated differently from the rest. Functions for which the domain is divided into two or more regions with different algebraic expressions in each are said to be defined piecewise (or "split functions"). The Federal Income Tax function is defined piecewise. The formula for your taxes in terms of your income depends upon your income bracket. But, no matter how complex its definition, any function is regarded as expressing *one* rule.

Natural Domain. In the definition of a particular function, the set of all numbers such that the defining expression yields a real number is called the natural domain. The two most common operations which may limit the natural domain are the taking of square roots (negative numbers are not in the natural domain because the square root of a negative number is not a real number) and division by 0 (we do not permit division by 0 since division by 0 is undefined).

Example 11: Let $f(x) = \sqrt{x}$. Find the natural domain.

The sentence "Let $f(x) = \sqrt{x}$," is shorthand for the more complete sentence, "Let f be the function with values defined by $f(x) = \sqrt{x}$, for all $x \geq 0$." Since the square root is not a real number for negative arguments, the natural domain is the set of real numbers greater than or equal to 0: $[0, \infty)$.

A graphics calculator can illustrate the concept of natural domain. If an x-value is not in the natural domain, the calculator will not be able to evaluate $f(x)$ as a real number and it will not be able to produce a point on the graph for that value of x.

Calculator Exercise 1: Evaluate $\sqrt{-2}$.

You will get an error message, or a complex number if your calculator is sophisticated enough. Negative numbers are not in the natural domain of the square root function.

Graph $f(x) = \sqrt{x}$ on your calculator. There are no points on the graph to the left of the y-axis because those x-values are not in the natural domain.

In this book, functions without explicit restrictions on their domains will be regarded as having their natural domains.

Example 12: Let $h(x) = \dfrac{x - 6}{x - 2}$. Find the natural domain.

When $x = 2$, the denominator is 0. Since division is not defined when the denominator is 0, the natural domain of h includes all real numbers *except* 2.

Example 13: If f is defined by $f(x) = \ln x$, the natural domain of f is the set of all numbers greater than 0.

Calculator Exercise 2: Evaluate ln(-2).

You will get an error message or a complex number. Negative numbers are not in the natural domain of the natural logarithm function.

Graph $f(x) = \ln x$ on your calculator. Note how the graph has no points on or to the left of the y-axis because the natural domain of *ln* does not include zero or negative numbers.

Example 13, continued: Suppose g is defined by $g(x) = \ln(2x - 6)$. Give the natural domain of g.

To evaluate $\ln(2x - 6)$ the argument of the *ln* function must be positive. Thus $2x - 6$ must be positive. Thus x must be greater than 3. The natural domain of g is the set of all real numbers greater than 3: $(3, \infty)$.

Calculator Exercise 3: Graph $f(x) = \ln(2x - 6)$ on your calculator. You can see that there are no points to the left of 3. If you use the *trace* feature and move the cursor to the left of $x = 3$, there will be no y-value for the x-values.

Calculator graphs have limitations. They can show the general behavior of graphs, but they use a limited number of columns of pixels and therefore a limited selection of x-values. An important x-value might not even be tried.

Calculator Exercise 4: Graph $f(x) = \sqrt{x - \pi}$ on your calculator.

From the graph alone it would not be possible to tell if π is in the natural domain or not. The interval to the right of $x = \pi$ visibly produces points and the interval to the left does not. But the calculator does not try $x = \pi$ (*trace* along and see), so the calculator's graph does not completely determine the natural domain. Since $\sqrt{0} = 0$ is a real number, π is in the natural domain which is $[\pi, \infty)$.

The Domain. The domain of a function is the set of all the argument values *under consideration*. Compare this term to "natural domain" which refers to all the argument values mathematically possible (in the sense that the image is a real number). Sometimes we do not want to consider *all* the x values for which the expression yields a real number.

Example 14: Suppose a farmer wants to make a rectangular animal pen with an existing long straight fence for one side and 100 feet of new fencing and gates for the other three sides (Figure 3). What dimensions of the pen would yield the maximum possible area of the pen?

To maximize the area, we need an expression for the area in terms of the dimensions of the pen. Let one side of the pen be "x" (Figure 3).

If the pen is x feet deep, $2x$ feet are used for the two sides pictured vertically, and $100 - 2x$ feet of fence remain for the other side. The area is length times width, so the area function is

$A(x) = x(100 - 2x)$.

What is the domain?

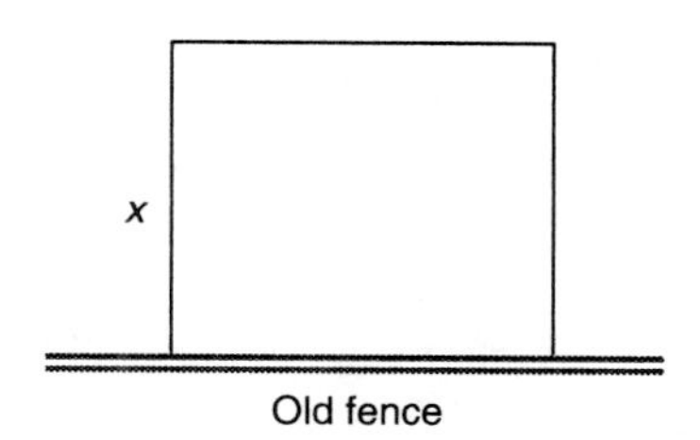

Figure 3: A pen made from 100 feet of new fence on three sides and an old fence on the fourth side.

Calculator Exercise 5: Graph this on your graphing calculator. You will see that the "standard" window does not display a representative graph. Change the window to show a representative graph.

Domains and representative graphs are related. In this example the standard window does not display all the relevant x- and y-values. To select a window, first think of which x's we want to consider. That is, determine the domain.

There is no point in using negative x's. Sides are not negative. Mathematically, negative x's make sense in the expression "$x(100 - 2x)$," but physically they do not. Similarly, that expression makes mathematical sense for large values of x, say $x = 300$, but it makes no physical sense. If $x > 50$, there is not enough fence for the sides. The domain is, by definition, the set of x's under consideration, which is not necessarily the "natural" domain.

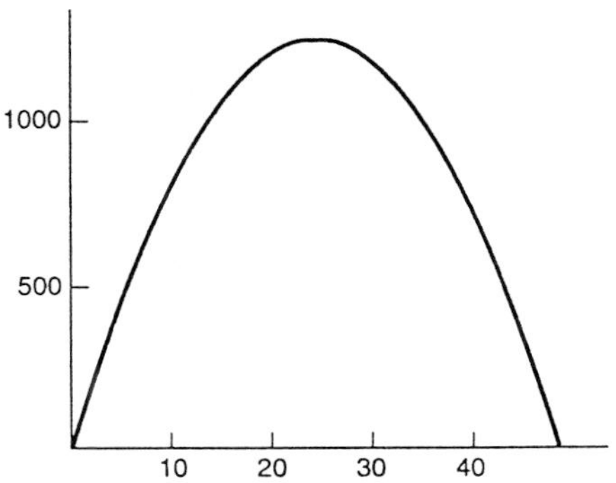

Figure 4: $x(100 - 2x)$ [0, 50] by [0, 1500].

So, in this problem, we would regard the domain as [0, 50], even though the "natural" domain of the expression is all real numbers.

To obtain a representative graph you will also have to select an appropriate "y" interval. Figure 4 uses [0, 1500].

<u>**Conclusion**</u>. In addition to learning how mathematical notation is used to express functions, you must grasp the concept of a function as a rule -- an operation or sequence of operations -- distinct from its image. You must distinguish the *function*, f (as a command, for example "Square it!"), from the *notation* for the function in which f is described by a statement, for example, "$f(x) = x^2$." The idea of a function as a rule or operation is fundamental to mathematics.

Terms: expression, argument, image, function, value of f at x, domain, natural domain, dummy variable, placeholder, formula notation, functional notation (as a statement), rule (as a command).

****

Exercises for Section 2.1, "Functions":

A1.* When you see the expression "$f(x)$" you should distinguish three things. Which three? In which order?

A2.* a) Define "natural domain."
b) Which are the two most commonly encountered operations that may cause the natural domain of a function to be restricted?

A3.* Put these three terms in their proper conceptual order: function, argument, image.

^ ^ ^ ^ Give the functions as commands [do not mention "x"]:
A4. a) $f(x) = 3x$. b) $f(x) = x - 9$.
A5. a) $f(x) = x/5$. b) $f(x) = x + 4$.
A6. a) $f(x) = 2(x + 1)$. b) $f(x) = x/2 - 5$.
A7. a) $f(x) = \log(x + 1)$ b) $f(x) = (x + 2)^2$.
A8. a) $A = \pi r^2$. b) $V = (4/3)\pi r^3$.
A9. a) $A = s^2$. b) $P = 4s$

^ ^ ^ ^ Give the standard "$f(x)$" definitions of the functions:
A10. a) "add 20 and then divide by 3" b) "Divide by 2 and then subtract 1.
A11. a) "subtract one and then take the absolute value." b) "square and then add one."
A12. a) "add 5 and then square." b) "subtract 9 and then divide by 5."

^ ^ ^ ^ Give the natural domain of f. In some cases, it may help to inspect a graph.

A13. a) $f(x) = 1/(x - 2)$. b) $f(x) = x^2 - 3$ c) $f(x) = \sqrt{x - 5}$

A14. a) $f(x) = 1/(x^2 - 4)$ b) $f(x) = 1/(x^2 + 3)$ c) $f(x) = \sqrt{x^2 - 9}$

^ ^ ^ ^ ^ ^ ^ ^ ^

B1.* What is a function?

B2.* Distinguish between "f" and "$f(x)$."

B3.* a) (Short essay) What is the relationship between commands and statements in mathematics?
b) Illustrate your answer with an example relating to the concept of "function."

B4.* In what sense is functional notation "backwards"?

B5.* What is abstract about a function?

B6.* Explain and illustrate with an example the meaning of "placeholder."

B7. Let $f(x) = x^2$. Express
a) $f(4)$ b) $f(z)$ c) $f(x + 1)$ d) $f(x/2)$

B8. Let $f(x) = 3x + 2$. Express
a) $f(4)$ b) $f(z)$ c) $f(x + 1)$ d) $f(x/2)$

B9. Suppose we "Let $y = 3x$." When $x = 4$, $y = 12$. When $x = 5$, $y = 15$. When x changes the numbers change, but something remains the same. What?

B10. Suppose we "Let $y = 2x^2$." When $x = 1$, $y = 2$. When $x = 10$, $y = 200$. When x changes, the numbers change, but something remains the same. What?

B11. Let $f(x) = 1/x$. Express a) $f(2x)$ b) $f(x + h)$ c) $f(1/x)$.

B12. Let $f(x) = x(x - 2)$. Express
a) $f(3)$ b) $f(x - 1)$ c) $f(x^2)$ d) $[f(x)]^2$. e) Why are the last two different?

B13. Let $f(x) = 2/x$. Express
a) $f(5)$ b) $f(x - 4)$ c) $f(1/x)$ d) $f(x^2)$ e) $f(c)$.

^ ^ ^ ^ Use the "empty parentheses" method to express the function.
B14. "Add 2 and then square." B15. "Divide by 3 and then add 5."

B16. Suppose you apply the function "multiply by 2" and then the function "multiply by 10" to the image. This is equivalent to applying one function to the original argument. Which function? [Express your answer in mathematical notation using "f"].

B17. Suppose you apply the function "add 12" and then the function "subtract 9" to the image. This is equivalent to applying one function to the original argument. Which function? [Express your answer in mathematical notation using "f"].

B18. Suppose you apply the function "multiply by 25" and then the function "divide by 5" to the image. This is equivalent to applying one function to the original argument. Which function? [Express your answer in mathematical notation using "f"].

B19. Suppose you apply the function "add 7" and then the function "subtract 7" to the image. This is equivalent to applying one function to the original argument. Which function? [Express your answer in mathematical notation using "f"].

^ ^ ^ ^ For each f, a) Give the rule, and
b) What would you do to solve "$f(x) = c$" for x? [Answer in English.]

B20. $f(x) = 3x + 7$.
B21. $f(x) = x/5 - 8$.
B22. $f(x) = (x - 4)/27$.
B23. $f(x) = 9(x - 5)$.

B24. Why might a mathematician object to the label "$\log x$" on a calculator key?

B25. The cost of photocopying a single original is 6 cents each for the first 10, and 5 cents each for each copy after the first 10. Express the function which gives the cost of n copies.

B26. The cost of printing many copies of a single page is 50 cents for the setup charge, plus 3 cents per copy for the first 200. Each copy after the first 200 costs only 2 cents. Express the function which gives the cost of n copies.

B27. (After B25 and B26) When is the printing method in B26 cheaper than the photocopy method in B25?

B28. There are two commonly-encountered operations that may cause the natural domain of a function to be restricted: square roots and division by zero. Give at least one more function with a naturally restricted domain.

B29. Suppose you apply the function "exp" and then the function "ln" to the image. This is equivalent to applying one function to the original argument. Which function?

B30. Suppose you apply the function "cube it" and then the function "take its cube root" to the image. This is equivalent to applying one function to the original argument. Which function?

B31. (Short essay) Why are functions interesting? Why don't we just talk about expressions, or equations such as "$y = 3x$"?

B32. a) What advantages does "formula notation" for functions have over "functional notation"?
b) What advantages does "functional notation" for functions have over "formula notation"?

Section 2.2. Functions, Graphs, and Tables

The previous section discussed the concept of *function* and algebraic notation for functions. This section discusses functions as represented by graphs and tables. It emphasizes the importance of the window of the graph.

Any function corresponds to a set of ordered pairs. For example, if f is defined by $f(x) = x^2$, the argument 3 corresponds to the image 9, which we can represent by the ordered pair (3, 9). The general argument x corresponds to the image x^2, represented by (x, x^2). So f (the function itself) corresponds to the set of all ordered pairs of the form (x, x^2). Ordered pairs can be visualized as points on an axis system, so we can visualize f with a graph (Figures 1A, B, C).

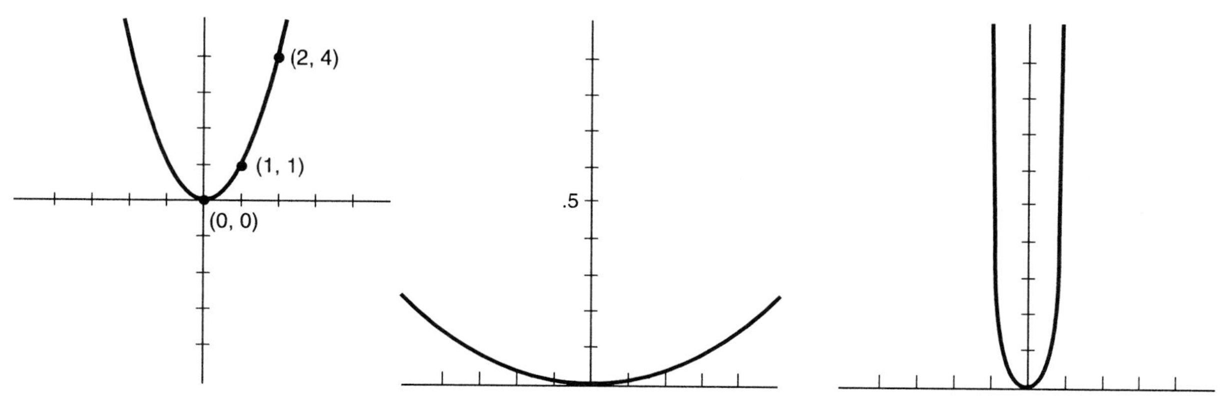

Figures 1A, 1B, and 1C: Three graphs of $f(x) = x^2$ in different windows.
[-5, 5] by [-5, 5] [-1/2, 1/2] by [0, 1] [-50, 50] by [0, 100]

Obviously, f given by $f(x) = x^2$ has more than one graph. The window strongly affects the appearance.

Calculator Exercise 1: Graph x^2 on the standard scale. Then change the x-interval to make it look wider. Which x-interval, [-20, 20] or [-5, 5], makes it look wider? Why? (Problem A1.)

__Graphs__. Functions and graphs are abstract. One rule that describes many number pairs is one function. One picture of many points is one graph.

Technically, by definition, the __graph__ of the function f is the set of ordered pairs $(x, f(x))$, for all x in the domain. But, most of us use the term "graph" to refer to a *picture* of that set of points, or, at least, of all the points that fit in the picture. Therefore, a second, less technical, definition of graph is a picture of a set of ordered pairs. We may use the term *graph* in the contexts of functions, expressions, or two-variable equations. The graph of the *function* "f" is the same as the graph of the *expression* "$f(x)$," which is the same as the graph of the *equation* "$y = f(x)$."

The domain of a function is the set of all argument-values under consideration (all x's). The range of a function is the set of all image values (all y's). Therefore, the range of f is the set of all y for which the equation "$f(x) = y$" has a solution.

Scale. Figure 1 shows that the scale of the axis system affects the appearance of a graph. How can you make the same function look different? It is easy on your graphics calculator -- just change the scale.

Calculator Exercise 2: Use the scale [-10, 10] by [0, 10] to graph $f(x) = x^2$ on your calculator (Figure 2A). What is the right most point on the graph that is still in the window? Now, adjust only the x-interval to make the graph go through the upper right corner of the window (Figure 2B).

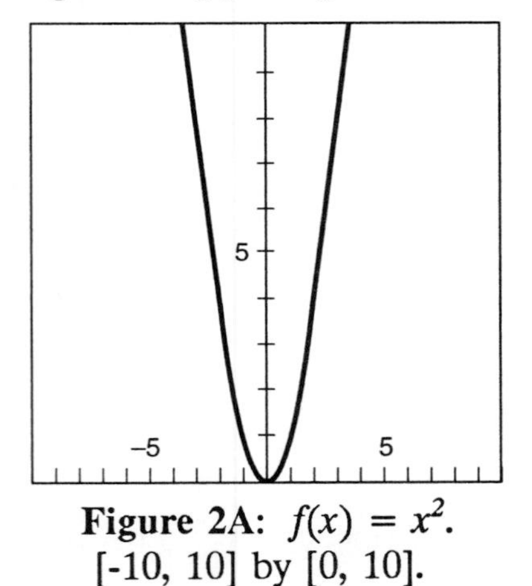

Figure 2A: $f(x) = x^2$.
[-10, 10] by [0, 10].

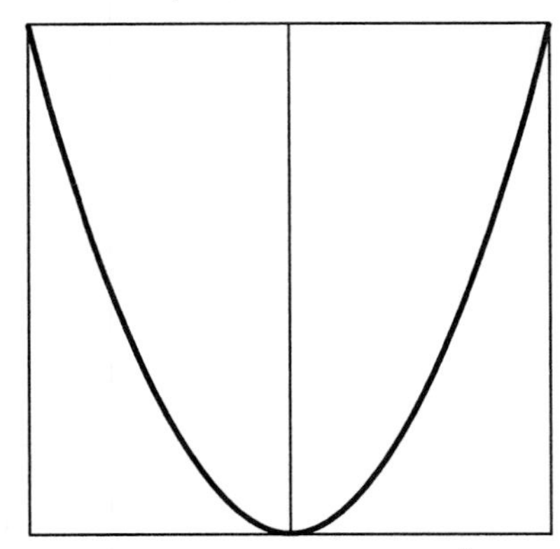

Figure 2B: $f(x) = x^2$.
What is the scale?

The curve exits the top of the window when x^2 is greater than 10, the largest possible y-value. So the right most point is when $y = 10$ and $x^2 = 10$, $(\sqrt{10}, 10)$ (Figure 2A). If we want that point to appear at the upper right, we can change the x-interval to $[-\sqrt{10}, \sqrt{10}]$ which is approximately [-3.16, 3.16].

Calculator Exercise 2, continued: Return to the original graph: x^2 on the scale, [-10, 10] by [0, 10] (Figure 2A). Adjust only the y-interval so that the graph goes through the upper right corner of the window (Figure 2B, again).

Now the key is to recognize that all points on the graph have the form (x, x^2) and the right edge is the line $x = 10$. The corresponding y-value is 100. Therefore, to show the point (10, 100) at the top, the y-scale needs to be adjusted to [0, 100].

Calculator Exercise 3: Learn how to change the scale to change the appearance of a graph. Figure 3A is a graph on the standard scale of $f(x) = 4x - x^3$. If you want to make the bumps the same width, but look taller in the window, how could you do it?

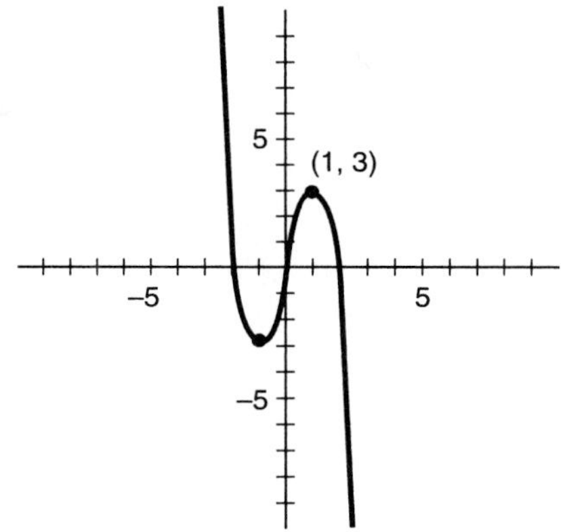

Figure 3A: $f(x) = 4x - x^3$. [-10, 10] by [-10, 10].

The width is determined by the x-values of the points, and the height by the y-values of the points. If we want a particular point to be higher in the window, we need to adjust the y-scale so that the same y-value looks higher. For example, the point (1, 3) is on the graph of f. Using the y-interval [-10, 10], that point is less than half way up in the first quadrant, since 3 is less than half of 10 (Figure 3A). However, if the y-interval were [-5, 5], then the same point would be more than half way up in the first quadrant, since 3 is more than half of 5. A shorter y-interval makes the same points look taller, since the same y-values are larger relative to the size of the interval (Figure 3B).

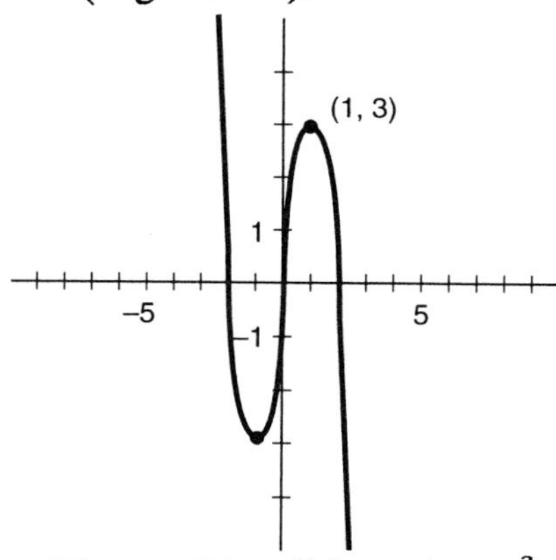

Figure 3B: $f(x) = 4x - x^3$. [-10, 10] by [-5, 5].

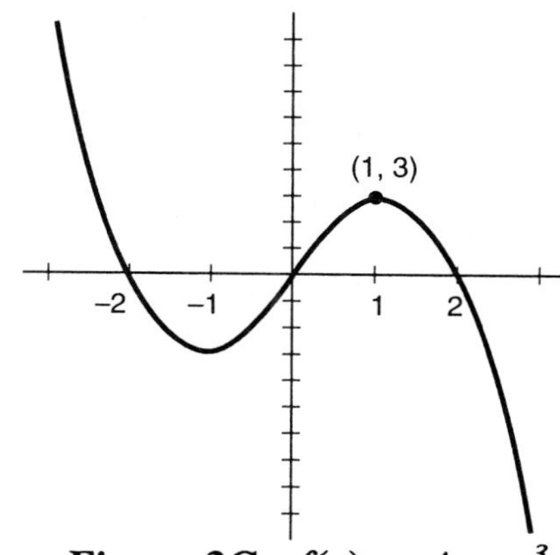

Figure 3C: $f(x) = 4x - x^3$. [-3, 3] by [-10, 10]

Calculator Exercise 3, continued: If you want to make the bumps the same height, but look wider in the window, how could you do it?

The point (2, 0) is on the graph. When the x-interval is [-10, 10], it is only two tenths the way from the center to the edge. The original graph is narrow (Figure 3A). If the x-interval were [-3, 3], that point would be two-thirds the way to the edge. The bumps would look wider (Figure 3C).

The points on a graph of a function are determined by the function, but the apparent location of those points is determined by the window.

Table I summarizes the affects of scale changes.

TABLE I

Interval change		**graph appears**
x-interval:	larger	narrower
	smaller	wider
y-interval:	larger	shorter
	smaller	taller

Example 1: Figure 4A plots the average during the year of the Dow Jones Industrial Stock price index for four years.

year	1990	1991	1992	1993
Dow average	2679	2929	3284	3754

The average gained 1075 points over those years, as the upward trend in the graph illustrates. Suppose you wanted to make the gain look more dramatic. How could you change the scale to do so?

The vertical interval in Figure 4A is 2000 to 5000. In Figure 4A, the gain was about one third the length of the y-interval. The gain will appear more dramatic if the gain is a larger fraction of the length of the y-interval. Figure 4B uses the interval [2500, 4000], which puts the low close to the bottom of the window and the high close to the top.

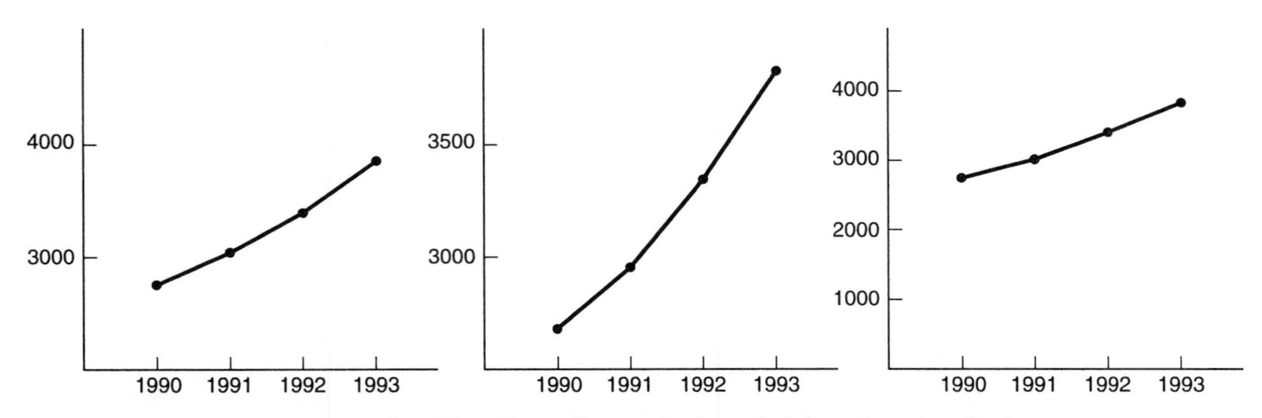

Figures 4A, B, and C: The Dow Jones Industrial Stock price index average, 1990 to 1993. Note how different vertical scales change the appearance.

Example 1, continued: Suppose you want to make the fluctuations in the Dow Jones average appear less dramatic. How could you adjust the scale to do so?

One way is to use a longer y-interval, so that the change in y is a smaller fraction of the window. Figure 4C uses 0 to 5000.

Sometimes we need to adjust the window so we can see the points of interest. For example, the minimum value might not be visible in the standard window.

Calculator Exercise 4: Let $f(x) = 10x^2 - 300x + 10000$. Use a graph to find x such that $f(x)$ is its minimum.

Some calculators have a "minimizer" that will automatically solve this problem. By repeatedly *evaluating* the expression and *comparing* the values to each other, it adjusts the x-value until the solution is determined with great accuracy.

However, many calculator minimizers require you to first select an x-interval in which the answer lies, so you would need to know something about the location of the minimum first. In that case, you still need to know how to find a representative graph.

A "minimizer" is a useful tool, but it is not a substitute for knowing how to analyze a function to find an suitable widow for its graph.

Do not expect the standard scale to yield a representative graph. Try it and see. You need to select a better window.

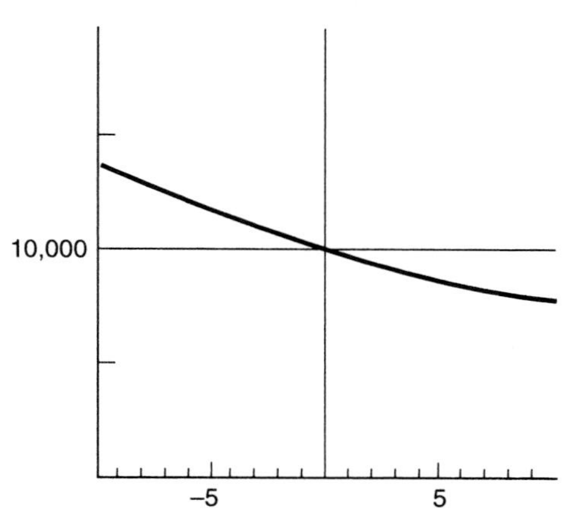

Figure 5: $10x^2 - 300x + 10,000$
[-10, 10] by [0, 20000].

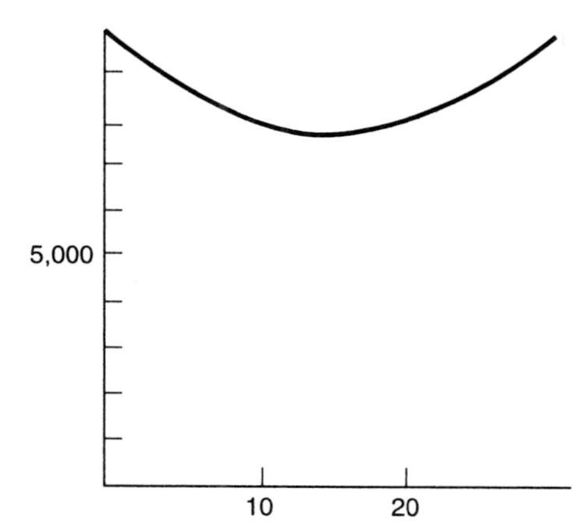

Figure 6: $10x^2 - 300x + 10,000$
[0, 30] by [0, 10000].

First try to determine an x-interval that contains the x-value of interest. It is not obvious which x-interval to choose. One option is to stick with the standard $-10 \leq x \leq 10$ and change it later if necessary. Now, determine the y-values that correspond to the given x-values. For example, when $x = 0$ (an easy case), $y = 10{,}000$. That is why the picture on the standard rectangle is blank -- the y-values are off the screen. There is no need to determine all the y-values accurately, just select any interval that is possible and modify it later. For example, try expanding the vertical scale a great deal, say, 0 to 20,000 (Figure 5).

Selecting a Custom Window
Step 1: Select an x-interval.
Step 2: Find y-values for those x-values (perhaps using the *trace* feature).
Step 3: Adjust the y-interval to display those y-values and look at the graph.
Step 4: Stop there, if you like the picture, or return to Step 1 if you don't.

The minimum y-value appears to be off the right edge of the graph and is certainly less than 10,000. So change the x-interval and, if you wish, the y-interval. Figure 6 uses the window [0, 30] by [0, 10000]. Now the general location of the minimum is clear. It can be found by zooming in (Problem A13). Use the *trace* function to display (x, y) pairs. Then the picture is no longer important -- the (x, y) pairs have the necessary information.

Calculator Exercise 5: Figure 7 pictures the graph of $f(x) = x^4(1 - x)^6$ in a certain window. Find the window.

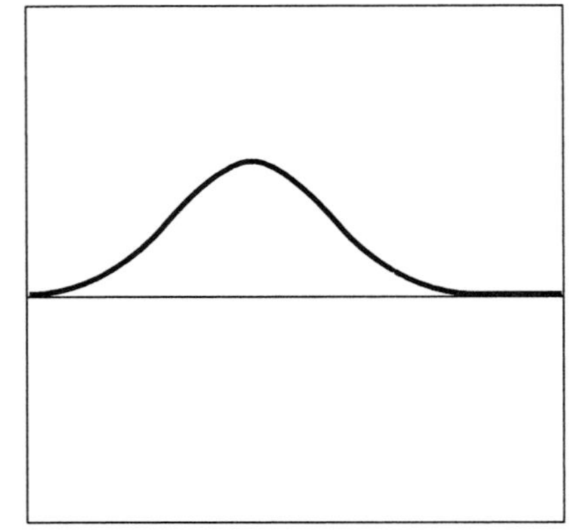

Figure 7: $f(x) = x^4(1 - x)^6$. What is the window?

The picture in the standard window is nothing like this. Try it and see. The only place on the graph in the standard window that appears horizontal is between $x = 0$ and $x = 1$, so you might use the *zoom* feature to expand that region. That does not work either. Try it.

The idea that the region is between $x = 0$ and $x = 1$ is right, but the *zoom* feature changes both scales simultaneously by the same factor and we need different factors on the two scales. Use $x = 0$ to $x = 1$ and select the y-values separately. The curve will lie along the x-axis and not show a bump unless small values of y are made to look larger. You can use the *trace* feature to exhibit (x, y) pairs and see which y-values occur. For example, when $x = 1/2$, $y = (1/2)^{10} = .00098$, so the y-interval needs to be very small to make small y-values look substantial. The window is [0, 1] by [-.002, .002].

Functions, Ordered Pairs, and Tables. Now you can see why functions and graphs are so closely related. Functions relate first numbers to second numbers. Points on graphs have first numbers and second numbers (coordinates). A function is one object consisting of many number pairs. One graph consists of many number pairs.

<u>Definition 2.2.1 (of "function")</u>: A <u>function</u> is a set of ordered pairs such that each first-coordinate value is paired with exactly one second-coordinate value.
The set of all first-coordinate values (arguments) is the <u>domain</u>. The set of all second-coordinate values (images) is the <u>range</u>.

In the previous section *function* is defined as a rule. That is the best definition to start with. However, in some cases this more abstract "ordered pairs" definition of *function* is useful because some numerical relationships cannot be expressed with convenient rules. Then we can regard the function as a set of ordered pairs, and graph the pairs or list them in a table.

You do not need to memorize anything about the particular example which follows; it is intended merely to illustrate how tables work.

Example 2: Table 2.2.2 gives values of the standard-normal cumulative-distribution function from statistics. It gives the probability that random "standard normal" numbers are less than or equal z. The function is usually denoted by the symbol Φ (the upper case Greek letter phi). This function is fundamental to statistics, but it has no convenient algebraic representation. Many calculators do not have a key for it. Consequently, statistics textbooks generally include a table at the back with a selection of argument-image pairs.

Table 2.2.2: The standard-normal cumulative-distribution function.

z	$\Phi(z)$	z	$\Phi(z)$	z	$\Phi(z)$	z	$\Phi(z)$
-3	.0014	-1.0	.1587	0.4	.6554	1.8	.9641
-2.5	.0062	-0.8	.2119	0.6	.7257	2	.9772
-2	.0228	-0.6	.2743	0.8	.7881	2.5	.9938
-1.8	.0359	-0.4	.3446	1	.8413	3	.9986
-1.6	.0548	-0.2	.4207	1.2	.8849		
-1.4	.0808	0	.5000	1.4	.9192		
-1.2	.1151	0.2	.5793	1.6	.9452		

From the table we see the standard-normal cumulative-distribution function value associated with argument 0 is 0.5. $\Phi(0) = 0.5$. $\Phi(1) = .8413$. $\Phi(-2) = .0228$.

Technically, the domain of Φ is all real numbers and Table 2.2.2 exhibits only a selection, but its value at unlisted arguments such as 1.32 and -2.8 can be approximated from the given entries. If more accuracy is desired, and occasionally it is, go find a better table! If you want to know the rule expressed as a command, the rule is, "Look it up in Table 2.2.2!"

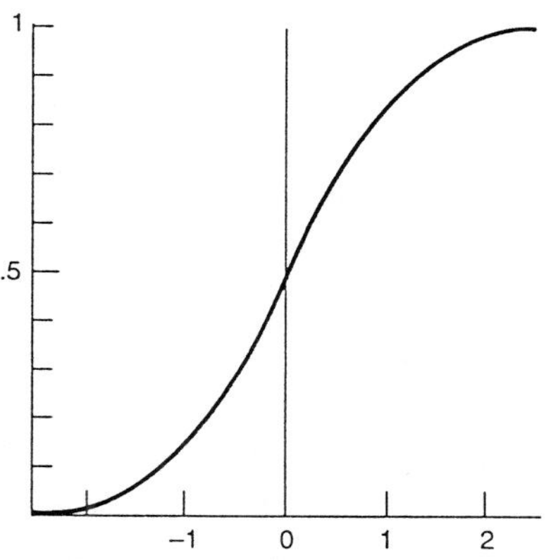

Figure 8: A graph of Φ. [-2.5, 2.5] by [0, 1].

Evaluating and Solving. With a table it is easy to see the distinction between "evaluating" and "solving." We evaluate expressions; we solve equations. To "evaluate" the expression $\Phi(x)$ for a particular x, look in the "x" column for that value of x and find the corresponding "y" value in the "$\Phi(x)$" column. For example, $\Phi(0.4) = .6554$, according to the table.

To solve the equation "$\Phi(x) = c$" means to find the value of x which yields the desired y value, in this case, c. For example, to solve "$\Phi(x) = .8849$," find the image .8849 in the "$\Phi(x)$" column of the table and report the corresponding x value. The solution is $x = 1.2$.

To evaluate is to go from argument to image. To solve is to go from image to argument.

To evaluate is to go from x to y.
To solve is to go from y to x.

Example 3: Figure 9 is a representative graph of f. Grid lines are one unit apart. a) Find $f(2)$. b) Solve $f(x) = 2$. c) Solve $f(x) < 0$.

The problem "Find $f(2)$" is a direct evaluation. Graphs are designed to be read from x to y. Recall that $x = 2$ is a vertical line. $f(2)$ is the y-value where that line intersects the graph. $f(2) = 2.4$

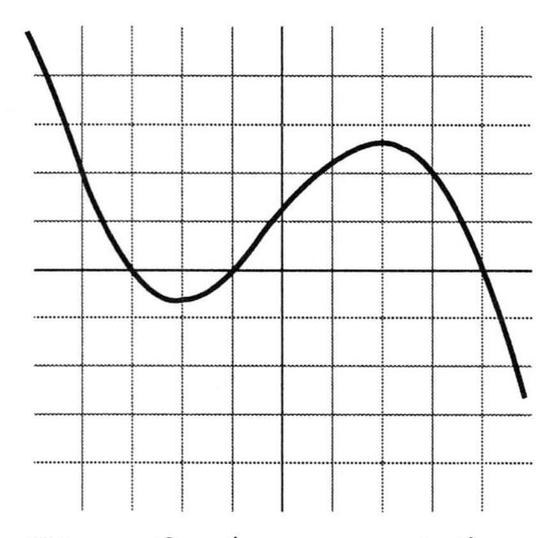

Figure 9: A representative graph of f.

The problem, "Solve $f(x) = 2$, is indirect. We have to read the graph "backwards"-- from y to x. Recall that $y = 2$ is a horizontal line. The question asks, "Which values of x correspond to points on the graph where $y = 2$?"

The solution is $x = -4$ or 1 or 3.

The problem "Solve $f(x) < 0$," asks you to find the x-values of the points on the graph where the y-value is less than zero. There are three things involved that must not be confused: x-values, y-values, and points. The relevant points are the points below the x-axis, since $f(x) = y < 0$. But the problem does not ask for points -- it asks for the x-values of those points.

The solution is $-3 < x < -1$ or $x > 4$.

The Conceptual Level of Functions. Functions are conceptual objects at a higher level then numbers. Here is an advanced example in which a function is treated as an thing (object) in its own right.

Example 4: Solve for f: $f(x + 1) = f(x) + 2$, for all x.

Pause now and solve it. At least give it some thought.

Most students in precalculus have no clue how to do this problem. Most expect to solve equations for "x". This equation does have an "x" in it, but it does not give the function f (it does not express a known command), so they do not know what to do.

In this problem the *function* is the unknown, not the numbers. f is the unknown object. The equation states "for all x," which tells us that x is not an unknown. It is a dummy variable.

You must (eventually) become comfortable with functions as objects that have "properties" such as the property "$f(x + 1) = f(x) + 2$" in this example.

One solution to this "functional equation" is f given by $f(x) = 2x$. There are other solutions (problem B38 and B43). Checking to see that the equation is satisfied by this f,

$$\begin{aligned} f(x + 1) &= 2(x + 1) && \text{[definition of } f\text{]} \\ &= 2x + 2 && \text{[Distributive Property]} \\ &= f(x) + 2 && \text{[definition of } f\text{]} \end{aligned}$$

as desired.

This example is at a much higher level of abstraction than typical equations in algebra. We do not expect you to think this abstractly yet. But, you are to strive to attain this level -- the level at which functions are things (objects) in their own right, with "properties" that do not depend upon the numbers involved. Many advanced functions are useful precisely because they have such properties.

Example 5: Suppose that, in a plant pathology laboratory, researchers note that a population of bacteria can be described as a function of time. They find that, "The population of bacteria doubles every 30 minutes." Describe that property of the function using functional notation.

To answer that, let time be the argument x and population be the image $f(x)$. The function is f. The problem says that when time (the argument) changes by 30 (minutes), the population (image) changes by a factor of 2. Mathematically stated, when the argument changes from x to $x+30$, the image changes from $f(x)$ to $2f(x)$. So the functional equation (the equation you solve for f, a function) is

$$f(x+30) = 2f(x), \text{ for all } x \geq 0.$$

Here "x" is a dummy variable and "f" is the unknown. This equation does not tell you f, the unknown function, but it does tell you a *property* of f.

Similarly, the numerical equation "$3x + 8 = x$" does not directly tell you x, but it tells you a property of x. You can find the number x after you learn how. The functional equation does not directly tell you f, but you will learn how to find f from that type of "functional" equation after you studied exponential functions in Chapter 5. By the way, one answer is $f(x) = 1000(2^{x/30})$. There are others (Problem B47).

Example 5 illustrates why we study exponential functions and other advanced functions: They can be used to describe real-world phenomena. They are important mathematical objects with properties that are interesting and useful.

Advantages and Disadvantages. A function can be described several ways:

1) as a rule
   a) in English (as a command), or
   b) in algebraic notation,
2) as a set of ordered pairs
   a) described in algebraic notation, or
   b) pictured as a graph, or
   c) listed as a table.

Each approach has advantages and disadvantages. The rule definition is the best to start with. Commands that express rules help you understand functions as distinct from numbers. But, English is not the right language in which to do algebra, so mastery of the algebraic notation for rules is essential.

Graphs help you understand functions because they provide a visual display of the argument-image relationship. A lot of learning is visual, and a graph of a function may be grasped more easily than the equivalent algebraic representation or table. For example, the "standard normal" information in Table 2 is easy to summarize in Figure 8. On the other hand, you cannot find images to four decimal places of accuracy from this graph, as you can in some tables.

Another disadvantage of graphs is that they can display only a limited domain, so most graphs do not exhibit the whole function. That is one reason we introduced the term "representative" graph (Section 1.5).

So what is it? Is a function a rule or a set of ordered pairs?

It's both. The two versions of the definition are perfectly compatible. Sometimes we want to emphasize the rule and the notation, sometimes we want to emphasize the ordered pairs in a graph or table. Take your pick.

Conclusion. The window strongly affects the appearance of a graph. For example, a narrower x-interval makes a graph appear wider. A shorter y-interval makes a graph appear taller. Learn how to find a window that yields a representative graph.

Functions are abstract objects (at a higher level than numbers) with properties of their own. They can even be solutions to "functional" equations.

There are two ways to define the term *function*: as a rule, or as a set of ordered pairs. Each definition says that the image must be determined by the argument.

Terms: graph, window, function, domain, range, table.

****

Exercises for Section 2.2, "Functions, Tables, and Graphs":

A1. Do Calculator Exercise 1.

A2.* The graph of the equation "$y = f(x)$" is the same as the graph of the expression ________ and the same as the graph of the function ________.

A3.* Which ordered pairs constitute the graph of the function f?

A4. Figure 10 locates (a, b) and (c, d) on an axis system. Locate (b, a) and (d, c).
A5. Figure 11 locates (a, b) and (c, d) on an axis system. Locate (b, a) and (d, c).

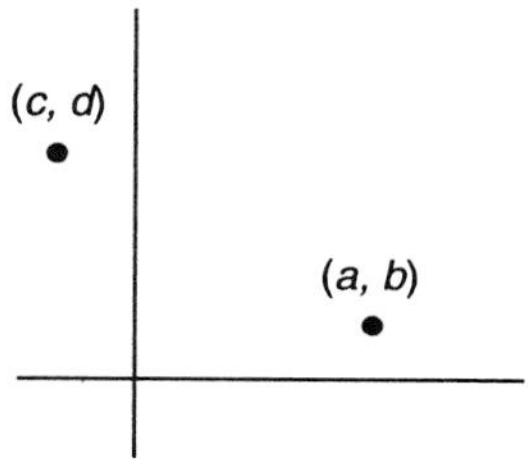

Figure 10

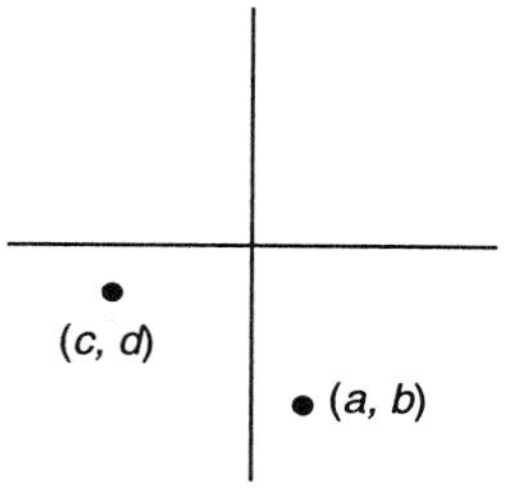

Figure 11

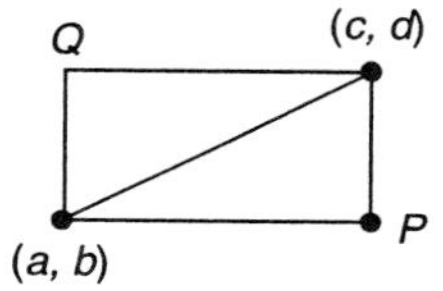

Figure 12

A6. Figure 12 gives (a, b) and (c, d). Find the coordinates of the point in the picture a) P b) Q.

^ ^ ^ ^The figure marks some points on the standard scale. Sketch where they would be in the given windows (and note if they would be outside the window).
Window A: [-5, 5] by [-10, 10]. Window B: [-20, 20] by [-10, 10].
Window C: [-10, 10] by [-5, 5]. Window D: [-10, 10] by [-20, 20].
Window E: [-2.5, 2.5] by [-2,5, 2.5]. Window F: [-40, 40] by [-40, 40]

A7. Figure 13: A, B, E.
A8. Figure 13: C, D, F
A9. Figure 14: A, B, E.
A10. Figure 14: C, D, F.
A11. Figure 15: A, B, E.
A12. Figure 15: C, D, F.

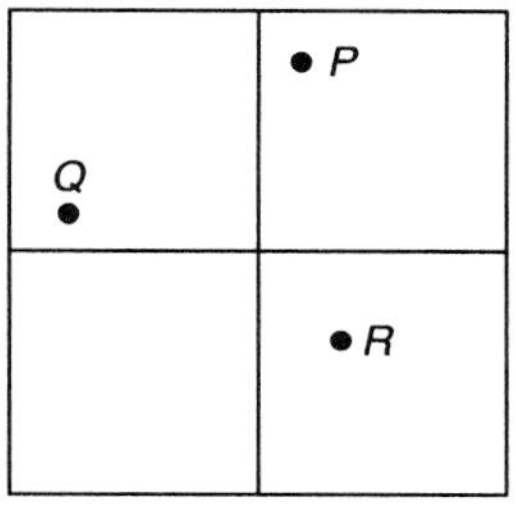

Figure 13

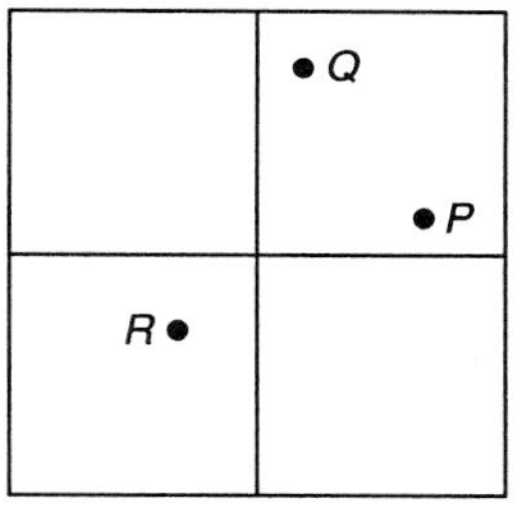

Figure 14

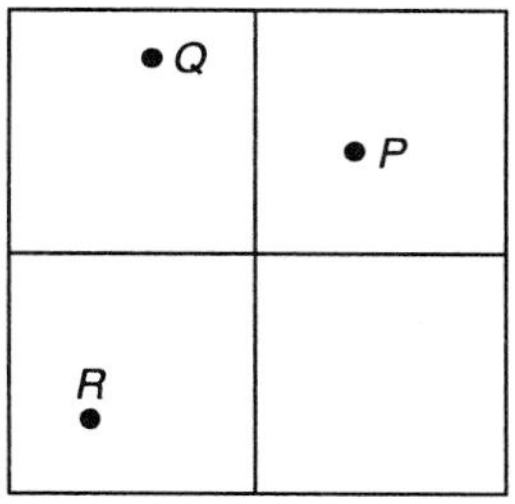

Figure 15

A13. Find the x-value that minimizes: $10x^2 - 300x + 10{,}000$.

A14.* a) In the context of functions, to evaluate is to go from ________ to ________.
b) In the equation "$f(x) = y$," to solve is to go from ________ to ________.

^ ^ ^ ^ Solve for x using Table 2.2.2.

A15. $\Phi(x) = .9772$. A16. $\Phi(x) = .6554$.

A17. Figure 16 gives a graph without giving its algebraic expression.
a) Evaluate $f(-1)$ b) Approximate the solution to $f(x) = 2$.

A18. Figure 17 gives a graph without giving its algebraic expression.
a) Evaluate $f(-2)$ b) Approximate the solution to $f(x) = 1$.

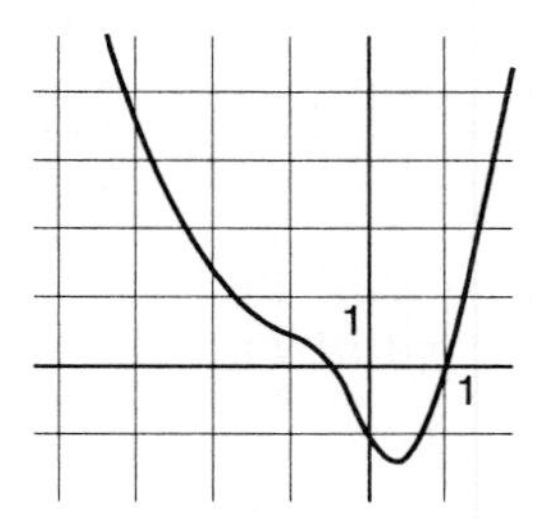

Figure 16.

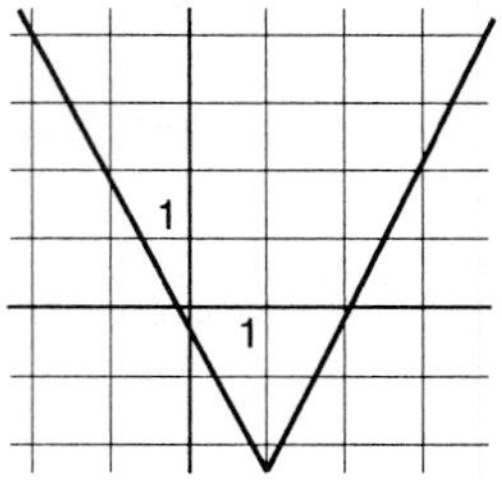

Figure 17.

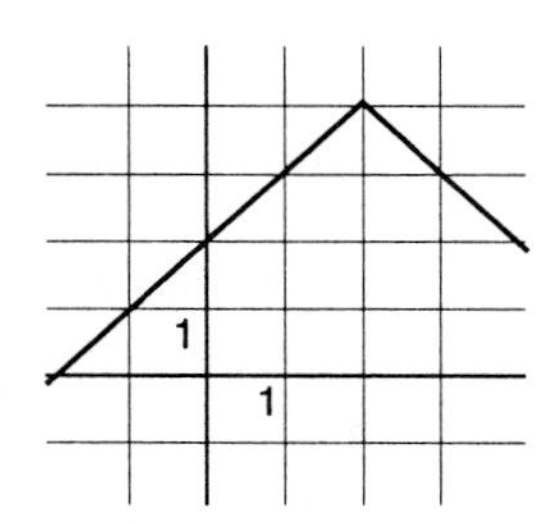

Figure 18.

A19. Figure 18 gives a graph without giving its algebraic expression.
a) Find $f(2)$ b) Solve $f(x) = 3$.

A20. Figure 19 gives a graph without giving its algebraic expression.
a) Find $f(2)$ b) Solve $f(x) = 3$.

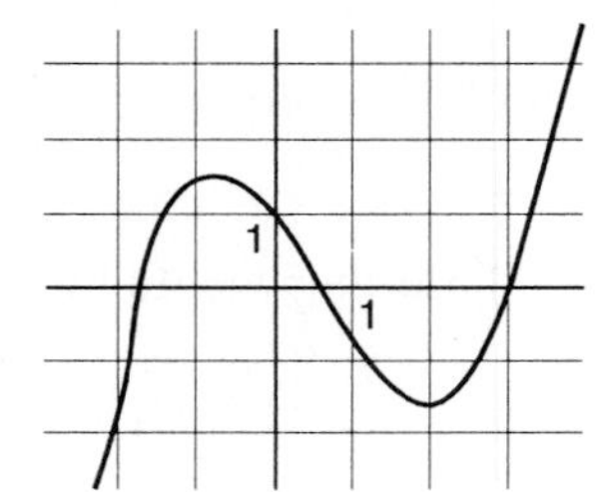

Figure 19.

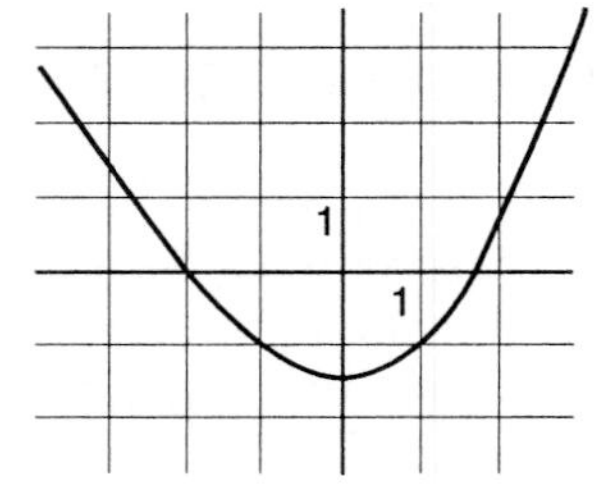

Figure 20.

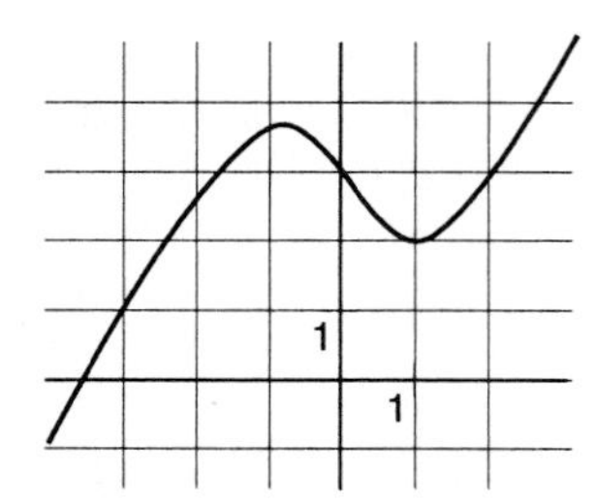

Figure 21.

A21. Figure 20 gives a graph of f, without giving its algebraic expression.
a) Find $f(3)$ b) Solve $f(x) = -1$.

A22. Figure 21 gives a graph of f, without giving its algebraic expression.
a) Find $f(1)$ b) Solve $f(x) = 1$.

^ ^ ^ ^ Let f be a set of ordered pairs. a) Is the given f a function? b) Give the domain as a set. c) Give the range as a set. d) Solve $f(x) = 5$.

A23. $f = \{(0, 10), (1, 5), (2, 4)\}$. A24. $f = \{(1, 5), (2, 5), (2, 10)\}$.

A25. $f = \{(-1, 3), (0, 12), (1, 5)\}$ A26. $f = \{(0, 4), (0, 6), (0, 5)\}$.

A27.* Define "representative" in the context of graphs.

A28.* Which of these have graphs that go through the origin?

a) x^2 b) x^3 c) $\sqrt{x}$ d) $1/x$

e) $|x|$ f) 10^x g) $\log x$.

A29.* Which of the expressions in the previous problem have graphs that go through the point (1, 1)?

A30. We will call the scale square if the vertical scale and the horizontal scales are the same, that is, one unit is the same distance on each scale. Square scales are important in calculus. a) What model of calculator do you have? b) Check to see if your calculator has a keystroke sequence which yields a "square" scale. If it does, graph something on a scale which is not square and then try that sequence to see what it does. Does it change the vertical scale or the horizontal scale?

A31.* The range of f is the set of all ____________ . Therefore, the range of f is the set of all c such that the equation __________ has a solution.

^ ^ ^ ^ Give the range of f. (You might want to look at the graph.)

A32. a) $f(x) = x^2$. b) $f(x) = x^3$. c) $f(x) = |x|$. d) $f(x) = 1/x$.

A33. a) $f(x) = x^2 - 7$. b) $f(x) = \sqrt{x}$. c) $f(x) = |x - 3|$. d) $f(x) = |x| + 6$.

^ ^ ^ ^ ^ ^ ^ ^

B1.* a) Which x-interval, [-10, 10] or [-20, 20], makes graphs look wider?
b) Which y-interval, [-10, 10] or [-5, 5], makes graphs look taller?

B2.* a) Which y-interval, [-10, 10] or [-20, 20], makes graphs look shorter?
b) Which x-interval, [-10, 10] or [-5, 5], makes graphs look narrower?

B3. Figure 22 gives a graph without giving its algebraic expression.
a) Evaluate $f(2) - f(0)$ b) Solve $f(x) < 0$.

B4. Figure 23 gives a graph without giving its algebraic expression.
a) Evaluate $f(1) - f(-1)$ b) Solve $f(x) > 0$.

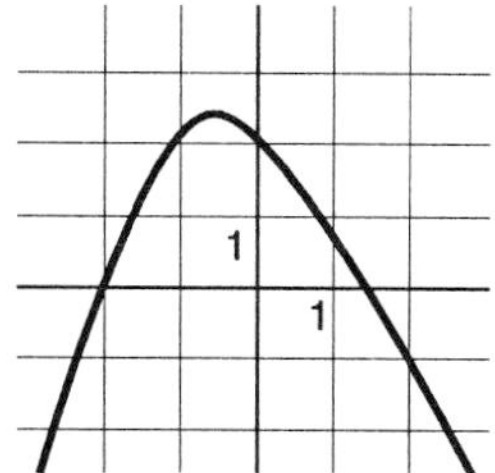

Figure 22.

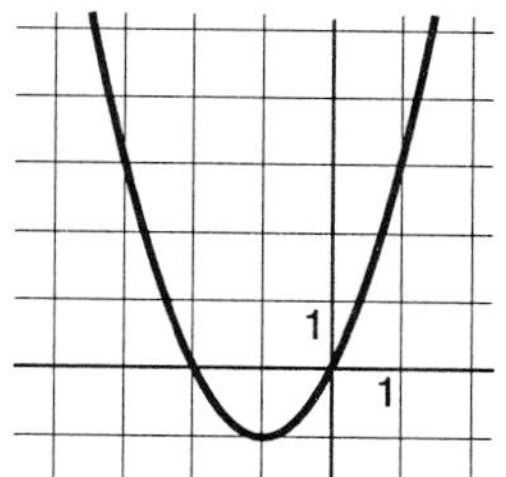

Figure 23.

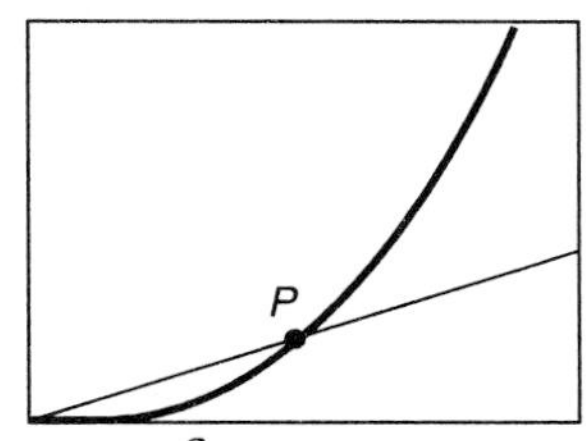

B5. x^3 and $5x$.

^ ^ ^ ^ Problems B5-B14 picture the given graphs in a certain window. a) Find the window (approximately). b) Find the coordinates of P. [Great accuracy is not required, just get it close enough to show you know how to do it.]

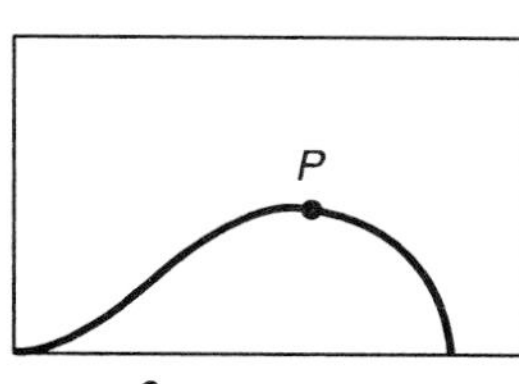

B6. $x^2(120 - 4x)$

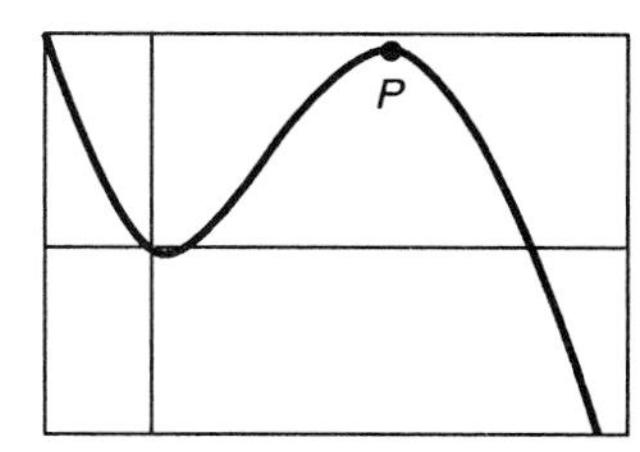

B7. $x(x - 20)(200 - x)$

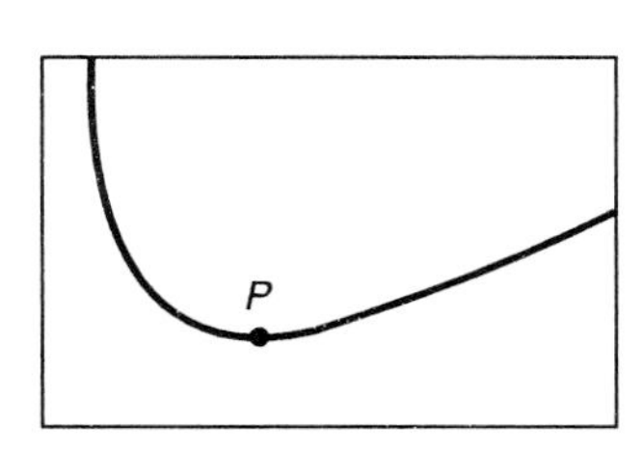

B8. $12x^2 + 12{,}000/x$.

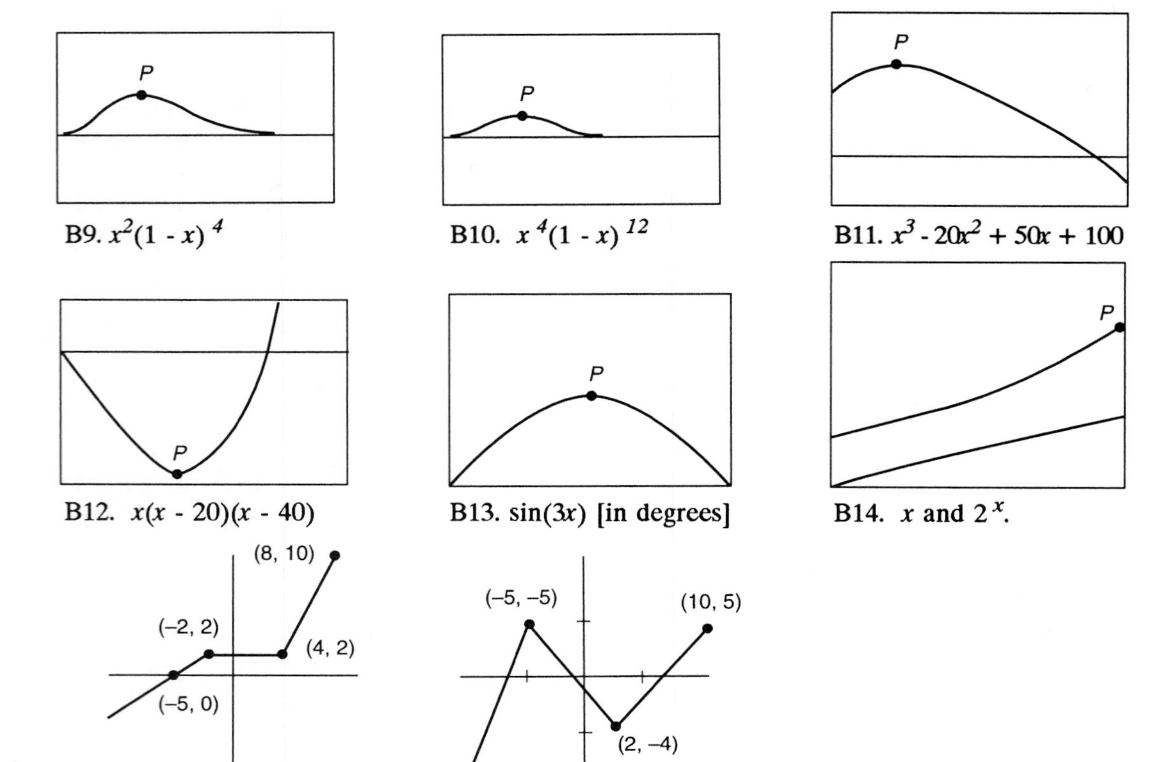

B9. $x^2(1 - x)^4$ B10. $x^4(1 - x)^{12}$ B11. $x^3 - 20x^2 + 50x + 100$

B12. $x(x - 20)(x - 40)$ B13. sin(3x) [in degrees] B14. x and 2^x.

Figure 24 **Figure 25**

^ ^ ^ ^Sketch the graph (given above in [-10, 10] by [-10, 10]) in the requested window.

B15. Figure 24 in [-5, 5] by [-10, 10].
B16. Figure 25 in [-5, 5] by [-10, 10].
B17. Figure 24 in [-10. 10] by [-5, 5].
B18. Figure 25 in [-10, 10] by [-5, 5].
B19. Figure 24 in [-2.5, 2.5] by [-2,5, 2.5].
B20. Figure 25 [-2.5, 2.5] by [-2,5, 2.5].

B21. Look at the graph of $f(x) = \sqrt{x}$ on [0, 10] by [0, 10].
a) Which new x-interval would make the graph exit the window at the upper right corner?
b) Which new y-interval would make the graph exit the window at the upper right corner?

B22. Look at the graph of $f(x) = x^3$ on [0, 8] by [0, 8].
a) Which new x-interval would make the graph exit the window at the upper right corner?
b) Which new y-interval would make the graph exit the window at the upper right corner?

B23. Look at the graph of $y = x$ on the standard scale. Then change the x-interval to make it look steeper. a) Which x-interval, [-20, 20] or [-5, 5], makes it look steeper? Why?
b) Now return to the standard x-interval and change the y-interval to make it look steeper. Which y-interval, [-20, 20] or [-5, 5], makes it look steeper? Why?

B24. Look at the graph of $y = |x|$ (the absolute value function) on the standard scale. Then change the x-interval to make it look steeper. a) Which x-interval, [-20, 20] or [-5, 5], makes it look steeper?
b) Now return to the standard x-interval and change the y-interval to make it look steeper. Which y-interval, [-20, 20] or [-5, 5], makes it look steeper?

B25. During a time period the price of gold rose from $320 per ounce to $385 per ounce and in between it fluctuated up and down. Suppose you must graph this. a) If you want to make the gain seem large, how should you pick the scale? b) If you want to make the price seem relatively stable, how should you pick the scale?

B26. During a time period the price of silver rose from $4.19 per ounce to $4.87 per ounce and in between it fluctuated up and down. Suppose you must graph this. a) If you want to make the gain seem large, how should you pick the scale? b) If you want to make the price seem relatively stable, how should you pick the scale?

B27. a) Sketch, in a rectangle you draw, a "representative" graph of $y = x^3 - 20x^2 + 150$. b) Give the x and y-intervals you used. c) For $x > 0$, find the minimum value of y [Great accuracy is not required, just get it close enough to show you know how to do it.]

B28. Suppose you wished to solve $f(x) = 4.32$. Inspect their graphs to determine which of the following functions would yield a unique solution.
a) x^2 b) x^3 c) $1/x$.

B29. Suppose you wished to solve $f(x) = 4.32$. Inspect their graphs to determine which of the following functions would yield a unique solution.
a) e^x b) e^{-x} c) $\ln x$ d) $\log x$

B30. Suppose you wished to solve $f(x) = 0.432$. Inspect their graphs to determine which of the following functions would yield a unique solution.
a) $\sin x$ b) $\cos x$ c) $\tan x$

B31.* Look at the graph of $1/(x - 1)$ on your calculator. Some graphics calculators include in the plot a vertical line near $x = 1$. Why?

B32. See if your graphics calculator has a keystroke sequence for a "square" scale. If it does, graph something on a non-square scale and use the sequence to put it on a square scale. What changes, the vertical or the horizontal scale?

^ ^ ^ ^ Write a "functional equation" to describe the given sentence. Use "x" for the argument and "f" for the function. [Do not solve them.]
B33. "In twice the time, the ball drops fours times as far." Argument: time. Image: How far the balls drops.
B34. "Every year the amount of money increases by a factor of 1.1." Argument: Time in years. Image: Amount of money.
B35. "Every 90 years the amount of the substance is halved." Argument" Time in years. Image: Amount of the substance.
B36. "When the earthquake wave size increases by a factor of 10, the Richter scale number goes up by 1." Argument: Earthquake wave size. Image: Richter scale number.
B37. "When the sound energy increases by a factor of 10, the decibel level goes up 10." Argument: Sound energy. Image: decibel level.

B38. Find other solutions for f to the functional equation "$f(x + 1) = f(x) + 2$, for all x," in Example 4.

B39. Solve for f: $f(x + 1) = f(x) + 3$, for all x.

B40. Solve for f: $f(x + 2) = f(x) + 1$, for all x.

B41. Solve for f: $f(2x) = 4f(x)$, for all x.

B42. Solve for f: $f(2x) = 8f(x)$, for all x.

B43. Example 4 gives a "functional" equation. Create a (simple) word problem which expresses that equation in a real-world context.

B44. Given the f in Figure 21, solve $f(x/3) = 1$.

B45. Given the f in Figure 22, solve $f(2x) = 3$.

B46.* There is a fundamental difference between the type of equation we "solve" and the type we "graph." What is it? Give an example of each type of equation to illustrate the difference.

B47. Verify that $f(x) = 1000(2^{x/30})$ is a solution to the functional equation in Example 5. Find another solution (not much different from the given solution).

B48.* Zoom in on the graph of $y = x^2$ near the origin. Use a square scale. Is the graph pointed there? How is it shaped at the origin? The graph looks almost like a line if you zoom in close enough. Which line? (Give the equation of that line.)

B49.* Zoom in on the graph of $y = x^3$ near the origin. Use a square scale. The graph looks almost like a line if you zoom in close enough. Which line? (Give the equation of that line.)

B50.* Zoom in on the graph of the square root function near the origin using a square scale. The graph looks almost like a line if you zoom in close enough. Which line? (Give its equation.)

B51.* How is a graph conceptually like a table?

B52. What are some advantages and disadvantages of graphs compared to other methods of representing functions?

B53. Which of these words and phrases comes closest to referring to functional relationships? "affects," "is correlated with," "is related to," "is determined by," "depends upon," "influences."

Section 2.3. Composition and Decomposition

The graphs we know by heart may number as few as a dozen. Obviously, there are far more than a dozen interesting relationships in the world. The purpose of this section is to expand our list to include relationships obtained by applying functions one after another.

Applying a second function to the image of a first is called composition of functions. The resulting two-stage function is said to be a composite function. Rewriting a composite function as a sequence of simpler functions applied one after the other is called decomposition.

After you understand that order matters it is not difficult to grasp the idea of composition. If you grasp the order in a composite function, you see the sequence of component functions. The difficulty, if any, is in using the "$f(x)$" style notation properly to express the component functions in order.

Example 1: The Celsius-to-Fahrenheit relationship is $F = 1.8C + 32$. It combines two operations, "Multiply by 1.8" and then "Add 32." Therefore it can be expressed as a composition of two simpler functions. Let g denote the "Multiply by 1.8" function: $g(x) = 1.8x$. Let f denote the "Add 32" function: $f(x) = x + 32$. In functional notation, g is applied first when it is on the right because it is inside the parentheses. Then $f(g(x)) = f(1.8x) = 1.8x + 32$, which is the abstract formulation of the Celsius-to-Fahrenheit relationship.

Order matters. This relationship says to multiply by 1.8 *first* and *then* add 32. In the other order you would get a different and incorrect relationship. In the other order, $g(f(x)) = g(x + 32) = 1.8(x + 32) = 1.8x + 57.6$, which is not the Celsius-to-Fahrenheit relationship. Order matters.

Example 2: The formula for the area of a circle given its radius is $A = \pi r^2$. Define two functions, f and g, such that the area function is a composition of them: $A(r) = f(g(r))$.

Solution: To evaluate A, we "Square and then multiply by π." The composition of the two components, "Square" and "Multiply by π," is evident. All we need to do is to express these functions in mathematical notation, with g first and f. $g(r) = r^2$ (or $g(x) = x^2$, the variable is a dummy variable). $f(x) = \pi x$. Then $f(g(r)) = \pi r^2$, the area function.

Algebra is about operations and order. A single *sequence* of operations can be regarded as one thing (for which the term *function* is appropriate), or as several things, several operations, one after another. Decomposition is essential for understanding complicated functions.

Example 3: Let $h(x) = (x - 3)^5$. Find $f(x)$ and $g(x)$ such that $h(x) = f(g(x))$.

h is "Subtract 3 and then take the fifth power." In "$f(g(x))$", g is first. So, let $g(x) = x - 3$ [g is "Subtract 3"], and $f(x) = x^5$ [f is "Take the fifth power"].

Then $f(g(x)) = f(x - 3) = (x - 3)^5$, as desired.

Example 4: Decompose $h(x) = \sqrt{x^2 + 9}$ into as many components as are appropriate.

Think functionally. The rule is "Square, then add 9, then take the square root." Let $f(x) = x^2$, $g(x) = x + 9$, and $k(x) = \sqrt{x}$. Then $h(x) = k(g(f(x)))$.

Example 5: Let $f(x) = 2x + 1$ and $g(x) = x^2$. Find $f(g(x))$.

$f(g(x)) = f(x^2) = 2(x^2) + 1 = 2x^2 + 1$. This is not the product of the two expressions: $(2x + 1)(x^2)$. Composition is not multiplication.

Order matters, so the other order would yield a different expression. $g(f(x)) = g(2x + 1) = (2x + 1)^2 = 4x^2 + 4x + 1$.

Example 6: Let $f(x) = \sqrt{x}$. Let $g(x) = x + 4$. Express $f(g(x))$.

f is "Take the square root." g is "Add 4." In the expression "$f(g(x))$," the first function is g, not f. Functional notation is read right-to-left (from inside to outside the parentheses). Therefore the composition is "Add 4 and then take the square root." The notation for that function is $f(g(x)) = \sqrt{x + 4}$. The extended square root symbol does the grouping of "$x + 4$" before the square root is taken.

The parentheses in "$f(g(x))$" do *not* refer to multiplication. Some students erroneously obtain the answer "$\sqrt{x}(x + 4)$" to Example 6. Similar notations can have different meanings. Note that "$f(...)$" signals the application of a function, but "3(...)" signals multiplication.

Example 7: Analyze $h(x) = (x + 4)\sqrt{x}$.

Composition refers to *successive* operations, beginning with the argument. $h(x)$ is a product, not a composition.

Composition for Calculus. Composition and decomposition are used almost every day in calculus. One of the important compositions occurs in the expression known as the "difference quotient" of f, $\dfrac{f(x+h) - f(x)}{h}$. The numerator gives the difference of the y-values corresponding to the two arguments x and $x+h$. The denominator gives the difference in those two x-

values. The quotient is, therefore, the "difference quotient."

Example 8: Express and simplify the difference quotient if $f(x) = x^2$.

$$\frac{f(x+h) - f(x)}{h} = \frac{(x+h)^2 - x^2}{h}$$
$$= \frac{x^2 + 2xh + h^2 - x^2}{h}$$
$$= \frac{2xh + h^2}{h}$$
$$= 2x + h, \textit{ if } h \neq 0.$$

In examples like this students sometimes make the mistake of treating $f(x + h)$ as $f(x) + h$. Don't! Order matters.

In calculus we then "Let h go to zero." As h goes to zero, the difference quotient goes to $2x$. Note that we could not just let h be zero in the original difference quotient, because the quotient would be the undefined expression "0/0." In calculus, this process defines the "derivative" of f. The derivative of x^2 is $2x$.

Composition and Graphs. We have just seen how algebraic representation of composition works. Now we consider graphical representation of composition with one or more of the four basic arithmetic operations.

The simplest graphical compositions occur when the image is affected.

Example 9: The graph of $f(x) = x^2$ is well-known (Figure 1, bold). How is the graph of $h(x) = x^2 + 3$ related to it?

The order in h has "Add 3" *after* the squaring function creates its images. So the "Add 3" function affects the images. All the images of h are 3 units greater than the images of f, so the graph of h would be 3 units above the graph of f (Figure 1). Similarly, the graph of "x^2 - 5" would be moved down five units.

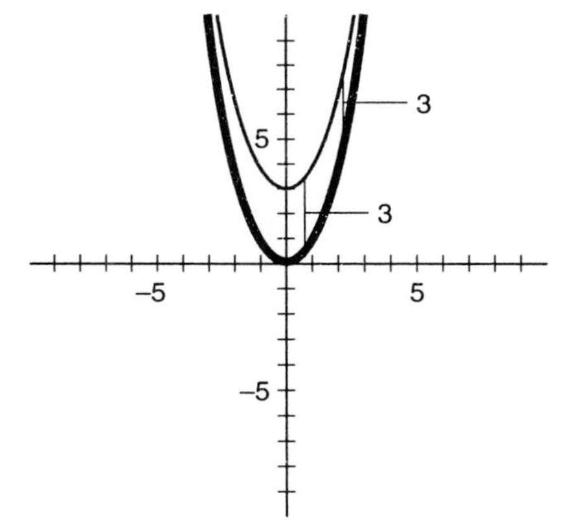

Figure 1: x^2 and $x^2 + 3$. [-10, 10] by [-10, 10].

Adding or subtracting *after* applying f changes the vertical location.

Example 10: Given the graph of $|x|$ (Figure 2, bold), graph $|x|$ - 6.

(The absolute value function is sometimes denoted "abs" on calculators.) The graph of "$|x|$ - 6" is simply down 6 units, since "Subtract 6" applies to the images on the vertical scale (Figure 2).

Calculator Exercise 1: Use your calculator to illustrate the graphical examples while you are reading. Calculators make it easy to draw and change graphs and get immediate feedback about whether you understand the idea. To test yourself, for each example, simply change the number and see if you can predict where the new graph will be.

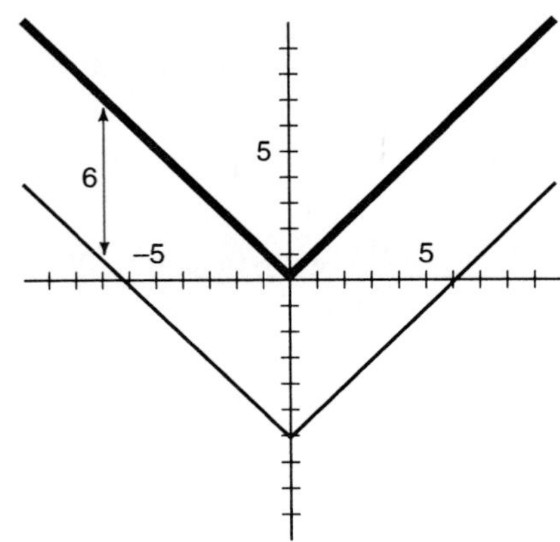

Figure 2: $|x|$ and $|x|$ - 6.
[-10, 10] by [-10, 10].

Example 11: Given the graph of sin x (Figure 3, bold), find the graph of $h(x) = 2 \sin x$.

The image of h is 2 times the image of sine, so the graph of h is magnified in the vertical direction by a factor of 2 (Figure 3).

Note that multiplication of the image by 2 expands the graph away from the horizontal axis. It does not just make it "taller," since multiplication of negative numbers by 2 makes them even more negative.

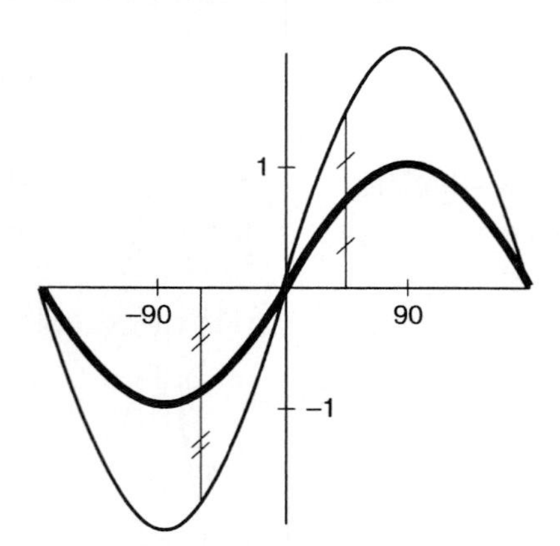

Figure 3: sin x [bold]
and 2 sin x.
[-180°, 180°] by [-2, 2].

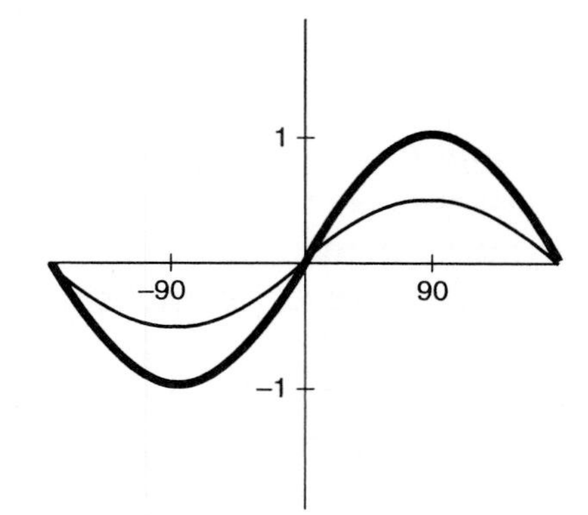

Figure 4: sin x [bold]
and (sin x)/2
[-180°, 180°] by [-2, 2].

Example 12: Define g by $g(x) = (\sin x)/2$. Graph g.

Division of images by 2, which is multiplication by 1/2, contracts vertical distances by a factor of 2. All points are closer to the horizontal axis (Figure 4).

Multiplying or dividing *after* applying *f* changes the vertical scale.

Example 13: Let $f(x) = x^2$. Graph $g(x) = -f(x)$.

$g(x) = -x^2$. The sign of the images is changed, so that positive y-values become negative. The graph of x^2 is turned upside down (Figure 5). It is reflected through the x-axis.

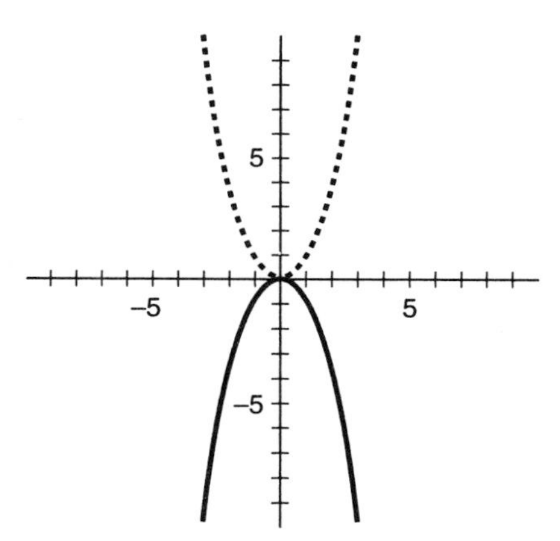

Figure 5: $f(x) = x^2$ [dashed] and $g(x) = -f(x) = -x^2$. [-10, 10] by [-10, 10].

When composition affects the image, only the vertical coordinate is affected and the new graphs are easy to understand relative to the original graphs (Examples 9-13). When composition affects the argument (as in $k(x)$ and $p(x)$ above), only the horizontal coordinate is affected and the relationship to the original graph is surprising.

Example 14: Given the graph of $|x|$, graph $|x - 6|$.

The graph of $|x|$ is well-known (Figure 6, bold). It has a corner (vertex) at (0, 0). When the argument is zero, the image is zero. Where is the corner on the graph of $|x - 6|$?

When $x = 6$, $x - 6 = 0$, and $|x - 6| = 0$. So the point (6, 0) is on the graph of $|x - 6|$. The vertex is 6 units to the *right* of the vertex of $|x|$ (Figure 6). The fact that subtraction *before* applying a function produces a shift *right* is surprising (since we usually think of subtraction as moving numbers left).

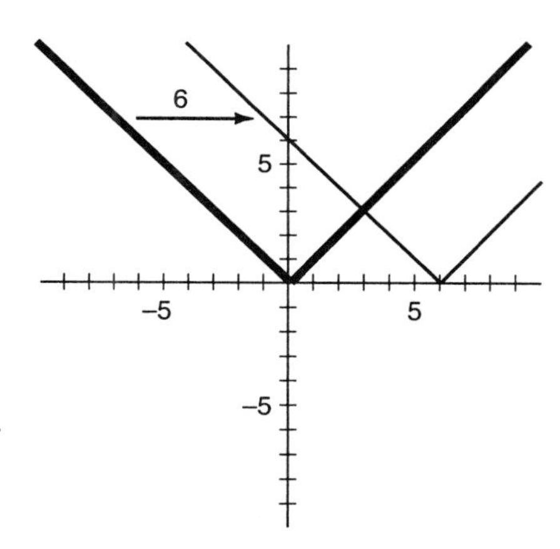

Figure 6: $|x|$ [bold] and $|x - 6|$ [-10, 10] by [-10, 10].

Calculator Exercise 2: Graph "$|x|$" and "$|x + 3|$." Why is the vertex of $|x + 3|$ where it is?

Adding or subtracting *before* applying f changes the horizontal location.

Subtraction before applying f shifts the graph of f *right*. Think of it this way: For the graph of $g(x) = f(x - c)$, the argument of f is zero when $x = c$, that is, the old zero is at the new c. So $g(c) = f(0)$. If c is positive, c is to the right of 0, so that produces a shift right. Subtraction produces a shift to the right. Similarly, for c positive, $h(x) = f(x + c)$ exhibits a shift left. Addition produces a shift to the left.

Example 15: Let $f(x) = x^2$. Graph $f(x + 4) = (x + 4)^2$.

A key point on the graph of $f(x) = x^2$ is (0, 0), the origin. $0^2 = 0$. In the expression "$f(x + 4)$" when will the argument of f be zero?

When $x = -4$, $(x + 4)^2 = 0^2$. Therefore, the (0, 0) point on x^2 occurs at (-4, 0) on $(x + 4)^2$. Therefore the graph is shifted *left* 4 units (Figure 7). The fact that *addition* before applying f is associated with a shift *left* is surprising (since we usually think of addition as moving numbers to the right).

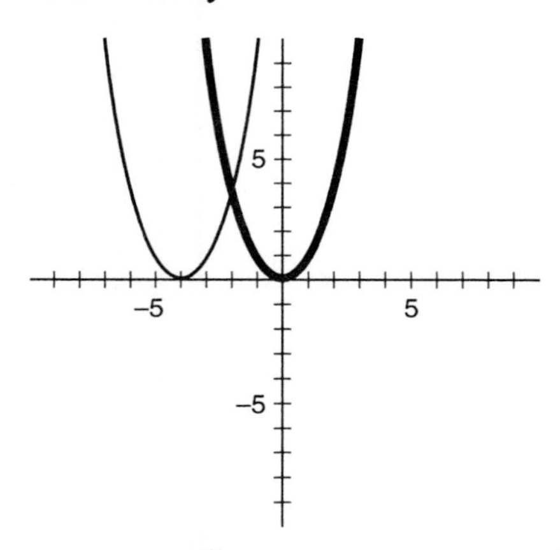

Figure 7: x^2 [bold] and $(x + 4)^2$. [-10, 10] by [-10, 10].

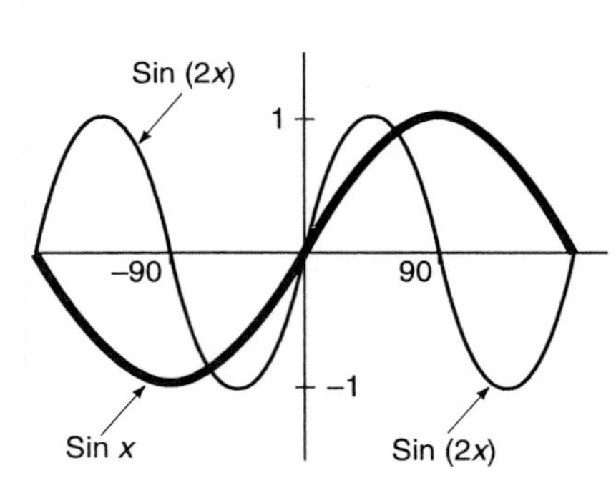

Figure 8: sin x [bold] and sin($2x$). $-180° \leq x \leq 180°$.

Example 16: Given the graph of sin x (Figure 8, bold), graph $g(x) = \sin(2x)$.

The argument of sine is affected. The images are the same, but they correspond to different arguments. For example, the image "sin 180" will occur not when $x = 180$, but when $2x = 180$, that is, $x = 90$. The image "sin 90" occurs when $2x = 90$, so $x = 45$. Every image that occurs on the "sin x" graph between $x = 0$ and $x = 180$ occurs on the graph of "sin($2x$)" between $x = 0$ and $x = 90$. The graph is compressed (contracted, shrunk) horizontally by a factor of two (Figure 8). (Actually the factor is one half, but, in the context of compression we can call the factor two.) The waves of sin($2x$) have twice the frequency of sin(x).

Multiplication before applying f changes the horizontal scale.

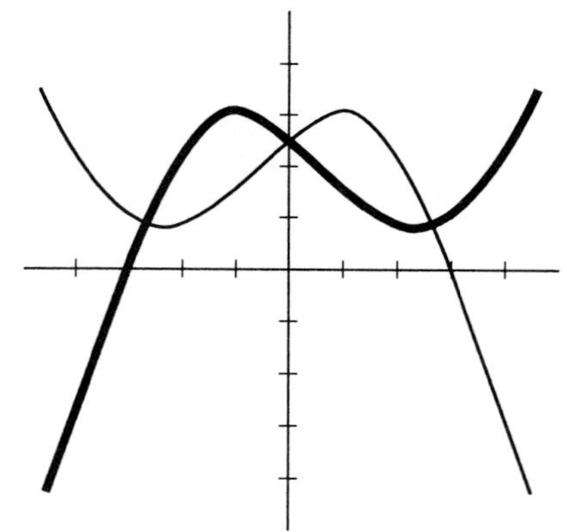

Figure 9: $f(x)$ [bold] and $f(-x)$.

Example 17: Figure 9 (bold) graphs $f(x)$. No algebraic representation is given. Graph $f(-x)$.

The argument is multiplied by negative one before the function is applied. Therefore, on the graph of $f(-x)$ the image "$f(2)$" is not plotted above $x = 2$, but above $x = -2$. All the images normally associated with positive x-values become associated with negative x-values, and vice versa. Therefore, the graph is flipped across the y-axis. It is a mirror image through the y-axis of the original graph (Figure 9).

<u>Composition With Two Arithmetic Functions</u>. Some composite functions consist of composition of *two* arithmetic operations with a function that has a well-known graph. The expressions for given graphs may be identified if they are simply shifts of well-known graphs.

Example 18: The graph in Figure 10 has the shape, but not the location, of the graph of x^2. Identify the expression graphed.

The graph in Figure 10 is the well-known graph of x^2 shifted to the right 1 and down 2. Its expression is, therefore, $(x - 1)^2 - 2$. Beginning with "x", subtracting 1 before squaring shifts it to the right, and subtracting 2 after squaring moves it down 2.

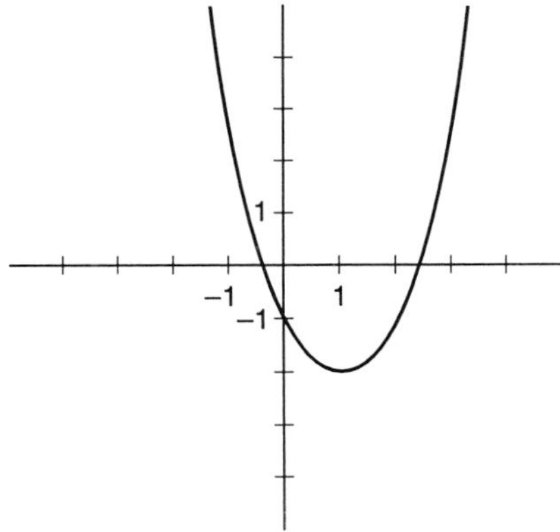

Figure 10: The graph of x^2, shifted: $(x - 1)^2 - 2$. [-5, 5] by [-5, 5].

We can also identify scale changes.

Example 19: The graph in Figure 11 has the shape, but not the scale, of the graph of sin x. Identify the expression graphed.

The vertical scale is twice that of the sine function, so the images are multiplied by 2. Also, the horizontal scale is compressed by a factor of three (changes happen three times as fast), so the argument is multiplied by 3. The expression is "2 sin(3x)."

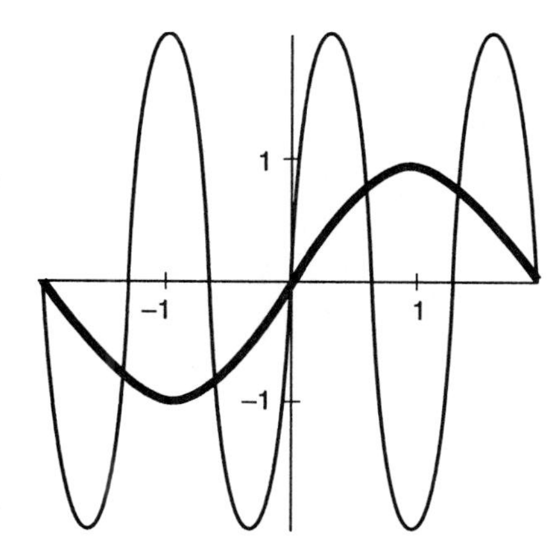

Figure 11: sin x [bold] and 2 sin(3x). [-180°, 180°] by [-2, 2].

Example 20: Consider the graph of $f(x)$ given in Figure 12. No algebraic representation is given. How does the graph of $f(x + 4)/2$ compare to it?

For $f(x + 4)/2$, note that "Add 4" applies to the argument, so it causes a shift left of 4 units (Figure 13, dashed curve). After f is applied, "Divide by 2" is applied to the images. The graph is compressed vertically by a factor of 2 (Figure 13, solid curve).

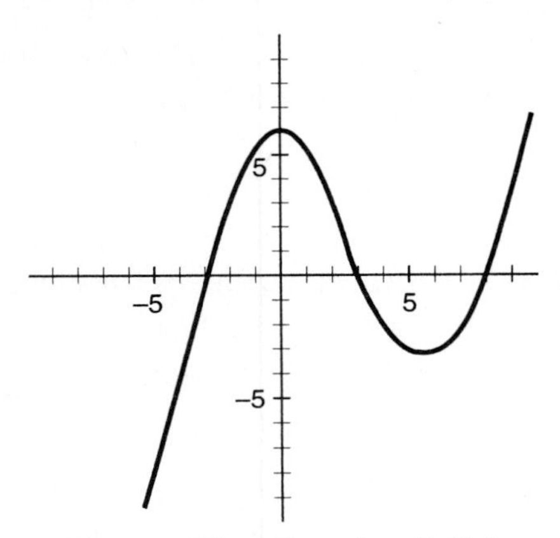

Figure 12: Graph of $f(x)$.
[-10, 10] by [-10, 10].

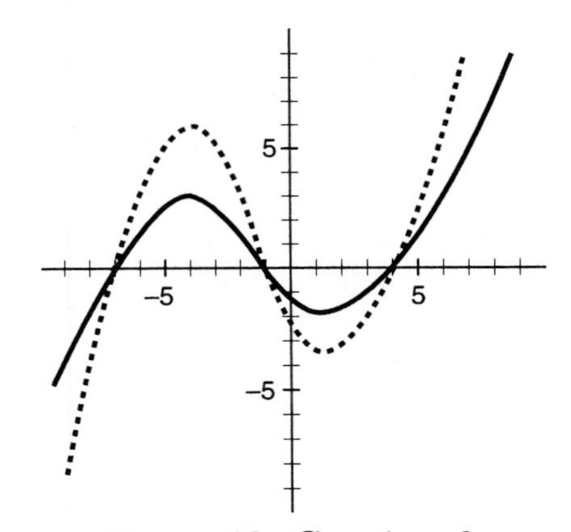

Figure 13: Graphs of
$f(x + 4)$ [dashed] and $f(x + 4)/2$.
[-10, 10] by [-10, 10].

Example 21: Suppose that all we know about f is that $f(4) = 7$. So we know the point (4, 7) is on the graph of $f(x)$. Find a point on the graphs of $f(x) + 2$, $-f(x)$, $f(x + 1)$, and $f(x/2)$.

function	point	
$f(x)$	(4, 7)	
$f(x) + 2$	(4, 9)	
$-f(x)$	(4, -7)	
$f(x+1)$	(3, 7)	[The argument of f is 4, so $x+1 = 4$ and $x = 3$.]
$f(x/2)$	(8, 7)	[The argument of f is 4, so $x/2 = 4$ and $x = 8$.]

Note how order matters. To graph $f(x+1)$, the points are $(x, f(x+1))$. We know $f(4) = 7$. To use $f(4)$, we need $x+1 = 4$, so $x = 3$. The point on the graph of $f(x+1)$ is (3, 7), one unit to the left (!) of the corresponding point on the graph of $f(x)$.

The points on the graph of $f(x/2)$ are $(x, f(x/2))$. To use $f(4) = 7$, we need $x/2 = 4$, so $x = 8$. The point on the graph of $f(x/2)$ is (8, 7), twice (!) as far from the y-axis as the corresponding point on the graph of $f(x)$.

Example 22: When radioactive substance decays, the amount of the substance remaining decreases as time elapses, as graphed in Figure 14. We use the notation $A(t)$ for the amount (in grams) at time t. Physics tells us that the shape of this graph is correct for all types of radioactive decay, rapid or slow, and including the decay of dangerous radioactive substances in nuclear reactors as well as the harmless decay of Carbon 14 used to date archeological materials. The *shape* is constant, but the *scale* varies from example to example.

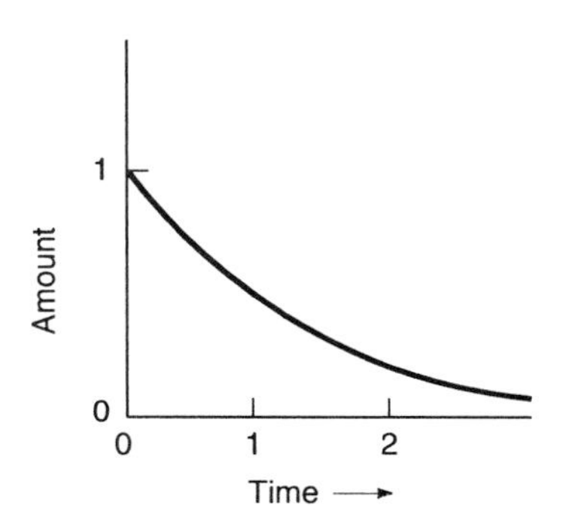

Figure 14: Radioactive decay. $A(t)$, the amount remaining at time t.

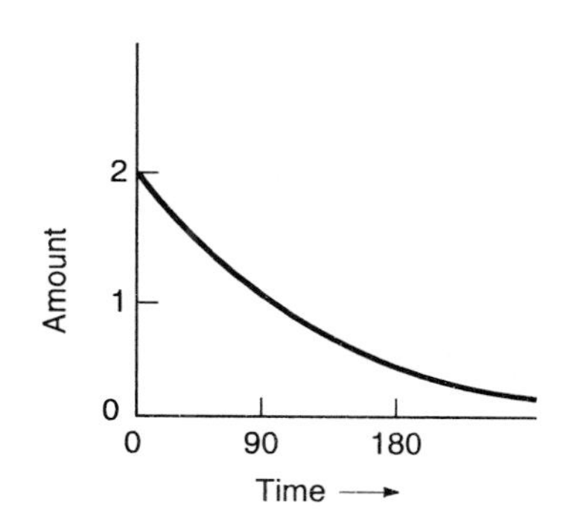

Figure 15: Radioactive decay. $A(0) = 2$ and $A(90) = 1 = (1/2)A(0)$.

How can we label the scale in Figure 14 for an example in which there are initially two grams of radioactive substance and half the original amount will remain after 90 years?

Figure 15 shows the same graph with the scales labeled to illustrate those two pieces of data.

Now, how can we find the formula for that graph, given that the basic shape in Figure 14 was graphed using $f(t) = (1/2)^t$?

We need to change the scales. For the basic shape function $f(t) = (1/2)^t$, $f(0) = 1$ (not 2) and $f(1) = 1/2$. That is, one-half the original amount remains after 1 unit of time (not 90 units of time). Because we want $A(0) = 2$, we want a vertical scale change by a factor of 2 (we want $A(t)$ to be twice as tall as $f(t)$) and a horizontal scale change by a factor of 90 (we want $A(t)$ to be 90 times as wide).

Therefore, the function we want is $A(t) = 2f(t/90)$.

$A(t) = 2(1/2)^{t/90}$. Graph it yourself and see (Problems B53-54).

The following tables summarize the effects of composition on graphs.

notation ($c > 0$)	change in graph	operation applied to the ...
$f(x) + c$	shift up c	image of f (changes are vertical)
$f(x) - c$	shift down c	
$cf(x)$ ($c > 1$)	expand vertically away from the x-axis by a factor of c	
$(f(x))/c$ ($c > 1$)	contract vertically toward the x-axis by a factor of c	
$-f(x)$	flip upside down (reflect) through the x-axis	
$f(x + c)$	shift *left* c	argument of f (changes are horizontal)
$f(x - c)$	shift *right* c	
$f(cx)$ ($c > 1$)	*compress* horizontally toward the y-axis with a factor of c	
$f(x/c)$ ($c > 1$)	*expand* horizontally away from the y axis by a factor of c.	
$f(-x)$	reflect through the y-axis	

type of arithmetic operation	applied to the	effect on the graph
add or subtract $c > 0$	argument	shift left or right
	image	shift up or down
multiply by $c > 1$	argument	*compress* horizontally
	image	expand vertically
divide by $c > 1$	argument	*expand* horizontally
	image	compress vertically

Conclusion. We can create new functions by composing simpler functions in sequence. The *argument-operation-image* concept of a function is extremely important for composition. The image of the first function becomes the argument of the second. **Order Matters!**

The examples show that composition with arithmetic functions produces a graphical location shift or a scale change. Arithmetic operations applied to the image seem to cause no problem. "Add c" and "up c" seem to go together.

But arithmetic operations applied to the arguments are harder to grasp. "Add c" and "*left* c" are a surprising combination. Similarly, it may be surprising that the graph of $f(2x)$ is *half* as wide as $f(x)$. While $2f(x)$ is expanded vertically, $f(2x)$ is compressed horizontally.

Terms: composition, composite, decompose, location, scale.

****

Exercises for Section 2.3, "Composition and Decomposition":

A1.* What makes a function "composite"?

A2. Do Calculator Exercise 2.

A3.* The notation "5(...)" signals multiplication, but the notation "g(...)" signals something else. What?

A4. Let $f(x) = 2x$. Find $f(f(f(3)))$.

A5. Let $f(x) = x^2$ and $g(x) = 2x + 7$.
a) Find $f(g(3))$. b) Find $g(g(4))$ c) Find $f(f(3))$.

A6. Let $f(x) = 3x - 1$ and $g(x) = (x + 1)/3$.
a) Find $f(g(5))$. b) Find $g(f(7))$. c) Find $f(f(2))$.

^^^^Express $h(x) = f(g(x))$ if

A7. $f(x) = x^2$ and $g(x) = x + 1$.
A8. $f(x) = 2x$ and $g(x) = x + 1$.
A9. $f(x) = x^2$ and $g(x) = \ln x$.
A10. $f(x) = \exp(x)$ and $g(x) = 4x$.
A11. $f(x) = \sin(2x)$ and $g(x) = x^2$.
A12. $f(x) = x^2$ and $g(x) = x + 5$.

^^^^Express $f(g(x))$ and then $g(f(x))$ if

A13. $f(x) = 2x + 1$ and $g(x) = (x - 1)/2$.
A14. $f(x) = x/5 + 3$ and $g(x) = 5(x - 3)$.

^^^^ Decompose $h(x)$ into $f(g(x))$ by identifying $f(x)$ and $g(x)$.

A15. $h(x) = |5x|$.
A16. $h(x) = 3x + 2$.
A17. $h(x) = (x - 2)/3$.
A18. $h(x) = \ln(2x)$.
A19. $h(x) = 3\exp(x)$.
A20. $h(x) = (5x)^2$.
A21. $h(x) = 2x^2$.
A22. $h(x) = \sqrt{(x + 7)}$.

^^^^Give the corner (vertex) on the graph.

A23. a) $|x - 2|$ b) $|x| + 2$ c) $|x + 5|$ d) $|x| - 7$.
A24. a) $x^2 - 9$ b) $(x - 8)^2$ c) $(x + 3)^2$ d) $x^2 + 4$.

^^^^Compare the graph to the graph of $|x|$:

A25. a) $|x| + 7$. b) $|x + 3|$.
A26. a) $|x - 4|$. b) $|x| - 2$.

^^^^Compare the graph to the graph of $\cos x$ (use radians):

A27. a) $3 \cos x$. b) $\cos(x - 3)$.
A28. a) $-\cos x$. b) $\cos(x + 2)$.

^ ^ ^ ^ ^ ^ ^ ^ ^

^ ^ ^ ^ Compare the graphs of the given expression to the graph of $f(x)$. [Assume $c > 1$].

B1.* $f(x) + c$ B2.* $cf(x)$ B3.* $f(x + c)$ B4.* $f(x) - c$.

B5.* $f(x - c)$ B6.* $(f(x))/c$. B7.* $f(cx)$ B8.* $f(-x)$.

B9.* $-f(x)$ B10.* $f(x/c)$.

B11.* Given the graph of $f(x)$, explain why the graph of $f(x + 4)$ is shifted 4 units *left*.

B12.* Given the graph of $f(x)$, explain why the graph of $f(x - 5)$ is shifted 4 units *right*.

B13.* Given the graph of $f(x)$, explain why the graph of $f(2x)$ is *half* as wide.

B14.* Given the graph of $f(x)$, explain why the graph of $f(x/2)$ is *twice* as wide.

B15. Use your calculator to help you sketch the graph of $y = 2^x$ (boldly, say on [-2, 2] by [0, 5]). On the same axis system, sketch the graph of $y = 2^{x/2}$. How do the two graphs compare? Pick a y-value between 1 and 5 and note how their x values compare. Pick a y-value between 0 and 1 and note how their x-values compare. Is the graph with "$x/2$" wider?

B16. Use your calculator to help you sketch the graph of $y = \cos x$ (in degrees, say on $[-360°, 360°]$ On the same axis system, sketch the graph of $y = \cos(2x)$. How do the two graphs compare? Pick a y-value between 0 and 1 and note how their x values compare. Pick a y-value between -1 and 0 and note how their x-values compare. Is the graph with "$2x$" wider?

^ ^ ^ ^ Give the vertex of the graph.

B17. $|x - 3| + 4$. B18. $|x + 2| - 5$.

B19. $(x - 8)^2 - 1$. B20. $(x + 3)^2 + 2$.

B21. If $f(x)$ is zero when $x = 7$, when is the following function zero?

a) $f(3x)$ b) $f(x - 4)$ c) $f(x/2)$

B22. If $f(x)$ is zero when $x = -3$, when is the following function zero?

a) $f(x + 2)$ b) $f(-x)$ c) $f(2x)$

B23. If $f(2) = 4$, find one point on the graph of

a) $f(x + 3)$ b) $f(x) + 2$ c) $f(2x)$

B24. If $f(-2) = 5$, find one point on the graph of

a) $f(x - 3)$ b) $f(x) - 4$ c) $f(x/2)$.

^ ^ ^ ^ Find and simplify the difference quotient. (See above Example 8.)

B25. Let $f(x) = 3x$. B26. Let $f(x) = 2x + 4$.

B27. Let $f(x) = (x + 1)^2$. B28. Let $f(x) = 1/x$.

^ ^ ^ ^ How should we modify "$\sin x$" to produce a graph with

B29. more waves of the same height? B30. fewer waves of the same height?

B31. more and taller waves? B32. fewer and shorter waves?

^ ^ ^ ^ How should we modify "x^2 " to produce a graph which is shifted

B33. left 5 units?

B34. up 3 units?

B35. down 2 units and to the left 5 units?

B36. left 4 units and turned upside down?

B37. The figures illustrate a graph which is the graph of $|x|$ shifted. Give the expressions graphed in Figures (a) and (b).

B38. The figures illustrate a graph which is the graph of $|x|$ shifted. Give the expressions graphed in Figures (c) and (d).

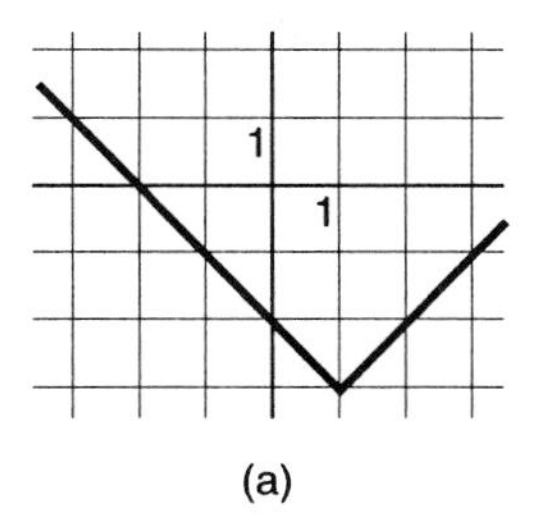

(a)

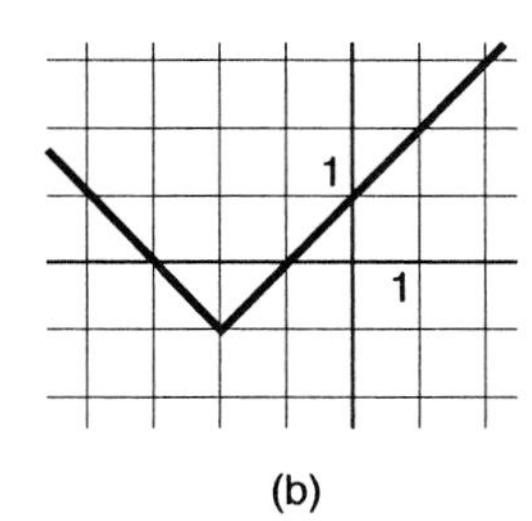

(b)

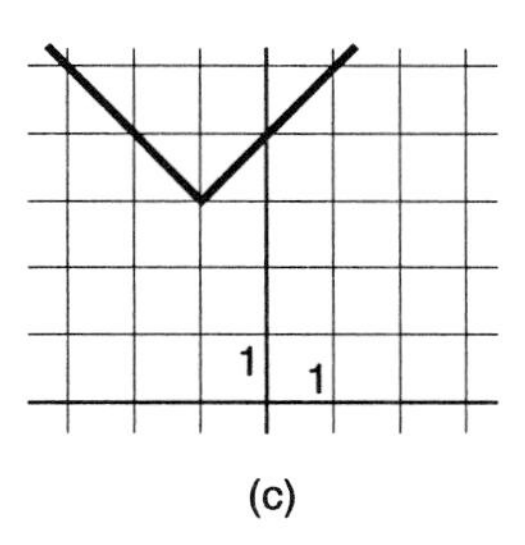

(c)

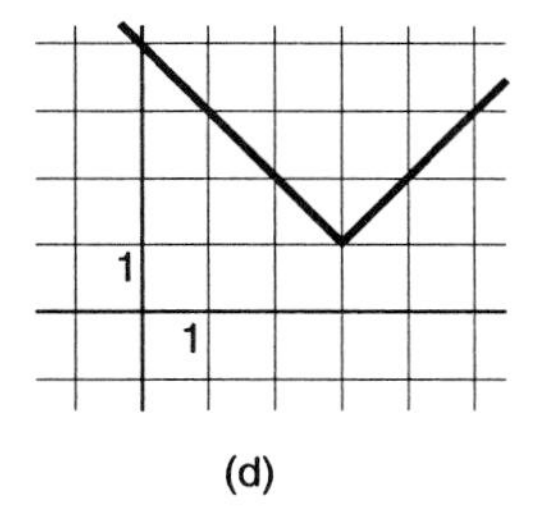

(d)

B39. Suppose $f(x)$ is graphed in Figure (a). Sketch the graph of $f(-x)$.

B40. Suppose $f(x)$ is graphed in Figure (b). Sketch the graph of $-f(x)$.

B41 Suppose $f(x)$ is graphed in Figure (c). Sketch the graph of $f(2x)$.

B42. Suppose $f(x)$ is graphed in Figure (d). Sketch the graph of $f(x/2)$.

B43. Suppose $f(x)$ is graphed in Figure (a). Sketch the graph of $f(x + 2)$.

B44. Suppose $f(x)$ is graphed in Figure (b). Sketch the graph of $f(x + 1) + 2$.

B45 Suppose $f(x)$ is graphed in Figure (c). Sketch the graph of $f(x - 2) - 3$.

B46. Suppose $f(x)$ is graphed in Figure (d). Sketch the graph of $f(x + 1) + 1$.

B47. How could we modify "cos x" to produce a graph which looks like the graph of "sin x"?

B48. How should we modify "sin x" to produce a graph which looks like the graph of "cos x"?

B49.* a) Define "composite" function.
b) Explain the idea behind decomposition of composite functions.

B50. Let $f(x) = \sqrt{x}$. Find and simplify its difference quotient. Let h "go to zero." If you "simplify" correctly, you will not obtain the undefined form "zero over zero." What do you think the "limit" would be?

B51. Let $f(x) = x^3$. Find and simplify its difference quotient.

B52. a) Suppose f is such that $f(x) = f(-x)$ for all x. Explain why the graph is symmetric about the y-axis.
b) Find a different condition on f that makes the graph symmetric about the line $x = c$.

^ ^ ^ ^ Find the function $g(t)$ with graph that fits Figure 14, however with

B53. $g(0) = 3$ and half remaining after 6 units of time.

B54. $g(0) = .05$ and half remaining after .3 units of time.

Section 2.4. Relations and Inverses

If a function is a set of ordered pairs and a graph exhibits a set of ordered pairs, how does a function differ from a graph?

Graphs may have more than one second number corresponding to a given first number.

Example 1: The equation of the graph in Figure 1 is:

$$x^2 + y^2 = 1.$$

This equation describes a circle centered at the origin with radius 1, which is called the "unit circle." Note that the equation does not express y as a function of x.

Any vertical line through the interior of the circle will cross the graph twice. Therefore the circle is not the graph of a function, because functions have exactly one image for each argument. The graph fails the "Vertical Line Test."

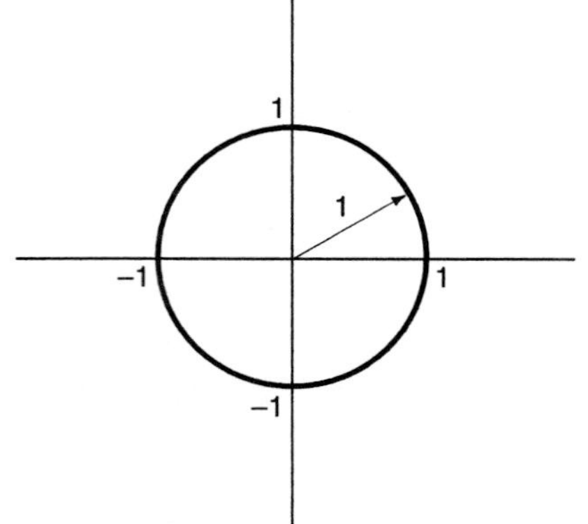

Figure 1: $x^2 + y^2 = 1$. The unit circle with radius 1 about the origin. [-2, 2] by [-2, 2].

The Vertical-Line Test (2.4.1): If there is a vertical line which intersects a graph twice, the graph is not the graph of a function.

Any graph, including a circle, determines a set of ordered pairs.

Definition 2.4.2: A relation is a set of ordered pairs.

Any set of ordered pairs determines a relation, even if some arguments have two or more images. Therefore, any graph determines a relation. Also, any equation with two variables (say, x and y) determines a relation given by the set of ordered pairs for which the equation is true. Furthermore, any function is a relation, but not all relations are functions.

Relations for Solving Equations. Relations are important for solving equations. Consider solving "$f(x) = c$." Because f is a function, for a given x there is only one value of c. But for a given c, there may be several values of x. Therefore solving may yield a relation.

Example 2: Suppose Figure 2 is a representative graph of the function f. Find $f(2)$. Solve $f(x) = 3$.

To evaluate $f(2)$, read the y-coordinate of the point intersected by the vertical line $x = 2$: $f(2) = 3$.

In the inverse direction, to solve the equation "$f(x) = 3$" read the x-coordinates of all the points intersected by the horizontal line $y = 3$. The solution is apparently "$x = -4$ or $x = 2$."

To evaluate you go from x to y; to solve you go from y to x.

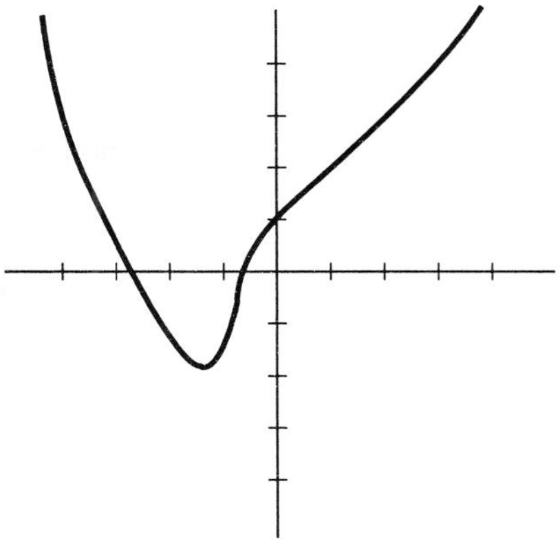

Figure 2: The graph of an arbitrary function, f. [-5, 5] by [-5, 5].

Evaluating and solving are inverse processes.

Many important functions have inverses which are relations, not functions.

Example 3: The absolute-value function is, of course, a function. But its inverse is not. There are two x's which solve "$|x| = 5$." Therefore, here we speak of an inverse *relation* as opposed to an inverse *function* (Figure 3).

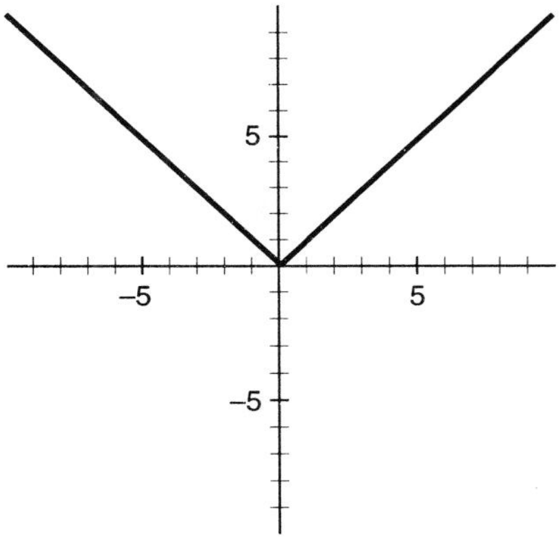

Figure 3: $|x|$. [-10,10] by [-10,10]

<u>Theorem 2.4.3, The Inverse Absolute-Value Relation</u>: $|x| = c$ iff $c \geq 0$ and ($x = c$ or $x = -c$).

In the equation "$|x| = c$" for each x there is one c, but if $c > 0$, for each c there are two x's. That is why the c-to-x direction is a relation, not a function. According to the theorem, the solution to "$|x| = 5$" is "$x = 5$ or $x = -5$."

Sometimes the functional nature of calculators leads students to mistakenly oversimplify relations into functions.

Example 4: Solve $x^2 = 20$.

Again we know that there is not a unique x which solves this equation. The solution is $x = \sqrt{20}$ or $x = -\sqrt{20}$. This may be conveniently abbreviated: $x = \pm\sqrt{20}$.

Calculator function keys are designed to operate as functions, not relations, because they display *one* result of each calculation. The square root symbol, $\sqrt{\ }$, denotes an inverse *function*. There are two solutions to $x^2 = 20$, but using the square root key ($\sqrt{20} = 4.472$) yields only one of them (on the right half of the graph, Figure 4). The other is easy to obtain with a simple sign change (if you remember to think of it).

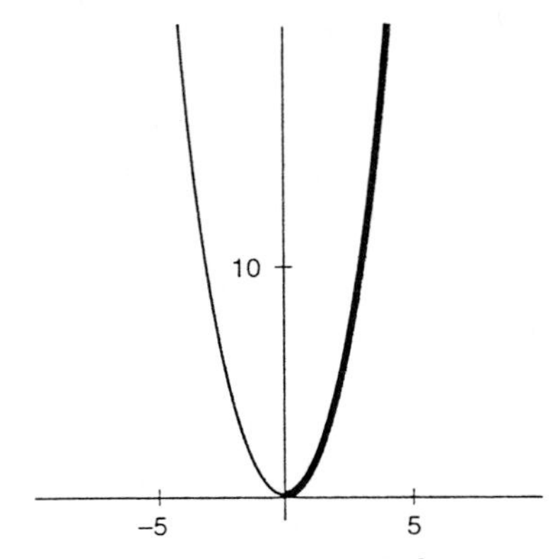

Figure 4: $y = x^2$ with the region where $x = \sqrt{y}$ emphasized. [-10, 10] by [0, 20].

Theorem 2.4.4 (The Inverse Square Relation):

$$x^2 = c \text{ iff } x = \sqrt{c} \text{ or } x = -\sqrt{c}.$$

Example 5: Solve $(x - 4)^2 = 7$.

By Relation 2.4.4 (where "x" can be "$x - 4$"), $x - 4 = \sqrt{7}$ or $x - 4 = -\sqrt{7}$. Thus $x = 4 + \sqrt{7}$ or $x = 4 - \sqrt{7}$. The only trick is to remember the second solution.

Some inverse relations are substantially more complex.

Example 6: *Sine* is a well-known trigonometric function (Figure 5). Given the value of x, your calculator will find the unique value of $\sin x$. But, given only the value of $\sin x$, you cannot be sure of the value of x. This is because there are many values of x which are related to a single value of $\sin x$.

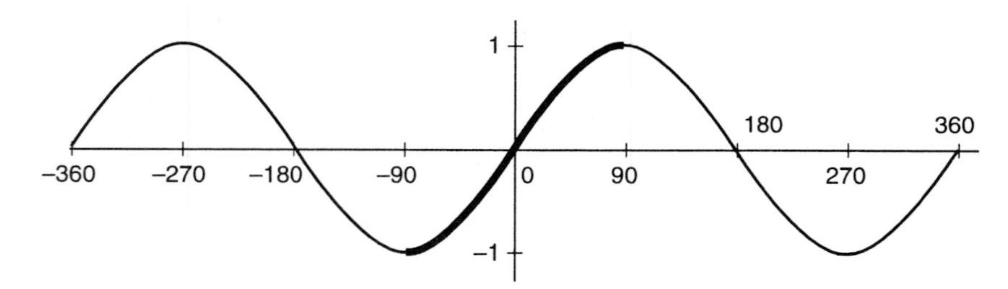

Figure 5: The graph of $y = \sin x$ with the points where $\sin^{-1} y = x$ emphasized. Also, $y = 1/2$ is graphed. The x-values of the points of intersection are the solutions to the equation "$\sin x = 1/2$." The two graphs intersect in many places, but only one of them is called "$\sin^{-1}(1/2)$." $-360° \leq x \leq 360°$, $-1 \leq \sin x \leq 1$.

Calculator Exercise 1. Set your calculator mode to degrees (instead of radians). Evaluate sin 30. You should obtain 0.5. Now find sin 150. Again the calculator displays 0.5. So 30 and 150 are *two* values of x related to the same value of $\sin x$ (Figure 5). When it comes time to solve the equation "$\sin x = 0.5$," there will be these two solutions, and many more. But your calculator yields only one of them. It has an inverse *function* keystroke-sequence which will find *one* solution, 30. The written inverse *function* is symbolized by "$\sin^{-1}$" or "arcsin", which are pronounced "inverse sine" or "arc sine." These are just two names for the same function.

Use your calculator to find $\sin^{-1}$ 0.5. It displays the *one* number, 30. It doesn't display 150, or 390, or -210, or any of the other solutions to "$\sin x = .5$" (Figure 5).

There is, of course, a way to find all solutions from the one your calculator yields. Here is a sample theorem to illustrate how complex inverse relations can be written. You are not expected to memorize it now, but you are expected to learn about inverses by reading it.

The superscript "-1" in the notation for the inverse sine function is not a power. $\sin^{-1} x$ is certainly not $1/(\sin x)$. The superscript "-1" indicates "inverse" and is commonly used to denote inverse functions. The inverse of the function f, if the inverse is a function, may be denoted by "f^{-1}," and read "f inverse."

<u>Theorem 2.4.5 (The Inverse Sine Relation)</u>: $\sin x = y$ iff

<u>A</u>) $x = \sin^{-1} y$
<u>B</u>) or $x = 180° - \sin^{-1} y$
<u>C</u>) or $x = \sin^{-1} y \pm 360n°$, for any integer n,
<u>D</u>) or $x = 180° - \sin^{-1} y \pm 360n°$, for any integer n.

Part A gives the solution your calculator displays. It is the one most frequently desired, but Part B is almost as important. Too many students simply accept the number their calculator provides without thinking that the number they really want may be one of the other alternatives.

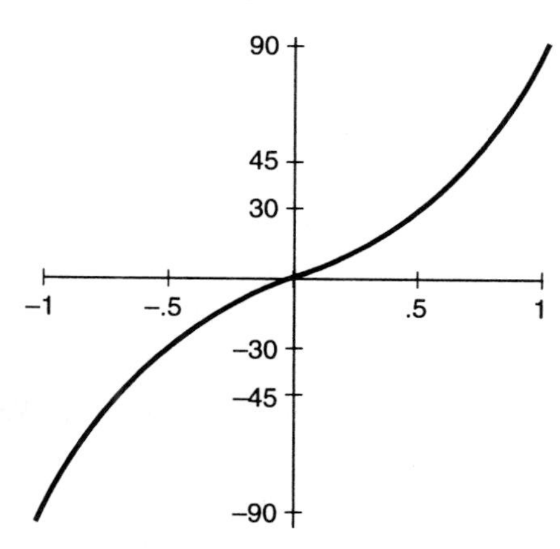

Figure 6: $\sin^{-1} x$ (= arcsin x) in degrees. [-1, 1] by [-90. 90].

Calculator Exercise 2: Try to evaluate $\sin^{-1}$ 2. What does your calculator say? Many return an error message and others give a complex number. Since the range of sine is [-1, 1], the natural domain of $\sin^{-1}$ is [1, 1], which does not include 2.

Calculator Exercise 3: Pick a number (say, 50) and evaluate sine of that number. Then evaluate $\sin^{-1}$ of that. Do you recover the original number?

You will *if* the original number was between -90 and 90 degrees. But you will not otherwise.

$\sin 50° = 0.7660444$. $\sin^{-1} 0.7660444 = 50°$.

$\sin 110° = 0.9396926$. $\sin^{-1} 0.9396926 = 70°$, not 110°. Note that $180° - 70° = 110°$, so 110° is the result of Part B.

Inverses: Theory. Inverses are important for solving equations. To evaluate f is to begin with argument x and find image y. To solve the equation "$f(x) = y$" is to begin with y and find all such x's.

Given a function which takes argument a and yields image b, the point of an inverse is to take argument b and return image a. If there is exactly one a for each b (which is the ideal case), then there is an inverse *function*. However, if there are two or more possible a's for some b's, then there is an inverse *relation*. Then, the point of an inverse *function* is to return a single a from which all the other values of a can be recovered. You want your calculator to select one of the a's and display it. Then (if possible) you generate the rest of the a's by some simple rule.

Definition 2.4.6. (Inverse Relations): Given a relation, R (a set of ordered pairs), its inverse relation, R^{-1}, is the set of all ordered pairs (b, a) such that (a, b) is in R.

Example 7: The first two columns give part of the Table 2.2.2, the standard-normal-cumulative-distribution function, Φ. The inverse function, Φ^{-1}, is given in the last two columns, or determined by reading the table backwards.

z	$\Phi(z)$	y	$\Phi^{-1}(y)$
-0.6	.2743	.2743	-0.6
-0.4	.3446	.3446	-0.4
-0.2	.4207	.4207	-0.2
0	.5000	.5000	0
0.2	.5793	.5793	0.2
0.4	.6554	.6554	0.4

The dummy variable, y, used to express the inverse, Φ^{-1}, is not important. The last two columns could be labeled "x" and "$\Phi^{-1}(x)$" without changing the meaning. In fact, the last two columns could be omitted. Given a table for a function, you do not need another table for its inverse function. Just read the original table in reverse.

Notation. Be careful. $f^{-1}(x)$ is not $1/f(x)$. The "-1" in "x^{-1}" means reciprocal. The "-1" in f^{-1} does not. It means *inverse*.

Many functions have inverses which are also functions. This is the case for the four arithmetic operations.

Theorem 2.4.7 (Inverses of Arithmetic Operations):
A) The inverse function of "Add c" is "Subtract c," and vice versa.
B) The inverse function of "Multiply by c" is "Divide by c", if c is not zero, and vice versa.
C) The inverse of "Change the sign" is "Change the sign."

These could be expressed in functional notation (problem B34).

Finding Inverses. Inverses express the "doing and undoing" method of solving equations, which I called the "inverse-reverse" method. The inverse function undoes what the function does (Figure 7).

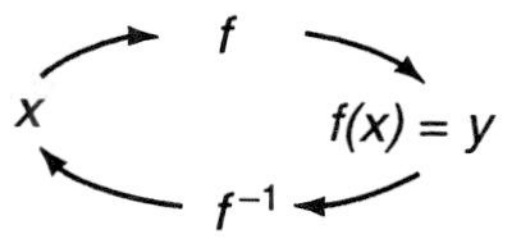

Figure 7: A function loop.

Example 8: Let $f(x) = x/2 + 3$. Find f^{-1}.

The function is "Divide by 2 and then add 3." By *inverse-reverse* thinking, f^{-1} is "Subtract 3 and multiply by 2." This could be expressed by "$f^{-1}(x) = 2(x - 3)$." An equivalent alternative is "$f^{-1}(y) = 2(y - 3)$." The variables "x" and "y" are just dummy variables anyway.

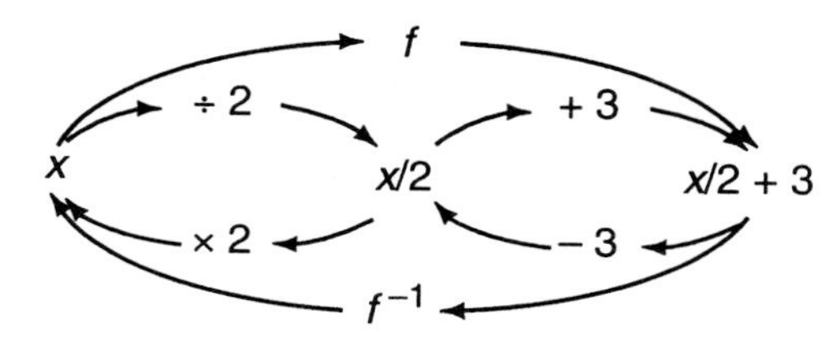

Figure 8: A "function loop" diagram for $f(x) = x/2 + 3$ and its inverse. f takes x to $f(x)$ along the top. f^{-1} returns x from $f(x)$ along the bottom.

Figure 8 illustrates why the method called *inverse-reverse* here is sometimes called "doing and undoing." The function f does something to x to yield the image y, and the solution process consists of undoing what f does to return from y back to x.

If "Solve $f(x) = 5$" is the problem and f^{-1} is known, $f^{-1}(5)$ is the solution. f^{-1} expresses the steps in the solution process for all problems of the form "Solve $f(x) = c$," for any c.

In the expression "$f(x)$", "f " expresses the *evaluation* process.
In the equation "$f(x) = y$", "f^{-1}" expresses the *solution* process.

> To find f^{-1}, solve $f(x) = y$ for x. The solution expresses $f^{-1}(y)$. You may, if you wish, switch dummy variables to express $f^{-1}(x)$.

Calculator Exercise 4: Graph examples of f and f^{-1} on the same *square* scale. (A scale is "square" when one unit is the same distance both vertically and horizontally.) The graph of f^{-1} is closely related to the graph of f. Can you see how?

For example, graph both $f(x) = x/2 + 3$ and $f^{-1}(x) = 2(x - 3)$ (from Example 8). Graph e^x and $\ln x$. Graph x^2 and $\sqrt{x}$. What is the relationship of the graphs (Problem B14)?

Example 9: Let $f(x) = 3 \log(x + 2)$. Find f^{-1}.

$3 \log(x + 2) = y$ iff $\log(x + 2) = y/3$

[now you need to know the inverse of the *log* function]

iff $x + 2 = 10^{y/3}$ [Theorem 1.4.2C]

iff $x = 10^{y/3} - 2$.

Therefore, $f^{-1}(y) = 10^{y/3} - 2$.

If you want to switch letters to obtain $f^{-1}(x)$, go ahead. The letters "x" and "y" are just dummy variables anyway.

Example 9, continued: Solve "$3 \log(x + 2) = 3.2$,"

In this equation, $f(x)$ from Example 9 is on the left and we already found f^{-1}. Therefore, we do not have to discover the solution steps again. f^{-1} tells us what to do. The answer is $f^{-1}(3.2)$. The answer is determined by doing f^{-1} to 3.2. Plug in 3.2 for y into "$f^{-1}(y) = 10^{y/3} - 2$."

Calculator Exercise 5: Use your calculator to solve that equation in exactly three steps using inverse-reverse thinking. f^{-1} expresses the steps. The solution is $x = 9.66$.

****

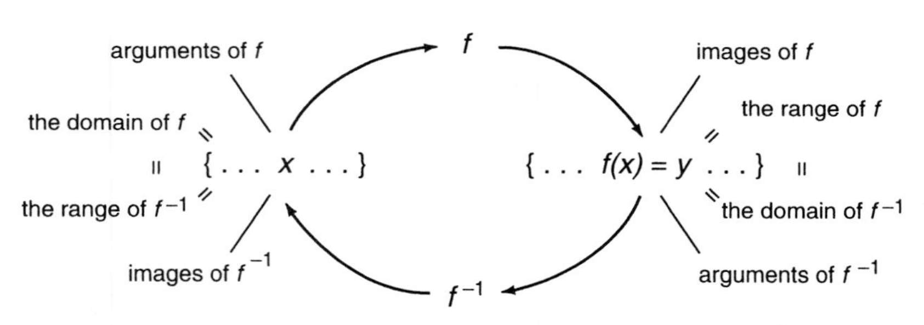

Figure 9: The relationship between f and f^{-1}.

****

Functions (such as "$f(x) = 2x + 7$") express their own *evaluation* process (as a command). To find, say, $f(4.1)$, you simply do what the function tells you to do ("Multiply by 2 and add 7."). On the other hand, equations do *not* express their own solution process. To *solve* the equation "$2x + 7 = 9.7$" you do *not* "Multiply by 2 and add 7." You *do* f^{-1}. f^{-1} is the rule of the solution process.

Example 10: Suppose you wish to rent a small truck for one day. The cost depends upon the number of miles you drive. If the cost function, C, is the rule for the cost in terms of the number of miles driven, what is the meaning of C^{-1}?

C^{-1} is the rule for the number of miles driven in terms of the cost.

Graphical Inverses: Graphically, inverse relations and functions are simple to see.

Example 11: Figure 10 pictures a representative graph of $f(x) = 2x + 3$. Every y-value which appears shows up only once. The function is one-to-one. There is a natural inverse function which can be read off the graph by reading it in reverse, that is, reading from y to x. $f(2) = 7$. $f^{-1}(7) = 2$.

A single graph is enough to read both the function and its inverse. To evaluate the function, read from horizontal to vertical (x to y). To evaluate the inverse, read from vertical to horizontal (y to x).

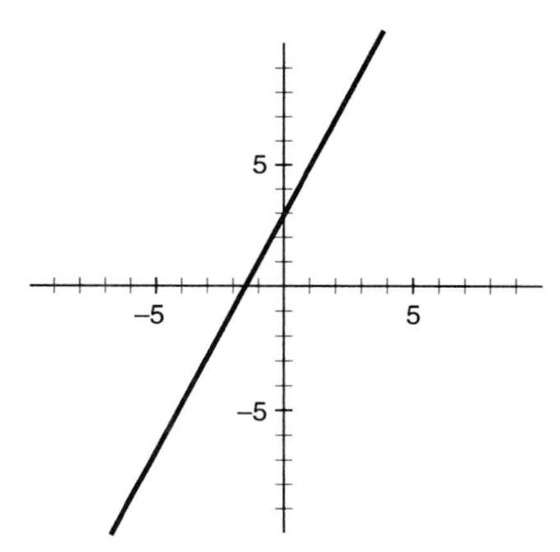

Figure 10: $f(x) = 2x + 3$. [-10,10] by [-10,10]

Definition 2.4.8: A function is said to be one-to-one if every image (y-value) which appears, appears exactly once.

This condition can be stated symbolically: "If $f(a) = f(b)$, then $a = b$." It is also equivalent to, "If $a \neq b$, then $f(a) \neq f(b)$."

Example 12: Consider this abbreviated table of $f(x) = x^2$.

x	$y = f(x) = x^2$
-2	4
-1	1
0	0
1	1
2	4

This function is not one-to-one. Some y-values appear twice. $f(2) = f(-2)$, but $2 \neq -2$. If we give the inverse, 4 will correspond to two values, both -2 and 2. The inverse is a relation containing both the ordered pairs (4,-2) and (4,2), so the inverse relation is not a function.

Theorem 2.4.9: If a function is one-to-one, then its inverse relation is a function.

The functions in Examples 7 through 11 are one-to-one and in each the inverse relation is also a function since for each y there is exactly one x.

Graphically it is easy to see if a function is one-to-one.

The Horizontal-Line Test (2.4.10): A function is one-to-one if and only if no horizontal line intersects its graph more than once. f is not one-to-one if and only if there is a horizontal line which intersects the graph of f more than once.

Here is why. If the horizontal line $y = c$ intersected the graph of f more than once at, say, $x = a$ and $x = b$, then there would be 2 x's for a single y. $f(a) = f(b) = c$, but $a \neq b$. Therefore, by definition, f would not be one-to-one.

In Example 11 (Figure 10), $f(x) = 2x + 3$ is one-to-one. Horizontal lines that cross the graph cross exactly once. No horizontal line intersects it more than once.

$f(x) = x^2$ is not one-to-one; horizontal lines above the x-axis intersect the graph twice (Figure 4).

Sine is not one-to-one if its domain is all real numbers. However, if we change the domain to [-90,90] (in degrees), the new, restricted, function is one-to-one (Figure 5).

Shortcuts. To solve "$f(x) = c$" using the inverse-reverse method "x" must appear only once. However, sometimes an expression for a function does not seem to begin with the argument. For example, "$5 - x$" may seem to begin with "5" and not with "x". If an expression does not begin with "x", reorder the sequence of operations so that it does. There are some handy ways to do this. Remember,

The purpose of any identity is to provide alternative sequences of operations.

Identity 2.4.11 (Alternative Ways to Subtract):

A) $c - x = (-x) + c$.

B) $c - x = -(x - c)$.

Example 13: $7 - 12 = -12 + 7 = -(12 - 7) = -5$.

$\pi - x = -x + \pi = -(x - \pi)$.

Example 14: Solve $5 - x = 14$ by the *inverse-reverse* method.

You might think of the left side as naturally beginning with "5". It does. But, for purposes of the *inverse-reverse* method we may regard "$5 - x$" as beginning with "x" as in "$(-x) + 5$." You can solve by subtracting 5 and changing the sign.

Of course, there are several ways to solve the original equation. The point is not just to know *one* way, but to be comfortable with various options.

Alternative Ways to Divide. Division sometimes appears to begin in an inconvenient place. For example, the expression "$5/x$" apparently begins with "5", but, for the argument-function-image sequence we want to regard it as beginning with x. We can do so with the proper identity.

The "reciprocal" function is given by $f(x) = x^{-1} = 1/x$. It is *not* appropriate to think of reciprocation as "1 divided by x," since the best way to regard a function is with the argument-operation-image sequence. It begins with argument x (not with 1) and ends with image $1/x$. Calculators have a key for it, labeled "x^{-1}" or "$1/x$".

Theorem 2.4.12 (The Inverse of the Reciprocal Function):

$$1/x = c \text{ iff } x = 1/c.$$

The reciprocal function has the interesting property that evaluating and solving are both done by the same process: reciprocation. It is its own inverse.

Example 15: $1/x = 5$ iff $x = 1/5$.

Note that this theorem yields a one-step solution to the first equation, instead of the two-step solution obtained by "Multiply by x and then divide by 5."

Calculator Exercise 6: Use your calculator to solve "$1/(x - 3) = \pi$" in as few steps as possible.

The function is, "Subtract three and take the reciprocal." Solve the equation in two steps after you enter π: Take the reciprocal. Add 3. Done.

The most common alternative method of solving the equation takes more steps (four, instead of two), and you might have to write down some work. However, using the idea of a reciprocal (instead of division), no writing is needed.

Symbolically, the steps could be written:

$1/(x - 3) = \pi$ iff $x - 3 = 1/\pi$ iff $x = 1/\pi + 3 = 3.318$.

When we want to regard division as beginning with the denominator, we use the next identity

Identity 2.4.13 (Division Beginning with the Denominator):

A) $c/x = x^{-1}c$.
B) $c/x = (x/c)^{-1}$.

Example 16: $2/10 = 10^{-1}(2)$ [= (1/10)2].
$9/4 = (4/9)^{-1}$ [= 1/(4/9)].

Calculator Exercise 7: See if you can reproduce these results using the reciprocal function.

Suppose you already have 1.234 on your calculator display and you wish to evaluate 7/1.234. One alternative is to use $1.234^{-1} \times 7$, which yields 5.67.

$$7^{1/5} = 7^{5^{-1}} = 1.476.$$

Calculator Exercise 8: Solve in exactly 3 steps: $4.3/(x - 3.2) = 12$.

Identify the three steps in the evaluation process: Subtract 3.2, take the reciprocal, and multiply by 4.3. Then use inverse-reverse on the image, 12.

$x = 3.56$.

Some advanced functions have famous inverses. For example, *ln*, the natural logarithm function, and *exp*, the exponential function, are inverses of one another (Figure 11.

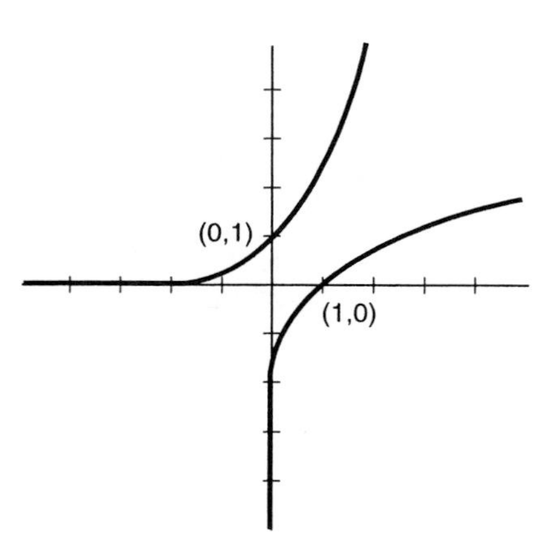

Figure 11: $y = e^x$.
$y = \ln x$ (lower).
[-5, 5] by [-5, 5].

Calculator Exercise 9: Solve $e^x = 7$.

You need to know how to "undo" exponentiation.

$e^x = 7$ iff $x = \ln 7$ (= 1.9459), according to the next theorem.

You may also think of e^x as applying a function to x to yield 7. To solve this, apply the inverse function, *ln*, to 7 to get back to x. This idea is expressed by the next result.

Theorem 2.4.14. $e^x = y$ iff $x = \ln y$.

Implicit is that $y > 0$, since $e^x > 0$ for all x, and *ln* has domain $x > 0$ (Figure 11). Because we usually solve for the variable "x", the theorem is apparently stated to solve problems employing the exponential function. However, the role of the dummy variables x and y may be switched to solve problems with "$\ln x$" (Problem B19).

Calculator Exercise 10: Solve $\ln x = 2.34$.

$\ln x = 2.34$ iff $x = e^{2.34}$ ($= 10.38$), by the theorem (with the roles of x and y interchanged).

Calculator Exercise 11: Select any value of x, say $x = 4.5$. Evaluate $e^{4.5}$. Then evaluate $\ln(e^{4.5})$. Your calculator should display the original number, 4.5.

Try evaluating "exp(ln x)" for some x. Note that you obtain the original value of x back *if* x is greater than zero, but not if x is zero or negative.

Now graph $y = e^{\ln x}$. What do you see?

Now graph $y = \ln(e^x)$. What do you see? Why is it not the same as the previous graph (problem B35)?

<u>**Geometry and Inverses**</u>. According to the definition of "inverse", (a, b) is on the graph of f if and only if (b, a) is on the graph of its inverse. There is a simple geometric relationship between these two points.

The graphs of f and f^{-1} are mirror images of each other through the line $y = x$.

Figure 12 illustrates several pairs of such points and the line $y = x$. Figures 14 through 16 illustrate three pairs of inverse functions.

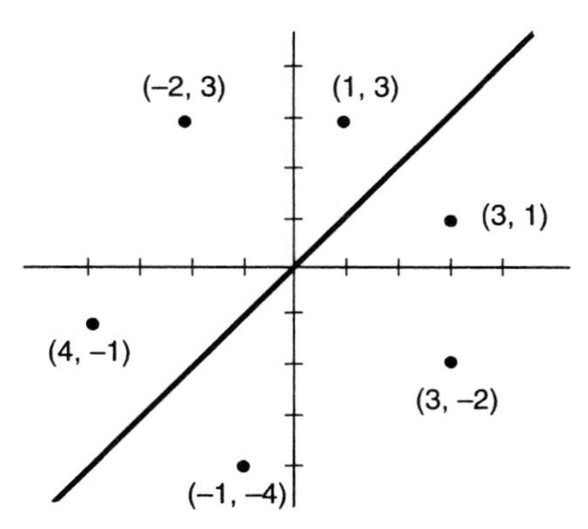

Figure 12: (a, b) and (b, a) are mirror images through the line $y = x$.

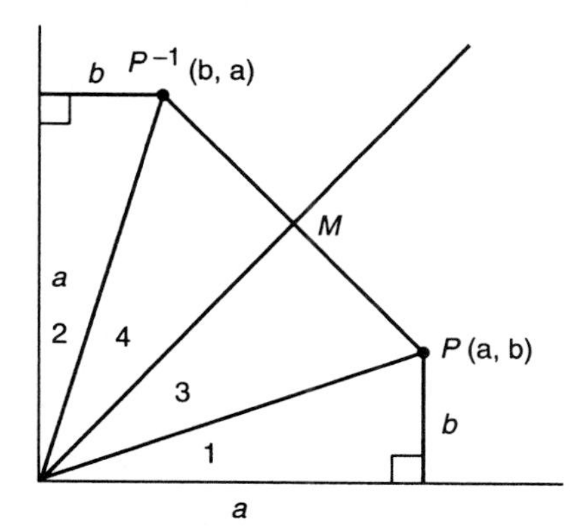

Figure 13: For a proof that (a, b) and (b, a) are mirror images of one another through the line $y = x$.

The two ordered pairs (a, b) and (b, a) are mirror images of each other through the 45 degree line $y = x$. Figure 13 includes the auxiliary lines for a geometric proof. The first point is labeled P and the second P^{-1}. A key idea is that angle 1 and angle 2 are congruent so angles 3 and 4 are also congruent. With a few more steps it can be shown that M is the midpoint of the line segment PP^{-1} which is perpendicular to the line $y = x$. Problem B41 asks for the details.

This mirror-image relationship can be seen in every example of the graphs of f and f^{-1}. Each graph is a reflection of the other through the line $y = x$. Note this in the graphs of $\ln(x)$ and e^x (Figure 11). The inverse of $f(x) = 2x + 7$ was given by $f^{-1}(x) = (x - 7)/2$. See the mirror images in Figure 14. The inverse of "add c" is "subtract c." See the mirror images in the graphs of "$x + c$" and "x - c" (Figure 15). The inverse of "multiply by 2" is "divide by 2." See the mirror images in the graphs of "$2x$" and "$x/2$" (Figure 16).

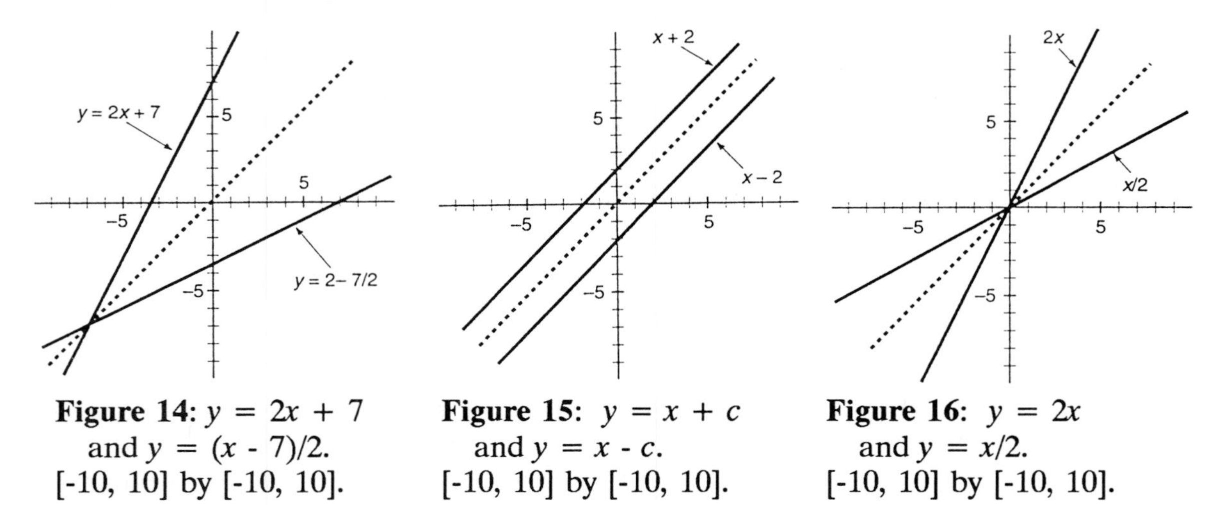

Figure 14: $y = 2x + 7$ and $y = (x - 7)/2$. [-10, 10] by [-10, 10].

Figure 15: $y = x + c$ and $y = x - c$. [-10, 10] by [-10, 10].

Figure 16: $y = 2x$ and $y = x/2$. [-10, 10] by [-10, 10].

<u>Functions Without Convenient Inverses</u>. Finding the inverse of f is equivalent to identifying the algebraic process for solving the equation "$f(x) = y$" for x. If f is simple enough, we can set up that equation and solve it to find the inverse function. For example, we could do this if $f(x) = 3(x + 5) = y$ to find $f^{-1}(y) = y/3 - 5$.

If f is not simple, but very important, we can *name* the inverse function and use its name to solve equations. For example, *sine* is a difficult function to evaluate by hand, but it is so important that we name its inverse and install both *sine* and *sin*$^{-1}$ on calculators. Then we can find *one* solution to "$\sin x = .4$" by using the name of a solution, "$\sin^{-1} .4$." But knowing its name does not mean it is easy to evaluate, unless you have a calculator.

There is one remaining case: functions which are not simple and not important, for example, "$f(x) = x^5 + x$." This function has an inverse, but it is not worth finding and naming. If we want to solve one equation with this function (for example, $x^5 + x = 8$), we use evaluate-and-compare. We expect to find and name inverse functions only when they are simple or important.

<u>Conclusion</u>. Evaluating and solving are inverse processes. A function can be regarded as a process. To evaluate $f(x)$ you apply the process named f to the number x. f^{-1} is the inverse process. To solve "$f(x) = c$" for x you apply the inverse process to c. Inverse relations are essential for equation-solving.

If a function takes argument a to image b, the point of an inverse relation is to take b and get back to a. If there is only one a for each b the function is said to be one-to-one and the inverse relation is a function. If there are two or more a's which yield the same b (as with squaring and *sine*), an inverse *relation* will take b and return *all* the corresponding a's. In that case the point of an inverse *function* is to take b and return *one* of the a's. This is what inverse functions on calculators do. Usually the other a's can be easily generated from the given a (for example, with squaring, by changing the sign).

Terms: Vertical Line Test, relation, inverse, one-to-one, Horizontal Line Test.

****

Exercises for Section 2.4, "Relations and Inverses":

A1.* a) What is another common name for the *inverse-reverse* method of solving equations? b) What is the format requirement for immediate application of the inverse-reverse method?

A2.* If (a, b) is on the graph of a function, then ________ is on the graph of its inverse.

A3.* a) In the expression "$f(x)$", "f" expresses the ________ process.
b) In the equation "$f(x) = y$", "f^{-1}" expresses the ________ process.

A4.* The range of f is the set of all _____________. Therefore, the range of f is the set of all c such that the equation ___________ has a solution.

A5.* If you solve a particular equation, say $f(x) = 25$, you find its solution. If you solve the general equation, $f(x) = y$, you discover ____________________.

A6.* Define "relation."

A7.* Define "one-to-one."

^ ^ ^ ^ State an identity giving an alternative way to
A8.* subtract. A9.* divide.

A10.* Which condition guarantees that a function will have an inverse function?

A11.* True or false? An equation with two variables always determines a relation.

A12. The first figure at the right locates (a, b) and (c, d) on an axis system. Locate (b, a) and (d, c).

A13. The second figure at the right locates (a, b) and (c, d) on an axis system. Locate (b, a) and (d, c).

^ ^ ^ ^ Make a function-loop diagram for f and its inverse.

A14. $f(x) = 5x + 7$. A15. $f(x) = 6 - x$. A16. $f(x) = (x - 3)/5$.

^ ^ ^ ^ Evaluate these expressions with a calculator. Give 3 significant digits. [A 2-digit answer is given.]

A17. a) e^7 [1100]. b) ln 23,000 [10].

A18. a) e^{-5} [.0067]. c) sin(-35°) [-.57].

^ ^ ^ ^ Solve these equations algebraically.

A19. $|x - 4| = 5$. A20. $(|x + 1| - 3)^2 = 4$ [four solutions].

A21. $(x^2 - 12)^2 = 100$ [four solutions].

A22. $|20 - |x|| = 12$ [four solutions].

A23. $\ln x = -1.23$ [.29]. A24. $\ln x = 5.6$ [270].

A25. $e^x = 20$ [3.0]. A26. $e^x = 0.1$ [-2.3].

^ ^ ^ ^ Use reciprocation to express these functions in English commands as a sequence of exactly three operations, beginning with x.

A27. $f(x) = 5.7/(x + 5)$. A28. $f(x) = 67/(x - 2)$.

A29. $f(x) = 1/|x - 4|$. A30. $f(x) = 5/x^2$.

A31. $f(x) = \ln(3/x)$. A32. $f(x) = (5/x)^2$.

^ ^ ^ ^ Use the sign-change operation to express these functions in English commands as a sequence of exactly three operations, beginning with x.

A33. $f(x) = 3 - |x|$ A34. $f(x) = (56 - x)^2$.

^ ^ ^ ^ Use reciprocation and the sign-change operation to express these functions in English commands as a sequence of exactly four operations, beginning with x.

A35. $f(x) = 45/(3 - x)$. A36. $f(x) = 6 - 5/x$.

A37. $f(x) = \ln(\pi - 1/x)$. A38. $f(x) = 1/|3 - x|$

^ ^ ^ ^ Solve algebraically in exactly three steps. Use your calculator to execute exactly three operations on the number on the right side. Do not write down intermediate steps.

A39. $2.1/(x - 5) = 34$ [5.1]. A40. $3.4/(x + 6) = .012$ [280].

A41. $\ln(3/x) = -4$ [160]. A42. $e^{2/x} = 1.05$ [41].

A43. $4.3\, e^{2x} = 10$ [.42]. A44. $\ln(5x - 4) = 2$ [2.3].

^ ^ ^ ^ Solve algebraically.

A45. $|x - 5|/4 = 10$. A46. $(|x| - 5)^2 = 16$.

A47. $144/(x - 2)^2 = 16$. A48. $24/|3 - x| = 8$.

A49. $3(|x/5| - 2) = 5$. A50. $[144/|x + 5|]^2 = 900$.

A51. $\ln(x^2 - 2) = -1.23$. A52. $e^{|x - 2|} = 20$.

^ ^ ^ ^ Find f^{-1} for the given f.

A53. $f(x) = x/2 + 7$. A54. $f(x) = (x - 5)/6$.

A55. $f(x) = 3(x + 4)$ A56. $f(x) = 5x - 2$.

^ ^ ^ ^ Here is f described as a set of ordered pairs. Does f^{-1} exist as a function? If so, give it. If not, why not?

A57. $f = \{(0, 1), (1, 2), (3, 10)\}$.

A58. $f = \{(2, 0), (3, 1), (4, 0)\}$

^ ^ ^ ^ Look at a graph of these functions and decide if f appears to be one-to-one.

A59. $f(x) = 1/x$.

A60. $f(x) = x^3 + x$.

A61. $f(x) = x^3 - x$.

A62. $f(x) = \ln x$.

^ ^ ^ ^ Give the range of f. (You may wish to look at a graph.)

A63. a) $f(x) = x^2$. b) $f(x) = x^3$.

A64. a) $f(x) = |x|$. b) $f(x) = 1/x$.

A65. a) $f(x) = x^2 + 1$. b) $f(x) = \sqrt{x} + 5$.

A66. a) $f(x) = |x - 3|$ b) $f(x) = x^2 - 5$.

^ ^ ^ ^ ^ ^ ^ ^ ^

B1.* a) Given $f(x)$, explain how to find f^{-1} (assuming it exists and is easy to find).
b) What is the essential difference between algebraically solving $f(x) = 5$ and finding f^{-1}?

^ ^ ^ ^ Find f^{-1} for the given f.

B2. $f(x) = 3 \log x$.

B3. $f(x) = e^{2x - 1}$.

B4. $f(x) = \log(10x) + 5$.

B5. $f(x) = 5/(x - \pi)$.

B6. $f(x) = 7/(9 - x)$.

B7. $f(x) = \ln(5/x)$.

B8. $f(x) = x/(x - 1)$.

B9. $f(x) = 2x/(x+3)$.

B10.* Distinguish between an inverse *function* and an inverse *relation*.

B11.* Explain the relationship between evaluating and solving. Be sure to mention the proper contexts.

B12.* Explain the inverse-reverse method of solving equations.

B13.* What is the relationship of the graphs of f and f^{-1}?

B14.* If a function is not one-to-one, how can an inverse function be defined for it?

B15. Compare and contrast the terms "function," "relation," and "graph" (the noun) in the context of ordered pairs of real numbers. a) Do all functions have graphs? b) Do all graphs correspond to functions? c) Do all relations have graphs? d) Do all graphs correspond to relations? e) Are all functions relations? f) Are all relations functions?

^ ^ ^ ^ From the given graph of f on a square scale, graph f^{-1}.

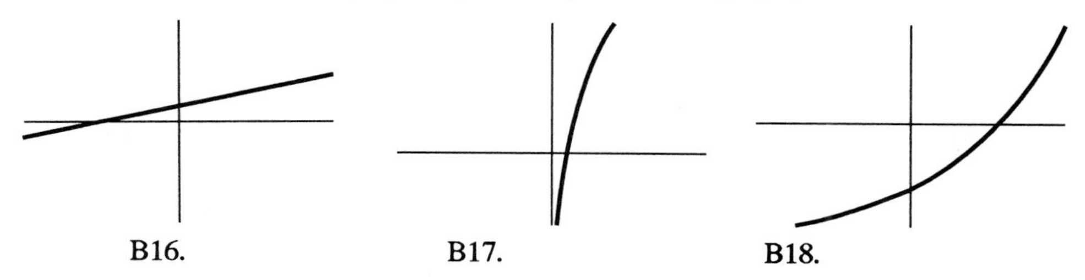

B16. B17. B18.

B19. a) Rewrite Theorem 2.4.14 with different uses of the dummy variables to emphasize solving logarithmic equations (such as "$\ln(x - 5) = 2$").
b) Rewrite Theorem 2.4.14 with dummy variables so that both exponential and logarithmic equations can be solved with it, but the notation does not emphasize either over the other.

^ ^ ^ ^ Read Theorem 2.4.5 to
B20. Find x such that $90° < x < 180°$ and $\sin x = .4$ [160°].
B21. Find x such that $90° < x < 180°$ and $\sin x = .6$ [140°].
B22. Find x such that $180° < x < 270°$ and $\sin x = -.4$ [200°].
B23. Find x such that $180° < x < 270°$ and $\sin x = -.6$ [220°].

^ ^ ^ ^ Find three angles x (in degrees) such that
B24. $\sin x = -.18$ [-10° is one].
B25. $\sin x = -.87$ [-60° is one].

^ ^ ^ ^ Solve $f(x) = 5$ when f^{-1} is as given.
B26. $f^{-1}(x) = x^3 + x$.
B27. $f^{-1}(x) = \log x + x$.

^ ^ ^ ^ How is it evident graphically that the given function is its own inverse?
B28. The reciprocal function.
B29. The "change sign" function.

B30. Consider $f(x) = \sqrt{x}$ and $g(x) = x^2$. Which of $f(g(x)) = x$ and $g(f(x)) = x$ hold? Are f and g inverse functions of one another?

B31. Give an example to show that this conjecture is false: $|x| = c$ iff $x = c$ or $x = -c$.

B32. "$\sqrt{(x^2)} = x$" is false as a generalization. Correct it.

B33. Discuss the truth of "$(\sqrt{x})^2 = x$."

B34. a) Express Theorem 2.4.7A in functional notation.
b) Express Theorem 2.4.7B in functional notation.

B35. a) Sketch the graph of $y = e^{\ln x}$. b) Sketch the graph of $y = \ln(e^x)$.
c) Why are they not the same if the functions are inverses of one another?

B36. a) Sketch the graph of $y = (\sqrt{x})^2$. b) Graph $\sqrt{x^2}$. c) Use the concepts of domain and range to explain why they are not the same.

B37.* Explain the idea of "inverse" functions using a "function loop."

B38.* If f has an inverse function, then the domain of its inverse is the ________ of f.

B39.* If f has an inverse function, then the range of its inverse is the ________ of f.

B40. a) Find three functions which are their own inverses.
b) Describe the features of the graph of a function which is its own inverse.

B41. Give a geometric proof that (a, b) and (b, a) in the first quadrant are mirror images of each other across the line $y = x$.

B42. For all x, $\cos x = \cos(-x)$. a) Interpret this graphically. For all x, $\cos x = \cos(x + 360°)$. b) Interpret this graphically. c) Use these facts to generate more solutions to $\cos x = c$ given the one solution, $x = \cos^{-1} c$.

B43. For all x, $\tan x = \tan(x + 180°)$. a) Interpret this graphically. b) Use this fact to generate more solutions to $\tan x = c$ given the one solution, $x = \tan^{-1} c$.

B44. a) Prove $|a| = |b|$ iff $a^2 = b^2$.
b) Use part a) to solve $|2x + 3| = |x - 1|$.

^ ^ ^ ^ Use B44 to solve:
B45. $|2x + 3| = |x|$. B46. $|2x| = |x + 1|$. B47. $|x - 2| = |2x|$.

B48. Use the fact that $\tan x$ repeats every 180 degrees to determine all solutions to $\tan x = c$ given the one solution $\tan^{-1} c$.

B49. Use the fact that $\cos x$ repeats every 360 degrees and the symmetry of its graph about the y-axis to determine all solutions to $\cos x = c$ given the one solution $\cos^{-1} c$.

B50. Distinguish between "inverses of functions" and "inverse functions."

CHAPTER 3

Fundamental Functions

Section 3.1. Lines, Distance, and Circles

Lines are everywhere in the real-world. This is a long section which connects lines to their algebraic, functional, expressions.

Points are located on rectangular-coordinate-system graphs by the intersection of vertical and horizontal lines.

In art, the term "lines" may include curves which are not straight, but, in Mathematics, lines are necessarily straight and geometric in nature.

Example 1: $x = 3$ denotes the vertical line through the point on the x-axis labeled 3 (Figure 1). In the plane we can interpret "$x = 3$" as "$x = 3$ and y = anything." Similarly, "$y = 2$" denotes the horizontal line through the point labeled "2" on the y-axis. The point (3, 2) is, by the definition of the axis system, at the intersection of these two perpendicular lines ($x = 3$ *and* $y = 2$) (Figure 1).

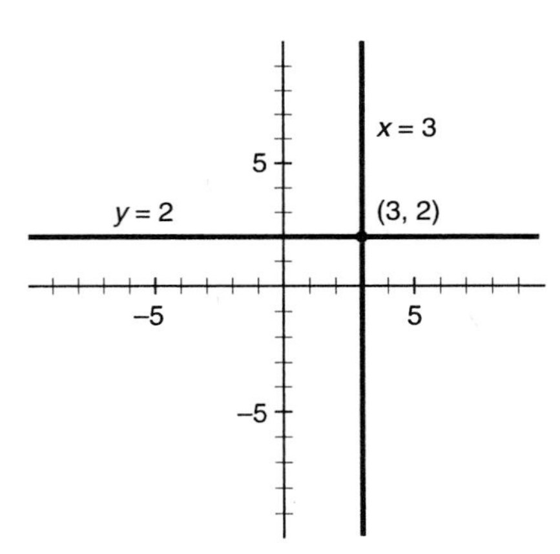

Figure 1: The lines $x = 3$ and $y = 2$ and the point (3, 2).

Example 2: Find equations of the vertical and horizontal lines through the point (π, 4.5).

By the definition of the rectangular coordinate system, that point is on the vertical line $x = \pi$ and the horizontal line $y = 4.5$.

The x-axis has equation $y = 0$, and the y-axis has equation $x = 0$. This switch of x's and y's is sometimes confusing.

horizontal line: $y = c$, for some c.
vertical line: $x = c$, for some c.
x-axis: $y = 0$.
y-axis: $x = 0$.

Slope. When a line is neither vertical nor horizontal, all the horizontal and vertical lines of the rectangular coordinate system intersect it and create similar right triangles (Figure 2). From geometry, we know that the ratios of corresponding sides of similar triangles are equal. Figure 3 illustrates two similar triangles (*ACB* and *ADP*). The ratio of the vertical side to the horizontal side is the same in both. It is characteristic of the line known as the slope.

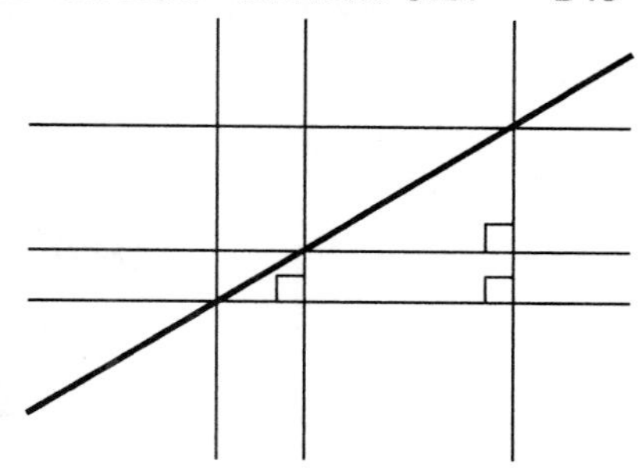

Figure 2: Three similar triangles formed by the intersection of a diagonal line with grid lines.

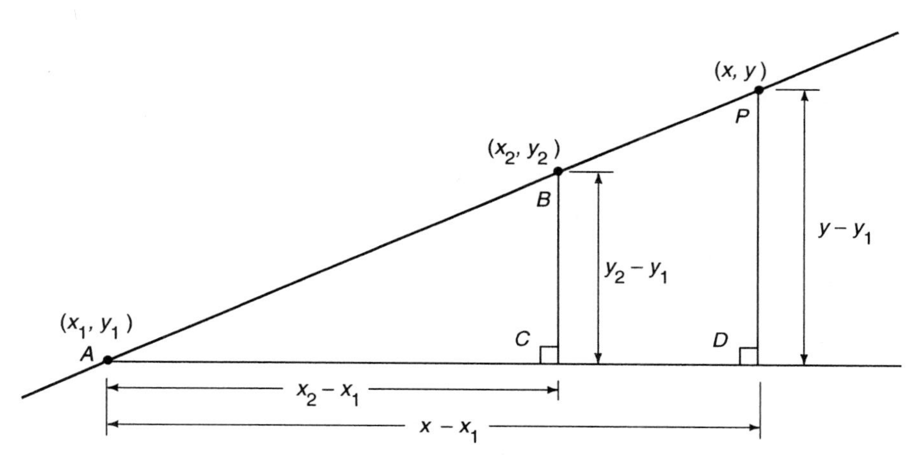

Figure 3: A line determined by two points, *A* and *B*, with another point, *P*, on it.

Learn the Two-Point Formula and Point-Slope Formula by understanding how they follow easily from this figure.

Suppose two given points on the line are A with coordinates (x_1, y_1) and B with coordinates (x_2, y_2) (Figure 3). The slope is given by

$$(3.1.1) \qquad m = \frac{y_2 - y_1}{x_2 - x_1} .$$

The pronunciation of "(x_1, y_1)" is "The point x sub one (pause) y sub one," or "x sub one (pause) y sub one," or, simply, "x one (pause) y one."

The slope is the difference in y's over the difference in x's.

Now, consider any other point on the line, P, with general coordinates given by (x, y). By similar triangles, PD is to DA as BC is to CA.

$$\frac{PD}{DA} = \frac{BC}{CA}.$$

Even as the location of P on the line changes, the ratio PD/DA remains constant. Algebraically,

$$\text{(3.1.2)} \qquad \frac{y - y_1}{x - x_1} = \frac{y_2 - y_1}{x_2 - x_1} = m \text{ , the slope.}$$

Multiplying through by "$x - x_1$" yields both the "two-point" and "point-slope" formulas.

<u>(3.1.3) Two-Point Formula</u>: $y - y_1 = \left(\frac{y_2 - y_1}{x_2 - x_1}\right)(x - x_1).$

or $y = \left(\frac{y_2 - y_1}{x_2 - x_1}\right)(x - x_1) + y_1.$

The second equation is a version of the first with "y_1" transferred to the right. The messy coefficient on x is just a fixed number, the slope. So this is essentially the "point-slope" formula for which the slope m and one point (x_1, y_1) are given (Figure 4).

<u>(3.1.4) Point-Slope Formula</u>:

$y - y_1 = m(x - x_1)$

or $y = m(x - x_1) + y_1$.

The point-slope form of a line is very important.

Remember both formulas by knowing how they follow from similar triangles.

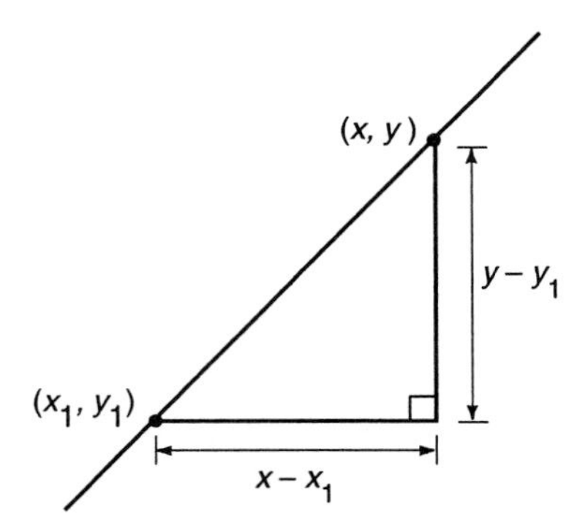

Figure 4: The point-slope formula: $y - y_1 = m(x - x_1)$. $(y - y_1)/(x - x_1) = m$.

Example 3: Find an equation of the line with slope 1/2 through the point (3, 1) (Figure 5).

Using the point-slope formula,

$y - 1 = (1/2)(x - 3)$, or $y = (1/2)(x - 3) + 1$.

The slope of a line, m, is often remembered as "the rise over the run" (Figure 6). Not all lines go upward to the right. The "rise" and slope can be negative (Figure 6).

(3.1.5)

$$\frac{y_2 - y_1}{x_2 - x_1} = \frac{rise}{run} = m = \frac{y_1 - y_2}{x_1 - x_2}.$$

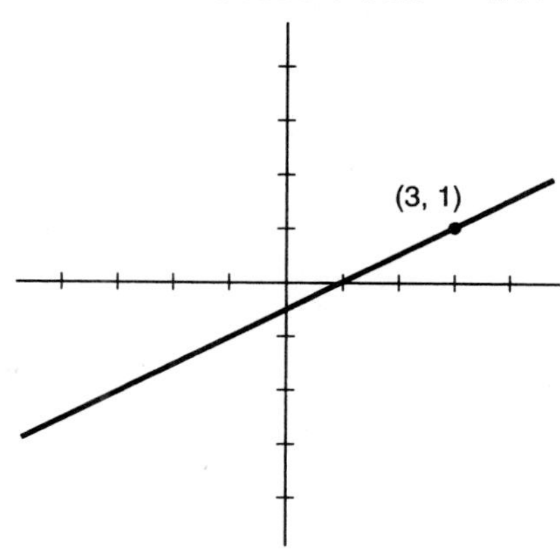

Figure 5: The line through (3, 1) with slope 1/2. $y - 1 = (1/2)(x - 3)$. [-5, 5] by [-5, 5].

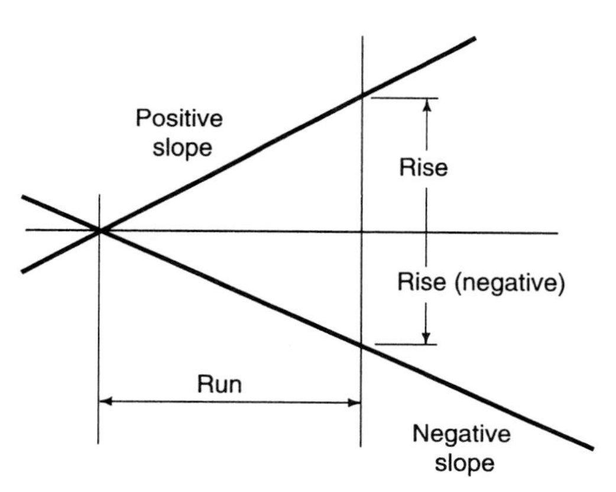

Figure 6: The slope is the rise over the run.

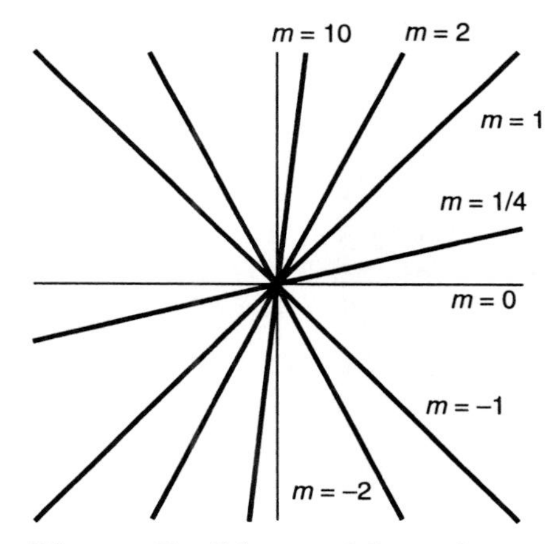

Figure 7: Lines with various slopes, graphed on a square scale.

Lines through the origin with various slopes are illustrated in Figure 7. A line goes up from left to right (is "increasing") if and only if its slope is positive. A line goes down from left to right (is "decreasing") if and only if its slope is negative.

Line	Slope
horizontal	zero
upward to the right	positive
downward to the right	negative
vertical	does not exist

The slope of any horizontal line is zero because its "rise" is zero. The point-slope formula yields "$y = 0(x - x_1) + y_1$. That is, $y = y_1$.

The slope of any vertical line does not exist, since the denominator (the "run") is zero. The point-slope formula does not work for vertical lines. The equation of a vertical line is "$x = c$" (and y = anything), for some c.

The slope is easy to interpret. When the change in x values is 1, the ratio is simply the change in y values. We can think of the slope as the amount the line goes up for each unit it goes to the right (Figure 8).

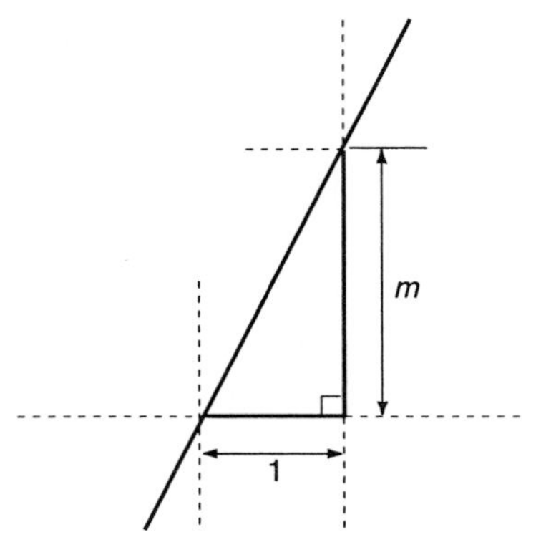

Figure 8: The slope of a line when $x - x_1 = 1$.

The slope is the change in *y*-value corresponding to a change of 1 in *x*-value.

Because "slope = rise over run," we also have

(3.1.4, again) rise = slope × run.

This is just the point-slope formula in words.
Inspect the symbolic version and Figure 4 to see that this is really the same.

Example 4: A line with slope 3 goes through (4, 2). What is the y-value when $x = 4.1$ (Figure 9)?

The x-value changes from 4 to 4.1, a change (run) of .1. The y-value changes 3 times as much (.3), from 2 to 2.3. $y = 2.3$.

What is the y-value when $x = 3.8$?

The x-value changes from 4 to 3.8, a change of -.2. The y-value changes 3 times as much (-.6), from 2 to 1.4. $y = 1.4$.

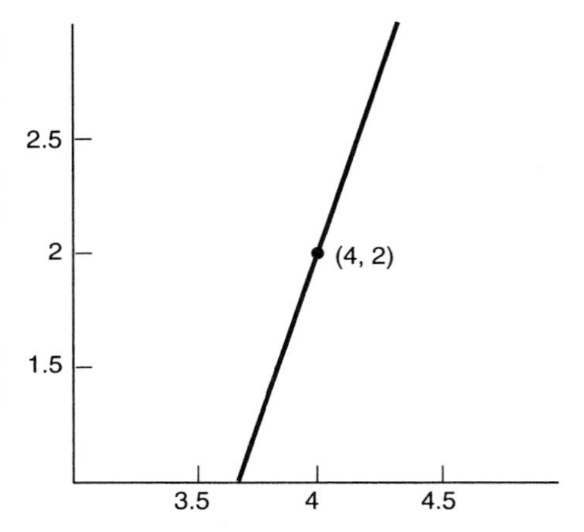

Figure 9: A line with slope 3 through the point (4, 2). [3,5] by [1, 3]

<u>Slope in the Real World</u>. The concept of slope as "rise over run" is abstract. In real life, the variable quantities represented by "x" and "y" would have units such as miles, hours, gallons, or dollars. Since slopes are quotients, particular examples of slopes would have the units of a quotient such as "miles per hour" or "dollars per gallon."

Example 5: Suppose drilling a well costs $500 to set up and $20 for each foot drilled. Express the cost of drilling a well x feet deep.

The total cost is $500 plus ($20 per foot) × (the number of feet). Abstractly, $y = 20x + 500$, where "y" has the units of dollars and "x" has the units of feet. And the slope, 20, is really "20 dollars per foot." It has the units of the quotient formula for slope (3.1.1).

Sometimes words can conceal the simple idea of putting a line through two points.

Example 6: A plane flies due east at a constant speed. At 1:30pm it is 100 miles west of Bozeman, and at 4:00pm it is 250 miles east of Bozeman. Express time in hours after noon and give the formula for the plane's location at any time (until the speed or direction changes).

Does this look like a line problem? It is. The "constant speed" in the problem corresponds to the constant slope of a line.

Let Bozeman be at location zero and east be the positive direction. "250 miles east" corresponds to +250, and "100 miles west" corresponds to -100. 1:30pm corresponds to $t = 1.5$ (hours after noon) and 4:00pm corresponds to $t = 4$. The formula for the plane's location is a line through two points, (1.5, -100) and (4, 250). Abstractly, according to Formula 3.1.3 (with "t" for "x"),

$$y = \frac{250 - (-100)}{4 - 1.5}(t - 1.5) - 100 .$$

This simplifies to $y = 140\,t - 310$. The slope, 140, is really the speed, "140 miles per hour," which has the units of a quotient.

The point is, we can study lines abstractly using "x" and "y", but they represent real formulas that have units that we omit in the abstract equations.

Example 7: There are 2.54 centimeters in an inch. To convert from inches to centimeters, multiply by 2.54 (centimeters per inch). The relationship is, abstractly, $y = 2.54x$. In reality, all of the terms in this equation have units. "x" has the units of inches. "y" has the units of centimeters. And "2.54," the slope, has the units of the quotient, "centimeters per inch." The slope gives the change in centimeters (change in y) corresponding to a change in one inch (change in x) (Figure 4). That is a real instance of "rise over run."

Proportional: Two variable quantities are said to be proportional (or "to vary directly") if one is a constant multiple of the other. If one is denoted by x and the other by y, y is proportional to x if and only if there exists a number, k ($\neq 0$), such that

(3.1.6) $$y = kx.$$

The constant k is called the constant of proportionality. The two quantities are functionally related by the function, "Multiply by k."

Example 7, revisited: Measurements in inches and centimeters are proportional. Measurements in centimeters are always 2.54 times measurements in inches. The constant of proportionality is $k = 2.54$ (centimeters per inch).

Example 5, revisited: In Example 5, the cost of a well is not proportional to its depth. With the formula "$y = 500 + 20x$," a one-foot well costs $520 and a two-foot well costs $540, not twice as much. The fixed set-up cost makes the total cost not proportional to the depth.

The graph of a proportional relationship will be a line *through the origin* with slope k; "k" plays the role of "m" in the usual notation. When y is proportional to x, x is also proportional to y, so they are proportional to each other (problem B70).

Example 8: Physics (Hooke's Law) tells us that the amount a spring stretches is proportional to the force applied to it. If a force of 5 pounds stretches a spring 3.2 inches, how far will a force of 7 pounds stretch it?

In school this is often done without bothering to obtain the whole function, since only one number is unknown. Call the stretch s.

s is to 7 as 3.2 is to 5. $s/7 = 3.2/5$ $s = 7(3.2)/5 = 4.48$ (inches).

Example 8, continued: Find the relationship between force and stretch for the spring.

In calculus we often want the entire relationship, rather than just the particular stretch for a particular force. We can obtain a general formula for the stretch, s, in terms of the applied force, x.

s is to x as 3.2 is to 5. $s/x = 3.2/5$. $s = (3.2/5)x$. $s = .64x$.

So the constant of proportionality, k, is .64. With this formula, we can evaluate s given x or solve for x given s, either way.

Here is another way to find k. This way uses the definition. Because they are "proportional," by definition, $s = kx$, where, at first, k remains to be determined. Using the given (x, s) pair,

$3.2 = k(5)$, so $k = 3.2/5 = .64$, as before.

Technically, the units of "s" are inches and the units of "x" are pounds. And, the units of ".64" the slope, are the units of the quotient 3.3.1, "inches per pound."

Example 9: Suppose photocopies cost 6 cents each. Is the total cost of n copies proportional to n?

Yes, the formula is "$C = 6n$." "C" has the units of cents. "n" has the units of photocopies. The slope (constant of proportionality) is a quotient, 6 (cents per photocopy).

Example 10: Is the area of a square proportional to its side?

This one fools some students. The term "proportional" does not mean simply "increasing together." It refers to a very particular type of relationship: Multiply by a constant. The area formula is $A = x^2$ in which x is not multiplied by a fixed constant. They are not proportional.

Parameters. The equation "$y = 2x$" is the equation of a line through the origin because (0, 0) satisfies the equation. Also, "$y = 3x$" is a line through the origin. We can consider all non-vertical lines through the origin at once by considering "$y = mx$." Different values of m yield different lines (Figure 8). In the equation "$y = mx$" there are three different letters, and the "m" plays a different kind of role than the "x" and the "y". For a given equation we imagine m to be fixed (say, 2.54) and the x and y to vary. However, for different lines there are different values of m, so m may vary too, but in a different context (there are different values of m in Examples 7, 8, and 9). Letters which are used to represent numbers which are constant in one context and variable in another, such as "m" here, are said to be parameters. In the context of a given line, m is fixed. In the context of a *family* of lines, m can vary to distinguish one line from another.

A Line Through Two Points. Together, the formula for the slope and the point-slope formula are sufficient to fit a line through two given points.

Example 11: Find an equation of the line through (1, 2) and (3, 8) (Figure 10).

The slope is $m = \frac{8 - 2}{3 - 1} = 3$. Now use either point in the point-slope formula:

$$y = 3(x - 1) + 2.$$

The slope, 3, and the point used, (1, 2), are clearly evident in this equation. Notice the problem requested "an" equation of "the" line. This is because there are many, equivalent, equations. Using the other point in the point-slope formula,

$$y = 3(x - 3) + 8.$$

This version clearly exhibits the slope and the second point, (3, 8). These are two equivalent equations of the same line. If you simplified, another equivalent equation in "slope-intercept" form would result. We discuss this form next.

It is common, although not necessary, to emphasize the point where the line crosses the y-axis. Its y-value is called the y-intercept. The term is "intercept" with "cept," not

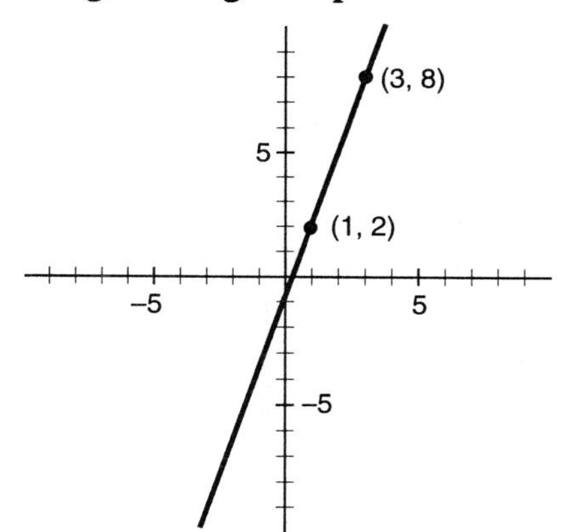

Figure 10: The line through (1, 2) and (3, 8). $y - 2 = 3(x - 1)$.

Equations in two variables may have *graphs* which are lines. The equation itself is not a line. Nevertheless, it is common, convenient, and not very misleading to speak as if equations could be lines. For example, the sentence, "Consider the line $y = mx + b$," is technically incorrect, but unlikely to be misunderstood.

"intersect," with "sect," as you might expect. The equation of the y-axis is $x = 0$, so that point is easily discovered by setting $x = 0$ (yes, $x = 0$, not $y = 0$) in any equation of the line.

Example 11, continued: Find the y-intercept of the line in Example 11: "$y = 3(x - 1) + 2$."

Set $x = 0$ in "$y = 3(x - 1) + 2$." Then $y = 3(-1) + 2 = -1$. The y-intercept is -1.

The slope-intercept form of a line emphasizes the slope and the y-intercept (Figure 11).

Slope-Intercept Form (3.1.7): $y = mx + b$.

This equation has two parameters: "m" for the slope and "b" for the y-intercept.

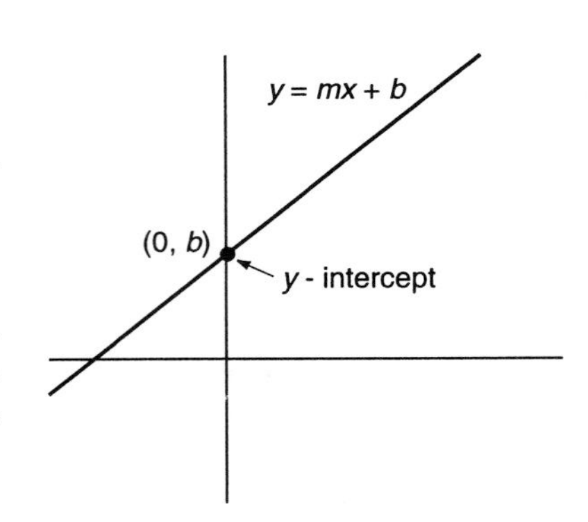

Figure 11: $y = mx + b$. b is the "y-intercept."

Example 11, revisited: The equation in Example 11, "$y = 3(x - 1) + 2$," can be rewritten in "slope-intercept form" by simplifying:

$$y = 3x - 1.$$

The slope, 3, and the y-intercept, -1, are apparent It crosses the y-axis at (0, -1).

Example 12: Find the slope and y-intercept of the line with equation "$y = -11x + 7$."

The slope is -11 and the y-intercept is 7.

The slope is readily visible on a graph when the scale is square (one unit is the same distance on both axes) as in Figure 10. But the appearance of the slope, as reflected in the steepness of the line, is dependent upon the scale. By changing the window we can make the same line look flatter or steeper.

More on Example 11: Figure 10 graphs the line "$y = 3x - 1$" in the standard window. The line is steep, corresponding to the slope, $m = 3$. Change the window so that the line looks much less steep.

The line will appear less steep if we make the change in y less impressive. We can do this by changing the window so that 2 or 5 or 10 times as many y-values are displayed in the same space. For example, change the window so that $-50 \le y \le 50$ (and still use $-10 \le x \le 10$). Then the slope of the line is unchanged, but its appearance is greatly changed (Figure 12).

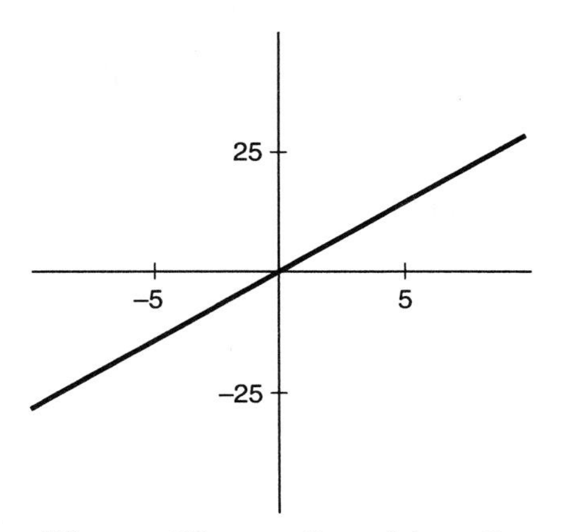

Figure 12: $y - 2 = 3(x - 1)$
[-10, 10] by [-50, 50].

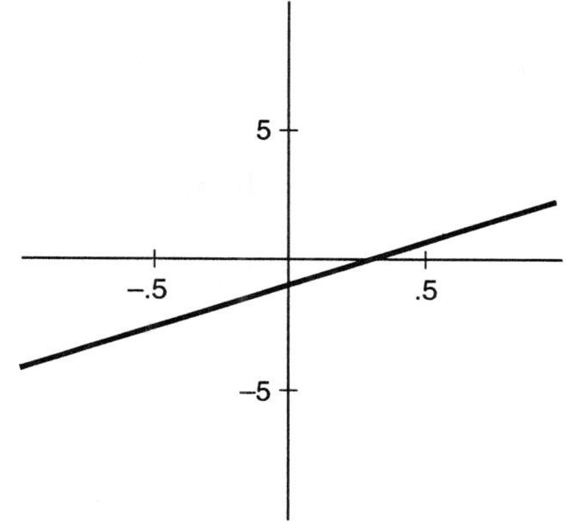

Figure 13: $y - 2 = 3(x - 1)$
[-1, 1] by [-10, 10].

Another way to make a given line appear less steep is to change the x-scale, leaving the y-scale as it was. Small changes in x make for small changes in y, so we can create the appearance of a small change in y by making the window express only a small change in x. For example, let $-1 \leq x \leq 1$, as in Figure 13. Scale is important.

Example 13: Find the line through the points on the graph of $f(x) = 1/x$ where $x = 2$ and $x = 3$ (Figure 14).

Points on the graph are determined by their x values (2 and 3). The two points are (2, 1/2) and (3, 1/3). The slope is $\dfrac{1/3 - 1/2}{3 - 2} = -\dfrac{1}{6}$. An equation of the line in point-slope form is

$$y = -(1/6)(x - 2) + 1/2.$$

There is no need to "simplify" to slope-intercept form.

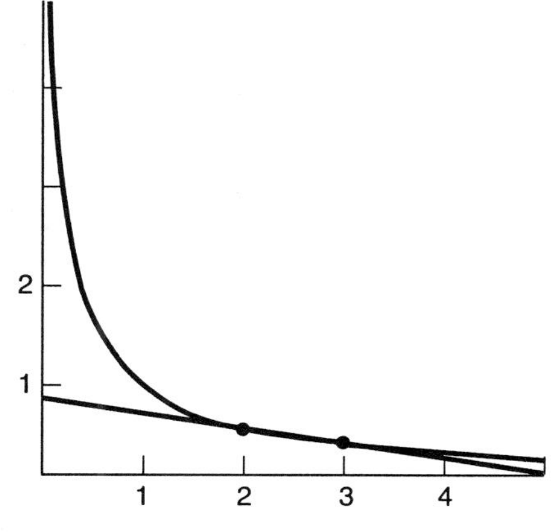

Figure 14: $f(x) = 1/x$ and a line through the points where $x = 2$ and $x = 3$.
[0, 5] by [0, 5].

<u>Solving Linear Equations</u>. An expression in "x" is said to be <u>linear</u> whenever it is equivalent to "$mx + b$" for some $m \neq 0$ and some b. (If m were zero, that expression would be a constant and would not have an "x" in it.) Linear equations can be easily solved using the inverse-reverse method.

Example 14: Solve $mx + b = c$ for the unknown, "x".

By the inverse-reverse method, simply subtract b from both sides and then divide by m.

If both sides of the equation have x's, you will want to "consolidate like terms" using the Distributive Law.

Example 15: Solve $x \tan 32° + 2.45 = \sin 32° - x \tan 55°$.

The appearance of trigonometric functions may, at first, be confusing. But "tan 32°" is just a number, not a variable. Put the terms with "x" on one side.

$$x \tan 32° + x \tan 55° = \sin 32° - 2.45.$$

Now factor, using the Distributive Law:

$$x(\tan 32° + \tan 55°) = \sin 32° - 2.45.$$

Now divide:

$$x = \frac{\sin 32° - 2.45}{\tan 32° + \tan 55°}.$$

Example 16: Solve for x: $\dfrac{x}{5 + x} = y$.

When solving for "x", "y" is treated just like a constant number. Again, the idea is to group like terms.

$$\frac{x}{5 + x} = y \quad \text{iff } x = (5 + x)y \text{ and } x \neq -5 \quad \text{[Theorem 1.4.3].}$$

As usual, we put all the terms with "x" on one side and all the remaining terms on the other. First multiply out the right side.

$$x = 5y + xy, \quad x - xy = 5y, \quad x(1 - y) = 5y,$$

$$x = \frac{5y}{1 - y}.$$

Example 16 solves the problem: "Find the inverse of f if $f(x) = x/(5 + x)$." The answer is: $f^{-1}(y) = 5y/(1 - y)$.

From the original equation we see that $x = -5$ is not in the domain of f, and from the last equation we see that $y = 1$ is not in the range of f or the domain of f^{-1}.

Part of the point of Examples 14 through 16 is that the same processes that apply to problems stated with simple numbers such as "3" and "5" also apply to problems with complicated expressions or letters. The expressions "sin 32°" and "y" can be treated just like any other number. To advance to calculus you will want to be just as capable of dealing with expressions and letters as with numbers.

Applications. There are many cases where it is natural to put a line through two points, even if we know that a line is not quite right. In the absence of a better model for the behavior of a function, the straight line fit often wins by default.

Example 17: Suppose market research in two sample markets shows The Bozeman Widget Company that it will probably sell about 2000 widgets per week if they are priced at \$50, but only 1500 widgets per week if they are priced at \$60. How many will they sell per week if widgets are priced at x dollars?

> In business examples, a <u>widget</u> is a mythical product manufactured by a mythical company. They can serve to illustrate the relationship between price and sales.

Who knows? Can we really expect a straight-line fit to be right? It would probably not be exactly right, but, in the absence of a better idea, let's try to answer the question by fitting a straight line.

We are given two points on the line, (50, 2000) and (60, 1500) (Figure 15). The slope is

$$\frac{1500 - 2000}{60 - 50} = -50 .$$

This may be interpreted as saying that for every dollar increase in price, 50 fewer widgets will be sold per week. Now use the point-slope formula with either point. With the point (50, 2000),

$$y = -50(x - 50) + 2000.$$

Now, for any price, x, you could estimate the sales, y.

It would not be appropriate to "simplify" this to Slope-Intercept form "$y = mx + b$." There is no point in emphasizing "b", which would be the number "sold" when the price is zero!

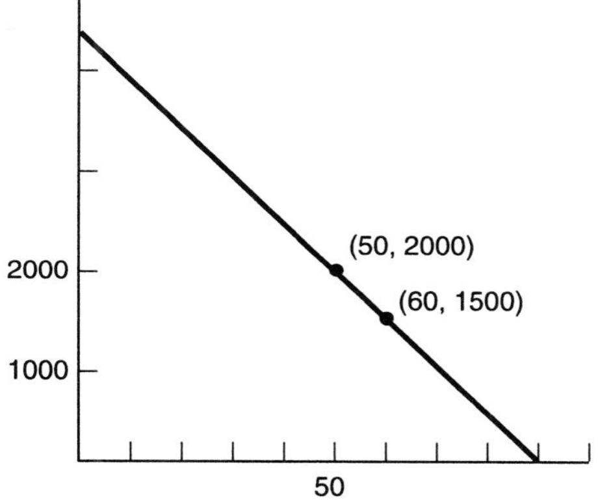

Figure 15: The line through (50, 2000) and (60, 1500).
$y - 2000 = -50(x - 50)$.
[0, 100] by [0, 5000].

Also, if we foolishly extend our line far enough to the right, we will eventually obtain negative numbers for the number sold at high prices. Also nonsense. It is common and reasonable to fit a line over a region where you have good data, but it is not reasonable to extrapolate that line far beyond the data.

If we use a line to estimate a functional value *between* x-values for which the y-values are known, the process is called <u>interpolation</u>. If we use a line to estimate functional values for x's *outside* that interval, the process is called <u>extrapolation</u>. There are so many important real-life problems where extrapolation has been grossly misused when the x-value was far from the known x-values that extrapolation has developed a bad reputation. Be very careful with extrapolation.

Basically, linear interpolation consists of putting a line through two points on a curve, and then using the line to approximate the curve.

Example 18: The standard-normal cumulative-distribution function, Φ, is very important in statistics, but difficult to compute. Use Table 2.2.2 to estimate $\Phi(1.65)$.

The table gives $\Phi(1.6) = .9452$ and $\Phi(1.8) = .9641$, but it does not give $\Phi(1.65)$. We need a better table. Or, we could try approximating the value on the curve by using the value on a line instead. According to the two-point formula (3.1.3), $\Phi(1.65)$ would be approximately

$$\frac{.9641-.9452}{1.8-1.6}(1.65 - 1.6) + .9452 = .9499.$$

Figure 16 displays the relevant part of the graph of Φ.

This approach is equivalent to recognizing that, on the line, the change in y is proportional to the change in x. The change in y between the two points we know is .9641 - .9452. The change in x for those two points is 1.8 - 1.6. The change in x we want to use is 1.65 - 1.6. The corresponding change in y will be proportional.

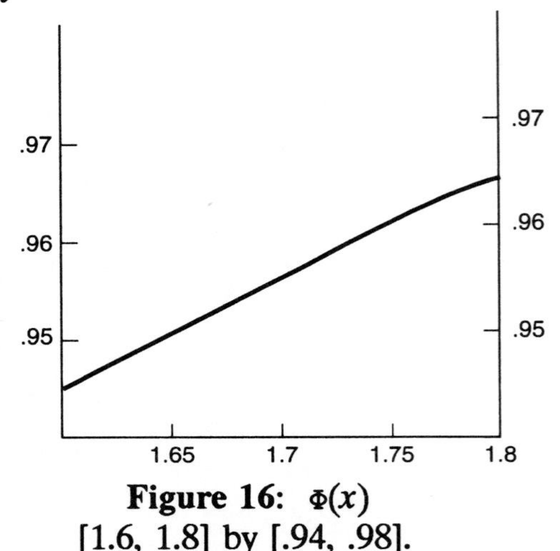

Figure 16: $\Phi(x)$
[1.6, 1.8] by [.94, .98].

$$\frac{y - .9452}{1.65 - 1.6} = \frac{.9641 - .9452}{1.8 - 1.6}.$$

The slope of the "linear interpolation" line appears on the right. After all, this approach almost derives the two-point formula (3.1.3) for a line again. The only difference is that "y" is computed for only the particular x-value 1.65, rather than for a general "x" value.

Solving Equations. Linear interpolation may also be used in reverse to solve equations.

Example 19: Suppose we want to solve "$f(x) = 0$," but $f(x)$ is hard to evaluate and we only know $f(.785) = -.00015$ (slightly too low) and $f(.786) = +.00071$ (slightly too high). Use linear interpolation to approximate x such that $f(x) = 0$ to four or more decimal places (Figure 17).

It is reasonable to assume that there is a solution between the two x-values and that the true curve is close to a straight line between the known points. If so, the point on the *line* where $y = 0$ will serve as a good approximation to the point on the *curve* where $y = 0$.

First fit a line through the two given points.

$$y - (-.00015) = \frac{.00071-(-.00015)}{.786 - .785}(x - .785)$$
$$= .86(x - .785)\ .$$

Then set $y = 0$ and solve for x.

$$.00015 = .86(x - .785)$$
$$\frac{.00015}{.86} + .785 = x$$
$$.78517 = x\ .$$

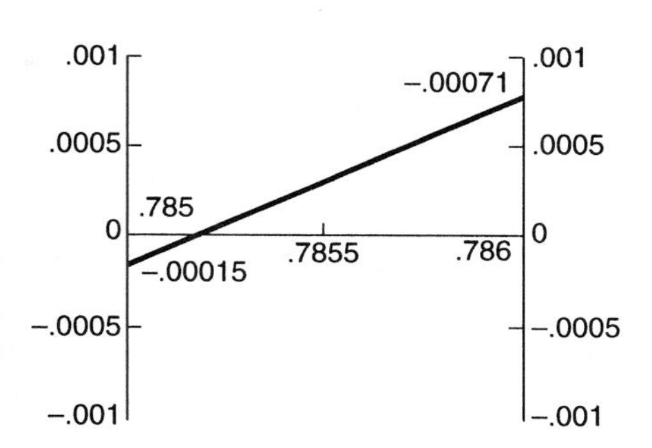

Figure 17:
$f(.785) = -.00015$
$f(.786) = .00071.$
[.785, .786] by [-.001, .001].

There is no guarantee that the extra two decimal places are both accurate, but this number is very likely to be substantially closer than either of the original two x-values which bracketed the solution.

The x-value at which the graph of $f(x)$ intersects the x-axis is called the x-intercept. To solve the equation "$f(x) = 0$" is equivalent to finding the x-intercepts of its graph.

A fundamental idea of calculus is that most curves that are not straight lines can, nevertheless, be approximated by straight lines over small regions.

Calculator Exercise 1: Use your calculator to graph any curved graph of your choice. Zoom in several times about any point on the curve to obtain the picture in a very small window. Does the graph look like a straight line? Usually, but not always, it will (Problems A5 and B68, where x^2 is used).

Example 20: Suppose that $f(x)$ is not a straight line, but can be approximated by a straight line. Suppose further that $f(35.4) = 0.0518$ and the slope of the tangent line to the graph at that point is -0.16. Approximate the solution to $f(x) = 0$ (Figure 18).

$f(35.4) = 0.0518$, which seems pretty close to 0, so maybe 35.4 is close to the solution we seek. However, we may be able to pick up another decimal place of accuracy, or even more, if the curve is closely approximated by its tangent line. The tangent line is

$$y - 0.0518 = -0.16(x - 35.4).$$

Set $y = 0$ on the line and find the corresponding x.

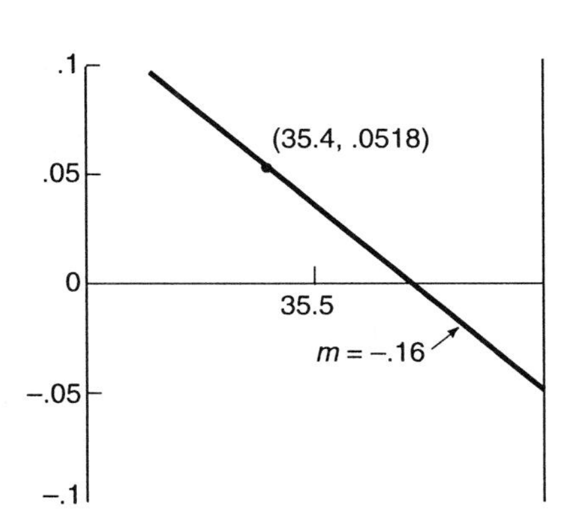

Figure 18: $f(35.4) = .0518$
$m = -0.16$ there.
[35, 36] by [-.1, .1].

$$0 - 0.0518 = -0.16(x - 35.4).$$

Dividing by -0.16 and then adding 35.4,

$$x = 0.32 + 35.4 = 35.72.$$

The approach of Example 20 can be generalized and expressed with dummy variables (Problem B87). Called "Newton's Method" after Sir Isaac Newton (the co-inventor of calculus), it is an important method in the subject of numerical analysis.

Parallel and Perpendicular Lines. Geometry can be used to determine the algebraic characteristics of parallel and perpendicular lines. Figure 21 illustrates why two lines are parallel if and only if they have equal slopes. Two lines crossing a single horizontal line will be parallel if and only if they form congruent angles with the horizontal, which will happen if and only if they have the same slopes.

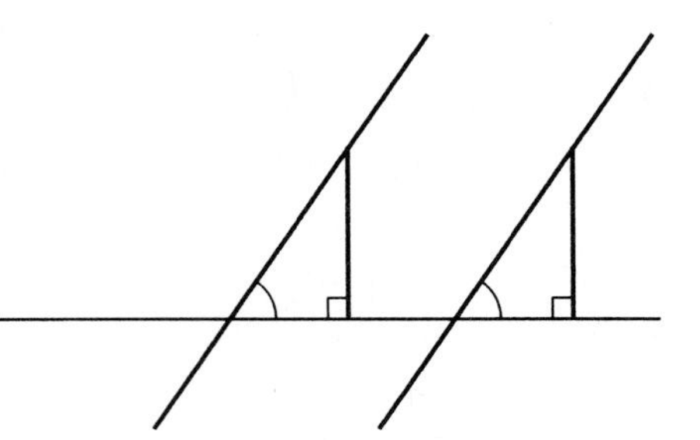

Figure 21: Parallel lines have the same slope.

Parallel Lines (3.1.8) A line has slope m if and only if any line parallel to it has slope m.

In their initial geometric definition, lines are determined by any *two* points they go through. Given two points, the algebraic "two-point form" is appropriate.

Then, Euclid's "parallel postulate" states, "Given a line and one point not on it, there exists exactly one line through the point parallel to the line." So a line is also determined by *one* point in conjunction with some way of describing parallel lines. Slopes provide that way. Given the slope and *one* point, the algebraic "point-slope form" is appropriate.

Example 21: Find an equation of the line through (6, -5) parallel to the line $y = -3x + 1$.

From the given equation we see $m = -3$, which is the key characteristic of all lines parallel to the given line. Therefore, using point-slope form,

$$y = -3(x - 6) - 5.$$

You may wish to "simplify" this, but it is fine as it is. It exhibits the slope and the point, which is all you need to know to graph the line.

Of course, vertical and horizontal lines are perpendicular. If a line is neither vertical nor horizontal, then the next theorem applies (Figure 22, Problem B90).

<u>Perpendicular Lines (3.1.9)</u>: A line has slope $m \neq 0$ if and only if any line perpendicular to it has slope $-1/m$.

Example 22: Find the line through (7, 3) perpendicular to the line with equation $y = 2x - 9$.

Because the slope of the original line is 2, the slope of the perpendicular line is -1/2. Now use the point-slope formula.

$$y = (-1/2)(x - 7) + 3.$$

This need not be "simplified." It emphasizes the key point as it is.

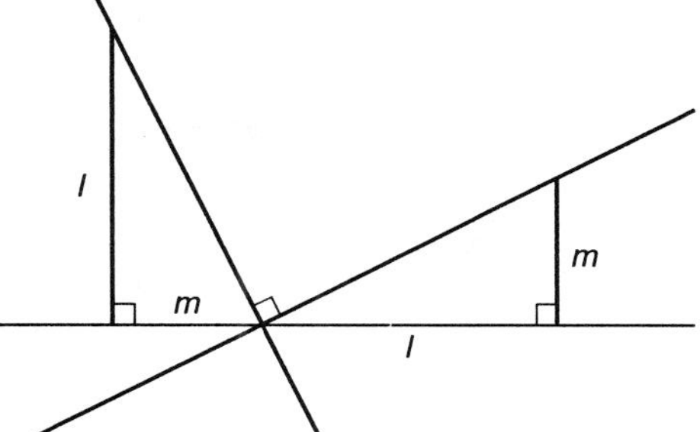

Figure 22: Slopes of perpendicular lines.

Example 23: Find the line through (3, 5) perpendicular to the line with equation $y = 4.3$.

The line $y = 4.3$ is horizontal and any line perpendicular to it is vertical. By the definition of the rectangular coordinate system, vertical lines all have the equation "$x = c$," for some constant c. Since the line we seek goes through (3, 5), c must be 3. The equation is simply "$x = 3$."

It is fair to say that two vertical lines have the same slope, in the sense that both slopes do not exist. There is a form for an equation of a line which handles all lines including vertical lines, which the formulas with slopes do not handle. The <u>general form</u> of a line is

(3.1.10) $ax + by + c = 0$, where a and b are not both zero.

The case $b = 0$ handles vertical lines. The "general" form is not particularly important -- it is just general. The point-slope and slope-intercept forms are far more important.

<u>Distance in the Plane</u>. The Pythagorean Theorem can be used to find the distance between any two points in the plane.

<u>Theorem 3.1.11 (The Pythagorean Theorem)</u>. Let triangle ABC have sides a, b, and c opposite vertices A, B, and C, as in Figure 23.

$c^2 = a^2 + b^2$ if and only if the angle at vertex C is a right angle.

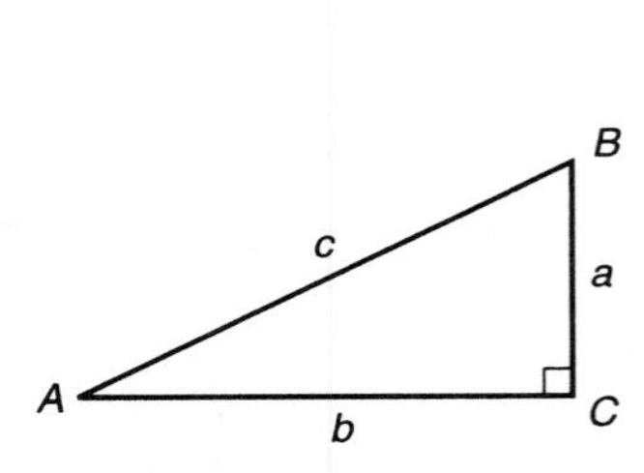

Figure 23: $a^2 + b^2 = c^2$ iff angle ACB is a right angle.

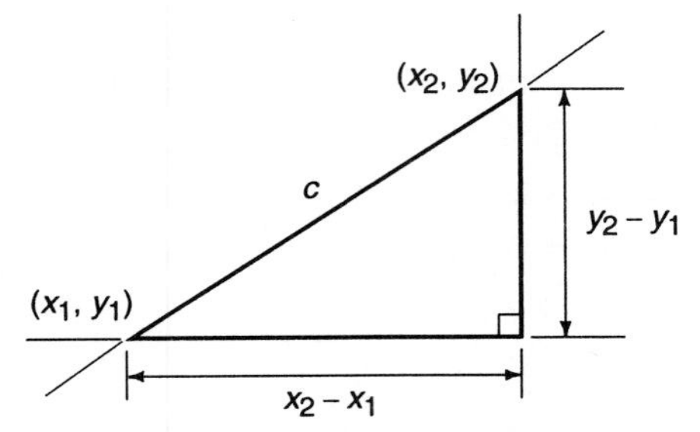

Figure 24: Distance in the plane. $c^2 = (x_2 - x_1)^2 + (y_2 - y_1)^2$.

To determine the formula for the distance between any two points in the coordinate plane, use the same picture we used to find the equation of a line through two points (Figure 24). The distance between the two points is the length of the hypotenuse.

Theorem 3.1.12 (The Distance Formula): The distance between points (x_1, y_1) and (x_2, y_2) is

$$\sqrt{(x_2 - x_1)^2 + (y_2 - y_1)^2} \; .$$

Use Figure 24 to help you remember this important formula.
It is the Pythagorean Theorem.

Example 24: Find the distance between the points (5, -2) and (-7, 3).
The distance is

$$\sqrt{(-7 - 5)^2 + (3 - -2)^2} = \sqrt{(-12)^2 + 5^2} = \sqrt{144 + 25} = \sqrt{169} = 13.$$

Example 25: Find the distance from (x, y) in the first quadrant to the y-axis.
The distance is simply x. The x-coordinate of any point is its directed distance from the y-axis. Similarly, the distance from (x, y) in the first quadrant to the x-axis is y. The distance formula 3.1.12 is not appropriate because it gives the distance between two points, not the distance from a point to a line.

Example 26: Find the expression for the distances from points on the line $y = 2x + 5$ to the origin (Figure 25).
In this problem the answer is not a number -- it is a formula. We have to create a new distance formula to fit the special conditions of the problem.

According to the distance formula 3.1.12, the distance from any point (x, y) (not necessarily on the line) to the origin is

$$\sqrt{(x - 0)^2 + (y - 0)^2} = \sqrt{x^2 + y^2} .$$

Points on the line satisfy $y = 2x + 5$. In the context of this problem this equation is called a "constraint" because it constrains the relationship between y and x. Replacing "y" in the distance formula with its equivalent in terms of x, the distance $d(x)$ satisfies

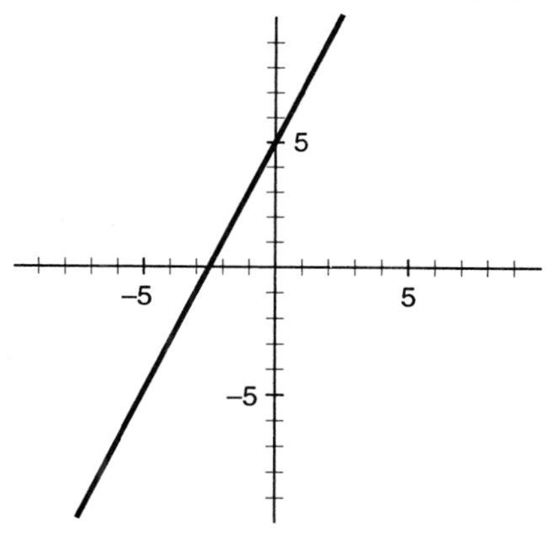

Figure 25: $y = 2x + 5$. [-10, 10] by [-10, 10].

$$d(x) = \sqrt{x^2 + (2x + 5)^2} .$$

This formula gives the distance from the point $(x, y) = (x, 2x + 5)$ on the line to the origin in terms of only the x-coordinate.

Example 26, continued: Use your graphics calculator to find the point on the line $y = 2x + 5$ closest to the origin.

What are we supposed to do? The word "closest" tells us we are to minimize the distance. For that we need the distance formula which we just obtained. Graph the function (x horizontal and distance $d(x)$ vertical) and find the x-value of the lowest (smallest distance) point on the graph (Figure 26). It appears to be about $x = -2$. From the original equation (not the distance formula), at the closest point, $y = 2x + 5 = 2(-2) + 5 = 1$. The closest point is $(-2, 1)$.

The vertical coordinate above $x = -2$ in Figure 26 is distance, *not* the y-value of the point we want. Be careful. There are *three* related numbers in this problem and graphs handle only two at a time. Figure 26 relates x to *distance* (from the origin to the point (x, y) on the line in terms of x). Figure 25 and the original equation relate x to y on the line.

Geometry and calculus both provide alternative ways to solve this problem (Problems B78-80).

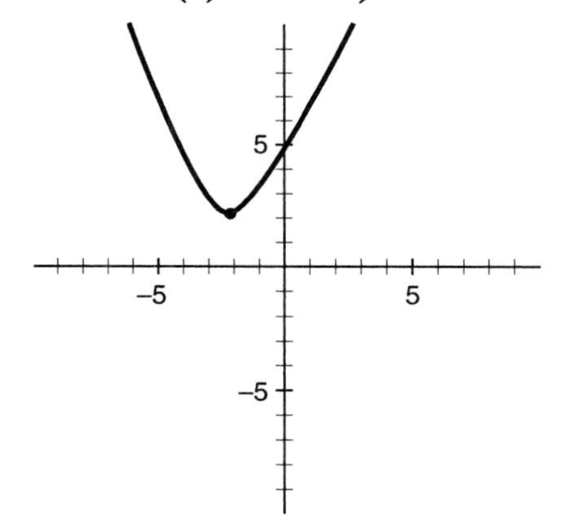

Figure 26: $d(x)$ from Example 26. The distance from the origin to $(x, 2x + 5)$ in terms of x. [-10, 10] by [-10, 10].

Circles. A circle of radius r is, by definition, the set of all points at distance r from a particular point called the center of the circle. Formula 3.1.12 for the distance between two points yields the equations of circles.

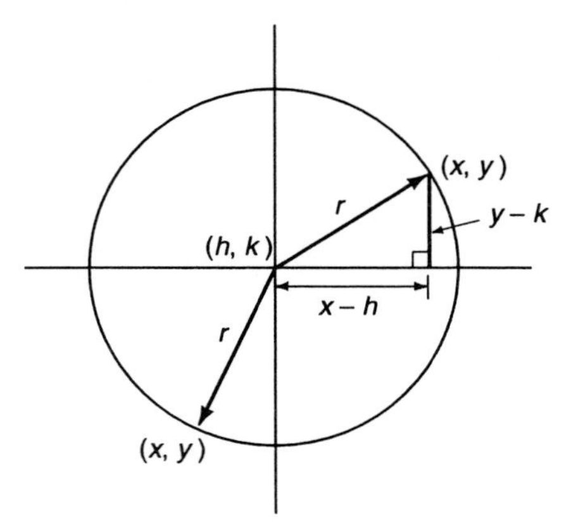

Figure 27: A circle. Center (h, k), radius r. $(x - h)^2 + (y - k)^2 = r^2$.

Circle (3.1.13): The standard form of the equation of the circle of radius r and center (h, k) is

$$(x - h)^2 + (y - k)^2 = r^2.$$

"x minus h, squared, plus y minus k, squared, equals r squared."

To describe this family of curves the standard form uses three parameters -- two for the location of the center and a third for the radius.

The standard form is just the square of the distance formula which gives r as the distance between (h, k) and the general point (x, y) on the circle (Figure 27).

Example 27: Find the equation of the unit circle, that is, the circle with center at the origin and radius 1.

$(x - 0)^2 + (y - 0)^2 = 1^2$, that is, $x^2 + y^2 = 1$.

This is the most famous and important circle.

Example 28: Find the equation of the circle with radius 5 and center (-4, 2).

Plugging in to 3.1.13, the equation is

$$(x + 4)^2 + (y - 2)^2 = 25.$$

Calculator Exercise 2: Graph the above equation on your calculator.

This equation describes a relation, not a function. If you want your graphing calculator to graph it using functions, you will need to solve for y and plot both of the two solutions on the same axis system.

Solving for y using the inverse-reverse method,

$$(y - 2)^2 = 25 - (x + 4)^2.$$

$$y - 2 = \pm\sqrt{25 - (x + 4)^2}. \qquad y = 2 \pm \sqrt{25 - (x + 4)^2}.$$

Graph these two equations to see the circle.

Conclusion. Lines are extremely important. Learn all about them.

Most line problems use point-slope form, which is simply the symbolic version of "Slope is rise over run." Certainly in calculus, the "point-slope" form is more important than the famous "$y = mx + b$" form. Linear interpolation uses the idea of approximating a curve by a line through two points on the curve.

Distance in the plane is given by the Pythagorean Theorem. Circles are determined as all points at a fixed distance (the radius) from the center.

Six Forms for Lines

Vertical: $x = c$, for some c.
Horizontal: $y = c$, for some c.

Two-Point Formula: $y - y_1 = \left(\frac{y_2 - y_1}{x_2 - x_1}\right)(x - x_1)$.

** **Point-Slope Formula:** $y - y_1 = m(x - x_1)$ or $y = m(x - x_1) + y_1$.
Slope-Intercept Form: $y = mx + b$.
General form: $ax + by + c = 0$.

Terms: slope, Two-point formula, point-slope formula, rise, run, proportional, parameter, y-intercept, Slope-Intercept form, widget, linear expression, parallel, perpendicular, general form (of a line), Pythagorean Theorem, distance, circle, standard form (of a circle), linear interpolation, x-intercept.

****

Exercises for Section 3.1, "Lines, Distance, and Circles":

A1.* a) Vertical lines have equation ____________________, for some c.
b) Horizontal lines have equation ____________________, for some c.
c) The equation of the x-axis is ____________________.
d) The equation of the y-axis is ____________________.

A2.* a) The slope of a horizontal line is __________.
b) The slope of a line which goes upward to the right is ______________.
c) The slope of a line which goes downward to the right is ______________.
d) The slope of a vertical line __________________.

A3.* Give the formula for the slope of a line through points (a, b) and (c, d).

A4.* Give the point-slope form of the equation of a line.

A5.* Look at the graph of x^2. Zoom in about the origin until the window is very small and the curve looks almost straight. What line does the graph resemble?

A6.* True or False: The idea behind linear interpolation is basically putting a line through two points.

A7.* Pick one. Linear interpolation can be used for
a) evaluating, b) solving, c) both of (a) and (b), d) Neither of (a) and (b).

A8.* Define "x-intercept."

A9.* Put these in their order of development on graphs:
a) the point (a, b) b) the line $x = a$. c) the x-axis.

A10.* Put these in their order of development on graphs:
a) the point (a, b) b) the line $y = b$. c) the y-axis.

A11.* The formula for the distance between two points in the plane is essentially just the ________________ theorem from geometry.

A12.* a) Give the "standard form" of the equation of a circle. b) How many parameters does it have?

^ ^ ^ ^ Find the equation of the
A13. a) vertical line through (6, 9). b) horizontal line through (123, 456).
A14. a) vertical line through (-4, 123). b) horizontal line through (19, 21).

^ ^ ^ ^ Find an equation of the line through the two points:
A15. (3, 9) and (2, 5) A16. (-1, 5) and (3, 7).
A17. (100, 50) and (300, 1000) A18. (5, 7) and (25, 47).

^ ^ ^ ^ Find an equation of the line through the given point with the given slope:
A19. (5, 7), $m = 3$. A20. (0, 2), $m = -3$
A21. (-2, -6), $m = 4$. A22. (100, 50), $m = 1/2$.

A23. a) Find an equation of the line through the point (50, 210) and parallel to the line $y = 5x + 12$. b) Find the slope of a line perpendicular to that line.

A24. a) Find an equation of the line through the point (65, 12) and parallel to the line $y = (1/4)x + 987$. b) Find the slope of a line perpendicular to that line.

^ ^ ^ ^ Use linear interpolation and Table 2.2.2 to approximate
A25. $\Phi(1.05)$. A26. $\Phi(.78)$

^ ^ ^ ^ Find the distance between
A27. (4, 6) and (2, -3). A28. (-1, 3) and (-9, 5).
A29. (5, 7) and (x, y). A30. $(x, 3)$ and $(2x, 7)$.

^ ^ ^ ^ Find the equation of the circle with given center and radius:
A31. Center (2, 6) and radius 3. A32. Center (-1, 4) and radius $\sqrt{5}$.

^ ^ ^ ^ Identify the center and radius of the circle:
A33. $x^2 + (y - 2)^2 = 16$. A34. $(x - 1)^2 + (y + 5)^2 = 10$.

A35. Suppose x and y are proportional. When $x = 4$, $y = 0.2$. Find the relationship.

A36. Suppose distance and time are proportional. In one hour and 12 minutes the distance is 30 miles. Find the relationship.

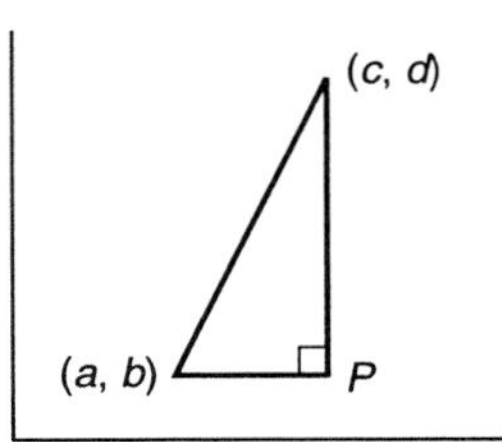

A37. The figure locates (a, b) and (c, d) on a line, and forms the right triangle with those two points as vertices. Give the coordinates of the vertex at the right angle.

A38. A spring stretches 1.2 inches when a force of 24 pounds is applied. Find the relationship between force and stretch.

A39. True or false?
a) If $y = 3x$, y is proportional to x.
b) If $y = 5x - 1$, y is proportional to x.
c) The area of a square is proportional to its side.
d) The perimeter of a square is proportional to its side.

A40. True or false?
a) If $C = 10n$, then C is proportional to n.
b) If $y = 2(x + 3)$, then y is proportional to x.
c) The area of a circle is proportional to its radius.
d) The circumference of a circle is proportional to its radius.

A41. True or false?
a) At a constant speed, the distance traveled is proportional to the time elapsed.
b) At a constant speed, the time elapsed is proportional to the distance traveled.
c) Over a fixed distance, the average speed is proportional to the time elapsed.
d) Over a fixed distance, the time elapsed is proportional to the average speed.

A42. True or false?
a) The volume of a cube is proportional to its side.
b) For a constant height, the area of a triangle is proportional to its base.
c) The area of a rectangle is proportional to its width.

A43. Find the distance from the point (1, 2) to
a) the line $x = -4$. b) the line $y = 10$.

A44. There are 45 calories in 1/4 cup. a) How many calories are in 1/3 cup?
b) Give the formula for the number of calories in any number of cups.

A45 There are 80 calories in 1/3 cup. a) How many calories are in 1/8 cup?
b) Give the formula for the number of calories in any number of cups.

^ ^ ^ ^ The graph of "$y = f(x)$" is a line. Solve $f(x) = 0$ if
A46. $m = .3$ and $f(1.5) = .04$ [1.4]. A47. $m = -.5$ and $f(12) = .1$ [12].
A48. $m = .8$ and $f(4.6) = -.5$ [5.2]. A49. $m = -.7$ and $f(.95) = -.2$ [.66].

^ ^ ^ ^ The slope of a line in a window depends upon the scale of the graph. Estimate the slope of the given line after noting the scale.

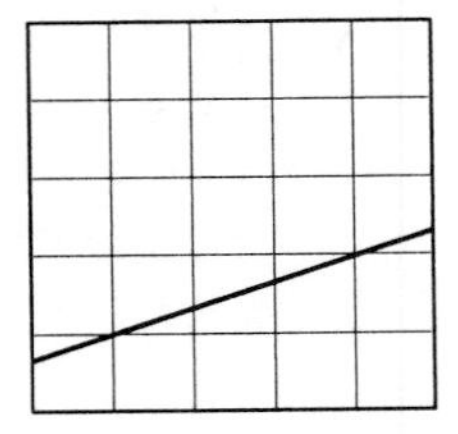

A50. [-5, 5] by [0, 20]

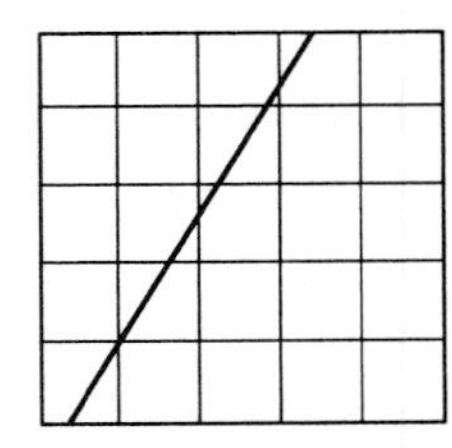

A51. [0, 100] by [0, 10]

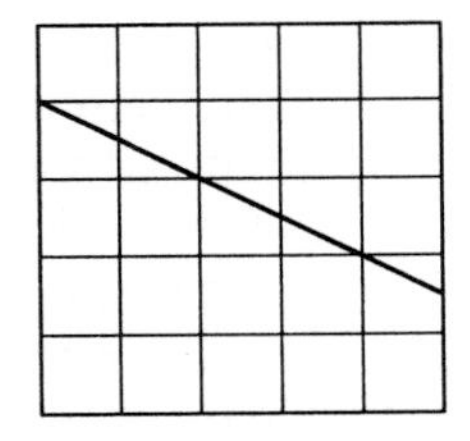

A52. [0, 10] by [0, 1000]

^ ^ ^ ^ ^ ^ ^ ^ ^

B1.* a) State the Two-Point Formula for a line. b) Sketch a (large) picture to illustrate *all* the points and distances in the formula. c) Use a bit of geometry to explain (or prove) why the formula in part (a) is, in fact, the right formula.

B2.* Explain how to find the equation of a line with a given slope through a given point.

B3.* Suppose a line with slope m goes through the point (x_1, y_1). a) State the formula for its equation. b) Sketch a picture to illustrate the formula and label on it the points and distances in the point-slope Formula. c) Use a bit of geometry to explain (or prove) why the formula in part (a) is, in fact, the right formula.

B4.* Suppose a line goes through the points (a, b) and (c, d). a) State the formula for its slope. b) Sketch a picture to illustrate the the formula and label on it the points and distances in your slope formula from part (a).

B5.* Sketch and label a picture and use it to explain how the distance formula is related to the Pythagorean Theorem.

B6.* Explain how to find the equation of a line through two given points.

B7.* Explain how to find the equation of a line through a given point and parallel to a line with a given equation. Write your answer in the format of a theorem. Make appropriate use of symbols.

B8.* Explain how to find the equation of a line through a given point and perpendicular to a line with a given equation. Write your answer in the format of a theorem. Make appropriate use of symbols.

B9.* Why might the phrase "*The* equation of a line" be misleading?

B10.* Which lines can *not* be written in the form "$y = mx + b$"?

B11.* a) State the "standard form" of the equation of a circle. b) Draw a (large) picture to illustrate it, labeling *all* its variables and expressions on the picture. c) Explain, using some geometry, why the standard form is right.

B12.* Why might point-slope form be preferred to slope-intercept form?

^ ^ ^ ^ Find the slope of the line through the points on the graph of

B13. x^2 where $x = 1$ and $x = 2$.

B14. $1/x$ where $x = 1$ and $x = 2$.

B15. x^3 where $x = 1$ and $x = 2$.

B16. $\sqrt{x}$ where $x = 1$ and $x = 4$.

B17. Consider the graph of $y = x^2$. Find the equation of the line through the points on the graph where $x = 2$ and where $x = 2.1$.

B18. Consider the graph of $y = \sqrt{x}$. Find the equation of the line through the points on the graph where $x = 2$ and where $x = 2.1$.

B19. If you want to rent a car and Firm A charges $40 a day plus 30 cents a mile and Firm B charges $30 a day and 34 cents a mile, when is Firm A better for you?

B20. Suppose you want to make many copies of a flyer and Photocopy World charges 7 cents per copy and PrintKing charges a $3 setup charge but after that only 4.5 cents per copy, when is PrintKing better for you?

B21. Suppose you want to use mayonaise with no more than 55 calories per tablespoon. You have both regular at 100 calories per tablespoon and Lite at 40 calories per tablespoon (but it's not as good). How many tablespoons of Lite should you mix in with 20 tablespoons of regular to make a mix with 55 calories per tablespoon?

B22. A cup of lentils has 646 calories of which 19 are from fat. Sausage is 50 calories per ounce of which 27 calories come from fat. Ervin wants to add sausage to his lentil soup, but does not want the calories from fat to be more than 15% of the calories. How many ounces of sausage can he add to a cup of lentils to have 15% of the calories from fat?

B23.* Explain what "linear interpolation" is.

B24.* Define "proportional." Be sure to give the proper context.

B25.* a) How can you tell, just by looking at their equations, if two lines are parallel? b) How can you tell, just by looking at their equations, if two lines are perpendicular?

B26.* a) Define "parameter." b) Give an example of a parameter.

B27. Find the equation of the circle with center (-2, 1) that goes through (3, 5).

B28. Find the equation of the circle with center (2, 5) that goes through (-1, -2).

B29. Find and simplify the slope of the line through the points on the graph of x^2 where $x = 1$ and where $x = 1 + h$.

B30. Find and simplify the slope of the line through the points on the graph of x^2 where $x = a$ and $x = b$. [The slope should simplify nicely.]

B31. Find and simplify the slope of the line through the points on the graph of $1/x$ where $x = 1$ and $x = 1 + h$.

B32. Find and simplify the slope of the line through the points on the graph of $1/x$ where $x = a$ and $x = b$.

^^^^Solve for x algebraically.

B33. $\dfrac{2 + x}{3 - x} = 4$. B34. $\dfrac{2x}{x - 5} = 3$.

B35. $\dfrac{x}{x + 3} = y$. B36. $\dfrac{2x + 1}{3x - 5} = y$.

^^^^Use linear interpolation to approximate a solution to $f(x) = 0$.

B37. $f(12.6) = .02, f(12.8) = -.01$. B38. $f(.78) = -.04, f(.81) = .017$.

B39. Suppose $f(3) = 10$ and we know that the slope of the tangent line at any point on the graph is given by $2x$. a) Find the tangent line at (3, 10). Then use the tangent line to approximate
b) $f(3.1)$. c) $f(2.8)$. d) The true f is given by $f(x) = x^2 + 1$. How inaccurate are these approximations in (b) and (c)?

B40. Consider the graph of $\sin x$ (in radians) near (0, 0). From calculus, we know the slope of the tangent line at $x = 0$ is 1. a) Find the tangent line at (0, 0). Use the tangent line to approximate b) sin (.1) c) sin (-.2) d) How inaccurate are these approximations?

B41.* Given the graph of $y = x$, interpret the graph of $y = mx + b$ as a scale change and then a location shift.

B42. Discuss the truth of "$a/b = c$ is equivalent to $a = bc$."

B43.* Prove that "$a = b$" is not equivalent to "$ca = cb$" by finding an example with three numbers where one equation is true and the other is not.

B44. Given the "general" form of a line, find the corresponding "slope-intercept" form using abstract letters for all constants.

B45. We can approximate the curve $y = x^2$ near $x = 5$ by a line of slope 10. a) Find the equation of that line. Approximate the value of the function by using the line, and find the error is so doing, at the points where b) $x = 5$; c) $x = 5.1$; d) $x = 4.95$; e) $x = 6$.
f) Comment on your findings.

B46. a) Identify and sketch the graph of "$x^2 + (y + 2)^2 = 25$.
b) What is the trick to getting your calculator to graph this?

B47. a) Identify and sketch the graph of "$(x + 1)^2 + (y - 3)^2 = 4$.
b) What is the trick to getting your calculator to graph this?

B48. Find the equation of the circle with center (2, 3) and tangent to the line $y = 7$.

B49. Find the equation of the circle with center (-1, 2) and tangent to the line $x = 5$.

B50. Solve for x: $x/(x - 2) = y$. B51. Solve for x: $3/x + 2 = y$.

^ ^ ^ ^ Find f^{-1} if f is as given.

B52. $f(x) = x/(x - a)$.

B53. $f(x) = a/x + b$.

^ ^ ^ ^ Use linear interpolation and Table 2.2.2 to approximate the solution to

B54. $\Phi(x) = 0.95$.

B55. $\Phi(x) = .8$.

B56. $\Phi(x) = .9$

B57. $\Phi(x) = .1$.

B58. What can the "general" form of a line do that "point-slope form" cannot?

^ ^ ^ ^ Solve for x:

B59. $x/a + b = x/c$.

B60. $ax + b = cx + d$.

B61. Robin walks and runs a total of 6 miles. She walks at 4 miles per hour and runs at 8 miles per hour. If she wants to finish in exactly one hour, how far should she run?

B62. The outside of a circle with equation $x^2 + y^2 = 1$ is mirrored so it reflects light. Light from point (2, 0) reflects off it through (0, 4). Where does it reflect off the mirror? [According to the science of optics, it reflects so that the total distance the light travels from (2, 0) to the point on the mirror to (0, 4) is the minimum possible.]

B63. Ms. Green, the president of Consolidated Widgets International, thinks that, for widgets at any price, an increase in price of $5 per widget would decrease sales by 400 widgets per week. The price is now $65 and sales are 5000 per week. Find a straight line fit for the price (x) to sales (y) relationship.

B64. Suppose you can rent a car from firm A at $30 a day plus 30 cents per mile. You can rent a comparable car from firm B for $40 a day plus 20 cents per mile. When is firm B the better deal for you?

B65. Suppose you can photocopy any number of copies of the same original for 6 cents each, or you can have the copies printed at 2.5 cents each after a setup charge of 50 cents. When is it cheaper to print copies than photocopy them?

B66. Suppose the interval [2,5] is subdivided into n intervals of equal length. Number the subintervals from left to right. a) What is the right endpoint of the 7th subinterval? b) What is the right endpoint of the kth subinterval?

B67. Suppose the interval $[a, b]$ is subdivided into n intervals of equal length. Number the subintervals from left to right. a) What is the right endpoint of the 5th subinterval? b) What is the right endpoint of the kth subinterval?

B68.* Look at the graph of x^2. Zoom in about the point (1, 1) until the window is very small and the curve looks nearly straight. What line does the graph resemble?

B69. Consider the line through points (a, b) and (c, d). Show that either point can be used in the point-slope form and equivalent equations result.

B70. Prove that, according to the definition, if y is proportional to x, then x is proportional to y. So we can say they are "proportional to each other."

B71. Two variable quantities x and y are said to be inversely proportional (to vary indirectly) iff there is a constant k ($\neq 0$) such that $xy = k$. For example, for rectangles of fixed area k, the length and width are inversely proportional. Give another example of quantities that are inversely proportional.

B72. Find f^{-1} if $f(x) = (ax + b)/(cx + d)$.

B73. Find f^{-1} if $f(x) = a/(x - b) + c$.

B74. Prove: If f is linear, then $f(f(x))$ is also linear.

B75. Prove that the composition of any two linear functions is a linear function.

B76. Lines are *not* good examples with which to graphically illustrate the effects of shifts due to composition with addition or subtraction. This is because shifts left or right cannot be distinguished graphically from shifts up or down. Explain, mathematically, how horizontal shifts of the graph of $f(x) = 3x - 1$ can be interpreted as vertical shifts of the same graph.

B77. $x^2 + y^2 = 1$ is the equation of a circle. a) Use ideas from Section 2.3 to explain the shape of the graph of $(2x)^2 + y^2 = 1$. b) Use ideas from Section 2.3 to explain the shape of the graph of $x^2 + (2y)^2 = 1$.

B78. Redraw a picture of the problem in "Example 26, continued" (Figure 25). Note that the shortest distance will be measured perpendicular to the line. Use Theorems 3.1.9 and 3.1.12 to determine the closest point.

B79. Redraw a picture of the problem in "Example 26, continued" (Figure 25). Note that the shortest distance will be measured perpendicular to the line. Use similar triangles and geometry to determine the distance to the closest point.

B80. The distance from the line $ax + by + c = 0$ to the point (x_0, y_0) is given by

$$\frac{|ax_0 + by_0 + c|}{\sqrt{a^2 + b^2}}.$$

Prove it. [There are several ways. One parallels problem B78, another parallels B79.]

B81. $y - y_1 = (\frac{y_2 - y_1}{x_2 - x_1})(x - x_1)$. Now set $y = 0$ and solve for x. [The result simplifies nicely. It is an important formula in the subject of "numerical analysis."]

B82. The midpoint of the line segment from (a, b) to (c, d) is given by $((a+c)/2, (b+d)/2)$. Illustrate why.

B83. Find the circle with center (2, 1) tangent to the graph of $y = 4x + 1$.

B84. A baseball park has the outfield fence 325 feet from home plate down both foul lines and 400 feet from home plate in the middle of center field. The fence is an arc of a circle. Where is the center of the circle located?

B85. Find the equation of the line tangent to the unit circle $x^2 + y^2 = 1$ at the point in the first quadrant on the circle where $x = .2$.

B86. Find and simplify the x-value of the point of intersection of the two lines $y = m_1x + b_1$ and $y = m_2x + b_2$.

B87. (Newton's Method) Some equations are difficult to solve, even by evaluate-and-compare, because the function is difficult to evaluate. Example 20 illustrates the idea of "Newton's Method" for solving such equations. Basically, the idea is to pick a potential solution, x, and to evaluate both $f(x)$ and the slope of the curve at $(x, f(x))$. Then use the method in Example 20 to improve the pick. Solve $x^2 - 10 = 0$, beginning with first guess $x = 3.16$ and using slope $2x$. Generate two improved approximations to the solution. How many decimal places of accuracy result?

^ ^ ^ ^These are functional equations. In each, a function is unknown. Solve for f.

B88. $f(x + 1) = f(x) + 3$, for all x.

B89. f: $f(x + 5) = f(x) - 3$, for all x.

B90. Prove 3.1.9 for positive slope m. Let the given line be L_1 with slope m and let the perpendicular line, L_2 intersect it at point P. On the horizontal line through P mark point Q one unit to the right. The vertical line through Q will intersect L_1 m units up at, say, R.

Locate point S m units to the left of P on the horizontal and then locate T at the intersection of L_2 and the vertical line through S. Show that point T is in fact one unit above S. Then from T to P on L_2 the rise is -1 and the run is m, so the slope of the perpendicular line is $-1/m$.

B91. In the subject of "differential equations," functions are described, not by giving an expression for $f(x)$ as usual, but by giving two pieces of information: 1) a point the graph goes through, and 2) an expression for the slope of the tangent line to the curve at any point.

Here is the problem: Suppose we do not know the expression for $f(x)$, but we know $f(25) = 5$ and we know that the slope of the tangent line at any point on the curve is given by $1/(2\sqrt{x})$.

a) Use this information to approximate $f(26)$.

b) In calculus you will learn how to solve for the unknown function f with the given slope formula. It turns out to be $f(x) = \sqrt{x}$. How much error was there in approximating $f(26)$? [Note that the approximation was pretty good!]

Section 3.2. Quadratics

Linear functions are both the simplest and most important functions. Quadratic functions (with an "x^2" term) are perhaps the next most important, largely because they are the next simplest. In physics, important examples of quadratic functions occur when an object is dropped or projected subject to the force of gravity. The shape of the graph of a quadratic function is called a parabola, and parabolic shapes have excellent reflective properties. Satellite dishes, microwave relay stations, and telescopes have parabolic reflectors to focus signals. In business, some simple profit functions are quadratics. And, of course, in mathematics, quadratic equations are very common, partly because they are simple enough to solve algebraically (by using the famous Quadratic Theorem).

Definition 3.2.1: Any expression equivalent to "$ax^2 + bx + c$" for some a, b, and c, where $a \neq 0$, is called a quadratic in "x". It is traditional to use "a" for the coefficient on "x^2", "b" for the coefficient on "x", and "c" for the constant term. "a", "b", and "c" are the parameters of this family of closely related expressions. With this notation, the quadratic is in "standard" form. a is sometimes called the leading coefficient. Many times a is simply 1.

Example 1: The basic quadratic is x^2 (Figure 1A). All the important properties of quadratics can be studied by studying x^2. The graph of x^2 has the shape known as a parabola. In an important sense, all parabolas are alike. For example, if a parabola seems "wide," it could be the graph of x^2 in a window which emphasizes points near the origin (Figure 1B, on [-.5, .5] by [0, 1]). "Wide" parabolas also could be the graph of x^2 with a large vertical interval (Figure 1C, on [-10, 10] by [0, 1000]).

If a parabola seems "narrow," it could be the graph of x^2 in a window which is wide (Figure 1D, on [-100, 100] by [0,100]). "Narrow" parabolas also could be the graph of x^2 with a small vertical interval (Figure 1E, on [-10, 10] by [0,4]). Figures 1A through 1E show that scale changes can have a dramatic affect on the appearance of a graph.

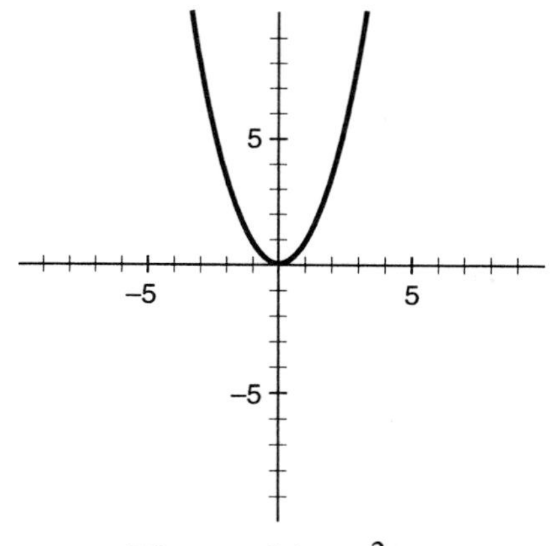

Figure 1A: x^2
[-10, 10] by [-10, 10].

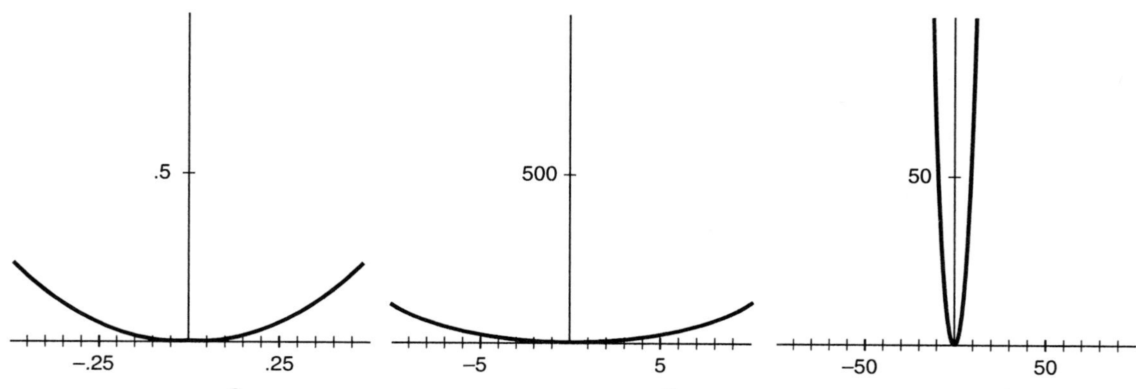

Figure 1B: x^2 near (0, 0).
[-.5, .5] by [0, 1]

Figure 1C: x^2
[-10, 10] by [0, 1000]

Figure 1D: x^2
[-100, 100] by [0, 100]

The most important point on the graph of a quadratic is its <u>vertex,</u> the point corresponding to (0, 0) on the graph of x^2. If the graph <u>opens up</u> like the graph of x^2 does, the vertex is the lowest point on the graph. If the graph is upside down, it <u>opens down,</u> and the vertex is the highest point on the graph (Figure 9).

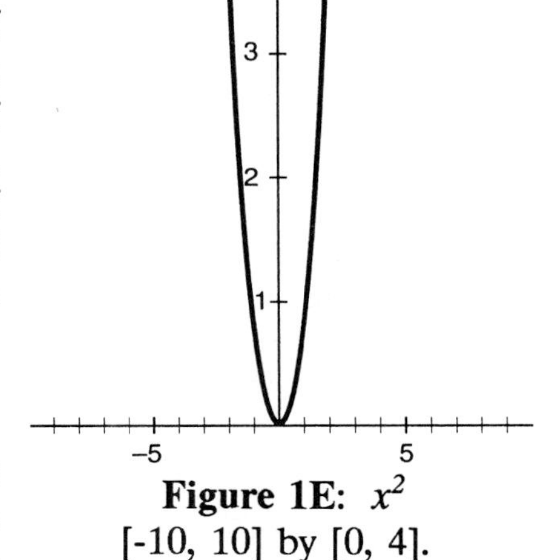

Figure 1E: x^2
[-10, 10] by [0, 4].

<u>Symmetry</u>. The graph of $f(x) = x^2$ is symmetric about its central axis, the y-axis. This is because the image of $-x$ is the same as the image of x:

$$\begin{aligned} f(-x) &= (-x)^2 \\ &= x^2 \\ &= f(x). \end{aligned}$$

Therefore the heights are the same on either side of the y-axis. The y-axis serves as an axis of symmetry.

If $c > 0$ there are two solutions to the equation "$x^2 = c$": $x = \sqrt{c}$ and $x = -\sqrt{c}$. They are at equal distances on either side of the axis of symmetry (Figure 2).

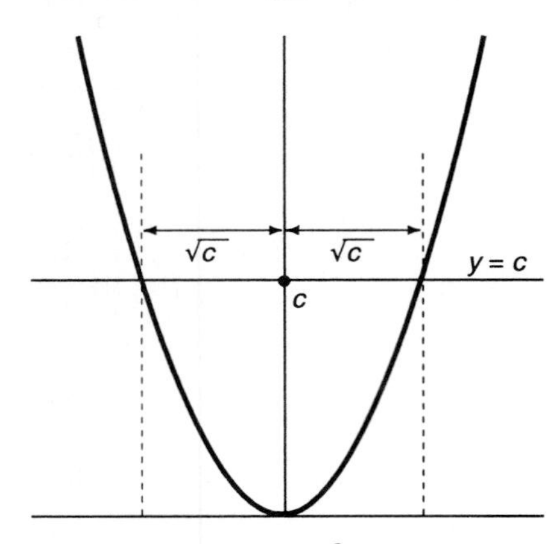

Figure 2: $y = x^2$ and $y = c$ intersect at distance $\sqrt{c}$ from the axis of symmetry.

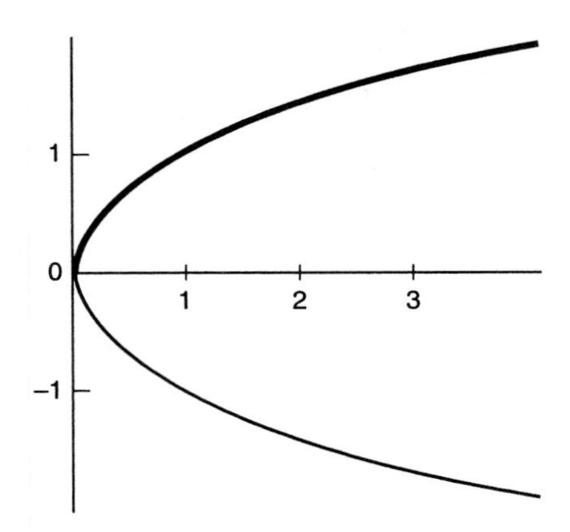

Figure 3: $y^2 = x$. $y = \sqrt{x}$ bold. $y = -\sqrt{x}$ on the bottom. [0, 4] by [-2, 2].

<u>The Square Root</u>: $y = \sqrt{x}$ implies $y^2 = x$. This is the reverse of the familiar "$x^2 = y$" relationship; we can graph it by reversing the axes. The graph of $y = \sqrt{x}$ (Figure 3) is the upper half (since $\sqrt{x} \geq 0$) of the horizontal version of the vertical parabola we have been discussing. The graph of x^2 is nearly horizontal near the origin. The graph of $\sqrt{x}$ switches the axes, so it is nearly vertical near the origin.

<u>Location Changes</u>. Consider shifting the graph left or right, and up or down. Shifts left or right shift the axis of symmetry.

Example 2: $y = x^2 - 6x$ defines a quadratic function. Graph it.

The Zero Product Rule yields the x-intercepts (where $y = 0$).

$x^2 - 6x = x(x - 6) = 0$ iff $x = 0$ or $x = 6$.

The graph crosses the x-axis at 0 and 6 (Figure 4). By the symmetry of the graph, the lowest point, the vertex, is half way between 0 and 6, at $x = 3$. The y-value there is $y = 3^2 - 6(3) = -9$.

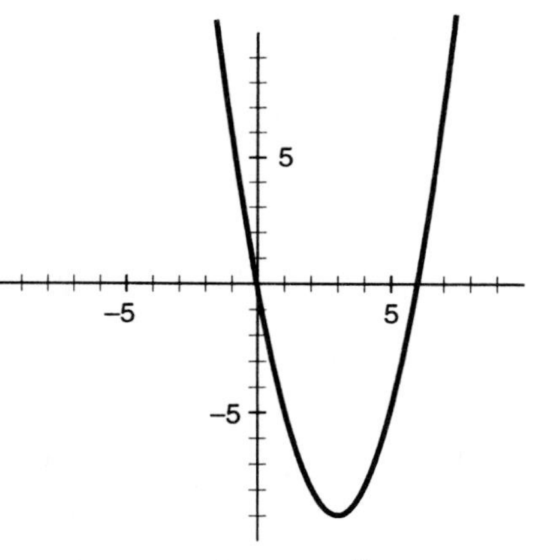

Figure 4: $y = x^2 - 6x$. [-10, 10] by [-10, 10].

The shape of the graph is the shape of $y = x^2$, but its location is different. In some important ways all quadratics are alike, so by studying x^2 you can learn about them all. We will see that the coefficient "a" on the "ax^2" term can change the scale, but first we note that all quadratics with $a = 1$ (as in Figure 4) have exactly the same shape (but not necessarily the same location).

The lowest point on the graph of x^2 occurs when $x = 0$, since $x^2 \geq 0$ for all x. The lowest point on the graph of "$(x - h)^2 + k$" will be when $x = h$ (so the argument of the squaring function is zero). Then y will be k. Therefore, the graph of $(x - h)^2 + k$ has its vertex at (h, k). That is:

Shifts (3.2.2) The graph of $(x - h)^2 + k$ differs from the graph of x^2 only in its location. It has the same shape. The vertex of the graph of "$(x - h)^2 + k$" is (h, k). The use of the letters "h" and "k" in this context is traditional.

Note that this expression is a perfect square plus or minus some constant.

Definition 3.2.3 (Complete the Square): To complete the square of a quadratic expression when $a = 1$ is to rewrite it as an equivalent expression in the form "$(x - h)^2 + k$," or "$(x + d)^2 + k$," in which case $h = -d$.

To complete the square we must discover h (or d) and k. We prefer to use the form with "h", since that is the x-value of the vertex. An expression such as "$(x + 5)^2$" is a perfect square and is fine to work with, but, in the end, we will treat it is as "$(x - -5)^2$," so we can see the "h".

We need the expanded form of perfect squares.

Theorem 3.2.4: $(x - h)^2 = x^2 - 2hx + h^2$, for all x and h.
$(x + d)^2 = x^2 + 2dx + d^2$, for all x and d.

Example 2, again: Complete the square of the quadratic $x^2 - 6x$.

The objective is to rewrite the given expression as a perfect square plus or minus some constant.

$$x^2 - 6x = x^2 - 6x + 9 - 9 = (x - 3)^2 - 9.$$

In the final expression the square is complete. $h = 3$ and $k = -9$. The vertex is $(3, -9)$ (Figure 4).

In this example we added and subtracted 9. How did we know to use 9? There is a general way to determine what to add and subtract.

Theorem 3.2.5 (Completing the Square):

$$x^2 + bx + c = x^2 + bx + (b/2)^2 - (b/2)^2 + c$$
$$= (x + b/2)^2 - (b/2)^2 + c.$$

The last expression "completes the square" of the first.

Therefore, if $a = 1$, to complete the square add and subtract the square of half the coefficient on x. Then rewrite the expression to exhibit the perfect square.

Example 2, reconsidered: Complete the square of $x^2 - 6x$.

Half the coefficient on "x" is -3. Add and subtract $(-3)^2 = 9$.

$$x^2 - 6x = x^2 - 6x + 9 - 9.$$

Now regroup, that is, reorder the operations to exhibit the perfect square.

$$x^2 - 6x = (x - 3)^2 - 9.$$

From this expression we can see that the vertex is (3, -9) (Figure 5).

Example 2, continued: Solve $x^2 - 6x = 15$.

Pretend we do not know the Quadratic Formula. Note that "$x^2 - 6x$" has two appearances of the variable "x" and does not fit the format requirement for the inverse-reverse method of solving equations, which is that there be only one appearance of "x". But the equivalent expression with the square completed will have only one appearance of "x" and will be ready for the inverse-reverse method:

$x^2 - 6x = 15$

iff $x^2 - 6x + 9 - 9 = 15$

iff $(x - 3)^2 - 9 = 15$ [this fits inverse-reverse format]

iff $(x - 3)^2 = 24$

iff $x - 3 = \pm\sqrt{24}$

iff $x = 3 \pm \sqrt{24}$.

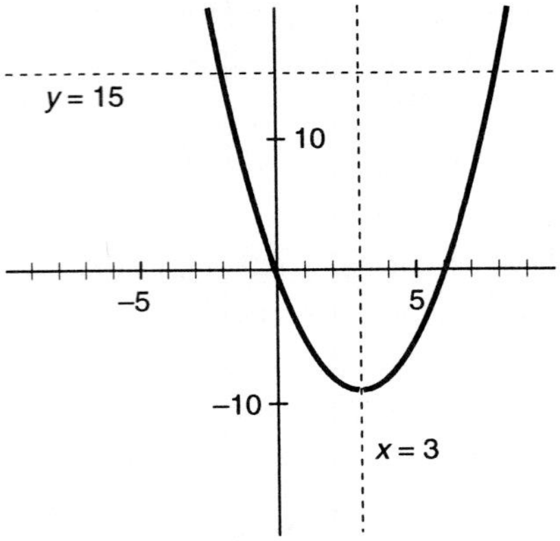

Figure 5: $y = x^2 - 6x$ and $y = 15$. [-10, 10] by [-20, 20]

The "3" is the "h", the x-coordinate of the vertex. Figure 5 shows that the two solutions are equal distances on either side of $x = 3$, which is the axis of symmetry.

This procedure can be generalized to handle any quadratic. When it is summarized and abbreviated to one step, the procedure yields the famous Quadratic Theorem, which we will prove as Theorem 3.2.6.

Why do we bother to "complete the square"? For one reason, the resulting expressions have only one appearance of "x" so the inverse-reverse method of solving equations applies (Example 2, continued). For another reason, the resulting expressions clearly exhibit location shifts (Examples 2 and 3).

Example 3: Identify the circle: $x^2 + 4x + y^2 - 6y = 5$.

A circle is identified by its center and radius. In Formula 3.1.13 we saw that

$$(x - h)^2 + (y - k)^2 = r^2$$

is the standard form of an equation of a circle with center (h, k) and radius r. To identify the circle means to identify h, k, and r. To do that, complete the square.

$$x^2 + 4x + y^2 - 6y = 5 \quad \text{iff}$$

$$x^2 + 4x + 4 + y^2 - 6y + 9 = 5 + 4 + 9 = 18,$$

$$\text{iff } (x + 2)^2 + (y - 3)^2 = 18 = (\sqrt{18})^2.$$

The center is (-2, 3) and the radius is $\sqrt{18}$ (Figure 6).

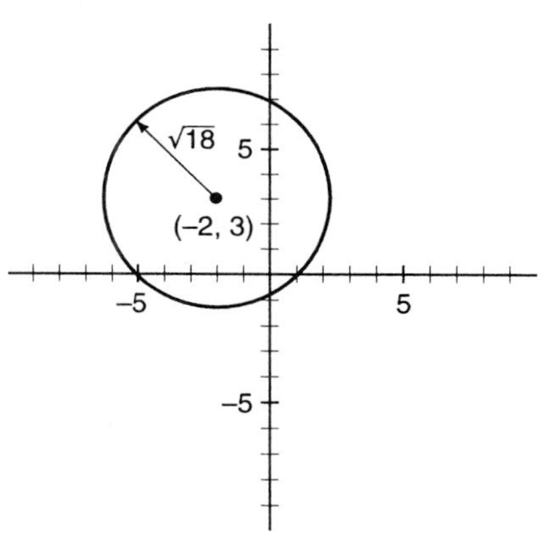

Figure 6: The circle with center (-2, 3) and radius $\sqrt{18}$. [-10, 10] by [-10, 10].

The Quadratic Theorem. By completing the square we can rewrite a quadratic in standard form ($ax^2 + bx + c$) as an equivalent expression in which "x" appears only once so that the inverse-reverse method applies. The famous "Quadratic Theorem" simply gives a one-step abbreviation of the process of first completing the square and then using the inverse-reverse method. Since the same process works for all quadratic equations, it can be expressed abstractly as a statement of fact.

Theorem 3.2.6 (The Quadratic Theorem). If $a \neq 0$, $ax^2 + bx + c = 0$ is equivalent to

$$x = \frac{-b \pm \sqrt{b^2 - 4ac}}{2a}.$$

"Negative b plus or minus the square root of (the quantity) b squared minus four a c, all over two a."

The idea of the proof is to use Theorem 3.2.5 **to complete the square and then to use the inverse-reverse equation-solving method.** Because Theorem 3.2.5 applies only when $a = 1$, the first step is to divide through by a.

Proof: Because $a \neq 0$, $ax^2 + bx + c = 0$ is equivalent to

$$x^2 + (\frac{b}{a})x + \frac{c}{a} = 0$$

$$x^2 + (\frac{b}{a})x = -(\frac{c}{a})$$

$$x^2 + (\frac{b}{a})x + (\frac{b}{2a})^2 = -(\frac{c}{a}) + (\frac{b}{2a})^2$$

$$(x + \frac{b}{2a})^2 = -(\frac{c}{a}) + \frac{b^2}{4a^2} \qquad \textit{[the square is complete]}$$

$$= \frac{b^2 - 4ac}{4a^2} \qquad \textit{[} 4a^2 \textit{ is a common denominator]}$$

Now there is only one appearance of "x". The inverse-reverse format requirement is satisfied.

$$x + \frac{b}{2a} = \pm\sqrt{\frac{b^2 - 4ac}{4a^2}}$$

$$x = -(\frac{b}{2a}) \pm \frac{\sqrt{b^2 - 4ac}}{2a} = \frac{-b \pm \sqrt{b^2 - 4ac}}{2a}.$$

This is the famous Quadratic Formula.

Example 4: Solve $4x + 2x^2 = 55$.

To use the Quadratic Theorem, you must identify a, b, and c of the usual form, and this is not in the usual form. This equation is equivalent to $2x^2 + 4x - 55 = 0$, in which $a = 2$, $b = 4$, and $c = -55$.

$$x = \frac{-4 \pm \sqrt{4^2 - 4(2)(-55)}}{2(2)} = 4.34 \textit{ or } -6.34.$$

Symmetry and the Quadratic Formula. The axis of symmetry of the graph of $ax^2 + bx + c$ is apparent in the Quadratic Formula. The "negative b over $2a$" gives it (Figure 7) -- the axis of symmetry is just the first term in the Quadratic Formula. The two solutions lie at equal distances on either side of that line of symmetry ("plus or minus the square root of...."). The axis of symmetry is also important since the extreme value of y occurs on the axis on symmetry.

$$x = \frac{-b}{2a} \pm \frac{\sqrt{b^2 - 4ac}}{2a}$$

axis of symmetry distance of solutions on either side.

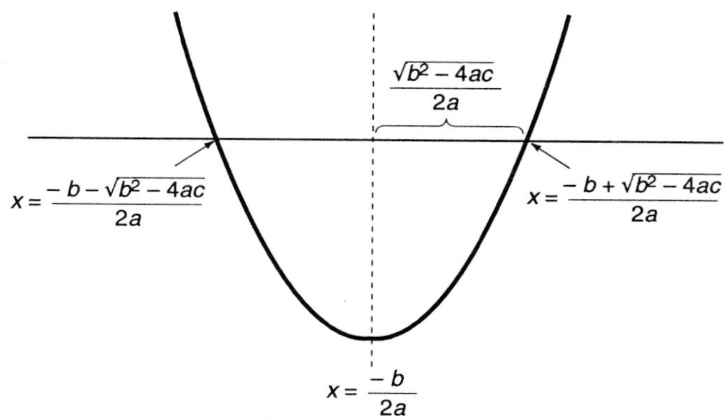

Figure 7: $ax^2 + bx + c$, labeled to show the terms of the Quadratic Formula.

The graph of "$ax^2 + bx + c$" is similar to the graph of x^2 with a scale change and a shift. By completing the square we can identify that shift, which is visible in Figure 7.

Theorem 3.2.7. The vertex of $ax^2 + bx + c$ is

$$x = -\frac{b}{2a} \text{ (again) and } y = c - \frac{b^2}{4a} = -\left(\frac{b^2 - 4ac}{4a}\right).$$

The central axis (the axis of symmetry) is $x = -\frac{b}{2a}$.

If $a > 0$, the graph opens up and the minimum value occurs at $x = -b/(2a)$.
If $a < 0$, the graph opens down and the maximum value occurs at $x = -b/(2a)$.

Use the Quadratic Formula to help you remember the x-value of the vertex in this theorem. Memorizing the y-value is not important; you can always find the y-value by substituting the x-value into the expression.

Example 5: Graph $x^2 + 5x$. Find x such that $x^2 + 5x$ is its minimum. Also, solve $x^2 + 5x = 3$ (Figure 8).

Since $a > 0$, the quadratic opens up and the vertex yields the minimum. So the minimum occurs at $x = -5/2$ [$= -b/2a$, by Theorem 3.2.7].

Then, the solutions to "$x^2 - 5x = 3$" must be equidistant on either side of $x = -2.5$. The symmetry tells us this, and so does the Quadratic Formula. The solution is problem A33.

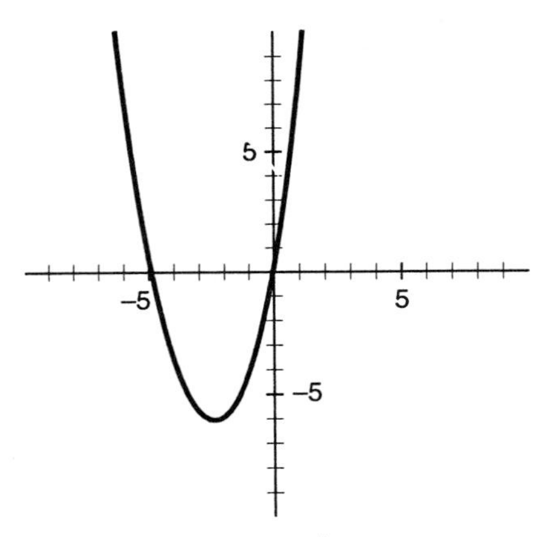

Figure 8: $x^2 + 5x$
[-10, 10] by [-10, 10].

Example 5, continued: Suppose the problem had been slightly different:
"Solve $x^2 + 5x = k$." Which values of k would give solutions?

This question determines the range of "$x^2 + 5x$."

Looking at the graph (Figure 8) and its vertex, we see that, for there to be solutions, k must be greater than or equal to the y-value of the vertex. Therefore, $k \geq -6.25$. The range of "$x^2 + 5x$" is $[-6.25, \infty)$.

Example 6: Graph $5 + 2x - x^2$ by completing the square and noting the location shift.

It is appropriate to factor out the minus sign, put the x^2 term first, and complete the square by 3.2.5 which only works for a = 1.

$$\begin{aligned} 5 + 2x - x^2 &= -(x^2 - 2x - 5) \\ &= -(x^2 - 2x + 1 - 1 - 5) \\ &= -[(x - 1)^2 - 6] \end{aligned}$$

We can graph this as the negative of the dashed curve in Figure 9, which is the graph of x^2 shifted right 1 and down 6. Then the "-" sign turns that graph upside down (Figure 9, solid curve).

To complete the square of $ax^2 + bx + c$, you may factor out a ("-1" in Example 7) and use the method of Theorem 3.2.5. Be careful to distribute a and not just discard it.

If the coefficient on x^2 is negative, the graph opens downward.

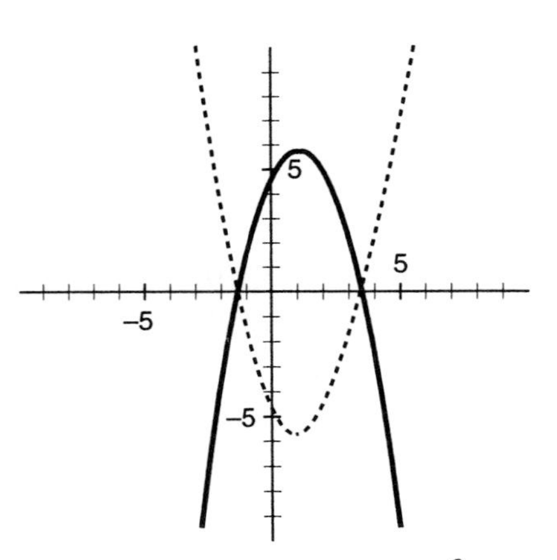

Figure 9: $5 + 2x - x^2$.
[-10, 10] by [-10, 10].

A quadratic equation has either two, one, or no real-valued solutions. To "Solve the equation $ax^2 + bx + c = 0$" is also to "Find the zeros of the expression $ax^2 + bx + c$," which is to "Find the x-intercepts of the graph of $ax^2 + bx + c$." It is easy to see how many x-intercepts the graph has.

Figure 10 illustrates how the location of the vertex of the graph affects the number of solutions when the graph opens up ($a > 0$). There are three cases. If the vertex is below the x-axis, there will be two solutions. If the vertex is on the x-axis, there will be one solution. Finally, if the vertex is above the x-axis, there will be no real-valued solutions (but there will be complex-valued solutions given by the same formula). If the graph opens down ($a < 0$), similar results hold with the position of the vertex reflected through the x-axis.

The Quadratic Formula always gives the solutions. The number of real-valued solutions depends upon whether the part under the square-root sign is positive.

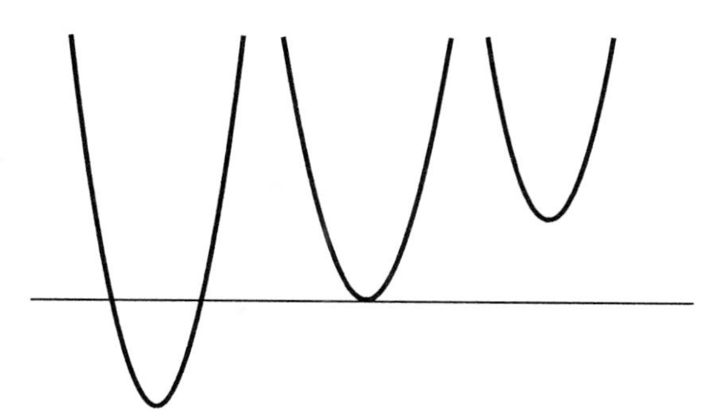

Figure 10: Quadratic equations have 2, 1, or no real-valued solutions.

Theorem 3.2.8: If $b^2 - 4ac > 0$, there are two distinct real-valued solutions.
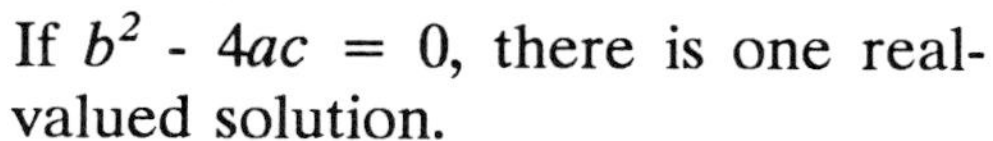
If $b^2 - 4ac = 0$, there is one real-valued solution.
If $b^2 - 4ac < 0$, there are no real-valued solutions (and two complex-valued solutions, because the square root of a negative number yields a complex number).

This is a direct result of the Quadratic Formula. This has nothing new to memorize. The expression "$b^2 - 4ac$" is called the discriminant because it discriminates between those three cases.

Complex Numbers. When the discriminant is less than zero, the Quadratic Formula expresses the square root of a negative number. The square root of a negative number cannot be a real number because $x^2 \geq 0$ for all real values of x. But it is possible to *define* a solution to $x^2 = -1$ and give it a name and properties. The solution will not be a "real" number, but, by the process of abstraction, we can treat this "made up" solution as an object with properties in the same way that we treat "7" and "-3" as objects with properties, even though they are not the sort of nouns you can touch. At first, and for most of school math, the idea of an "imaginary" number, called "i" which satisfies "$i^2 = -1$" seems silly. It seems to be nothing, or just something invented to serve as a symbol for the square root of negative one, which really doesn't have a square root. But, at a higher conceptual level, negative one does have a square root -- and a very useful one at that. Unfortunately, the use and value of "imaginary" and "complex" numbers in electronics, physics, and mathematics is not easily explained in basic algebra. We will delve into them later, but for now, we merely assert they exist and are important.

Definition 3.2.9: "i" is a number such that $i^2 = -1$. The number i can be added to or multiplied by a real number. For a and b real numbers, a number of the form "$a + bi$" is called a complex number, and "a" is its real part and bi is its imaginary part. All the usual arithmetic operations applied to complex numbers yield complex numbers and all the usual rules of arithmetic operations (such as the commutative, associative, and distributive properties) apply to complex numbers. Also, for real-valued $c \geq 0$,

$$\sqrt{-c} = i\sqrt{c} .$$

With this definition the Quadratic Formula always yields solutions to a quadratic equation -- but they might be complex-valued instead of real-valued.

Example 7: Solve $x^2 + 2x + 5 = 0$.

The Quadratic Formula yields

$$x = \frac{-2 \pm \sqrt{2^2 - 4(1)(5)}}{2} = \frac{-2 \pm \sqrt{-16}}{2} = \frac{-2 \pm 4i}{2} = -1 \pm 2i .$$

For now, this is all you need to know about complex numbers. They exist. As abstract as they are, if you work with them enough they will become "concrete." In the sciences they certainly are a valuable type of object. Because their value is most clearly understood after trigonometry, we will study them then.

<u>**Dummy Variables**</u>. The variables in the Quadratic Formula, as in other theorems, are dummy variables. The coefficients could be letters other than "a", "b", and "c" and the unknown does not have to be "x". The key to the meaning of the formula is not in the letters, but, of course, in the places they hold. It can be used to solve an equation with the same letters in different places.

Example 8: Solve $x^2 + cx + 2b = 0$.

Now "c" and "b" do not play their usual roles. Read the Quadratic Theorem by position, not letter.

$$x = \frac{-c \pm \sqrt{c^2 - 4(1)(2b)}}{2} .$$

Calculator Exercise 1: Graph $x^2 + 3xy + y^2 - 14 = 0$.

Try it. This is not trivial.

Most graphics calculators only graph equations in the form "y = ..." Can you convert this equation to that form?

Solve for "y" using the Quadratic Theorem. The theorem is about solving equations that express squaring, even if it is "y" squared instead of "x" squared. The coefficients need not be numbers; they can be expressions. Reorganize it to identify the coefficients on "y^2" and "y":

$$(1)y^2 + (3x)y + (x^2 - 14) = 0.$$

$$y = \frac{-3x \pm \sqrt{(3x)^2 - 4(1)(x^2 - 14)}}{2(1)}.$$

Graph these two equations (Figure 11).

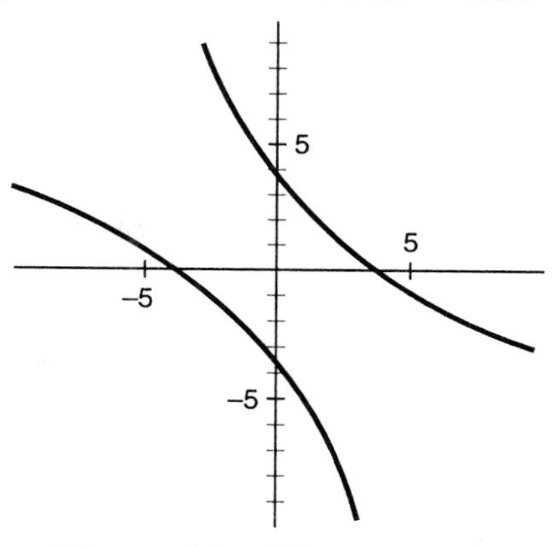

Figure 11: The graph of $x^2 + 3xy + y^2 - 14 = 0$. A hyperbola. [-10, 10] by [-10, 10]

__Ellipses__. Equations of circles have terms with x squared and y squared. Scale changes covert circles to ellipses.

Example 9: In Section 3.1 we noted that the equation of the unit circle is "$x^2 + y^2 = 1$." Graphed on a square scale, it looks like a circle, as it should. However, with a different window it can look different. Graph it with the window [-1, 1] by [-2, 2].

Your calculator wants y given functionally. So, solve for y.

$$y^2 = 1 - x^2,$$

$$y = \pm\sqrt{1 - x^2}.$$

Plot two graphs together, one for the plus sign and one for the minus sign.

Try different windows. Figure 12 is [-1, 1] by [-2, 2] and the shapes of the graph is an ellipse. Equations of circles yield circles *only on square scales*.

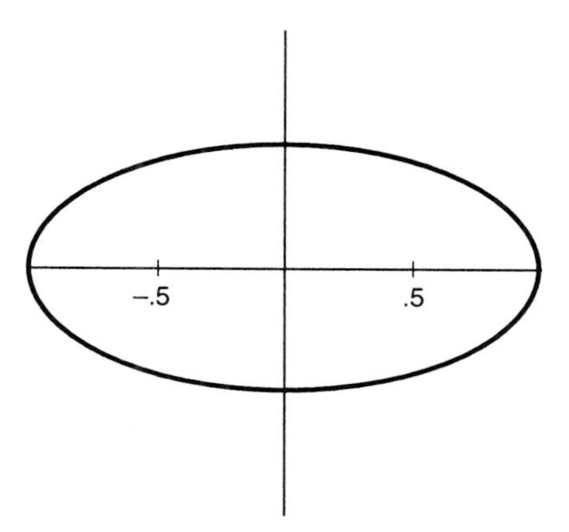

Figure 12: $x^2 + y^2 = 1$ on [-1, 1] by [-2, 2].

Example 10: Reconsider "$x^2 + y^2 = 1$." What would be the difference if "x" were replaced by "$x/2$"?

Graph "$(x/2)^2 + y^2 = 1$" on a square scale.

Compare the two equations:

$$x^2 + y^2 = 1 \qquad \text{[unit circle]}$$
$$(x/2)^2 + y^2 = 1.$$

The only difference is that the x-value of the second is divided by two before it is squared. So $x = 1$ in the first corresponds to $x = 2$ in the second. $x = 1/2$ in the first corresponds to $x = 1$ in the second. For a given y-value, the x-value of the second is doubled compared to the first. Therefore, the graph of the second is the graph of the first expanded horizontally by a

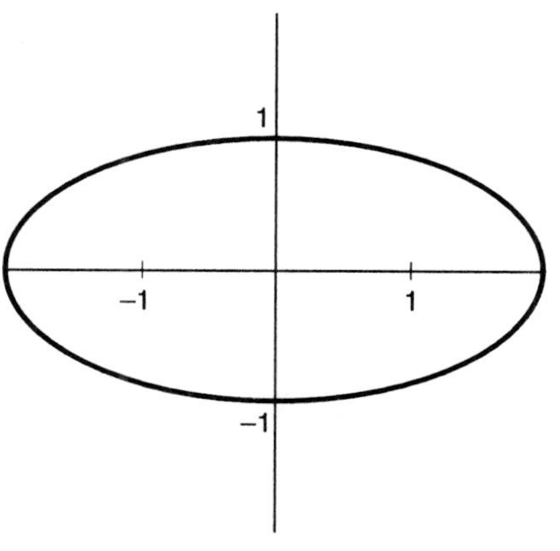

Figure 13: $(x/2)^2 + y^2 = 1$. [-2, 2] by [-2, 2].

factor of 2 (Figure 13). This shape is called an ellipse.

This fits the result from Section 2.3 about composition which states that dividing x by a before applying the function yields a graph which is expanded horizontally by a factor of a.

Example 11: Graph $x^2 + (y/2)^2 = 1$.

Compare this equation to the equation of the unit circle:

$$x^2 + y^2 = 1 \quad \text{[unit circle]}$$
$$x^2 + (y/2)^2 = 1.$$

For any given value of x, the y-value in the second equation would have to be twice the y-value in the first equation. The graph is the unit circle expanded vertically by a factor of two (Figure 14).

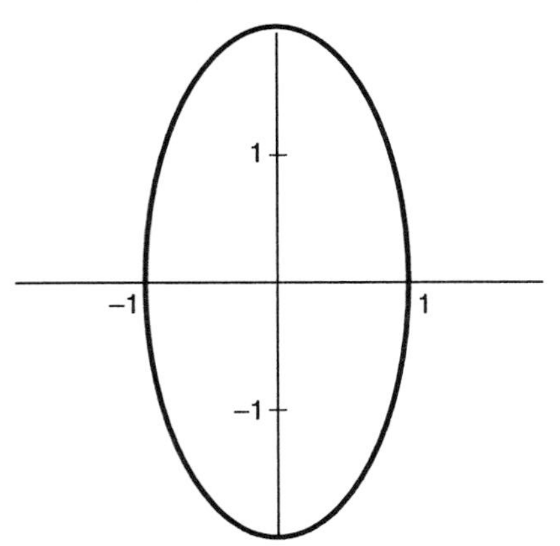

Figure 14: $x^2 + (y/2)^2 = 1$.
[-2, 2] by [-2, 2].

Did you expect the graph to be shorter, since "y" is divided by 2? Be careful, it is not the values of y in the first equation which are divided by 2. The first equation did *not* have an expression for y in terms of x, say $g(x)$, which was divided by 2 to obtain "$y = g(x)/2$." On the contrary, we replaced "y" with "$y/2$." Then, replacing "y" by "$y/2$" in "$y = g(x)$" yields

$$y/2 = g(x).$$

Therefore,

$$y = 2g(x).$$

The new graph is taller, as we found.

<u>Ellipse (3.2.10)</u>: The equation $\frac{x^2}{a^2} + \frac{y^2}{b^2} = 1$ is the "standard form" of the equation of an ellipse centered at the origin. Its graph can be obtained from the graph of the unit circle by expanding the graph horizontally by a factor of a and vertically by a factor of b.

Of course, an ellipse centered at (h, k) can be represented by replacing "x" by "$x - h$" and "k" by "$y - k$," according to the ideas about location shifts in Section 2.3.

<u>Ellipse (3.2.11)</u>: The equation $\frac{(x - h)^2}{a^2} + \frac{(y - k)^2}{b^2} = 1$ is the "standard form" of the equation of an ellipse centered at (h, k). It extends a units to the left and right of (h, k) and b units above and below (h, k). To describe this family of curves the standard form uses four parameters -- two for the location of the center and two more for the size.

So ellipses are squashed or expanded circles. They have many important properties which will be studied later. "Completing the square" plays an important role in determining the center if the equation is not already in "standard form."

Example 12: Graph the ellipse "$x^2 + 6x + 4y^2 - 32y + 37 = 0$" by putting the equation in standard form.

First, complete the squares on both "y" and "x".

$$x^2 + 6x + 4y^2 - 32y + 37 = 0 \quad \text{iff}$$
$$x^2 + 6x + 4(y^2 - 8y) = -37 \quad \text{iff}$$
$$x^2 + 6x + 9 - 9 + 4(y^2 - 8y + 16 - 16) = -37 \quad \text{iff}$$
$$(x + 3)^2 + 4(y - 4)^2 = -37 + 9 + 64 = 36$$

Now, divide by 36 to obtain the "1" on the right required for standard form.

$$\frac{(x + 3)^2}{6^2} + \frac{(y - 4)^2}{3^2} = 1 .$$

The center is (-3, 4). It stretches 6 units to the right and to the left and 3 units up and 3 units down (Figure 15).

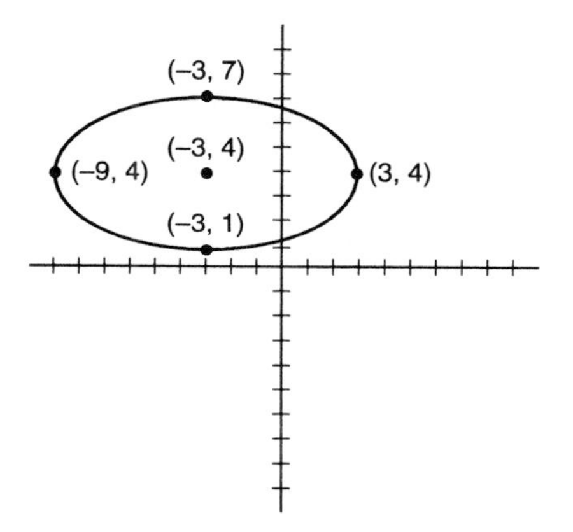

Figure 15: An ellipse centered at (-3, 4) with $a = 6$ and $b = 3$.

<u>**Applications**</u>: There are many applications of quadratics. Some arise in physics.

Example 13: Suppose a rock is thrown from an initial height s_0 (in feet) with an initial upward velocity of v_0 (in feet per second). Dropping the rock corresponds to $v_0 = 0$. Throwing it downward corresponds to a negative v_0. Suppose the projectile is subject to the usual force of gravity, but other forces such as wind resistance are neglected. Then the formula for the vertical position, $s(t)$, at time t (in seconds) is

$$s(t) = -16t^2 + v_0 t + s(0).$$

This is a quadratic "in t" instead of "in x." It opens downward, since the coefficient on t^2 is negative. The coefficient "-16" is the value determined by the force of gravity (and the units we use, which, here, are "feet per second squared").

If a rock is thrown upward from ground level with an initial velocity of 80 feet per second, how high will it go?

Let the ground level be 0 so $s(0) = 0$. The formula is $s(t) = -16t^2 + 80t$. The highest point on a parabola which opens downward is the y-value of its vertex. You can use Theorem 3.2.7 and read off the y-value in terms of "a", "b", and "c". Another way is to recall that the vertex occurs at "$t = -b/(2a)$" and substitute in to the formula to find y. $t_{max} = 80/32$.

$$s(t_{max}) = -16(80/32)^2 + 80(80/32) = 100.$$

When will the rock be 60 feet above the ground?

This merely asks you to solve the equation: $-16t^2 + 80t = 60$. There will be two solutions, one on the way up, and the other on the way down (Problem A34).

Example 14: In business, profit is revenue minus cost. Suppose the cost of producing and selling each widget is \$200. Suppose that market research tells the firm that, if they are priced at \$300, 4000 will be sold. However, if they are priced at \$400, only 2200 will be sold. Assume that the number sold at other prices would be on a straight line through those two points. What price will maximize profits?

This is a complex problem. Draw a picture of the price to number sold relationship (Figure 16). The idea behind the business problem is that if the price is higher, you make more profit per widget, but sell fewer widgets. If the price is lower, you sell more widgets, but make less profit per widget. Where is the optimum level?

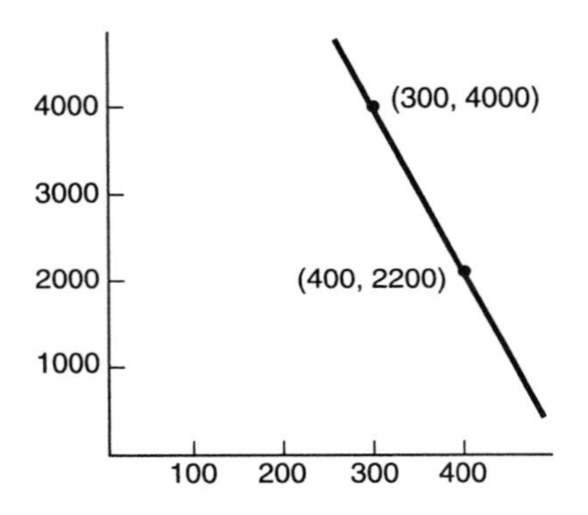

Figure 16: Price (x) to number sold (y) given the two points (300, 4000) and (400, 2200).

Let x be the price. If we know how many will be sold, s, we can calculate the profit from the revenue and cost. If every widget costs the same amount, the total cost is the number of widgets times the cost of each. Given s, the cost is easy: cost = $200s$ (\$200 each). Also, if every widget is sold at the same price, the revenue is the number sold times the price of each: revenue = xs (because they cost \$$x$ each). The trick is to find the relationship between x and s.

The number sold, s, is a function of x, and we are told to assume a straight line fit of the given points. So put a line through the two given points which give the price and corresponding number sold. They are: (300, 4000) and (400, 2200). The slope is

$$\frac{2200 - 4000}{400 - 300} = -\frac{1800}{100} = -18. .$$

The line is (using the point-slope form)

$$s = -18(x - 300) + 4000.$$
$$= -18x + 9400$$

Therefore,

$$\text{profit} = \text{revenue} - \text{cost}$$
$$= xs - 200s$$
$$= (x - 200)s \qquad [x - 200 \text{ is profit per widget}]$$
$$= (x - 200)(-18x + 9400)$$
$$= -18x^2 + 13000x - 1{,}880{,}000.$$

This is a quadratic. We know how to maximize quadratics.

The parabola opens downward, since the leading coefficient is negative. The maximum profit occurs at the vertex, where $x = -b/(2a) = 13{,}000/36 = 361$ (dollars).

Calculator Exercise 2: Graph the price-to-profit function.

The trick is to select an appropriate window. Graphing this in the "standard" window will show nothing.

Learn to analyze the function to find the domain and range. The word problem itself mentions prices of \$300 and \$400, so the window should include those x-values. What about y? A quick calculation shows that at \$300 the 4000 widgets sold make \$100 each for a profit of $100 \times 4000 = 400{,}000$. So y must go up to at least that. Figure 17 graphs it on a window [0, 500] by [0, 1000000].

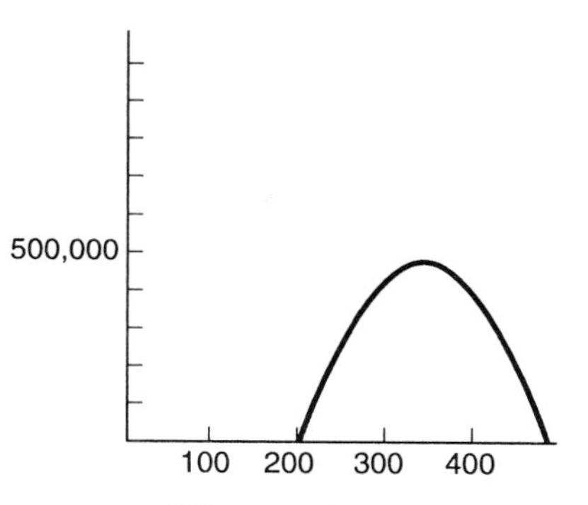

Figure 17:
$-18p^2 + 13{,}000p - 1{,}880{,}000$
[0, 500] by [0, 1000000].

<u>**Conclusion**</u>. Memorize the Quadratic Theorem. Of course, you must be able to solve equations by "plugging in" to the Quadratic Formula. However, you must strive for more understanding of quadratics than just that. Why are quadratics symmetric? How can their locations be shifted? How can their scales be changed? In what sense are they all similar? What is "completing the square" good for? What if the letters in the problem are non-standard?

Terms: quadratic, coefficient, leading coefficient, standard form, complete the square, vertex, open up, open down, Quadratic Theorem, axis of symmetry, discriminant, ellipse.

****

Exercises for Section 3.2, "Quadratics":

A1.* Zoom in on the graph of x^2 near the origin using a square scale. What does it look like there if you zoom in a lot?

A2.* Zoom in on the graph of $\sqrt{x}$ near the origin using a square scale. What does it look like there if you zoom in a lot?

A3.* What does it mean to "complete the square"?

A4. Simplify $\frac{f(x + h) - f(x)}{h}$ given $f(x) = x^2$.

A5.* If you have the Quadratic Formula memorized, how can you use it to remember the axis of symmetry of the graph of a quadratic?

A6. a) Give the "discriminant" of a quadratic.
b) What does it discriminate? c) Where is it in the Quadratic Formula?

^ ^ ^ ^ Which x-value yields the minimum value of the expression?
A7. $x^2 - 9x + 3$. A8. $x^2 + 7x + 12$.
A9. $5x + 3 + 2x^2$. A10. $3x + 14 + 4x^2$.

^ ^ ^ ^ Which x-value yields the maximum value of the expression?
A11. $50 + 500x - x^2$. A12. $12 - 10x - x^2$.

A13. Find a window [0, 4] by [0, d], such that the graph of x^2 goes through the upper right corner of the window.

A14. Find a window [-10, 10] by [0, d] such that the graph of x^2 exits the right side of the window half way up.

^ ^ ^ ^ Give the equation of the graph which has the same shape as the graph of x^2, but is
A15. a) 2 units to the left. b) 4 units up.
A16. a) 3 units to the right. b) 7 units down.
A17. a) twice as tall b) twice as wide.
A18. a) half as tall. b) half as wide.

^ ^ ^ ^ Use the Quadratic Formula and your calculator to solve. As usual, give at least three significant digits.
A19. $x^2 - 3.2x + 1 = 0$ [2.8,...]. A20. $2.3x - 1.6x^2 - 4 = 0$ [1.7,...].
A21. $4 + 3.5x^2 = 12x$ [3.1,...]. A22. $-1.2x^2 + 10x - 2.3 = 0$ [8.1,...].

^ ^ ^ ^ Complete the square of the expression:
A23. $x^2 - 2x$ A24. $x^2 + 7x + 2$
A25. $x^2 + 3x + 1$ A26. $x^2 - 10x + 31$.

^ ^ ^ ^ Use the Quadratic Formula to solve the following equations. Write your answer in "$a + bi$" form:
A27. $x^2 = -16$. A28. $x^2 + x + 1 = 0$.
A29. $x^2 + 5x + 5 = 0$ A30. $3x^2 + 1 = 0$.

A31. Indicate on a sketch of $x^2 + 5x$ the solutions to $x^2 + 5x = 3$. Label the sketch with the two terms of the Quadratic Formula. Solve the equation.

A32. Indicate on a sketch of $2x^2 + 7x$ the solutions to $2x^2 + 7x = 8$. Label the sketch with the two terms of the Quadratic Formula. Solve the equation.

A33. Solve the equation in Example 5: $x^2 - 5x = 3$ [5.5,...]

A34. Solve $-16t^2 + 80t = 60$.

A35. Use the Quadratic Formula to give the solutions to "$dx^2 + ex + f = 0$."

^ ^ ^ ^ ^ ^ ^ ^ ^ ^

B1.* If $a > 0$, illustrate the three possibilities for the number of real-valued solutions to a quadratic equation (Give three graphs and the corresponding number of solutions).

B2.* If $a < 0$, illustrate the three possibilities for the number of real-valued solutions to a quadratic equation (Give three graphs and the corresponding number of solutions).

B3.* Suppose "$ax^2 + bx + c = 0$" has two solutions and $a > 0$. Draw a sketch of the relevant graph and label it to illustrate the two major terms of the Quadratic Formula.

B4.* a) How can you tell which quadratic expressions have minima and which have maxima? b) Where is the x-value at which the extreme value is attained?

B5.* How can you tell from the graph of a quadratic expression $f(x)$ how many complex-valued solutions the equation "$f(x) = 0$" has?

B6.* The key ideas to deriving the Quadratic Formula were to first divide through by a to make the coefficient on x^2 be 1 and then to ________________ and then to use the ____________________ method of solving equations.

B7.* If $a > 0$, the minimum value of $ax^2 + bx + c$ occurs when $x =$ ________.

B8.* How are ellipses related to circles?

^ ^ ^ ^ Give the equation of the graph which has the same shape as the graph of x^2, but is

B9. a) two units down and three units to the right.
b) 4 units to the left and 7 units up.

B10. a) 5 units down and 3 units to the left.
b) 6 units to the right and 9 units up.

B11. Find x such that the distance from (1, 2) to (x, 5) is 6 units.

B12. Find y such that the distance from (-5, -3) to (2, y) is 10 units.

^ ^ ^ ^ Give the center and radius of the circle:

B13. $x^2 + 6x + y^2 - 10y = 0$.

B14. $x^2 + y^2 + 4y = 6x - 12$.

B15. $4x^2 + 6x + 4y^2 - 8y = 12$.

B16. $x^2 - 5x + y^2 = 7$.

^ ^ ^ ^ Complete the square:

B17. $2x^2 + 5x$

B18. $4x - 2x^2$

B19. $3x^2 + x + 1$

B20. $10x^2 + 12x + 7$

^ ^ ^ ^ Complete the squares to find the standard form of the ellipse with the given equation.

B21. $x^2 - 2x + 3y^2 - 12y = 50.$

B22. $4x^2 + 8x + y^2 - 6y = 90.$

B23. Suppose the cost of producing and selling each widget is $100. Suppose that market research tells your firm that, if they are priced at $200, 50,000 will be sold. However, if they are priced at $175, only 80,000 will be sold. Assume that the number sold at other prices would be on a straight line through those two points. What price will maximize profits?

B24. Alvin flies his fighter plane at low altitude from his base at (0, 0) into the first quadrant. His flight path has the equation $y = x(8 - x)$. How close does he come to an enemy with a Stinger missile at location (5, 17)? [Distances are in miles.]

B25. Find the point on the line $y = 3x - 7$ closest to the origin. [$x = ...$ and $y = -.70$]

B26. Find the point on the curve $y = x^2$ closest to the point (2, 1). [$x = ...$ and $y = 1.36$]

^ ^ ^ ^ In the picture the graph of x^2 is shifted. Identify the expression graphed.

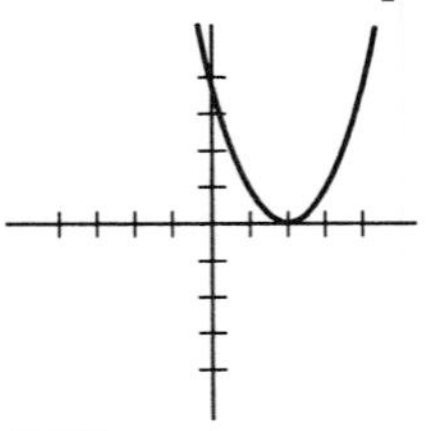

B27.

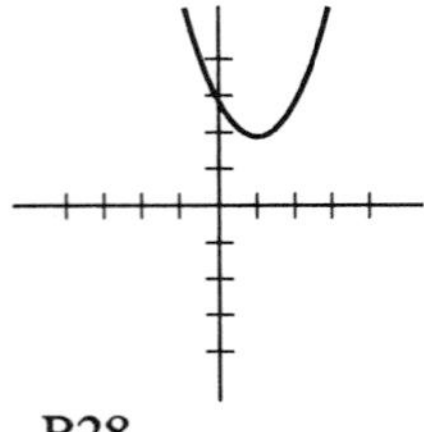

B28.

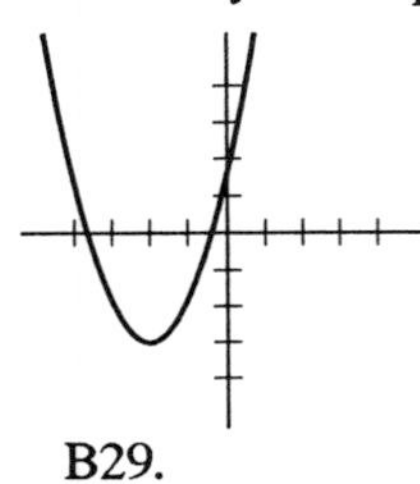

B29.

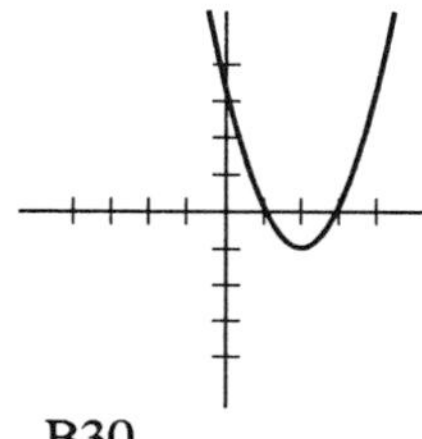

B30.

^ ^ ^ ^ Use the Quadratic Theorem to solve

B31. $16^2 = 20^2 + b^2 - 40b \cos 39°$, for b. [25,...]

B32. $3^2 = a^2 + 4^2 - 7a \cos 20°$, for a. [5.2,...]

B33. $bx^2 + 2ax + 3c = 0$, for x.

B34. $y^2 + 2by + a = 0$, for y.

B35. $cx^2 + bx + 1 = 0$, for x.

B36. Solve "$(\sin x)^2 + 3\sin x = 2$" for x, with the help of Theorem 2.4.5 on the inverse sine function. [Give two solutions.]

^ ^ ^ ^ Use the Quadratic Theorem to help you sketch the graph.

B37. $x^2 + xy + 2y^2 = 20.$

B38. $x^2 + 3xy + y^2 = 17.$

B39. Suppose a punt in football has a "hang time" of 3.6 seconds. About how high did it go?

B40. Suppose you drop a rock off a high bridge and count off the seconds until it hits the water below. Give a formula for the height of the bridge in terms of the elapsed time.

^ ^ ^ ^ Consider the graph of $x^2 + y^2 = 1$ on [-2, 2] by [-2, 2].

B41. How could you choose a new y-interval to make it look half as high?

B42. How could you choose a new y-interval to make it look twice as high?

B43. How could you choose a new x-interval to make it look twice as wide?

B44. How could you choose a new x-interval to make it look one quarter as wide?

B45. The Worldwide Widget Company calculates total profit as profit-per-widget-sold times the number-of-widgets-sold. If the profit-per-widget is $100, they sell 2000 widgets. If the profit-per-widget is $150, they sell only 1000 widgets. Assume a straight line fit for the profit-per-widget to number-of-widgets-sold relationship. a) Find a formula for total profit in terms of the profit-per-widget. b) What profit-per-widget should the company use to maximize total profit?

B46. The area of a circle is πr^2. a) Determine the area of the ellipse $(x/2)^2 + y^2 = 1$ by considering scale changes. b) Determine the area of the ellipse $(x/a)^2 + (y/b)^2 = 1$ by considering scale changes.

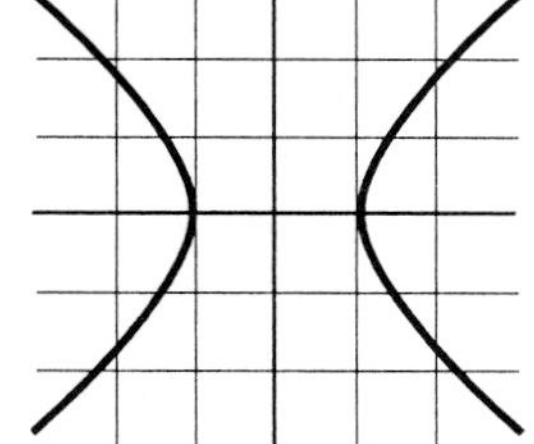

B47. Here is the graph of $x^2 - y^2 = 1$ (a hyperbola). Use it and a scale change to find the graph of
a) $(x/2)^2 - y^2 = 1$.
b) $x^2 - (y/2)^2 = 1$.

B48. Find the sum and product of the solutions to: $ax^2 + bx + c = 0$. [The result should simplify nicely.]

B49.* For the graph of x^2, a vertical scale change is much like a horizontal scale change. Why?

B50. Prove that the location of the vertex is as stated in Theorem 3.2.7.

B51. Prove the Quadratic Theorem (without looking).

Section 3.3. Graphical Factoring

Students spend a lot of time learning to factor quadratics. You should know *why* we want to factor quadratics. The primary motivation for factoring expressions is to be able to solve equations using the Zero Product Rule.

The Zero Product Rule takes factors of an expression and yields solutions to an equation. The "Factor Theorem" (3.3.5, below) reverses this. It takes solutions to an equation and yields factors of an expression. It facilitates factoring by allowing us to, instead, solve an equation. Then the Quadratic Formula and graphical techniques of solving equations make factoring much easier than using the old "guess and check" method. After a review of basic ideas about factoring, this section emphasizes these advanced factoring methods.

To <u>factor</u> an expression means to write it as a product. To factor a quadratic expression means to write it as an equivalent product of linear terms, usually in one of the forms:

$(x - b)(x - c)$	[if the leading coefficient is 1] or
$k(x - b)(x - c)$ or $(kx - d)(x - c)$	[the leading coefficient is k] or
$(ax - b)(cx - d)$	[the leading coefficient is ac].

The Zero Product Rule makes solving some quadratic equations extremely simple -- if the expression is already factored.

Example 1: Solve $(x - 2)(x + 3) = 0$.

This is trivial because it is already factored. The solution is: $x = 2$ or $x = -3$. Figure 1 graphs the expression. To solve the equation is to determine the x-values where the value of the expression is zero.

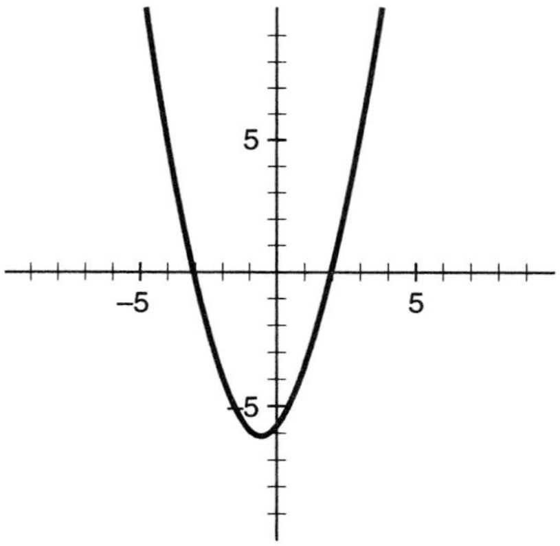

Figure 1: $(x - 2)(x + 3)$
[-10, 10] by [-10, 10].

Factored form can be useful; however, there are many times when we do *not* want a quadratic expression factored.

Example 2: Solve $(x - 5)(3x + 4) = 5$.

In this equation factored form is useless. The Zero Product Rule only works for *zero* products, and this product is 5. We will need to multiply it out and use a different method (Problem A21).

Of course, any quadratic equation can be solved with the Quadratic Theorem, so we never need to use the Zero Product Rule to solve quadratic equations. Therefore, there is much less reason to factor expressions after you understand the Quadratic Theorem. Nevertheless, there are some other uses of factoring (see Section 4.3 on higher-degree polynomial equations and Section 4.5 on rational functions) and you are supposed to know how to factor quadratics. Because factoring is the reverse of "multiplying out," it is appropriate to multiply out a few products first.

The Distributive Property. In previous sections we have had reasons to "consolidate like terms," "factor," and "multiply out" expressions. All of these processes rely on the Distributive Property (of multiplication over addition).

The Distributive Property 3.3.1: $a(b + c) = ab + ac$.

Example 3: $3(102) = 3(100 + 2) = 3(100) + 3(2) = 306$.

You use the Distributive Property every time you multiply a two-digit number times a one-digit number by hand.

Example 4:

$$\begin{array}{r} 24 \\ \underline{\times 3} \\ 72 \end{array} \qquad \begin{array}{r} 20 + 4 \\ \underline{\times 3} \\ 12 \\ \underline{60} \\ 72 \end{array} \qquad \begin{array}{r} b + c \\ \underline{a} \\ ac \\ \underline{ab} \\ ab + ac \end{array}$$

You may have learned to write your work all on one line (as in the leftmost version) by "carrying." 4 times 3 equals 12, put down the 2 and carry the 1. Then 3 times 2 equals 6 plus the carried 1 is 7. Put down the 7.

Really, there is no "3 times 2" in this problem. There is "3 times 20." The algorithm (method) shortens the problem by dealing with the 20 as 2 tens by working with the tens digit. Also, there is no "1" to "carry"; it is really 10.

The Distributive Property also works for subtraction, which can be regarded as addition of the negative.

Example 5: $3(98) = 3(100 - 2) = 3(100) - 3(2) = 300 - 6 = 294$.

This subtraction version comes in handy when you are shopping. When you want to know how much 2 compact discs priced at \$14.99 each cost, use the Distributive Property: $2(15) - 2(0.01) = 30 - .02 = 29.98$.

The Extended Distributive Property deals with products of two sums.

Theorem 3.3.2 (Extended Distributive Property): For all a, b, c, and d,

$$(a + b)(c + d) = ac + ad + bc + bd.$$

This theorem follows from using the Distributive Property three times, once with "$c + d$" as a single factor which multiplies both "a" and "b", and then with "a" and "b" as factors which multiply both "c" and "d" in the sum "$c + d$."

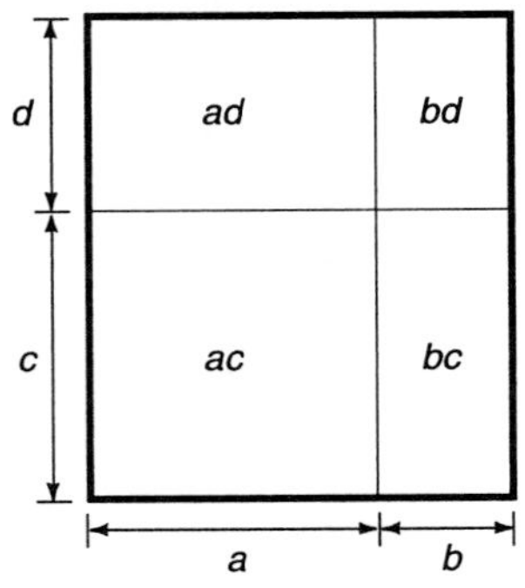

Figure 2: The extended Distributive Property.

Proof:

$(a + b)(c + d) = a(c + d) + b(c + d)$

[using the Distributive Property with "$c + d$" as the factor and "$a + b$" as the sum]

$= ac + ad + bc + bd$

Many people remember this using the acronym FOIL: First, Outer, Inner, Last.

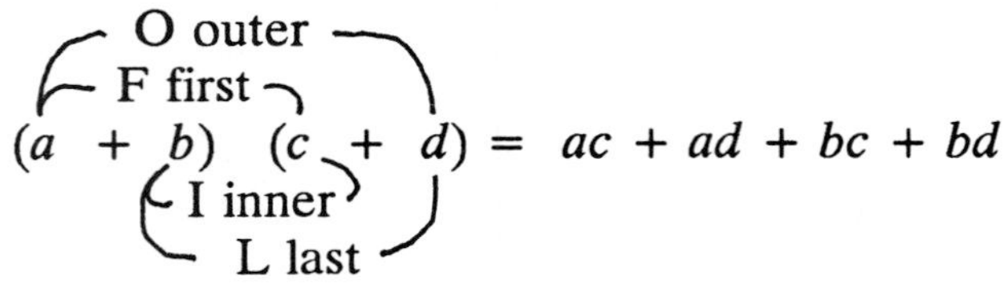

O outer
F first
$(a + b)\ (c + d) = ac + ad + bc + bd$
I inner
L last

Figure 3: FOIL

This pattern is exactly the same pattern we use to multiply two-digit numbers by each other:

Example 6: Compute 27×34 by hand.

abbreviated	expanded	more expanded		with variables
34	34	$30 + 4$		$c + d$
27	27	$20 + 7$		$a + b$
238	28	7×4		bd
68	210	7×30		bc
918	80	20×4		ad
	600	20×30		ac
	918	918	total	$ac + ad + bc + bd$

Figure 4: Multiplication of two-digit numbers.

The method you learned in school is used on the left of Figure 4. The "expanded" version in the second column shows the real steps without so much abbreviation. The role of the Distributive Property is emphasized in the "more expanded" version in the third column. The version with variables is on the right. But the idea is identical, whether we use numbers or variables.

By definition, a corollary to a theorem is another result so closely related to the first result that it needs little or no further proof.

Corollary 3.3.3. Common Products: For all a, b, and x,

A) $x^2 + (a + b)x + ab = (x + a)(x + b)$.
B) $x^2 + 2ax + a^2 = (x + a)^2$.
C) $x^2 - 2ax + a^2 = (x - a)^2$.
D) $x^2 - a^2 = (x + a)(x - a)$.
E) $x^2 + ax = x(x + a)$.

This theorem is used left-to-right for factoring. Reading right-to-left, each product is "multiplied out" or "expanded."

Figure 5 illustrates Part A. The term "$(a + b)x$" is called the "cross product" term. Perhaps this term comes from the "cross" in the following image of multiplication with variables:

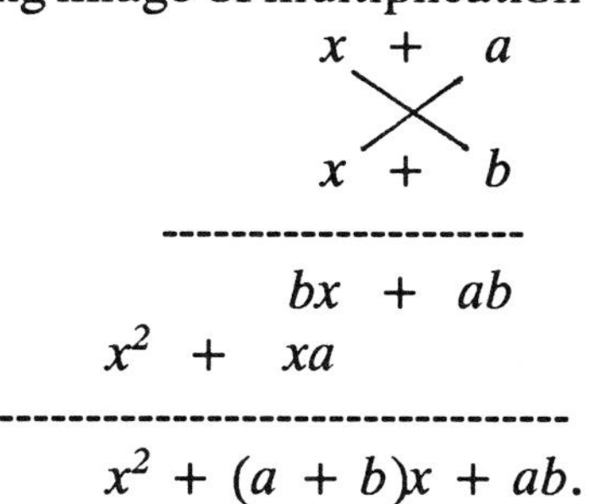

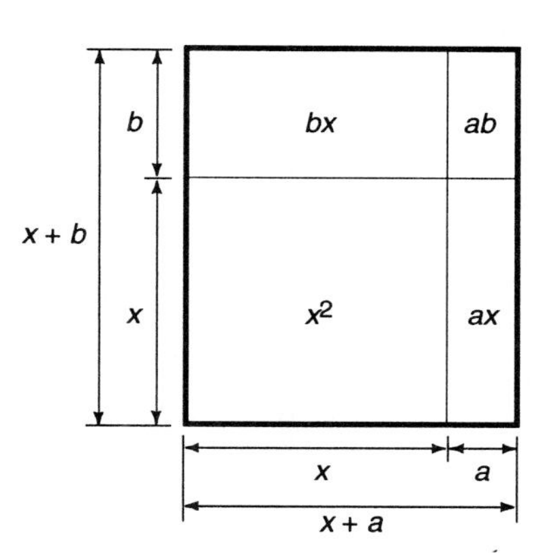

Figure 5: $(x + a)(x + b) = x^2 + ax + bx + ab$.

Each term in each line multiplies each term in the other line. The "bx" and "xa" terms in the product consolidate into the cross product term.

In 3.3.3 part B note that

the square of a sum is not the sum of the squares.

Order matters. In part C note that the square of a difference is not the difference of the squares. Order Matters. The difference of the squares is factored in part D. The expression "$x^2 + a^2$" does not appear. It does not generally factor using real numbers.

Usually students do not immediately learn to factor general quadratics, but only special cases that factor *using integers*.

Example 7: Factor "$x^2 - 2x - 15$" using integers.

If it factors into $(x + a)(x + b)$, then, according to Corollary 3.3.3, Part A, by matching corresponding coefficients, $ab = -15$ and $a + b = -2$. If a is to be integer-valued and $ab = -15$, then a must be 1, 3, 5 or 15, or the negative of one of these. The problem becomes one of trial-and-error. Experience helps, but the method is simply to run through all the limited number of integer-valued possibilities until you find the right one.

By inspecting the possible integer-valued a and b such that $ab = -15$, we see that $a = -5$ and $b = 3$ will work.

$$x^2 - 2x - 15 = (x - 5)(x + 3).$$

The factoring ideas in Example 7 can be summarized in a theorem.

<u>Corollary 3.3.4 on Factoring</u>: $x^2 + cx + d$ factors into $(x + a)(x + b)$ iff

$$ab = d \text{ and } a + b = c.$$

This follows immediately from Corollary 3.3.3A by matching the coefficients on "x" and the constants.

$$\begin{aligned} & x^2 + \quad cx + \quad d \\ = (x + a)(x + b) = {} & x^2 + (a + b)x + ab. \end{aligned}$$

So $a + b = c$ and $ab = d$.

<u>**Advanced Factoring Methods**</u>. Usually we are not interested in factoring for its own sake (except on school homework). Usually we want to factor expressions so we can use the Zero Product Rule to solve related equations. The upcoming Factor Theorem tells us we can also reverse the process; we can solve an equation to factor an expression.

For this factoring method you must keep the distinction between expressions and equations in mind.

Example 7, revisited: "$x^2 - 2x - 15$" is a quadratic *expression*. Call it "$P(x)$".
"$x^2 - 2x - 15 = 0$" is a quadratic *equation*, "$P(x) = 0$."

Suppose we want to solve the equation "$P(x) = 0$." One way is to factor the expression to obtain the equivalent equation:

$$\text{"}(x + 3)(x - 5) = 0.\text{"}$$

Here the solutions "-3" and "5" are evident. Because "$x + 3$" is a factor, "-3" is a solution. Because "$x - 5$" is a factor, "5" is a solution.

The Factor Theorem allows us to reverse both the problem and the steps. It says that, because "-3" is a solution, "$x + 3$" is a factor. Because "5" is a solution, "$x - 5$" is a factor. So the Factor Theorem says that we can factor a polynomial expression, "$P(x)$", by, instead, solving an equation, "$P(x) = 0$," whenever we can find some other way besides factoring to solve the equation. The Quadratic Theorem and graphing provide other ways.

Example 7, continued: Factor "$x^2 - 2x - 15$."

Pretend we don't see its factors. The upcoming Factor Theorem says solving "$x^2 - 2x - 15 = 0$" helps.

By the Quadratic Formula, $x = 5$ or $x = -3$. The factors are therefore "$x - 5$" and "$x - -3$."

$$x^2 - 2x - 15 = (x - 5)(x + 3).$$

Part of the point of this method is that it does not use guess-and-check. **A quadratic expression can always be factored without any guessing** by using the Factor Theorem and the Quadratic Theorem.

Example 7, graphically: Factor "$x^2 - 2x - 15$."

We may graph the expression to solve the equation "$x^2 - 2x - 15 = 0$" (Figure 6). The solutions appear to be 5 and -3. Evaluating the expression at those values confirms they are solutions. Therefore, by the Factor Theorem, "$x - 5$" and "$x + 3$" are factors.
Again, $x^2 - 2x - 15 = (x - 5)(x + 3)$.

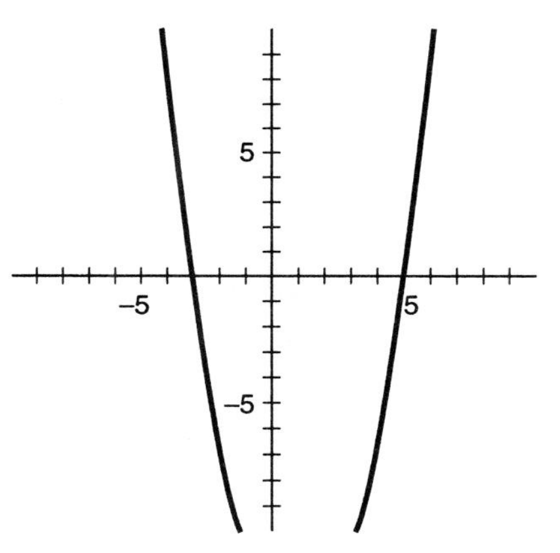

Figure 6: $x^2 - 2x - 15$. [-10, 10] by [-10, 10].

Here is the formal statement of the Factor Theorem for quadratics.

<u>The Factor Theorem (3.3.5)</u>: "$x - c$" is a factor of the quadratic expression $P(x)$ if and only if c is a solution to the quadratic equation "$P(x) = 0$" (that is, iff $P(c) = 0$).

Therefore, b and c are the solutions to the quadratic equation "$P(x) = 0$," if and only if $P(x) = k(x - b)(x - c)$, for some constant $k \neq 0$. (Note that the Factor Theorem does *not* determine the leading coefficient, k, because the zeros of a polynomial do not determine its vertical scale.)

Every solution to the polynomial *equation* "$P(x) = 0$"
yields a factor of the *expression* "$P(x)$".

From the Zero Product Rule we already knew the reverse -- that factors of "$P(x)$" yield solutions to "$P(x) = 0$." The solutions to the equation "$P(x) = 0$" are also known as the <u>zeros</u> of the expression $P(x)$. Sometimes zeros are also called "roots".

Example 8: Factor $x^2 - 2x - 24$ using integers.

One way is to use the method of Corollary 3.3.4 and try the integer-valued factors of 24. Another way is to solve the equation "$x^2 - 2x - 24 = 0$" (which is probably the reason you would want to factor the expression in the first place). According to the Factor Theorem, if you solve the equation you have found the factors. You may use the Quadratic Theorem to find the solutions: $x = 6$ and $x = -4$. These give the variable factors, but not k, the constant factor:

$$x^2 - 2x - 24 = k(x - 6)(x + 4).$$

Since the leading coefficient is 1, $k = 1$. Therefore,

$$x^2 - 2x - 24 = (x - 6)(x + 4).$$

Example 8, graphically: Another way to factor "$x^2 - 2x - 24$" is to find its zeros using its graph (Figure 7). The zeros appear to be 6 and -4. Evaluating the expression at these values confirms they are solutions to the equation "$x^2 - 2x - 24 = 0$," so the Factor Theorem then gives the desired linear factors, "$x - 6$" and "$x + 4$."

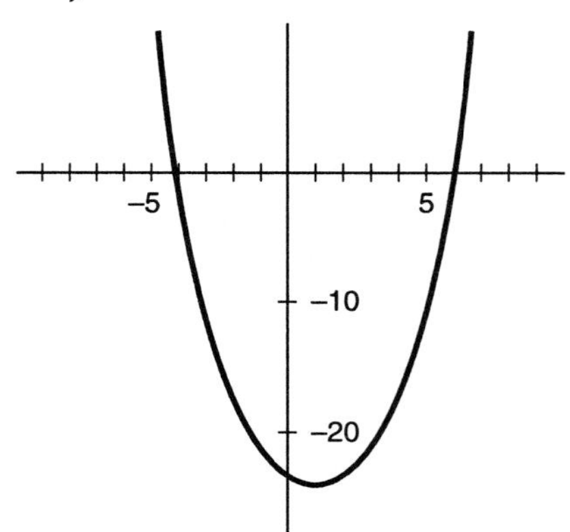

Figure 7: $x^2 - 2x - 24$. [-10, 10] by [-30, 10].

If a quadratic expression has leading coefficient 1 and it factors in integers, a graph and the Factor Theorem make finding the factors easy.

The Factor Theorem does not, by itself, determine the constant factor.

Example 9: Find the quadratic $P(x)$ given $P(4) = 0$, $P(-2) = 0$, and $P(1) = 7$ (Figure 8).

The zeros of P are given as 4 and -2. They tell us that $P(x)$ factors into

$$P(x) = k(x - 4)(x + 2),$$

where k remains to be determined. We can use the one other value, $P(1) = 7$, to determine k. From above, substituting 1 for x,

$$P(1) = k(1 - 4)(1 + 2) = -9k.$$

But $P(1) = 7$ was given.

Equating these two,

$$-9k = 7.$$

Therefore $k = -7/9$. $P(x) = (-7/9)(x - 4)(x + 2)$.

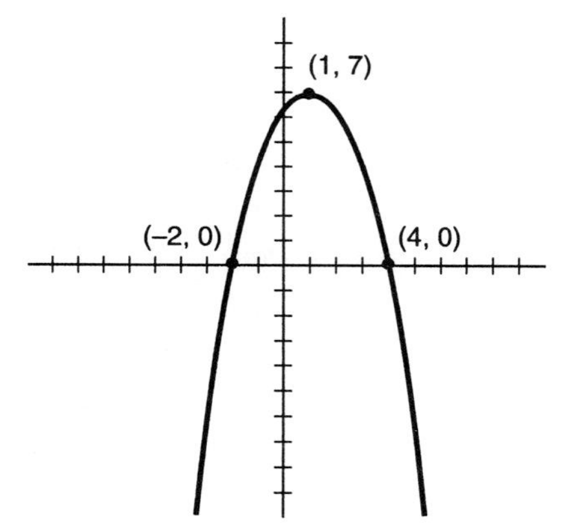

Figure 8: A quadratic with $P(4) = 0$, $P(-2) = 0$, and $P(1) = 7$. [-10, 10] by [-10, 10].

A quadratic is determined by three parameters. In "standard" form they are the "a", "b", and "c" in "$ax^2 + bx + c$." In factored form they are the "k", "b", and "c" of "$k(x - b)(x - c)$." (The uses of the letters differ between the two forms). *Three* parameters can be determined by *three* facts. The *two* solutions to the equation in the Factor Theorem are not enough by themselves. The leading coefficient served as the third fact in Example 8, and "$P(1) = 7$" served as the third fact in Example 9.

<u>Factoring using Real or Complex Numbers</u>. In previous math classes, when you were asked to factor a quadratic, chances were good it would factor *in integers*. If a quadratic does not factor *in integers*, you wonder if you want to bother to factor it at all. Usually you don't. This subsection considers those rare cases when you do.

It is interesting to note that there are many quadratics which do not factor using only integers, but which do factor if you use real numbers. The ones that do not factor using real numbers do factor when complex numbers are used. So, when the instructions say, "Factor such-and-such a polynomial," you must first decide what the ground rules are. Probably the rules are to factor it using factors with *integer* coefficients. However, it is possible (and easy) to factor any quadratic into two linear factors using the Quadratic Theorem, even if the result does not have integer coefficients.

Example 10: Factor $x^2 - 2x - 14$.

What are the ground rules? Are we to use integers? Or real numbers?

If it factors as in Corollary 3.3.4 using *integer-valued* a and b, then $ab = -14$ and a must be 1, 2, 7, or 14, or the negative of one of these. If you try every possibility, you discover that none work. This does not mean that it does not factor. It just means it does not factor *nicely* (that is, using integers). It does factor using real numbers, but you will not guess the factors easily. Use the Factor Theorem and the Quadratic Theorem to find the factors.

The solutions to $x^2 - 2x - 14 = 0$ yield the factors. Using the Quadratic Theorem we discover one solution is $1 + \sqrt{15}$ and the other is $1 - \sqrt{15}$. Treating these solutions as a and b, it factors into $(x - a)(x - b)$.

$$x^2 - 2x - 14 = [x - (1 + \sqrt{15})][(x - (1 - \sqrt{15})].$$

Another approach to finding the factors would be to solve the equations in Corollary 3.3.4 for a and b using two equations and two unknowns (and not expecting the answer to be integers): $ab = -14$ and $a + b = -2$ (Problem A35).

According to the Factor Theorem, quadratics can *always* be factored. That is not to say they always factor using integers. Real-valued solutions to the equation yield real-valued factors, and solutions that use complex numbers yield complex-valued factors.

Example 11: Factor $x^2 - 2x + 5$ over the complex numbers.

By the Quadratic Theorem, the solutions to $x^2 - 2x + 5 = 0$ are $1 \pm 2i$. Therefore, because the leading coefficient is 1, by the Factor Theorem,

$$x^2 - 2x + 5 = [x - (1 + 2i)][x - (1 - 2i)].$$

Example 12: Does "$x^2 + x - 1$" factor using integers?

This can be answered by trial-and-error using Corollary 3.3.4. Another approach is to graph it (Figure 9). If it factors into $(x - b)(x - c)$ using integers, b and c would have to be integer-valued zeros. Since the graph does not intersect the x-axis at integer values, it does not factor using integers. It will, however, factor using real numbers (Problem A17), as we can see from the fact the graph crosses the x-axis twice.

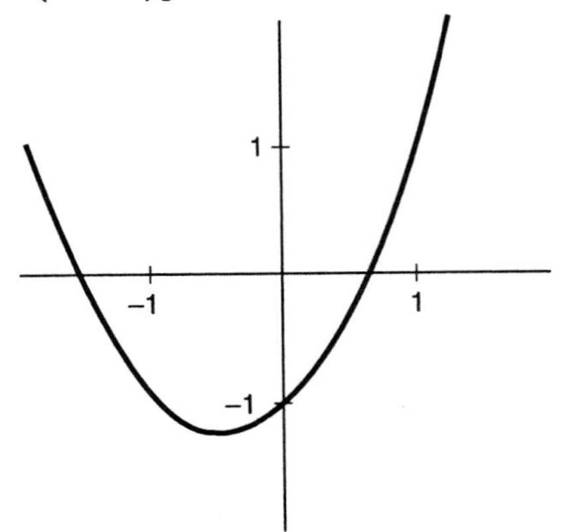

Figure 9: $x^2 + x - 1$. [-2, 2] by [-2, 2].

<u>Conclusion</u>. The Zero Product Rule gives solutions from factors. The Factor Theorem gives factors from solutions.

The primary motivation for factoring is the Zero Product Rule for solving equations. Therefore, the incentive to factor quadratics is much reduced after the Quadratic Theorem has been learned. However, if factoring is required (and occasionally it is), the Quadratic Theorem or graphing can be used with the Factor Theorem to determine factors.

Terms: Distributive Property, factor (verb and noun), extended Distributive Property, cross product, Factor Theorem.

****

Exercises for Section 3.3, "Factoring":

A1.* Why do we want to be able to factor quadratic expressions?

A2.* True or False?
a) Any quadratic equation can be solved using the Quadratic Theorem.
b) Any quadratic expression can be factored using integers.
c) Factored form is always more useful than expanded (multiplied out) form.
d) Expanded form is always more useful than factored form.

A3.* Let $P(x)$ be a quadratic. The zeros of P are the solutions of __________________.

A4.* Let $P(x)$ be a quadratic. If c is a zero of P, then ________ is a factor of $P(x)$.

A5.* a) Define "product" (the noun). b) Define "factor" (the verb).

A6.* Draw a figure to illustrate the Extended Distributive Property.

A7.* a) What does the acronym "FOIL" mean?
b) State, using variables, the theorem FOIL is intended to help you remember.

A8.* True or false?
a) $a^2 + b^2 = (a + b)^2$ b) $a^2 - b^2 = (a - b)^2$ c) $(a + b)(a - b) = a^2 - b^2$.

A9.* a) If a quadratic factors into "$(x - a)(x - b)$" using integers, what can we say about the graph of the quadratic? b) Which theorem tells you that?

A10.* Restate the Distributive Property using some other letters.

A11. Can "$x^2 + 4x - 13 = 0$" be solved
a) by completing the square and using the inverse-reverse method?
b) by the Quadratic Theorem?
c) by factoring using integers?
d) by evaluate-and-compare?

A12. Can "$x^2 - 5x - 14 = 0$" be solved
a) by completing the square and using the inverse-reverse method?
b) by the Quadratic Theorem?
c) by factoring using integers?
d) by evaluate-and-compare?

^ ^ ^ ^ A13-A16 factor using integers. Factor the expression by using a graph (and the Factor Theorem) to discover the factors.
A13. $x^2 - x - 12$ A14. $x^2 - 10x + 16$.
A15. $2x^2 + 10x - 48$. A16. $3x^2 + 6x - 144$.

^ ^ ^ ^ A17-A20 do not factor using integers. Factor the expression by using the Quadratic Theorem (and the Factor Theorem) to discover the factors.
A17. $x^2 + x - 1$. A18. $x^2 - 3x + 1$.
A19. $2x^2 + x - 4$. A20. $5x^2 + x - 2$.

A21. Solve $(x - 5)(3x + 4) = 5$. A22. Solve $(x - 1)(x + 2) = 19$.

^ ^ ^ ^ Factor the following expressions using traditional methods such as Corollaries 3.3.3 and 3.3.4:
A23. $x^2 - 4$. A24. $x^2 + 2x + 1$. A25. $x^2 + 6x + 8$.

^ ^ ^ ^ Multiply out ("expand") the following products:
A26. $(x - 3)(x - 2)$. A27. $(x - 3)^2$. A28. $(x - \sqrt{5})^2$.

A29. If "$3x - 5$" is a factor of a quadratic, $P(x)$, give a solution to $P(x) = 0$.

A30. If "$2x + 7$" is a factor of a quadratic, $P(x)$, give a solution to $P(x) = 0$.

^ ^ ^ ^ Consolidate like terms:
A31. $3x + x \tan 25 + 3\sqrt{x} + (\sqrt{x})\ln 5$. A32. $\ln(x) + 3x + 2\ln(x) - x \sin 36°$.
A33. $x \ln 5 + 5\ln(x) + 7x - \ln(x)$.

A34. Restate Corollary 3.3.4 using some other letters.

A35. Solve $ab = -14$ and $a + b = -2$ (for Example 10).

^ ^ ^ ^ ^ ^ ^ ^

B1.* Draw a figure to illustrate $(x + a)(x + b) = x^2 + (a + b)x + ab$. Label all the distances and areas mentioned in the identity.

B2.* Draw a figure to illustrate $(x + a)^2 = x^2 + 2ax + a^2$. Label all the distances and areas mentioned in the identity.

B3.* How can its graph help you find the factors of a quadratic expression?

B4.* The Factor Theorem and the Zero Product Rule are related. How?

B5.* a) Which of the "four ways to solve an equation" (Section 1.6) could be used to solve a quadratic which has two real-valued solutions?
b) (Medium-length essay) Since there is more than one possible way, how do you choose between them?

B6.* a) Explain how the Quadratic Theorem can be used to factor quadratics.
b) Will it always work?

B7.* We do not always prefer quadratic expressions to be factored rather than multiplied out. Give a reason why we might rather have a quadratic multiplied out. Give an example with a factored quadratic that we would rather have multiplied out.

B8.* If a quadratic expression "$x^2 + kx + p$" factors into "$(x + b)(x + c)$," what is the relationship between the letters in the two expressions?

B9.* Draw a figure to illustrate $(x - a)^2 = x^2 - 2ax + a^2$. Explain where the "$+a^2$" comes from.

B10.* You can tell if a quadratic does not factor using real numbers just by looking at its graph. How?

B11.* Explain how to do problems like "$3.98 times 4" and "$7.95 times 2" in your head.

B12.* Multiply out "$(a + b + c)^2$."

^ ^ ^ ^ Give a linear factor of $P(x)$ *using integers*.

B13. If $x = 3/2$ solves $P(x) = 0$.

B14. If $x = -5/7$ solves $P(x) = 0$.

^ ^ ^ ^ With the aid of a graph, factor these expressions using integers.

B15. $4x^2 + 4x - 3$.

B16. $4x^2 - 5x - 6$.

B17. Give the correspondence of letters between Theorem 3.3.2 and 3.3.3A.

B18. Give the correspondence of letters between Theorem 3.3.2 and 3.3.3B.

^ ^ ^ ^ Draw a sketch to illustrate the product and how it relates to the extended Distributive Property in Theorem 3.3.2 and Figure 2.

B19. 23×37.

B20. 12×26.

^ ^ ^ ^ B21-B23. Use the figure and the labeled points to find the expression graphed (factored form is recommended).

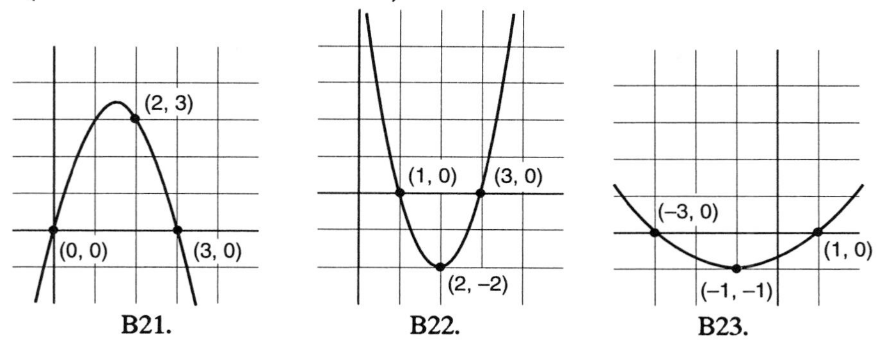

B21. B22. B23.

B24. Find a quadratic expression $P(x)$ with zeros at 2 and 5 such that $P(1) = 12$.

B25. Find a quadratic expression $P(x)$ with zeros at -3 and 4 such that $P(0) = 2$.

^ ^ ^ ^ Factor

B26. $3x^2 - 5x - 1$ using real numbers.

B27. $2x^2 + x - 3$ using real numbers.

B28. $x^2 + x + 4$ using the complex numbers.

B29. $3x^2 + 2x + 2$ using complex numbers.

B30. $3x^2 - 5x + 1$.

B31. $4x^2 + 2x - 7$.

B32. Factor as much as possible using only real numbers: $x^4 - 16$.

^ ^ ^ ^ Solve algebraically:

B33. $(x - 2)^2(x + 1) + (x - 2)(x + 1)^2 = 0$.

B34. $(x^2 - 4)(x + 3) + (x^2 + x - 2)(x - 4) = 0$.

B35. $3(x - 2)^5(x + 7)^2 - 5(x - 2)^4(x + 7)^3 = 0$

B36. $x^2(2x - 5)^3 - x^3(2x - 5)^2 = 0$.

B37. The "Remainder Theorem": If a polynomial $P(x)$ is divided by $x - c$, then the remainder is $P(c)$. Let $P(x) = x^2 + 3x + 5$. Verify this for $c = 0$ and 2.

B38. If $P(x) = cx^2 + dx + e$ is a quadratic with integer-valued c, d, and e and $P(a) = 0$ for some integer a, then a divides e (with no remainder). Explain why.

B39. Suppose $P(x) = cx^2 + dx + e$ is a quadratic with integer-valued c, d, and e. Suppose further that for some rational number, r, $P(r) = 0$. Write $r = n/m$ in lowest terms, where n and m are integers. Prove that the other solution also must be rational. What can we deduce about n and m in terms of c, d, and e?

B40. Someone might say "$x^2 - 3x - 17$ does not factor." Is he right? Discuss this technically.

Section 3.4. Word Problems

Word problems, also known as story problems, often strike fear into the hearts of calculus students. But it does not have to be that way. Use this section to help you improve your word-problem skills. You may even grow to like them!

Algebra is supposed to be applicable to real-world problems, not just to homework problems with x's. However, with only the background material in Chapters 1 and 2 we are nowhere near ready to use algebra to discuss important mathematical problems such as the elliptical orbits of satellites or the statistical analysis of relationship between cigarette smoking and lung cancer. But interesting questions and methods for answering them are down the road in front of you, if only you travel far enough. Word problems, also known as story problems, are an essential part of the trip.

Math educators have studied why students find word problems difficult. This section discusses some of the reasons. Identify which apply to you so that you can improve your ability to do word problems.

Elementary Problems. To begin, I will discuss simple word problems of the kind seen in grade school and beginning algebra. Our interest is in determining what to **do** given the words in the problem.

Words that suggest mathematical operations (for example, "difference" may indicate subtraction) are called cue words. The degree of difficulty of a grade-school word problem depends upon how the cue words are used. Some math educators use the term direct to categorize word problems in which the information and language cues indicate the action required to do the problem.

Example 1: Sue sold 40 raffle tickets. John sold 25. How many less than Sue did John sell?

The phrase "less than" indicates subtraction and the order of the numbers to be subtracted is as given in the problem. $40 - 25 = 15$. The problem is "direct."

Example 2: The theater has 20 rows of seats with 14 seats in each row. What is the total number of seats in the theater?

In this problem the words "each" and "total" indicate multiplication. The numbers are the right numbers to multiply. $20 \times 14 = 280$. The problem is "direct."

To do direct word problems correctly you only need to know the correspondence between the "cue" words in English and the mathematical operations. *Direct problems are basically computational tasks expressed in words.* Very few students have difficulty with direct word problems. *If* the cue words in a word problem actually indicate the required operations, the problem is easy. But that's a big "if".

An indirect word problem is one in which the language suggests operations that are not the operations required to do the problem.

Example 3: John sold 12 fewer tickets than Bill. John sold 35. How many did Bill sell?

The word "fewer" may suggest subtraction, but the appropriate operation to solve the problem is addition. The problem is indirect. The answer is $35 + 12 = 47$.

Look at the words as expressing a formula. Let "J" denote the number John sold and "B" denote the number Bill sold. In mathematical notation, the sentence, "John sold 12 fewer tickets than Bill," is given by the formula: "$J = B - 12$," which does mention subtraction.

Given this formula, the problem is easy. Just "plug in" 35 for J to obtain "$35 = B - 12$." $B = 47$.

In this example the cue word "fewer" does *not* suggest the right operation to *do*, but it does suggest the right operation to *express* in the formula ($J = B - 12$). Indirect word problems use cue words to build formulas.

The major difference between the easy "direct" word problems and the harder "indirect" word problems is that in direct problems the cue words express operations you actually *do* to the given number or numbers, whereas in indirect problems the cue words express operations you are supposed to *represent* in an algebraic formula, without actually doing them.

Students who think they "just can't do word problems" usually can do direct problems, which are not really algebraic. Direct problems use operations to generate new numbers, and most students are comfortable with numbers (the conceptual level of a direct problem remains at the number level). However, in indirect problems, students often become stuck because they must deal with (that is, express or represent) operations (a higher-level concept than numbers), without actually doing them! The key to doing indirect problems is to represent the operations algebraically in a formula.

Formulas. Let's examine the formula route to success. Students find problems easy when they know the formula.

Example 4: A circular oak table is 48 inches across. How far is it around?

Figure 1 illustrates the information. Recall the formula, "$C = \pi d$," for the circumference of a circle, C, in terms of its diameter, d. If a problem gives you the diameter, the formula tells you what to do to find the circumference, "multiply by π."

Not only is "$C = \pi d$" the right formula, but its in the right order to do the problem directly because you are given an argument, $d = 48$, and asked to find an image. That's the way functions operate. So, $C = 48\pi = 150.8$ (inches). The table is 150.8 inches around.

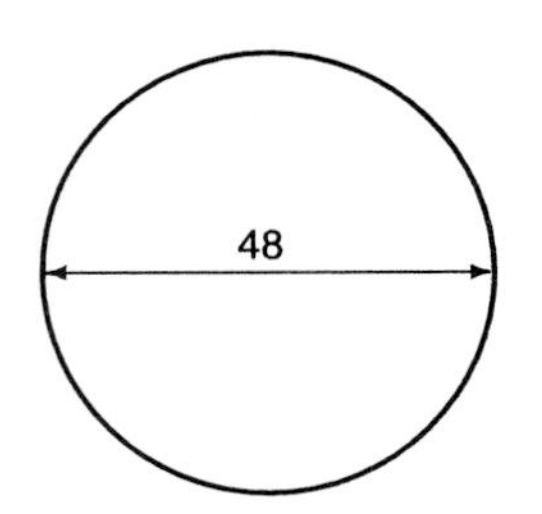

Figure 1: A circular table.

Sometimes the appropriate formula must be used "backwards".

Example 5: Measured 1.5 meters above the ground, a large fir tree is 4.9 meters around. What is its diameter?

Solution Process: First make the reasonable decision that a cross-section of the tree can be regarded as circular. Then the formula from Example 4 applies. This time we are given the image and asked for the argument.

Plug in to "$C = \pi d$": $4.9 = \pi d$. $d = 4.9/\pi = 1.56$ (meters).
The problem was indirect, but still easy because the formula is well-known.

Many students can do these two examples easily, but still do not regard themselves as "good at word problems." That is because there is something particularly simple about these two problems: There is a well-known formula. Word problems are relatively easy if you know -- or can create -- the formula.

Example 6. A pyramid is 20 feet high and it has a square base. Its volume is 6,000 cubic feet. How long is a side of its base?

You can draw a picture (Figure 2), but to do this problem you must know the formula. A pyramid with height h and a square base with side b has volume given by the formula $V = (1/3)b^2h$. With this formula the problem is easy. "Plug in" to obtain the equation:

$$60{,}000 = (1/3)b^2(20).$$

Now the "word" part of the word problem is over. Solving,

$$6{,}000 = (1/3)b^2(20)$$

iff $900 = b^2$

iff $b = 30$ or $b = -30$.

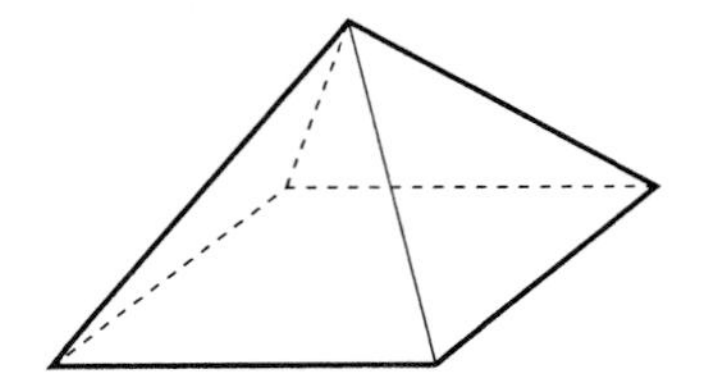

Figure 2: A pyramid.

Since sides are not negative, the solution $b = -30$ is extraneous and the actual solution is $b = 30$. The sides are 30 feet long.

The point of this example is *not* how to do pyramid problems. The point is to see that

Knowing the formula makes the problem possible.

Problems with formulas are relatively easy. Therefore, if a problem does not seem to fit a well-known formula, you may want to build the relevant formula yourself.

Example 7: A rectangle has one side three inches longer than the other. The total area is 75 square inches. How long are the sides?

Don't rush to the answer. Forget about the "75 square inches" for now. First find the relevant formula.

This problem has cue words. The phrase "three inches longer" suggests adding 3, and the word "area" suggests using the well-known "area = base times height" formula. But the given number, 75, is an area and the suggested operations apply to sides, not area. This problem is indirect. Therefore we *represent*, in a formula, the operations suggested by the cue words.

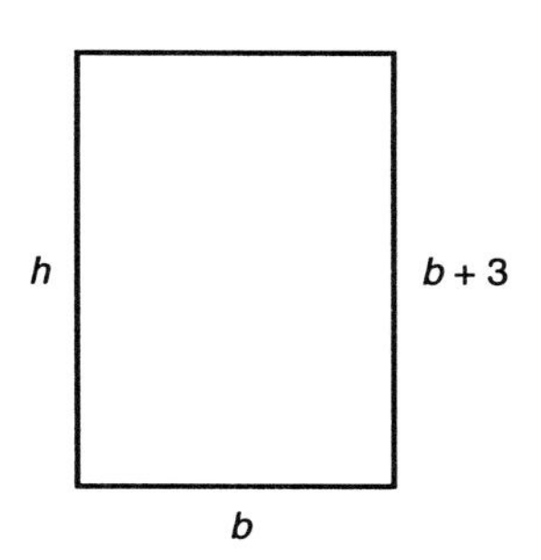

Figure 3: A rectangle with one side 3 inches longer than the other.

To find the formula, take several steps. Draw a picture (Figure 3). According to the cue words, if the shorter side is b, the longer side is $b + 3$. To find the area, we would multiply the sides together, according to the well-known building-block formula,

$$A = bh.$$

So, substituting,

$$A = b(b + 3).$$

This is the formula we want. It is not famous, but it is the right formula for this problem. Now we can "plug in" the particular value, 75, for the area:

$$b(b + 3) = 75.$$

Now the "word" part of the word problem is over. The rest is algebraic manipulation (problem A7).

Example 7, revisited: Was the answer to the word problem an argument or an image of the formula we built?

The answer was "b", which was an argument of the formula. The image, A, was a given of the problem ("area is 75 square inches"). Therefore, the problem did not require us to execute the suggested operations. On the contrary, we were supposed to use algebraic notation to *represent* the operations in a formula. Only after we got the formula did we use the given particular number "75," because the problem was indirect.

The advice of this section is: When the quantities in the word problem do not already fit a well-known formula, use the cue words and well-known formulas to

Build your own formula.

<u>**Direct Evaluation and Indirect Problems**</u>. Algebraic notation is designed to express operations. When a problem, such as Example 7, uses a relationship between an argument (side) and image (area), there are two closely related possible problems: 1) Given the argument (side), find the image (area) (This would be a *direct* evaluation problem requiring you to *do* the suggested operations), and 2) Given the image (area), find the argument (side) (This would be the *indirect* inverse problem requiring you to *represent* the suggested operations in a formula).

Many students are able to do direct problems and yet are unable to do very similar indirect problems. Their difficulty is not in solving the equation, it's in obtaining the equation to solve.

How do you obtain the equation to solve?

The answer illustrates the very purpose of algebraic notation.

The purpose of algebraic notation is to represent operations and order,

even if the number to which the operations apply is unknown. In an indirect word problem, algebraic notation is used to *represent* the sequence of operations in the cue words and relevant well-known formulas, even if the argument is unknown. Then the given image is plugged into the formula, which yields the an equation to solve. Then the "word" part is over.

"Solving" an indirect word problem is far more complex than "solving" an equation, because in a word problem first you must create the equation to solve.

Example 8: A cup of lentils has 646 calories of which 19 are from fat. Sausage is 50 calories per ounce of which 27 calories are from fat. Jon wants to add sausage to his lentil soup, but does not want the calories from fat to exceed 10% of the calories. How many ounces of sausage can Jon add to a cup of lentils to have 10% of the total calories from fat?

This problem gives the percentage of calories from fat and asks you to solve for the corresponding number of ounces of sausage. The key to the problem is to determine how to directly evaluate the percentage of calories from fat given the (unknown) number of ounces of sausage. That is, treat the answer as the *argument* of some formula which we will determine.

If Jon uses x ounces of sausage, the total number of calories will be

$646 + 50x$ [646 from the lentils and $50x$ from the sausage].

The number of calories from fat will be

$19 + 27x$ [19 from the lentils and $27x$ from the sausage].

Therefore, the fraction from fat will be

$$\frac{19 + 27x}{646 + 50x}.$$

This is the formula we want.

The problem asks for the value of x that makes this 10%. Plugging in to the formula we built:

$$\frac{19 + 27x}{646 + 50x} = .1.$$

Now the "word" part is over.

The hard part of this problem is not solving this equation, it is using algebra to express the direct evaluation process in a formula. Once you have the formula, the problem is easy (problem A11).

The key to many word problems is to use the cue words and well-known formulas to express the direct evaluation process. If the formula is unknown,

Build your own formula.

This suggests a way to discover equations in word problems. **Express, in symbols, the direct evaluation process** as it would apply to the unknown quantity, if you knew the quantity, rather than concentrating on dealing with given numbers.

Example 9: A plane flying due east at 120 miles per hour leaves Bozeman at noon. Another plane leaves at 1:00pm and flies due south at 150 miles per hour. When will they be 200 miles apart?

First, try to understand the problem. A picture may help (Figure 4).

The question asks "When?" and gives a distance. Therefore we must relate time to distance.

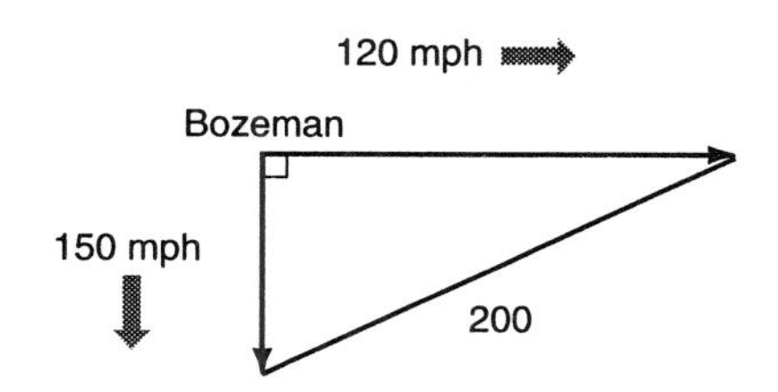

Figure 4: Two planes flying from Bozeman.

The basic formula is "distance equals rate times time" for motion of a single plane. But this problem has two planes and requires a custom formula for distance given time. The formula should *express* the distance apart with time (the answer) as the argument. Then, setting the distance equal to the given "200 miles" will yield the equation.

Evaluating distance once for a particular time can expose the evaluation process that we want to represent in a formula. For example, if the time were two hours after noon the eastbound plane would have flown 240 miles (= 120 miles per hour × 2 hours). The southbound plane would have flown 150 miles (= 150 miles per hour × 1 hour). Their distance apart, c, would be given by the Pythagorean Theorem.

$$240^2 + 150^2 = c^2, \quad c = 283.0 \text{ (miles)}.$$

This calculation uses a sequence of operations. If you can see the sequence of operations, instead of the particular numbers, you can write the formula we need.

Use algebraic notation to express that evaluation process. Let t express time in hours after noon. To find the position of the eastbound plane take the elapsed time and "Multiply by 120"; at time t the eastbound plane is $120t$ miles east of Bozeman. The southbound plane does not fly as long. To find the position of the southbound plane take the time since noon, subtract one hour, and multiply by 150; at time t it is $150(t - 1)$ miles from Bozeman. These express the legs of the triangle. The distance apart at time t, $c(t)$, is therefore,

$$c(t) = \sqrt{(120t)^2 + [150(t - 1)]^2}\ .$$

This is the formula we wanted to build. With the formula, the problem is straightforward, if not short. Now, and only now, we can use the "200 miles" in the problem. "Plug in" the 200 and solve for t.

$$\sqrt{(120t)^2 + [150(t - 1)]^2} = 200\ .$$

Solve it by squaring both sides and consolidating like terms to obtain the usual form for the Quadratic Theorem (Problem A8).

Evaluate-and-Compare. "Evaluate-and-Compare" was my name for one of the four ways to solve an equation. It is sometimes called "guess-and-check." It is not one of the three traditional high-school methods, but it is of increasing importance in school math now that calculators and computers have made it so easy to evaluate expressions. And it has long been an important method in higher mathematics. This section shows how evaluation can help with word problems.

As we have seen, indirect word problems have two distinct stages:

1) Find a general formula that expresses the relationship of the quantities in the problem (this is often the hard part, and may require building your own formula), and
2) Plug a known quantity into the formula to get an equation to solve, and then solve it.

This section concerns only the first stage, finding the relationship. If you can see the relationships in a word problem well enough to directly evaluate (calculate) the image given the argument, it is only a short step to expressing the correct relationship in symbolic notation. The idea is to apply the same *operations* to "x" that you would apply to any particular number.

Example 10: We all know that walking a hypotenuse saves distance compared to walking the legs of a right triangle. Suppose the only sidewalk from building A to building B has a right angle at point C, as pictured in Figure 5. Let AC be 400 yards and CB be 200 yards. Suppose you begin to walk from A toward C, but at point D, you leave the sidewalk and head straight for B. If this saves you 100 yards of walking, where is point D?

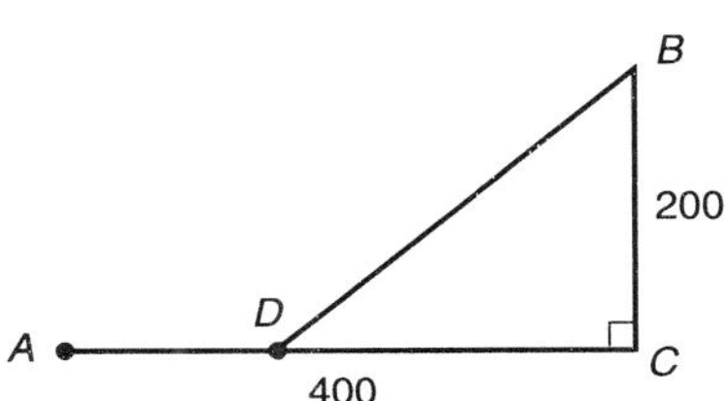

Figure 5: A shortcut off the sidewalk from A to B.

Solution process: Ask yourself how the position of D is related to the distance saved. Clearly, the location of point D is related to the distance walked and therefore to the distance saved.

Next, find the formula which relates those quantities. Decide how to directly *evaluate* the distance saved if you knew the location of D. For example, let D be 50 yards from A. Then the distance walked would be 50 yards plus the distance from D to B. That we can evaluate from the Pythagorean Theorem. From D to C is 350 yards, so from D to B along the hypotenuse is

$\sqrt{350^2 + 200^2}$ - 403.11 *yards*. . The total distance is 403.11 + 50 = 453.11 yards cutting across the grass. The long way would be 400+200 = 600 yards. The savings is 600 - 453.11 = 146.89 yards. This "guess" did not "check," but we did not expect it to.

The point of the guess was to expose the operations involved in the direct evaluation.

How did we get the numbers which resulted from our guess? What were the operations in which order? How could we express them using "x" as the argument?

Let D be x yards from A. Our trial calculation was with $x = 50$. We got 350 by subtracting from 400. That operation is expressed by "$400 - x$." We got the length of DB using the Pythagorean Theorem:

$$DB = \sqrt{(400 - x)^2 + 200^2}\ .$$

Then we added 50 to get the total distance walked:

$$x + \sqrt{(400 - x)^2 + 200^2}\ .$$

Then we subtracted from 600 to find the distance saved, s:

$$s = 600 - [x + \sqrt{(400 - x)^2 + 200^2}\ ]\ .$$

This is the formula we wanted. The rest is relatively easy. Plug in 100 yards for s and the "word" part is over. Then solve the equation (Problem B6).

Sometimes you may wish to do more than one evaluation to be sure you see the pattern. You might want to use a table to exhibit your results.

Example 11: An open-topped box is formed from a 10 by 15 inch rectangular sheet of metal by cutting a square from each corner, folding up the sides, and sealing the seams. How big are the cutout squares which yield the maximum possible volume?

Solution Process: First try to understand the problem. Reread it several times, if necessary. Draw a picture (Figure 6).

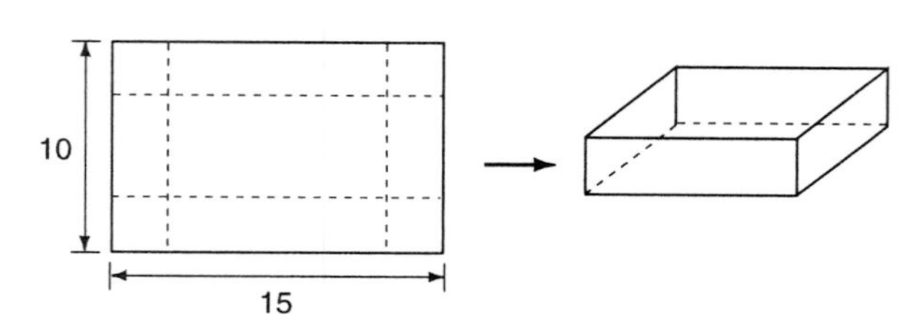

Figure 6: An open-topped box formed from a sheet of metal by cutting out the corners and folding up the sides.

Now determine how the cutout size is related to the volume. We want the argument which maximizes the image. We can find it using a graphics calculator, *if* we have the formula.

To find the formula which relates side of the cutout square to the volume of the box, analyze the evaluation process. For example, consider any particular case.

Suppose the cutout were 2 inches square. What would the volume of the box be? Well, the base would be 6 inches by 11 inches (why?), and the depth would be 2 inches. Therefore, the volume would be $6 \times 11 \times 2 = 132$ (cubic inches).

The key is not the numbers. Where did the "6", "11, "2" and "132" come from? The key is to see *how* the numbers were obtained. It is the operations we want to discover. How did we find the short base side? The longer base side? The depth? The volume?

From the picture we can see that the sides are reduced by two times the amount folded up.

cutout size	shorter base side	longer base side	depth	volume
2	6	11	2	$6 \times 11 \times 2 = 132$
3	4	9	3	$4 \times 9 \times 3 = 108$
1	8	13	1	$8 \times 13 \times 1 = 104$
x	$10 - 2x$	$15 - 2x$	x	$(10 - 2x)(15 - 2x)x$

The last line expresses how the numbers were obtained. It expresses the evaluation process. It expresses the proper formula:

$$V = (10 - 2x)(15 - 2x)x.$$

To solve the problem is easy given the formula. Use evaluate-and-compare by graphing the expression and locating the x-value which yields the maximum.

To graph this, consider the domain. What are reasonable values for x? Negative values are out. Also, you cannot cut out a square greater than 5 inches on a side (why?). So the appropriate domain is [0,5]. If you try to graph the function on this domain, you may not get a good picture unless you choose the y-interval well. Inspecting the table, we see that y-values can be at least as large as 132. So, try a y-interval with larger y-value, say [0, 150] (Figure 7).

To actually find the maximum given the graph is easy (Problem A9).

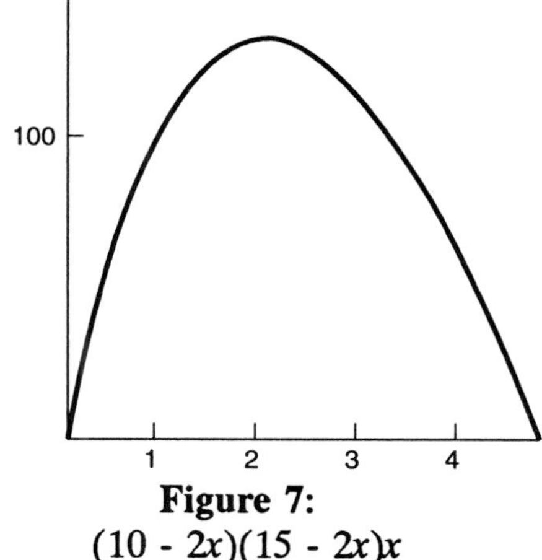

Figure 7:
$(10 - 2x)(15 - 2x)x$
[0, 5] by [0, 150].

Evaluation Patterns. Mathematical expressions express sequences of operations (Have you read this 100 times yet?). A modern method of teaching young students to set up word problems correctly is to use the evaluation-pattern approach used in Example 11. That is, use an evaluate-and-compare type of procedure several times until you see the pattern of operations. Then express that pattern with an expression using a letter, usually "x". Then you have the formula you need.

Here is one more example of the evaluation-pattern approach.

Example 12: A pair of animal pens are constructed from 100 linear feet of fence (including two gates) in the rectangular configuration in Figure 8. The total area is 416 square feet. What are the dimensions of the construction?

Solution process: Try to understand the problem. Read it several times if necessary.

The problem asks about dimensions and gives the area. So, determine how the dimensions are related to the area. Also, the length and width of the pens are related because the total length of the fence is 100 feet.

Suppose the construction is w feet wide. Then we could determine the length of the pens and then the area.

To find the relationship in terms of the unknown, w, the "evaluation pattern" approach suggests you first try a few particular numbers. For example, if the width were 30 feet, what would the length and area be?

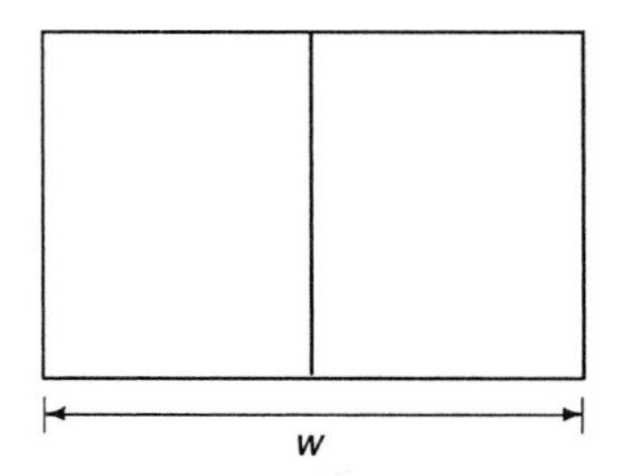

Figure 8: Animal pens constructed from 100 linear feet of fence.

If the width were 30 feet, in the picture the top and bottom together would be 60 feet, and, out of the 100 total feet, the remaining 40 would be split among three equal lengths. So the length would be 40/3 and the area $30 \times 40/3 = 400$ (square feet).

width	fencing for widths	length	area
30	60	40/3	$30 \times 40/3 = 400$
20	40	20	$20 \times 20 = 400$
40	80	20/3	$40 \times 20/3 = 266.67$
w	$2w$	$(100 - 2w)/3$	$w(100 - 2w)/3$

The operations have been abstracted in the final formula. Now, returning to the original problem, we see that the equation is:

$$w(100 - 2w)/3 = 416.$$

This is a quadratic which can be solved with the quadratic formula (problem A10).

Patience and the Philosophy of Word Problems. Formulas are abstract. *One* formula can apply to many problems. *One* function expresses *many* number pairs. If you only want one number as the answer to a problem, why bother with the abstract formula? There are two good reasons. One is, the formula helps you get it right. The formulas approach works. The other is, in higher-level courses, including calculus, the relationships are more important than particular numbers. It is a good idea to learn to deal with relationships in this course before you get to calculus.

Impatient students who skip the formula stage often cannot do word problems. Then they wonder why they are "not good at word problems." Take the time to build your own formula. Have a little patience.

Guidelines. To use math to solve an indirect problem which is already accurately stated in words, the words must be translated into mathematical notation. The translation from words to math can be difficult. There are some simple guidelines you can follow which will help you with word problems.

1) Understand the problem. Read it closely for meaning. Reread it several times if necessary. Good mathematicians do.
2) Draw a picture, if appropriate. Pictures can illustrate relationships between the components of the problem.
3) Build your own formula. Write down, and expect to use as components of your formula, well-known formulas about the quantities in the problem. Cue words suggest operations to be expressed in algebraic notation in your custom-built formula.

   [This is the stage where evaluate-and-compare may help you find the formula.]
4) Plug in the given amount for the known quantity. Now you have the information in the problem expressed in mathematical notation.
5) Solve for the unknown quantity.

In this outline the "solving" part is only one part out of five, and only the last part at that. Learn to have the patience to complete steps 1 through 4 first.

Conclusion: Indirect word problems are relatively easy if the relevant formula is known, or can be created. Formulas express relationships, and relationships (not numbers) are the key to word problems. The process of direct evaluation can be used to discover the relevant formula, because the formula expresses the direct evaluation process in mathematical notation. The purpose of algebraic notation is to express processes (operations and order), even if (*especially* if) the numbers to which they apply are unknown. Once you have the formula, plug in the given value and the rest is just solving, for which we already have numerous techniques.

Terms: direct (word problem), cue word, evaluation pattern.

****

Exercises for Section 3.4, "Word Problems":

Note: This text occasionally supplies solutions with two significant digits in brackets. These are given so you can immediately determine whether you used the correct solution process. On the homework, prove that you know the correct process by giving a more accurate answer with three or more significant digits.

A1. a)* Define "cue" word [in the context of a word problem].
b) Define "direct" in the context of a word problem.

A2.* Word problems are relatively easy if you know the ______________.

A3.* If you don't know the relevant formula, the advice of this section is to ________ ________ ________ ________ (fill in exactly four words).

A4.* What stage of doing indirect word problems is the evaluate-and-compare method supposed to help? What is the point of doing the calculations with a guess that is probably not correct?

A5. Compare the word problem to the usual building-block formula to determine whether the answer would usually be regarded as an argument (as opposed to an image).
a) Example 7. b) Example 8. c) Example 9.

A6. Compare the word problem to the usual building-block formula to determine whether the answer would usually be regarded as an argument (as opposed to an image).
a) Example 10. b) Example 11. c) Example 12.

A7. Use the quadratic formula to solve the equation in Example 7, "$75 = b(b + 3)$."

A8. Solve the equation in Example 9. [1.5]

A9. Find the x-value at which the expression in Example 11 is a maximum. [2.0]

A10. Find the width which yields the maximum possible area of the pen construction in Example 12 with the given constraint that the fence totals 100 feet long.

A11. Solve the equation in Example 8. [2.1]

^^^^^^^^^

B1.* Explain the role of formulas in indirect word problems.

B2.* How is "evaluate-and-compare" supposed to help with indirect word problems?

B3.* a) What is the purpose of algebraic notation?
b) What does this have to do with indirect word problems?

B4.* Algebraic expressions (such as "$b(b + 3)$") can be interpreted at two conceptual levels. What are the two levels? Explain which is appropriate for indirect word problems and why.

B5.* Explain what is meant by saying that solving an equation is the inverse of directly evaluating an expression.

B6. Solve the equation in Example 10.

B7. An open-topped box is formed from a 8½" by 11" sheet of cardboard by cutting squares out of the four corners and folding up the sides. a) Obtain the formula for its volume. b) Find the size of the cutout that maximizes the volume. [1.6]

B8. If the price of a WidgetCar is $12,000, then 100,000 will be sold. If the price is $13,000, then 70,000 will be sold. Assume the price-to-number sold relationship is a straight line.
a) Find the price to number-sold relationship.
b) Find the revenue if the price is "x" dollars.
c) If the cost of selling each is $9000, find the selling price that would maximize profit (where profit is revenue minus cost). [12,000]

B9. A rain gutter is formed from a sheet of metal 9 inches wide which is bent to form a trough. Three inches are mounted on the house, three inches form the bottom of the trough, and three inches form the lip (see the figure). You wonder if you bend the lip out a bit whether the cross-sectional area would be greater. The trough would be wider, but not so deep. How far out should the lip be bent (x) to maximize the area of the cross section (shaded area)?

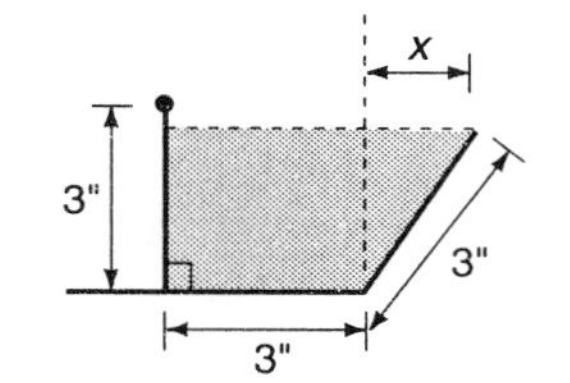

B10. A picture is 24 inches by 16 inches. It is mounted with a border of uniform width on all four sides. If the surface area of the border is half the surface area of the picture, how wide is the border? [2.2]

B11. A picture is 12 inches by 20 inches, and is mounted in a frame of similar shape ("similar" is a technical term from geometry). If the total area of the picture and frame is twice the area of the picture, what are the dimensions of the frame? [17]

B12. A field is in the shape of a square with a semicircular cap on one side (see the figure). Its area is 10,000 square feet. What is the length of a side of the square? [85]

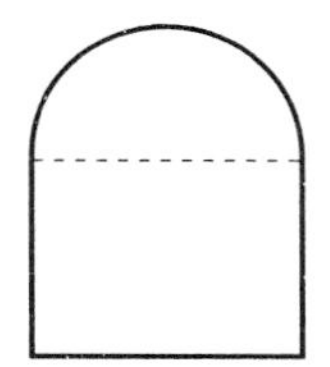

B13. A right triangle has hypotenuse 5 inches long. Its area is 4 square inches. How long are the other sides? [1.7 and ...]

B14. A semicircular enclosure has perimeter 20 centimeters. What is its area? [24]

B15. Find the equation of the circle with center (3, 2) that is tangent to the unit circle ($x^2 + y^2 = 1$).

B16. A pyramid has height half of the side of its square base and a volume of 10 cubic centimeters. How long is a side? [3.9]

B17. A figure consists of a rectangle with a triangle on one end as in the figure. If the rectangle is 6 units high, the entire width of the bottom edge is 10 units, and the entire area is 40 square units, how wide is the rectangle? [3.3]

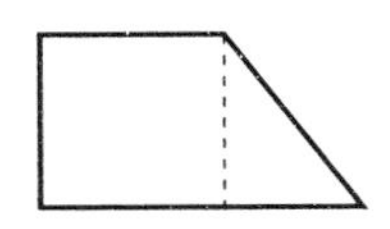

B18. A given length of string can be made into either the circumference of a circle or the perimeter of a square. If the area of the circle is 10 square inches more than the area of the square, how long is the string? [24]

B19. A firm in Japan orders wheat with 15% protein content from your firm and they will take all you have. You have 100 tons of wheat on hand, but it has only 14.4% protein content, so you will buy some wheat with 16% protein content to mix in with your wheat to create wheat with exactly 15% protein content. How much wheat should you buy? [60]

B20. Suppose 100 linear feet of fence made three (3) animal pens as in Example 12 and Figure 8, except with three pens side by side instead of only two. What dimensions would maximize the area? [12]

B21. Gold weighs 19.3 grams per cubic centimeter, and silver weighs 10.5 grams per cubic centimeter. a) Develop a formula for the density in grams per cubic centimeter of a mixture of silver and gold. b) If a mixture weighs 12.7 grams per cubic centimeter, what is the amount by weight of gold in it? [25%]

B22. A jet fighter shoots cannon shells at 4000 feet per second. An enemy plane flies a straight-line course perpendicular to the course of the fighter at 1000 feet per second. The enemy is 6000 feet directly in front of the fighter when the cannon is fired. Where should the cannon be aimed? (Express the answer in terms of how far along the projected straight-line path of the enemy plane the cannon should be aimed.) See the figure. [1500]

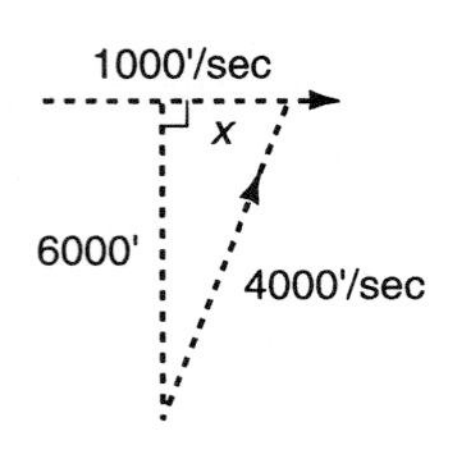

B23. The length plus girth of packages shipped by Universal Package Service must not exceed 150 inches. Your company wishes you to design the cylinder-shaped package with the largest possible volume that they can ship. The "length" is the height of the cylinder and the "girth" is the circumference of the circle. SET up all the relevant formulas and then find an expression, with one variable, to maximize. Make it very clear which expression you intend to maximize. Last (and least) find the radius of the cylinder of maximum volume. [16]

B24. At this instant a jet fighter is flying north at 1000 miles per hour straight toward a site 200 miles away. Also at this instant an enemy plane 100 miles west of that site is flying east straight toward the site at 600 miles per hour. The jet's missile radar will lock on when they are 60 miles apart. SET UP (but do <u>not</u> bother to solve) the equation to solve the question "How long from now (in hours) until the jet's radar locks on?" [Just give a clear and correct equation.]

B25. Janet walks and runs a total of 5 miles. She walks at 4 miles per hour and runs at 8 miles per hour. She wants to finish in exactly 50 minutes. How far should she run?

B26. Alvin runs 3 miles and walks 2 miles. He runs twice as fast as he walks and he finishes in exactly one hour. How fast does he walk? [3.5]

B27. Sue likes to walk in the desert. She walks at 2 miles per hour in the desert, and at 4 miles per hour on roads. She wants to get to the parking lot as quickly as possible when she is 1 mile from the road and the parking lot is 2 miles down the road, as pictured. Rather than walking straight toward her car, or straight toward the road, she will walk at an angle toward the road so she can spend more time walking faster on the road. Where should she aim to get to her car the quickest?

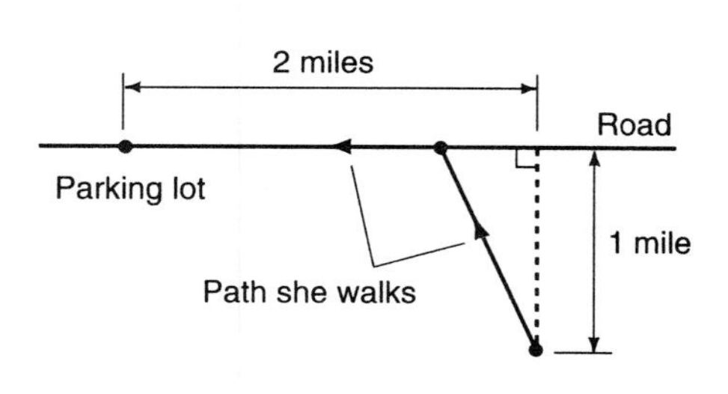

B28. A rancher builds three pens using a long existing straight fence for one side. She uses 200 feet of new fencing and gates in an arrangement as pictured. What are the dimensions that maximize the total area of the pens?

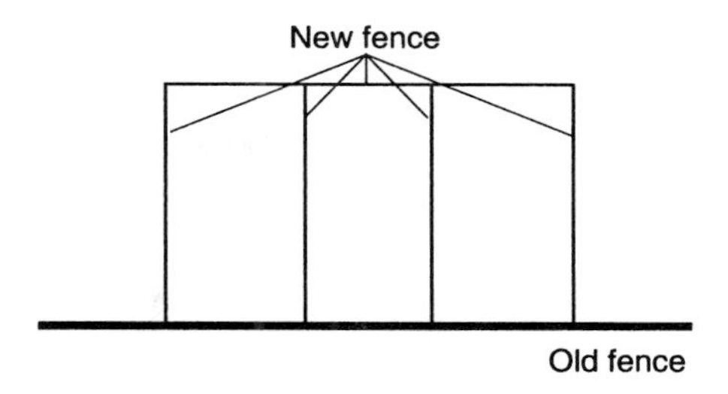

B29. Find the equation satisfied by all points (x, y) that are equidistant from the point (0, 1) and the line $y = -1$.

B30. Find the equation satisfied by all points (x, y) such that the sum of the distances from (1, 0) and (-1, 0) to (x, y) is d, where $d > 2$.

B31. A warship is sailing due east at 25 miles per hour. When it is 3 miles north of a submarine, the submarine fires a torpedo which travels at 70 miles per hour (Figure 12). The torpedo is aimed so that it will meet the ship without changing course (if the ship does not change course). Where will it meet the ship? [1.1 miles east]

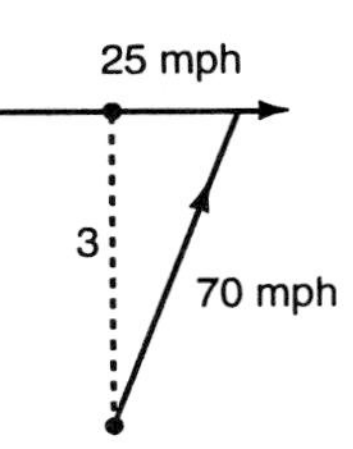

B32. The formula for the volume of a cone is $(1/3)Bh$, where "B" is the area of the base, which is the area of a circle, and h is the height. If you take a circular sheet of paper 10 inches in diameter and cut out a sector as in the picture, it can be shaped into a cone by taping together the two edges of the cutout. How much of the circumference should be cut out to make the cone of maximum volume?

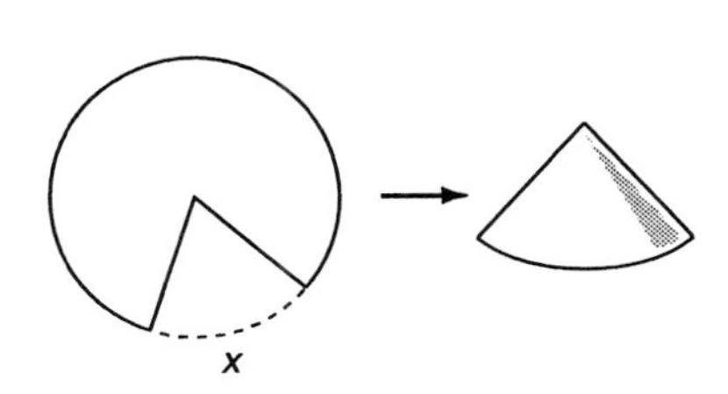

B33. A plane flies northeast from Bozeman. It leaves at noon and flies at a constant speed of 170 miles per hour. Assume Billings is 140 miles east of Bozeman. The plane will appear on the Billings radar when it is 110 miles from Billings. When will the plane be 110 miles from Billings?

B34. Suppose there are no taxes on the first \$14,000 of income and any income above \$14,000 is taxed at a 15% rate. A tax-law change is proposed that there should be no taxes on the first \$16,000 of income and the rate should be 17% on any income above \$16,000. People with which incomes will pay less under the proposed changes?

Section 3.5. Word Problems with Constraints

Many formulas have two independent variables and one dependent variable. For example, the area of a rectangle is base times height. In this formula base and height are variables that can take on any positive values independent of one another. However, in some problems another relationship may be known which determines the base as a function of the height or the height as a function of the base, so they are not, in that particular problem, independent. Relationships between variables which are normally independent are called constraints.

Example 1: Express the area of a rectangle given its perimeter is 40 inches.

The usual formula for the area of a rectangle is $A = bh$ (Figure 1). There are two independent variables. However, if the perimeter is given, b and h are no longer independent. We can rewrite the area in terms of one variable, say, b.

The formula for the perimeter is

$$P = 2b + 2h.$$

The problem states the perimeter is 40. That is a constraint.

$$40 = 2b + 2h,$$
$$20 = b + h,$$
$$h = 20 - b.$$

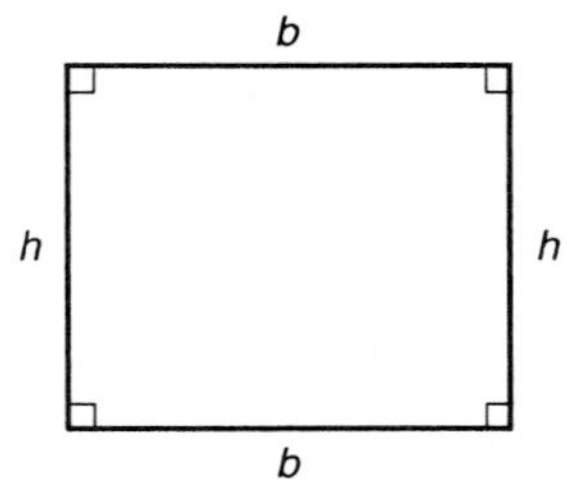

For rectangles with perimeter 40, the area formula can be rewritten:

$$A = bh = b(20 - b).$$

Figure 1: A rectangle with base b and height h.

Now there is only one independent variable, b. The constraint converts a two-variable formula for area into a one-variable formula. This can be graphed, and when set equal to a constant, it can be solved. With this new area formula, many questions can be answered.

Example 1, continued: The area of a rectangle with perimeter 40 inches is 60 square inches. What are the sides?

Use the formula we built. Solve $60 = b(20 - b)$, a quadratic equation (problem A1).

More from Example 1: Find the dimensions of the rectangle with maximum possible area, given its perimeter is 40 inches.

The area formula is $A = b(20 - b)$. With the formula, the "word" part of the word problem is over. There are various ways to find the extreme value of a quadratic. One is to determine the x-value of the vertex. If the quadratic is "$ax^2 + bx + c$" the vertex is at $x = -b/(2a)$ (Theorem 3.2.7). The letters are different in the given area formula ("b" plays the role of "x"). In this example,

$$b(20 - b) = -b^2 + 20b.$$

The vertex occurs when the base, b, is $-20/(-2) = 10$. The other side is then also 10.

As you may already know, the maximum area of a rectangle of fixed perimeter occurs when the rectangle is a square (problem B4).

Another way to maximize $b(20 - b)$ is to change "b" to "x", graph it, and maximize graphically (Figure 2). To select a good window, note the reasonable x-values and y-values. Since "x" represents the side of a rectangle, there is no point in considering negative values of x. Since the perimeter is 40, no side can be longer than 20. So use $0 \leq x \leq 20$. The "y" value is the area. It is zero at the endpoint of the x-interval. In the middle, $x = 10$ and the area is 100. So pick a y-interval wide enough to display both $y = 0$ and $y = 100$. Figure 2 uses $0 \leq y \leq 125$.

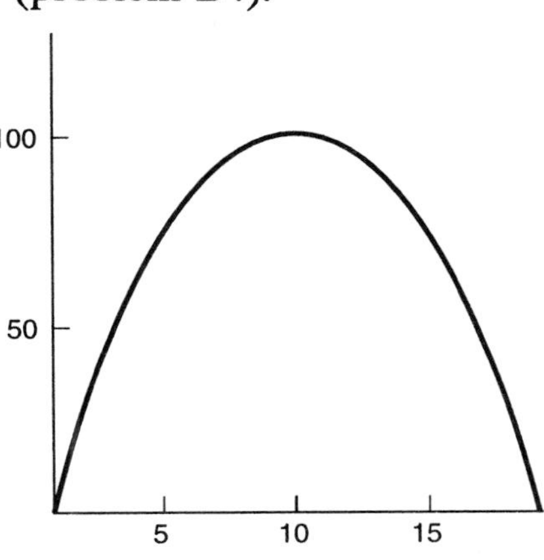

Figure 2: $x(20 - x)$
[0, 20] by [0, 125].

Example 2: Suppose the area of a rectangle is 100 square inches. Find its perimeter if one side is 8 inches.

The two relevant formulas are: $A = bh$ and $P = 2b + 2h$. You are given $b = 8$ and $A = 100 = bh$. Solve for h and plug both b and h into the perimeter formula (Problem A2).

Example 2, continued: Suppose the area of a rectangle is 100 square inches. Find the dimensions which yield the minimum possible perimeter.

Last time we used a particular number (8) for a side, instead of a general number, "b". This time we want a general formula for perimeter.

Again, $A = bh = 100$. This is the constraint. Also,

$$P = 2b + 2h.$$

Using the constraint, we can solve for either b or h, say h. $h = 100/b$. Then

$$P = 2b + 2(100/b).$$

This is the formula we want.

To minimize P, switch "b" to "x", graph it, and use evaluate-and-compare (Problem A3).

Now the meaning of "constraint" is clearer. The context requires two formulas with variables in common. Then, information from one of the formulas (the constraint) can be used to rewrite the other formula with fewer variables.

Example 3: Suppose the length of the hypotenuse of a right triangle is fixed at 10 units. Find the base if the area is 20 square units.

Solution Process: Draw a picture (Figure 3). Think about well-known relationships. From the picture, the area formula and the Pythagorean Theorem leap to mind.

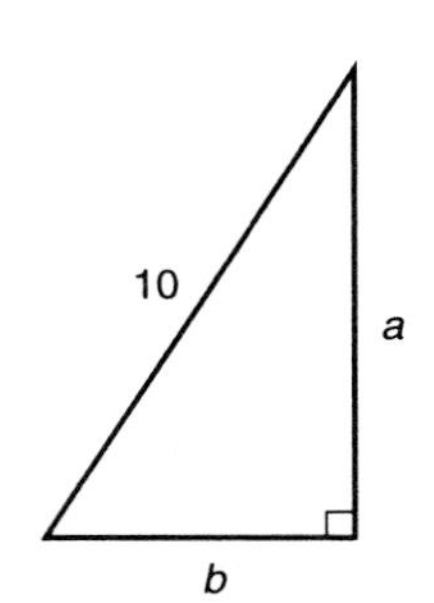

Figure 3: A right triangle with hypothenuse 10.

The question gives area and asks about the base. Therefore, we need a formula for the area in terms of the base. To *solve* the problem we must determine how to *evaluate* the area and then express that process in a formula.

The usual area formula is

$$A = (1/2)bh = (1/2)ba.$$

In a right triangle, one leg is the base and the other is the height. Even if we knew the base, we could not evaluate this directly because the height is not given. But the problem statement gives the hypotenuse, which is a constraint that relates the base and the height. From the base and the hypotenuse we can evaluate the height.

If you know how to directly build the proper formula and set up the equation, you may skip right to it. If not, consider the following "evaluation-pattern" approach.

Try a particular base, any base, and compute the area. Compute it with the intention of noticing the operations you employ, not just the numbers you obtain.

For example, suppose the base is 3. If $b = 3$ and the hypotenuse is 10, we can use the Pythagorean Theorem to compute the length of the other side.

$$a^2 + b^2 = c^2,$$
$$a^2 + 3^2 = 10^2$$
$$a^2 = 10^2 - 3^2$$

$$a = \sqrt{10^2 - 3^2} = \sqrt{91}\ .$$

Now, using the area formula, the area associated with our guess of "3" would be

$$A = (1/2)ba = (1/2)3\sqrt{91} = 14.31.$$

We wanted an area of 20, not 14.31. But we did not expect to guess right. *The purpose of the calculation was to illustrate the evaluation process* so we could express it in terms of "*b*", or "*x*", if we prefer to use "*x*" for a dummy variable.

If we begin with "x" for the length of the base instead of "3", instead of

$$a = \sqrt{10^2 - 3^2} = \sqrt{91} .$$

we would obtain
$$a = \sqrt{10^2 - x^2} ,$$

and instead of
$$A = (1/2)3\sqrt{10^2 - 3^2}$$
the area formula would yield

$$A = (1/2)x\sqrt{10^2 - x^2} .$$

This is the desired formula.

Now take stock of the situation and reread the original problem. The equation is

$$(1/2)x\sqrt{10^2 - x^2} = 20 .$$

To solve it, use evaluate-and-compare (Problem A4).

There are many types of constraints. For example, general points can be anywhere in the plane, but a constraint may force points to be on a line.

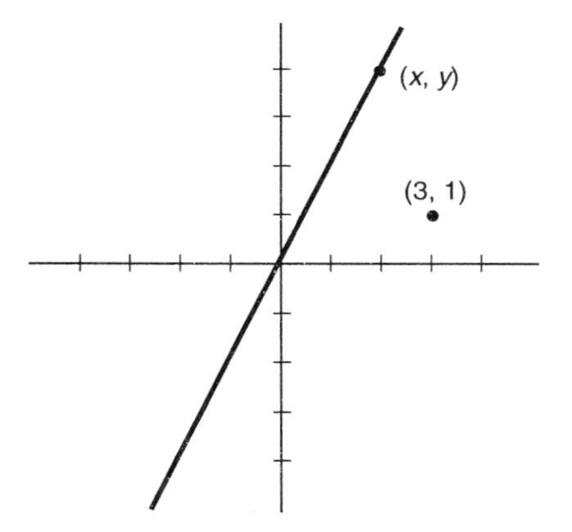

Figure 4: The point (3, 1) and the line $y = 2x$. [-5, 5] by [-5, 5].

Example 4: Find a formula for the distance from the point (3, 1) in the plane to points on the line $y = 2x$.

You wouldn't forget to draw a picture, would you (Figure 4)?

The two relevant formulas are the distance formula for distances between any two points in the plane (Theorem 3.1.14), and the given formula ("$y = 2x$") for points on the line.

Recall that the distance between two points (x_1, y_1) and (x_2, y_2) is given by

$$d = \sqrt{(x_2 - x_1)^2 + (y_2 - y_1)^2}$$

One point is (3, 1). What is the other?

The distance from (3, 1) to any point (x, y) is given by

$$d(x,y) = \sqrt{(x - 3)^2 + (y - 1)^2} .$$

Now the use of the same two variables in two formulas is evident. The given formula "$y = 2x$" is a constraint. We may use it to plug in for "y" in the distance formula:

$$d(x) = \sqrt{(x - 3)^2 + (2x - 1)^2} .$$

This is the formula we want. It has only one variable.

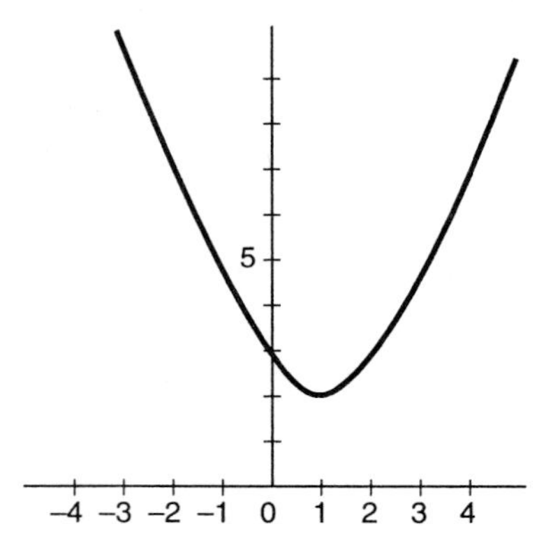

Figure 5: $d(x)$. The distance from (3, 1) to $(x, 2x)$ in terms of x. [-5, 5] by [0, 10].

Example 4, continued: Find the points on the line $y = 2x$ which are 4 units away from (3, 1).

Plug "4" in for "d" in the formula we built (problem A5).

More from Example 4: Find the point on the line $y = 2x$ closest to (3, 1).

The word "closest" tells us to minimize distance. The distance formula we built is the formula we want. Graph it to find the minimum (Figure 5, problem A6). A convenient way to minimize the expression is to square it and find the minimum of its square, since the x-value which minimizes the square of the distance also minimizes the distance (problem B30). There are also geometric ways to do this problem.

Some constraints are so simple there is little need to emphasize their separate nature.

Example 5: A wire 100 inches long will be cut into two lengths and each will be bent to form the perimeter of a square. Express the total area of the two squares in terms of where the wire is cut.

Draw a picture (Figure 6). The relevant formulas are the formula for the area of a square, the formula for the perimeter of a square, and given fact that the perimeters of the two squares sum to 100 inches.

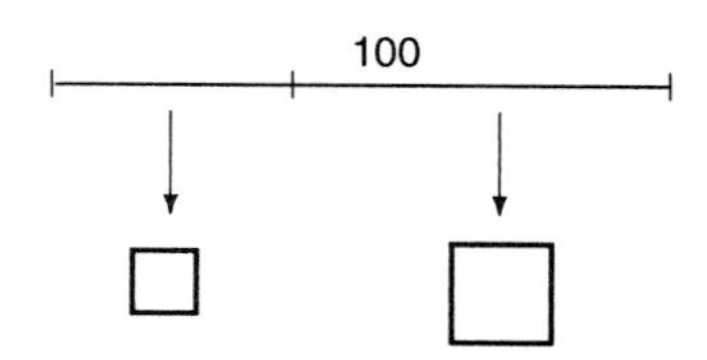

Figure 6: A wire 100 inches long cut and bent to form squares.

It is often advisable, in longer problems, to check how the computations would go in a specific case. This is the idea of an "evaluation pattern." Perhaps even imagining the physical process of cutting the wire and bending it may be helpful. For example, if the wire were cut 40 inches from one end, there would be two pieces of lengths 40 inches and 60 inches. From them we could form two squares, of sides 10 inches (10 = 1/4 of 40) and 15 inches (15 = 1/4 of 60). The total area would then be $10^2 + 15^2$ (square inches). Now, if you are comfortable with those calculations, we can do them again with a cut at a general distance, x.

After the cut there will be two lengths, say x and y. The constraint is that $x + y = 100$. The formula for the area of a square, A, in terms of its side, s, is simple ($A = s^2$), but we are given the perimeter, not the side. However, $P = 4s$, so $s = P/4$. Thus $A = (P/4)^2$, by composition of functions. Thus, for any two lengths of wire, x and y, the total of the two square areas, T, would be

$$T = (x/4)^2 + (y/4)^2$$

as a function of x and y. But our constraint tells us that x and y are related: $y = 100 - x$. Thus, under the constraint, T can be expressed as a function of x alone:

$$T = (x/4)^2 + [(100 - x)/4]^2.$$

You might wish to rewrite this as

$$T = x^2/16 + (x^2 - 200x + 10{,}000)/16 = (2x^2 - 200x + 10{,}000)/16.$$

Example 5, continued: A wire 100 inches long was cut into two lengths and each was bent to form the perimeter of a square. The total area of the two squares is 350 square inches. Where was the wire cut?

Now that we have the formula, the rest is relatively easy. The particular fact that the total area is 350 square inches can be used to set up an equation:

$$T = (2x^2 - 200x + 10{,}000)/16 = 350.$$

This can be solved for x using the Quadratic Theorem (problem A7).

More from Example 5: A wire 100 inches long will be cut into two lengths and each piece bent to form the perimeter of a square. Where should it be cut to minimize the total area of the two squares?

With the formula, this is easy. The minimum will occur at the vertex of the parabola, since the coefficient on "x^2" is positive (Problem A8).

The next example illustrates that you may need to name and utilize an intermediate variable to express the relationships. A variable is <u>intermediate</u> when it is neither the argument nor the image of the given relationship, but serves to relate the two.

Example 6: The perimeter of a semicircular enclosure is 10 meters (Figure 7). What is its area?

The question mentions "perimeter" and "area," both of which are usually discussed in term of the radius (or diameter). The radius is a useful "intermediate" variable.

One formula is well-known: $A = \pi r^2$. The other can be built from half a circumference of a circle plus a diameter:

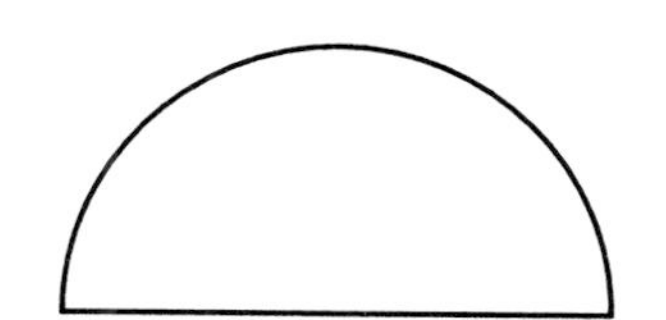

Figure 7: A semicircular enclosure.

$$P = (2\pi r)/2 + 2r = (\pi + 2)r.$$

Now use this formula to solve for r and then use the area formula (Problem A9).

For the next example the formula for the relation between total cost, cost per unit area, and area is relevant.

$$\text{cost} = (\text{cost per unit area}) \times (\text{units of area}).$$

Example 7: Suppose a box with a square bottom is to contain 1000 cubic inches. The top costs 7 cents per square inch, the 4 sides cost 3 cents per square inch, and the bottom costs 5 cents per square inch. Assume there are no other costs. Find the dimensions of the box with minimum cost.

Solution process: Read the problem again. Draw a picture (Figure 8). Decide what is related to what.

The relevant formulas concern volume, area, and cost. Using l for length, w for width, and h for height, the general formula for the volume, V, of a box is

$$V = lwh.$$

In this problem, the box has a square base (which is a constraint), so "l" and "w" are equal. Let's use x for both l and w.

$$V = x^2h.$$

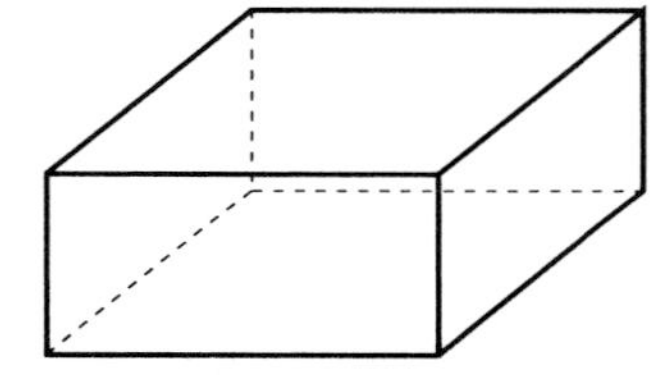

Figure 8: A box with a square bottom.

The problem did not mention the height, h, but we need to use it. It is an important variable which is "intermediate" between the given information and the formula we want. The volume is given as 1000, which is a constraint on x and h:

$$1000 = x^2h.$$

Now build the cost formula. There are 6 sides. The top and bottom each have area x^2. The 4 other sides each have area xh.

$$C = 7x^2 + 5x^2 + 3(4xh) = 12x^2 + 12xh.$$

We can use the constraint to replace either x or h in the formula for cost. It is simpler to replace h. From the constraint,

$$h = 1000/x^2, \text{ so}$$

$$C = 12x^2 + 12x(1000/x^2) = 12x^2 + 12{,}000/x.$$

This is the formula we want. It has only one variable; "h" is gone. To minimize the image, graph it and use evaluate-and-compare. Calculus will provide other techniques. The only trick to graphing it is finding an appropriate window (Problem A10).

Example 8: The length plus girth of packages shipped by the Universal Package Service must not exceed 120 inches (Figure 9). The "length" is the length of the longest edge, and the "girth" is the perimeter around the side with the shorter two edges. Suppose the smaller side is a square. Find the box of maximum volume that can be shipped with the Universal Package Service.

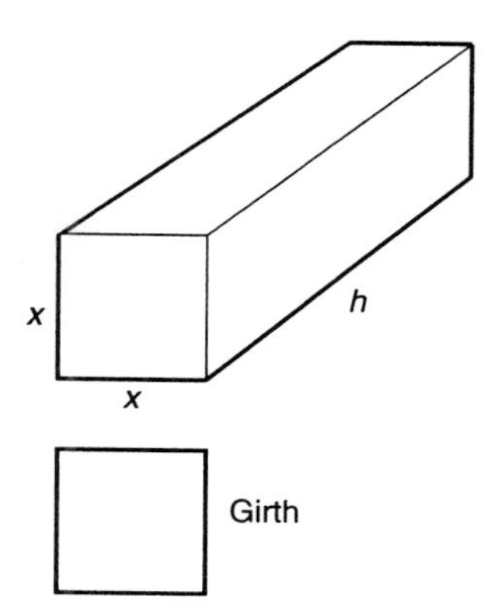

Figure 9: A box with a square side. The distance around the square is the girth.

Reread the problem. Relevant formulas are the volume of a box, and the given constraints. Let "x" be the side of the square. The volume of the box is

$$V = x^2h,$$

where "h" is the length. (Avoid the letter "l" because it looks too much like the numeral "1".)

The perimeter of the square side (the "girth") is given by $P = 4x$. The constraint is

$$4x + h = 120.$$

Therefore, $h = 120 - 4x$. Substituting for h in the volume formula,

$$V = x^2(120 - 4x).$$

This is the formula we want. It can be graphed and maximized using evaluate-and-compare (Problem A11). Calculus techniques also work.

__Conclusion__. Constraints are equations which relate two or more variables. Word problems with constraints are usually best solved by obtaining general formulas first and using particular facts later on in the solution process. Whenever possible, draw a picture. Think functionally. Name argument and image variables, and intermediate variables when needed.

Terms: constraint, intermediate variable.

****

Exercises for Section 3.5, "Word Problems with Constraints":

Note: As usual, report solutions with three significant digits.

A1. Solve the equation in "Example 1, continued." [3.7, ...]

A2. Find the perimeter in Example 2. [41]

A3. Find b that minimizes $P(b)$ in "Example 2, continued." [10]

A4. Solve the equation in Example 3. [4.5,...]

A5. Solve the equation in "Example 4, continued." [2.5,...]

A6. Find the x-value of the minimum of the distance in "More from Example 4." [1.0]

A7. Solve the equation in "Example 5, continued." [67,...]

A8. Find the minimum of the expression in "More from Example 5." [310]

A9. Solve for "A" in Example 6. [1.9]

A10. Minimize the expression in Example 7. [x = ..., cost is 2268 cents]

A11. Maximize the expression for the volume in Example 8. [x = ..., V = 16,000]

^ ^ ^ ^ ^ ^ ^ ^ ^

B1.* Explain what a "constraint" is. Be sure to give the proper context.

B2.* Explain what the "evaluation patterns" of the previous section are supposed to be good for in the context of a word problem.

B3.* Explain what an "intermediate" variable is.

B4. Prove that the maximum area of a rectangle with a fixed perimeter occurs when the rectangle is a square. Use a general variable, P, to represent the fixed perimeter.

B5. In "Example 5, continued" there are two solutions. Explain why, both algebraically (from the appearance of the equation) and geometrically.

B6. Suppose you wish to mix 10% acid solution with 24% acid solution to make 10 liters of 14% acid solution. a) What is the constraint?
b) How much of each should you use? [7.1 liters of 10% solution]

B7. 100 centimeters of wire are cut into two lengths, one of which is bent to form a square and the other a circle. Express the total area of the two figures and find where the cut should be to minimize that total. [56 for the square]

B8. Find the point on the parabola $y = x^2$ which is closest to the point (5, 1). [(1.5,...)]

B9. Find the point on the parabola $y = x(x - 1)$ which is closest to the point (0, 1). [(...,.74)]

B10. A sports club wishes to make an indoor 200 meter track in the shape of two straight parallel lines connected by semicircles on each end. The lines form two sides of a rectangle. Find the rectangle of largest possible area. [straight sides 32]

B11. A jogger runs out at a constant speed and runs back 2 miles per hour faster. She averages 8 miles per hour over the 6 mile round trip. How fast does she run out? [7.1]

B12. It costs \$10 per foot to lay cable underwater, and \$4 per foot to lay it over land along the shore. A lighthouse is 100 feet off shore (Figure 10, L). To connect a terminal (T) 200 feet down the shore to the lighthouse with cable, it is cheaper to lay the cable from the lighthouse to a point on shore somewhat toward the terminal, rather than lay it perpendicular to the shoreline. Where should the cable meet the shore to minimize the cost? [44 feet toward the terminal]

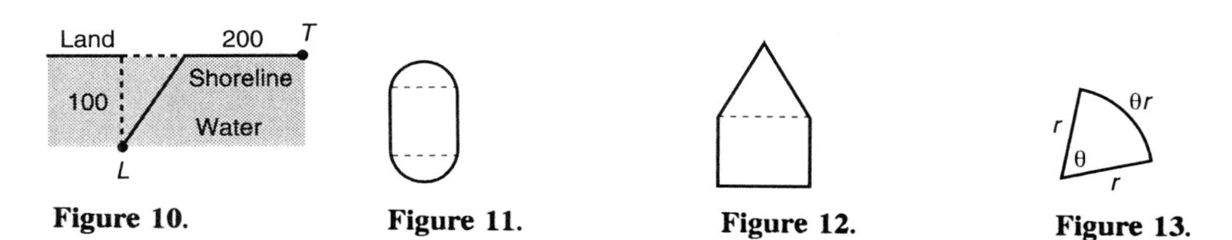

Figure 10. **Figure 11.** **Figure 12.** **Figure 13.**

B13. 100 inches of wire are to be cut into two lengths. One is bent to form a rectangle with one side 10 inches long. The other forms a square. Find the cut which produces the minimum possible total area of the two figures. [60 for the rectangle]

B14. A figure is in the shape of a square with two semicircular caps on opposite sides (Figure 11). Its perimeter is 100. Find its side. [19]

B15. The side of a large outdoor playhouse has the shape pictured in Figure 12. It is 13 feet high and 8 feet wide. The area of the side is 76 square feet. How tall are the vertical sides? [6.0]

B16. The area of a sector of a circle of radius r and central angle θ (in radians) is $A = \theta r^2/2$. The perimeter of the sector is $2r + \theta r$ (two radii plus an arc) (Figure 13). Suppose the perimeter is fixed. Find the value of r which maximizes the area.

B17. Suppose the hypotenuse of a right triangle is twice the length of a leg and the area is 100. Find the leg. [11]

B18. Suppose a semicircle is constructed inside a square using a side of the square as its diameter. The area in the square that is not in the semicircle is 45. Find the side of the square. [8.6]

B19. Consider a 3-4-5 right triangle. Consider all possible rectangles inside the triangle that have two sides along the triangle's sides of length 3 and 4 (Figure 14). Which one has the largest area? [Use algebra.]

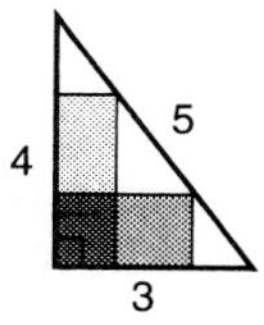

4 5 3

Figure 14.

B20. Let point P be somewhere on the x-axis. Consider the total of the distance from (0, 5) to P and the distance from P to (9, 3). Where should P be located to minimize that total?

B21. Find the largest rectangle (in area) that fits in the first quadrant inside the curve $y = 10 - \sqrt{x}$.

B22. Find the largest rectangle (in area) that fits in the first quadrant inside the line that goes through (5, 1) and (1, 4). [$x = 3.2$]

B23. Jane walks and runs a total of 4 miles. She wants to finish in exactly 40 minutes. If she walks at 4 miles per hour and runs at 8 miles per hour, how many miles should she run?

B24. Find the set of all points (x, y) such that the distance from (x, y) to $(1, 0)$ is the same as the distance from (x, y) to the line $x = -1$. Simplify.

B25. Find the set of all points (x, y) such that the distance from (x, y) to $(1, 0)$ is the same as the distance from (x, y) to $(-1, 0)$. Simplify.

B26. An isosceles triangle has two equal sides of length 20 and area equal to 120 square units. How long is the other side?

B27. The WidgetCan Company wishes to make the cheapest cylindrical metal can which can hold 2000 cubic centimeters. They reckon the cost of the metal as $1 per square meter (with no waste) and the cost of sealing the seams (around the top and bottom, and one up the side) as 1 cent per 10 centimeters. All other costs are considered the same, no matter what the shape of the can is. What are the dimensions of the cheapest can?

B28. In "Example 1, continued" can the fact that the area is 60 square inches be regarded as a constraint?

B29. A warship is sailing due west at 30 miles per hour. When it is 4 miles north of a submarine, the submarine fires a torpedo which travels at 90 miles per hour. The torpedo is aimed so that it will meet the ship without changing course (if the ship does not change course). Where will it meet the ship?

B30. In "More on Example 4," it states "Another way to minimize the expression is to square it and find the minimum of its square, since the x-value which minimizes the square of the distance also minimizes the distance." State this as a concise theorem (with abstract letters) and explain why the theorem is true.

B31. A fisher stands with her eye level six feet above the water, and a fish ten feet out is two feet below the water. Light travels 1.33 times as fast in air as in water. This causes refraction. That is, the light she sees will not come straight from the fish to her eye, but will appear to bend at the surface of the water. A principle of physics says that the light she sees will have traveled the path of least time. In Figure 15, find x such that the time for light to travel from F to P to A is least.

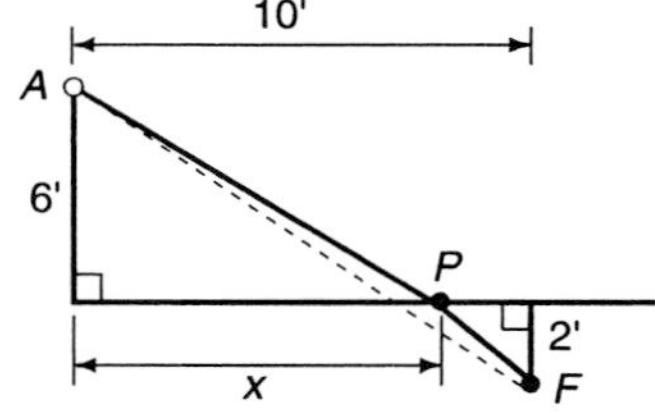

Figure 15:

B32. Suppose the bottom and sides of an open-topped (no top) cubical box cost 0.3 cents per square inch (neglect other contributions to the cost). Find the volume of a box costing 75 cents. [350]

CHAPTER 4

Powers

Section 4.1. Powers and Polynomials

Power functions have images given by powers of x such as x squared (x^2) or x cubed (x^3). In the expression "b^p", "b" is called the base and p is called the power or exponent. If p is regarded as fixed and b as the variable argument (as in "x^2"), we call this a power function. If b is regarded as fixed and p as the variable argument (as in "2^x," "two to the x"), we call this an exponential function. Power and exponential functions are very closely related. This section concentrates on power functions; exponential functions are the subject of the next chapter.

When a power is a positive integer, it is appropriate to think of *repeated multiplication.* The power gives the number of repeated factors in the product. For example, denote

$$2 \times 2 \times 2 \times 2 \text{ by } 2^4 \; (= 16), \text{ and}$$
$$b \times b \times b \text{ by } b^3.$$

Let $b^p = b \times b \times ... \times b$ [p factors]. Clearly, for any p,

$$1^p = 1. \tag{4.1.0}$$

The properties of integer powers follow from the repeated-multiplication interpretation. For example, a product of integer powers is easy to express in an alternative form.

$$b^2b^3 = (b \times b) \times (b \times b \times b) = b^5.$$

Evidently, the power (5) of a product ($b^2 \times b^3 = b^5$) is the sum of the powers ($2 + 3 = 5$). This generalizes to:

$$b^pb^r = b^{p+r}. \tag{4.1.1}$$

Many students find this identity hard to grasp. The left side displays *powers and multiplication*, in that order. But the right side displays *addition and a power*, in that order. Not only the order changes, but also the operations change.

Example 1: $x^2x^4 = x^6$ [not x^8].
$(2x + 1)^3(2x + 1)^4 = (2x + 1)^7$ [not $(2x + 1)^{12}$].

A quotient of different powers is easy to simplify if the bases are the same and the power in the numerator greater than the power in the denominator. For example,

$$\frac{b^5}{b^2} = \frac{b \times b \times b \times b \times b}{b \times b}$$
$$= b \times b \times b = b^3 .$$

Evidently, the power (3) of a quotient ($b^5/b^2 = b^3$) is the difference of the powers (5 - 2 = 3). This generalizes to (if $b \neq 0$):

(4.1.2A)
$$\frac{b^p}{b^r} = b^{p-r}.$$

Example 2: $\frac{x^6}{x^2} = x^4$ [*not* x^3].

This example simplified a quotient of powers when the top power is greater than the bottom power. But what if the top power isn't greater? If it isn't, we can use this property to define what we mean by the 0 power and negative powers.

Here are some powers of 2.

power	power of 2	
4	$16 = 2^4$	
3	$8 = 2^3$	
2	$4 = 2^2$	
1	$2 = 2^1$	
0	1	so $1 = 2^0$
-1	1/2	so $1/2 = 1/(2^1) = 2^{-1}$
-2	1/4	so $1/4 = 1/(2^2) = 2^{-2}$
-3	1/8	so $1/8 = 1/(2^3) = 2^{-3}$

The results for *positive* powers fit the idea of repeated multiplication. There is a pattern. When the power is increased by 1, the image is multiplied by 2. When the power is decreased by 1, the image is divided by 2. Continue that pattern to determine the zero power and negative powers.

The power-function pattern illustrated for base 2 holds for any base b. To *add* 1 to the power is to *multiply* the image by b, the base. To *subtract* 1 from the power is to *divide* the image by b. Therefore,

(4.1.2B) $$b^0 = 1.$$

(4.1.2C) $$b^{-p} = \frac{1}{b^p} = (\frac{1}{b})^p .$$

Example 3: $$3^{-2} = \frac{1}{3^2} = \frac{1}{3 \times 3} = \frac{1}{3} \times \frac{1}{3} = (\frac{1}{3})^2 .$$

$$x^{-1} = \frac{1}{x} . \qquad x^{-2} = \frac{1}{x^2} .$$

$$x^3 x^{-4} = x^{-1} = 1/x.$$

$$\frac{x^2}{x^{-1}} = x^{2 - -1} = x^3 .$$

Example 4: Find $P(x)$ that satisfies the given equation.
$x^4 P(x) = x^6$. Then $P(x) = x^2$.
$(x - 1)^3 P(x) = (x - 1)^7$. Then $P(x) = (x - 1)^4$.
$x^4 P(x) = x^{-3}$. Then $P(x) = x^{-7}$.
$x^{-3} P(x) = x^{-2}$. Then $P(x) = x$.

Example 5: Factor $3x^2(1 - x)^5 - 5x^3(1 - x)^4$.
Both terms have "x" to a power and "$1 - x$" to a power. Factor out the largest powers the terms have in common, that is, the lesser power of each: x^2 and $(1 - x)^4$. It equals

$$x^2(1 - x)^4[3(1 - x) - 5x]$$
$$= x^2(1 - x)^4[3 - 8x].$$

Example 6: Factor $5x^{-2}(1 - x)^4 + 2x^{-3}(1 - x)^5$.
The two terms have factors in common. Factor out the *lesser* power of each. It equals

$$x^{-3}(1 - x)^4[5x + 2(1 - x)]$$
$$= x^{-3}(1 - x)^4[3x + 2].$$

Now consider powers of powers. If both powers are integers, the idea of repeated multiplication shows, for example,

$$(b^2)^3 = b^2 \times b^2 \times b^2 = (b \times b) \times (b \times b) \times (b \times b) = b^6.$$

In general,

(4.1.3) $$(b^r)^p = b^{rp}.$$

Example 7: $(2^3)^4 = 2^{12}$.

$(x^2)^5 = x^{10}$.

$((1/2)^3)^2 = (1/2)^6 = 2^{-6}$, using 4.1.3 and then 4.1.2C. You may obtain the same result another way: $((1/2)^3)^2 = (2^{-3})^2 = 2^{-6}$, using 4.1.2C and then 4.1.3.

Powers distribute over multiplication. For example,

$$\begin{aligned}(2 \times 5)^3 &= (2 \times 5)(2 \times 5)(2 \times 5) \\ &= (2 \times 2 \times 2)(5 \times 5 \times 5) \\ &= 2^3 5^3\end{aligned}$$

You know what the word "distribute" means in English: to deal out or allot. The same meaning holds in Mathematics. The power in the expression $(ab)^2$ is distributed to each of the two factors, a and b, to form a^2b^2. Note the parallel with the Distributive Property of multiplication over addition: $a(b + c) = ab + ac$. The factor (a) which multiplies the sum $(b + c)$ is distributed to both terms to form $ab + ac$.

This generalizes to:

(4.1.4A) $$(ab)^p = a^p b^p.$$

Similarly (Problem B9),

(4.1.4B) $$\left(\frac{a}{b}\right)^p = \frac{a^p}{b^p}.$$

Example 8: $(5x)^2 = 5^2x^2 = 25x^2$.

$(x/3)^2 = x^2/3^2 = x^2/9$.

$(2/x)^3 = 2^3/x^3 = 8/x^3 = 8x^{-3}$.

Negative numbers to *integer* powers are positive or negative depending upon the number of factors. For example,

$$(-7)(-7) = 7^2.$$

$$(-7)(-7)(-7) = -7^3.$$

Be careful with the order conventions. -7^2 is negative. It is $-(7^2)$. -7^2 is not $(-7)^2$. Powers are executed before multiplication, and the minus sign is treated like multiplication by -1. So $-7^4 = -(7^4)$, not $(-7)^4$.

These examples generalize to:

(4.1.5) If n is even, $(-b)^n = b^n$.
If n is odd, $(-b)^n = -b^n$.

Here are the properties repeated with base "x" replacing "b".

Table 4.1.6. Properties of Powers

For $x > 0$, $a > 0$, and any p and r,

(4.1.1). $x^p x^r = x^{p+r}$.

(4.1.2A) $\dfrac{x^p}{x^r} = x^{p-r}$.

(4.1.2B) $x^0 = 1$.

(4.1.2C) $x^{-p} = \dfrac{1}{x^p} = (\dfrac{1}{x})^p$.

(4.1.3) $(x^r)^p = x^{rp}$.

(4.1.4A) $(ax)^p = a^p x^p$.

(4.1.4B) $(\dfrac{x}{a})^p = \dfrac{x^p}{a^p}$.

(4.1.5) If n is even, $(-x)^n = x^n$.
If n is odd, $(-x)^n = -x^n$.

Earlier in this section these properties were motivated with *integer* powers. Nevertheless, they hold for all *real-valued* powers if b or x is *positive*.

Example 9: For positive values of x,

$x^\pi x^2 = x^{\pi+2}$, by 4.1.1
$x^2/x^{1.5} = x^{0.5} = \sqrt{x}$, by 4.1.2A
$(x^{1.3})^2 = x^{2.6}$, by 4.1.3
$(3x)^{.7} = (3^{.7})x^{.7}$, by 4.1.4A
$(x/5)^{1/4} = x^{1/4}/5^{1/4}$, by 4.1.4B

These fractional powers of x do not yield real numbers for negative values of x. If x is negative, the properties in the table still hold for *integer* powers. However, for fractional powers of negative numbers there are complications which will be discussed in Section 4.3.

Polynomials. Because integer powers come from repeated multiplication, polynomials can be evaluated using methods from arithmetic. Logarithmic, exponential, and trigonometric expressions cannot. We can exactly evaluate "2.3^4," but there is no way to exactly evaluate "log 2.3" or "$2^{2.3}$ " or "sin 2.3."

Definition 4.1.7: A monomial is any expression of the form "cx^n ", where n is a non-negative integer. If $c \neq 0$, n is its degree and c is its coefficient.

Discovery 1: Graph x^p for various positive integer values of p, including $p = 1, 2, 3, 4,$ and 5 (Problem A1). Which resemble each other?
Which are one-to-one and therefore have unique solutions to "$x^p = c$"? State your result as a theorem.

Example 10: "$4.5x^3$ " is a monomial of degree 3 with coefficient 4.5.

"$-x^2$ " is a monomial of degree 2 with coefficient -1.

"7" is a monomial of degree zero, because it can be regarded as "$7x^0$," since $x^0 = 1$. The coefficient is 7. All numbers are monomials.

"$x^5/3$" is a monomial of degree 5 with coefficient 1/3 (not 3, coefficients are *multiplicative* constants).

Definition 4.1.8: A polynomial is either a monomial, or a sum or difference of two or more monomial terms. The degree of a polynomial is the highest power of any of its terms. The leading coefficient is the coefficient on the highest-power monomial term.

Example 11: "$x^2 + 2x + 1$" is a polynomial of degree 2 with leading coefficient 1. All quadratics are polynomials of degree 2.

"$3x - 7$" is a polynomial of degree 1 with leading coefficient 3. All linear expressions of the form $mx + b$ where m is not zero are of degree 1.

"$7x^2 - 5x^3$" is a polynomial of degree 3 with leading coefficient -5. The terms "degree" and "leading coefficient" refer to the term with the highest power, which is not necessarily on the left.

"$3x^2(5x^4 + 2x + 12)$" is a sixth degree polynomial written in factored form, with leading coefficient 15.

Graphs of Polynomials. This section concentrates on cubic and higher-degree polynomials because we have already thoroughly discussed polynomials of degree 2 (quadratics) and degree 1 (lines).

Example 12: Cubic polynomials ("cubics") are polynomials of degree 3. Unlike quadratics, they come in a variety of shapes. The basic cubic, x^3, is familiar (Figure 1). Images in the first quadrant correspond to images in the third quadrant. This is because

$$(-x)^3 = -(x^3),$$

by 4.1.5. For positive x, the point (x, x^3) is on the graph in the first quadrant. Then $(-x, -x^3)$ is on the graph in the third quadrant. Every point in the first quadrant corresponds to another point in the third quadrant. The graph of x^3 is symmetric about the origin.

Discovery 2: What slope does the graph of x^p have at the origin for $p \geq 2$? Graph x^3, x^4, and x^5 on [-2, 2] by [-2, 2] and see.

Which two points do all these graphs go through (Problem A2)? State your results as theorems.

Definition 4.1.9. Functions (such as f defined by $f(x) = x^3$) with the property that $f(-x) = -f(x)$ for all x are said to be odd functions. The term "odd" is used because all odd-power monomials have this property (4.1.5).

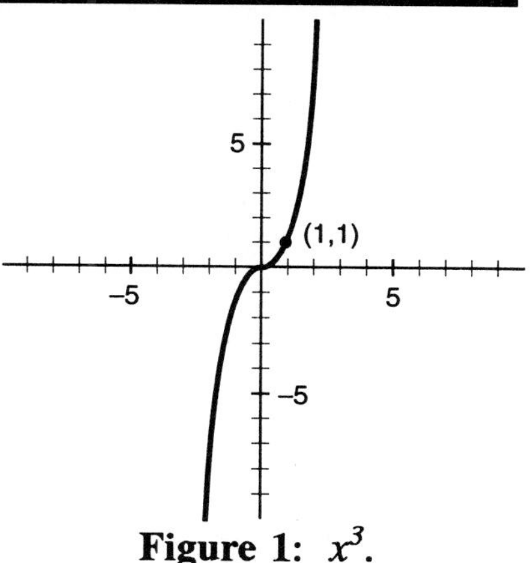

Figure 1: x^3.
[-10, 10] by [-10, 10].

The graphs of odd functions are said to be point-symmetric about the origin. That is, if (a, b) is on the graph, so is $(-a, -b)$, which is equidistant from the origin and directly opposite (a, b) through the origin.

The graph of "$y = x$" ($x = x^1$, a first degree monomial) has this point symmetry. Note that the graphs of odd functions are not limited to first and third quadrant points -- they may exhibit point symmetry about the origin between second and fourth quadrant points (Figure 2).

The graph of $x^3 - 5x$ exhibits two local extrema (Figure 2. "Extremum" is singular. "Extrema" is plural). We will say a graph has a "local extremum" where it has either a "local maximum" or a "local minimum." A graph has a "local maximum at $x = x_0$" when there is an interval surrounding x_0 for which $f(x) \leq f(x_0)$ for all values of x in the interval. That is, the graph is locally highest there. In Figure 2, the graph has a local maximum at $x = -1.29$, where the graph is higher than at nearby x values. This is not the overall highest point on the graph (which is why the maximum is only "local"). The key to a "local maximum at $x = x_0$" is that $f(x_0)$ is at least as great as $f(x)$ for all *nearby* x-values. Similarly, a graph has a local minimum at $x = x_0$ if $f(x_0) \leq f(x)$ for all values of x in some interval surrounding x_0. Figure 2 exhibits a local minimum at $x = 1.29$.

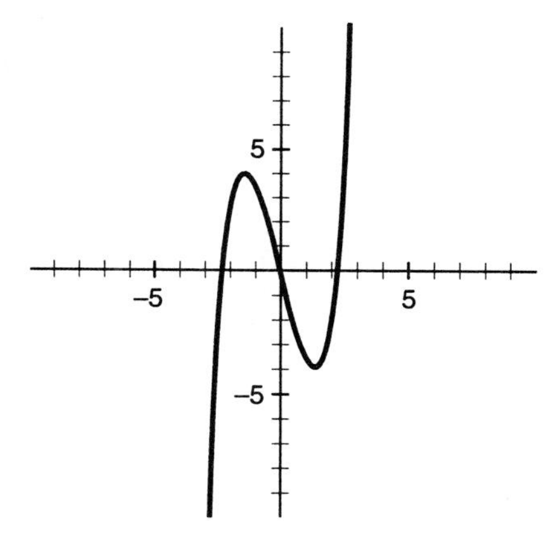

Figure 2: $x^3 - 5x$.
[-10, 10] by [-10, 10].

Quadratics have one local extremum. For example, the well-known graph of x^2 exhibits a local minimum at $x = 0$ (Figure 3.2.1), and the graph of $-x^2$ exhibits a local maximum at $x = 0$.

Cubics may have two local extrema, or no local extrema, but they cannot have exactly one local extremum or three or more local extrema. It takes a higher degree polynomial to have three or more local extrema.

Example 13: The graph of x^4 (Figure 3) is somewhat similar to the graph of x^2. All even power monomials have some features in common. They are symmetric about the y-axis. If p is even, from 4.1.5,

$$(-x)^p = x^p,$$

which says that the images to the left of the y-axis (images of $-x$) duplicate the images to the right of the y-axis (images of x).

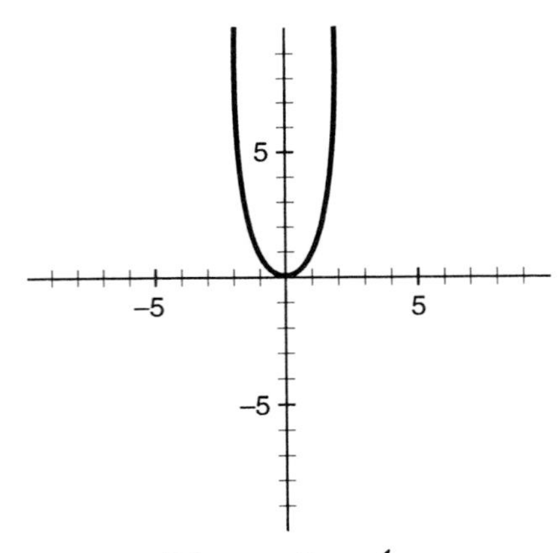

Figure 3: x^4.
[-10, 10] by [-10, 10].

Definition 4.1.10. Functions (such as f defined by $f(x) = x^2$) with the property that $f(-x) = f(x)$ for all x are called even functions. The term "even" is used because all even-power monomials have this property (4.1.5). Graphs of even functions are symmetric about the y-axis.

The extreme case is the constant function, $f(x) = c$, which is graphed as a horizontal line, which is clearly symmetric about the y-axis. Recall that $c = cx^0$, so it can be regarded as a constant times an even power of x (0 is an even number).

The graph of x^4 has only one local extremum (Figure 3). The graph of $x^4 - 5x^2$, on the other hand, has three (Figure 4).

In general, a polynomial of degree n can have at most $n - 1$ local extrema.

A graph is symmetric...

about the y-axis iff $(-x, y)$ is on the graph whenever (x, y) is.
about the origin iff $(-x, -y)$ is on the graph whenever (x, y) is.
about the x-axis iff $(x, -y)$ is on the graph whenever (x, y) is. Because functions have a unique y-value for each x-value, graphs of functions generally are not symmetric about the x-axis.

Odd degree polynomials have even numbers of local extrema.

For example, lines -- of degree 1 -- have 0 local extrema. Cubics may have 2 (Figure 2) or zero (Figure 1).

Even degree polynomials have odd numbers of local extrema.

For example, quadratics (of degree 2) have 1 local extremum, and polynomials of degree 4 have either 1 or 3.

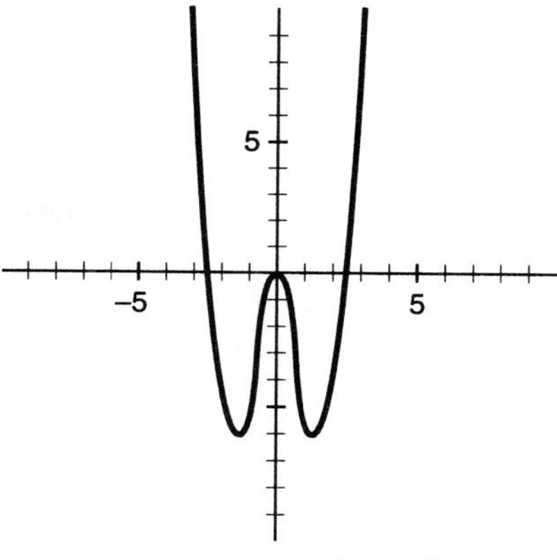

Figure 4: $x^4 - 5x^2$.
[-10, 10] by [-10, 10].

Example 14: Graph $x^3(x - 2)$.

This is a polynomial of degree 4. Instead of having 3 local extrema, it has 1 (Figure 5).

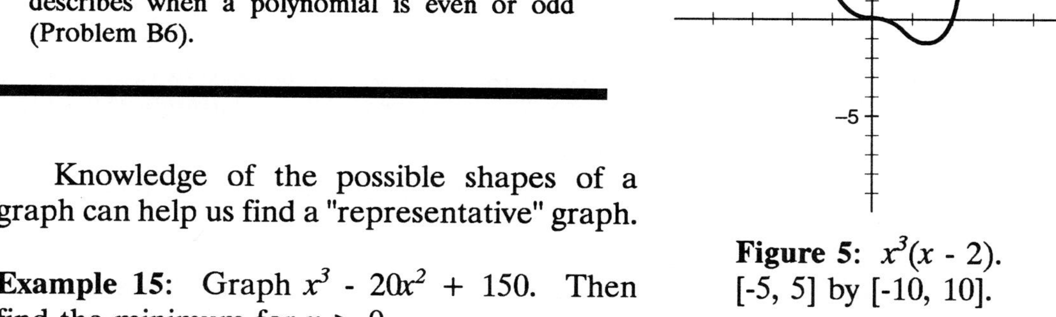

Discovery 3: In Example 14, $x^3(x - 2)$ is a polynomial of degree 4, but it is not an "even" function. Why not?

Look at the graphs of $x^2 + x$, $x^2 + x + 1$, and $x^2 + 1$. Which are even? Look at the graphs of $x^3 + x^2$, $x^3 + x$, and $x^3 + 1$. Which are odd? Generalize. That is, state a theorem that describes when a polynomial is even or odd (Problem B6).

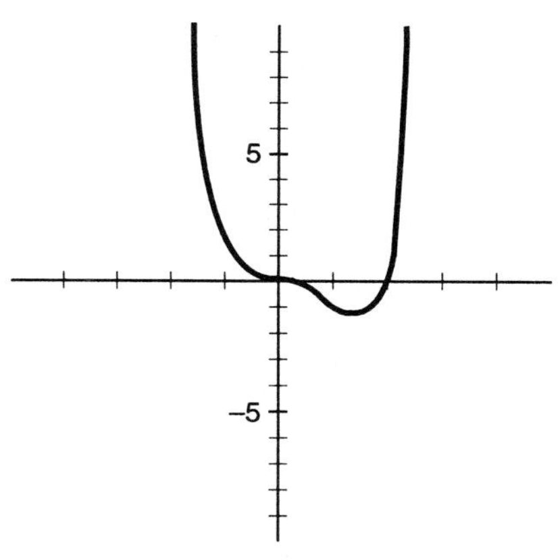

Figure 5: $x^3(x - 2)$.
[-5, 5] by [-10, 10].

Knowledge of the possible shapes of a graph can help us find a "representative" graph.

Example 15: Graph $x^3 - 20x^2 + 150$. Then find the minimum for $x > 0$.

The picture on the standard scale is not very illuminating; try it yourself. The picture consists of two nearly vertical slashes. The window is wrong. A glance at the expression tells us that y-values can be as large as 150, so we must change the vertical scale. Try, say $-500 \leq y \leq 500$ (Figure 6).

If you didn't know what cubics look like, you might think Figure 6 is a representative graph. It's not.

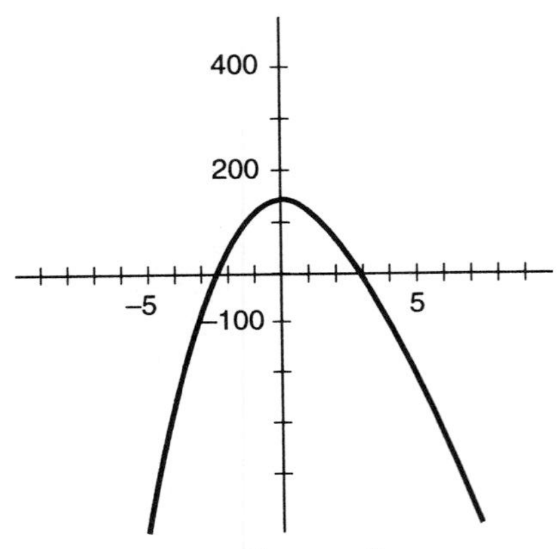

Figure 6: $x^3 - 20x^2 + 150$.
[-10, 10] by [-500, 500].

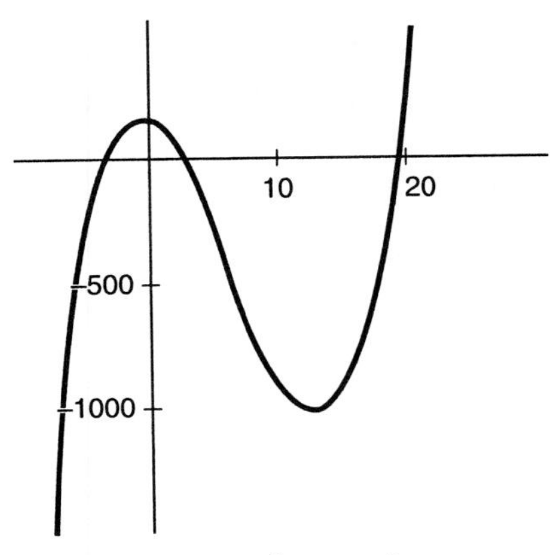

Figure 7: $x^3 - 20x^2 + 150$
[-10, 30] by [-1500, 500].

Cubics can not have exactly one local extremum. There must be a second. The window is still not right. Try again. In Figure 7 the scale is [-10, 30] by [-1500, 500]. The picture looks like a cubic. It must be representative. The minimum, for $x > 0$, is easy to find (Problem A27).

<u>**End Behavior**</u>. Some features of the graph of a polynomial are determined by its degree and its leading coefficient. If you "zoom out" far enough, the graph of any polynomial looks much like the graph of its highest-degree monomial. To express this clearly, we assign a name to the highest-degree monomial in a polynomial.

<u>Definition 4.1.11</u>. The <u>end-behavior model</u> of a polynomial is its highest-degree monomial term.

Example 16: The end-behavior model of "$x^2 + 2x + 1$" is "x^2".
The end-behavior model of "$7 - 5x^3$" is "$-5x^3$."
The end-behavior model of "$2x^{10} + 165x^8 + 942$" is "$2x^{10}$."
The end-behavior model of "$(x^2 - 4)(3x - 5)$" is "$3x^3$."

The point of the end-behavior model is that for large x, the polynomial behaves much like its leading term. If, as in Figure 6, your picture does not look like the end-behavior model for large x, you know you do not have a representative graph.

Example 17: The end-behavior model of the polynomial $2x^3 - 7x - 3$ is $2x^3$. For large values of x, the other part, $-7x - 3$, is large, but $2x^3$ is far larger. The behavior of the polynomial for large absolute values of x is determined by the behavior of $2x^3$. For small values of x the value of "$-7x + 3$" is substantial relative to the value of "$2x^3$". On a small scale the graphs of the polynomial and its end-behavior model need not be alike (Figure 8). But on a large scale, their "end-behaviors" are alike (Figure 9).

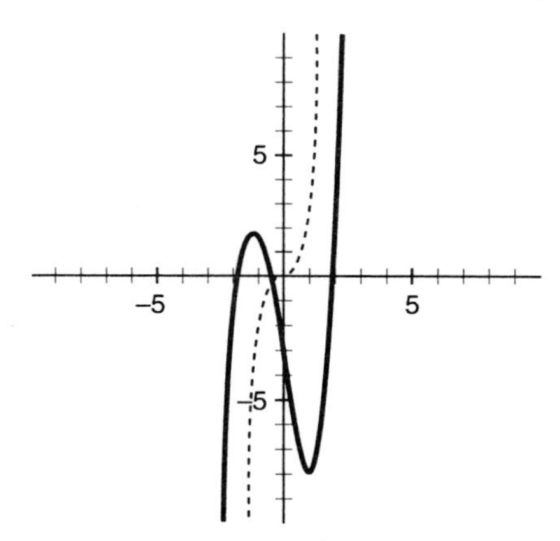

Figure 8: $2x^3$ - $7x$ - 3, and $2x^3$ (dotted). [-10, 10] by [-10, 10].

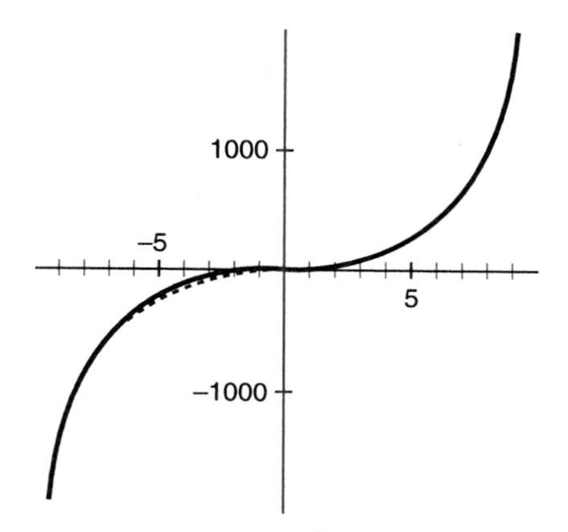

Figure 9: $2x^3$ - $7x$ - 3 and $2x^3$. [-10, 10] by [-2000, 2000]

Example 18: A fourth-degree polynomial may have three local extrema. For small x, the graph of x^4 - $5x^2$ (Figure 4) is quite unlike the graph of x^4 (Figure 3), its end-behavior model. But for large x, they are very similar (Figure 10).

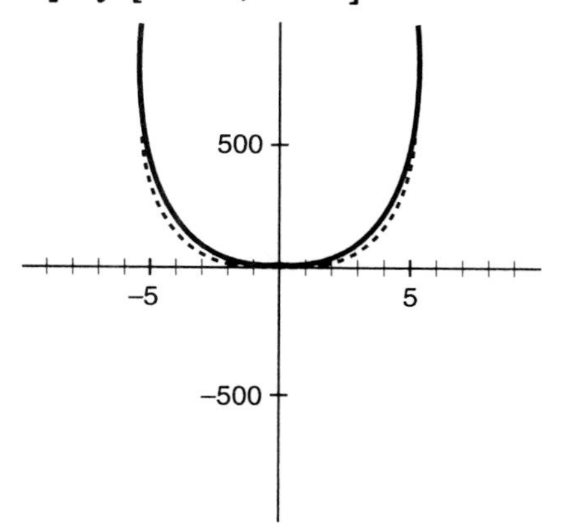

Figure 10: x^4 - $5x^2$ and x^4 (slightly higher). [-10, 10] by [-1000, 1000].

A crude indication of the behavior of the graph of a polynomial for large absolute values of x can be given with two arrows, one to indicate its behavior on the far left and one for the far right. For example, the end-behavior of the graph of x^2 might be described by "↑↑" and the graph of x^3 by "↓↑". The graph of $-x^2$ would be described by "↓↓" and the graph of $-x^3$ by "↑↓".

<u>**The Use of Polynomials**</u>. Polynomials can express real-world functions such as the relationships from physics and business discussed in the section on quadratic functions. Cubics also serve for three-dimensional volume problems. There are applications of polynomials in many different areas.

Example 19: Suppose voters are chosen at random from a huge pool of voters and asked whom they prefer, candidate A or candidate B. Probability theory gives the probability of any sequence of responses. For example, the

probability of 5 responses being in favor of A, then B, then B, then A, and then A is approximately

$$x^3(1 - x)^2,$$

where x is the true fraction ($0 \le x \le 1$) of voters who would select candidate A (Figure 11). The power "3" is because 3 voters preferred A, and the power "2" is because 2 voters preferred B. This is a fifth-degree polynomial in factored form.

Here is an example of an important type of problem in statistics: Find x which maximizes that expression over the domain $0 \le x \le 1$.

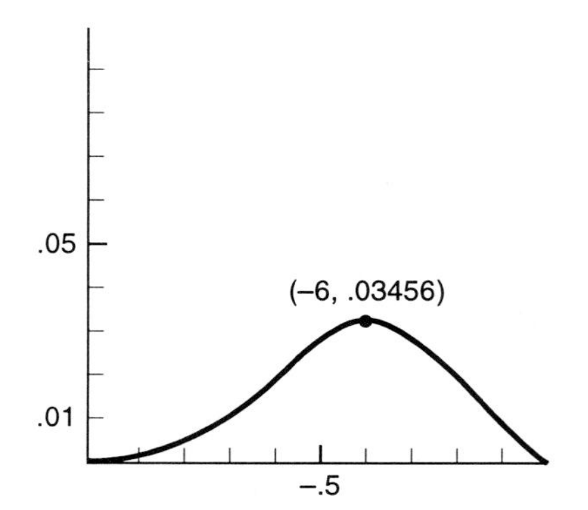

Figure 11: $x^3(1 - x)^2$. [0, 1] by [0, .1].

Calculus will provide an exact answer. But we do not have to wait for calculus -- we can obtain an answer graphically. As usual, the scale is important. If you use the standard scale the tiny region of interest with x between 0 and 1 will be hard to see. Figure 11 uses a viewing rectangle [0, 1] by [0, .1]. Note the small y-interval. The maximum occurs when $x = .6$ (Problem B23). Using calculus, it can be proved that this numerical result is exact.

Approximation. Your calculator will evaluate "sin x" for you to numerous decimal places of accuracy. How does it do it?

It used to be, a mere 30 years ago, that people looked up $\sin x$ in printed tables. We may, instead, approximate $\sin x$ with a polynomial, which can be evaluated using only arithmetic operations that your calculator can rapidly execute.

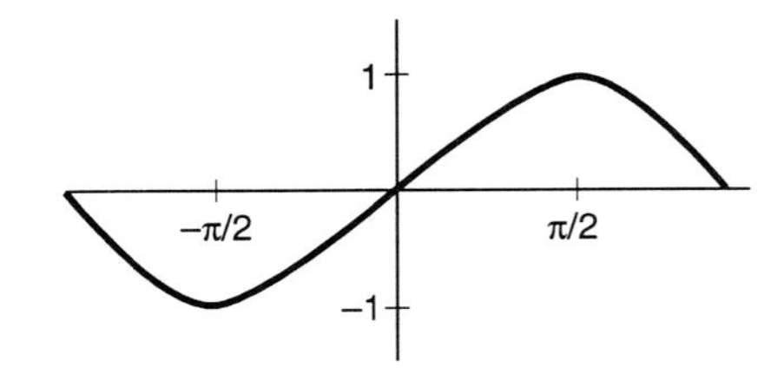

Figure 12: $\sin x$. $-\pi \le x \le \pi$. The range is $-1 \le y \le 1$.

When the sine curve is graphed on the domain in Figure 12, $-\pi \le x \le \pi$ (equivalent to -180° to 180°, where the angle is measured in radians), it almost looks like a polynomial -- perhaps a cubic with two local extrema.

Example 20: Consider $\sin x$, where x is in radians. Near the origin the graph of $\sin x$ is almost a straight line, $y = x$ (Figure 12). For small values of x, $\sin x$ is nearly x. $\sin .1 = 0.0998$, only 0.0002 off. $\sin .2 = 0.1987$, only 0.0013 off. Therefore, the linear polynomial "x" can approximate $\sin x$ for x near 0.

The graphs of $\sin x$ and x together show that the linear approximation rapidly becomes worse as x moves further from 0. $\sin 1 = .84$, not very close to 1. However, a cubic approximation can do better:

For small x,

sin x is approximately $x - x^3/6$.

x	0.1	0.2	0.5	1.0	$\pi/2 = 1.5708$
$\sin x$ $x - \sin x$	0.099833417 0.000166583	0.198669331 0.001330669	0.4794255 0.0205745	0.84147 0.15853	1 0.5708
$x - x^3/6$ $\sin x - (x - x^3/6)$	0.099833333 0.000000084	0.198666667 0.000002664	0.4791667 0.0002588	0.83333 0.00814	0.9248 0.0752

Higher degree polynomials can give better approximations (problem B35). A polynomial that is accurate to within 2 digits in the ninth decimal place over the interval $[0, \pi/2]$ is:

$$\sin x = x(1 + a_2x^2 + a_4x^4 + a_6x^6 + a_8x^8 + a_{10}x^{10}),$$

where $a_2 = -.4999999963$, $a_4 = .0416666418$, $a_6 = -.0013888397$,
$a_8 = .0000247609$, and $a_{10} = -.0000002605$.

No! You don't have to memorize this! If you want to know how someone figured that out, you will have to take more math!

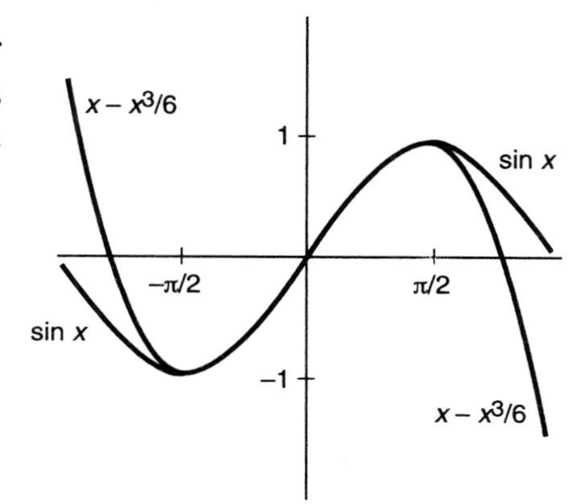

Figure 13: sin x and $x - x^3/6$. $[-\pi, \pi]$ by $[-2, 2]$.

<u>**Conclusion**</u>. Integer powers can be interpreted as repeated multiplication. If the base is positive, all the properties of powers can be understood by considering repeated multiplication, even if the powers are not integers. If the base is negative, the usual properties of integer powers hold, but there are complications for fractional and irrational powers.

Polynomials are very important because they can be evaluated using only arithmetic operations. Monomials are an important special case of polynomials. Monomials serve as end-behavior models for all polynomials, and the graphs of monomials are fundamental.

Many special functions which would otherwise be very difficult to evaluate can be closely approximated by polynomials, at least over a restricted domain.

Terms: base, power, exponent, monomial, polynomial, odd function, point-symmetric about the origin, local extremum, local maximum, local minimum, even function, symmetric about the y-axis, end-behavior model.

****

Exercises for Section 4.1, "Powers":

A1.* Do "Discovery 1." a) Sketch the graphs of x^3, x^4, and x^5. b) Which two look alike? c) Which have unique solutions to "$x^p = c$," for any c? State your result as a theorem.

A2.* Do "Discovery 2." a) All graphs of x^p for positive values of p go through the two points ________ and ________. b) For all $p \geq 2$, the slope of the graph of x^p at the origin appears to be ________.

A3.* a) Discuss whether all quadratics have similar shapes.
b) Do all cubics have similar shapes?

A4.* Comment on these assertions: a) x^2 and x^4 and x^6 have graphs with somewhat similar shapes. b) x^3 and x^5 and x^7 have graphs with somewhat similar shapes.

^^^^Simplify to an equivalent expression with only one appearance of "x":

A5. a) x^5/x^2. b) x^6x^4 c) $(x^3)^7$.
A6. a) $(2x)^5/(8x)$. b) x^{-3}/x c) $(x^{-4})x^5$.
A7. a) $(4x)^3/(2x)^2$. b) $(\sqrt{x})^6$. c) $x/\sqrt{x}$.
A8. a) $\sqrt{x}/x$. b) x^2x^5. c) x^2/x^7.

^^^^Find $P(x)$ given

A9. $x^5P(x) = x^{15}$.
A10. $x^4P(x) = x^7$.
A11. $(x - 2)^2 P(x) = (x - 2)^3$.
A12. $(\log x)^2 P(x) = (\log x)^5$.

^^^^Factor

A13. $4x^3(1 - x)^2 - x^4(1 - x)$.
A14. $5x^3(1 - x)^3 - 3x^5(1 - x)^2$.
A15. $4(2x + 1)^3(x - 2)^2 + 2(x - 2)(2x + 1)^4$.
A16. $2(x + 3)(x - 2)^3 + 3(x + 3)^2(x - 2)^2$.

^^^^Give the end-behavior model of the polynomial.

A17. $2x^3 - 6x^2$.
A18. $4x^5 - 20x^2 + 14$.
A19. $5 - x^2$.
A20. $5x + 13 - 3x^3$.
A21. $(x - 3)(5x^2 + 4)$.
A22. $(4x - 1)(x + 2)$.
A23. $(2x + 1)(x - 3)(x - 5)$.
A24. $(x^2 + 3)(5x^2 + 7)$.

^^^^Which of the following can not be simplified to only one appearance of "x":

A25. a) $x^7 - x^4$ b) $x^3(1 - x^2)$ c) $5x^2(x^6)$
A26. a) $x\sqrt{x}$ b) $x(x^2 + 2)$ c) $x^2/\sqrt{x}$

A27. Find, for $x > 0$, the minimum of $x^3 - 20x^2 + 150$. [-1000]

A28. a) Give the first five powers of 2. b) Approximately what magnitude is 2^{10}? [Powers of 2 are numbers you will see in articles about computers.]
c) Approximately what magnitude is 2^{20}?

^^^^Decide if the given expression is "odd" (4.1.9), "even" (4.1.10), or neither.

A29. x^4.
A30. x^5.
A31. $x^6 + 2x^2$.
A32. $x^2 + 7$.
A33. $x^3 + 5$.
A34. $x^3 + 6x$.
A35. $x^3 + 3x + 5$.
A36. $x^3 + 3x^2$.
A37. $x^4 + 3x^2 + 3$.
A38. $x^4 + 4x^3$.

A39.* Distinguish between "monomial" and "polynomial."

^^^^^^^^^

B1.* Sketch, roughly, all the shapes that graphs of monomials can have.

B2.* a) How many different types of shapes can cubics have?
b) Sketch an example of each shape.
c) Do you think that turning a graph upside down makes a new shape?

B3.* Discuss how many local extrema the graph of a polynomial can have.

B4.* a) Define "end-behavior model."
b) Comment on this remark of a student (short essay): "The graph a polynomial looks like the graph of its end-behavior model."

B5.* Explain, as if to a younger student who does not know, why any number (except possibly 0) to the 0 power is 1.

B6. Do "Discovery 3." In Example 14, $x^3(x - 2)$ is a polynomial of degree 4, but it is not an "even" function. Why not? Look at the graphs of $x^2 + x$, $x^2 + x + 1$, and $x^2 + 1$. Which are even? Look at the graphs of $x^3 + x^2$, $x^3 + x$, and $x^3 + 1$. Which are odd? Generalize. That is, state a result that describes when a polynomial is even or odd.

^ ^ ^ ^ Explain (include an example) why
B7.* $a^n a^m = a^{n+m}$. B8.* $(ab)^n = a^n b^n$. B9.* $(a/b)^n = a^n/b^n$.

^ ^ ^ ^ Note which can <u>not</u> be rewritten directly using properties of powers:
B10. a) a^2b^3. b) a^3b^3. c) $(a^2 + b^2)^{1/2}$.
B11. a) $5^3 6^4$. b) $(x^2 + 4)^{1/2}$. c) $(xy^2)^2$.
B12. a) $x^{1/3} + y^{1/3}$. b) $(a/b)^3$. c) $x^2 + y^2$.

^ ^ ^ ^ Simplify:
B13. $(1/9)^n\, 3^{2n+1}$. B14. $5^{n+1}/25^{n/2}$.
B15. $2^n(1/2)^{2n+1}$. B16. $4^n/2^{3n}$.

^ ^ ^ ^ Suppose a polynomial has the following characteristics. What degree is it?
B17. 2 local maxima and one local minimum. B18. 1 local maximum.
B19. 3 local minima and 2 local maxima. B20. 2 local extrema.

B21. Consider a sheet of metal 12 inches by 20 inches. Cut squares out of each corner and fold up the sides to make an open-topped box. Find the size of the cutout squares that maximizes the volume of the box. [2.4]

B22. Consider a sheet of metal in the form of a right triangle with base 36 inches and height 20 inches. What are the dimensions of the rectangle of largest area that can be cut from that sheet?

B23. Find a window so that the expression $x^3(1 - x)^2$ in Example 19 exhibits the behavior you would expect of a fifth-degree polynomial (Figure 11 does not). Then sketch it and label the viewing rectangle.

B24. Near 0, cos x (in radians) can be approximated by $1 - x^2/2 + x^4/24$. What is the maximum error of the approximation on the interval [0, .5]? [.000022]

B25. Sketch the possible shapes of the graph of $ax^4 + bx^3 + cx^2 + dx + e$.

B26. Sketch the possible shapes of the graph of $ax^5 + bx^4 + cx^3 + dx^2 + ex + f$.

B27. What are the characteristics of the graph of an even function?
B28. What are the characteristics of the graph of an odd function?

B29. State the property of e^x that parallels:
a) 4.1.1. b) 4.1.2A. c) 4.1.2B. d) 4.1.2C.

B30. a) Prove that a sum of two odd functions is an odd function.
b) Prove that a product of two odd functions is an even function.

B31. a) Prove that a sum of two even functions is an even function.
b) Prove that a product of two even functions is an even function.
c) True or false? The product of and odd and an even function is an even function.

B32. The definition of "monomial" was given both in English and symbolically in 4.1.7, but the definition of "polynomial" in 4.1.8 was given only in English. Give it symbolically.

B33. See Example 12. The probability of a sequence of n randomly chosen voters responding to form a particular string of a choices for candidate A and the rest of the choices for B is given approximately by $P(x,a,n) = x^a(1 - x)^{n-a}$.
Find the x-value which maximizes it over $0 \leq x \leq 1$ for $a = 6$ and $n = 17$.

B34. In calculus the idea of end-behavior is formalized after introducing the concept of a limit. In essence, the idea is that the values of the ratio of the polynomial to its end-behavior model approach 1 as x becomes large in absolute value. Let $P(x)$ be the polynomial and $E(x)$ be its end-behavior model. *As x goes to* $\pm\infty$, $\dfrac{P(x)}{E(x)}$ *goes to* 1. *

Therefore, for large absolute values of x, the ratio of $P(x)$ to $E(x)$ is approximately 1. Justify the starred assertion.

B35. $x - x^3/6 + x^5/120$ is approximately $\sin x$, for small x (in radians). Find the largest error used in approximating $\sin x$ by it when $0 \leq x \leq \pi/2$.

B36. A polynomial approximation of degree 2 to e^x near $x = 0$ is given by "$1 + x + x^2/2$" (Figure 15). How accurate is it a) For $x = 0$? b) For $x = .1$?
c) Find the largest value of x such that the error of the approximation is no more than 1% of the true value.

Section 4.2. Polynomial Equations

Polynomial expressions can be *evaluated* using only arithmetic operations. But polynomial equations cannot necessarily be *solved* using only arithmetic operations. Polynomial equations are often difficult to solve, and frequently we must be satisfied with approximate solutions. This section distinguishes the types that can be solved algebraically using roots or factoring. Of course, the evaluate-and-compare method always works, so it is an important alternative even when algebraic methods would work, and it also works when algebraic methods do not.

Let $P(x)$ denote a polynomial expression. The problem in this section is to solve the equation "$P(x) = c$," for constant c. If the polynomial is linear, the equation can be solved using only arithmetic operations.

Example 1: Solve $mx + b = c$ for x, for $m \neq 0$.

In this problem "m" and "b" are parameters of a line, and "x" is the unknown. By inverse-reverse thinking, $x = (c - b)/m$. The solution uses only arithmetic operations.

If $P(x)$ is not linear, the solution to "$P(x) = c$" is much more complex. Even such a simple quadratic as "$x^2 = 15$" cannot be solved using only arithmetic operations. We are so accustomed to using the square root symbol that we may take square roots for granted, but they are not trivial to compute. The first electronic calculator with square roots came out in 1965 and cost $1850 then!

In Mathematics we use the square root symbol to *name* the *non-negative* solution to "$x^2 = c$," for non-negative c. The idea of naming the solution to the equation "$x^n = c$" for integer values of n leads to the concept of an "n^{th} root."

<u>**Solving Monomial Equations**</u>. Let $P(x)$ be an nth degree monomial [ax^n], for $n \geq 2$. Then we can solve "$P(x) = c$" using the nth-root function. For example, the inverse of the third-power (cubing) function is commonly called the cube root function and sometimes given a special notation like the square root function, except with a tiny "3" in the crook of the radical symbol:

$$\sqrt[3]{x} = x^{1/3} .$$

The latter notation with a fractional power is preferable in calculus. We will use it. This is not to say that the cube root is easy to evaluate by hand -- it is not. But, at least, the cube root function is on your calculator.

Calculator Exercise 1: Check the keystrokes required by your calculator to obtain these results.

$1000^{1/5} = 3.981$. [This can be evaluated with at least two keystroke sequences, one of which uses 5^{-1} in place of 1/5.]

$(-300)^{1/3} = -6.694$.

Evaluate $(-300)^{1/4}$.

For this one some calculators return an error message. Even-degree roots of negative numbers are not real numbers.

Calculator Exercise 2: Graph x^n for any odd n, say, $n = 3$ or 5. Is it one-to-one? Does it have an inverse? Graph $x^{1/3}$ or $x^{1/5}$. What is its domain?

Now graph x^n for any even n, say $n = 4$. Is it one-to-one? When does $x^4 = c$ have two solutions? No solutions? Does x^4 have an inverse? Graph $x^{1/4}$. What is its domain?

We define the nth-root function to be the inverse function of the nth-power function over a region where the nth-power function is one-to-one. It turns out that the nth-root function is the $1/n$-power function.

Theorem 4.2.1: For odd integer values of n, $x^n = c$ iff $x = c^{1/n}$.
For even integer values of n, $x^n = c$ iff $c \geq 0$ and $x = \pm c^{1/n}$.

This theorem has two distinct cases, depending on whether n is odd or even. If n is odd, x^n can be negative and the nth-root function can be defined for negative arguments. If n is even, x^n is non-negative, and the nth-root function is, therefore, not defined for negative arguments (however, it can be defined using complex numbers).

Example 3: Solve $5x^3 = 12$.

By inverse-reverse thinking, first divide by 5. Then "uncube" with the one-third power function.

$x^3 = 12/5 = 2.4$ iff $x = 2.4^{1/3} = 1.339$.

Example 4: Solve $x^6 = 1200$.

Since 6 is even and 1200 is greater than 0, there will be two solutions, much like "$x^2 = 15$" has two solutions. $1200^{1/6} = 3.2598$. The two solutions are $x = \pm 3.2598$.

The reason the nth root is the $1/n$ power can be seen by inspecting the result about a power of powers (4.1.3): $(x^r)^p = x^{rp}$. Therefore, for $x > 0$,

$$(x^n)^{1/n} = x^{n(1/n)} = x^1 = x.$$

This is why the square root function, $\surd$, is also the one-half power function. All the properties of power functions hold for square roots. Problem B4 asks you to rewrite the properties of powers from Table 4.1.6 using square-root notation.

Solving Polynomial Equations. Four types of polynomial equations are easy to solve -- those with lines, quadratics, or monomials, and those already factored or nearly factored. All the other types of polynomial equations are substantially more difficult to solve algebraically.

There are formulas for solving polynomial equations of degree 3 (cubics) and degree 4 in terms of roots, but they are so long and messy that very few mathematics professors have bothered to learn them. They are available in some computer-software programs, but yield solutions that are a full line long and virtually useless until converted into decimal notation.

So, if the cubic and quartic formulas are too long to use, how does a mathematician solve a cubic equation? There are two basic ways, selected from the "Four Ways to Solve an Equation." One is to factor and use the Zero Product Rule. Much of this section discusses clever methods of factoring cubic and higher-degree polynomials. The other way is to use evaluate-and-compare.

Evaluate-and-Compare. To solve the equation "$P(x) = 0$" is to find the values of x such that the value of $P(x)$ is zero. The terms for this process depend upon whether the context is an equation or an expression. Any value of x such that $P(x) = 0$ is called zero (or "root") of the polynomial P. To "solve" the equation "$P(x) = 0$" is to "find the zeros" of the expression $P(x)$.

The easiest way to solve a polynomial equation of degree 3 or higher is to graph it and use evaluate-and-compare.

Example 5: Solve $x^3 - 4x - 1 = 0$.

Graph the cubic (Figure 1). Apparently, there are three solutions. We can discover each of the three solutions to any desired degree of accuracy. $x = -1.861$ and $x = -.254$ are two solutions. The third is left to the reader (Problem A11).

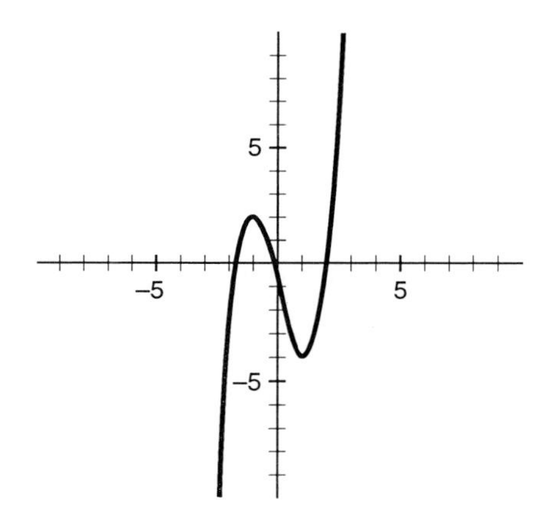

Figure 1: $x^3 - 4x - 1$
[-10, 10] by [-10, 10].

The only trick to evaluate-and-compare is to remember the concept of a "representative" graph (which exhibits all relevant features of the function) and the fact that a polynomial of degree n may have up to n zeros. We cannot assume that all the zeros will be exhibited on the standard scale.

Example 6: Solve $(x^3 + 20x^2 - 16x - 320)/100 = 0$.

From the graph (Figure 2, on the standard scale) it may appear that there are two solutions. But that graph is not a representative graph and changing the window gives a different picture (Figure 3). When you see a cubic polynomial you know from its end-behavior model that a representative graph of it cannot look like Figure 2. To the left of the window in Figure 2 the graph must come back down and cross the x-axis again.

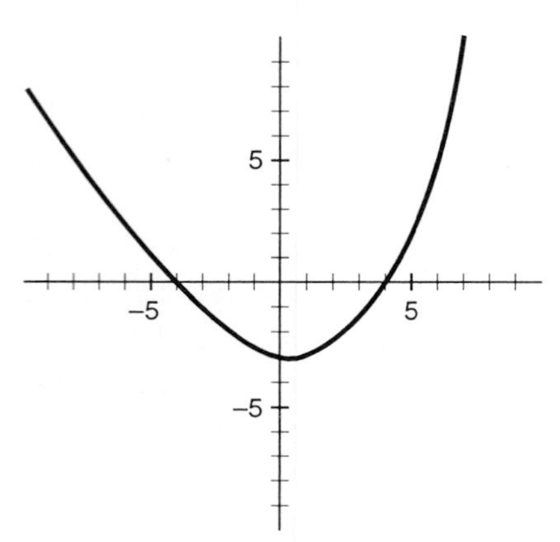

Figure 2:
$(x^3 + 20x^2 - 16x - 320)/100$
[-10, 10] by [-10, 10].

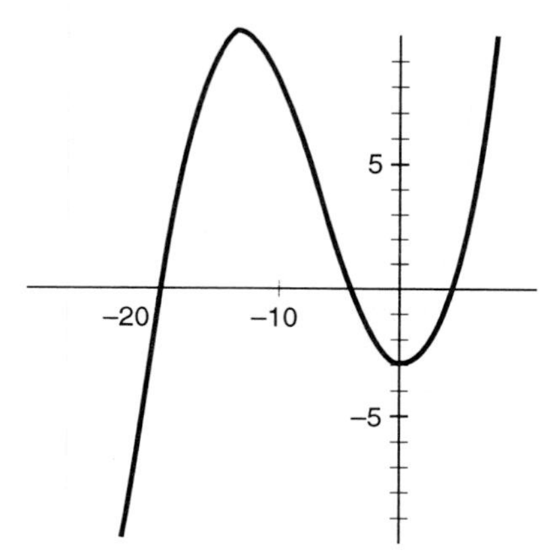

Figure 3:
$(x^3 + 20x^2 - 16x - 320)/100$
[-30, 10] by [-10, 10].

In Figure 3 we see there will be three solutions, the maximum number possible for a third degree polynomial (Problems A13 and 14).

Look at Figure 4. It shows that the number of possible solutions to a polynomial equation $P(x) = c$ is related to the number of local extrema of the graph of the expression $P(x)$. For example, a quartic can have at most three local extrema, and can yield 4 solutions in certain cases (for example, when $c = c_4$), two in others (when $c = c_2$), and none in others (still other c_0). Also one or three solutions are possible as special cases when the line $y = c$ is tangent to the curve ($c = c_1$ or $c = c_3$).

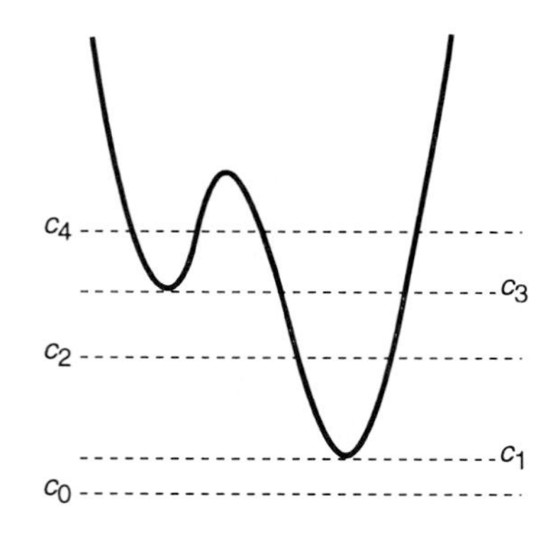

Figure 4: A quartic (polynomial of degree 4). $f(x) = c$ can have 4, 2, or no solutions. It can have 1 or 3 if the horizontal line $y = c$ is tangent to a bump.

Cubic equations always have at least one

real-valued solution. The range of every cubic expression is all real numbers, including zero. Their behavior is either ↓↑ or ↑↓, so their graphs must cross the horizontal line $y = c$ for every c, so there will be at least one solution to $P(x) = c$. This generalizes to all odd-degree polynomials: All odd-degree polynomial equations have at least one real-valued solution. But there is no parallel result for even-degree polynomials. For example, we already knew that quadratic equations do not necessarily have any real-valued solutions.

A cubic equation may have at most three real-valued solutions (Example 5, Figure 1). In general,

A polynomial equation of degree *n* may have at most *n* real-valued solutions.

The Zero Product Rule. The first way to solve a high-degree polynomial is to use evaluate-and-compare. The only simple remaining way to solve a polynomial of degree 3 or higher is to factor it and use the Zero Product Rule. The Zero Product Rule is the primary reason you spent so much time learning to factor in school.

The Zero Product Rule (4.2.2): $ab = 0$ iff $a = 0$ or $b = 0$, for all a and b.

The Zero Product Rule has a natural extension to three or more factors.

Example 7: Solve $(x - 2)(x + 3)(x - 7) = 0$.

The solution is obtained by setting each factor equal to zero. $x = 2$ or $x = -3$ or $x = 7$ (Figure 5).

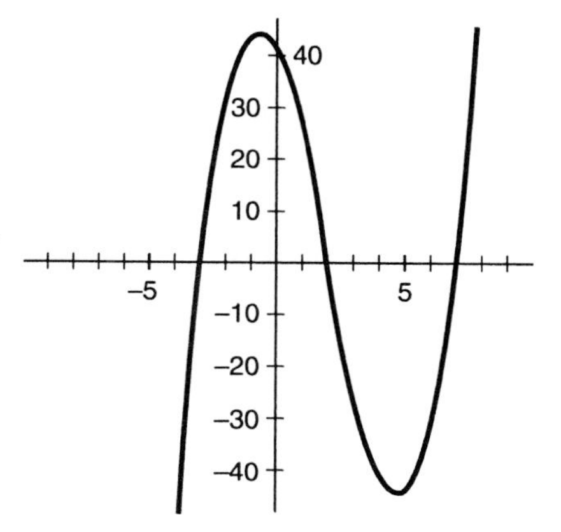

Figure 5:
$(x - 2)(x + 3)(x - 7)$
[-10, 10] by [-50, 50].

The next example illustrates a type of problem that appears frequently in calculus. Sometimes factoring is facilitated because the terms of a sum are already factored.

Example 8: Solve $-2x^3(1 - x) + 3x^2(1 - x)^2 = 0$.

First factor out the common factors. Each has an "x^2" and each has a "$1 - x$".

$$-2x^3(1 - x) + 3x^2(1 - x)^2$$
$$= x^2(1 - x)[-2x + 3(1 - x)]$$
$$= x^2(1 - x)[3 - 5x].$$

Therefore, by the Zero Product Rule, the original equation is equivalent to

$$x^2 = 0 \text{ or } 1 - x = 0 \text{ or } 3 - 5x = 0.$$

The solution is $x = 0$ or $x = 1$ or $x = 3/5$.

Calculator-Aided Factoring Techniques. Occasionally, but only occasionally, polynomials of degree three or higher can be factored using integers. Mostly this occurs when the textbook author or exam author has carefully selected the polynomial to be one of the rare ones that does factor using integers. The Factor Theorem can be very helpful in these cases.

The Factor Theorem (4.2.3): "$x - c$" is a factor of the polynomial expression $P(x)$ iff c is a solution to $P(x) = 0$ (that is, iff $P(c) = 0$).

To factor a cubic or higher-power polynomial is difficult unless one of the factors is "$x - c$" where c is an integer. The Factor Theorem makes these factors easy to find. But, there is no easy algebraic way to factor general high-degree polynomials unless you get lucky and stumble across an integer-valued factor by using evaluate-and-compare.

Equivalent Statements

c is a solution to $P(x) = 0$.
c is a root of $P(x)$.
c is a zero of $P(x)$.
$P(c) = 0$.
c is an x-intercept of the graph of $P(x)$.
$x - c$ is a factor of $P(x)$.

Example 9: Part 1: Factor $x^3 - 3x^2 - 5x + 15$.
Part 2: Solve $x^3 - 3x^2 - 5x + 15 = 0$.

These are two very closely related problems. Let's see if we get lucky. Graph the expression (Figure 6). There appears to be a zero at, or at least near, $x = 3$. By evaluating the expression for $x = 3$, we find 3 is a zero. That means that "$x - 3$" is a factor. The other factor must be a quadratic. We can find it by long division.

$$\begin{array}{r|l} & x^2 \qquad - 5 \\ \hline x - 3 & x^3 - 3x^2 - 5x + 15. \\ & \underline{x^3 - 3x^2} \\ & \quad 0 \; - 5x + 15 \\ & \qquad \underline{- 5x + 15} \\ & \qquad\qquad 0 \end{array}$$

So, $x^3 - 3x^2 - 5x + 15 = (x - 3)(x^2 - 5)$.

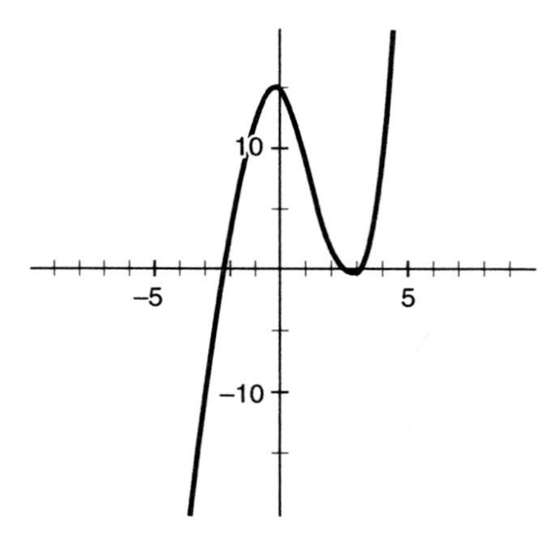

Figure 6:
$x^3 - 3x^2 - 5x + 15$.
[-10, 10] by [-20, 20].

The process of polynomial long division parallels regular long division of integers. The layout is the same and the steps are the same. Also, the remainder will be zero if and only if the divisor is truly a factor.

The equation in Part 2 is equivalent to $x = 3$ or $x^2 - 5 = 0$. The solutions to "$x^2 - 5 = 0$" are $\sqrt{5}$ and $-\sqrt{5}$. $\sqrt{5}$ is not an integer, so the factor theorem tells us it cannot be factored further using only integers. If you are willing to use real numbers, it factors into

$$(x - 3)(x - \sqrt{5})(x + \sqrt{5}).$$

Example 9, another way: Factor $x^3 - 3x^2 - 5x + 15$ using integers.

Another method of factoring third-degree polynomials is called "grouping." This method requires specially designed examples that work (most don't), takes a lot of practice, and uses the Distributive Law (what else?).

By inspection, $x^3 - 3x^2 - 5x + 15 = x^2(x - 3) - 5(x - 3)$. Regrouping, this $= (x^2 - 5)(x - 3)$.

It is very rare for this "grouping" method to find any factors that will not be found more easily by graphing and using the Factor Theorem. I mention it only because school texts (prior to graphics calculators) thought it was worthwhile because it can be used to factor some selected cubics and quartics.

Example 10: Solve $x^3 - 2x^2 - 5x - 12 = 0$ for all solutions, real and complex.

Without using the cubic formula, we have only two ways to go. Factor it, or use evaluate-and-compare. You may try to factor this by guess-and-check, which is a fine method when it works. However, there is an intelligent way to guess.

A graph can show us what to expect (Figure 7). Cubic equations may have 1, 2, or 3 real-valued solutions, and this one appears to have 1. By the Factor Theorem, there will be a corresponding linear factor. Then the other factor must be a quadratic. Furthermore, that quadratic will not factor using real numbers.

Evaluate-and-compare confirms that "$x = 4$" is a solution: $P(4) = 0$. Therefore "$x - 4$" is a factor. The other factor can be determined by polynomial long division:

$$\begin{array}{r|l} & x^2 + 2x + 3 \\ \hline x - 4 & x^3 - 2x^2 - 5x - 12 \\ & \underline{x^3 - 4x^2} \\ & \quad 2x^2 - 5x \\ & \quad \underline{2x^2 - 8x} \\ & \qquad 3x - 12 \\ & \qquad \underline{3x - 12} \\ & \qquad\qquad 0 \end{array}$$

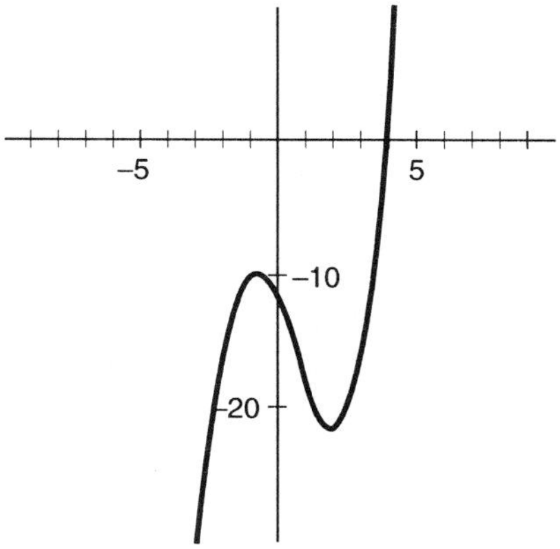

Figure 7:
$x^3 - 2x^2 - 5x - 12$.
[-10, 10] by [-30, 10].

So it factors into $(x - 4)(x^2 + 2x + 3)$. The solution is then the "$x = 4$" we already knew, together with the solutions to the quadratic equation $x^2 + 2x + 3 = 0$, which are easy to obtain:

$$x = 4 \quad or \quad x = \frac{-2 \pm \sqrt{4 - 12}}{2}.$$

$$x = 4 \quad or \quad x = -1 \pm i\sqrt{2} .$$

The last step uses $\sqrt{-8} = i\sqrt{8} = i\sqrt{4(2)} = 2i\sqrt{2}$.

Example 11: Find a cubic polynomial $P(x)$ that goes through (2, 0), (4, 0), (-3, 0), and (1, 4) (Figure 8).

The Factor Theorem and the zeros tell us the cubic is, in factored form,

$$k(x - 2)(x - 4)(x + 3),$$

where k remains to be determined. The zeros alone do not determine the constant factor. We use the point (1, 4) to determine k. Substitute 1 for x.

$$4 = P(1) = k(1 - 2)(1 - 4)(1 + 3) = 12k.$$

Therefore $k = 1/3$ and the cubic is

$$(1/3)(x - 2)(x - 3)(x + 4).$$

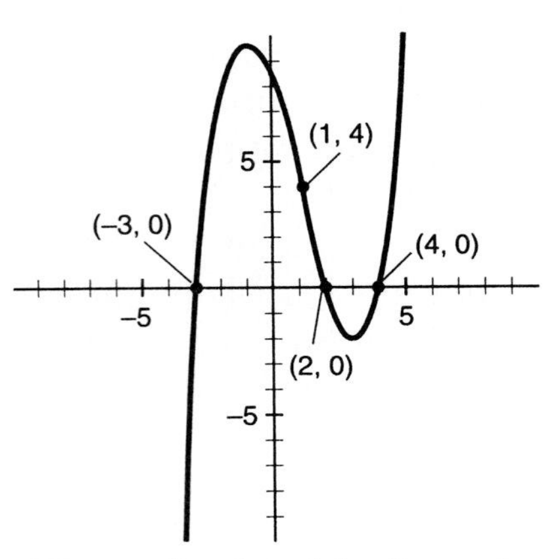

Figure 8: A cubic through 4 given points. [-10, 10] by [-10, 10].

__Identities__. The properties of powers given in 4.1.1 to 4.1.3 (see Table 4.1.6) are useful for reordering expressions.

Example 12: Solve $x^2x^3 = 1000$.

To use inverse-reverse the expression on the left must be rewritten to exhibit only one appearance of "x". Property 4.1.1 ($x^px^r = x^{p+r}$) will do it.

$x^5 = 1000$. $x = 1000^{1/5} = 3.981$.

Example 13: Solve $x^7/x^4 = .04$.

By Property 4.1.2 ($x^p/x^r = x^{p-r}$), $x^7/x^4 = x^3 = .04$. $x = .04^{1/3} = .342$.

Example 14: Solve $(x^2)^3 = 12$.

By Property 4.1.3 [$(x^r)^p = x^{rp}$], $(x^2)^3 = x^6 = 12$ and $x = \pm 12^{1/6} = \pm 1.513$.

Example 15: Solve $x^3 + 2x^2 = 7$.

A sum of terms with different powers (here, 3 and 2) does not simplify into a single power function. There is no shortcut to solving this equation. Use evaluate-and-compare (Problem A12).

Example 16: Solve $x^3 + 2x^2 = 0$.

This equation is much different from the previous equation, despite having the same polynomial on the left, because the right side is zero. Factor out x^2 and use the Zero Product Rule.

$x^3 + 2x^2 = 0$ iff $x^2(x + 2) = 0$ iff $x^2 = 0$ or $x + 2 = 0$.
The solution is $x = 0$ or $x = -2$.

Factoring in Integers. The problem is still to solve "$P(x) = 0$" where P is a cubic or higher-degree polynomial. Remember that any polynomial equation can be solved for its real-valued solutions using evaluate-and-compare, so think of that possibility first. This subsection concerns finding algebraic, exact, solutions and complex-valued solutions when the leading coefficient is an integer that is not 1. The techniques are based on the Factor Theorem.

Example 17: $6x^2 - x - 35 = (3x + 7)(2x - 5)$.

Note that the "3" and the "2" in the factors divide the leading coefficient "6." Also, the "7" and the "-5" divide the constant, "-35." The next theorem says that this type of behavior always happens when you are dealing with *integer* coefficients. The leading coefficients of the factors must divide the leading coefficient of the original polynomial. The constant terms of the factors must divide the constant of the polynomial.

Example 17, continued: Solve $6x^2 - x - 35 = 0$.

For the previous part, $6x^2 - x - 35 = (3x + 7)(2x - 5)$. Therefore, by the Zero Product Rule, $x = -7/3$ or $x = 5/2$.

Note that the solutions to $P(x) = 0$ have denominators that divide the leading coefficient, 6, and numerators that divide the constant, -35. This type of behavior always happens when you are dealing with a polynomial with *integer* coefficients. The upcoming "Rational Zeros" theorem states this idea in general.

Theorem on Factoring Polynomials with Integer Coefficients (4.2.4): Suppose $P(x)$ is a polynomial with integer coefficients, leading coefficient a, and constant term c. ["j" and "k" denote integers in this subsection.]

If $P(x)$ factors into factors with integer coefficients and "$jx - k$" is one of the factors, then j divides the leading coefficient, a, and k divides the constant, c.

This theorem describes all the potential factors with integer coefficients (problem B31). If none are actually factors, then $P(x)$ has no linear factors with integer coefficients.

Example 18: Factor $2x^3 - x^2 - x - 3$ using integers.

According to the theorem, any linear factor "$jx - k$" must have j that divides 2 (that is, $j = \pm 1$ or ± 2) and k that divides -3 (that is, $k = \pm 1$ or ± 3). Therefore, any linear factor must be on this list:

$x - 1, x + 1, x - 3, x + 3, 2x - 1, 2x + 1, 2x - 3$, or $2x + 3$.

Now the problem can be solved using trial-and-error among these possibilities. [We do not need to try the negatives of all of these expressions because we can treat any negative sign as on the other factor, since $(-a)b = a(-b)$].

That looks like a lot of long division to try. Can't the Factor Theorem help? (We will continue this problem shortly.)

A theorem very much like the Factor Theorem can help. Of course, by the Zero Product Rule, a linear factor "$jx - k$" yields the rational zero $x = k/j$. The "Rational Zeros Theorem" (also called the "Rational Root Theorem") tells us that we can find the rational zeros exactly where we can find the factors with integer coefficients.

Theorem 4.2.5 -- The Rational Zeros Theorem. Suppose $P(x)$ is a polynomial with integer coefficients, leading coefficient a, and constant term c.
A) Suppose $x = k/j$ is a *rational* solution in lowest terms to the equation "$P(x) = 0$." Then the denominator j divides the leading coefficient a and the numerator k divides the constant term c.
B) Every rational solution corresponds to a linear factor with integer coefficients, and vice versa.

Therefore, we can find the factors by looking for zeros at the rational possibilities. By graphing the polynomial we can rapidly locate any likely rational possibilities, so, instead of trying the whole list of possible linear factors, all we need to try is the linear factors which would yield zeros where they appear to be on the graph.

Example 18, continued: We want to factor $2x^3 - x^2 - x - 3$ using integers, but there are more possible factors than we are happy to try. Use the Rational Zeros Theorem to narrow the search.

Graph the expression (Figure 9) and look for rational zeros with denominator 1 or 2, (because they are the only possibilities that divide the leading coefficient, 2). The numerator of the zero must be ±1 or ±3. Figure 9 shows the only zero is between 1 and 2. Therefore, the only viable candidate for a zero is 3/2 and the only viable factor is the one with a zero at 3/2: $2x - 3$. If that expression is not a factor, it does not have a linear factor with integer coefficients.

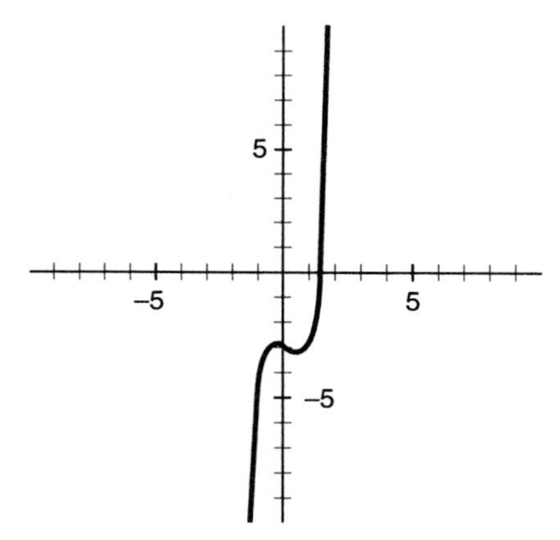

Figure 9: $2x^3 - x^2 - x - 3$. [-10, 10] by [-10, 10].

Long division shows it is a factor. We obtain $2x^3 - x^2 - x - 3 = (2x - 3)(x^2 + x + 1)$.

Now we can solve for all zeros, real and complex. $x = 3/2$ is one zero. From the quadratic factor (using the Quadratic Theorem), the other two zeros are:

$$\frac{-1 \pm \sqrt{1^2 - 4(1)(1)}}{2} = -\frac{1}{2} \pm \frac{i\sqrt{3}}{2}.$$

Example 19: Solve $2x^5 - 5x + 5 = 0$.

The first idea is to graph it (Figure 10). The graph tells us to expect only one real-valued solution. We can use evaluate-and-compare to find it to any desired degree of accuracy. However, if we want an algebraic solution, we can hope this factors in integers. If it does not factor in integers, we will be stuck (No one said that every problem has an algebraic solution).

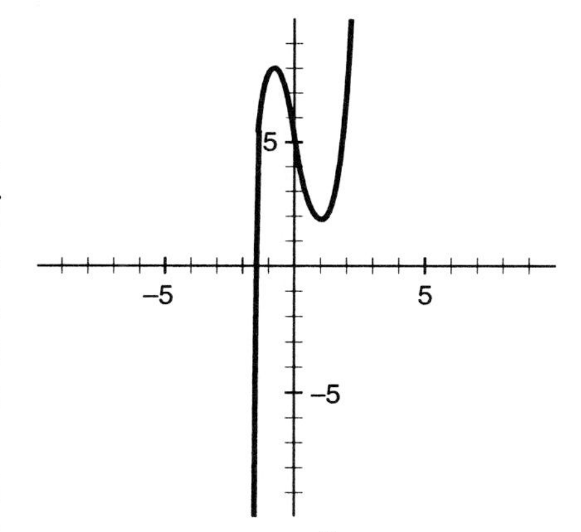

Figure 10: $2x^5 - 5x + 5$. [-10, 10] by [-10, 10].

To factor it in integers, look for a rational solution with denominator ±1 or ±2 (the only divisors of the leading coefficient, 2). Also, its numerator must be ±1 or ±5 (the only divisors of the constant term, 5). We can check all of these candidates using evaluate-and-compare, or, cleverly, we can narrow the search by checking only the candidates near the solution we can see on the graph. The solution is between -2 and -1. There are no candidates there -- so we conclude it does not factor in integers. We are stuck. Be happy with the one solution you can find using evaluate-and-compare (Problem B30).

Corollary 4.2.6 to Theorem 4.2.5A: If $P(x)$ is a polynomial with integer-valued coefficients and its leading coefficient is 1, then the only *rational* solutions to "$P(x) = 0$" are integer-valued solutions; there are no rational solutions that are not integers (that is, there are no solutions such as 1/2 or -5/3, but there might be solutions such as $\sqrt{2}$).

Example 20: Factor $x^3 + 3x^2 - 4x - 8$ in integers.

If it factors in integers, there will be an integer zero. Graph it and see (Figure 11).

You might want to zoom in to assure yourself that the three zeros are definitely not integers. Since they are not, the cubic does not factor in integers. The problem cannot be done.

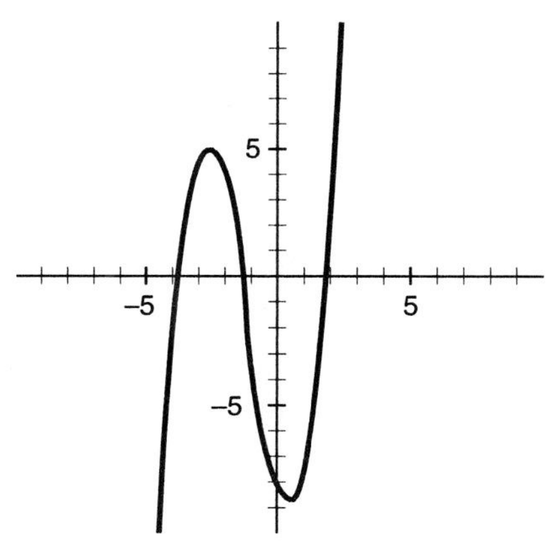

Figure 11: $x^3 + 3x^2 - 4x - 8$ [-10, 10] by [-10, 10].

There is a theorem, the "Fundamental Theorem of Algebra," which says that any polynomial of degree n with real-valued coefficients is equivalent to a product of a constant and n linear factors of the form "$x - c$". Unfortunately, the c's may be complex numbers and the theorem does not say how to find them. We will not deal with complex numbers until after we have studied trigonometry, which, perhaps surprisingly, is important for understanding complex numbers.

Conclusion. Polynomial equations that are linear or quadratic are easy to solve algebraically. The only other easy cases occur when the polynomial is a monomial, is factored or nearly factored, or has an integer or rational solution.

Any polynomial equation can be solved for all its real-valued solutions using evaluate-and-compare. The Factor Theorem helps solve cubic and higher-degree polynomials algebraically whenever one solution is easily found. The Rational Zeros Theorem can help find "easy" algebraic solutions.

Terms: n^{th}-root function, factor (verb and noun), zero (of an expression), Factor Theorem, Rational Zeros Theorem.

****

Exercises for Section 4.2, "Polynomial Equations":

A1.* State the Zero Product Rule.

A2.* Which equation-solving method can you always use to solve a polynomial equation?

A3.* Solve $mx + b = c$ for x.

A4.* True or false? The cubic formula is easy to memorize.

A5.* The Factor Theorem is closely related to a rule we have used a lot. Which one?

A6. There is another place in Figure 4 where the right value of "c" would yield three solutions. Where?

^ ^ ^ ^ Solve graphically.

A7. $x^3 + 2x = 8$. [1.7]

A8. $3x^3 + x = 1$. [.54]

A9. $x^4 - 2x^3 = 5$. [2.4,...]

A10. $x^4 + 5x^3 = 6$. [1.0, ...]

A11. Find the third solution in Example 5.

A12. $x^3 + 2x^2 = 7$ (Example 15).

A13. Find the leftmost solution to the equation in Example 6 (Figure 3).

A14. Find the rightmost solution to the equation in Example 6 (Figure 3).

^ ^ ^ ^ Use the Quadratic Theorem or a graph to factor the expression using integers.

A15. $6x^2 - 11x - 10$

A16. $8x^2 + 30x - 27$

^ ^ ^ ^ Factor into two factors.

A17. $x^3 - 5x^2 + 8x - 16$.

A18. $x^3 - x - 6$.

A19. $x^3 - x^2 - 3x + 6$.

A20. $x^3 - 5x^2 - 4x - 12$.

^ ^ ^ ^ Solve algebraically.

A21. $3x^2(1 - x)^4 - 4x^3(1 - x)^3 = 0$.

A22. $5x^4(1 - x)^3 - 3x^5(1 - x)^2 = 0$.

A23. $4x^3(2x + 1)^3 + 6x^4(2x + 1)^2 = 0$.

A24. $2(x + 1)(x - 2)^3 + 3(x + 1)^2(x - 2)^2 = 0$.

A25. True or false: Constant factors are determined by the Factor Theorem.

A26. a) Suppose $P(x)$ has a factor of $5x + 17$. Where will it have a zero?
b) Suppose $P(x)$ has integer coefficients and 5/2 is a zero. Give a factor with integer coefficients.

A27. a) Suppose $P(x)$ has a factor of $3x - 4$. Where will it have a zero?
b) Suppose $P(x)$ has integer coefficients and 13/4 is a zero. Give a factor with integer coefficients.

^ ^ ^ ^ ^ ^ ^ ^

B1.* Some types of polynomial equations are relatively easy to solve, compared to all the rest. Which ones? Which ones had their own sections (which sections?) prior to this section in this text?

B2.* Let n denote a positive integer and $c \neq 0$.
a) $x^n = c$ has two real-valued solutions if ________________.
b) It has one solution if ________________.
c) It has no real-valued solutions if ________________.

B3.* a) Discuss the number of local extrema the graph of a polynomial can have.
b) How does the number of local extrema affect the number of possible solutions to a polynomial equation?

B4.* Use the square root symbol to rewrite properties 4.1.3, 4.1.4A, and 4.1.4B (when $p = 1/2$, as in Table 4.1.6).

B5.* Use fractional notation for the nth root (as the $1/n$ power in place of p) and rewrite the properties 4.1.3, 4.1.4A, and 4.1.4B (from Table 4.1.6).

B6.* Why is factoring so important in solving equations?

B7.* "A quotient is zero if and only if the top is zero and the bottom is not."
State that result abstractly using dummy variables.

B8. Why is there an essential difference between the methods of solving "$x^5 + 5x^3 = 1$" and "$x^5 + 5x^3 = 0$"?

^ ^ ^ ^ Algebraically solve for all solutions:

B9. $x^3 - x^2 - 11x - 10 = 0$. [4.2,...,...]

B10. $x^3 + 5x^2 - x - 20 = 0$. [1.8,...,...]

B11. $x^3 - 5x^2 + x + 12 = 0$. [2.3,...,...]

B12. $x^3 - 21x^2 + 17x + 60 = 0$.

B13. $x^4 + x^2 - 3 = 0$. [1.1,...]

B14. $x^4 + 2x^2 - 5 = 0$.

B15. $x + \sqrt{x} = 5$. [3.2]

B16. $x - 2\sqrt{x} = 7$. [15]

^ ^ ^ ^ Find a polynomial P with the given properties.
B17. P is quadratic, with zeros at 4 and -2 such that $P(1) = 3$.
B18. P is quadratic, with zeros at -1 and 2 such that $P(0) = 4$.
B19. P is cubic, with zeros at 2, 3, and -1 such that $P(0) = 5$.
B20. P is cubic, with zeros at -3, 1, and 2 such that $P(-1) = 6$.
B21. P is cubic with integer coefficients, with zeros at 3, 4, 1/2.
B22. P is cubic with integer coefficients, with zeros at 1, 5, and 3/4.

B23. Explain why, if $P(1.5) = 0$, then $2x - 3$ is a factor of the polynomial $P(x)$.

B24. Someone might say "$x^2 - 3x - 17$ does not factor." Is he right? Discuss this technically.

^ ^ ^ ^ Here are graphs of quadratics. Find the quadratic by noting its zeros and one other value.

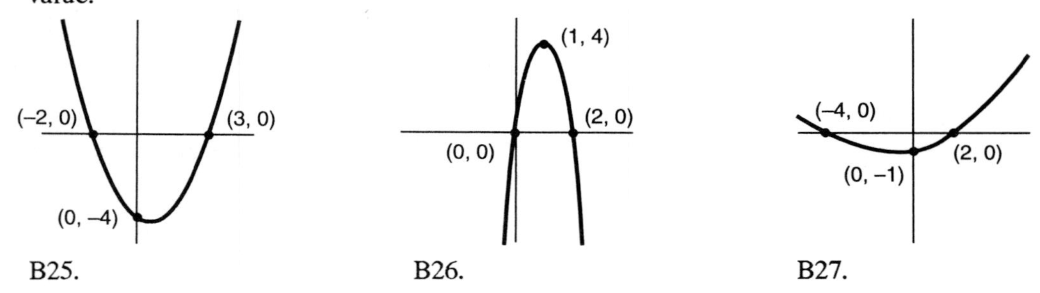

B25. B26. B27.

B28. Algebraically solve for all solutions: $x^4 + x^3 + 2x + 2 = 0$.

B29. a) Solve $nx^{n-1}(1 - x)^m - mx^n(1 - x)^{m-1} = 0$. b) This equation can be used to find the extreme of the expression $x^n(1 - x)^m$, where m and n are integers. Which solution corresponds to the maximum of the expression for x in the interval [0,1]?

B30. (A hard problem) In Example 19 there were no integer-valued solutions, so we could not factor the expression using the methods of this section. That means there will be some complex-valued solutions we cannot find using the methods of this section. But there must be a way to find the complex-valued solutions. Think of one. [You need not actually use it; just explain how it would work, in theory.]

B31. Prove Theorem 4.2.4 by assuming $P(x) = (jx - k)Q(x)$, for some polynomial $Q(x)$ with integer coefficients and multiplying the right hand side out.

B32. The Remainder Theorem states, "When a polynomial $P(x)$ is divided by $x - c$, the remainder is $P(c)$." The remainder, $R(x)$, satisfies

$$P(x) = (x - c)Q(x) + R(x),$$

for some polynomial $Q(x)$. a) Verify it for $x^3 - 2x^2 - 14$ divided by $x - 3$.
b) Prove the Remainder Theorem.

Section 4.3. Fractional Powers

The properties of powers given in Section 4.1, "Powers," also apply to fractional powers, including square roots. Fractional powers commonly arise in distance problems and whenever different integer powers are combined in the same equation.

A common technique for solving equations with square roots is to "square both sides."

Example 1: Solve $x - 1 = \sqrt{x + 11}$.

Squaring both sides, $x^2 - 2x + 1 = x + 11$, which is a quadratic.

$x^2 - 3x - 10 = 0$ iff $(x - 5)(x + 2) = 0$ iff $x = 5$ or $x = -2$.

But this is not the answer. This is the right process, and the steps are correct, but the answer is not yet right. We have created a sequence of equations, but not a sequence of *equivalent* equations. Whenever you employ the procedure "square both sides," you will find all solutions, but there is no guarantee that a solution has not been unintentionally added. The procedure may introduce an extraneous solution -- one that satisfies the terminal equation that does not solve the original equation. Extraneous solutions can be eliminated by checking all the solutions to the terminal equation back in the original equation and discarding those that do not really work. In this example, 5 works ($5 - 1 = \sqrt{5+11}$), but -2 does not ($-2 - 1 \neq \sqrt{-2+11}$). "-2" is extraneous. The solution is $x = 5$.

A simpler example shows why this can happen. Consider the pair of equations "$x = 3$" and "$x^2 = 9$." Obviously, "$x = 3$" has only one solution. But, if you square both sides to obtain "$x^2 = 9$," the new equation has two solutions (including $x = -3$, which does not solve the original equation). Squaring introduced an extraneous solution.

The graph of the two expressions in Example 1 shows what happened (Figure 1). The graphs intersect only once, although they look as if they would intersect again if the parabola were complete, instead of just the part with non-negative y-values. The

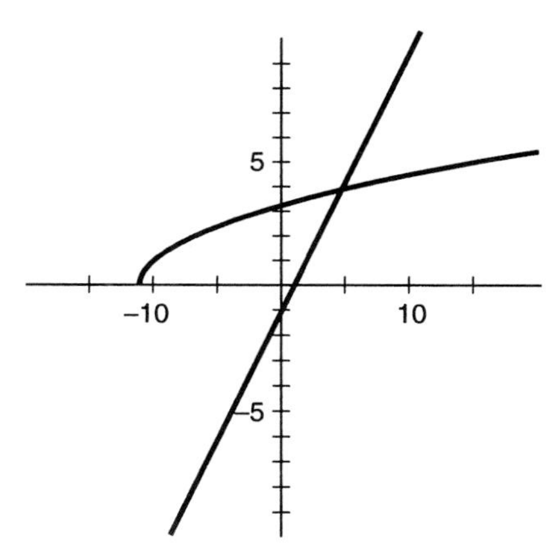

Figure 1: $x - 1$ and $\sqrt{x + 11}$. [-20, 20] by [-10, 10].

extraneous solution is the one that would have been there if the negative part of the parabola had been included.

Squaring both sides does not always introduce an extraneous solution.

Example 2: Solve $4\sqrt{x-2} = x$.

Square both sides to obtain: $16(x-2) = x^2$, and then $x^2 - 16x + 32 = 0$. According to the Quadratic Theorem,

$$x = \frac{16 \pm \sqrt{16^2 - 4(1)(32)}}{2} = 2.34 \quad or \quad 13.65.$$

Neither solution is extraneous. Figure 2 shows the two original expressions on the standard scale. Only one solution is evident, but the graph is not representative. On the scale [0, 20] by [0, 20] (Figure 3) the two intersections can be seen.

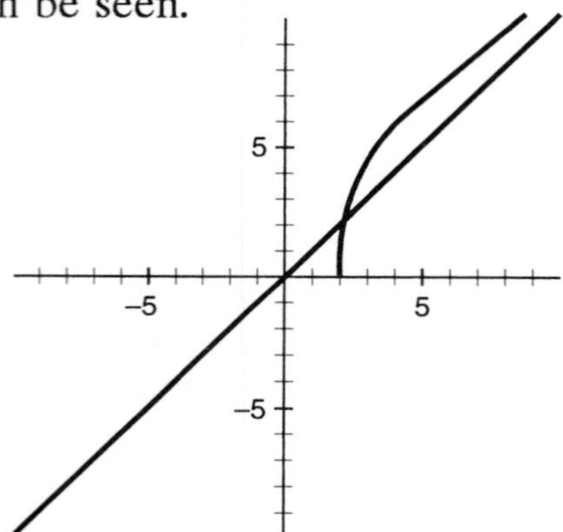

Figure 2: $4\sqrt{x-2}$ and x
[-10, 10] by [-10, 10].

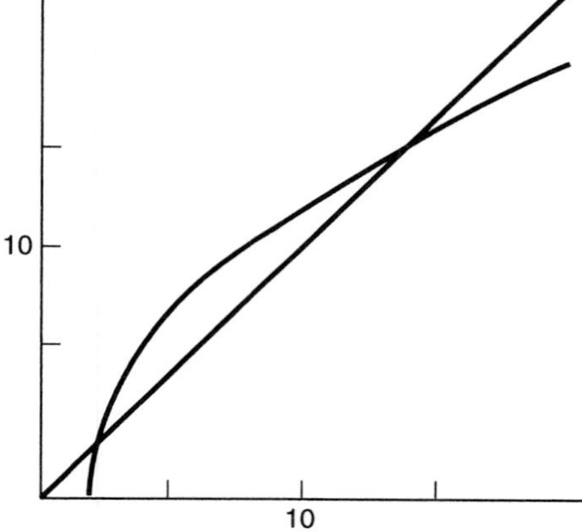

Figure 3: $4\sqrt{x-2}$ and x
[0, 20] by [0, 20].

The relevant theorems are next.

<u>Theorem 4.3.1 (on squaring)</u>: If $a = b$, then $a^2 = b^2$.
If $\sqrt{a} = b$, then $a = b^2$.

<u>Reading Mathematics</u>. Note that this theorem is not stated with "iff." It does not assert equivalence. On the other hand, the two equations might be equivalent, but we don't know whether they are or not from this theorem. We know by checking.

When a theorem (such as the one above) describes a process for altering an equation (such as squaring both sides) there are two possibilities. If it is stated with "if and only if," the process always yields an equivalent equation which has the same solutions. However, if the theorem uses "if..., then...," the process will find all the solutions, but it may yield extraneous solutions as well.

Example 3: Here is a theorem. What process is it about?

Theorem 4.3.2: A) If $c \neq 0$, then $a = b$ iff $ca = cb$.
B) If $a = b$, then $ca = cb$.

This theorem concerns the process of multiplying both sides of an equation by an expression.

Now, use the theorem to solve $\dfrac{x^2 + x - 6}{x - 2} = 2x - 1$.

You may "multiply through by "$x - 2$," (using Part B).

$$x^2 + x - 6 = (2x - 1)(x - 2)$$
$$= 2x^2 - 5x + 2.$$
$$0 = x^2 - 6x + 8 = (x - 2)(x - 4).$$
$$x = 2 \text{ or } x = 4, \text{ by the Zero Product Rule.}$$

The solution may appear to be "$x = 2$ or $x = 4$," but it is not. In the initial equation, the expression on the left is undefined at $x = 2$ because of division by zero. Thus $x = 2$ cannot be a solution; it is extraneous. The solution is "$x = 4$."

The theorem warns us about the possibility of extraneous solutions by using the connective "if..., then...."

Multiplying through by an expression such as "$x - 2$" will preserve solutions in most examples, but not in this one. When $x = 2$, multiplying by "$x - 2$" is multiplying by 0, and multiplying by 0 will convert any equation (even a false one) into the true equation "$0 = 0$."

Sometimes, to eliminate a square root you may wish to reorganize the equation first.

Example 4: Solve $2 + \sqrt{x} = x$.

Squaring immediately does not remove the square root. If you square both sides as is, there will be cross product term on the left that will still have the square root of x in it. Isolate the square root first.

$$2 + \sqrt{x} = x \text{ iff } \sqrt{x} = x - 2.$$

Now square.

$$x = x^2 - 4x + 4. \quad 0 = x^2 - 5x + 4. \quad 0 = (x - 4)(x - 1).$$
$$x = 4 \text{ or } x = 1.$$

If you quit here you will have made a mistake. Whenever you square, you must check for extraneous solutions. Check $x = 4$. It works. Check $x = 1$. It does not. The solution is $x = 4$.

Why did the extraneous solution $x = 1$ appear? The equation we squared was "$\sqrt{x} = x - 2$." Checking $x = 1$ in that one, it says "$\sqrt{1} = 1 - 2$," that is,

"1 = -1," which is false. It is false, but its square is true. Squaring loses track of signs, which is how extraneous solutions can appear.

The distance formula employs the square root function.

Example 5: Where are the points on the line $y = x$ that are 4 units from the point (5, 2)? See Figure 4.

According to the distance formula (3.1.12), the distance from (x, y) to (5, 2) is given by

$$\sqrt{(x-5)^2 + (y-2)^2} .$$

The constraint is that the point must be on the line $y = x$, so, the distance from (5, 2) to the point, in terms of x, is

$$\sqrt{(x-5)^2 + (x-2)^2} = \sqrt{x^2 - 10x + 25 + x^2 - 4x + 4}$$
$$= \sqrt{2x^2 - 14x + 29} .$$

Set this equal to 4 and solve. Squaring both sides,

$$2x^2 - 14x + 29 = 16.$$

This can be solved using the Quadratic Theorem (Problem A59), which will give two solutions. By Figure 4 and geometry, we can see that there should be two solutions, so they are right.

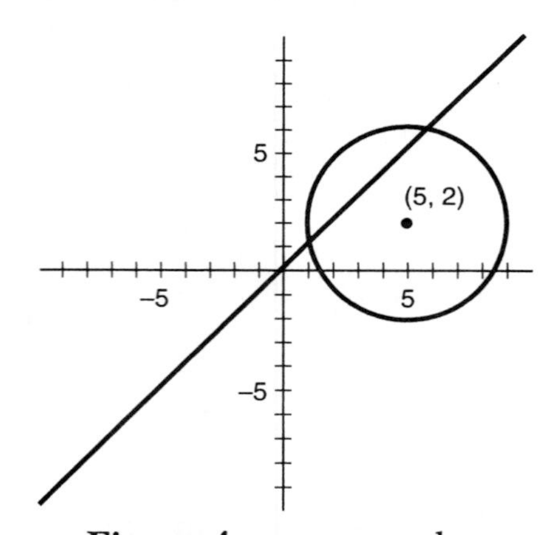

Figure 4: $y = x$, and a circle of radius 4 centered at (5, 2). [-10, 10] by [-10, 10].

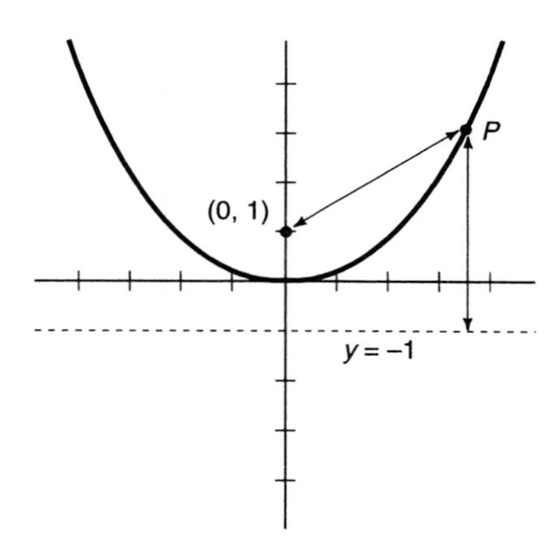

Figure 5: The point (0, 1), the line $y = -1$, and P, a point equidistant from both. [-5, 5] by [-5, 5].

Example 6: Find all points P equidistant from (0, 1) and the line $y = -1$ (Figure 5).

The problem uses the term "equidistant" to assert two distances are equal. We first need two formulas for distance. Let P be denoted by (x, y).

The distance from (x, y) to the line $y = -1$ is $|y + 1|$, which is simply $y + 1$ for y above the line. The distance from (x, y) to the point $(0, 1)$ is $\sqrt{(x^2 + (y - 1)^2)}$. So

$$|y + 1| = \sqrt{x^2 + (y - 1)^2}$$

Squaring,

$$y^2 + 2y + 1 = x^2 + (y - 1)^2 = x^2 + y^2 - 2y + 1.$$
$$4y = x^2. \quad y = x^2/4.$$

This is a parabola. The set of points equidistant from any line (the "directrix") and any point (the "focus") not on the line is a parabola.

Sometimes equations have x's inside two square root symbols. For these we may need to square twice (problems B13, 14, 28, and 30).

Example 7: Solve $1 + \sqrt{2x} = \sqrt{x + 7}$.

Square both sides: $1 + 2\sqrt{2x} + 2x = x + 7$.
Isolate the square root: $2\sqrt{2x} = 6 - x$.
Now square again. $8x = 36 - 12x + x^2$.
Solve that quadratic: $x = 2$ or $x = 18$.

Checking the original equation for extraneous solutions, we see $x = 2$ works and $x = 18$ is extraneous. Therefore, $x = 2$ is the solution.

Fractional Powers. The $1/n$ power is sometimes called the nth root. We have already seen properties of the nth root function. Theorem 4.3.3 repeats Theorem 4.2.1.

Theorem 4.3.3: For odd integer values of n, $x^n = c$ iff $x = c^{1/n}$.
For even integer values of n, $x^n = c$ iff $c \geq 0$ and $x = \pm c^{1/n}$.

We have already used this result to solve monomial equations.

Example 8: Solve $x^5 = 100$.
Take the fifth root. $x = 100^{1/5} = 2.51$.

The same idea works for non-integer powers if the unknown is positive.

Theorem 4.3.4. For $x > 0$ and $p \neq 0$,
A) "$x^p = c$" is equivalent to "$x = c^{1/p}$ and $c > 0$."
B) x^p and x^r are inverse functions iff $pr = 1$, that is, iff $p = 1/r$.

The domain is given as $x > 0$ because complex numbers are needed to define negative numbers to general powers. Fortunately, the answers we seek are usually positive numbers, so restricting consideration to positive values of x is usually acceptable.

Powers do not have to be integers or reciprocals of integers.

Discovery 1: Graph x^p for various values of p, including fractions and irrational numbers. Try p = 1/2, 1/3, 2/3, 3/4, -1/5, and .41. Which yield no points to the left of the y-axis (Problem #B4)?

Example 9: Solve $x^{4.7} = 1000$.

Take the 1/4.7 power. $x = 1000^{1/4.7} = 4.348$.

Powers do not have to be positive.

Example 10: Solve $x^{-1.2} = 5$.

Take the 1/(-1.2) power. Theorem 4.3.4 permits negative powers.
$x = 5^{1/(-1.2)} = .2615$.

Example 11: Solve $\sqrt[3]{x}$ - 7.8.

The cube root is the 1/3 power. The equation is $x^{1/3} = 7.8$. The reciprocal of 1/3 is 3. Take the third power:
$x^{1/3} = 7.8$ iff $x = 7.8^3 = 474.55$.

Integer powers are easy to understand as repeated multiplication. But how can we grasp a fractional power such as $x^{3.7}$? Any rational power can be expressed as the sum of integer powers and roots. Then Property 4.1.1 gives the image as a product of components. Here is how.

To evaluate x^{n+p}, where n is an integer and $0 \leq p < 1$, Property 4.1.1 says:

$$x^{n+p} = x^n x^p.$$

If we knew x^p for every p, $0 \leq p < 1$, we could use this to evaluate x^{n+p} because the "x^n" part is easy for integer n.

Example 12: To evaluate $10^{3.7}$, use a calculator. To understand how $10^{3.7}$ is determined, think of 3.7 as 3 + .7, which is 3 + .5 + .2 = 3 + 1/2 + 1/5. Now use 4.1.1.

$10^3 = 1000$.
$10^{.5} = 10^{1/2} = \sqrt{10} = 3.16228$.
$10^{.2} = 10^{1/5} = 1.58489$.

Both of these roots require special calculation and are determined by 4.2.1 as inverses of integer power functions. Now,

$10^{.7} = 10^{.5}10^{.2}$ [by property 4.1.1]
$= 3.16228(1.58489) = 5.01187$.

Then, $10^{3.7} = 10^3 10^{.7} = 1000(5.01187) = 5011.87$.

Property 4.1.3 yields equivalent alternative approaches. For one,

$10^{.7} = 10^{7/10} = (10^{1/10})^7$ or $(10^7)^{1/10}$. Then multiply by 10^3.

For another, treat $10^{3.7}$ as $10^{37/10} = (10^{1/10})^{37}$ or $(10^{37})^{1/10}$.

This example shows how x to any rational exponent can be built from component factors using integer powers and integer roots. Irrational exponents are discussed in Problem B38.

Graphs. Root functions are also called "radical" functions. The graphs of the most important root functions, the square root (Figure 6) and the cube root (Figure 7), resemble the graphs of x^2 and x^3, except the x- and y-axes are switched and the square root is only half a parabola, since the square root is defined to be non-negative. The graph of "$y = x$" is included for comparison in each case.

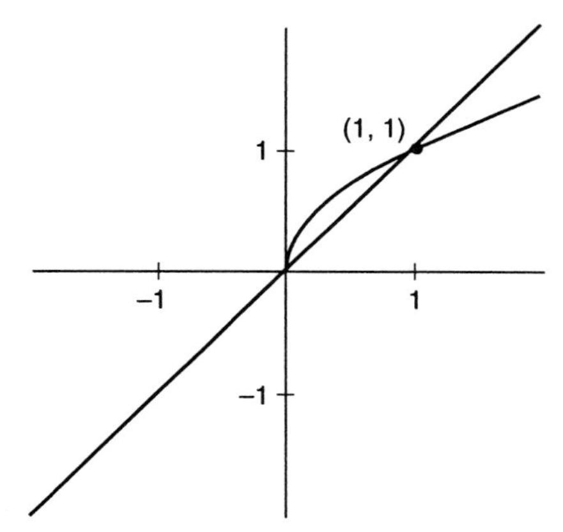

Figure 6: $\sqrt{x} = x^{1/2}$ and x.
[-2, 2] by [-2, 2].

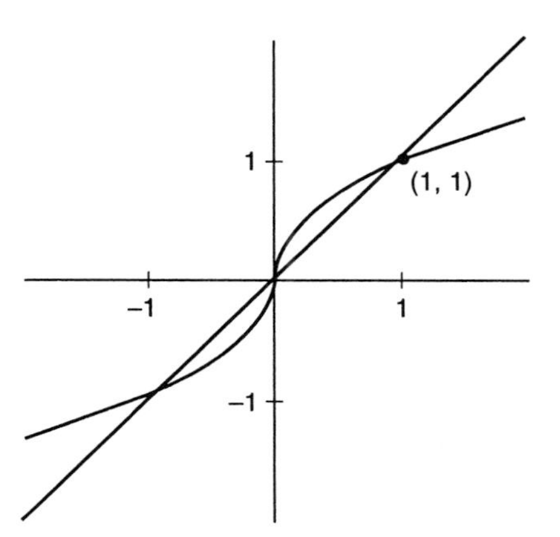

Figure 7: $x^{1/3}$ and x.
[-2, 2] by [-2, 2].

The graphs of all even root functions resemble the graph of $x^{1/2}$, and the graphs of all odd root functions resemble the graph of $x^{1/3}$ (Figures 8 and 9).

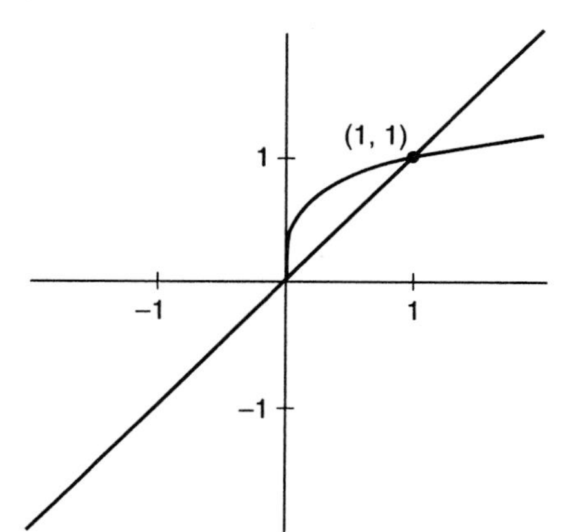

Figure 8: $x^{1/4}$ and x.
[-2, 2] by [-2, 2].

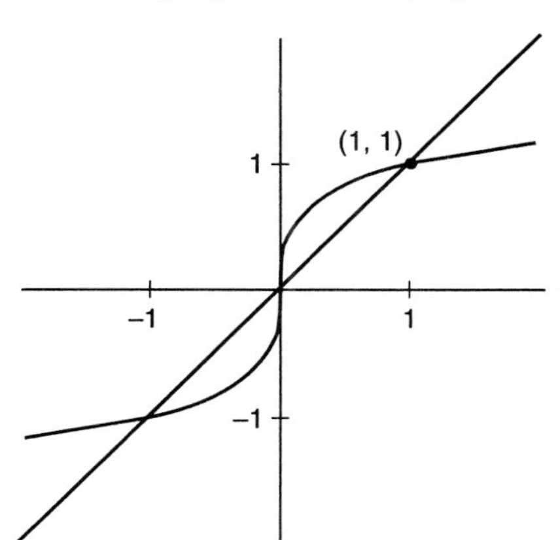

Figure 9: $x^{1/5}$ and x.
[-2, 2] by [-2, 2].

There is a big difference between even root functions and odd root functions: the domain of even root functions does not include negative numbers (unless complex numbers are permitted as images). Therefore, when discussing real-valued powers it is natural to exclude negative numbers so that all the powers will exist. For positive numbers, the shape of the power curve is easy to anticipate from the approximate size of the power. Powers between 0 and 1 are concave down like the square root curve (for example, the 0.41 power in Figure 10). Powers greater than 1 are concave up like the x^2 curve (Figure 11). Negative powers have a vertical asymptote at $x = 0$ and are concave up like the reciprocal function, $1/x$ (Figure 12). By the way, there is nothing special about the powers 0.41, 1.41, and -.41, they are just arbitrary examples.

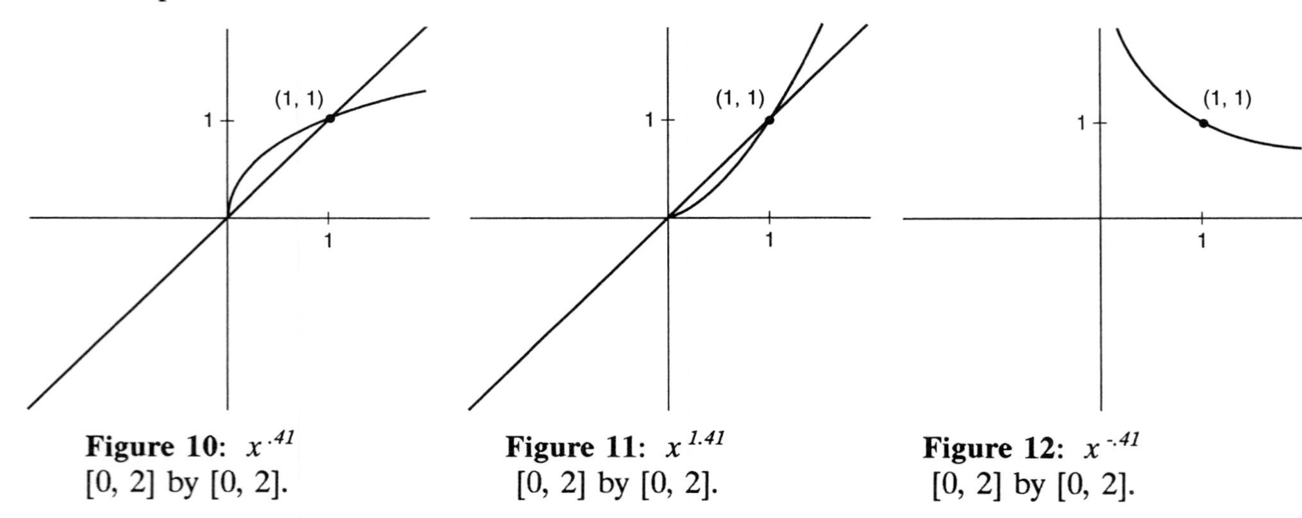

Figure 10: $x^{.41}$
[0, 2] by [0, 2].

Figure 11: $x^{1.41}$
[0, 2] by [0, 2].

Figure 12: $x^{-.41}$
[0, 2] by [0, 2].

Rational powers arise naturally when different integer powers are combined in the same equation.

Example 13: Kepler's Third Law of planetary motion says: $T^2 = ca^3$, where T is the time required for a planet to orbit the sun and a is the average distance of the planet from the sun. c is a constant which takes into account the force of gravity and the units employed to express time and distance.

Solve for T in terms of a.

All variables in the equation are positive, so we do not need to worry about negative solutions or square roots of negative numbers.

$$T = \sqrt{ca^3} = (ca^3)^{1/2} = c^{1/2}a^{3/2} .$$

The finals steps used properties of powers which also apply to n^{th} roots when $p = 1/n$. Reformulated for roots:

Theorem 4.3.5: For $a > 0$, $b > 0$, and any $p \neq 0$,

$$(ab)^{1/p} = a^{1/p}b^{1/p}, \qquad \text{[see 4.1.4A] and}$$

$$\left(\frac{a}{b}\right)^{1/p} = \frac{a^{1/p}}{b^{1/p}}. \qquad \text{[see 4.1.4B]}$$

$$(b^r)^{1/p} = b^{r/p} = (b^{1/p})^r. \qquad \text{[see 4.1.3].}$$

Theorem 4.3.5 is *not new*. It restates Theorems 4.1.4 and 4.1.3 using letters differently to emphasize different applications.

Example 13, continued: Solve for a in terms of T: $T^2 = ca^3$.
The expression "ca^3" is primarily a product. So, first divide.

$$T^2/c = a^3, \text{ and } \quad a = \left(\frac{T^2}{c}\right)^{1/3} = \frac{(T^2)^{1/3}}{c^{1/3}} = \frac{T^{2/3}}{c^{1/3}}.$$

Example 14: Give the surface area, A, of a cubical box in terms of its volume, V.

We usually express area and volume in terms of the length of an edge, s. Because a cube has 6 sides, $A = 6s^2$. For the volume, $V = s^3$. Use this equation to isolate s: $s = V^{1/3}$. Substituting this in for s in the area formula,

$$A = 6(V^{1/3})^2, \text{ which is}$$
$$A = 6V^{2/3}, \text{ (Figure 13).}$$

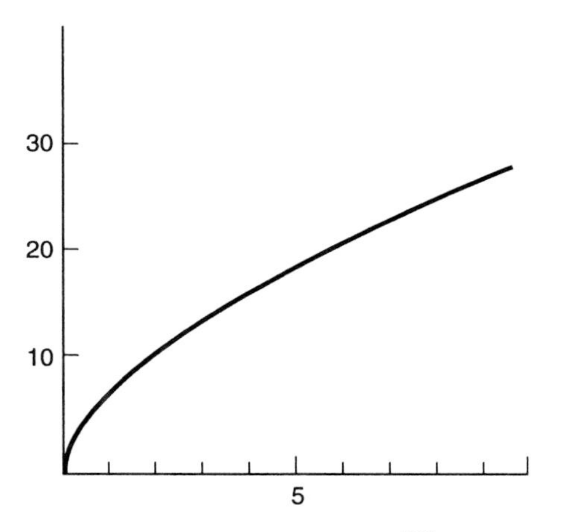

Figure 13: $A = 6V^{2/3}$.
[0, 10] by [0, 40].

Complications. There are complications when negative bases are considered. The next example shows what can go wrong, even in an expression as simple as "x".

Example 15: $x = x^1 = x^{(1/2)2}$, for all x.

This much is clear. But we cannot go one step further and say this is always $(x^{1/2})^2$ (misapplying Theorem 4.3.5 about a power of powers to negative values of x). Here is why.

$x^{(1/2)2}$ is not always x because the square root of x is not real if x is negative, so the operations cannot be executed in that order for negative x. x is defined for all x, but $(\sqrt{x})^2$ is not. Anytime a fractional power has an even denominator, this type of difficulty will occur with negative bases.

Example 16: Use your calculator to graph $x^{2/3}$.

Did you get what you expected?

Fractional powers are tricky. We can think of this as $(x^2)^{1/3}$ or $(x^{1/3})^2$. My calculator graphs these two for all x including negative x (Figure 14). But, for some reason (studied in complex analysis), it does not consider negative values of x to be in the domain of $x^{2/3}$ and therefore gives only the right side of the graph in Figure 14. This problem has to do with the calculator's method of evaluating negative numbers to fractional powers. Fortunately, real-world problems with fractional powers almost always require positive values of x, so the difficulty with negative values of x is of little concern.

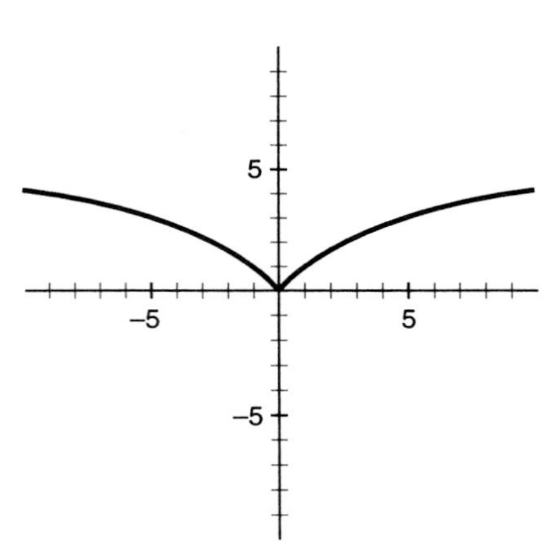

Figure 14: $(x^2)^{1/3}$ or $(x^{1/3})^2$. [-10, 10] by [-10, 10].

Identities. Identities can be used to simplify expressions.

Example 17: Simplify $\dfrac{\sqrt{x^3}}{(2x)^2}$ until only one "x" appears.

$$\frac{\sqrt{x^3}}{(2x)^2} = \frac{x^{3/2}}{(2x)^2} = \frac{x^{3/2}}{4x^2} = \frac{x^{-1/2}}{4}.$$

Other "simplified" expressions are possible: $\dfrac{x^{-1/2}}{4} = \dfrac{1}{4x^{1/2}} = \dfrac{1}{4\sqrt{x}}$.

Example 17, continued: Solve $\dfrac{\sqrt{x^3}}{(2x)^2} = 4.3.$

The previous simplification did most of the work. The equation is equivalent to

$$x^{-1/2}/4 = 4.3; \; x^{-1/2} = 17.2.$$

Now apply the -2 power (because -2 is the reciprocal of -1/2) to find $x = 17.2^{-2}$ [by Theorem 4.3.4A]. In decimal form, $x = .00338$. Do not be afraid of negative powers.

There is another way to solve this which avoids negative powers. Convert the equation to $1/\sqrt{x} = 17.2$ and multiply through by $\sqrt{x}$ to make the negative power disappear. But we are discussing powers -- including negative powers -- so learn the new way too.

The next three examples concern a type of problem that occurs in calculus when finding maxima and minima of expressions.

Example 18: Factor "$\sqrt{x}(1 + x)^3 + x^{3/2}(1 + x)^2$."

Both have a factor of $(1 + x)^2$. Also, $\sqrt{x} = x^{1/2}$, so both have a factor of $x^{1/2}$ ($x^{3/2} = x^{1/2}x^1$, by 4.1.1).

$$\sqrt{x}(1 + x)^3 + x^{3/2}(1 + x)^2 = x^{1/2}(1 + x)^2[(1 + x) + x]$$
$$= x^{1/2}(1 + x)^2(1 + 2x).$$

Negative powers can be treated the same way.

Example 19: Factor "$(-1/3)x^{1/2}(1 - x)^{-2/3} + (1/2)x^{-1/2}(1 - x)^{1/3}$."

Pull out a factor with the lesser of the powers on the similar factors. For "$x^{1/2}$" and "$x^{-1/2}$," the lesser power is -1/2. Then $x^{-1/2}x = x^{1/2}$. For "$(1 - x)^{-2/3}$" and "$(1 - x)^{1/3}$," the lesser power is -2/3. Note $(1 - x)^{-2/3}(1 - x) = (1 - x)^{1/3}$. Therefore,

$$(-1/3)x^{1/2}(1 - x)^{-2/3} + (1/2)x^{-1/2}(1 - x)^{1/3}$$
$$= x^{-1/2}(1 - x)^{-2/3}[(-1/3)x + (1/2)(1 - x)]$$
$$= x^{-1/2}(1 - x)^{-2/3}[1/2 - 5x/6].$$

Now, if the problem had been to solve the equation in which the original expression is set equal to zero, the Zero Product Rule would apply. However, only the last factor would yield a solution, because variables to negative powers are never zero. Think of negative powers as being a type of division (because $x^{-p} = 1/(x^p)$, which is never zero).

Example 20: Solve algebraically: $\dfrac{(4/3)x^{1/3}(1 + x)^2 - 2x^{4/3}(1 + x)}{(1 + x)^4} = 0$.

Recall that a quotient is zero if and only if the top is zero and the bottom is not. Therefore, the idea is to set just the top equal to zero and solve, remembering that "$x = -1$" can not be a solution because the bottom cannot be zero. So, factor the top.

$$(4/3)x^{1/3}(1 + x)^2 - 2x^{4/3}(1 + x) = (1 + x)[(4/3)x^{1/3}(1 + x) - 2x^{1/3}x^{3/3}]$$
$$= (1 + x)x^{1/3}[(4/3)(1 + x) - 2x]$$
$$= (1 + x)x^{1/3}[(4/3) - (2/3)x] .$$

Now the Zero Product Rule applies: $x^{1/3} = 0$ or $(4/3) - (2/3)x = 0$ or $1 + x = 0$. The last case is ruled out because the bottom would be zero. The solution is $x = 0$ or $x = 2$.

Conclusion. Squaring may introduce extraneous solutions. Also, multiplying by an expression with a variable may introduce extraneous solutions. When the relevant theorem is stated with "if..., then..." (as opposed to "iff"), extraneous solutions may arise. They can be eliminated by checking.

For positive bases power functions with real-valued powers have properties which parallel the properties of power functions with integer powers. Not all the parallels hold for negative bases, but we rarely need to use negative bases.

Terms: extraneous solution, iff, if...then..., n^{th} root, concave up, concave down.

****

Exercises for Section 4.3, "Fractional Powers":

A1.* a) To solve an equation is to find the values of the variable that make the equation ________. b) An extraneous solution is a value of the variable that makes the terminal equation ________ but makes the original equation ________. c) True or false: Extraneous solutions never arise unless you make a mistake.

A2. Which graph, $\sqrt{x}$, x^2, or $1/x$, do the following resemble the most for $x > 0$?
a) $x^{.68}$ b) $x^{1.2}$ c) $x^{.2}$ d) $x^{-.5}$.

A3.* Expand the square: $(a + b + c)^2$ [You should obtain 6 terms.]

A4. Expand the square: $(x - \sqrt{a})^2$.

^ ^ ^ ^ Evaluate these expressions and report at least three significant digits. To help you check your keystrokes, the value is given with two significant digits in brackets.

A5. a) $4^{2.5}$ [32] b) $4^{-2.5}$ [.031]
A6. a) $2^{3.4}$ [11] b) $2^{-3.4}$ [.095].
A7. a) $.6^3$ [.22] b) $.6^{-3}$ [4.6]
A8. a) $.3^{2.1}$ [.080] b) $.3^{-2.1}$ [13].
A9. a) $100^{1/5}$ [2.5] b) $100^{-1/5}$ [.40]
A10. a) $1000^{1/7}$ [2.7] b) $1000^{-1/7}$ [.37]
A11. a) $.02^5$ [3.2×10^{-9}] b) $.02^{-5}$ [310,000,000]
A12. a) $.15^3$ [.0034] b) $.15^{-3}$ [300]

^ ^ ^ ^ Solve algebraically for a positive solution and evaluate the solution as a decimal with at least three significant digits. To help you check your work, the solution is given with two significant digits in brackets.

A13. $x^{3.4} = 16$ [2.3].
A14. $x^{4.5} = 300$ [3.6].
A15. $x^{-2.3} = 20$ [.27].
A16. $x^{-3.2} = 50$ [.29].
A17. $x^{1/5} = 1.43$ [6.0].
A18. $x^{1/4} = 4.2$ [310].
A19. $x^{1/3} = .4$ [.064].
A20. $x^{1/5} = .2$ [.00032].
A21. $x^{2/3} = 10$ [32]
A22. $x^{3/4} = 100$ [460].
A23. $(1 + r)^{12} = 1.23$ [.017].
A24. $(1 + r)^{25} = 8$ [.087].
A25. $(1 + r/2)^4 = 1.18$ [.084].
A26. $(1 + r/12)^{36} = 1.6$ [.16].

^ ^ ^ ^ Simplify to only one appearance of "x" and one power, if possible. Assume $x > 0$.

A27. $(x^{1/2})x$
A28. $(x^{2/3})x$.
A29. $x^{1/2}/x$
A30. $x/x^{1/2}$.
A31. $(x^{1/3})^2$.
A32. $(x^{3/2})^{1/2}$.
A33. $(x^6)^{1/2}$.
A34. $(x^2)^{1/3}$.
A35. $(x^{1/3})^{1/2}$.
A36. $(x^{1/2})^{1/4}$.

^ ^ ^ ^ Over the interval $1 < x$, which is larger?

A37. x or x^{-1}.
A38. x^2 or x^{-2}.
A39. $x^{1/2}$ or x.
A40. $x^{1/3}$ or x.
A41. x^2 or x.
A42. x^2 or x^3.
A43. $x^{1/2}$ or $x^{1/4}$.
A44. $x^{1/3}$ or $x^{2/3}$.
A45. If $p > 0$, x^p or x^{-p}.

A46. Generalize from Exercises A37-A45 to determine a simple way to decide which is larger over the interval $1 < x$: x^p or x^r.

^ ^ ^ ^ Over the interval $0 < x < 1$, which is larger?

A47. x or x^{-1}. A48. x^2 or x^{-2}. A49. $x^{1/2}$ or x.

A50. $x^{1/3}$ or x. A51. x^2 or x. A52. x^2 or x^3.

A53. $x^{1/2}$ or $x^{1/4}$. A54. $x^{1/3}$ or $x^{2/3}$. A55. If $p > 0$, x^p or x^{-p}.

A56. Generalize from Exercises A47-A55 to determine a simple way to decide which is larger over the interval $0 < x < 1$: x^p or x^r.

A57. Find the equation of the circle with center (1, 3) through the point (2, 6).

A58. Find the equation of the circle with center (-2, 5) through the point (0, 0).

A59. Solve the equation in Example 5: $2x^2 - 14x + 29 = 16$.

A60. a)* How can you solve "$x^{-p} = c$" in one step?

b) Solve $x^{-3.7} = 5$ using that approach.

A61. In the equation in Example 3, identify the expressions corresponding to the letters "a", "b", and "c" of the theorem.

^ ^ ^ ^ ^ ^ ^ ^

B1.* a) Define "extraneous" solution. b) How can one arise?

c) Do you have to make a mistake for an extraneous solution to arise?

B2.* a) Name two processes for solving equations that may lead to extraneous solutions.

b) Do they always create extraneous solutions?

c) How can you eliminate extraneous solutions?

B3.* To solve an equation, multiplying both sides by an expression may introduce extraneous solutions. a) How? b) Create a very simple illuminating example. c) State the relevant theorem from this section.

B4. Do Discovery 1: Use your graphics calculator to look at the graph of x^p for various values of p, including fractions and irrational numbers. Try p = 1/2, 1/3, 2/3, 3/4, -1/5, and .41. Which yield no points to the left of the y-axis. Generalize.

^ ^ ^ ^ Give the domain.

B5. a) $x^{1.5}$ b) $x^{1/3}$ c) x^{-2} d) $x^{-1/2}$.

B6. a) x^{-4} b) $x^{2.3}$ c) $x^{1/4}$ d) $x^{1/5}$.

^ ^ ^ ^ Solve algebraically.

B7. $2\sqrt{x} + 3 = x$. B8. $2\sqrt{x} + 1 = x$. [5.8]

B9. $\sqrt{x + 5} = 7 - x$. B10. $\sqrt{10 - x} = x + 10$. [23]

B11. Solve $\sqrt{20 - x} = 14 - x$. [16] B12. Solve $\sqrt{6x - 5} = 16 - x$.

B13. $\sqrt{x} + \sqrt{x + 5} = 10$. [23] B14. $\sqrt{x} + \sqrt{x + 10} = 9$. [16]

^ ^ ^ ^ Factor.
B15. $x^{1/2}(2 + x)^{5/2} + 2x^{3/2}(2 + x)^{3/2}$.
B16. $4x^{11/6}(3 + x)^{3/4} + x^{5/6}(3 + x)^{7/4}$.
B17. $x^{-1/2}(1 - x)^{7/2} + 3x^{-3/2}(1 - x)^{9/2}$.
B18. $5x^{-1/3} + 2x^{2/3}$.

^ ^ ^ ^ Solve algebraically.
B19. $-6\sqrt{x}(1 - 2x)^2 + (1/2)(1 - 2x)^3x^{-1/2} = 0$.
B20. $(5/2)(1 - x)^2x^{3/2} - 2x^{5/2}(1 - x) = 0$.
B21. $(1/3)x^{1/4}(x + 3)^{-2/3} - (1/4)x^{-3/4}(x + 3)^{1/3} = 0$.
B22. $(5/3)x^{2/3}(1 - x)^{3/4} - (3/4)x^{5/3}(1 - x)^{-1/4} = 0$.

B23. Use your calculator to graph $x^{1/3}$. Now use it to graph $x^{2/6}$ (entering the operations in the given order). Are they different graphs? Use the following idea to explain why. $x^{2/6} = (x^2)^{1/6}$ for all x, but $(x^2)^{1/6}$ is not $(x^{1/6})^2$ for all x.

^ ^ ^ ^ The surface area of a sphere is $4\pi r^2$. The volume of a sphere is $(4/3)\pi r^3$.
B24. Express the surface area in terms of the volume.
B25. Express the volume in terms of the surface area.

B26. The moon and satellites obey Kepler's Third Law (Example 13). If the moon orbits the Earth in 27.3 days and is 58 times as far away from the center of the earth as a satellite, how long does it take the satellite to orbit the earth?

B27. Set up the equation. Let A (left) and B (right) be fixed points on a line. Let M be on the line and vary between A and B. Let P be on the perpendicular to the line at M such that $MP^2 = AM(MB)$. Find the equation for all such points P.

B28. Set up the equation. Let A (left) and B (right) be fixed points on a line. Let M vary to the right of B. Let P be on the perpendicular to the line at M such that $MP^2 = AM(MB)$. Find the equation for all such points P.

B29. Let point A be the point (-3, 0) and point B be (3, 0). Find all points P such that the sum of the distances from P to A and B is 10 [Generalized in B30].

B30. [Generalizes B29.] Let A be $(-c, 0)$ and B be $(c, 0)$. Set up the equation for all points P such that the sum of the distances AP and BP is $2a$, where $2a > 2c > 0$. [This yields an ellipse which can be written in the form 3.2.10. Continued in B31.]

B31. Simplify the equation obtained in B30. Let $b^2 = a^2 - c^2$ and obtain the form of an ellipse in 3.2.10: $x^2/a^2 + y^2/b^2 = 1$.

B32. Let A be $(-c, 0)$ and B be $(c, 0)$. Set up the equation for all points P such that the difference of the distances AP and BP is $\pm 2a$, where $0 < 2a < 2c$. [This yields a hyperbola. Continued in B33]

B33. Simplify the equation obtained in B32. Let $b^2 = a^2 - c^2$ and obtain the form:
$x^2/a^2 - y^2/b^2 = 1$.

B34. Find an equation for the set of all points that are twice as far from the y-axis as from (3, 0). Simplify.

B35. a) Sketch, on the standard scale, the graphs of the two expressions in Example 4: "$2 + \sqrt{x}$" and "x".
b) Mark (perhaps with dashes) the rest of the parabola of which only a part is visible.
c) Where, on your sketch, is the point corresponding to the extraneous solution?

B36. Consider the equation $\sqrt{x} = mx + b$.
Sketch the two expressions (for various m and b) to illustrate that there can be one solution, two solutions, or no solutions.

B37. Show that Theorem 4.3.3 would be false without the condition "$c > 0$" in the second half. Do so by giving an example such that "$x^p = c$" and "$x = c^{1/p}$" are not equivalent.

B38. Prove, for non-negative a and b: $(a + b)/2 \geq \sqrt{(ab)}$.

B39. x^c can be defined for irrational c by taking the "limit" of the values of x^r where r varies over rational values approaching c. Give a sequence of rational r's approaching π.

B40. Which power functions have negative numbers in their domain (without using complex numbers)?

Section 4.4. Percents, Money, and Compounding

The term "percent" occurs in the context of multiplication. When you see or hear the word "percent," think *multiplication*. For example, the number of students described by the phrase "70 percent of the students" is obtained by multiplying the number of students by .70. Do not be fooled by other words into thinking about addition or subtraction. For example, to "add 10 percent" yields a final amount which is obtained by *multiplying* by 1.10. To "take 20 percent off of the price" yields a price obtained by *multiplying* the price by .80.

Repeated multiplication by the same factor produces powers. For example, when money grows at 8% per year for 5 years there will 5 factors of 1.08 and the total growth factor will be 1.08^5. The idea of such "compound" interest makes the connection with powers, which are the subject of this chapter.

<u>Review of Percents.</u> One way to compare two amounts is to compute their difference. For example, to compare 60 to 50 we might say "60 is 10 more than 50." The difference, 10, expresses a comparison. Another way to compare two amounts is to compute a multiplicative factor. For example, we could say "60 is 1.2 times 50," where the factor, 1.2, expresses the comparison. There are many occasions when a comparison using multiplication is more appropriate than a comparison using addition. In this section comparisons will be determined using multiplication. Percents provide a convenient way to express those comparisons.

"Percent" means, in its Latin derivation, "by the hundred." It is a way of comparing two quantities where the standard of comparison is on a scale of 100. The standard is "100 percent." Amounts less than the standard are less than 100 percent, and amounts greater than the standard are greater than 100 percent.

(4.4.1) To say "b is p percent of a" is to say $b = (\frac{p}{100})a$.

Think of percents as expressing a functional relationship using multiplication. The standard plays the role of the argument, a.

Mathematicians usually work percentage problems using decimals. They convert fractions like "$p/100$" to "p percent" expressed in decimals.

Example 1: The function signaled by the phrase "25 percent of" is "Multiply by .25," because 25/100 = .25.

25 percent of 600 is .25(600) = 150.

The function signaled by "72 percent of" is "Multiply by .72."

The function signaled by "220 percent of" is "Multiply by 2.20."

To compare b to the standard a, the percent p is given by

$$\frac{b}{a} = \frac{p}{100}. \qquad \text{Therefore,} \quad p = 100\left(\frac{b}{a}\right). \tag{4.4.2}$$

The standard is in the denominator.

Example 2: Compare the area of a circle inscribed in a square to the area of the square (Figure 1).

The area of the circle is given by $A = \pi r^2$. The side of the surrounding square is $2r$, so the area of the square is $(2r)^2 = 4r^2$. That is the standard.

$$\frac{\pi r^2}{4r^2} = \frac{\pi}{4} = .785.$$

The circle has 78.5% of the area of the square.

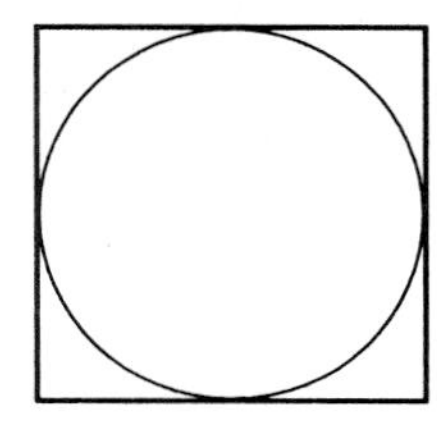

Figure 1: A circle inscribed in a square.

<u>Incorporating Change</u>. Often percents are used in the context of change. "Sales went *up* 12 percent." "During the sale prices are knocked *down* 35 percent." "The index of leading indicators *gained* 0.3 percent last month." "The Dow Jones stock market index *lost* 0.4 percent last week."

Example 3: A 9 percent increase corresponds to a final amount that is "109 percent of" the original. The standard is automatically 100%, so the increase of 9% brings the total up to 109%. The final amount compares to the original amount by the function, "Multiply by 1.09."

Example 4: The phrase "80 percent of" corresponds to a 20 percent decrease, which is equivalent to "20 percent off." The words "off" and "of" can be confounded. Their spelling differs by only one letter, but they have radically different interpretations. "y is 20 percent of x" means $y = .2x$. "y is 20 percent off x," means $y = .8x$. "Off" refers to the percent being a decrease. That second "f" in "off" makes a big difference.

Example 5: Express as a rule the function which relates the new amount to the standard.

phrase	rule
"gain 20%"	multiply by 1.2
"down 10%"	multiply by .9
"70% off"	multiply by .3
"up 6%"	multiply by 1.06
"20% discount"	multiply by .8

(4.4.3A) To "increase," "gain," or "go up," p percent means the final amount is the standard multiplied by a factor of $1 + p/100$.

(4.4.3B) To "decrease," "lose," or "go down," p percent means the final amount is the standard multiplied by a factor of $1 - p/100$.

Example 6: Express as a formula: "B is 30% more than A."

Decide what is the standard. B is being compared to A, so A is the standard. A is therefore regarded as 100%, so B is 130% of A: $B = 1.3A$.

Example 7: Skirts originally priced $80 are now $60. How much off are they?

There are two kinds of answers to this problem. Thinking additively, since $80 is $20 more than $60, they are $20 off.

But it is common and often important to think multiplicatively. Think of $60 as some multiple of $80.

$$60 = 80x. \quad x = 60/80 = .75.$$

Therefore the skirts are "25% off."

Another, equivalent, approach is to think of "$20 off" compared to a standard of $80. The factor is obtained from the equation: $20 = 80x$. So $x = .25$, which is a second way to obtain the answer, "25 percent off."

Example 8: At 20 percent off a jacket costs $100. What did it cost originally?

Many people would erroneously reply, "$120." They have reversed the image and the argument.

The phrase "20 percent off" refers to the original price as argument (the standard). The cost is .80 times the original cost. $100 = 0.8c$ yields $c = 125$ (dollars).

This problem illustrates the value of the functional, formula, approach to word problems.

Sometimes it is nice not to have to have a fixed standard.

Example 9: Suppose sidewalks are laid out perpendicular to each other to form a square grid (Figure 2). If you walk across the grass on the diagonal of a square rather than on the sidewalks, how much distance do you save?

Obviously, the distance you save depends upon the size of the square.

But, since all squares are similar, you can express the distance you save as a fraction of the standard distance. Percents are good for this.

If the sidewalks are 1 unit apart, then the diagonal is $\sqrt{2}$ units. Instead of walking 2 units (the standard) you could walk $\sqrt{2}$ units. The standard goes in the denominator: $\sqrt{2}/2 = .707 = 70.7$ percent. The savings is 29.3 percent.

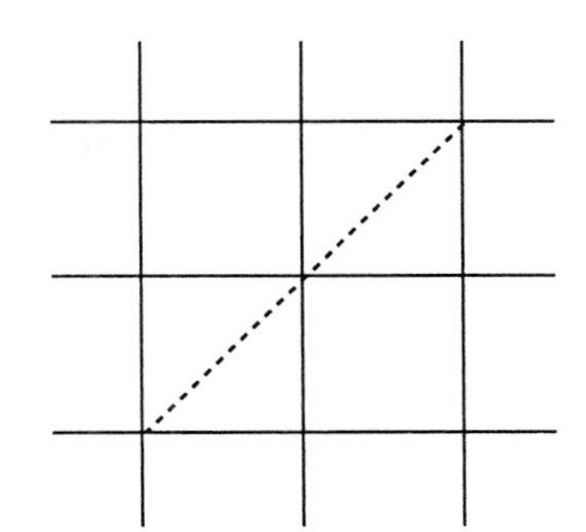

Figure 2: Sidewalks laid on a square grid and a diagonal shortcut.

Example 9, continued: How much further is it to walk on the sidewalks rather than cut across on the grass?

Now the standard has changed. The standard is the length of the diagonal, $\sqrt{2}$ units. The standard goes in the denominator: $2/\sqrt{2} = 1.414$. It is 41.4 percent further.

Note that an increase of p percent does not cancel a decrease of p percent. These concepts are not additive. Percent is not an additive concept!

<u>**Composition of Functions**</u>. One trick to percent problems is that the standard can change within a single problem.

Example 10: Sales of widgets went up 30 percent in 1992 and down 20 percent in 1993. How did they change over the two years?

Do not add or subtract. The context for percents is *multiplication*, not addition or subtraction. If you think it went "up 10 percent," because 30 - 20 = 10, you must change your thoughts about how percents are calculated.

The factor from 1991 to 1993 is: (1.30)(0.80) = 1.04. Sales went up 4 percent (not 10 percent).

Example 11: Which was a better investment, stocks which went up 11.5 percent and were sold for a 1.5 percent commission or antiques which went up 30 percent and were sold for a 20 percent commission? Note that the difference is 10 percentage points (not 10 percent) in each case.

We are comparing two growth factors. Build your own formula for each.

The factor for stocks is 1.115(.985) = 1.098275, for a 9.8275 percent return. The factor for antiques is 1.30(.80) = 1.04, for a 4 percent return. Stocks were the better investment.

Example 12: B is 30% more than A. How much less than B is A?

The "30% more than" is used to compare B to A. The question asks us to compare A to B.

The first sentence is expressed by:

$$B = 1.3A.$$

Therefore, solving for A,

$$A = B/1.3.$$

To express the answer we must convert this division to multiplication:

$$A = .769B.$$

So A is 76.9% of B, so A is 23.1% less than B (because 100 - 76.9 = 23.1).

The cumulative effect of successive percentage changes is *not* additive. But, *if two* successive changes are small (this is a big *if*), say, less than 10% each, they are *nearly* additive (problem B38). The cumulative effect of successive changes is obtained by (surprise!) *multiplication*.

Example 13: An increase of 20% was followed by a decrease of 20%. What was the cumulative effect?

Do not add to get 0. Percentage changes are multiplicative, not additive. The factor is given by

$$1.2(.8) = .96,$$

which corresponds to a decrease of 4%.

For many word problems it is a good idea to

Build your own formula.

Example 14: Suppose you add x ounces of an 80 percent acid solution to 30 ounces of 40 percent acid solution. How many ounces should you add to make a 70 percent acid solution?

Note that each of the three percentages mentioned has a different standard. The amount of acid in the final solution will be 40 percent of 30 ounces plus 80 percent of x ounces. The amount of solution will be $30 + x$ ounces, which is the standard for the 70 percent. Thus the relevant equation is:

$$.4(30) + .8x = .7(30 + x).$$

Solving, $12 + .8x = 21 + .7x$. $\quad .1x = 9$. $\quad x = 90$.

Money and Compounding. Percents are often used in the context of money and investments. The "rate of growth" or "rate of return" on invested money is expressed in percent *per year*. For example, if the rate of return on $100 is 10 percent per year, it would grow to $110 in one year. However, the growth of money over time periods of duration greater than or less than one year is not so simple.

To "compound" money means to compute percentage increases periodically based on the current amount of money rather than on the original amount. The original amount of money invested is called the principal. As the money grows in value, investors want percentage increases calculated based on the larger current amount and not on the smaller principal. This is the idea of compounding -- periodic readjusting of the standard on which the percentages are calculated.

Example 15: Suppose $1000 is invested at 8% rate of interest, compounded yearly. How much will there be after 3 years?

The term "compounded yearly" means that each year the amount is multiplied by factor of 1.08. Three years yield three such factors.

1000 [original]
1000(1.08) [after 1 year]
1000(1.08)(1.08) [after 2 years]
1000(1.08)(1.08)(1.08) [after 3 years]

Of course, repeated multiplication can be expressed with powers. The final amount is $\$1000(1.08)^3$. This comes to $1259.71.

Note that the money grew by a factor of $1.08^3 = 1.25971$. It grew 25.971 percent, which is more than 3 times 8 percent = 24 percent. Over a short time period the difference between compounding and not compounding is not very great. However, over a longer time period the effect of compounding can be phenomenal.

Example 16: In 1626 Manhattan Island (the site of the future New York City) was purchased from the natives for the equivalent of $24. This is often considered to be one of the greatest bargains in history. Was it?

Suppose the natives had been able to invest the $24 they got at 8% compounded yearly. What would it be worth in 1996?

The time, $t = 1996 - 1626 = 370$ (years). There would be 370 factors of 1.08. The value would be

$$\$24(1.08)^{370} = \$5.5 \times 10^{13} = 55 \text{ trillion dollars!}$$

This makes the point that compounding over a long period of time can produce dramatic results, although, of course, it would be rare for an investment to be committed for even one-tenth that long.

Let an amount P (for principal) be invested at an interest rate of i per time period and compounded for n time periods (here "i" stands for "interest rate," not the square root of -1). At the end of n time periods its amount, A, satisfies

$$A = P(1 + i)^n. \tag{4.4.4}$$

Money is usually compounded more frequently than once a year. If so, the cited percentage rate *per year* (the so-called "annual percentage rate" or "APR") must be converted to a percentage rate *per time period*. The rate per time period, i, is defined to be the rate per year, r, divided by the number of time periods per year.

Example 17: Suppose an investment yields 12% per year compounded monthly. How does it grow?

The money will be compounded 12 times per year. For the calculations, the rate per year is converted to a rate per month and time is reckoned in months. "12% per year compounded monthly" is computed as "1% per month compounded monthly." Each month the investment will grow by 12%/12 = 1%, that is, it will change by a factor of 1.01. Let the initial amount be P and the integer number of months be m. The formula for the amount after m months is $P(1.01)^m$.

After 1 year the amount is 1.01^{12} times the original amount. $1.01^{12} = 1.1268$, which corresponds to a growth of 12.68%, not 12%.

Note that there is no factor corresponding to the "12%" mentioned in the problem. To facilitate comparison, interest rates are quoted "per year," even when they are not computed that way. "12%" is the "annual percentage rate," which must be used in combination with the frequency of compounding to express how the money grows. By itself, the annual percentage rate but does not fully describe the growth rate. The actual growth in one year is known as the effective rate. Banks take pains to distinguish between the name of the interest rate (the APR, annual percentage rate) and the "effective" rate, which is usually somewhat higher. In this example, the APR is 12% and the effective rate is 12.68%.

Compound Interest Formula. Suppose an amount of money, P (for "principal") is invested at an annual rate, r, and is compounded k times per year. The formula for the amount of money A, after t years (where t is a multiple of $1/k$) is given by

$$A = P(1 + \frac{r}{k})^{kt} . \tag{4.4.5}$$

Example 18: Find the future value of $10,000 invested at 8% compounded quarterly for 10 years. To compound money "quarterly" means to compound it four times per year. The rate per time period is one-fourth the annual rate.

$P = 10{,}000$, $k = 4$, $r = .08$, and $t = 10$.
$A = 10{,}000(1.02)^{40} = \$22{,}080.40$

From 4.4.5, after one year, the amount A is

$$P(1 + \frac{r}{k})^k .$$

Therefore, After one year, the increase in amount is

$$P(1 + \frac{r}{k})^k - P = P\,[(1 + \frac{r}{k})^k - 1]\ .$$

The factor by which an amount increases in one year is called the effective rate. Therefore, the effective rate corresponding to an annual percentage rate of r compounded k times per year is

(4.4.6) $$r_{effective} = (1 + \frac{r}{k})^k - 1\ .$$

Example 18, continued: What is the effective rate corresponding to 8% APR compounded quarterly?

The phrase "8% compounded quarterly" means 2% per quarter, for four quarters a year. The effective rate is $1.02^4 - 1 = 1.0824 - 1 = .0824 = 8.24\%$. This is somewhat higher than the APR of 8%.

Example 19: Some credit cards charge interest at 1½% per month. What is their effective rate?

$(1 + .015)^{12} = 1.1956$. Their effective rate is 19.56% (per year).

Example 20: Suppose the yearly growth rates of an investment over the past three years were 10%, 43%, and 15%. What was the average (compound) growth rate?

In this context the word "average" does not mean to add them and divide by three. The intention is to find a constant annual growth rate that, over three years, would have produced the same return as the given varying growth rates. The actual growth was by a factor of

$$1.10(1.43)(1.15) = 1.80895.$$

Growth over three years at the annual rate r would produce growth by a factor of

$$(1 + r)(1 + r)(1 + r) = (1 + r)^3.$$

Set these equal and solve for r.

$$(1 + r)^3 = 1.80895,$$

$$1 + r = (1.80895)^{1/3} = 1.2185,$$

so $r = 21.85\%$. Money invested at the fixed annual rate of 21.85% would have produced the same three-year return.

The average (compound) rate of n successive rates, $r_1, r_2, ..., r_n$, is the solution for r to

(4.4.7) $$(1 + r)^n = (1 + r_1)(1 + r_2) \dots (1 + r_n).$$

Example 21: Suppose the yearly growth rates of an investment were 10%, 45%, -35%, and 12%. What is the average yearly growth rate?

$$(1 + r)^4 = 1.10(1.45)(.65)(1.12) = 1.16116.$$
$$1 + r = 1.16116^{1/4} = 1.0381.$$
$$r = 3.81\%.$$

The loss of 35 percent in the third year erased most of the growth. In spite of a big gain of 45 percent in the second year, a fixed annual growth rate of 3.81% would have produced the same growth over four years.

Example 22: Money invested at a fixed rate per year tripled in 11 years. What was the annual rate?

The growth factor is $(1 + r)^{11}$. The problem tells us this was 3. So solve

$$(1 + r)^{11} = 3,$$
$$1 + r = 3^{1/11} = 1.10503, \text{ and } r = 10.503\%.$$

The annual compound rate was 10.503 percent.

In Formula 4.4.5 for the growth of money, if time begins now ($t = 0$ now) we may think of the principal "P" as the "present value" of the investment and "A" as its "future value" in t years. A particular dollar value of money in the future is worth less than the same dollar value now. Everyone would rather win $1000 cash now than win a promise of a check for $1000 to be paid in the year 2013. If an amount is to be paid in the future, we can determine its present value by using Formula 4.4.5: $A = P(1 + r/k)^{kt}$. If we want, we can rewrite it to emphasize P.

Let P be the present value and A be the future value after t years of an amount of money invested at an annual rate, r, compounded k times per year.

(4.4.9) $$P = A(1 + r/k)^{-kt}.$$

Example 23: Suppose a retirement fund has $200,000 invested at 10% per year and inflation is projected to average 6% per year. Find the amount after 20 years and then find the present value of that amount, discounted for inflation.

The dollar amount will grow substantially over 20 years. The growth rate is 10%.

$$A = \$200{,}000(1.10)^{20} = \$1{,}345{,}500.$$

This number is pretty impressive. But a dollar will not buy as much in 20 years as it does now. What is the present value of that amount? The inflation rate is 6%.

$$P = \$1{,}345{,}500(1.06)^{-20} = \$419{,}533.26.$$

Still a nice amount, but not so impressive.

<u>**Annuities**</u>. An <u>annuity</u> is of a sequence of payments or deposits.

Example 24: John wins a $1,000,000 lottery and finds out he does not really get $1,000,000 now. He gets $50,000 now and payments of $50,000 each year for 19 more years. John actually wins an annuity.

Suppose payments, each of amount R, are made at equal time intervals beginning one time period from now and continuing for n payments. The <u>future value</u>, S, of the annuity (the value just as the last payment has been made at time n, at rate i per time period) is given by

(4.9.10) $$S = R[(1 + i)^n - 1]/i.$$

The <u>present value</u>, A, of the annuity (one time period before the first payment) is given by

(4.9.11) $$A = R[1 - (1 + i)^{-n}]/i.$$

Example 24, continued: John, who won the "million dollar" lottery, wants all his money now. So he agrees to sell his annuity for a lump sum payment. How much is the lump sum?

The value of his "million" dollars depends upon the going interest rate. If the interest rate is high, payments promised in the distant future are worth less than money now. Suppose the rate is 9.5% per year. He gets $50,000 now and 19 more payments in an annuity. According to 4.9.11, the total is

$$\$50,000 + \$50,000[1 - (1 + .095)^{-19}]/.095 = \$482,477.92.$$

That amount of money invested now at 9.5% could yield the same sequence of payments. I expect that John will be disappointed that his "million" is worth less than half of that.

<u>**Conclusion**</u>. Percents are a convenient way to express the functional relationship "Multiply by a constant." Because the context for percents is multiplication and not addition, successive percentage changes are not totaled by adding and not averaged by dividing.

Compounding of money invested at a fixed rate produces amounts given by power functions that are effectively repeated multiplication.

Terms: percent, discount, split function, percentage point, principal, annual percentage rate, compound interest, effective rate, present value, average growth rate, annuity.

****

Exercises for Section 4.4, "Percents":

A1.* What does "percent" mean, in its Latin derivation?

A2.* The context for "percent" is (pick one)
a) addition b) multiplication c) subtraction d) powers

A3.* Which percents refer to less than the standard?

A4.* Write the numbers as percents (assuming the context is multiplication).
a) 1 b) 1/2 c) 2 d) 1/4.

A5.* Rachel likes to give 15% tips when she eats in a restaurant. Explain, in English, how she can approximate in her head 15% of amounts such as $8 or $20.

^ ^ ^ ^ Express, *as a rule* (for example, "Multiply by 1.3") the function which relates the original amount to the final amount, as expressed by the given phrase:

A6. "87 percent of"
A7. "up 15 percent"
A8. "down 5%"
A9. "40 percent off"
A10. "gained 33 percent"
A11. "12% of"
A12. "25% discount"
A13. "14% more than"
A14. "70 percent less than"

^ ^ ^ ^ Give the functional relationship expressed by using appropriate letters for the variables.
A15. "Sales of video games in 1993 were 1300 percent of what they were in 1980."
A16. "55 percent of the students in Freshman math classes are female."
A17. "54 percent of eligible voters voted,"
A18. "Liquidation sale: All hand tools priced at 20 percent of manufacturer's suggested retail price."

A19. The market index went from 415.7 to 401.8. What was the percent change?

A20. Gina's salary just went up from $21,125 per year to $22,456 per year. What percentage raise did she get? [6.3%]

A21. State employees just got an across-the-board $970 per year raise. For Sally, that works out to a 4.2% raise. How much did she earn before the raise? [$23,000]

A22. a) "Sales this year are 250% of last year's" How much are sales up?
b) "Sales this year are 96% of last year's" How much are sales down?

A23. What was the original price of a jacket that sold for $114 at "40% off"?

A24. Including a 6% sales tax a car cost $13,668.70. What was the pre-tax price?

^ ^ ^ ^ What is the cumulative change of
A25. an increase of an increase of 10% followed by an increase of 15%?
A26. a loss of 70% followed by an increase of 100%?
A27. a loss of 10% followed by an increase of 10%?
A28. a gain of 70% followed by an loss of 50%?

A29. Suppose a country has a population of 15,000,000 and the population grows at a rate of 3% per year. How many people will it have in 20 years? [27,000,000]

A30. Suppose $10,000 is invested at 10% compounded monthly. What will be its value after ten years? [$27,000]

A31. Money invested at a fixed rate per year doubled in 9 years. What was the annual rate?

A32. Solve $5000(1 + r/2)^{24} = 5600$. [.0095]

^ ^ ^ ^ What is the effective rate of

A33. 6% compounded monthly?

A34. 10% compounded quarterly?

A35. Which is larger after a year? Money compounded at 10.2% monthly or 10.3% semiannually?

A36. Over four years an investment went up 8%, down 12%, up 22%, and up 15%. What was the average (compound) growth rate? [up 7.5%]

A37. Find the average annual percentage increase corresponding to three annual increases of 10%, 15%, and 50%. [24%]

A38. Which yields the greater cumulative increase: 1) an increase of 10% followed by an increase of 20%, or 2) an increase of 20% followed by an increase of 10%?

A39. a) An investment went up 12% one year and 42% the next. How much did it go up over the two years? b) What was its average (annual) growth rate? [59, 26]

A40. a) The cost of computer memory went down 30% one year and 60% the next. How much did it go down over the two years? b) What was its average (annual) rate of change? [72, 47% down]

A41. a) An investment went up 34% one year and down 12% the next. How much did it change over the two years? b) What was its average (annual) rate of change? [18% up, 8.6%]

^ ^ ^ ^ ^ ^ ^ ^

B1. a) What operation is usually signaled by "percent"?
b) If there are two successive percents in a problem, what operation is required to find their cumulative effect?
c) What operation is sometimes erroneously used to find the cumulative effect of two operations expressed in percents?

B2.* Explain the difference between "200 percent more than" and "200 percent of".

B3.* In the equation "$c = b(1 + d)^p$" there are four letters. Suppose we are given the values of three and one is originally unknown.
a) If d is not given, how can we find it?
b) If b is not given, how can we find it?
c) If c is not given, how can we find it?
d) If one letter is not given, arithmetic operations, powers, and roots do not suffice to find it. Which one?

B4. a) What percent of the side is the perimeter of a square?
b) What percent of the perimeter is the side of a square?

B5. a) What percent of the diameter is the circumference of a circle?
b) What percent of the circumference is the diameter of a circle? [32%]

B6. Box A is 40% heavier than Box B. How much lighter is Box B than Box A? [29%]

B7. One year the second quarter earnings of IBM fell 92%. How much would they have to go up to reach the previous level?

B8. Profits are down 50% this year. What increase will bring profits back up to last year's level?

B9. Over the last two years the Widget Mutual Fund went up 60%. Last year alone it went up 35%. How much did it go up the year before last? [19%]

B10. Worldwide Widget stock went down 10% in the last 5 years, and in the first of those 5 years it went up 20%. How much has it gone down in the last 4 years?

B11. Kristi invested in two mutual funds. She put some money in one and it went up 8% and, at the same time, she put twice as much money in another and it went up 12.6%. How much did her money go up, overall? [11%]

B12. The July 29- August 4, 1991, issue of the *Washington Post Weekly* said that the Sahara desert grew by 16% in the early 1980's and has shrunk 9% since then. What is the overall percentage change?

B13. Ricky has some round logs that are four times as long as their diameter. To make them burn better, he decides to split them lengthwise in quarters. How much does that increase their surface area? [110%]

B14. The cost of that computer has dropped 20% twice since it was introduced at $1000. What does it cost now?

B15. Wheat cost 200 times as much in 300 A.D. as it did in 150 A.D. What was the average annual rate of inflation of the cost? [3.6%]

B16. A path is in the shape of a quarter circle from point A to point B. How much shorter is it straight from A to B than around the path? [10%]

B17. "Profits were down 40 percent last year, but they are up 55 percent this year." How are company profits doing compared with two years ago?

B18. Income and taxable income are not the same. Suppose the first $12,500 of income is not taxable, and income in excess of that amount is taxable at 15%. At what income level would taxes amount to 10% of income?

B19. Suppose that under the current tax system the first $15,000 of income is not taxed, and any income over that is taxed at 15%. A possible revision has the first $16,000 not taxed, but any income over that taxed at 18%. People with which incomes will pay less under the proposed revision?

B20. Ervin has $100,000 to invest and is comparing tax-free municipal bonds at 6% to other bonds at 8%. Which has a higher net return? Assume that taxes take 28% of the interest on the other bonds.

B21. Add pure acid to 10 ounces of 30% acid solution. How much should be added to form a 50% acid solution?

B22. Add an amount of 5% salt solution to 10 ounces of 8% salt solution. How much does it take to make a 6% salt solution?

B23. Jane's house has 120° hot water and 45° cold water. How should they be mixed to make 95° water? [67% hot]

B24. Suppose your investments make 10%, inflation is 4%, and taxes take 28% of the increase in value of your investments. Adjusted for inflation, how do your investments increase over one year? [See also the next problem and B39.] [up 3.1%]

B25. Suppose your investments make 20%, inflation is 14%, and taxes take 28% of the increase in value of your investments. Adjusted for inflation, how do your investments increase over one year? [See also the previous problem and B39.] [up .35%]

B26. A course grade is based on 3 unit exams of 100 points each and a final of 200 points. If a student averages 86% on the first three exams, give the formula for how her 200-point final will affect her average.

B27. Crystal's employer offered to pay her either of two ways. On Plan A she would get $40 a day plus 2 percent of all her sales. On Plan B she would get $20 a day plus 5 percent of all her sales. When is Plan B the better option for her?

B28. Rose invested in several mutual funds. 15% of her investment was in a fund that went down 65%. But, overall, her investments went up 40%. How did her other funds (the other 85%) do? [up 59%]

B29. Suppose a retirement fund has $200,000 invested at 10% per year and inflation is projected to average 6% per year. Find the amount after 20 years and then find the present value of that amount, discounted for inflation. [$1,300,000, $420,000]

B30. Box A is $p\%$ heavier than Box B. How much lighter is Box B than Box A?

B31. On Sept. 1, 1993 you could buy 104 Brazilian cruzeiros for a dollar. On Nov. 23, you could buy 217. The time interval is 83 days. What was the daily inflation rate? If that rate of inflation kept up for 365 days, what would be the annual effective inflation rate?

^ ^ ^ ^ See Example 24.
B32. If the interest rate is 12%, what is the present value of John's "million" dollars?
B33. If the interest rate is 5%, what is the present value of John's "million" dollars?

B34. Erin invests, beginning next month, $50 a month for the next 30 years (30×12 payments). a) What is the sum of his payments?
b) If the yield is a constant 8% per year, what will be the value of the annuity at the end of 30 years? [$75,000]
c) If the inflation rate is 5%, what would be the present purchasing power of that amount? [$17,000] [See also B43.]

B35. Here is a problem which helps explain why bond prices move in the opposite direction of interest rates. Suppose a $1000, 8% bond really gives you interest payments, calculated at 8% per year, every year for 30 years beginning one year from when you buy it, and, in addition, at the end of the 30 years you get $1000. If the current interest rate drops to 7% after two years (and there are still 28 interest payments left, plus the $1000), what is the present value of the bond then? [$1100]

B36. Here is a another problem which helps explain why bond prices move in the opposite direction of interest rates. Suppose a $1000, 10% bond really gives you interest payments, calculated at 10% per year, every year for 20 years beginning one year from when you buy it, and, in addition, at the end of the 20 years you get $1000. If the current interest rate goes up to 12% after three years (and there are still 17 interest payments left, plus the $1000), what is the present value of the bond then? [$860]

B37. Suppose the first $12,000 of taxable income is taxed at 15%, and that any taxable income above that is taxed at a 28% rate. At what income level would taxes be 20% of taxable income?

B38. a) Let an amount change p percent and then change q percent. What is the total change? The actual percentage change is $p+q+(pq/100)$. Prove it.
b) Find the cumulative effect of a 5% change followed by a 10% change two ways: using part (a), and by directly computing the cumulative multiplicative factor.

B39. [After B23 and 24] Compare the results of problems B23 and B24. If your investments make a constant $p\%$ over the inflation rate, are you better off with low or high inflation?

B40. What percent of the volume of a cube is filled by the largest sphere that fits in it?

B41. (Leverage) Suppose you can control an investment without paying for it all. For example, suppose you buy stock by immediately paying 20 percent and agreeing to pay interest on the remaining amount at 1 percent per month. Of course, when you sell it you will pay off the balance. Suppose in one month the stock is up 5 percent, and you sell it for a 1.5 percent commission. How did you do?

B42. [Compare with B34.] Erin invests, beginning one year from now, $600 a year for the next 30 years. a) What is the sum of his payments?
b) If the yield is a constant 8% per year and all increases are taxed each year at a 28% federal rate and a 10% state rate, what will be the value at the end of 30 years?
c) If the inflation rate is 5%, what would be the current purchasing power of that amount?

B43. Perhaps you have heard that some paintings have sold for phenomenal values (up to $80 million!) to Japanese collectors. A major part of the reason is the Japanese tax system which encourages the purchase of art. Here is why.

Suppose you are old and rich and wish to pass your estate down to your heirs. But the death-tax rate is over 50 percent, so, without some scheme, your heirs will get less than half. However, the value of art (as opposed to other assets such as land) is generally determined by the owner, not the government, and the government seems to accept art valued at only 10% of its actual cost. Thus, by borrowing against other assets and buying art with the money, the apparent value of the estate can be decreased and thus death taxes decreased. Even after the heirs pay a 20 percent commission to have the art sold, they should come out well ahead.

Find out how well this scheme works by making the following assumptions and determining the percentage of the value of the estate that the heirs can retain. Suppose the death-tax rate is 55 percent and 70 percent of the value of the estate is art.

Section 4.5. Rational Functions

Polynomial expressions can be evaluated using only the four arithmetic operations. But the distinction of being the most complex type of expression which can be evaluated using only the four arithmetic operations belongs to the "rational" functions, that is, functions which are quotients of polynomials.

Definition 4.5.1: An expression is rational if and only if it is a quotient of polynomials.

Example 1: $\dfrac{x^2 - 1}{2x - 3}$ is a rational expression.

$1/x$ and $1/x^2$ are rational expressions.

$\dfrac{\sqrt{x}}{x + 1}$ is not a rational expression -- the numerator is not a polynomial.

Solving Rational Equations. A primary concern with division is to avoid division by zero. Numbers for which the denominator is zero are not in the domain of a rational function.

The most important features of the rational function $P(x)/Q(x)$ occur when $P(x) = 0$ or $Q(x) = 0$.

When the bottom is zero the expression is undefined. Near those x-values the behavior of the graph is remarkable and will be discussed shortly. Of course, the zeros of a function are always important.

Theorem 4.5.2 (Zero Quotient Rule):

$$\frac{P(x)}{Q(x)} = 0 \quad \textit{iff} \quad P(x) = 0 \quad \textit{and} \quad Q(x) \neq 0 \; .$$

Example 2: Solve $\dfrac{x - 2}{x - 1} = 0$.

The quotient is zero when the top is zero and the bottom is not. The solution is simply $x = 2$, from the top. The "$Q(x) \neq 0$" part of the theorem tells us to check $x = 2$ to make sure the bottom is not zero there. It is not.

In many problems forgetting to check when the denominator is zero causes no mistakes, because the solutions ruled out were not found anyway. But, sometimes, we do need to worry about the denominator.

Example 3: Solve $\dfrac{x^2 - 4}{x - 2} = 0$.

The top is zero when $x = 2$ or $x = -2$. But $x = 2$ is not in the domain, because it would make the bottom zero. Therefore, the complete solution is $x = -2$. "$x = 2$" is ruled out by the part of Theorem 4.5.2 that says that "$Q(x) \neq 0$."

You might think to factor "$x^2 - 4$" into "$(x - 2)(x + 2)$" and cancel the "$x - 2$" factors in the top and the bottom, but this approach requires caution. Often this causes no difficulty, but sometimes it does. Canceling may change the domain and make solutions that should be ruled out look legal.

Example 4: Solve $\dfrac{(x - 2)^2(x - 3)}{(x - 2)} = 0$.

In the original expression it is clear that $x = 2$ is illegal. But, after canceling, the equation would read "$(x - 2)(x - 3) = 0$," in which it is not clear that $x = 2$ is illegal. This new equation has solution $x = 2$ or $x = 3$, which is *not* equivalent to the original equation.

It is easy to misuse the near-identity, $a/a = 1$. This is, of course, only true *if* $a \neq 0$. In rational function problems where the denominator has a variable it may not be so easy to see when it is zero. Be careful.

Example 5: In calculus, problems like the following occur: Solve

$$\frac{3(x - 1)^2(x - 2)^2 - 2(x - 2)^3(x - 1)}{(x - 1)^4} = 0 .$$

According to Theorem 4.5.2 the first step is to set the numerator equal to zero and solve, remembering to exclude any resulting solutions for which the denominator would also be zero.

$$\begin{aligned} 3(x - 1)^2(x - 2)^2 - 2(x - 2)^3(x - 1) &= 0 \\ \textit{iff} \quad (x - 1)(x - 2)^2[3(x - 1) - 2(x - 2)] &= 0 \\ \textit{iff} \quad (x - 1)(x - 2)^2[3x - 3 - 2x + 4] &= 0 \\ \textit{iff} \quad (x - 1)(x - 2)^2(x + 1) &= 0 \\ \textit{iff} \quad x = 1 \ \textit{or} \ x = 2 \ \textit{or} \ x &= -1 . \end{aligned}$$

The solution $x = 1$ is ruled out by the denominator, so the solution to the original equation is $x = 2$ or $x = -1$.

Difference Quotients. Examples in which the top and the bottom are zero for the same values of x may, at first, seem like contrived special cases, but similar concerns with quotients occur in the definition of the derivative in calculus. Derivatives of functions are defined in terms of "difference quotients."

Definition 4.5.3. The difference quotient of f at x and a is

$$\frac{f(x) - f(a)}{x - a}.$$

If $f(x)$ is a polynomial or a rational function, this difference quotient is a rational function.

Example 6: Let $f(x) = x^2$. Find and simplify the difference quotient.

$$\begin{aligned} \frac{f(x) - f(a)}{x - a} &= \frac{x^2 - a^2}{x - a} \\ &= \frac{(x - a)(x + a)}{x - a} \\ &= x + a \qquad [\textit{ if } x \neq a \text{ }]. \end{aligned}$$

It is critical that x cannot equal a. Otherwise the original difference quotient would be zero over zero, which is undefined. Everywhere except at a the difference quotient has the relatively nice form "$x + a$." In calculus, we are interested in the difference quotient "as x goes to a," in which case "$x + a$" goes to $2a$. This limit calculation yields the derivative of x^2 at a -- which is $2a$.

Example 7: Let $f(x) = 1/x$. Find and simplify the difference quotient.

$$\begin{aligned} \frac{f(x) - f(a)}{x - a} &= \frac{\frac{1}{x} - \frac{1}{a}}{x - a} \\ &= \frac{\left(\frac{a - x}{xa}\right)}{x - a} = \frac{\frac{-(x - a)}{xa}}{x - a} \\ &= \frac{-1}{xa} \qquad [\textit{ if } x \neq a \text{ }]. \end{aligned}$$

Again, it is easy to see that x cannot equal a in any difference quotient, but x could equal a in the last expression, if we did not remember to rule it out.

When the Quotient is not Zero. To solve a rational equation when the quotient is not zero it is common to "Multiply through by the denominator." Again, we must be careful that the denominator is not zero.

Example 8: Solve $\dfrac{3x - 5}{x + 1} = 2$.

By multiplying through by the denominator a rational equation is converted into a polynomial equation.

$$\begin{aligned} 3x - 5 &= 2(x + 1) \\ &= 2x + 2. \\ x &= 7. \end{aligned}$$

Technically, we must remember to check that this potential solution does not make the original denominator zero. It doesn't, so it is right. The relevant theorem is next.

Theorem 4.5.4. $a/b = c$ iff $a = bc$ and $b \neq 0$.

The "$b \neq 0$" part tells us that any solution to the "$a = bc$" part that makes $b = 0$ must be ruled out.

Example 9: Solve $\dfrac{x^2 - x - 2}{x - 2} = 5$.

Multiplying through by "x - 2," we obtain

$$x^2 - x - 2 = 5(x - 2),$$

but this is not guaranteed to be equivalent to the original equation. The right approach is to proceed as usual and to remember to check the potential solutions to see that "$x = 2$" is ruled out, if it occurs. The equation is now a quadratic.

$$\begin{aligned} &x^2 - x - 2 = 5x - 10, && \\ &x^2 - 6x + 8 = 0, && \text{[consolidating like terms]} \\ &(x - 2)(x - 4) = 0, && \text{[factoring]} \\ &x = 2 \text{ or } x = 4 && \text{[by the Zero Product Rule]} \end{aligned}$$

But "$x = 2$" is extraneous and must be omitted, according to Theorem 4.5.4. The complete solution is "$x = 4$."

Of course, examples like this where extraneous solutions arise from multiplying through by the denominator are not common.

If two rational functions are combined in one equation, we may treat them like fractions and multiply through by a common denominator.

Example 10: Solve $\dfrac{1}{x - 2} + \dfrac{2}{x - 1} = 3$.

The product of the denominators is a common denominator. Multiplying through,

$$1(x - 1) + 2(x - 2) = 3(x - 1)(x - 2).$$

Now this is a quadratic. It remains to consolidate like terms, solve, and check the solutions to make sure they do not include "$x = 1$" or "$x = 2$" (Problem A24).

Graphs. Rational functions can have complicated graphs. That is part of the reason we study them -- they can represent a wide variety of relationships that cannot be represented by polynomials. For example, the graphs of rational functions can display vertical asymptotes which cannot be a feature of the graphs of polynomials.

Definition 4.5.5. A vertical line $x = a$ is said to be a vertical asymptote of the graph of $f(x)$ if, as x approaches a (from either the left or right), $f(x)$ is positive and becomes arbitrarily large or $f(x)$ is negative and becomes arbitrarily large in absolute value.

The reciprocal function exhibits the classic case of a vertical asymptote.

Example 11: $1/x$ has a vertical asymptote at $x = 0$ (Figure 1). As x nears 0 from the right, $1/x$ is positive and becomes arbitrarily large. For example, when $x = 1/100 > 0$, $1/x = 100$, which is large. Similarly, as x nears 0 from the left, $1/x$ is negative and becomes arbitrarily large in absolute value. For example, when $x = -1/1000 < 0$, $1/x = -1000$ which is a very large negative number. $1/x$ is undefined at $x = 0$; 0 is not in the domain.

The reciprocal function is an odd function (-1 is an odd number):

$$f(-x) = 1/(-x) = -1/x = -f(x).$$

Its graph exhibits point symmetry about the origin (Figure 1).

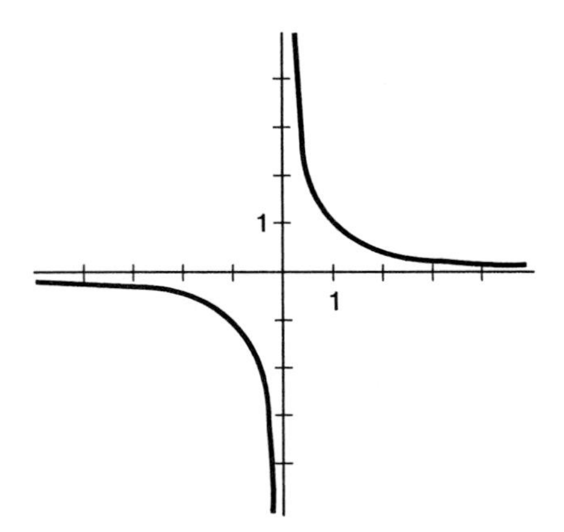

Figure 1: $1/x$.
[-5, 5] by [-5, 5].

Example 12: $1/x^2$ has a vertical asymptote at $x = 0$, which is a zero of the denominator (Figure 2). $1/x^2$ is undefined at $x = 0$. As x nears 0 from either side, $1/x^2$ is positive and becomes arbitrarily large.

$f(x) = 1/x^2$ is an even function (-2 is an even number): $f(-x) = f(x)$. Its graph exhibits symmetry about the y-axis.

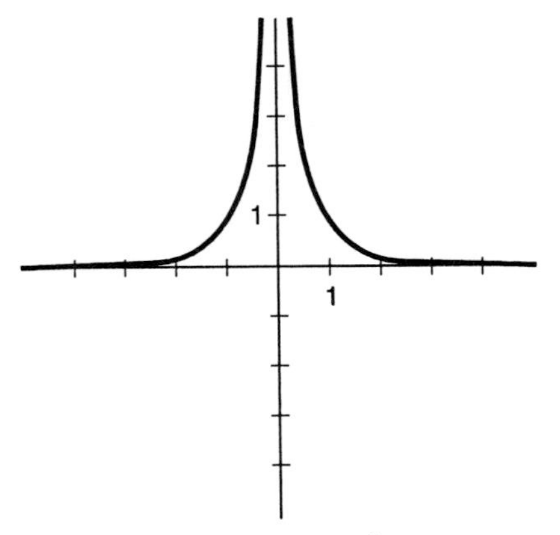

Figure 2: $1/x^2$.
[-5, 5] by [-5, 5].

In rational functions, vertical asymptotes usually occur where the denominator would be zero. Suppose $x = a$ makes the denominator

0. If the numerator is not also zero at the same place, then, as x nears a, the top approaches some number ($P(a)$) and the bottom approaches 0. The quotient of a number and a very small number is a number with large absolute value, either positive or negative depending upon its sign.

Example 13: Graph $2/(x + 3)$.

Expect a vertical asymptote at $x = -3$. There are no zeros.

Think of this as a shift and a scale change applied to the reciprocal function. Let $f(x) = 1/x$. This is $2f(x + 3)$. Therefore its graph is much like the graph of $1/x$, but three units to the left and twice as high (Figure 3).

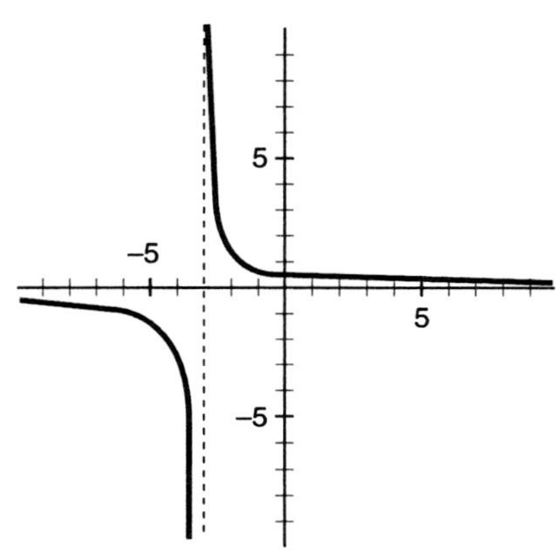

Figure 3: $2/(x + 3)$.
[-10, 10] by [-10, 10].

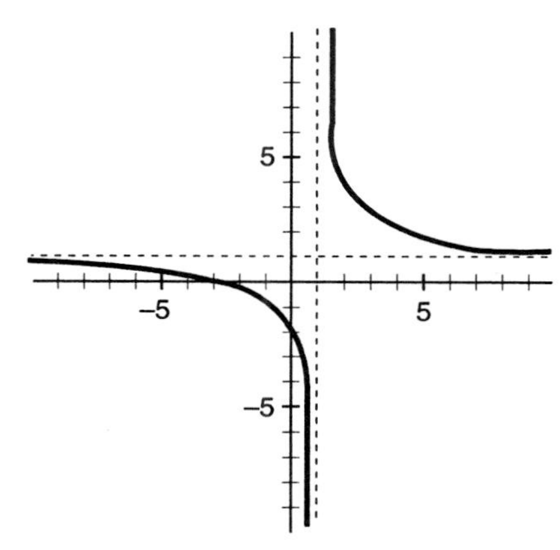

Figure 4: $(x + 3)/(x - 1)$
[-10, 10] by [-10, 10].

Example 14: Graph $(x + 3)/(x - 1)$.

Expect a vertical asymptote at $x = 1$. The zero is $x = -3$.

The graph of a quotient of linear polynomials usually resembles the graph of $1/x$ (Problem B31). By long division,

$$\begin{array}{r|l} & 1 \;\; + 4/(x - 1) \\ x - 1 & x + 3 \\ & \underline{x - 1} \\ & \quad 4 \end{array}$$

Therefore $(x + 3)/(x - 1) = 1 + 4/(x - 1)$, which has a graph that is the graph of $1/x$ after a shift and a scale change (Figure 4).

The only cases where there is not necessarily a vertical asymptote when the denominator equals 0 is when the numerator also equals 0 at the same value. Then the quotient approaches "0 over 0," which is an undefined form studied in calculus. By the factor theorem, if both the top and the bottom of a rational function are zero at the same value of x, the Factor Theorem tells us they both have a factor of "$x - a$," which can be factored out and canceled, if $x \neq a$. The original rational function is then identical to a reduced rational function (except possibly at $x = a$, where the original function was not defined, but the new function might be) which can be studied and graphed instead.

Expect a vertical asymptote when the denominator is zero. (There are some exceptions when the numerator is zero for the same value of x.)

Example 15: Graph $\dfrac{(x - 2)(x - 5)}{x - 5}$.

At first, we expect a vertical asymptote at $x = 5$. But $x = 5$ is also a zero of the numerator, so the top and the bottom have a common factor (which is obvious here, but which would not be so obvious if the top were multiplied out). Therefore, the quotient should be reduced first, keeping in mind that $x = 5$ is not in the domain.

$$\frac{(x - 2)(x - 5)}{x - 5} = x - 2 \text{ , } \textit{if } x \neq 5 \text{ .}$$

Therefore, the rational expression is nearly identical to "x - 2," a line. The difference is only that there is a tiny hole in the line at $x = 5$ where the rational function is not defined. Holes that are only one point wide are too small to see, and graphing calculators usually do not notice them, but, on a man-made graph, we can put a tiny open circle to indicate the existence of a hole (Figure 5).

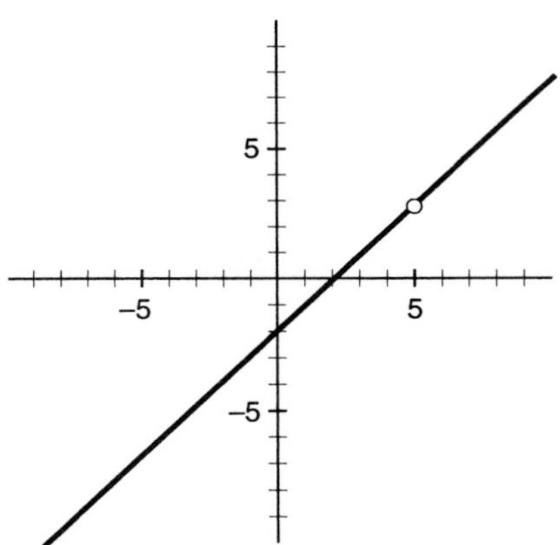

Figure 5:
$(x - 2)(x - 5)/(x - 5)$.
[-10, 10] by [-10, 10].

The most important features of a rational function are its zeros and asymptotes. To find the zeros, set the numerator equal to zero and solve, and check to make sure the denominator is not zero for the same x-values. To find the vertical asymptotes, set the denominator equal to zero and solve, and check to make sure the numerator is not zero at the same x-values.

If the numerator and denominator are higher-degree polynomials, the graph can be quite complex. We separate out for special consideration "end behavior" and behavior near asymptotes.

End-Behavior. The behavior of a rational function for large absolute values of x is easy to determine by looking at the quotient of the leading terms in the numerator and denominator.

Definition 4.5.6. Let a rational function, $R(x)$, be the quotient of polynomials $P(x)$ and $Q(x)$. Suppose $P(x)$ has leading term $p(x) = ax^n$ of degree n and $Q(x)$ has leading term $q(x) = bx^m$ of degree m. Then the rational function

$$\frac{P(x)}{Q(x)} \text{ has end-behavior model } \frac{ax^n}{bx^m} = \left(\frac{a}{b}\right) x^{n-m} .$$

Example 16: Determine the end-behavior models of selected functions.

$$\frac{3x^2 + 2x + 5}{x - 4} \text{ has end-behavior model } \frac{3x^2}{x} = 3x .$$

$$\frac{x^5 - 2x + 1}{4x^2 + 5} \text{ has end-behavior model } \frac{x^5}{4x^2} = \frac{x^3}{4} .$$

$$\frac{1 - x}{x^2 + 5} \text{ has end behavior model } \frac{-x}{x^2} = -1/x .$$

$$\frac{(2x + 3)(x - 7)}{x^2 - 2x - 5} \text{ has end-behavior model } \frac{2x^2}{x^2} = 2 .$$

For large absolute values of x, the graph of a rational function resembles the graph of its end-behavior model.

Example 17: Find the end-behavior model of

$$\frac{(x - 2)(x + 3)}{x - 1} \text{ and compare its graph to the}$$

graph of its end-behavior model.

The end-behavior model is simply $x^2/x = x$, the graph of which is a diagonal line. The rational function has zeros at 2 and -3, and a vertical asymptote at $x = 1$. For small x we do not expect the graph of the rational function to be like the graph of the line that is its end-behavior model, but for large x the similarity is evident (Figure 6).

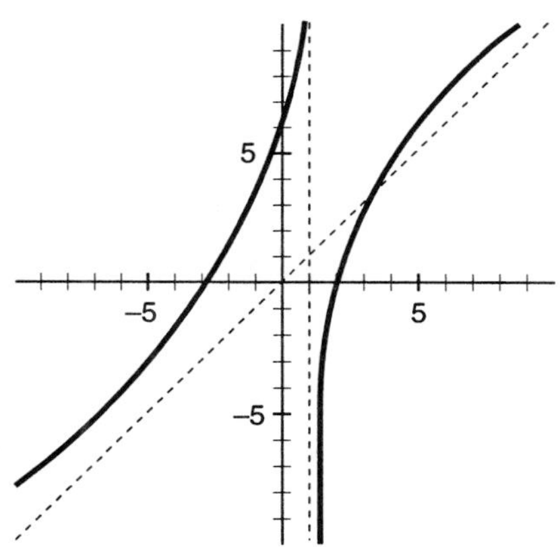

Figure 6: x and $(x - 2)(x + 3)/(x - 1)$. [-10, 10] by [-10, 10].

End-behavior models are helpful for large absolute values of x, but are not intended to help for small values of x. However, they can sometimes indicate when a graph is not "representative" because the window is too small.

Example 18: The end-behavior model of $\dfrac{2x^3 + 12x - 20}{6 - x}$ is

$\dfrac{2x^3}{-x} = -2x^2$. On a moderate-scale graph its end-behavior is not clear (Figure 7). It almost looks like a cubic in that window. But that cannot be a representative graph, because the end-behavior is not indicated correctly. If we zoom out far enough, we should be able to see the end-behavior. The end-behavior refers only to large absolute values of x, and what is "large" depends on the particular rational function. For large x, but not for small x, the graph resembles the graph of $-2x^2$ (Figure 8).

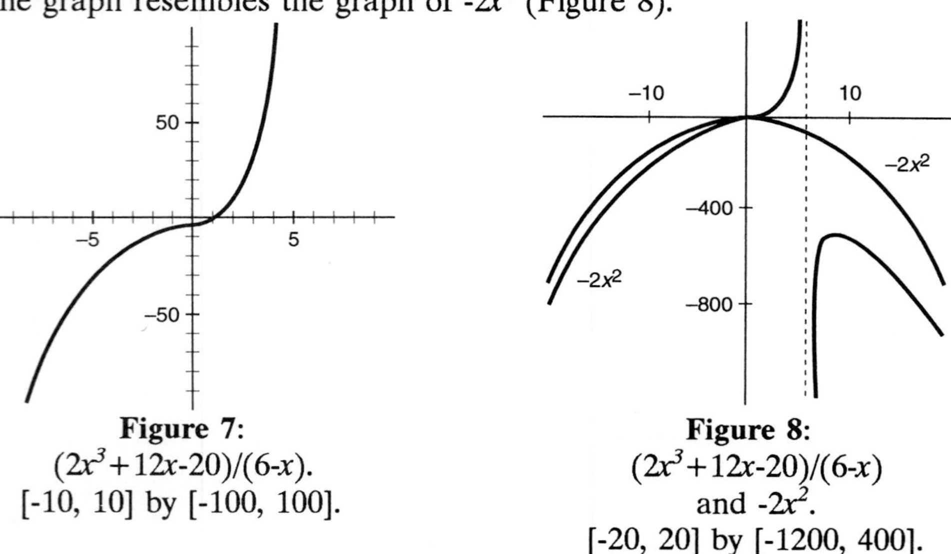

Figure 7:
$(2x^3+12x-20)/(6-x)$.
[-10, 10] by [-100, 100].

Figure 8:
$(2x^3+12x-20)/(6-x)$
and $-2x^2$.
[-20, 20] by [-1200, 400].

Sometimes the graph of a rational function approaches a horizontal line.

Definition 4.5.7. If, as x becomes large, or as x becomes large and negative, the graph of a function approaches a horizontal line, that horizontal line is said to be a horizontal asymptote of the graph.

A horizontal asymptote indicates a special type of end-behavior. It is easy to tell from the degrees of the component polynomials whether there will be a horizontal asymptote, and if there is, it is easy to tell the equation of the line.

Example 19: The graphs of $1/x$ (Figure 1), $1/x^2$ (Figure 2) and $2/(x + 3)$ (Figure 3) have the horizontal asymptote $y = 0$. If the denominator has higher degree than the numerator, $y = 0$ will be a horizontal asymptote.

If the numerator and denominator have the same degree, $y = c$ will be a horizontal asymptote where c is the quotient of the leading coefficients. For example, the rational function "$(x + 3)/(x - 1)$" (Example 10, Figure 4) is a quotient terms of degree 1 with leading coefficients 1 and 1, so their quotient is $1/1 = 1$, and the horizontal asymptote is $y = 1$.

Example 20: Determine the major features of the graph of $\dfrac{2x^2 + x - 7}{x^2 - 2x - 5}$.

The zeros are perhaps the most important features. To find them, solve $2x^2 + x - 7 = 0$. It does not factor easily, so use the Quadratic Theorem to obtain $x = (-1 \pm \sqrt{57})/4 = 1.63$ or -2.14.

The vertical asymptotes are perhaps the next most important feature. To find them, solve $x^2 - 2x - 5 = 0$. It does not factor easily, so use the Quadratic Theorem. $x = 1 \pm \sqrt{6}$. The two equations have no solutions in common -- the rational function cannot be reduced. It has vertical asymptotes at $x = 1 \pm \sqrt{6}$.

Now consider the end-behavior model: $2x^2/x^2 = 2$. It has a horizontal asymptote: $y = 2$. These ideas isolate only major features of the graph (Figure 9). Evaluation of particular values provides the details.

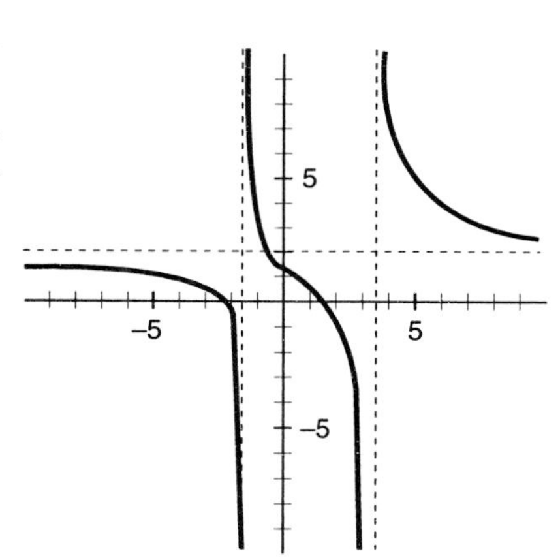

Figure 9: $\dfrac{2x^2 + x - 7}{x^2 - 2x - 5}$. $[-10, 10]$ by $[-10, 10]$.

Horizontal Asymptotes

For a rational function, if the degree of the numerator is less than or equal to the degree of the denominator there will be a horizontal asymptote. If the degree of the numerator is higher, there will not be a horizontal asymptote.

A rational function has a horizontal asymptote if and only if the end-behavior model is a constant or c/x^n for some $n = 1, 2, 3, \ldots$.

If the end-behavior model is a constant, c, then $y = c$ is a horizontal asymptote (Figures 4 and 9).

If the end-behavior model is c/x^n for $n \geq 1$, then $y = 0$ is a horizontal asymptote (Figures 1, 2, and 3).

The "major" features of a graph of a rational function include zeros and asymptotes.

Example 21: Graph $\dfrac{(x - 2)(x + 5)}{2x + 3}$ and identify its major features.

The zeros occur when $x = 2$ or $x = -5$ (neither makes the denominator 0). The denominator is zero when $2x + 3 = 0$, that is, when $x = -3/2$. Because the top is not zero for the same x, $x = -3/2$ is a vertical asymptote and the zeros of the numerator are the zeros of the rational function (Figure 10). The end-behavior model is $x^2/(2x) = x/2$, a line (Figure 10, dashes).

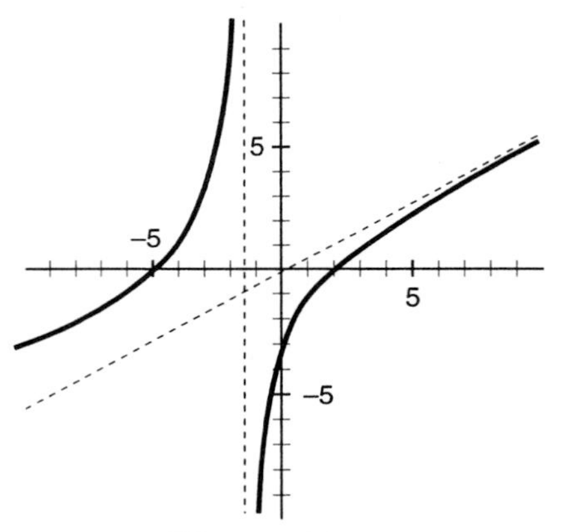

Figure 10:
$(x - 2)(x + 5)/(2x + 3)$.
[-10, 10] by [-10, 10].

Calculator Exercise 1 : Graphics calculators may draw graphs with artifacts. When graphing Figure 10, mine produces a vertical line at $x = -3/2$ that should not be there. Does yours? (Problem A33.)

Such artifacts are produced because the calculator fills in the graph by connecting dot to dot. A dot (pixel) high to the left of the asymptote is connected to a dot low to the right. For most functions, connecting dot to dot is the right thing to do, but not for this rational function.

<u>**Behavior near Vertical Asymptotes.**</u> Once a vertical asymptote has been discovered, the next question is whether the graph goes up or down along it on either side. Of course, graphic calculators have taken most of the work out of graphing functions, but you should try to understand *why* rational functions behave near asymptotes as they do.

Example 22: Let the rational function be simply the reciprocal function, $1/x$. There is a vertical asymptote at $x = 0$ (Figure 1). As x nears 0 from the left, x is negative and small and the numerator is positive, so the quotient will be negative and large. Therefore the graph goes down along the asymptote from the left. As x nears 0 from the right, x is positive and the quotient is positive and large, so the graph goes up along the asymptote.

The graph of $1/x^2$ was different (Figure 2). It went up along both sides of the asymptote because the denominator x^2 is positive and the quotient is positive on both sides of the asymptote.

Even without a graphics calculator, we can distinguish the ups from the downs by the sign of the quotient as x approaches the asymptotic x-value.

Example 23: Graph $\dfrac{(x - 3)(x + 1)}{x^2 - 4}$. Explain the behavior of the graph near the vertical asymptotes.

The two zeros of the numerator are $x = 3$ and $x = -1$. The denominator is zero at different places, $x = 2$ and at $x = -2$, so the graph has vertical asymptotes at $x = 2$ and $x = -2$. The end-behavior model is $x^2/x^2 = 1$, so it has a horizontal asymptote at $y = 1$ (Figure 11).

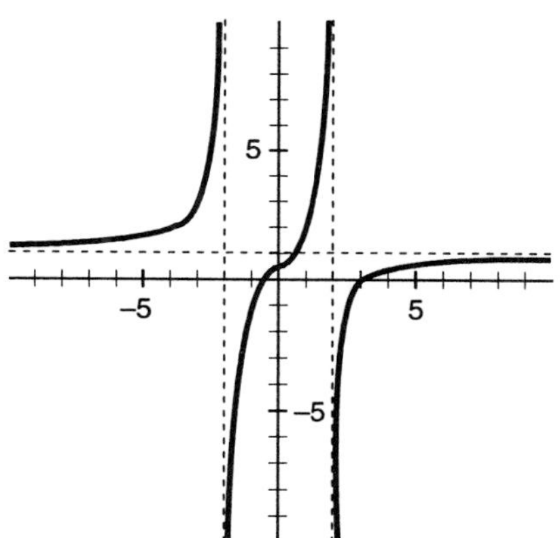

Figure 11: $\dfrac{(x - 3)(x + 1)}{x^2 - 4}$

[-10, 10] by [-10, 10].

How does the graph behave near the vertical asymptotes?

Near $x = 2$, the numerator is near its value *at* $x = 2$: $(2 - 3)(2 + 1) = -3$. The numerator is negative. Near $x = 2$, but to the right, the denominator is small and positive, so the quotient is large and negative; the graph goes down along the asymptote (Figure 11).

Near $x = 2$, but to its left, the denominator is small and negative, so the quotient is large and positive. Therefore, the graph goes up along the asymptote. The behavior at the other asymptote is determined in a similar manner.

To determine whether a graph goes up or down along an asymptote $x = a$, determine the sign of the quotient near $x = a$. *Near a* the numerator will have the same sign it has *at a*, and the sign of the denominator can be determined by considering the signs of its factors in the two cases, $x < a$ and $x > a$.

<u>**Uses of Rational Functions**</u>. Rational functions are very useful in higher mathematics. They express sums, help evaluate special functions by approximating them, and are used to define special functions in calculus.

Example 24: (4.5.8) $1 + x + x^2 + ... + x^n = (1 - x^{n+1})/(1 - x)$, for $|x| \neq 1$.

The left side gives a (possibly long) sum of $n+1$ terms, whereas the right side gives a simple rational function with only four terms. This is a result from the study of series (a series is a sum of terms) (Problem B34).

Of particular value is the famous sum of the unending infinite series,

(4.5.9) $$1 + x + x^2 + ... + x^n + ... = 1/(1 - x), \text{ for } |x| < 1.$$

For example, when $x = 1/2$, $1 + 1/2 + 1/4 + 1/8 + 1/16 + ... = 1/(1 - \frac{1}{2}) = 2$.

On the left the individual terms go to zero as n increases. There are more and more terms, and the individual terms become smaller and smaller, if $|x| < 1$. This is a fascinating case in which a rational function can be used to express a sum of an infinite number of terms (Problem B26).

Polynomials can closely approximate many functions, but not functions that have a vertical asymptote. Rational functions can be used to approximate functions with vertical asymptotes. The point of finding such approximations is to obtain a method of evaluating hard-to-evaluate functions with nearby functions which can be evaluated using only arithmetic operations. For example, how does your calculator know the value of tan x? This is a very difficult problem and the next example does not answer it, but it does give a glimpse of how rational functions can be used.

Example 25: Figure 12 displays the graph of tan x on $0 \leq x \leq \pi$ (angle x in radians). $x = \pi/2$ is a vertical asymptote. Because no polynomial has a vertical asymptote, it is not possible to closely approximate tan x near $\pi/2$ with a polynomial. However, advanced mathematics tells us that the rational function $1/(\pi/2 - x)$ is close to tan x for x close to $\pi/2$.

x	tan x	$1/(\pi/2 - x)$
$\pi/2 - .1$	9.96664	10
$\pi/2 - .01$	99.9666	100

Although tan x is hard to evaluate, the rational approximation near $\pi/2$ is easy to evaluate.

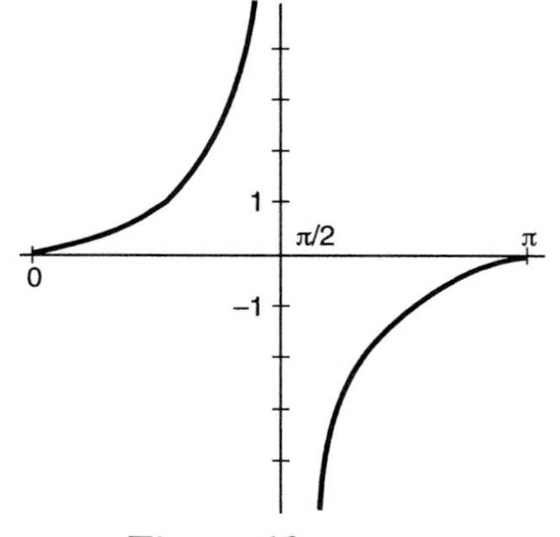

Figure 12: tan x. $[0, \pi]$ by $[-5, 5]$.

Example 26: Sue runs four miles every day, two miles out and two miles back. She likes to speed up on the return trip, so she runs back three miles per hour faster than she runs out. If her run takes 30 minutes, how fast does she run out?

The problem gives the distance and elapsed time, and asks for the rate.

So distance, time, and rate are related. How are they related?

Of course, the basic formula is "distance equals rate times time" ($d = rt$). The difficulty is in adapting it to this problem. Let her rate out (in miles per hour) be r. Then her rate back is $r + 3$. The total time it takes is her time out plus her time back. So we need the basic formula rewritten as $t = d/r$. Then the formula for the total time it takes given r is:

$$T = \frac{2}{r} + \frac{2}{r + 3}.$$

Building this rational formula is the hard part. Now the word problem tells us the total time is also 1/2 hour (do not use "30," because rates are usually measured per hour, not per minute). Set these two expressions equal and solve for r (Problem B14).

Conclusion. Rational functions are the most complex type of function that can be evaluated using only arithmetic operations. Since their behavior can be so complex, before solving rational equations it is a good idea to use your calculator to graph them and see their behavior. Be sure to choose a window that yields a representative graph. If you use algebraic methods to solve a rational equation you can check your answer with the graph. The most important features of the rational function $P(x)/Q(x)$ occur when $P(x) = 0$ or $Q(x) = 0$.

Terms: rational expression, vertical asymptote, end-behavior model, horizontal asymptote.

****

Exercises for Section 4.5, "Rational Functions":

A1.* a) How do you solve an equation of the form "$a/b = 0$"?
b) What part can cause trouble?
c) State the relevant theorem (using variables)

A2.* Define "rational" expression.

A3.* The major features of the graph of a rational function include its ________ and its ____________________.

^ ^ ^ ^Solve algebraically

A4. $1/(x - 1) = 3/x$. A5. $(2x - 5)/(4 - x) = 12$. [3.8]

A6. $(x^2 - 1)/(x + 2) = 1$. [2.3,...]

^ ^ ^ ^ Identify all zeros and asymptotes (including horizontal asymptotes).

A7. $(x - 5)/(x + 1)$.

A8. $(x + 2)/(2x - 3)$.

A9. $(3x + 1)/(x - 3)$.

A10. $3 + x/(x - 4)$.

A11. $(x - 3)(x - 10)/(x + 2)$.

A12. $(x - 9)/[(x + 1)(x + 2)]$.

A13. $(x^2 + 5x + 6)/x$.

A14. $(x + 1)/(x^2 + x)$.

A15. $(x^2 - 3x + 1)/(2x^2 - 5)$.

A16. $(x^2 + 3)/(x^2 - x - 6)$.

A17. $2/x + 3(x - 4)$.

A18. $x + 1/x$.

A19. $\dfrac{3}{2 + \dfrac{1}{x}}$.

A20. $\dfrac{x - 1}{x + \dfrac{1}{x}}$.

A21. Solve $1/x - 1/(x - 2) = 3$. [1.6,...]

A22. Solve $(3 - x)/(x^2 - 2) = 4$. [1.5,...]

A23. A car travels 100 miles at an average speed of 40 miles per hour and then another 100 miles at an average speed of 60 miles per hour. What is the average speed over the 200 miles? [Hint: It is not 50 miles per hour.]

A24. Solve "$1(x - 1) + 2(x - 2) = 3(x - 1)(x - 2)$" from Example 10.

^ ^ ^ ^ Give the end-behavior models:

A25. $x^3/(x - 2)$

A26. $(5 - x)/x^2$

A27. $x/(x + 3)$

A28. $(x - 2)/x$.

A29. $(2 - x - 3x^2)/[(x - 1)(x + 5)]$.

A30. $(2x - 3)(x + 1)/[(5x)(x + 12)]$.

A31. a) Do the long division of $x - 2$ into $x + 1$.
b) Use your result from part (a) to explain the appearance of the graph of $(x + 1)/(x - 2)$.

A32. a) Do the long division of $x - 1$ into $2x - 5$.
b) Use your result from part (a) to explain the appearance of the graph of $(2x - 5)/(x - 1)$.

A33. a) What model graphics calculator do you use? b) When it graphs Figure 10, does it produce an artifact by making a vertical line at $x = -3/2$?

^ ^ ^ ^ ^ ^ ^ ^

B1.* Where are the most interesting places on the graph of a rational function?

B2.* a) Looking at the expression for a rational function, how can you tell where the vertical asymptotes are likely to be?
b) Will there always be a vertical asymptote where you expected (from your answer to part (a))?
c) Looking at the expression for a rational function, how can you tell where to expect zeros?
d) Will there always be zeros where you expected them (from your answer to part (c)?

B3.* a) Looking at the expression for a rational function, how can you tell whether there are horizontal asymptotes and where they will be?
b) Will there always be a horizontal asymptote where you expected (from your answer to part (a))?

B4.* a) State (with variables) the theorem about solving a rational equation by multiplying through by the denominator. b) What part can cause trouble?

B5.* The text says to "expect" a vertical asymptote of a rational function where there is a zero of the denominator. Why doesn't it say "There will be a vertical asymptote where there is a zero of the denominator"?

B6. Give the simplest rational function with a zero at 2, a vertical asymptote at -3, and a horizontal asymptote at 4.

B7. Give the simplest rational function with a zero at 0, a vertical asymptote at 1, and a horizontal asymptote at -1.

B8. Give the simplest rational function with zeros at 2 and 5 and a vertical asymptote at 1.

B9. Give the simplest rational function with a zero at -3 and vertical asymptotes at 0 and 4.

B10. Give the simplest rational function with a zeros at 2 and 0, vertical asymptotes at 1 and 3, and a horizontal asymptote at 5.

B11. Give a simple rational function that has vertical asymptotes at 0 and 2 and it goes up along both sides of $x = 0$ and down along both sides of $x = 2$ [You need not "simplify" your answer].

B12. a) Do the long division of $x + 1$ into $x^2 + 3x + 3$.
b) Explain, from your result in part (a), the appearance of the graph of $(x^2 + 3x +3)/(x + 1)$.

B13. a) Do the long division of $2x - 1$ into $6x^2 + x - 4$.
b) Explain, from your result in part (a), the appearance of the graph of $(6x^2 + x - 4)/(2x - 1)$.

B14. Solve the rational equation in Example 26.

B15. Find and simplify the difference quotient (at x and a) of $f(x) = x^3$.

B16. Use 4.5.9 to rewrite $1/(1 + x^2)$ as an infinite sum for $x^2 < 1$.

B17. Solve $(x - 3)/(x + 2) + (x - 5)/(x + 1) = 7$. [-3.8,...]

B18. Find the slope of the line through the points on the graph of $1/x$ where $x = 1$ and $x = 1 + h$. Simplify.

B19. Find the slope of the line through the points on the graph of $1/x$ where $x = a$ and $x = b$. Simplify.

B20. The formula $d = rt$ holds for a *constant* rate over a distance, or for an *average* rate over a distance. The average rate over a two-part trip may not be the numerical average of the rates of the two parts. Suppose a round trip consists of traveling a distance d out at and average rate of r and then the same distance back at an average rate of s. What is the average rate for the whole trip [Hint: It is not $(r + s)/2$].

B21. Teresa runs 8 miles round trip each day. She runs out at a constant rate and runs back 2 miles per hour faster. If she averages 7 miles per hour, how fast does she run out? Before doing the work, make a reasonable guess. Then do the detailed work and calculations. How close was your guess?

B22. Give the simplest rational function with a vertical asymptote at 3, zeros at 0 and 5, and a horizontal asymptote at 2. Assume it has no other zeros or vertical asymptotes.

B23. Give the simplest rational function with a vertical asymptotes at 3 and -2, a zero at 5, and a horizontal asymptote at -1. Assume it has no other zeros or vertical asymptotes.

^ ^ ^ ^ Solve for a and b [This way to rewrite a fraction is called "partial fractions." See problem B37.] [Hint: Find a common denominator and match coefficients in the numerators.]
B24. "$(5x - 7)/(x^2 - 3x + 2) = a/(x - 1) + b/(x - 2)$, for all x."
B25. "$(x + 1)/(x^2 - x - 6) = a/(x + 2) + b/(x - 3)$, for all x."

B26. [See 4.5.9] Graph $1 + x + x^2 + x^3 + x^4$ and $1/(1 - x)$ on [0, 1] by [0, 10]. a) Where are they within .1 of each other? b) What changes if we add $x^5 + x^6$ to the sum?

^ ^ ^ ^ The given graph is of $P(x)$. Grid lines are one unit apart. Sketch the corresponding graph of its reciprocal, $1/P(x)$.

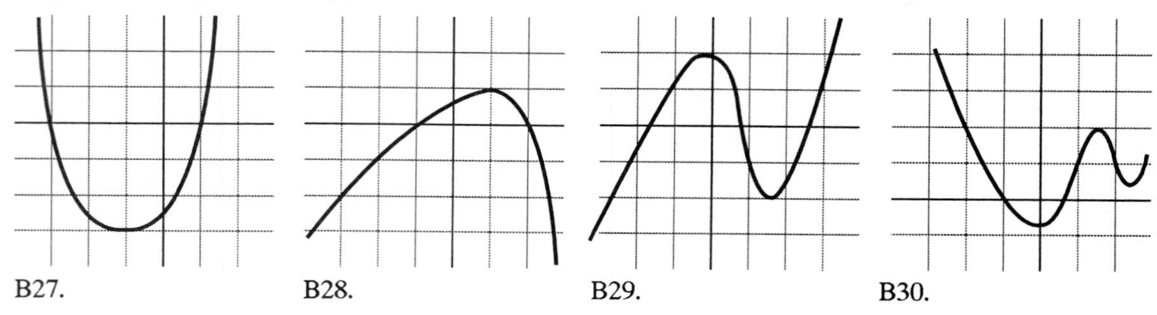

B27. B28. B29. B30.

^ ^ ^ ^ ^ ^ ^ ^ ^

B31. a) Show that the graph of a quotient of linear polynomials usually resembles the graph of $1/x$ (shifted or rescaled) (Use a general quotient of linear polynomials).
b) When wouldn't it?

B32. When the end-behavior model is a diagonal line (such as $y = x$ in Example 17), the graph will parallel the line, but not necessarily come closer and closer to it. There will be a slant asymptote (also called an "oblique asymptote") that the graph does come closer and closer to, but that line (asymptote) is not quite the same as the end-behavior model. Find the line that the graph in Example 17 does come closer and closer to as x gets large.

B33. All powers (including irrational powers) can be defined using $b^p = (e^{\ln b})^p = e^{p \ln b}$ (using Property 4.1.3). Thus b^p can be evaluated as $e^{p \ln b}$, using the exponential function without needing the "repeated multiplication" idea which motivated the definition of powers but which really worked only for integer powers. Suppose we know e^x for all x.
a) How would $4.3^{2.8}$ be evaluated with this approach?
b) Define $x^{2.3}$ with this approach.

B34. Prove: $1 + x + x^2 + ... + x^n = (1 - x^{n+1})/(1 - x)$, for $|x| \neq 1$.

B35. Use 4.5.9 to rewrite $1/x$ as an infinite sum for $0 < x < 2$.

B36. Use 4.5.9 to rewrite $1/(1 + x^2)$ as an infinite sum for $|x| < 1$.

B37. (Partial Fractions) To add fractions we use a common denominator. For example,

$$\frac{2}{x-1} + \frac{3}{x-2} = \frac{2(x-2) + 3(x-1)}{(x-1)(x-2)} = \frac{5x-7}{(x-1)(x-2)}.$$

In this example a sum of terms with constant numerators and linear denominators is turned into a quotient with a linear numerator and a quadratic denominator. Sometimes we wish to reverse the process. (The reverse process is called "partial fractions.") a) Find A and B

such that $$\frac{x-4}{(x-1)(x-2)} = \frac{A}{x-1} + \frac{B}{x-2}.$$

b) Solve for A and B in terms of a, b, m, and k in

$$\frac{mx+k}{(x-a)(x-b)} = \frac{A}{x-a} + \frac{B}{x-b}.$$ c) Is there always a solution?

Section 4.6. Inequalities

This section discusses the various ways to solve an inequality. It is important to remember that some of the processes which work for solving equations do not necessarily work for solving inequalities. In particular, multiplying or dividing both sides by a non-zero number or by an expression may not preserve the solutions to an inequality.

Interval Notation. To solve an inequality means to find the values of the variable (often "x") that make the inequality *true*. Solutions may be expressed in two ways. One way is to give an equivalent inequality which exhibits the x-values which make the original inequality true. The other is to consolidate those x-values into a single set.

Example 1: The solution to "$2x < 10$" can be expressed as the equivalent inequality, "$x < 5$," which exhibits the solutions. The solution can also be expressed as a set which, in "interval notation," can be written "$(-\infty, 5)$."

The inequality describing an interval of numbers and the corresponding interval notation for the corresponding set are given next. If the endpoints are included, square brackets are used. If they are excluded, parentheses are used.

$a < x < b$	(a, b)	$x < b$	$(-\infty, b)$
$a \leq x \leq b$	$[a, b]$	$x \leq b$	$(-\infty, b]$
$a < x \leq b$	$(a, b]$	$x > a$	(a, ∞)
$a \leq x < b$	$[a, b)$	$x \geq a$	$[a, \infty)$.

The symbols "∞" (called "infinity") and "$-\infty$" are not real numbers. The symbols "∞" and "$-\infty$" are used to take the place of the other (non-existent) endpoint when the interval has only one real-number endpoint (as in the right column).

Sentences such as "$a < x < b$" express two inequalities connected by "and": "$a < x$" and "$x < b$." The word "and" means both inequalities apply to x, so x's in the region must satisfy both. Figure 1 shades the three regions.

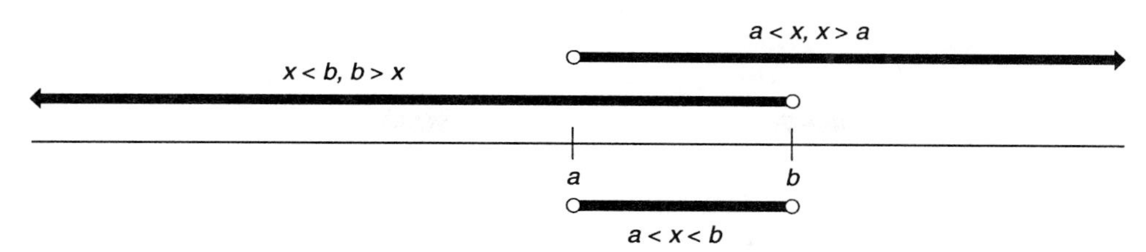

Figure 1: "$a < x$", "$x < b$", and "$a < x < b$."

Combining two inequalities into one string creates a single interval as in Figure 1. If we want to say that x belongs to either of two intervals the proper connective to use is "or", not "and". For example, Figure 2 illustrates the region of the number line in which "$|x| > 2$."

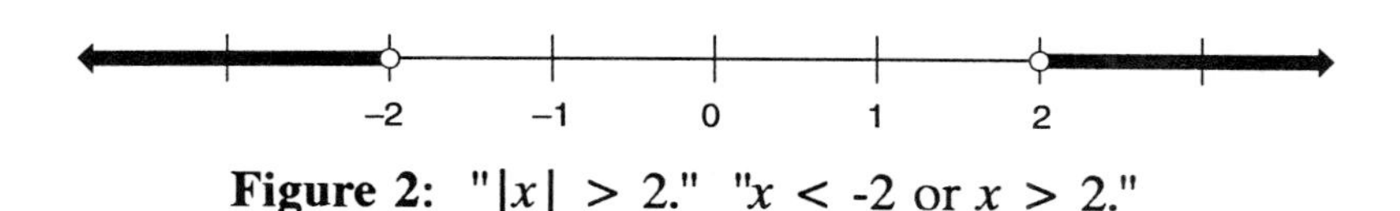

Figure 2: "$|x| > 2$." "$x < -2$ or $x > 2$."

The region consists of two intervals. A particular "x" cannot be in both. It can be in one *or* in the other. The way to express this is "$x < -2$ or $2 < x$," using "or".

This region is *not* described by "$x < -2$ *and* $x > 2$" -- there are no such x's. Be careful with the usage of "and," because English is ambiguous in a way that Mathematics is not. If we want "and" to apply only to the region the two inequalities have in common in sentences such as "$a < x$ and $x < b$" (the single interval (a, b) in Figure 1), we must be consistent and always use it to refer to the region two inequalities have in common. So we must use a different word for regions consisting of two separate intervals as in Figure 2. That word is "or."

A string of inequalities is always interpreted as inequalities connected by "and," never "or." Therefore it is not permitted to state "$2 < x < -2$." There are no such x's.

The concept of set "union," denoted by "$\cup$", can be used to combine separate intervals into one set. For example, the set in Figure 2 is $(-\infty, -2)\cup(2, \infty)$. For another example, the set described by the sentence, "$x < -2$ or $1 \leq x < 3$" is $(-\infty, -2)\cup[1, 3)$.

Graphical Solutions. The solution to an inequality can be approximated graphically. Every point on a graph is associated with *two* numbers, an "x" value and a "y" value. The inequality "$f(x) > 0$" asserts that the "y" values are greater than zero. So, graphically, the problem is to "Find the x-values of the points that are above the x-axis."

The solution to an *equation* is usually one or more numbers. The solution to an *inequality* is frequently one or more *intervals* of numbers.

Example 2: Solve "$f(x) > 0$" for the function graphed in Figure 3. Assume it is a representative graph.

Locate all the points that are above the x-axis. Determine and report the x-values of those points as the solution.

Points on the far left and far right are above the x-axis. They have x-values less than (to the left of) -3 or greater than (to the right of) 2. The solution is: $x < -3$ or $x > 2$.

In "interval notation," the solution is expressed as a set: $(-\infty, -3) \cup (2, \infty)$.

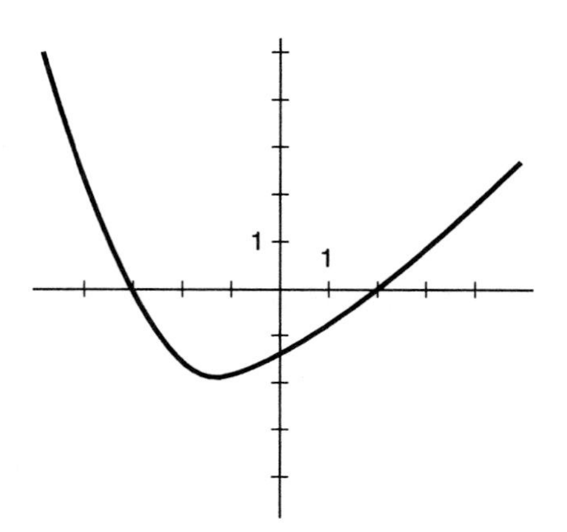

Figure 3: $f(x)$, without a given algebraic representation. [-5, 5] by [-5, 5].

Methods of solving inequalities can be divided into the various types of traditional, algebraic methods and the modern, guaranteed-to-work (but not algebraic), "evaluate-and-compare" method.

Evaluate-and-Compare. A graphics calculator makes solving inequalities easy.

Example 3: Solve "$x^2 < 2x + 3$."

One approach (not the only approach) is to graph both expressions "x^2" and "$2x + 3$" (Figure 4). Look along vertical columns of pixels for points on the graph of "x^2" which are below the corresponding points on the graph of "$2x + 3$" (because "below" corresponds to "less than" for images). Then, read off the x-values of those points.

The solution is $-1 < x < 3$.

Be very clear about one thing: The *points* themselves are not solutions, it is *the x-values of the points* that are solutions.

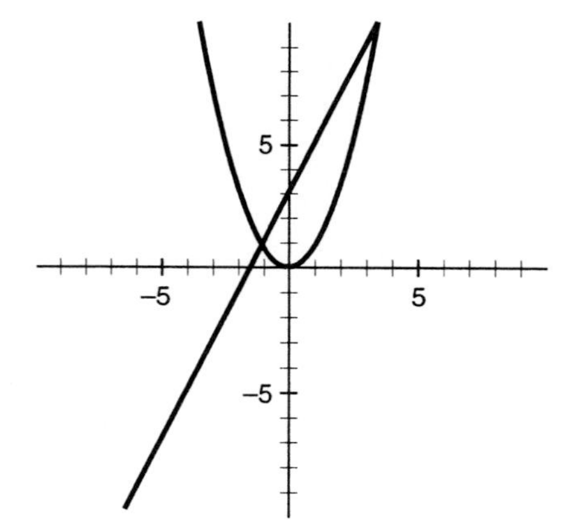

Figure 4: x^2 and $2x + 3$. [-10, 10] by [-10, 10].

Calculator Exercise 1: Graph two graphs simultaneously. Learn how to use the *trace* feature of your calculator and the arrow keys to move the cursor from one graph to the other. As the cursor skips from one graph to the other only the y-value should change while the x-value remains the same. The inequality compares the y-values, but the solutions will be x-values.

Rather than compare two y-values to each other, it is easier to compare their difference to zero, which is justified by this theorem.

Theorem 4.6.1: For all a, b, and c,

A) $a < b$ iff $a + c < b + c$.

B) $a < b$ iff $a - b < 0$.

This tells us that, for solving inequalities, adding and subtracting work just like they do for equations. But, beware, multiplying and dividing do not (Examples 13 and 14, later).

Example 3, revisited: Solve "$x^2 < 2x + 3$."

Use Theorem 4.6.1 to restate the inequality as the equivalent inequality,

$$"x^2 - 2x - 3 < 0."$$

Now graph the one expression "$x^2 - 2x - 3$" which can be compared to 0 (Figure 5). The solution corresponds to the x-values of the points below the x-axis: $-1 < x < 3$.

There are still more ways to solve this inequality that we will discuss after Theorem 4.6.5.

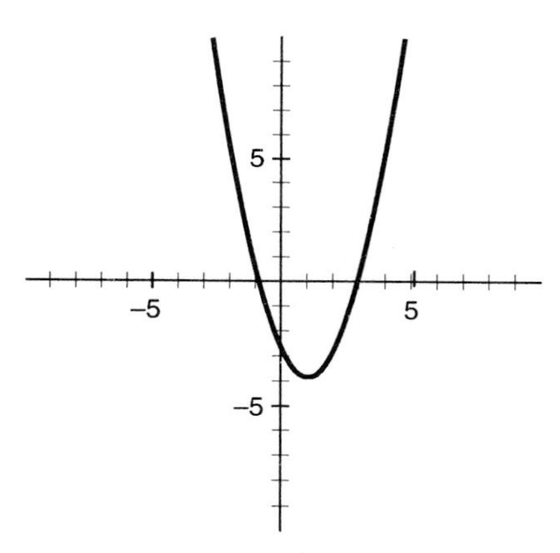

Figure 5: $x^2 - 2x - 3$. [-10, 10] by [-10, 10].

The "evaluate-and-compare" method will work for many inequalities for which no algebraic method works.

Example 4: Solve "tan $x < 2.1x$" for x in $[0, \pi/2)$ (in radians).

"tan x" is a trigonometric expression. "$2.1x$" is a polynomial expression. Combinations of trigonometric and polynomial expressions can rarely be solved algebraically. But evaluate-and-compare will always work.

First put everything on one side to obtain "tan $x - 2.1x < 0$." Then graph the expression "tan $x - 2.1x$" and compare it to 0 (Figure 6). The points on the graph are below 0 for x in the interval $0 < x < 1.190$.

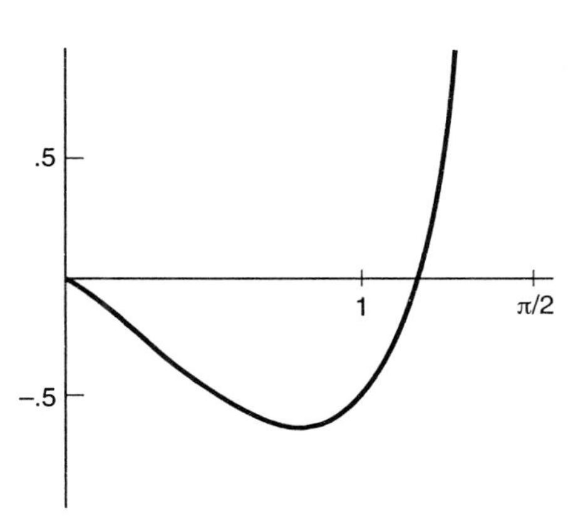

Figure 6: tan $x - 2.1x$. $[0, \pi/2]$ by [-1, 1].

Absolute Values. Absolute values are used in one-dimension to express size and distance. -1,000,000 is less than 5, but we would not usually call it small. What is a "big" or "small" number is usually determined by its absolute value, and the absolute value of -1,000,000 is 1,000,000, a "big" number by most standards (unless you have been thinking about the national debt).

Definition: The absolute value of a real number, x, is denoted by $|x|$, which is read "the absolute value of x." It is defined by $|x| = x$ if $x \geq 0$ and $|x| = -x$ if $x < 0$.

Note that $-x$ is positive if $x < 0$, so the absolute value of x is always non-negative. On the real number line, the absolute value of x can be interpreted as its distance from the origin, 0. The locations of $-x$ and x are the same distance from the origin, which reflects the fact that $|-x| = |x|$.

The absolute value of "$x - a$" is the distance from x to a on the number line. So we may read "$|x - a|$" aloud as "the distance from x to a," which is the same as "$|a - x|$", "the distance from a to x."

Example 5: $|7.6| = 7.6$. $|-7.6| = 7.6$. $|0| = 0$. $|7 - 5| = 2$. $|5 - 7| = 2$.

In calculus, absolute values are used to express the idea of numbers being "close to" one another.

Example 6: Which numbers are "close" to 5?

The distance between x and 5 is $|x - 5|$. The inequality "$|x - 5| < \delta$" ("δ" is the lower-case Greek letter delta) describes the numbers that are within δ units of 5. The smaller the δ, the closer x is forced to be.

Theorem on Absolute Values (4.6.2):

A) For all x and c, $|x| < c$ iff $-c < x < c$ [Figure 7A].

B) For all x and c, $|x| > c$ iff $x < -c$ or $x > c$ [Figure 7B].

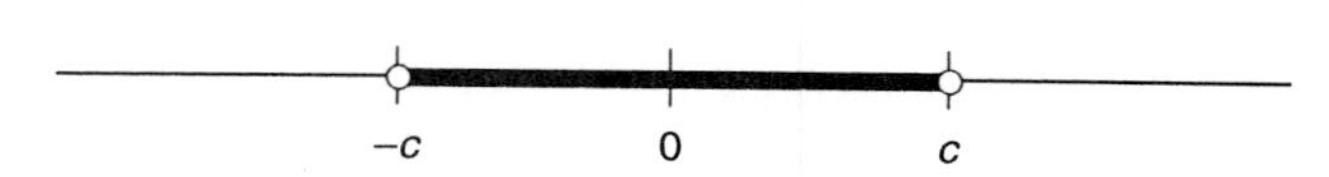

Figure 7A: "$|x| < c$." "$-c < x < c$."

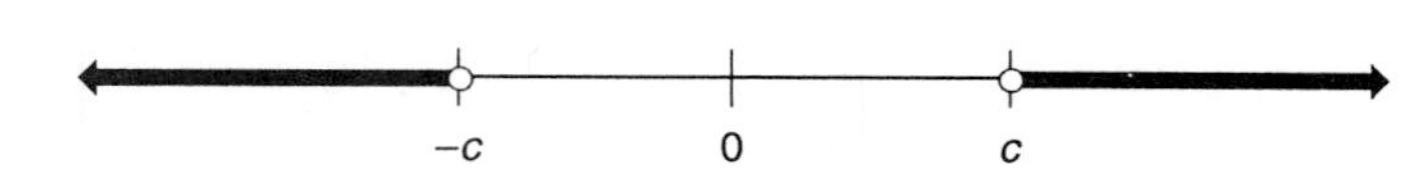

Figure 7B: "$|x| > c$." "$x < -c$ or $x > c$."

Read "$-c < x < c$" as "negative c is less than x is less than c," or "negative c is less than x *and* x is less than c." The connective "and" is not written in "$-c < x < c$," but it is really there.

Example 6, continued: Solve $|x - 5| < 1$. Shade the interval on a number line.

$|x - 5| < 1$ iff $-1 < x - 5 < 1$, [by the Theorem on Absolute Values, 4.6.2]

iff $4 < x < 6$ [by Theorem 4.6.2A twice, adding 5 to all sides of both inequalities.]

The solution is the interval $4 < x < 6$, all numbers within 1 unit of 5. Figure 8 illustrates the solution as an interval on the real number line (problem A43).

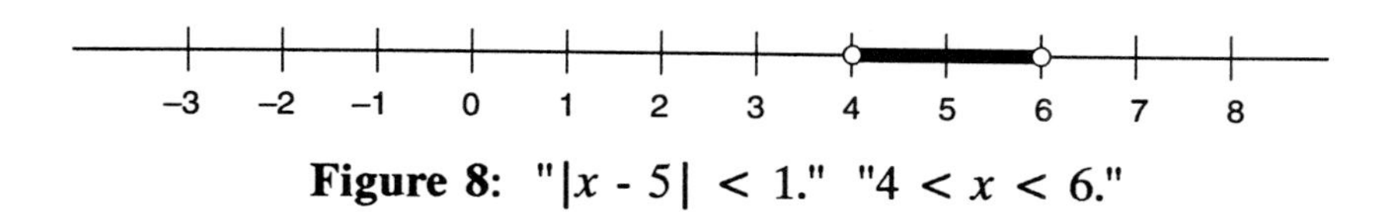

Figure 8: "$|x - 5| < 1$." "$4 < x < 6$."

Example 6, continued further: Solve $|x - 5| < \delta$.

$|x - 5| < \delta$ iff $-\delta < x - 5 < \delta$ [by the Theorem on Absolute Values]

iff $5 - \delta < x < 5 + \delta$ [by Theorem 4.6.1, twice.]

The inequality describes an interval on the number line of width 2δ about a central point. In this case, the center of the interval is 5.

A major purpose of absolute values in calculus is to describe intervals centered about a point.

<u>Theorem 4.6.3</u>: $|x - a| < \delta$ iff $a - \delta < x < a + \delta$.

This theorem gives two alternatives for expressing "All the numbers within δ units of a." The interval is of width 2δ, so δ is its "half-width." Its center is a (Figure 9).

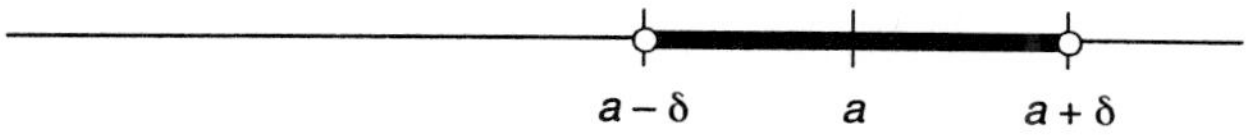

Figure 9: "$|x - a| < \delta$." "$a - \delta < x < a + \delta$."

<u>Proof</u>: $|x - a| < \delta$ iff $-\delta < x - a < \delta$ iff $a - \delta < x < a + \delta$.

Example 7: Use absolute value notation to express all numbers within 2 units of 3.4. Then find the endpoints of the interval.

Directly we can write down the answer to the first part, "$|x - 3.4| < 2$." By Theorem 4.6.3, the answer to the second part is determined by $3.4 - 2 < x < 3.4 + 2$, so the endpoints are 1.4 and 5.4.

Example 8: Express the interval (2.1, 2.2) using absolute values to emphasize its center and width. Use a single inequality of the form "$|x - a| < b$."

The center is $a = 2.15$, and the half-width is $b = .05$. The interval could be expressed by "$|x - 2.15| < .05$."

Example 9: Solve $|x - 3| > 2$.

Since absolute values are applied last, use Theorem 4.6.2B (the part for "is greater than").

$|x - 3| > 2$ iff $x - 3 < -2$ or $x - 3 > 2$

iff $x < 1$ or $x > 5$ (Figure 10).

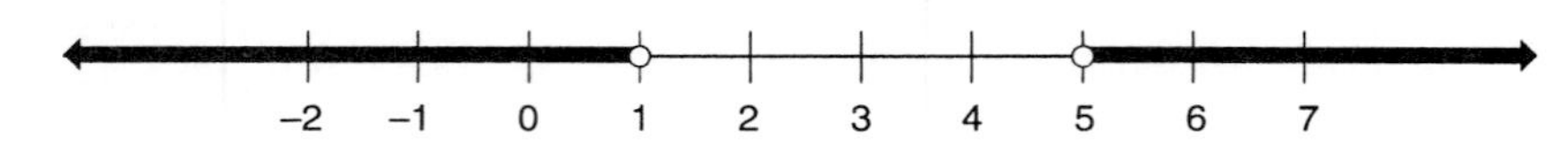

Figure 10: "$|x - 3| > 2$." "$x < 1$ or $x > 5$."

This expresses all the points on the number line that are greater than 2 units away from the location of "3". There are two separate intervals, which is expressed with "or". In interval notation, the set of solutions is: $(-\infty, 1) \cup (5, \infty)$.

<u>Solving Linear Inequalities</u>. The processes which can be used to solve a linear inequality are *not* perfectly parallel to the processes that we use to solve equations. Inequalities are more complex than equalities. Students who fail to acknowledge the difference often make mistakes.

Adding to or subtracting from both sides of an inequality works just like it does for equations. But multiplying or dividing or squaring both sides of an inequality does *not* work just like it does for equations. Multiplying both sides of an inequality by a number may or may not produce an equivalent inequality. It does only if the number is positive.

Example 10: Sue's employer offered to pay her either of two ways. On Plan A she would get \$25 a day plus 4 percent of her sales. On Plan B she would get \$10 a day plus 7 percent of her sales. When is Plan B the better option for her?

This is a "word" problem. Remember our methods of solving word problems, and remember that inequalities consist of a relation between two expressions. You need the two expressions first. So, before we set up the equation, build a formula for her pay under Plan A, and a separate formula for her pay under Plan B. Let "x" denote the dollar value of her sales in a day. Under Plan A, her pay is:

$$25 + .04x \text{ (dollars)}.$$

Under plan B her pay is:

$$10 + .07x \text{ (dollars).}$$

Plan B is better for her when her pay under Plan B is greater:

$$10 + .07x > 25 + .04x.$$

To solve this, subtract 10 from both sides and subtract $.04x$ from both sides (which is justified by Theorem 4.6.1):

$$.03x > 15.$$

Now divide by .03, which is a positive number.

$$x > 15/.03 = 500 \text{ (dollars).}$$

Plan B is better if she sells more than $500 per day.

In the last step, it is critical that the number that was used to divide (.03) was *positive*. If a number is *negative*, division or multiplication by it *reverses* the direction of the inequality. Figure 11 illustrates the fact that multiplying both sides of an inequality by -1 reverses the orientation of the inequality.

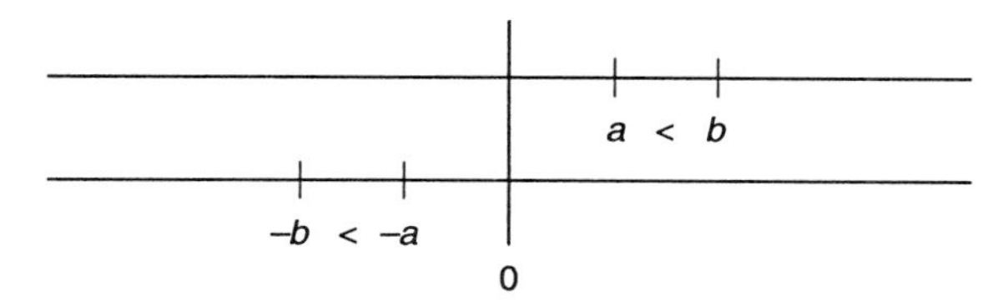

Figure 11: On the upper line, $a < b$.
On the lower line, $-b < -a$. Equivalently, $-a > -b$.

Example 11: Solve $8 - 4x > 10$.

$8 - 4x > 10$ iff $-4x > 2$ [subtracting 8, Theorem 4.6.1]
iff $x < 2/(-4) = -1/2$. [changing the direction]

Example 12: Solve $|3 - x| < 1$.

$|3 - x| < 1$ iff $-1 < 3 - x < 1$ [the Theorem on Absolute Values]
iff $-4 < -x < -2$ [subtracting 3 from all sides]
iff $4 > x > 2$ [dividing by -1, a negative number].

Do not forget to change the direction of the inequality for the last line. If you wish to rewrite the final interval as "$2 < x < 4$," you may.

Using the interpretation of absolute values of a difference as the distance between the two numbers on the number line, we can see that $|3 - x|$ expresses the distance between 3 and x, which will be less than 1 if x is between 2 and 4.

<u>Theorem 4.6.4 (on multiplying both sides of an inequality by a number)</u>:
<u>A</u>) Let $c > 0$. Then $a < b$ iff $ca < cb$, and $a < b$ iff $a/c < b/c$.
<u>B</u>) Let $c < 0$. Then $a < b$ iff $ca > cb$, and $a < b$ iff $a/c > b/c$.

Part B handles the negative numbers in Examples 11 and 12. The only trick to this theorem is part B. If you forget to worry about the direction of an inequality and you multiply or divide both sides by a number and your number is positive, Part A says things turn out right. However, if you multiply or divide by a variable which could be negative, things usually go wrong.

Example 13: Solve $2/x < 1$.

Your first thought to simplify this might be to "multiply through by x" to obtain "$2 < x$." This new inequality would be simpler, but it wouldn't be *equivalent*. That is, multiplying through by x would be a big blunder. We have no result that says you may multiply both sides of an *in*equality by "x", *unless you already know whether "x" is positive or negative.*

To do this problem requires a new approach. One approach that works is to separately consider two cases as in Theorem 4.6.4 above.
<u>Case I</u> ($x > 0$): If $x > 0$, we may "multiply through by x" to convert "$2/x < 1$" to "$2 < x$." (The direction of the inequality is preserved because x is positive.) The numbers that satisfy both conditions ($x > 0$ and $2 < x$) solve the inequality, so "$2 < x$" is part of the solution.
<u>Case II</u> ($x < 0$): If $x < 0$, we may "multiply through by x," but we must remember to change the direction of the inequality (as in Part B of the theorem) to obtain "$2 > x$." The numbers that satisfy both conditions ($x < 0$ and $2 > x$) solve the inequality, so "$x < 0$" is part of the solution.

The solution is "$x < 0$ or $2 < x$."

Graphing the difference of the two expressions in Example 13 is illuminating (Figure 12, for solving "$2/x - 1 < 0$."). The graph of "$2/x - 1$" is below the x-axis in two distinct regions. But, if we illegally "multiply through by x" to convert "$2/x < 1$" into "$2 < x$" we lose track of all the negative solutions (because multiplying through by a negative number should have changed the direction of the inequality by Theorem 4.6.4B).

Dividing both sides of an inequality by an expression with a variable is also dangerous.

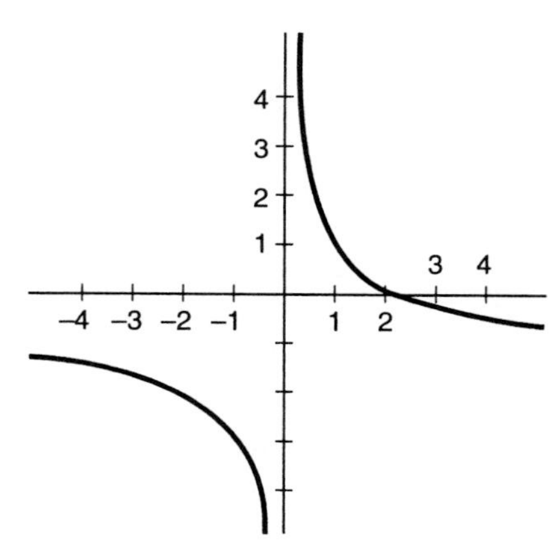

Figure 12: $2/x - 1$.
[-5, 5] by [-5, 5].

Example 14: Solve "$x^2 < 2x$."

Your first thought might be to simplify this by canceling the factor of x on both sides. The resulting inequality ($x < 2$) would then be simpler, but it wouldn't be *equivalent*. That is, canceling "x" would be a big blunder. We have no result that says you may divide both sides of an inequality by an expression, *unless you already know whether the expression is positive or negative*.

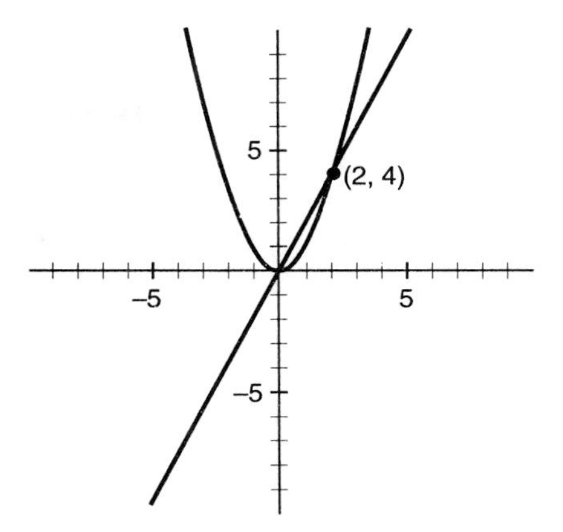

Figure 13: x^2 and $2x$.
[-10, 10] by [-10, 10].

Graph the two expressions (Figure 13). The graph of x^2 is below the graph of $2x$ when x is between 0 and 2. That is, the correct solution to "$x^2 < 2x$" is "$0 < x < 2$. But, suppose a novice (incorrectly) cancels "x" to convert "$x^2 < 2x$" into "$x < 2$." The new inequality includes negative values of x that do not solve the original equation. Such canceling is illegal.

Example 14, again: Solve "$x^2 < 2x$."

$x^2 < 2x$ iff $x^2 - 2x < 0$.

iff $x(x - 2) < 0$.

The expression on the left is a polynomial (Figure 14). Now the problem is easy because we can simply look at the graph and read off the answer.

The zeros of $x(x - 2)$ are $x = 0$ and $x = 2$. Only the middle interval yields negative images. The solution is $0 < x < 2$.

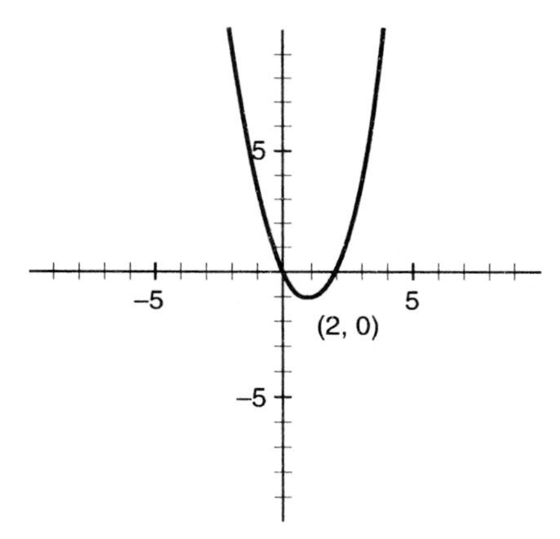

Figure 14: $x^2 - 2x$
$= x(x - 2)$.
[-10, 10] by [-10, 10].

Zeros and Inequalities. The zeros of a polynomial expression divide the x-axis into intervals. Inside each of these intervals the polynomial can not change sign, because it would have to cross the axis at a zero to do so. For example, in Figure 15, the three zeros divide the line into four intervals. Inside any one of those intervals the polynomial is either positive everywhere or negative everywhere.

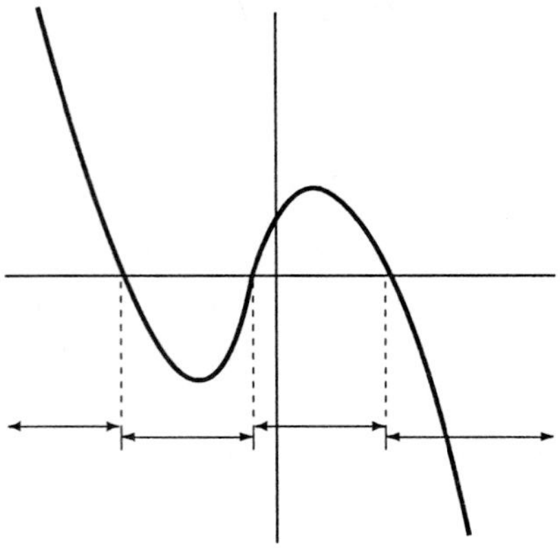

Figure 15: The zeros of a polynomial divide the line into intervals and, in the interior of each interval, it is either positive for every x-value or negative for every x-value.

Theorem on Zeros and Signs (4.6.5):

A) The zeros of a polynomial $P(x)$ divide the number line into intervals. The sign of $P(x)$ at any point in the interior of an interval is the sign of $P(x)$ for all points in the interior of that interval.

B) Let $R(x)$ be a rational expression, $P(x)/Q(x)$. The zeros of P together with the zeros of Q divide the number line into intervals. The sign of $R(x)$ at any point in the interior of an interval is the sign of $R(x)$ for all points in the interior of that interval.

Use this theorem with a graph. This is my favorite way to solve polynomial or rational function inequalities. Step one is to put everything on one side so an expression can be compared to zero. Step two is to graph it and find the zeros and vertical asymptotes, which are endpoints of intervals. Step three is to note from the graph which of those intervals yield solutions to the inequality. The final step is to check the endpoints separately.

Example 15: Solve $\dfrac{x+3}{x-1} \le 0$.

Graph it (Figure 16). The most important features of the graph are the zero at $x = -3$ and the vertical asymptote at $x = 1$. The interiors of the three intervals of the theorem are $(-\infty, -3)$, $(-3, 1)$ and $(1, \infty)$.

It is easy to see from the graph that the negative images occur in the interval from $x = -3$ to $x = 1$. According to the Theorem on Zeros and Signs, the solution includes the interior of that interval. The endpoints must be checked separately. Try -3. It is a zero, and, because the inequality is "$\le$", the zero is a

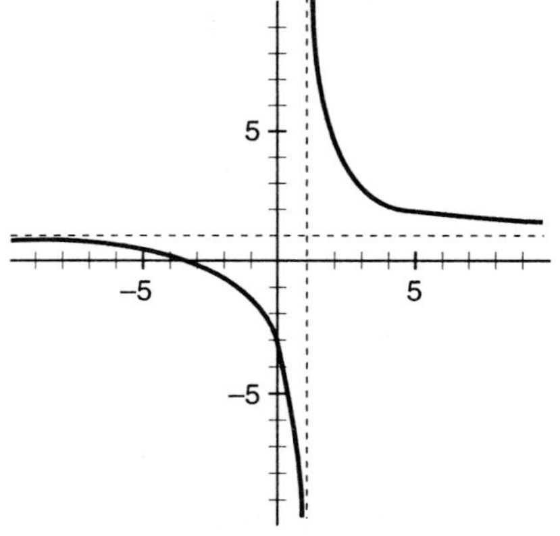

Figure 16: $(x + 3)/(x - 1)$ [-10, 10] by [-10, 10].

solution. Try 1. It is not in the domain and does not solve the inequality. So the complete solution is: $-3 \le x < 1$. As a set, it could be expressed in interval notation: $[-3, 1)$.

In Example 15, wouldn't it be easier to just "multiply through by $x - 1$" and eliminate the quotient? It would be *easier*, but it would be *wrong*. The expression "$x - 1$" can be either positive or negative, and when it is negative it would be wrong to leave the direction of the inequality unchanged. By the way, you would erroneously get "$x + 3 < 0$" and then "$x < -3$," which is completely wrong.

The idea of "Zeros and Signs" works only for comparing expressions to *zero*.

Example 16: Solve $(x - 1)(x + 3) \ge 1$.

In many examples we are pleased to see factored form, but in this one the factoring is useless. Factored form is useful if and only if the expression is compared to *zero*.

$(x - 1)(x + 3) \ge 1$ iff $x^2 + 2x - 3 \ge 1$ [multiplying out]

iff $x^2 + 2x - 4 \ge 0$ [subtracting 1].

So we solve this inequality instead. Now the zeros of $x^2 + 2x - 4$ are relevant. It factors, but not using integers. Use the quadratic formula to find the zeros: $x = 1 \pm \sqrt{5}$. In decimal form, the zeros are approximately -3.236 and 1.236.

Graph it (Figure 17). From the picture, the middle interval gives the negative values, and the two outer intervals give the positive values. Because the inequality has this expression greater than or equal to zero, we want the positive values. The solution is

$$x \le -1 - \sqrt{5} \text{ or } x \ge -1 + \sqrt{5},$$

where the endpoints are included because the inequality uses "≥ 0," which includes the zeros.

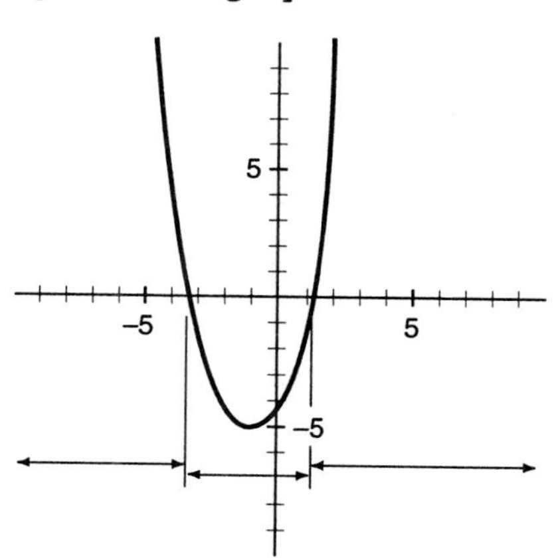

Figure 17: $x^2 + 2x - 4$
$[-10, 10]$ by $[-10, 10]$.

Example 18: Solve $\dfrac{x^2 + 3x - 5}{(x + 1)(x - 3)} \ge 0$.

Graph it (Figure 18). We see the zeros of the top as zeros, and the zeros of the bottom as vertical asymptotes.

The zeros of the denominator are $x = -1$ and $x = 3$. The zeros of the numerator can be determined by the Quadratic Formula.

$$x = \frac{-3 \pm \sqrt{29}}{2} = 1.19 \ \textit{or} \ -4.19 .$$

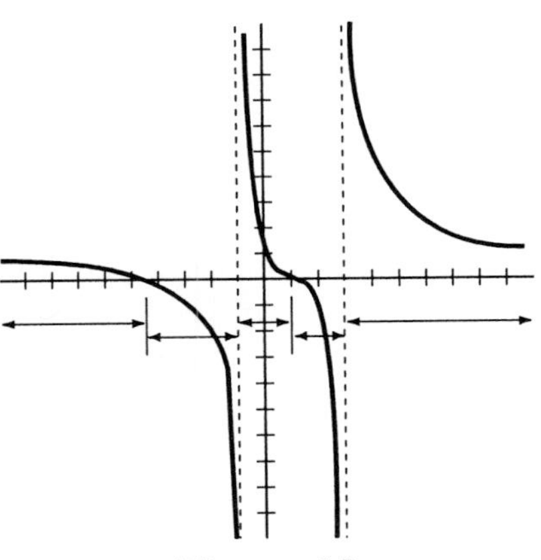

Figure 18: $\dfrac{x^2 + 3x - 5}{(x + 1)(x - 3)}$ [-10, 10] by [-10, 10].

The four zeros divide the line into 5 intervals. The graph shows where the expression is positive. Checking the endpoints separately, we see that the zeros of the numerator are included and the zeros of the denominator are excluded. The solution is
$x \le (-3 - \sqrt{29})/2$ or $-1 < x \le (-3 + \sqrt{29})/2$ or $x \ge 3$.
This is an algebraic solution because we found the endpoints of the intervals algebraically.

Conclusion. Inequalities are different than equations. Some of the processes that work for solving equations do not necessarily work for solving inequalities. Multiplying by a negative number reverses the direction of the inequality. Also, multiplying by an expression is not correct unless you know in advance whether the expression is positive or negative. Otherwise you cannot tell the direction of the resulting inequality. Also, squaring both sides is not justified.

To solve an inequality, evaluate-and-compare will always work. However, it is not an algebraic method. The advice, "Put everything on one side," is usually appropriate. The Theorem on Zeros and Signs is helpful when the zeros and asymptotes can be found. If they are found algebraically, the theorem yields an algebraic solution, even if a graph is used to help see where the expression is positive or negative.

Terms: inequality, solution, interval notation, and, or, union, $\cup$, absolute values.

****

Exercises for Section 4.6, "Inequalities":

A1.* To solve an inequality, you may multiply both sides by a ________ number and preserve the direction of the inequality. However, if the number you multiply by is ________, you must remember to switch the __________________.

A2.* Suppose you apply the following operations to both sides of an inequality and leave the direction of the inequality unchanged. Which operations preserve the solutions?

a) add 2. b) subtract 2. c) add -2. d) subtract -2.
e) multiply by 2. f) divide by 2. g) multiply by -2. h) divide by -2.
i) square. j) add x. k) multiply by x. l) divide by x.

A3.* Sentences such as "$a < x < b$" abbreviate two inequalities connected by the word ________.

^ ^ ^ ^ Describe in English the solution to

A4.* "$|x - a| < d$."

A5.* "$|x - a| > d$."

A6.* The solution to an equation is likely to be one or more points (numbers). The solution to an inequality is likely to be one or more ________________.

A7.* Give a single inequality describing the points within d units of a.

A8.* Give a single inequality describing the points further than d units from a.

A9. Here is a "sign pattern" theorem: $ab > 0$ iff $[(a > 0$ and $b > 0)$ or $(a < 0$ and $b < 0)]$. Rephrase it in English beginning, "A product is positive if and only if ..."

^ ^ ^ ^ Solve algebraically:

A10. $3x - 5 < 13$.

A11. $-4x < 20$.

A12. $10 - 2x < 40$.

A13. $5x < 8 - 2x$.

^ ^ ^ ^ Solve algebraically:

A14. $|x - 9| < 1$.

A15. $|x - 4| \leq 3$.

A16. $|x + 2| > 1$.

A17. $|2x - 1| < 7$.

A18. $|5 - x| < 1$.

A19. $|20 - 4x| < 1$.

A20. $|x - 20| > .01$.

A21. $0 < |x - 2|$.

A22. $0 < |x + 3|$.

^ ^ ^ ^ Write in the form "$|x - a| < \delta$":

A23. $2.4 < x < 2.8$.

A24. $6.8 < x < 7.0$.

A25. $40 < x < 42$.

A26. $-5 < x < -4$.

^ ^ ^ ^ Solve algebraically:

A27. $3/x > 4$.

A28. $5/x < 15$.

A29. $12/(x - 3) \leq 2$.

A30. $1/(x - 1) < .2$.

^ ^ ^ ^ Use the Theorem on Zeros and Signs and a graph to algebraically solve:

A31. $(x + 5)/(x - 4) > 0$.

A32. $(x - 6)/(3 - x) \leq 0$.

A33. $(x - 3)/(x + 4) \leq 0$.

A34. $(x - 1)(x + 3)(x - 5) > 0$.

A35. $(x - 1)(x - 2)/(x - 3) \geq 0$.

A36. $(x + 2)/[(x - 1)(x - 5)] \leq 0$.

^ ^ ^ ^ Find the natural domain of f.

A37. $f(x) - \sqrt{5 - 2x}$.

A38. $f(x) - \sqrt{1 - \frac{3}{x}}$.

A39. $f(x) - \sqrt{16 - x^2}$.

A40. $f(x) - \sqrt{x^2 - 25}$.

A41. $f(x) - \sqrt{x(x + 5)}$.

A42. $f(x) - \sqrt{\frac{x - 1}{x + 1}}$.

A43. [See Example 6 and Figure 8.] Graph the two expressions in "$|x - 5| < 1$." How is the solution visible?

A44. a) Which is easier to solve algebraically? "$(x - 5)(x + 7) > 0$" or "$(x - 5)(x + 7) > 2$"? b) Why?

^ ^ ^ ^ ^ ^ ^ ^ ^

B1.* Equations and inequalities are different. What are some important differences in their solution processes?

B2.* a) To solve an inequality, why is it dangerous to multiply or divide both sides of an inequality by an expression with a variable? b) What can you do to avoid the danger?

B3.* What is the difference between the uses of "or" and "and" to connect inequalities such as "$x < a$" and "$x > b$"?

B4.* True or False: a) "$x < a$" is equivalent to "$x^2 < a^2$."
b) "$xf(x) < xg(x)$" is equivalent to "$f(x) < g(x)$."
c) "$f(x)/x < g(x)/x$" is equivalent to "$f(x) < g(x)$."
d) "$a < b$" is equivalent to "$ax < bx$."

B5.* True or False: a) $|x - a| = |a - x|$ b) $|x| = -|x|$
c) $|x| = |-x|$ d) $|x - a| = |x| - a$. e) $|b| = |-b|$
f) $|kx| = k|x|$ g) $|x + a| = |x| + |a|$.

B6.* To solve the inequality "$f(x)/g(x) > h(x)$" a student might multiply both sides by $g(x)$ to obtain "$f(x) > h(x)g(x)$." Explain what can go wrong.

B7.* To solve the inequality "$f(x)g(x) > f(x)h(x)$" a student might divide both sides by $f(x)$ to obtain "$g(x) > h(x)$." Explain what can go wrong.

B8. a) State a "sign pattern" theorem for solving "$a/b > 0$" that resembles the theorem in problem A9. b) Use it to solve "$x/(x - 2) > 0$."

B9. Suppose a student says the solution to "$x^2 > 9$" is "$x > 3$ and $x < -3$." The use of "and" is wrong. a) Why? b) What word should be used in place of "and"?

B10. Inequalities can be solved in the form "$f(x) < g(x)$" by evaluate-and-compare. Why bother, even when using evaluate-and-compare, to "put everything on one side"?

^ ^ ^ ^ Solve algebraically:

B11. $|1 - 3x| < 5$.
B12. $|x^2 - 4| < 1$.
B13. $|9 - x^2| < 1$.
B14. $|4 - 2x| < 1$.

^ ^ ^ ^ Solve algebraically:

B15. $2/(x - 1) < 6$.
B16. $10/(x + 3) < 2$.
B17. $(x - 1)/(x - 4) < 2$.
B18. $(2x + 1)/(x - 1) < 1$.

^ ^ ^ ^ Use the Theorem on Zeros and Signs and a graph to algebraically solve:

B19. $(x - 2.3)(x + 4)(2x + 3) \geq 0$.

B20. $(x^2 + 3)(x - 20)/(x + 100) \leq 0$.

B21. $x^2/(x - 5) > -3$.

B22. $3/x > 2 - x$.

B23. $\dfrac{x^2 - x - 7}{(x - 3)(x + 6)} > 0$.

B24. $\dfrac{x^2 + 2x - 13}{(x + 5)(x - 1)} > 0$.

B25. $\dfrac{x}{x - 5} < 2$.

B26. $\dfrac{x^2}{x + 3} < 1$.

B27. $1/(x - 1) \leq x$.

B28. $1/x + 1/(x - 2) \leq 3$.

^ ^ ^ ^Here is a theorem: $|a| < |b|$ iff $a^2 < b^2$ (See also B31). Use it to solve these.

B29. $|x| < |x - 3|$.

B30. $|x - 2| < |2x|$.

B31. Prove "$a < b$" is not equivalent to "$a^2 < b^2$." a) What does this tell you about solving an inequality by squaring both sides? b) Give an illuminating example.

B32. Charles' employer gave him a choice of pay plans. Under Plan A he gets \$30 a day plus 4% of his sales. Under Plan B he gets \$16 a day plus 6% of his sales. When is Plan B better for Charles?

B33. You are is asked to advise your firm whether they should rent a photocopy machine on Plan A or Plan B. On Plan A they would pay \$200 per month plus 3 cents per copy. On Plan B they would pay \$300 per month but only 2.2 cents per copy. When is Plan B better for your firm?

B34. Use symbolic notation and inequalities with absolute values to restate: "If x is within d units of a, then $f(x)$ is within c units of $f(a)$."

B35. A student "solved" the inequality "$6/x > 5$" by multiplying through by x to obtain "$6 > 5x$" and then "$6/5 > x$." What went wrong and why?

B36. A student "solved" the inequality "$x(x - 4) < 3(x - 4)$" by dividing through by "$x - 4$" to obtain "$x < 3$." What went wrong and why?

B37. Solve $2 + x \geq 2/(2 - x)$

B38. $1 + 1/x \leq 5 + x$

B39. *State* a theorem which allows us to do the first step in solving $(2x - 5)^2 \leq 17$. [No proof or solution required.]

B40. *State* a theorem which allows us to do the first step in solving $(2x - 5)^2 > 17$. [No proof or solution required.]

B41. a) Solve algebraically: $x^5 > 20$.
b) *State* a theorem which justifies your step [we have stated no theorems about powers and *inequalities*, but you can guess one].

CHAPTER 5

Exponential and Logarithmic Functions

Section 5.1. Exponents and Logarithms

Exponential functions are closely related to power functions. We can regard the expression "b^p" in two ways. If b is the variable and p is constant (x^p, for example, x^2 or x^3), it expresses a power function. If p is the variable and b is a constant (b^x, for example 2^x or 10^x), it expresses an exponential function. This section introduces exponential functions and their inverses, logarithmic functions.

Example 1: $10^4 = 10{,}000$. Is this a power-function fact about x^4 when $x = 10$, or an exponential fact about 10^x when $x = 4$?

It's both. Properties of power functions and exponential functions are perfectly parallel.

$10^3 10^2 = 10^5$. This is an example of the power-function fact $x^3 x^2 = x^5$, for $x = 10$. On the other hand, it is also an example of the exponential-function fact: $10^x 10^y = 10^{x+y}$, for $x = 3$ and $y = 2$.

Properties of 10^x. Some basic facts are: $10^1 = 10$, $10^2 = 100$, $10^3 = 1000$, $10^6 = 1{,}000{,}000$ (one million), $10^9 = 1{,}000{,}000{,}000$ (one billion), and $10^{12} = 1{,}000{,}000{,}000{,}000$ (one trillion).

Other basic facts are: $10^0 = 1$, $10^{-1} = 1/10 = .1$, $10^{-2} = 1/100 = .01$, and $10^{-3} = 1/1000 = .001$.

Read "10^6" as "ten to the 6th," which is short for "ten to the 6th power." Read "10^{-3}" as "ten to the negative 3" or "ten to the minus 3."

The argument 9 is associated with a huge image. The argument -3 is associated with a tiny image. We cannot draw a wonderful graph of 10^x because no vertical scale can handle the rapid changes in y-values. For example, if the window were our "standard" window (Figure 1, [-10, 10] by [-10, 10]) and the actual picture were an inch and a half high, the image of $x = 10$, 10^{10}, would be 12,000 miles above the top of the screen! Expanding the vertical interval helps, but not much. Figure 2, with window [-4, 4] by

[0, 1000] is legible only for x-values between 1 and 3.

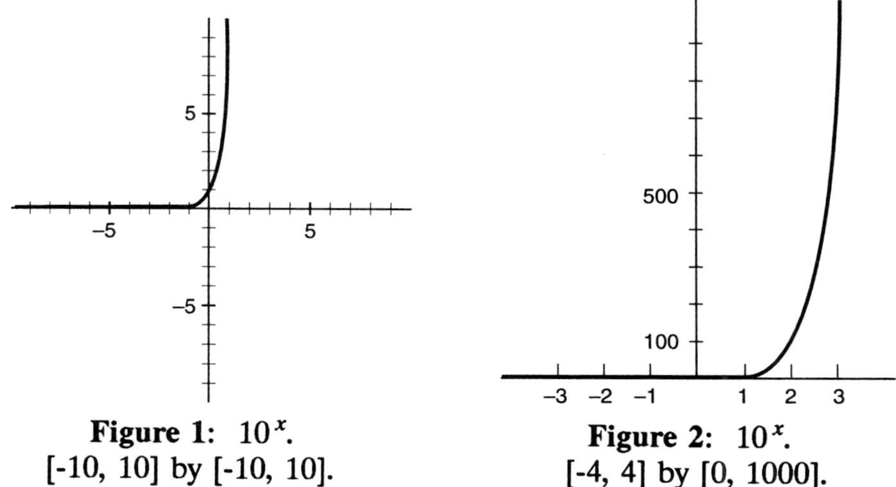

Figure 1: 10^x.
[-10, 10] by [-10, 10].

Figure 2: 10^x.
[-4, 4] by [0, 1000].

Note that the images are never negative. The range of the exponential function is all positive real numbers. For example, 10^{-5} is not negative; it is positive and small.

Scientific Notation. Base 10 is used for "scientific" notation for numbers -- particularly very large or very small numbers where the number of zeros to the left or right of the decimal place becomes unwieldy. The idea is to rewrite a number as a number between 1 and 10, times 10 to the appropriate power.

Example 2: $123 = 1.23 \times 10^2$.
$1{,}200{,}000 = 1.2 \times 10^6$.
$.00000005 = 5 \times 10^{-8}$.
In chemistry, Avogadro's number (the number of atoms in a gram-molecule) $= 6.023 \times 10^{23}$. No one wants to write, or read, a 23 digit number.
In astronomy, a "light year" is a unit of length equivalent to approximately 5.88×10^{12} miles, which is approximately 9.47×10^{12} kilometers.

Properties of Exponential Functions. All the important properties of exponential functions hold regardless of the base.

Discovery 1: Look at your calculator's graph of b^x for various values of b, including $b = e$, 2, 1.1, .9, and 1/2. Which ones resemble each other, except possibly for the scale (Problem B5)? Which ones are increasing? Decreasing?

Different letters can be used to express the same facts, and the choice of letter may influence the interpretation of a theorem (Section 1.4). We can rewrite the properties of power functions as exponential-function theorems merely by switching letters. We give them new numbers to go with the new letters so they can be cited with "5.1" numbers from this section.

Table of 5.1.1 through 5.1.3

	Power Functions with General Variables	Variables Emphasizing Exponential Functions	
	For $x > 0$ and for any p and r,	For $b > 0$ and for any x, y, and p,	
(4.1.1)	$x^p x^r = x^{p+r}$.	$b^x b^y = b^{x+y}$.	(5.1.1)
(4.1.2)	$\dfrac{x^p}{x^r} = x^{p-r}$.	$\dfrac{b^x}{b^y} = b^{x-y}$.	(5.1.2)
(4.1.2B)	$x^0 = 1$.	$b^0 = 1$.	(5.1.2B)
(4.1.2C)	$x^{-p} = \dfrac{1}{x^p} = (\dfrac{1}{x})^p$.	$b^{-x} = \dfrac{1}{b^x} = (\dfrac{1}{b})^x$.	(5.1.2C)
(4.1.3)	$(x^r)^p = x^{rp}$.	$(b^p)^x = b^{px}$.	(5.1.3)

In this table b could represent 10 or e or any other positive number.

Evaluating 10^x. To evaluate 10^x, use a calculator. But, how was it evaluated before calculators?

To find $10^{2.4}$, 2.4 could be treated as 2+.4, so $10^{2.4} = 10^{2+.4} = (10^2)(10^{.4})$ [by 4.1.1, or 5.1.1]. The 10^2 factor is easy. So we only need to know $10^{.4}$. Before calculators, students looked this part up in tables of 10^p for p between 0 and 1. Then, multiplication by 10^2 shifts the decimal point 2 places to the right, but does not change the digits.

The base, b, of an exponential function is always a positive number. Regardless of the choice of $b > 0$,

The **domain** of b^x is all real numbers

The **range** of b^x is all positive numbers.

If $b > 1$, the graph is increasing and resembles the graphs for bases 10, e, and 2 (Figures 1, 2, 3, and 4).

If $b < 1$, the graph is decreasing and resembles the graph for $(1/2)^x$ (Figure 5).

Discovery 2: Use your calculator to evaluate $10^{.4}$.

Now try $10^{1.4}$, $10^{2.4}$, etc. Do you see a pattern?

Now try $10^{-.6}$ and $10^{-1.6}$. Why do they fit in the same pattern (Problem B19)?

Logarithms. Logarithms are exponents. When $10^x = y$, the log of the number y is the exponent x. The next example gives pairs of facts. The first is in "exponential form" and the second is the same fact in "logarithmic form."

Example 3: Inspect these facts closely to see that "logs are exponents."

$10^1 = 10$. $\log 10 = 1$.
$10^2 = 100$. $\log 100 = 2$.
$10^3 = 1000$. $\log 1000 = 3$. log of one thousand equals three.
$10^{-3} = .001$. $\log .001 = -3$.
$10^6 = 1{,}000{,}000$. $\log 1{,}000{,}000 = 6$. log of one million equals six.
$\sqrt{10} = 10^{1/2} = 3.162$. $\log 3.162 = 1/2$.
$10^0 = 1$. $\log 1 = 0$.

Note that 10^3 has 3 zeros to the left of the decimal point, but 10^{-3} has only two zeros to the right of the decimal point.

This example also shows that logs are inverses of exponentials.

For base 10 and all other bases greater than one, exponential functions are increasing and have inverses (Figures 1 through 4). The inverse functions are called a logarithmic functions. Logarithms evaluated using base 10 are sometimes called "common" logarithms. ("Natural" logarithms are base $e = 2.718....$) Here are the inverse relationships.

Definition 5.1.4: $10^x = y$ iff $\log y = x$.
$e^x = y$ iff $\ln y = x$.
In general, $b^x = y$ iff $\log_b y = x$.

Inspect this closely to see why my favorite sound bite about logs is

Logs are exponents.

Read "e^x" as "e to the x." Read "$\ln y$" as "$\log y$." Yes, "ln" is usually pronounced the same way as "log". If you are reading them, the difference is clear. If you are hearing them, it is not. Be careful. If you need to be very clear, you could say "the natural log of y."

Example 4: Solve $10^x = 1000$.

$10^x = 1000$ iff $x = \log 1000 = 3$.

Solve $10^x = 50$.

$10^x = 50$ iff $x = \log 50 = 1.699$ (from a calculator). The ".699" part of the answer is not obvious, but it is easy to see that log 50 is somewhere between 1 and 2 because 50 is between $10 = 10^1$ and $100 = 10^2$.

Discovery 3: Evaluate log 2 and log(1/2) and note their signs. Try to evaluate log 0 and log(-2). Then, use your calculator to graph "$\log x$" and "$\ln x$". Note the domain and range. Are there points to the left of the y-axis? Are there points below the x-axis? What does all this tell you about the domain and range of "log"?

Graphs. Given the graph of 10^x, its inverse, log, can be determined by reading from y to x (instead of from x to y). Any inverse can be graphed by recalling that (b, a) is on a graph of the inverse if and only if (a, b) is on the graph of the function (Section 2.4, "Inverses"). The graph of log is the reflection of the graph of 10^x through the line $y = x$ (Figure 3). Both are increasing functions. 10^x grows extremely rapidly; log x grows extremely slowly.

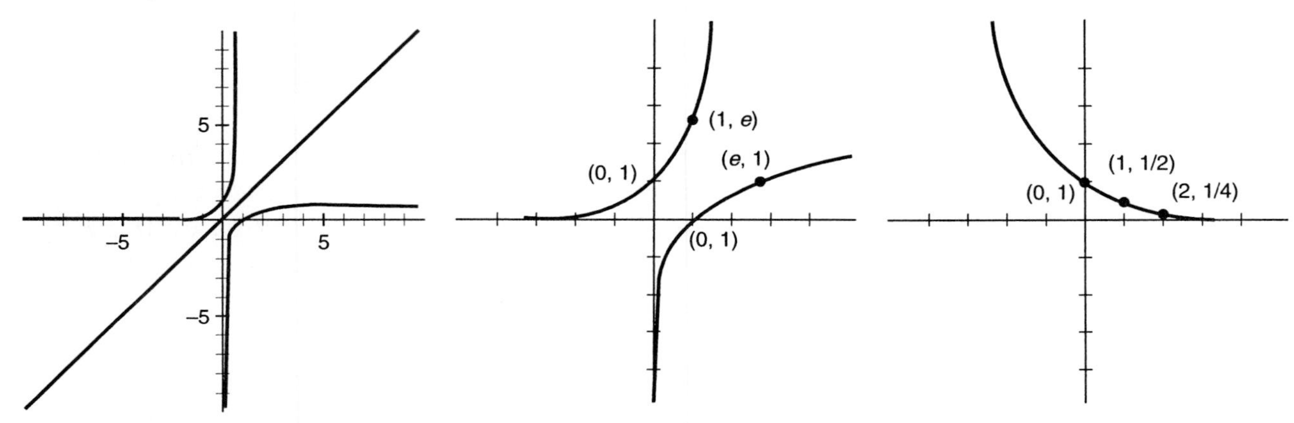

Figure 3: 10^x, $\log x$ and $y = x$. [-10, 10] by [-10, 10].

Figure 4: e^x and $\ln x$. [-5, 5] by [-5, 5].

Figure 5: $(1/2)^x$. [-5, 5] by [-5, 5].

Corollary 5.1.4A (inverse function version):

For all x, $\log(10^x) = x$. For all $x > 0$, $10^{\log x} = x$.

For all x, $\ln(e^x) = x$. For all $x > 0$, $e^{\ln x} = x$.

For any $b > 0$, for all x, $\log_b(b^x) = x$. For all $x > 0$, $b^{\log_b x} = x$.

Calculator Exercise 1: Use your calculator to illuminate the inverse relationship stated in this corollary.

For example, evaluate 10 to any power, say 3.4. $10^{3.4} = 2511.886$. Then take log of that number to recover your original number: $\log 2511.886 = 3.4$.

Conversely, evaluate log of any positive number, say 55. $\log 55 = 1.74036$. Then take 10 to that power to recover your original number: $10^{1.74036} = 55$.

Graph $\log(10^x)$. Graph $10^{\log x}$. Why is it not always x?

Definition 5.1.4, "$10^x = y$ iff $\log y = x$," is expressed with two dummy variables, "x" and "y". Of course, we can rewrite it with different letters.

Version B addresses equations in which *log* is applied to the unknown.

Definition 5.1.4B: $\log x = y$ iff $x = 10^y$.

Example 5: Solve $\log x = 5$.

$\log x = 5$ iff $x = 10^5 = 100{,}000$.

$\log x = 3.7$ iff $x = 10^{3.7} = 5011.8$. Obtaining this exact value requires a calculator, but the order of magnitude of the solution is clear from properties of powers. Because 3.7 is between 3 and 4, the answer is between $10^3 = 1000$ and $10^4 = 10000$.

Solve $\log x = -2.3$.

$\log x = -2.3$ iff $x = 10^{-2.3} = .00501$. Because -2.3 is between -2 and -3, the answer is between $10^{-2} = .01$ and $10^{-3} = .001$.

Discovery 4: Use your calculator to evaluate log 5.

Now try log 50, log 500, log 5000, etc. Do you see a pattern?

Now try log .5 and log .05. Why do they fit in the same pattern (Problem B20)?

Properties of Logarithms. Because all log facts are exponential facts (in another form), we can extract log facts from properties 5.1.1 through 5.1.3 about exponentials. The key idea is that

Logs are exponents.

Here we derive the properties of logarithms with base 10, but all the properties hold for any base. Exponential facts and logarithmic facts come in pairs, and we use the definition

(5.1.4, again) $$10^x = y \text{ iff } x = \log y$$

to reformulate exponential facts as log facts. Recall 5.1.1.

(5.1.1, again) $$10^p\, 10^r = 10^{p+r}.$$

The left side is a product (10^p times 10^r) and the log of the product is visible as the exponent on the right side: $p+r$. The logs of the factors are also visible: p and r.

"The log of a product is the sum of the logs of the factors."

Using dummy variables c and d to represent the factors 10^p and 10^r, we see

(5.1.1L, log version) $$\log(cd) = \log c + \log d.$$

Example 6: $\log(7x) = \log 7 + \log x$.
$\log(x(1 - x)) = \log x + \log(1 - x)$.

The log of a quotient can be rewritten. Recall

(5.1.2, again) $$\frac{10^p}{10^r} = 10^{p-r}.$$

The left side is a quotient and its log is visible on the right side: $p - r$. This is the log of the numerator minus the log of the denominator.

(5.1.2L, log version) $\log(c/d) = \log c - \log d$.

"The log of a quotient is the difference of the logs."

Example 7: $\log(x/3) = \log x - \log 3$.
$\log(1/x) = \log 1 - \log x = -\log x$.

$$\log\left(\frac{x}{1-x}\right) = \log x - \log(1 - x).$$

Property 5.1.2B says $10^0 = 1$. In logarithmic form, this is

(5.1.2BL, log version) $\log 1 = 0$.

(5.1.2CL, log version) $\log(1/c) = -\log c$.

"The log of a reciprocal is the negative of the log."

Example 8: $\log(1/2) = -\log 2 = -.301$.

The most useful result about logarithms comes from 5.1.3.

(5.1.3, again) $(10^r)^p = 10^{rp}$.

Let $c = 10^r$, so $\log c = r$ and the left side can be regarded as c^p. Then, $\log(c^p)$ is visible as the exponent on the right: rp, which is $p \log c$.

(5.1.3L, log version) $\log(c^p) = p \log c$.
$\log(c^x) = x \log c$.

"The log of an exponential is the exponent times the log of the base."

The second line repeats the first, but emphasizes solving for unknown exponents by using "x" as the exponent.

Example 9: $\log 10^3 = 3 \log 10$ [$= 3(1) = 3$].
$\log(7^{2.4}) = 2.4 \log 7$.
$\log(1.08^x) = x \log 1.08$.
$\log(x^3) = 3 \log x$.

Logs are useful for solving for unknown exponents.

Example 10: Solve $2^x = 20$.
Take logs.
$2^x = 20$ iff $\log(2^x) = \log 20$ [log is one-to-one]
iff $x \log 2 = \log 20$ [5.1.3L]
iff $x = (\log 20)/(\log 2)$.
It is interesting that it does not matter which base you use to evaluate this quotient. Try common logarithms:
$(\log 20)/(\log 2) = 1.301/.301 = 4.322$.
Try natural logarithms:
$(\ln 20)/(\ln 2) = 2.996/.693 = 4.322$. The result is the same either way.

Theoretically, we could try to do this problem by taking logs base 2, but they are not available on most calculators. If the base is not 10 or e, 5.1.3L is more valuable than 5.1.4.

Example 10 fits the pattern: $b^x = c$. Taking logs, $x \log b = \log c$. Therefore, for $b > 0$ and $c > 0$,

(5.1.5) $$b^x = c \text{ iff } x = (\log c)/(\log b).$$

Example 11: Solve $1000(1.006)^x = 1200$.
$1000(1.006)^x = 1200$ iff $1.006^x = 1.2$.
Taking logs of both sides, $\log(1.006^x) = \log 1.2$ [log is one-to-one]
$x \log 1.006 = \log 1.2$ [by 5.1.3L]
$x = (\log 1.2)/(\log 1.006) = 30.48$.

Example 12: Solve for r: $2 = (1 + r)^{4.5}$.
We could use logs, but using inverse powers is more to the point. Use inverse-reverse and undo the 4.5 power:
$2^{1/4.5} = 1 + r$. Therefore, $r = 2^{1/4.5} - 1$. In decimal form, $r = .1665$.

Logs are useful for solving for unknown exponents. In this example, the exponent is not unknown. Use roots.

Table 5.1.6

	Exponential Form (base 10)	Logarithmic Form (base 10)	
(5.1.4)	$10^x = y$	$\log y = x.$	(5.1.4)
(5.1.4inv)	For $x > 0$, $10^{\log x} = x$.	$\log (10^x) = x.$	(5.1.4invL)
(5.1.1)	$10^x 10^y = 10^{x+y}.$	$\log(cd) = \log c + \log d.$	(5.1.1L)
(5.1.2)	$\dfrac{10^x}{10^y} = 10^{x-y}.$	$\log(c/d) = \log c - \log d.$	(5.1.2L)
(5.1.2B)	$10^0 = 1.$	$\log 1 = 0.$	(5.1.2BL)
(5.1.2C)	$10^{-x} = \dfrac{1}{10^x} = (\dfrac{1}{10})^x.$	$\log(1/c) = -\log c.$	(5.1.2CL)
(5.1.3)	$(10^p)^x = 10^{px}.$	$\log(c^x) = x \log c.$	(5.1.3L)
	10^x has domain all real numbers.	log has range all real numbers.	
	10^x has range all positive numbers.	log has domain all positive numbers.	

The properties in this table also hold for base e and any other positive base b in place of base 10.

Properties of exponentials and logarithms may be used to solve equations.

Example 13: Solve for d: $2^{3/d} = 5$.

The unknown is in the exponent. Take logs.

$\log(2^{3/d}) = \log 5$

$(3/d)\log 2 = \log 5$ [by 5.1.3L]

$(3 \log 2)/(\log 5) = d$ [isolating d]

$d = 1.29.$

Example 14: Solve $1.06^x = .7(1.09)^x$.

Because the unknown is an exponent, take logs. The right hand side is a product, so its log is a sum. By 5.1.3L and 5.1.1L:

$$x \log 1.06 = \log .7 + x \log 1.09.$$

$$x(\log 1.06 - \log 1.09) = \log .7$$

$$x = (\log .7)/(\log 1.06 - \log 1.09) = 12.78.$$

Example 14, another way: Solve $1.06^x = .7(1.09)^x$.

Divide both sides by 1.09^x to obtain, using 4.1.4B,

$$\frac{1.06^x}{1.09^x} = .7$$

$$\left(\frac{1.06}{1.09}\right)^x = .7 .$$

Taking logs,

$x \log(1.06/1.09) = \log .7.$

$x = 12.78.$

Example 15: How many years does it take for \$10,000 invested at 11% compounded quarterly to become \$30,000?

From 4.4.5, the formula for compounding money is $A = P(1 + r/k)^{kt}$, where in this problem $A = 30{,}000$, $P = 10{,}000$, $r = .11$, $k = 4$, and t is unknown. So the problem is to solve for t in:

$$30{,}000 = 10{,}000(1 + .11/4)^{4t}.$$

First divide by 10,000 to obtain

$$3 = (1 + .11/4)^{4t}.$$

Now take logs:

$$\log 3 = 4t \log(1 + .11/4).$$

$$t = 10.12 \text{ (years)}.$$

Example 16: Use properties of logarithms to rewrite: $\log[x^5(1 - x)^7]$. [Without logs of powers, logs of products, or logs of quotients.]

The order of operations is important. As usual, work with the last operation first. Inside the log expression, multiplication is last, so use the log of a product from 5.1.1L.

$$\log[x^5(1 - x)^7] = \log(x^5) + \log[(1 - x)^7]$$

$$= 5 \log x + 7 \log(1 - x), \text{ by 5.1.3L.}$$

There is no simplification of the logarithm of a difference, so this is the final expression.

Example 17: Use properties of logs to rewrite $\log\left[\frac{x^2(1 - x)^5}{(1 + x)^3}\right]$.

Inside the log expression, the division is last, so use 5.1.2L.

$$\log\left[\frac{x^2(1 - x)^5}{(1 + x)^3}\right] = \log[x^2(1 - x)^5] - \log[(1 + x)^3]$$

In the first term on the right multiplication is last, so use 5.1.1L.

$$= \log(x^2) + \log[(1 - x)^5] - \log[(1 + x)^3] .$$

Now use 5.1.3L to deal with powers.

$$- 2 \log x + 5 \log(1 - x) - 3 \log(1 + x) .$$

There is no simplification of the logarithm of a sum or a difference, so this is the final expression.

Conclusion. Power functions and exponential functions are intimately related. Properties of power functions can be rewritten as properties of exponential functions just by changing letters. Logarithmic functions are inverses of exponential functions. Definition 5.1.4 gives the basic corresponding exponential and logarithmic forms. For solving equations, the most important property of logs is 5.1.3L: $\log(c^x) = x \log c$. It is useful for solving for unknown exponents.

Terms: base, power, exponent, scientific notation, exponential function, logarithmic function, exponential form, logarithmic form.

****

Exercises for Section 5.1, "Exponents and Logs":

A1.* Properties of exponential functions are just properties of __________ functions with the letters switched.

A2.* In exactly three words, what are logs?

A3.* Write this in "logarithmic form": $b^a = c$.

A4.* There are only two commonly-used types of logarithms. Which two?

A5.* The log of a product is the ______________ of the logs.

A6.* The log of a quotient is the ______________ of the logs.

A7.* True or false?: There are properties of logarithms in this section for expressing logs of a) products b) sums c) quotients d) differences e) powers

A8.* Note the correct alternatives: There are properties of logarithms for rewriting a) differences of logs b) products of logs c) sums of logs d) quotients of logs.

A9.* a) 10^n has _______ zeros to the left of the decimal point.
b) 10^{-n} has _______ zeros to the right of the decimal point.

A10.* Which of the following are *not* properties of logs?
a) $\log(x/4) = (\log x)/4$. b) $\log(x/4) = (\log x)/(\log 4)$. c) $\log(x - 2) = \log x - \log 2$.
d) $\log(\sqrt{x}) = \sqrt{(\log x)}$. e) $\log(x^2) = (\log x)^2$.

A11.* In each part of A10 that is false, correct the right side, if possible, to make it true.

A12. Write in logarithmic form: a) $10^4 = 10{,}000$. b) $10^{-2} = .01$.
A13. Write in logarithmic form: a) $10^{2.5} = 316.2$ b) $10^{.3} = 1.995$.

A14. Write in exponential form: a) $\log 100 = 2$. b) $\log(.1) = -1$.
A15. Write in exponential form: a) $\log 50 = 1.699$. b) $6 = \log 1{,}000{,}000$.
A16. Write in exponential form: a) $\log 150 = 2.176$ b) $\log(.015) = -1.824$.

^ ^ ^ ^ Solve

A17. $(1 + r/2)^8 = 1.35$ [.076].
A18. $(1 + r)^{20} = 5$ [.084].
A19. $(1.005)^t = 2$ [140].
A20. $(1.03)^t = 1.5$ [14]
A21. $1000(1.005)^{12t} = 1500$ [6.8].
A22. $500(1.03)^{2t} = 1000$ [12].
A23. $\log(x^2 - 1) = 2$ [10,...].
A24. $\log(x/5) = 2.5$ [1600].
A25. $\log[x/(1 - x)] = -1$ [.091].
A26. $\log(3^x) = 2$ [4.2].
A27. $10^{x+1} = 1000$ [2.0].
A28. $10^{x/3} = 30$ [4.4].

^ ^ ^ ^ Use properties of logs to rewrite the expression as an equivalent expression without logs of products, logs of quotients, or logs of powers.

A29. $\log(3x)$.
A30. $\log(1.05^x)$.
A31. $\log(x^5)$.
A32. $\log[x(2x + 1)^2]$
A33. $\log[x/(5x + 1)]$.
A34. $\log(7/x^2)$.

A35. You should be able to answer these in your head:
a) The common name for 10^6 is ________.
b) $\log 1000 =$ ________.
c) $\log 10 =$ ________.
d) $\log .001 =$ ________.
e) $10^3\, 10^6 =$ ________.
f) If $10^x = 1$, $x =$ ________.

A36. If $x < y$, factor "$b^x + b^y$."

A37. Do Discovery 3. That is, use your calculator to graph "$\log x$" and "$\ln x$". Note the domain and range. a) Are there points to the left of the y-axis? b) Are there points below the x-axis? c) Note the signs of log 2 and log(1/2). d) Try to evaluate log 0 and log(-2). What does all this tell you about the domain and range of "log"? Look at the graph of "10^x" and "e^x". e) Are there points to the left of the y-axis? f) Are there points below the x-axis? g) Note the signs of 10^{-2} and 10^0. h) What does all this tell you about the domain and range of "10^x"?

^ ^ ^ ^ ^ ^ ^ ^ ^

B1.* Explain the relationship between power functions and exponential functions.

B2.* a) Write in exponential form: $\log a = b$. b) Write in logarithmic form: $10^c = d$.

B3.* Give three important power-function facts and the associated exponential-function facts.

B4.* Give three important exponential-function facts and the associated logarithmic-function facts.

B5. Do Discovery 1. That is, look at a calculator's graph of b^x for various values of b, including $b = e$, 2, 1.1, .9, and 1/2. a) Which ones resemble each other, except possibly for the scale? b) Which ones are increasing? Decreasing? c) Find a window $[-c, c]$ by $[-10, 10]$ such that the graph of 1.1^x in that widow looks like the graph of 2^x on $[-10, 10]$ by $[-10, 10]$. What is the value of c?

B6. Determine and state a theorem for solving: $Ac^x = B$ (state it using logs base 10, and assume A, B, and c are greater than 0).

^ ^ ^ ^ Solve algebraically:

B7. $1.06^x/1.04^x = 5$ [84].

B8. $2^x/3^x = .5$ [1.7].

B9. $2^x/3^{2x} = .01$ [3.1].

B10. $10^x = 2^{x+1}$ [.43].

B11. $(1/2)^{5/h} = .3$ [2.9]

B12. $(1.03)^t = .01(1.07)^t$ [120].

B13. $2^{t/30} = 10(2^{t/40})$ [400].

^ ^ ^ ^ Use properties of logs to rewrite the expression as an equivalent expression without logs of products, logs of quotients, or logs of powers.

B14. $\log[x^3(1 - x)^6]$.

B15. $\log[x^2/(1 - x)^3]$.

B16. $\log[20x^{-5}(10 - x^2)]$.

B17. $\log[(1 + x)/(2 + x)^3]$

B18. Look at a calculator's graph of b^{-x} for some $b > 1$. How does the graph of b^{-x} compare to the graph of b^x ?

B19. Do Discovery 3. a) State, symbolically, the pattern which relates $10^{.4}$ to $10^{1.4}$, etc.. b) Why does -.6 fit in with .4?

B20. Do Discovery 4. a) State, symbolically, the pattern which relates log 5 to log 50, etc.. b) Why does .5 fit in with 5?

B21. Find the approximate doubling time of money invested at 7% compounded monthly. [9.9 years]

B22. Find the approximate doubling time of a population that grows 2% per year. [35]

B23. Investment Plan A returns 10% compounded monthly. Investment Plan B returns 12% compounded monthly, but it costs 5% in broker's fees (so your principal is only 95% as much). After how many years will Plan B be better than Plan A? [2.6]

B24. Investment Plan A returns 8% compounded quarterly. Investment Plan B returns 9.2% compounded monthly, but there is a 10% penalty for taking the money out before you retire. Suppose you will want your money before you retire. How long do you need to wait before the amount under Plan B is better, even after the 10% penalty? [8.5]

B25. Population A is growing at 3% per year (compounded yearly). Population B starts out 40% larger, but is only growing at 1% per year (compounded yearly). If these rates remain the same, in how many years will the Population A equal Population B? [17]

B26. Scientific notation and exponential notation have some similarities and differences. Discuss them.

^ ^ ^ ^ Solve algebraically:

B27. $\log(x^5) + \log(x^2) = 3$ [2.7].

B28. $\log x + \log(3x) = 2$ [5.8].

B29. *State* the identity which expresses how to evaluate b^p in terms of the natural exponential function.

Section 5.2. Base 2 and Base *e*

The previous section introduced exponential functions and logarithmic functions with special emphasis on base 10. This section emphasizes other bases. All exponential functions have much in common, and all logarithmic functions have much in common. Various minor reasons cause us to use one base rather than another. Base 2 is convenient for certain applications in which an amount doubles in a given amount of time. The number e (= 2.718...) may appear to be awkward, but it turns out to be the most convenient base in calculus.

Base 2. It is hard to graph the exponential function with base 10 because the images change so fast that no uniform vertical scale can illustrate them over a wide domain. If we want to graph an exponential function, it is more convenient to use a smaller base, say, base 2, which is important in computer science. Table 5.2.1 gives values of 2^x for integer values of x.

Table 5.2.1

x	-3	-2	-1	0	1	2	3	4	5	6	7	8	9	10
2^x	1/8	1/4	1/2	1	2	4	8	16	32	64	128	256	512	1024

This table can be used to illustrate the properties of exponential functions.

Example 1: $2^3 \times 2^4 = 2^{3+4} = 2^7$.
That is, $8 \times 16 = 128$.

$$2^7 \times 2^{-3} = 2^{7+-3} = 2^4.$$
That is, $128 \times 1/8 = 16$.

$$2^9/2^4 = 2^{9-4} = 2^5.$$
That is, $512/16 = 32$.

$$(2^3)^2 = 2^{2\times 3} = 2^6.$$
That is, $8^2 = 64$.

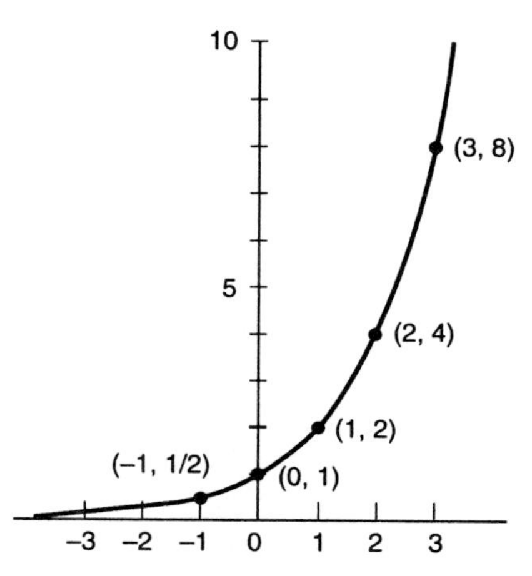

Figure 1: 2^x.
[-4, 4] by [0, 10].

For the function 2^x, to add 1 to the argument is to multiply the image by 2. To add 3 to the argument is to multiply the image by 8. The graph of 2^x interpolates between the values given in Table 5.2.1 (Figure 1). It looks just like the graph of 10^x or e^x, only with a horizontal scale change (We will see why later in Example 17, Figure 7).

Exponential Growth. In the real world it is common for one quantity to be related to another by an exponential function. Often the independent variable is time, so exponential functions are often expressed with argument "t" instead of "x".

Example 2: Let $f(t) = 2^t$. Compare $f(t+1)$ to $f(t)$.

$$\frac{f(t+1)}{f(t)} = \frac{2^{t+1}}{2^t} = \frac{2(2^t)}{2^t} = 2 .$$

The ratio is 2 regardless of t. Therefore, regardless of its value, $f(t)$ will double in one more unit of time.

If the "doubling time" is not *one* unit of time, then a scale change is in order. Remember that scale changes of the argument arise by multiplying or dividing the argument by a constant *before* applying the function (Section 2.3).

Example 3: Suppose a bacterial population doubles every thirty minutes. Change the horizontal scale of the expression $P\,2^t$ so that it describes doubling every 30 units of time instead of every one unit of time.

When $t = 0$, $P\,2^t = P$, the initial population. When the exponent becomes 1, the population will have doubled. We want this to happen when $t = 30$, so use $P\,2^{t/30}$.

Now $t = 30$ yields exponent 1 and the population doubles in 30 units of time. $t = 60$ yields exponent 2 and the population quadruples in 60 units of time.

Dividing the argument by 30 *before* applying the function makes the graph 30 times as wide. Effectively, this just relabels the horizontal scale (Figure 2).

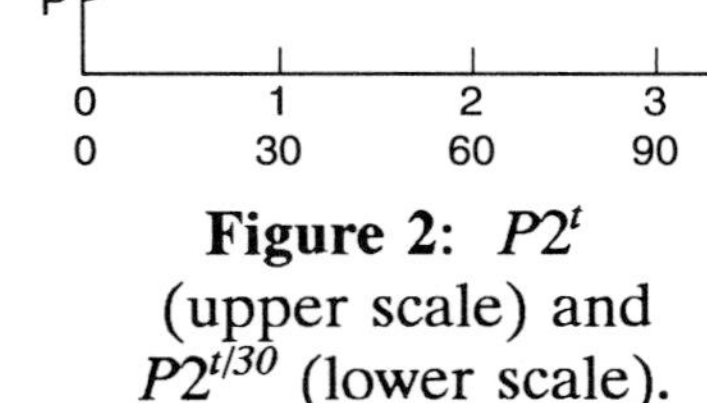

Figure 2: $P2^t$ (upper scale) and $P2^{t/30}$ (lower scale).

Example 3, continued: Determine the population after 80 minutes if the initial population is 10,000.

The formula is $P(t) = P\,2^{t/30}$.

This is a straightforward "plug in" evaluation problem.

$$P(80) = 10{,}000(2^{80/30}) = 63{,}496 \text{ (about 63,500)}.$$

Example 3, continued further: How long will it take for this population to triple?

The question asks when $P(t) = 3P$, or $P(t)/P = 3$. It asks when the growth factor will be 3. Now, at any time, $P(t)/P = 2^{t/30}$. So the problem is to solve:

$$2^{t/30} = 3.$$

Taking logs (any base, say, base 10) of both sides and using 5.1.3L:

$$\log(2^{t/30}) = \log 3,$$
$$(t/30)\log 2 = \log 3,$$
$$t = (30 \log 3)/\log 2 = 47.55 \text{ (minutes)}.$$

The population will triple every 47.55 minutes. Of course, in the real world the exponential model of population growth eventually breaks down as the bacteria become crowded and the food supply becomes inadequate to support continued exponential growth.

If the doubling time of an amount undergoing exponential growth is given, the exponential function base 2 is convenient. The doubling-time model generalizes Example 3. The term mathematical model is used to mean a mathematical formula intended to quantify some real-world phenomenon.

Doubling-time Model: Let a population, $P(t)$, undergo exponential growth such that it doubles every d units of time. Then

$$P(t) = P(0)\, 2^{t/d}. \tag{5.2.2}$$

The growth factor is "$2^{t/d}$."

Time:	0	d	$2d$	$3d$	$4d$
Factor:	1	2	4	8	16

Example 4: A bacterial population doubles every 35 minutes and is 10,000 at 3:00pm. How many bacteria will there be at 4:10pm?

4:10pm is 70 minutes after 3:00 -- exactly 2 doubling times later. Therefore the population will have grown by a factor of 4. It will be 40,000.

How many will there be at 5:00pm?

This is much like "Example 3, continued," where we calculated a population after 80 minutes. The difference is that here we must convert clock time to elapsed time. In exponential problems you can choose time 0 to be any convenient time. Here, let 3:00pm be time 0 and compute elapsed time in minutes, so 5:00pm, which is two hours later, corresponds to $t = 120$. Now this is a "plug in" evaluation.

$$P(120) = 10{,}000(2^{120/35}) = 107{,}672 \text{ (about 108,000)}.$$

Take logs when you want to solve for an unknown exponent.

Example 5: Suppose a population undergoing exponential growth grows 10 percent in three hours. What is its doubling time?

Plug in to 5.2.2, using $t = 3$ and $P(3) = 1.1\ P(0)$.

$P(3) = 1.1\ P(0) = P(0)\ 2^{3/d}$.

$1.1 = 2^{3/d}$. Taking logs,

$\log 1.1 = (3/d)\log 2$. $d = (3 \log 2)/\log 1.1 = 21.8$ (hours).

Base 1/2. Base 1/2 is convenient for "half-life" models, where, instead of a constant doubling time, there is a constant halving time. That is, half the amount remains after a certain amount of time elapses. For this model it is convenient to use $(1/2)^t$. Table 5.2.3 gives $(1/2)^t$ for selected integer values of t.

Table 5.2.3

t	0	1	2	3	4	5
$(1/2)^t$	1	1/2	1/4	1/8	1/16	1/32

Of course, $(1/2)^t$ is just 2^{-t}, so the only virtue of dealing with the fraction 1/2 instead of the integer 2 is that it avoids negative numbers in the exponent. Instead of talking about a "halving time," we use the term "half-life." As is, the expression $A(1/2)^t$ has a half-life of one unit. For any other half-life, a scale change is in order.

Example 6: The half-life of Carbon 14, used to date archaeological materials, is 5730 years. Change the horizontal scale of the expression $A(1/2)^t$ so that it describes a half-life of 5730 units of time instead of one unit of time.

The initial amount at time $t = 0$ is A. Half will remain when the exponent is 1, so we want the exponent to be 1 when $t = 5730$. So divide by 5730 before applying the exponential function. The function $A(1/2)^{t/5730}$ has the desired property. Division by 5730 before applying the function makes the graph 5730 times as wide (Section 2.3), or, what is the same, relabels the horizontal scale (Figure 3).

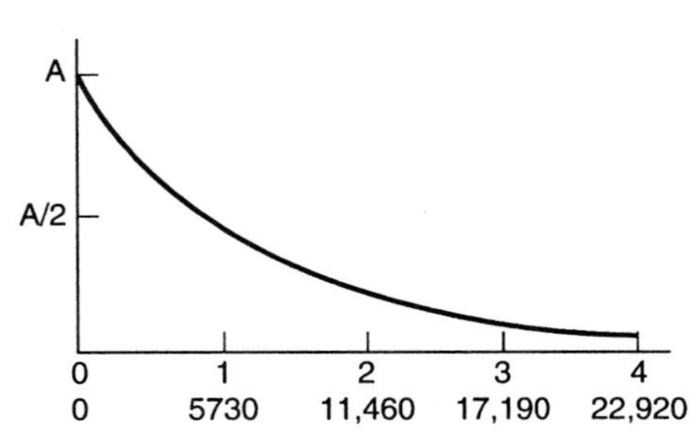

Figure 3: $A(1/2)^t$ (upper scale) and $A(1/2)^{t/5730}$ (lower scale).

Example 6, continued: Switch to functional notation. Let $A(t)$ be the amount at time t. The initial amount is then denoted by $A(0)$. If the initial amount is 0.012 grams, how much will remain after 11,460 years?

11,460 years is 2 × 5730 years -- exactly two half-lives. Therefore, one quarter of the original amount will remain: .003 grams.

How much will remain after 4000 years?

This is a "plug in" evaluation problem:

$A(4000) = 0.012(1/2)^{4000/5730} = 0.0074$ (grams).

The half-life model generalizes Example 6.

Half-Life Model: Let $A(t)$ denote the amount of a substance at time t. Let it undergo exponential decay such that half remains after every h units of time. Then

$$A(t) = A(0)(1/2)^{t/h}. \qquad (5.2.4)$$

"$(1/2)^{t/h}$" is the decay factor.

Time:	0	h	$2h$	$3h$	$4h$
Factor:	1	1/2	1/4	1/8	1/16.

Example 6, continued further: Suppose the half life of a substance is 5730 years and 0.00036 grams remain after 2000 years. How many grams existed originally?

Given the formula, word problems are easy. Just plug in for the given quantities.

$$.00036 = A(0)(1/2)^{2000/5730}.$$

We want to know the original amount, $A(0)$.

$$A(0) = \frac{.00036}{(\frac{1}{2})^{2000/5730}} = .00046 \, .$$

Originally, there existed 0.00046 grams.

Example 7: Suppose the half life of a substance is 90 years. When will only 1/1000 of the original amount remain?

The factor "$(1/2)^{t/h}$" is to be 1/1000. h is given as 90. So solve

$$(1/2)^{t/90} = 1/1000.$$

The unknown is in an exponent, so take logarithms. Use 5.1.3L.

$$(t/90)\log(1/2) = \log(1/1000).$$

$$t = \frac{90 \log(1/1000)}{\log(1/2)} = 897 \textit{ years}.$$

One one-thousandth of the original amount will remain after 897 years.

Example 8: Suppose a substance undergoes rapid radioactive decay. Only 1/10 the original amount remains after 20 seconds. What is its half life?

This is another "plug in" problem.

$A(20)/A(0) = 1/10 = (1/2)^{20/h}$.

$\log(1/10) = (20/h)\log(1/2)$. $h = [20\log(1/2)]/[\log(1/10)] = 6.02$ (seconds).

Compound Interest. Money problems serve to illustrate the close relationship between exponential functions and power functions.

The formula for compound interest was given in Section 4.4 on Percents. Here it is again.

(5.2.5) $$A = P(1 + \frac{r}{k})^{kt}.$$

"P" is the principal, the original amount of money. "A" is the amount at time t, where "t" is expressed in years. "r" is the annual percentage rate, and "k" is the number of times per year the money is compounded.

To solve for t in the compound interest model, take logarithms of both sides using any convenient base (base 10 or base e).

Example 9: How long does it take \$1000 to triple when invested at 12% compounded monthly?

Plugging into Formula 5.2.5, the problem becomes:

Solve $3000 = 1000(1.01)^{12t}$ for t (in years).

$3 = 1.01^{12t}$.

$\log 3 = \log(1.01^{12t}) = 12t \log 1.01$.

$t = (\log 3)/(12 \log 1.01) = 9.2$ years.

Actually, the formula is only valid when t expresses an integer number of months, and 9.2 years is not precisely an integer number of months. But it is close enough to answer the spirit of the question.

Example 10: How long does it take money to double when invested at 8 percent compounded yearly?

This problem asks us to solve: $2 = 1.08^t$.

Taking logs, $\log 2 = t \log 1.08$, and $t = (\log 2)/(\log 1.08) = 9.01$ years.

Example 11: Money compounded monthly grew from 10,000 to 14,000 in 3 years. What was the annual percentage rate?

This problem asks us to solve for r in: $14{,}000 = 10{,}000(1 + r/12)^{36}$.

The first step is to divide by 10,000.

$$1.4 = (1 + r/12)^{36}.$$

There is no need to take logs. Logs are great for solving for unknowns that are in the exponent. This unknown is not. Use inverse-reverse and take roots.

$$1.4^{1/36} = 1 + r/12.$$

$1.00939 = 1 + r/12.$ $.00939 = r/12.$ $r = 12(.00939) = .1127.$
The rate was 11.27%.

Base *e*. In calculus, base e (= 2.718...) is the preferred base for exponential and logarithmic functions. This may seem like an awkward number, but it really does have some wonderful properties that make it "natural" in calculus.

Logarithms base e are called "natural" logarithms and the function is denoted by "ln" to distinguish it from all other logarithmic functions which use the spelling "log." Nevertheless, "ln" is usually pronounced "log," not "linn" or "el en." In calculus, logs will almost always be natural logs, so the identical pronunciation of two different functions ("ln" and "log") will rarely cause confusion.

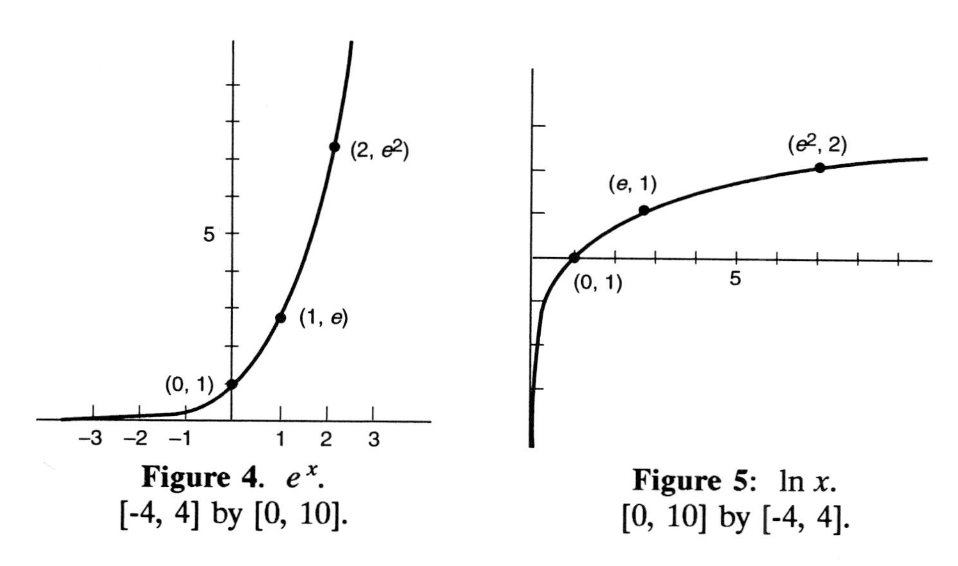

Figure 4. e^x.
[-4, 4] by [0, 10].

Figure 5: ln x.
[0, 10] by [-4, 4].

All the general facts about exponential functions and logarithms hold for base e. They are repeated here for convenience. The same reference numbers are used again.

Table 5.2.7

	Exponential Form (base e)		Logarithmic Form (base e)	
(5.1.4)	$e^{a} = c$	iff	$\ln c = a.$	
(5.1.4inv)	For all $c > 0$, $e^{\ln c} = c.$		For all a, $\ln(e^{a}) = a.$	(5.1.4invL)
(5.1.1)	$e^{x}e^{y} = e^{x+y}.$		$\ln(cd) = \ln c + \ln d.$	(5.1.1L)
(5.1.2)	$\dfrac{e^{x}}{e^{y}} = e^{x-y}.$		$\ln(c/d) = \ln c - \ln d.$	(5.1.2L)
(5.1.2A)	$e^{0} = 1.$		$\ln 1 = 0.$	(5.1.2BL)
(5.1.2B)	$e^{-x} = \dfrac{1}{e^{x}} = (\dfrac{1}{e})^{x}.$		$\ln(1/c) = -\ln c.$	(5.1.2CL)
(5.1.3)	$(e^{x})^{p} = e^{xp}.$		$\ln(c^{p}) = p \ln c.$	(5.1.3L)
	e^{x} has domain all real numbers.		*ln* has range all real numbers.	
	e^{x} has range all positive numbers.		*ln* has domain all positive numbers.	

It takes a background in calculus to fully understand why the preferred base for logarithms is base e. The upcoming examples will give a glimpse of the reasons.

The Limit of Compounding. An important reason for using base e is that it arises naturally as a limit of compounding. Reconsider the compounding factor in Formula 5.2.5: $(1 + r/k)^{kt}$. Suppose the annual percentage rate, r, and the time, t, are fixed, but we compound more and more frequently. For example, first consider compounding yearly, then semiannually (twice a year), then quarterly, monthly, daily, and even more frequently. What happens to the effective yield?

Calculator Exercise 1: Let $t = 1$ and $r = 1$. Evaluate $(1 + 1/k)^{k}$ for various values of k, including large k.

k	1	2	4	12	365	$\to \infty$
$(1 + 1/k)^{k}$	2	2.25	2.44	2.61	2.714	$\to$ 2.718...= e.

As k goes to ∞, $(1 + 1/k)^{k}$ goes to e (This is the definition of e.)
Graph $(1 + 1/x)^{x}$. As x increases, this increases to e (Problem A26)

(5.2.8) As k goes to ∞, $(1 + r/k)^{kt}$ goes to e^{rt}.

Calculator Exercise 2: Pick any time and any rate. Say, $t = 3$ and $r = .1$. 5.2.8 tells us that, if k is large enough, the compounding factor will be near $e^{.1(3)} = e^{.3}$. Try various values of k:

k	12	100	365	$\to$	∞
$(1 + r/k)^{kt} = (1 + .1/k)^{3k}$	1.34818	1.34966	1.34980	$\to$	$1.34986 = e^{rt}$.

This result justifies the use what is called "continuous" compounding. Note that the number resulting from compounding 365 times a year (1.34980) is very nearly the "limiting" result expressed using *e* (1.34986). Compounding more frequently than that will hardly make any difference. If money were compounded every second (3 million times a year; that's a large k!), the result would be indistinguishable from "continuous" compounding.

Continuous-Compounding Model: As before, let A be the future amount, P be the principal (the original amount), and r be the annual percentage rate (not to be confused with the "effective" rate), and t be time in years. Then

$$A = Pe^{rt}, \tag{5.2.9}$$

Example 12: \$5,000 is deposited in an account. It earns an annual percentage rate of 7.65% compounded continuously. How much will it be in 4 years?

Plug in. $A = 5{,}000\, e^{.0765(4)} = 6{,}789.91$ (dollars).

Example 13: How long will it take money to double if it is invested at an annual rate of 10% compounded continuously?

This problem asks us to solve: $2 = e^{0.1t}$ for t. By 5.1.4 (taking natural logarithms), $\ln 2 = .1t$, and $t = (\ln 2)/.1 = 6.93$ years.

Example 13, continued: What is the effective rate of money compounded continuously at a 10 percent annual percentage rate?

In one year, the multiplicative factor will be $e^{0.1} = 1.10517$, for an effective rate of 10.517 percent.

As illustrated by this example, effective rates are generally slightly greater than annual percentage rates.

Polynomial Approximations. A number like *e*, 2.718..., seems like such an unnatural base for a "natural" exponential function and "natural" logs. What could be natural about such a base? For one thing, it yields very simple polynomial approximations.

The image e^x can be closely approximated by "$1 + x$" for small x. No other base yields such a simple approximation.

Calculator Exercise 3: Compare "e^x" to "1 + x" for various small values of x (Figure 6).

x	0	.01	.1	.2	-.02	-.1
e^x	1	1.01005	1.10517	1.2214	.98020	.90484
$1 + x$	1	1.01	1.1	1.2	.98	.9
$e^x - (1 + x)$	0	.00005	.00517	.0214	.00020	.00484

There are better approximations which use higher-degree polynomials (see Section 4.1, Example 21).

Both e^x and $1 + x$ go through (0, 1). When $x = 0$, the slope of the e^x curve is 1, which is matched by the slope of the line "$y = 1 + x$," which line turns out to be the tangent line to the curve e^x (Figure 6). That is a fact from calculus, and it explains why $1 + x$ is so close to e^x for small x.

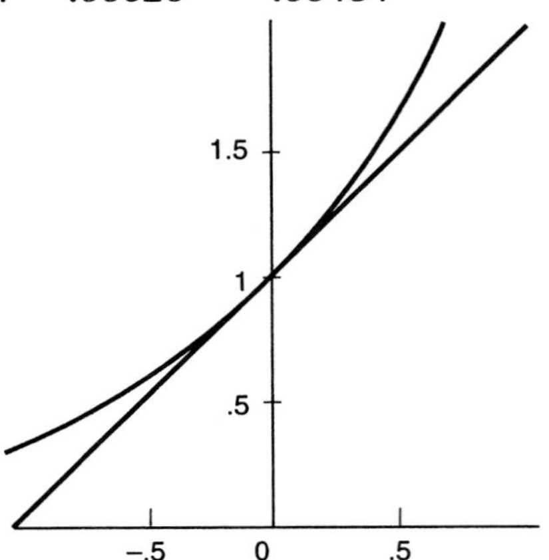

Figure 6: e^x and $1 + x$.
[-1, 1] by [0, 2].

Similarly, natural logarithms have a simple polynomial approximation when the argument is near 1: $\ln(x)$ is near $x - 1$, for x near 1. Graph them both and see. Try a few values yourself (problem B23).

The point is, polynomial approximations of b^x and $\log_b x$ are particularly simple when $b = e$.

<u>Exponential Model</u>: Using "P" (for population, or principal), "t" for time, and "k" for the growth rate parameter, the exponential model is:

(5.2.10) $$P(t) = P(0)e^{kt}.$$

This model relates four things: the initial amount, the growth rate, the elapsed time, and amount after the time has elapsed. Given any three of these, the fourth can be determined.

The growth factor is "e^{kt}," which is also "$(e^k)^t$." Because e^k can be any positive number including 2 or 1/2, this model includes the doubling-time and half-life models.

If $k > 0$, there will be growth (the population will increase), because e^k is greater than 1 when k is positive. Each unit of time the population will increase by a factor of $e^k > 1$. For example, if $k = .2$, $e^{.2} = 1.221$ and the population increases 22.1% each unit of time.

If $k < 0$, there will be decay (the population will decline), because e^k is less than 1 when k is negative. For example, if $k = -.05$, $e^{-.05} = .951$ and the population decreases 4.9% each unit of time.

Example 14: Suppose a bacterial population doubles every 30 minutes and there are initially 10,000 bacteria. Give a formula for the population growth. When will there be 60,000 bacteria?

This can be solved using base 2, as in Example 3. But here, to illustrate that base e is versatile, we do the problem using base e.

The model is: $P(t) = P(0)e^{kt}$,

The problem states that $P(0) = 10{,}000$. Furthermore, the doubling time tells us that when $t = 30$ minutes, $e^{kt} = 2$. Solving $e^{30k} = 2$ for k, by taking natural logarithms,

$$30k = \ln 2 \text{ and } k = (\ln 2)/30 = .0231.$$

Now the formula has been identified. It is

$$P(t) = 10{,}000e^{0.0231t}.$$

Now the second question, "When will there be 60,000 bacteria?" is easy to answer. Solve

$$60{,}000 = 10{,}000e^{0.0231t}.$$

$$6 = e^{0.0231t};\ \ln 6 = 0.0231t;\ t = (\ln 6)/(0.0231) = 77.6 \text{ (minutes)}.$$

Example 15: Suppose the amount of a substance undergoing radioactive decay has a half-life of 90 years. Give the formula for the amount at any given time in terms of the original amount, $P(0)$.

This problem is easy using base 1/2, but, here we show it can be done with base e. The model, in terms of base e, is: $P(t) = P(0)e^{kt}$. The only unknown is k, since we are letting the initial amount be general, $P(0)$. The half-life tells us that $P(90) = (1/2)P(0)$, that is, the multiplicative factor is 1/2 at time 90. This can be used to solve for k.

$$1/2 = e^{90k},\quad \ln(1/2) = 90k,\quad k = \ln(1/2)/90 = -.693/90 = -.0077.$$

k is negative because the amount is declining. The model is

$$P(t) = P(0)e^{-.0077t}.$$

Example 16: Suppose the population of a country is growing at a fixed rate of 3% per year. Model its population using the exponential model.

To "model" it means to give a mathematical formula for it.

The most straightforward model is: $P(t) = P(0)(1.03)^t$, where t is the number of years.

This can be interpreted as an exponential model, 5.2.10, where $e^k = 1.03$ because

$$P(0)e^{kt} = P(0)(e^k)^t = P(0)1.03^t,$$

which is the straightforward model. The required value of k is $\ln 1.03 = .0296$. So, $P(t) = P(0)\, e^{.0296t}$.

Examples 14, 15, and 16 show that the "exponential model" 5.2.10 is very general and models several different types of problems. Because of the advantages of base e in calculus, many mathematicians emphasize this general model over the other particular models (Problems B41 and B42).

<u>**Change-of-Base**</u>. Examples 14, 15, and 16 show that base e can be used to do exponential problems of types that we had previously done in either base 2, base 1/2, or some other base. All exponentials are much alike, and all logarithmic functions are much alike. In fact, it is simple to change back-and-forth between different bases. The base we actually use is not a matter of necessity -- it is merely a matter of convenience.

Note that the graph of 2^x looks just like the graph of 10^x and the graph of e^x, only with a horizontal scale change -- the graph of 2^x is wider (Figure 7).

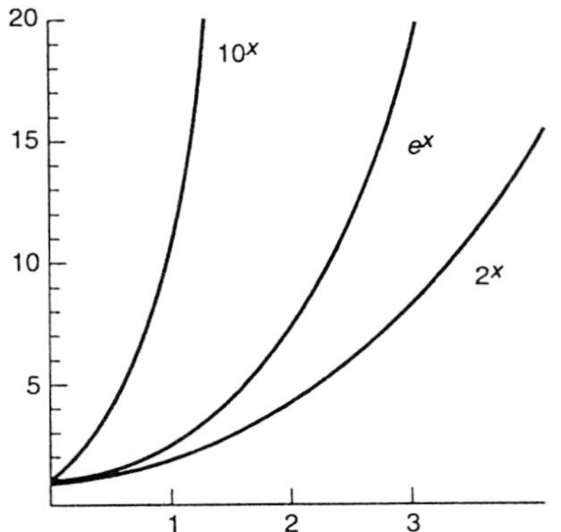

Figure 7: 10^x (left), e^x, and 2^x (right). [0, 4] by [0, 20].

Example 17: Compare the graph of e^x to the graph of 10^x.

Recall $e = 10^{\log e}$. Therefore,

$$e^x = (10^{\log e})^x = 10^{x \log e} = 10^{0.434x}$$
$$= 10^{x/2.3}.$$

Therefore, the graph of e^x is approximately 2.3 times as wide as the graph of 10^x (Figure 7).

Similarly, $b^x = (e^{\ln b})x = e^{x \ln b}$. Thus any exponential can be written as an exponential base e (Problems B26 and B27).

<u>Logarithms</u>. Your calculator will evaluate 2 to any power using the general power key. However, most calculators do not have a key for logarithms base 2. But logs base 2 are just a constant multiple of logs base 10 or base e, and both those logarithmic functions are on your calculator.

Example 18: Find $\log_2 x$ in terms of logs base 10.

$$2^{\log_2 x} = x.$$

$\log_2 x\,(\log 2) = \log x$ [by 5.1.3L1]

$\log_2 x = (\log x)/(\log 2)$.

By the way, $\log 2 = .301$, so

$\log_2 x = (\log x)/.301 = 3.32 \log x$.

The graph of $\log_2 x$ is about 3.32 times as high as the graph of $\log x$ (Figure 8).

This argument could be repeated using logarithms with base e ("ln") or any other base (Problem B29 gives the general result).

The method of Example 18 works to evaluate logs to unusual bases (bases for which your calculator does not have a key).

Calculator Exercise 5: Obtain the following logs base 2 using base 10.

$\log_2 256 = (\log 256)/(\log 2) = 8.$

$\log_2 100 = (\log 100)/(\log 2) = 6.644$

[approximately 3.32 times 2].

Using "ln" twice instead of "log" yields the same results.

Graph "$\log_2 x$" by graphing "$(\log x)/(\log 2)$." It will be the graph of "$\log x$" with a vertical scale change (Figure 8).

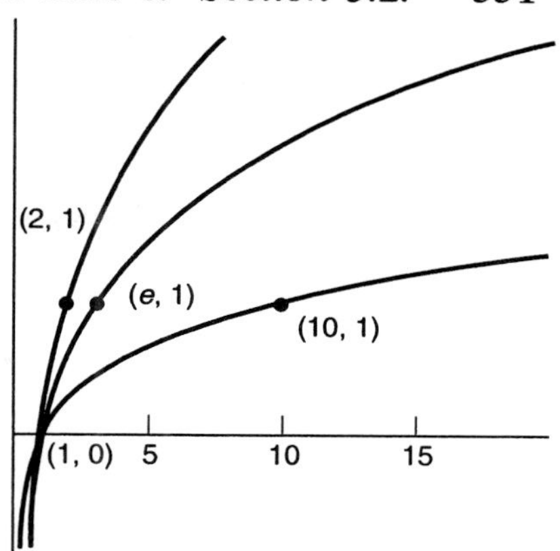

Figure 8: $\log_2 x$ (upper), $\ln x$ (middle), $\log x$ (lower). They are constant (vertical) multiples of each other. [0, 20] by [-1, 3].

To solve equations with logs to unusual bases, convert to exponential form using Definition 5.1.4

Example 19: Solve $\log_2 x = 2.7$.

The given problem is in logarithmic form. In exponential form it is:

$$x = 2^{2.7} = 6.498.$$

<u>**Conclusion**</u>: Exponential growth and decay are common real-world phenomena that are modeled using exponential functions. Although base 10 has substantial advantages, there are often good reasons to use base 2 and base *e*. Any exponential model with any base can be rewritten in terms of base *e*, which is the most important base in calculus. Logarithms undo exponentiation, so they are used for solving equations in which the unknown is in an exponent.

All exponential functions with base greater than 1 are much alike -- they differ only in their horizontal scale. Similarly, all logarithmic functions with base greater than 1 are much alike -- they differ only in vertical scale.

Terms: exponential growth, mathematical model, doubling time, half-life, compound interest, *e*, continuous compounding, exponential models.

****

Exercises for Section 5.2, "Base 2 and Base *e*":

A1. a)* Give the first 5 integer powers of 2.
b) Give round numbers close to 2^{10} and 2^{20}.

A2.* State the logarithmic fact corresponding to the given exponential fact.
a) $e^0 = 1$. b) the domain of e^x is all real numbers.
c) the range of e^x is all positive numbers. d) $e^a = c$.

A3.* State the exponential fact corresponding to the given logarithmic fact.
a) $\ln 1 = 0$. b) The range of $\ln x$ is all real numbers.
c) the domain of $\ln x$ is all positive numbers. d) $\ln c = a$.

A4.* In the exponential growth model, if the doubling time is d, what is the quadrupling time (time until there are four times as many)?

^^^^Solve algebraically:

A5. $2^x = 5$ [2.3].
A6. $1700 = 1000(2^{t/5})$ [3.8].
A7. $3 = 2^{40/d}$ [25].
A8. $(1/2)^{50/h} = .01$ [7.5].
A9. $20 = 100(1/2)^{t/30}$ [70].
A10. $1.3 = 2^{6/d}$ [16].
A11. $2 = e^{5k}$ [.14].
A12. $3600 = 2000(e^{t/20})$ [12].
A13. $1/2 = e^{-k}$ [.69].

A14. Using the doubling-time model, when will there be 10^6 bacteria if there are now 1000 bacteria and their doubling time is 35 minutes? [5.8 hours]

A15. Using the doubling-time model, if there were 10^4 bacteria four hours ago and there are now 5×10^6 bacteria, what is the doubling time? [.45 hours]

A16. Using the half-life model, if 1/15 of the original amount remains after 12 seconds, what is the half life? [3.1]

A17. Using the half-life model, if the half-life is 5730 years and .762 of the original amount remains, how much time has elapsed? [2200]

A18. Complete the identity: $\ln(1/x) =$ ___________.

^^^^Write in exponential form:

A19. $\log_8 x = 3.4$
A20. $\log_2 8 = 3$.
A21. $\log_2(1/4) = -2$.
A22. $\log_2 1024 = 10$.
A23. $\log_2(1/64) = -6$.

A24. True or false (for positive a and b)?
a) $\log(a/b) = (\log a)/(\log b)$. b) $\sqrt{(ab)} = (\sqrt{a})(\sqrt{b})$. c) $e^{a+b} = e^a + e^b$.

A25. True or false (for positive x and y)?
a) $(xy)^2 = x^2y^2$. b) $\log(x - y) = \log x - \log y$. c) $1/e^x = e^{-x}$.

A26. Graph $(1 + 1/x)^x$ and find out what happens as x goes to infinity.

^^^^^^^^^

B1.* Solve the following equation: $d = cb^a$ for the unknown variable, assuming the other values are known. If logarithms are required, use only logarithms base e or base 10.
a) a is unknown b) b is unknown c) c is unknown.

B2.* What type of problem can logs solve that arithmetic operations and taking powers cannot?

B3.* State the logarithmic fact corresponding to the given exponential fact.
a) $e^{a+b} = e^a e^b$. b) $e^{a-b} = (e^a)/(e^b)$.
c) $(e^a)^b = e^{ab}$. d) $e^{-b} = 1/(e^b)$.

B4.* State the exponential fact corresponding to the given logarithmic fact.
a) $\log(a^x) = x \log a$.
b) $\log a + \log b = \log(ab)$
c) $\log a - \log b = \log(a/b)$.
d) $\log(1/x) = -\log x$.

^ ^ ^ ^ Solve algebraically:

B5. $3(2^{t/20}) = 2^{t/19}$ [600].

B6. $.95(1.08)^t = 1.06^t$ [2.7].

B7. $1000(2^{t/30}) = 2^{t/29}$ [8700].

B8. $1.2(1.02)^{4t} = 1.008^{12t}$ [11].

B9.* $f(kx)$ has a graph which is $1/k$ times as wide as the graph of $f(x)$. Explain why (as if to a student who had not mastered Section 2.3 on composition).

B10. Population A is 40% larger than Population B, but Population B is growing at a 3% rate (compounded continuously), whereas Population A is growing at only a 1% rate (compounded continuously). When will the two populations will be equal? [17]

B11. There is now ten times as much of substance C as of substance D, but substance C is decaying with a half-life of 120 minutes and substance D is decaying with a half-life of 200 minutes. When will the two amounts be equal? [1000 minutes]

B12. Solve for b: $b^{x+1} = 5\,b^x$.

B13. In the exponential growth model, if the doubling time is d, what is the time until there are k times as many?

B14. In the half-life model, if the half life is h, how long until there are $1/k$ times as many?

B15. Suppose you are considering two alternative investments. One is to invest in a "cash equivalent" fund offering 8% compounded continuously. The other is to invest in collectable art, which you are willing to guess will appreciate at 12% per year compounded continuously. But there is a 20% commission to sell the art you buy. That is, you realize only 80% of its "value." Since 12% is greater than 8%, you figure art will be the better investment in the long run. How long do you have to hold on to the art before it begins to outperform the cash equivalent investment? [5.6]

B16. Bacteria of type A double every 30 minutes. Bacteria of type B double every 29 minutes (that is more frequently), but they begin with only 1 for every 1000 of type A. How long before they are 1000 times as common as type A? [Express your answer in the most comprehensible units: seconds, minutes, hours, days, or years.] [12]

B17. In 1993 Japan's economy was 70% the size of the U.S. economy. If Japan's economy grew at an exponential rate of 4% and the U.S. economy grew at 2%, in how many years would Japan's economy overtake the economy of the U.S.? [18]

B18. In the middle of the second century AD a modius of wheat (about 18½ quarts) cost 1/2 denarius. In 301 it cost 100 denarii. What was the average inflation rate over those 150 years? [3.6%]

B19. During the 30 years of the reign of the Roman emperor Constantine the Great (307-337 AD), the intrinsic value of the copper coin denomination dropped to 1/6 its original value. What was the average yearly rate of inflation over that period? [6.2%]

B20. As a linear (not constant) approximation, if x is near 0, then e^x is near ________.

B21. The graph of e^x is approximately ________ times as ________ as the graph of 10^x.

B22. The graph of $\ln x$ is approximately ________ times as ________ as the graph of $\log x$.

B23. a) Graph both "$\ln x$" and "$x - 1$" and compare them near (1, 0) by zooming in. What do you see? b) Graph $\ln x - (x - 1)$ near $x = 1$. What degree polynomial does it look like? c) Find the maximum of the absolute value of the difference between $\ln x$ and $x - 1$ on $1 \le x \le 1.2$.

B24. Simplify $\log(cx) - \log x$. What does this tell you about the graph of $\log x$?

B25. Solve for c: $\log(cx) - \log x = 1$.

B26. Write as an exponential function base e: $2^{t/d}$.

B27. Write as an exponential function base e: $(1/2)^{t/h}$.

B28. Interpret this in logarithmic form: $2^a = (10^{\log 2})^a = 10^{a \log 2}$. Which theorem results?

B29. Prove the general change-of base result for logarithms: $\log_c x = \dfrac{\log_b x}{\log_b c}$.

B30. Prove $\log_b c = 1/\log_c b$.

B31. a) Prove the change-of-base result for exponentials: $b^x = (c^{\log_c b})^x = c^{(\log_c b)x}$
b) Use it to explain how the graph of b^x compares to the graph of 10^x.

B32. In the exponential growth model, what is the relationship between doubling time and tripling time?

B33. Let $y = cb^x$. Take logs of both sides to obtain a linear relationship between $Y = \ln y$ and x. If $Y = mx + b$, identify m and b.

B34. Let $y = cx^p$. Take logs of both sides to obtain a linear relationship between $Y = \ln y$ and $X = \ln x$. If $Y = mX + b$, identify m and b.

B35. How many digits does 950^{950} have?

B36. Cari was thinking of investing her money at $p\%$ per year compounded continuously, but she calculated that if she could get a 1% higher rate, her money would double exactly 1/2 year sooner. What was p?

B37. $(\log a)/(\log b) = (\ln a)/(\ln b)$, for all positive a and b. Prove it.

B38. In banking there is a "Rule of 72" that approximates the doubling time of money invested at various rates. It says that if the interest rate is p percent per year, the doubling time is about $72/p$ years. For example, if the interest rate is 8%, the doubling time is about 9 years ($72/8 = 9$). a) Compute the doubling time in terms of p.
b) Compare the result of the "Rule of 72" to the actual doubling time for money compounded yearly at 4%.
c) Use part (a) to explain why the Rule of 72 works. [Use B23. Note that "72" is used because it is evenly divisible by so many likely interest rates: 1,2,3,4,6,8,9,12,15,18, and 24.]

B39. Let $y = x^2$, $Y = \ln y$ and $X = \ln x$. Now plot Y against X. What does the graph look like?

B40. Let $y = 2^x$ and $Y = \ln y$. Now plot Y against x. What does the graph look like?

B41. Example 14 is a particular example of how the exponential model 5.2.10 includes the doubling-time model 5.2.2. Prove that, if d is the doubling time, then $k = (\ln 2)/d$.

B42. Example 15 is a particular example of how the exponential model includes the half-life model 5.2.4. Prove that, if h is the half-life, then $k = (-\ln 2)/h$.

Section 5.3. Applications

In this section we introduce a new way to express and graph relationships between quantities. It is called "change-of-variable" or "change-of-scale." It is appropriate in many contexts where multiplicative factors are more relevant than additive terms. We will discuss what it means for multiplicative factors to be more relevant than additive terms, and we will give several real-world examples. Changing variables can be particularly valuable when data fits the exponential model. The usual sort of graph (the kind used for every previous graph in this book) is inappropriate and misleading in certain contexts, so we develop a new type of graph (on a logarithmic scale) to correctly illustrate the information.

Example 1: How can you report the power of an earthquake? Many alternatives have been proposed. In the United States the Richter scale is used. It is a method of reporting the energy released as indicated by the amplitude of the seismic waves produced. The amplitude of these waves can vary widely. A weak earthquake might have waves with amplitude one-tenth millimeter or less and a very strong earthquake might have waves with amplitude one million times as great. Rather than report the amplitude itself, the Richter scale reports a function of the amplitude. Let a be the amplitude of the earthquake (measured in microns, where 1000 microns equal one millimeter, on a standard seismograph located 100 kilometers from the epicenter). Define

$$R(a) = \log a. \tag{5.3.1}$$

Then report $R(a)$ rather than a. For example, a moderately weak earthquake with waves of amplitude 10 millimeters would be reported as a 4 on the Richter scale because 10 millimeters is 10,000 microns and $\log 10{,}000 = 4$. An earthquake 10 times as strong with waves of amplitude 100 millimeters would be reported as a 5. A very strong earthquake with waves of amplitude 10 meters (= 10,000 millimeters = 10,000,000 microns) would be reported as a 7 ($\log 10{,}000{,}000 = 7$).

An increase in amplitude by a factor of 10 corresponds to an additive increase of 1 on the Richter scale. An additive increase of 2 on the Richer scale corresponds to an increase in amplitude by a factor of 100.

Let b be the amplitude of a stronger earthquake and a be the amplitude of a weaker earthquake. How do their Richter-scale values differ?

Suppose they differ by d, that is,

let $d = R(b) - R(a)$ [d for difference]

$= \log b - \log a$ [by 5.3.1]

$= \log(b/a)$ [by Property 5.1.2L of logs].

Therefore,

$b/a = 10^d$ and $b = (10^d)a$.

The stronger has 10^d times the amplitude of the weaker. This proves the first half of the next theorem. The proof of the second half is problem B17.

Theorem 5.3.2: A difference of d on the Richter scale corresponds to the stronger having 10^d times the amplitude of the weaker.

A stronger earthquake having amplitude k times that of a weaker earthquake corresponds to a difference of $\log k$ on the Richter scale.

Example 1, continued: How much stronger is an earthquake that is .5 higher on the Richter scale?

By the first part of the theorem, the stronger is $10^{.5} = 3.1$ times as strong.

For example, a 6.7 earthquake is about 3.1 times as strong as a 6.2 earthquake.

More about Example 1: If a second earthquake is 20 times as strong as a first, how do their Richter scale values compare?

By the second part of the theorem, the stronger has a Richter scale value that is $\log 20 = 1.3$ higher.

For example, if the first was 3.2, the second was 4.5. If the first was 5.1, the second was 6.4.

Example 1, continued further: Suppose you were writing an article on earthquakes and you wished to illustrate, with a graph, the magnitudes of several famous earthquakes. How would you do it?

Plotting the amplitude of the seismic waves, a, will not work well. Some famous earthquakes are 100 times as strong as others, so plotting them with a on the vertical scale would make the smaller ones look puny. Plot R instead.

Example 2: The loudness of sound varies widely. In a quiet forest we can hear leaves rustling. At the other extreme, the loud sounds at a rock concert may be 10,000,000,000 (ten billion) times as loud as leaves rustling. To report the volume of sound we do not want a scale with 10 billion different levels of volume! The decibel scale was devised to take this wide range of values and convert it into a manageable range. The idea is to avoid reporting the actual volume in terms of the energy of sound, but to report a logarithmic function of the volume. Let v be the volume (energy of the sound) and $D(v)$ be the corresponding number of decibels.

(5.3.3) $$D(v) = 10 \log(v/v_0),$$

where v_0 is chosen to be the volume of the faintest sound that can be heard by healthy young people.

Table of Sound Volumes

cause	multiple of the faintest volume	decibel level
faintest sound	1	0
leaves rustling	30	15
whisper	1000	30
normal conversation	1,000,000	60
vacuum cleaner	10,000,000	70
jackhammer	10,000,000,000	100
painful sound	≥ 1,000,000,000,000	≥ 120

It is impossible to graph all the numbers in the "multiple" column on the same uniform scale. If you want to graph these, use the numbers in the decibel column instead. That switch from using a quantity (volume = energy) to a function of it (the log of the volume) instead is called a "change of variable." The new numbers are on a new scale, so this produces a "change of scale" to a logarithmic scale.

Let w be the volume of a louder sound and v be the volume of a quieter sound. How do they differ on the decibel scale?

$$D(w) - D(v) = 10 \log(w/v_0) - 10 \log(v/v_0) \quad \text{[by 5.3.3]}$$
$$= 10[\log(w/v_0) - \log(v/v_0)] \quad \text{[DistributiveProperty]}$$

$$= 10 \log\left(\frac{\left(\frac{w}{v_0}\right)}{\left(\frac{v}{v_0}\right)}\right) = 10 \log\left(\frac{w}{v}\right).$$

Therefore, the difference on the decibel scale is 10 times the log of the ratio of their volumes. This proves the first half of the next theorem. The second half is Problem B18.

Theorem 5.3.4: If a louder sound is k times as loud as a quieter sound, they differ on the decibel scale by $10 \log k$.

If two sounds differ on the decibel scale by d, the louder is $10^{d/10}$ times as loud as the quieter.

Example 2, continued: When the volume dial is adjusted, my car radio displays the volume in decibels. The smallest increase it records is 2 decibels. What change in volume does that correspond to?

According to the second part of the theorem, the change is by a factor of $10^{2/10} = 1.58$. The smallest increase it records is by a factor of 1.58, a 58% increase. This type of increase in volume is noticeable, but rather minimal and certainly not dramatic.

More about Example 2: If one sound is 500 times as loud as a second sound, how do their decibel levels compare?

By the first part of the theorem, their decibel levels differ by 10 log 500 = 27 decibels.

In chemistry, the measure of acidity called "pH" is on a similar logarithmic scale (problems B11-13.)

When Changes are Multiplicative. In many applications it is appropriate to record changes by giving the factor that produces the change, rather than by giving the difference of the values.

Example 3: Suppose Mr. Black invests some money and it earns $10,000 in one year, and Mrs. Smith invests some money and it earns $500 in that same year. Who did better?

What is the standard of comparison for "better"? Usually it is not the amount earned, rather the percent earned. If Mr. Black invested $1,000,000 and earned only $10,000 on it, his return is a mere 1 percent. If Mrs. Smith invested $1000 and earned $500 on it, her return is an amazing 50 percent. It is more useful to know the percentage return (which is multiplicative) than the actual return (which is additive).

Addition is not the right way to express the return. The proper way to express returns on money is by giving the *multiplicative* factor.

Reporting Stock Market Gains. Invested money tends to grow over time. How can the value be graphed to exhibit useful information about the growth rate of the money over the years?

As Example 3 shows, we are interested in percentage growth. Invested money will compound -- multiply -- not just add. Its growth should resemble exponential growth, which is hard to interpret on the usual type of scale because even a *constant* growth rate produces a rapidly *changing* graph (Figures 5.2.1 and 2). It is very difficult to detect changes in growth rates if information is plotted on the usual type of scale.

We want a graph to show the *factor* by which the money grows, not the *amount* by which it grows. In the context of money, a change from $10,000 to $20,000 should be equivalent to a change from $100,000 to $200,000, because in both cases the money doubled. The usual type of graph will show the second change 10 times as large. A change-of-scale to a logarithmic scale can fix that.

Example 4: Consider a new type of graph. Let the horizontal scale be time (say, in years) labeled as usual. But label the vertical axis differently. Mark off equidistant parallel lines above the horizontal axis. Label the first line above the horizontal axis "$10,000," the second "$20,000," and the third "$40,000" (not "$30,000"). That is, label each line double the preceding line (Figure 1). The fourth is $80,000, the fifth $160,000, etc. Then a change of one vertical unit means the money doubled, regardless of the amount. This makes it fair to compare the growth in 1980 to the growth in 1990.

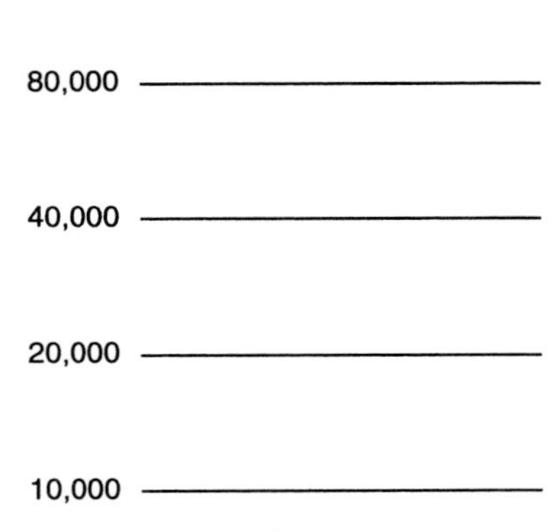

Figure 1: A logarithmic vertical scale.

The method of generating the points is as follows. Let t be the year (1954 to 1990). The Templeton Fund opened in late 1954. Imagine an initial investment of $10,000. Due to startup fees (its so-called "load") it had barely regained its initial value by the end of 1955, but then it began to take off. Let $v(t)$ be the value at the end of the year t. We want to compare $v(t)$ to the initial $10,000 cost. Use $\log_2[v(t)/10{,}000]$.

t	value, $v(t)$	$\log_2[v(t)/10{,}000] = L(t)$.
1954	10,000	0
1955	10,000	0
1960	18,000	.85
1965	25,000	1.32
1970	46,000	2.20
1975	102,000	3.35
1980	345,000	5.1
1985	660,000	6.05
1990	1,531,000	7.26

Let the first line above the horizontal axis correspond to "0" in the $L(t)$ column. Now here is an interesting idea: Plot L but label it "v" anyway. We are still interested in the value of the money, not the log (Figure 2).

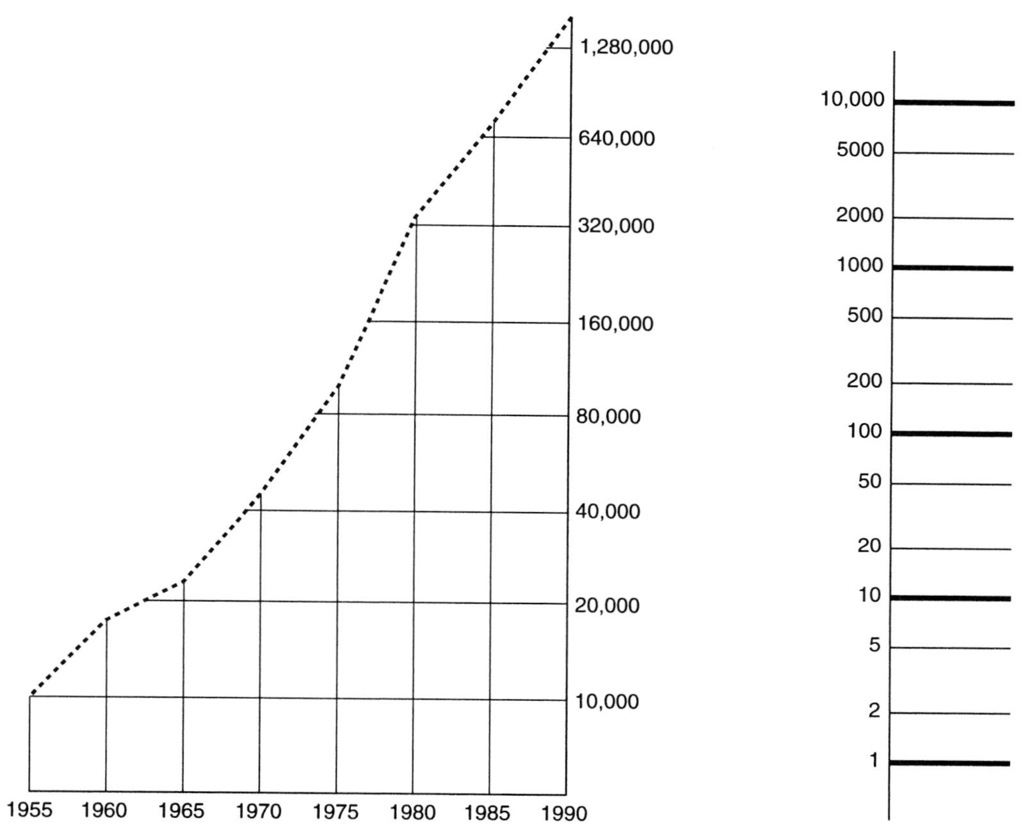

Figure 2: Templeton Fund Growth.

Figure 3: A logarithmic vertical scale, base 10.

Figure 2 does not use a "uniform" scale, it use a "logarithmic" scale. It is a "semi-log" graph. "Semi" means "half." The horizontal scale is uniform, but the vertical half is logarithmic.

Money graphs tend to emphasize growth by a factor of 2. In many other applications growth by a factor of 10 is emphasized. You can buy graph paper lined and ready to go for such growth. Figure 3 illustrates "semi-log" graph paper on which the bold horizontal lines differ by a factor of 10.

Exponential Model. Money compounded continuously at a constant rate of growth fits the Exponential Model of growth. Of course, the growth rate of money does not remain constant over 35 years; rates go up and down. But, for comparison, the Exponential Model is appropriate. Let's see how a constant growth rate would look on a logarithmic scale such as in Figures 1, 2, and 3.

The exponential model can be converted to a linear model by a change of variable. Here is the exponential model 5.2.10 again:

$$P(t) = P_0\, e^{kt}. \tag{5.3.5}$$

The parameters are k and P_0. The variables are t and $P(t)$. Watch what happens if we use $\ln(P(t))$ as the second variable. Taking the natural logarithm,

$$\begin{aligned} \ln[P(t)] &= \ln(P_0\, e^{kt}) \\ &= \ln(P_0) + \ln(e^{kt}) && \text{[log of a product, 5.1.1L]} \\ &= \ln(P_0) + kt && \text{[inverses, 5.1.4L].} \end{aligned}$$

Now call $\ln[P(t)] = y(t)$ and $\ln(P_0) = y_0$. Then

$$y(t) = y_0 + kt. \tag{5.3.6}$$

This is the Slope-Intercept form of a line with variable t (3.1.7).

This tells us that graphing $\ln[P(t)]$ instead of $P(t)$ itself yields a line (which has *constant* slope instead of the ever-increasing slope of an exponential growth curve.

If the Templeton Fund growth rate had been constant, the graph in Figure 2 would be a straight line. As you can see, in the long run it appears to tend upward like a straight line, but there are many irregularities over the short term. That's the way it is with money.

Example 5: Figure 4 illustrates the growth of a bacterial colony on a logarithmic scale. Time is given in hours. The horizontal lines are one logarithmic unit (base 2) apart, so they represent a doubling in numbers of bacteria. What is the doubling time?

The graph is a straight line, which means the exponential growth is at a constant rate. The doubling time is visible on the graph because it is the time between crossing one horizontal line and the next. It appears to be about 1/2 hour. The semi-log graph makes it easy.

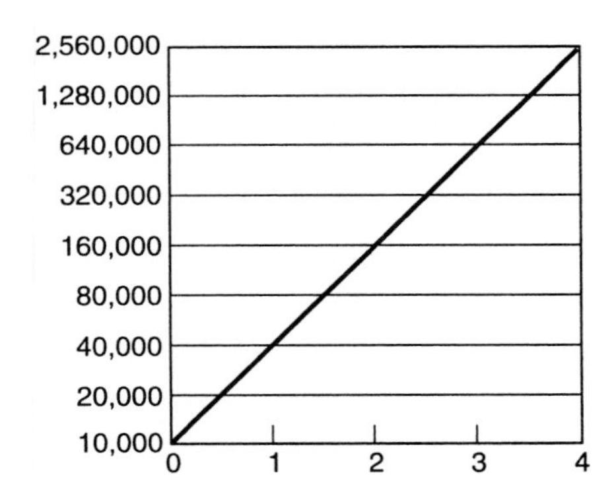

Figure 4: Bacterial growth on a semi-log scale.

<u>**Putting an Exponential Curve Through Two Points**</u>. Many mathematical models use the exponential function. The homework problems have a few, and we will explore one here to illustrate the remarkable way that time can

be shifted in exponential models.

The exponential model 5.2.10 and 5.3.5 is repeated here:

The Exponential Model: $P(t) = P_0 e^{kt}$.

This is a family of models with two parameters, P_0 (the initial population), and k (the growth rate). Usually with two facts we can determine two parameters.

Example 6: After 2 hours a population undergoing exponential growth is 100,000 and after 3.7 hours it is 980,000. Find the growth rate and the initial number.

The question asks us to find the specific model from the family of exponential models. If we knew P_0 and k we would know the exact formula. It gives two facts -- two points we want the curve to go through.

Plug in the given information:

$$P(2) = 100{,}000 = P_0 e^{2k}.$$
$$P(3.7) = 980{,}000 = P_0 e^{3.7k}.$$

This is a system of two equations with two unknowns. Since both expressions on the right have P_0 as a factor, we can eliminate P_0 by dividing.

$$\frac{980{,}000}{100{,}000} = \frac{P_0 e^{3.7k}}{P_0 e^{2k}} = \frac{e^{3.7k}}{e^{2k}}$$
$$= e^{3.7k - 2k} = e^{1.7k}.$$

So, $9.8 = e^{1.7k}$. Taking the natural logarithm of both sides,

$$1.7k = \ln 9.8. \quad k = (\ln 9.8)/1.7 = 1.343.$$

Now, plug this back into either original equation to find P_0.

$$100{,}000 = P_0 e^{2(1.343)}. \quad P_0 = 100{,}000/e^{2(1.343)} = 6821.$$

So the specific model is

$$P(t) = 6281e^{1.343t}.$$

This formula can be used to answer any question about what the population will be at any time (as long as exponential growth continues at the same rate).

Linear models ($y = mx + b$) are so important that you study how to put a line through two points. Of course, the slope turns out to be a critical concept. Exponential models are important too. Example 6 puts an exponential curve through two points. Next, we analyze that example to see how, in general, to put an exponential curve through two points. The growth rate turns out to be a critical concept.

The equation for the growth rate k (obtained by dividing the original two equations) ended up using the *difference* in times (3.7 - 2 = 1.7) and the *factor* by which the population grew (980,000/100,000 = 9.8). The initial population canceled out. This always happens. We deduce that time zero plays no special role in these problems. We can slide the time scale back and forth without changing anything essential. Example 6 mentioned a population of 100,000 "after 2 hours." Naturally we called that time 2, that is $t = 2$. But, in exponential model problems, we can choose time zero wherever we like. We could let $t = 0$ refer to the time when there were 100,000, in which case two hours earlier would have been time -2.

Example 6, revisited: Choose the time scale such that time 0 is when there was 100,000. According to the example, there were 980,000 "after 3.7 hours," which would be at $t = 1.7$. Then the question asks about the number two hours before there were 100,000. That would be at time $t = -2$. Here is the model:

$$P(t) = P_0 e^{kt}.$$

We know P_0 because we choose time 0 to correspond to 100,000. So the model is

$$P(t) = 100{,}000\, e^{kt},$$

where only k remains to be discovered. Use the fact that $P(1.7) = 980{,}000$.

$$980{,}000 = 100{,}000\, e^{1.7k}.$$

$$9.8 = e^{1.7k}. \quad \ln 9.8 = 1.7k. \quad k = (\ln 9.8)/1.7 = 1.343,$$

as before.

Now, to find the "initial" number, plug in $t = -2$:

$$P(-2) = 100{,}000\, e^{1.343(-2)} = 6821,$$

as before.

This second approach to putting an exponential curve through two points is particularly simple because, by locating $t = 0$ at one of the points, P_0 is given. Then the other point is used to create one equation for the growth rate, k. The only trick is to remember where the horizontal scale begins so that times mentioned as, say, "3.7 hours later," are put in the proper position relative to time $t = 0$.

If you do not want to adjust time zero, you can always use the method of Example 6. Here it is in abstract form.

Example 7: Fit the exponential model through the two points (s, u) and (t, v).

The two equations are

$$u = P_0 e^{ks} \text{ and } v = P_0 e^{kt}.$$

Dividing,

$$v/u = e^{kt}/e^{ks} = e^{k(t - s)}.$$

Solving for k by takings natural logs,

$$\ln(v/u) = k(t - s). \qquad k = \ln(v/u)/(t - s).$$

Now use either of the original two equations to solve for P_0.

Conclusion. When we are young, we study addition before we study multiplication. In algebra, we study additive models before we study multiplicative models. This is because we need exponentials and logarithms to master multiplicative models. In exponential models multiplication by a factor is more relevant than adding an amount.

To graph rapid growth it may be useful to change variables or change scales. The exponential model can be converted into a linear model with a change of variable. The growth rate constant is the slope of the line.

There is a straightforward method of putting an exponential curve through two points.

Terms: exponential model, change of variable, change of scale.

****

Exercises for Section 5.3, "Applications":

A1. On the Richter scale, a difference of 1 corresponds to what difference in energy of the earthquake?

A2. If an earthquake is 100 times as strong as a 4.3 earthquake, what does it register on the Richter scale?

A3. If an earthquake is twice as strong as a 4.3 earthquake, what does it register on the Richter scale? [4.6]

A4. How much stronger is a 5.2 than a 4.6 earthquake?

A5. How much stronger is a 7.8 earthquake than a 5.2 earthquake? [400 times]

A6. If one noise is 10 times as loud as another, how do they differ on the decibel scale?

A7. How much louder is an 80 decibel noise than a 60 decibel noise?

A8. How much louder is a 90 decibel noise than a 55 decibel noise? [3200 times]

A9. If a noise is twice as loud as a 50 decibel noise, how many decibels is it? [53]

A10. If a noise is 1000 times as loud as a 50 decibel noise, how many decibels is it?

^ ^ ^ ^ ^ ^ ^ ^ ^

B1. a) Logarithms convert multiplication to ________________.
b) It may be appropriate to graph data on a logarithmic vertical scale. Why?

B2. The amount of light that penetrates water is related to the clarity of the water and the distance the light must penetrate. The model is $I = I_0 e^{-kd}$, where d is the distance the light penetrates the water, I is the intensity of light, and k is a parameter that describes how clear the water is.
a) If half the light reaches a depth of 10 feet, find k. [.069]
b) How deep will it be where only 1% of the light reaches? [66]

B3. (See B2) If 1/3 the light reaches a depth of 20 feet, how much light will reach a depth of 90 feet? [.0071]

B4. (See B2) a) If $I_0 = 20$ and $I = 5$ when $d = 10$, find k. [.14]
b) Find I when $d = 15$. [2.5]

B5. (See B2.) If $I_0 = 100$ and $k = .4$, find d such that $I = 60$.

B6. One type of sheet glass lets through 96% of the light that hits it. Give a mathematical model for the amount of light passing through any number of sheets of this glass.

B7. On Jan. 1, 1980 Sally invested some money which was compounded continuously at a constant rate until Jan. 1, 1990. On that terminal date it was $77,217.82. On Jan. 1, 1986 it was $48,746.34. How much did she invest originally and at what rate? [$24,000 ...]

B8. Newton's Law of Cooling yields a model for changes in temperature of small objects moved to cooler or warmer surroundings.
Newton's Law of Cooling Model: $T(t) = T_m + (T_0 - T_m)e^{-kt}$,
where t is elapsed time, $T(t)$ is the temperature of the small object at time t, T_0 is its initial temperature, and T_m is the temperature of the surrounding medium. The rate of cooling is described by k. a) How many parameters does the model have?
b) How many pieces of information will be needed to specify the exact model?

B9. (See B8 for the model) Suppose a small metal ball at 70° is put in boiling water at 212°. Suppose further that $k = 3$ when time is measured in minutes. Find the elapsed time until the ball reaches 150°. [.28]

B10. (See B8 for the model) A can of soda at 70° is immersed in ice water at 32°. In two minutes it is 55°. Find the formula for its temperature at any time.

B11. (pH in chemistry.) Acidity is measured in terms of pH, which is defined as the negative of the common logarithm of the molar hydronium-ion concentration. That is, pH = $-\log[H_3O^+]$. Pure water has pH = 7, which is considered neutral, neither acidic nor basic. Acids have lower pH numbers.
a) Suppose orange juice has $[H_3O^+] = 2.80 \times 10^{-4}$. Give its pH.
b) Suppose milk has pH 6.4. What is its ion-concentration?

B12. [See B11] "Normal" rain has a pH of 5.6. If an acid rain has pH 2.6, how much more acidic than normal is it?

B13. [See B11] If a rain is 50 times as acidic as normal rain, what pH is it?

B14. Star Magnitude. In ancient times people classified stars by how bright they appeared. A star of the first magnitude was distinctly brighter than a star of the second magnitude, etc. Modern astronomers put that intuitive scale on a mathematical basis. They decided that a difference of 5 magnitudes would be equivalent to a 100-fold difference in brightness. What does a difference of 1.00 magnitude correspond to?

B15. Solve for m: $m^2e^{-m}/2 = .1$. [.61]

B16. The chance of making n baskets in a row is approximately p^n where p is the probability (between 0 and 1) of making one basket. Find the highest value of n such that the chance is still above 1/2.

B17. Prove the second half of Theorem 5.3.2.

B18. Prove the second half of Theorem 5.3.4.

B19. How does the model in B2 compare to the model in B6? What is the relationship?

B20. In exponential-model problems we can let time zero be wherever we like. State and prove some mathematical result that, at least partly, explains why that is so.

Section 5.4. More Applications

Power functions such as x^2 have curved graphs. However, the logarithm of a power function has a graph that is a straight line. Because we know so much about lines, we sometimes choose to make this change-of-variable.

Power Model. In the exponential model, graphing on a vertical logarithmic scale may be useful (as in Examples 5 and 6 from the previous section). Such graphs have a uniform horizontal scale as usual and the vertical scale is changed. However, for some models it is sometimes useful to change both scales. The power model is one. If two variables are related by a power function, graphing on a "log-log" scale may be useful.

Let the variables be x and $y(x)$.

(5.4.1) The Power Model: $y(x) = c\, x^p$, for some p.

Example 1: One example of a power model we have seen is Kepler's Third Law (Example 4.3.13). The time required for a planet to orbit the sun, y, depends upon the mean distance of that planet from the sun, x (these were called by different letters in Example 4.3.13, but that does not matter). Kepler's law is

(5.4.2) $$y = c\, x^{3/2}, \text{ for some } c.$$

When he started, Kepler did not know the right mathematical description of the relationship between x and y. He did not know a power model was right, and he certainly did not know the right power was 3/2. It took him years to figure out the relationship 5.4.2 from the data he had. Here is how we might do it nowadays with good data and statistical methods that use the upcoming "log-log" idea. The "time" and "distance" columns in the following table give the primary data (Figure 1).

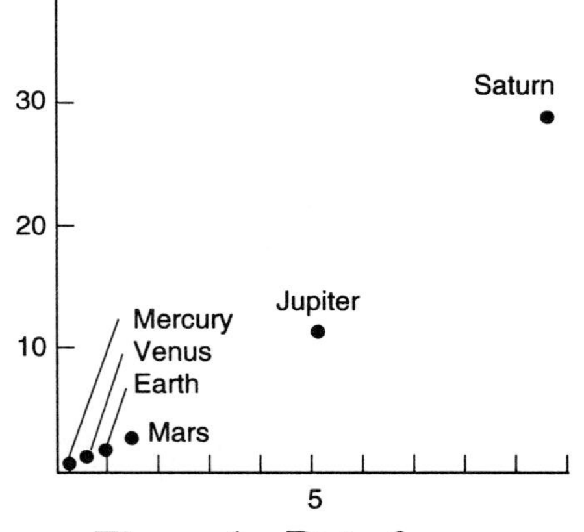

Figure 1: Data for Kepler's Third Law. distance [0, 10] by time [0, 40].

Data for Kepler's Law

planet	distance (astronomical units)	time	log(distance)	log(time)
	x	y	$\log x$	$\log y$
Mercury	.387	.241	-.412	-.618
Venus	.723	.615	-.141	-.211
Earth	1	1	0	0
Mars	1.524	1.881	.183	.274
Jupiter	5.202	11.862	.716	1.074
Saturn	9.359	29.457	.980	1.469

Distance is measured in "astronomical units" in which the distance of the Earth to the Sun is 1 unit. Plotting that data gives a curve that looks like some power of x, but it is difficult to identify the precise power (Figure 1). Also, too many planets are crammed in the bottom left corner near the origin. The usual uniform scale is awkward.

The table also includes two more columns giving the logs of the distance and time. These are the "change-of-variable" columns.

To see what the two "log" columns can do, take logs of both sides of the Power Model:

$$y = c\,x^{p} \qquad [5.4.1]$$
$$\log y = \log(c\,x^{p}) \qquad \text{[taking logs]}$$
$$= \log c + \log(x^{p}) \qquad [5.1.1L]$$
$$= \log c + p \log x. \qquad [5.1.3L]$$

Now change variables. Let $\log y = Y$ and $\log x = X$. This is the change of variable. Also, $\log c$ is just a constant, so call it b. Now relate (capital) Y to (capital) X. So, with this change of variable the power model is equivalent to

(5.4.3) $$Y = pX + b.$$

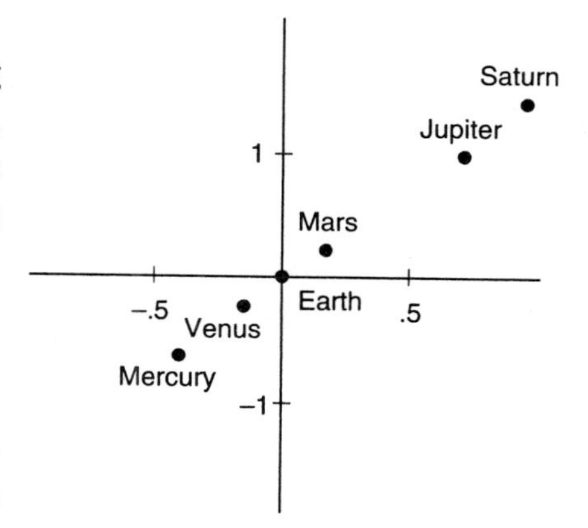

Figure 2: The planetary data on a log-log scale.

You have seen this before. This is the Slope-Intercept form of a line, with "p" denoting the slope. So, plot the X's and Y's (instead of the x's and y's) and a line will fit through the points (if a power model is correct). Figure 2 plots the data about the planets as (X, Y) pairs. The slope is the desired value of p (problem A1). If the log-log data fit a straight line, then the relationship of the original data is given by the Power Model 5.4.1.

Special graph paper called "log-log" graph paper allows us to plot the original values on logarithmic scales directly. Both scales are labeled like the logarithmic vertical scale in Figure 3.

Original Applications. The original application of logarithms was to facilitate multiplication. The values of logs and exponentials were looked up in tables. Now we have calculators which are far more convenient. Logarithms convert multiplication to addition, using properties 5.1.4 and 5.1.1. Here is the theory:

$$a \times b = 10^{\log a} \times 10^{\log b} = 10^{\log a + \log b}.$$

This is an identity and, like other identities, it expresses an alternative procedure. Therefore, to multiply a times b (expressed on the left of the identity) we may instead do the procedure expressed on the right of the identity:

1) Find log a
2) Find log b
3) *Add* log a to log b to obtain the sum. Call it c.
4) Find 10^c, the answer, which is the product ab.

Line 3 exhibits the step where multiplication is converted to addition.

In the old days, the log and exponential functions would have been evaluated with a huge table of values.

Does this seem long? Actually, with practice, people in the pre-calculator age found it a relatively fast way to multiply many-digit numbers by each other. These properties are the principle on which the old engineer's "slide rule" was based (Problem B5). Now the principles are still interesting, but slide rules are obsolete and multiplication by using logarithms is positively medieval. Nevertheless, try the next example. with your calculator replacing a table of logarithms.

Calculator Exercise 1: Find 56×78 using the above approach, with your calculator replacing a table of logarithms.

Find log 56. Find log 78. Add them. Find 10 to that power.

Now check your answer by multiplying the usual way.

Logarithms can be used to evaluate powers. Here is an identity which expresses the theory:

$$b^p = (10^{\log b})^p = 10^{p \log b}.$$

As with any identity, the right side expresses an alternative procedure to evaluate the left side. To find b^p for any $b > 0$ and any p,

1) Find log b
2) *Multiply* it by p, to obtain their product. Call it c.
3) Find 10^c, the answer, which is the power b^p.

Prior to the invention of computers this was the only practical way to evaluate non-integer powers.

Calculator Exercise 2: Find $2^{4.6}$ using this approach, with your calculator replacing a table of logarithms.

Find log 2. Multiply it by 4.6. Find 10 to that power. Now check your answer against the usual direct way to evaluate $2^{4.6}$.

Of course, we do not need logs to multiply, but your calculator does evaluate general powers with a method like the one just expressed (although usually using natural logarithms with base e). These exercises illustrate the remarkable properties of exponential functions and logarithms (Problem B4).

<u>Conclusion</u>. Taking logs converts a power model into a straight-line model. This can be seen by graphing the logs of the data (or using a log-log scale). The slope of the resulting line is the power.

Terms: power model, change of variable, change of scale.

****

Exercises for Section 5.4, "More Applications":

A1. Use the log data from Mercury and Saturn to compute the slope of the line in Figure 6. Is your answer near 1.5?

A2. Use the logarithmic method of Calculator Exercise 1 to multiply 42×36. Show the results of the four steps.

^ ^ ^ ^ ^ ^ ^ ^

B1. The following data come very close to fitting a power model. Graph the data on a log-log scale. Then find the model.

$f(1.1) = 0.8, f(1.6) = 2.04, f(2.1) = 4.0, f(3.3) = 12.46.$

B2. The following data come very close to fitting a power model. Graph the data on a log-log scale. Then find the model.

$f(0.4) = 0.77, f(1.1) = 2.58, f(1.8) = 4.66, f(2.4) = 6.58.$

B3. Kepler's Third Law gives the periods of planets as a function of their distances from the sun. Assuming circular orbits (not correct, but not far wrong), derive speed in orbit as a function of distance from the sun.

B4. State an identity in the spirit of the ones used in Calculator Exercises 1 and 2 that expresses a way to divide using logarithms.

B5. Suppose two meter sticks are marked with "2" at position log 2, that is, with "2" at 30.1 centimeters from the left end. Similarly, suppose they are labeled "3" at log 3, "4" at log 4, etc. Now, slide one along the other until its left end is above position "2" of the other (see the Figure).

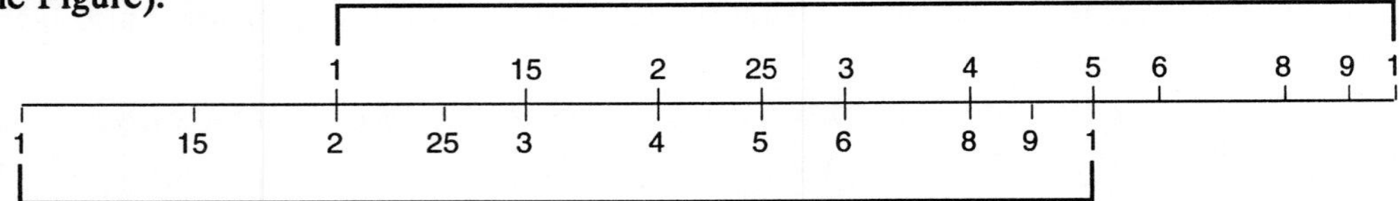

Two similar rules, the upper one slid to the right.

Explain how this "slide rule" can facilitate multiplication by any number. Which property is the key? How can you multiply 25×30 with it?

CHAPTER 6

Trigonometry

Section 6.1. Geometry for Trigonometry

Trigonometry is the branch of mathematics which concerns the relationships of the sides and the angles of triangles. Geometry is also concerned with triangles, so trigonometry and geometry are closely related. There are, however, other applications of trigonometry to circular motion and other types of periodic motion such as waves (including sound and light waves) and the swing of a pendulum.

Motion produces change, and calculus is the mathematical subject which studies change. So the aspect of trigonometry which regards trigonometric functions as "circular" functions in the context of motion is important in calculus. The functions of triangle trigonometry are also used as circular functions; it is really the context in which the functions are applied that distinguishes triangle trig from circular trig. As is traditional, we will begin with triangle trigonometry.

To solve a triangle in trigonometry means to find (the measures of) all the sides and angles given some of them. Given three parts in a particular arrangement, geometry tells us how many solutions to expect when we solve for the missing parts numerically using trigonometry. The upcoming trigonometric "Law of Sines" and "Law of Cosines" (6.3.3 and 6.3.4) both yield unique answers when the geometric case determines a unique triangle, and both yield *two* answers when the geometric case yields *two* triangles (Angle-Side-Side is the trickiest case). Therefore, understanding the various geometric cases will help you understand the Law of Sines and the Law of Cosines.

Solving a Triangle. In trigonometry a triangle is regarded as having six parts -- three sides and three angles. From now on we are more interested in the measure of the angles and sides than in their location, so the word "side" will refer to a length, and "angle" to a measure (at first, in "degrees").

In our notation (Figure 1), side a is opposite vertex (VER tex) A (the plural of *vertex* is *vertices* ("VER teh SEES", not "ver TEX is"). Angle A is at vertex A (using the same letter for two different things, a point and an angle). Angle A is formed between sides b and c. Side a is between angles B and C. Sometimes an angle is denoted by a lowercase Greek letter such as α (alpha), β (beta), γ (gamma), or θ (theta) placed inside the angle (such as α in Figure 1, which is also angle A and angle BAC).

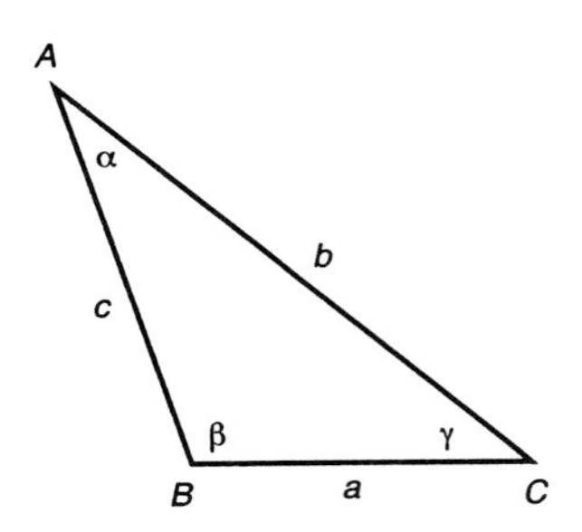

Figure 1: A triangle, with typical labeling.

Geometry for Trigonometry. This section reviews geometric constructions that you learned in geometry to remind you **how** and **why** a triangle is or is not uniquely determined by three of its parts. Most of all, it is intended to **warn you when not to expect a unique answer** when using trigonometry to solve for unknown parts of a triangle.

(6.1.1) Side-Side-Side (SSS): A triangle is determined by its three sides. That is, its three angles are determined by the three sides, so both its size and shape are determined (but its location is not).

The geometric construction is illuminating. Begin with three sides (Figure 2). In geometry, "sides" are line segments of the desired lengths, but not necessarily yet in the desired locations. In trigonometry, "sides" are the measures of the sides, that is, their lengths.

Figure 2: The parts for a SSS construction.

Now construct a triangle with those sides by the following steps.

1) Lay out a line anywhere (often horizontal) and reproduce length a on it. Label the endpoints B and C (Figure 3).

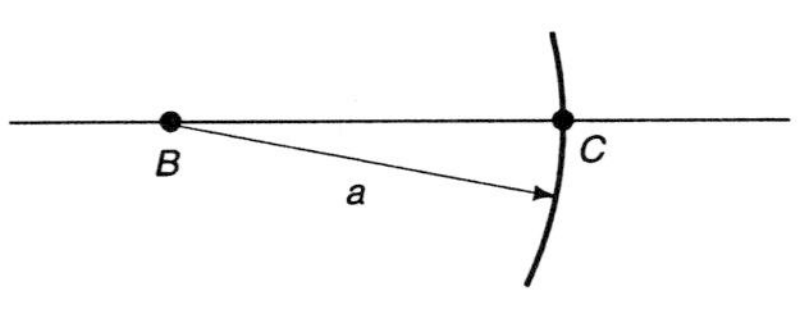

Figure 3: Step 1 of SSS. A line segment BC of length a.

2) From B draw a circle (part of it, an arc, will do) of radius c (Figure 4).

3) From C draw a circle (an arc will do) of radius b.

4) Locate point A at the intersection of the two circles. (There will be two candidates for A, labeled "A" and "A' " which is pronounced "A prime," Figure 4.)

5) Connect A to B and C with the straightedge, thus forming the sides of the triangle (Figure 5).

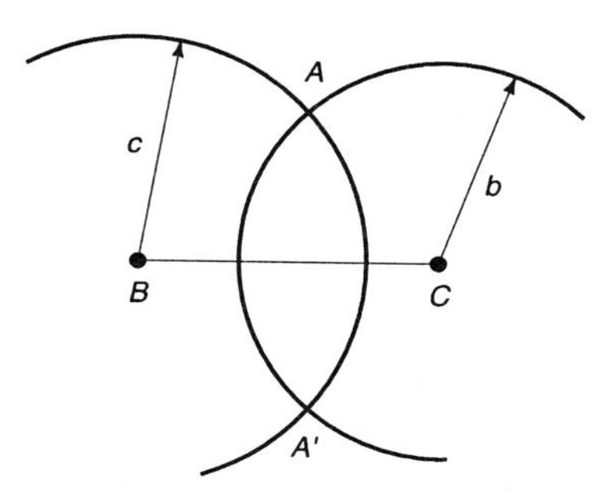

Figure 4: Further into the SSS construction. The vertices are located. (Steps 2, 3, and 4)

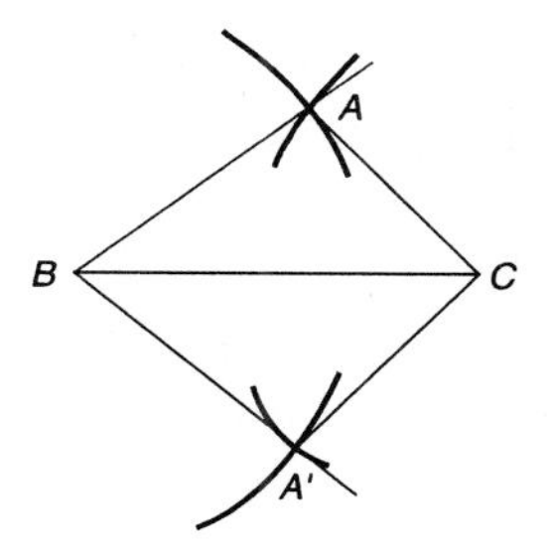

Figure 5: The result of the SSS construction. Triangles ABC and $A'BC$.

In Figure 5, the angles in triangle ABC have the same measures as the angles in triangle $A'BC$. So in trigonometry we do not care which one you construct since corresponding parts have equal measures in all six cases.

<u>(6.1.2) Side-Angle-Side (SAS)</u>. Two sides and the included angle determine a triangle. By "included" angle we mean the angle formed by the two sides. This case is called Side-Angle-Side (SAS) because the angle is between the sides.

Begin with sides a and b and the angle, C, which will be between them (Figure 6). Note the angle cannot be labeled "A" or "B"; angles and *opposite* sides are labeled with corresponding letters. Since we have already been given sides a and b, the third side will be c, so the angle between a and b, which is opposite c, must be at vertex C.

1) Lay out a line anywhere and reproduce length a on it. Label the endpoints C and B (Figure 7).

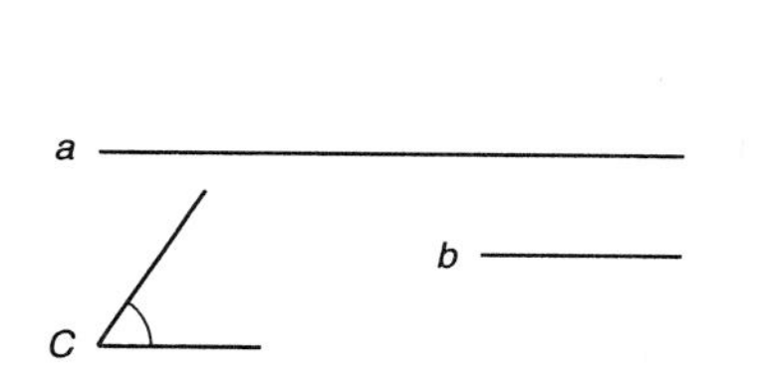

Figure 6: SAS. The parts.

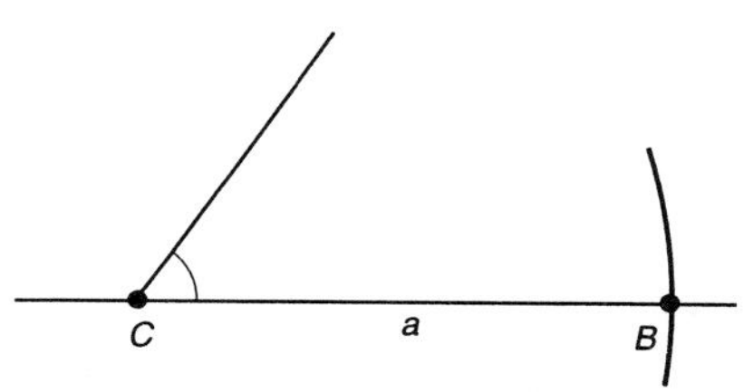

Figure 7: Part way through the SAS construction.

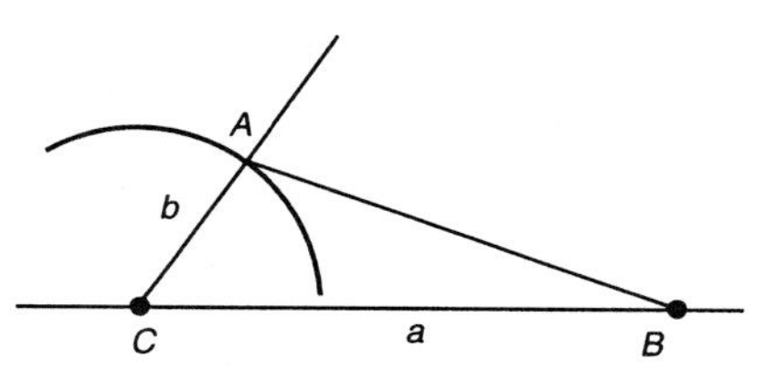

Figure 8: The result of the SAS construction.

2) Reproduce angle C at point C, creating a ray (half a line) extending from point C (How to reproduce an angle is the subject of problem B10).

3) With one end at point C, reproduce length b on the ray. Call the other end A (Figure 8).

4) Draw the line segment through A and B.

Steps 2 and 3 could be interchanged, and, in step 1, length b could have been used instead of length a. This would merely change the orientation of the triangle, but not the measures of the sides or angles.

(6.1.3) Angle-Side-Angle (ASA). Two angles and the included side determine a triangle. By "included" we mean the side between the two angles.

Begin with two angles, B and C, and the included side, a (Figure 9).

1) Draw any line and duplicate side a on it. Call the points at the ends B and C.

2) Duplicate angle B at point B, creating a ray from point B (Figure 9).

3) Duplicate angle C at point C, creating a ray from point C (Figure 10).

4) Locate A where the two rays intersect.

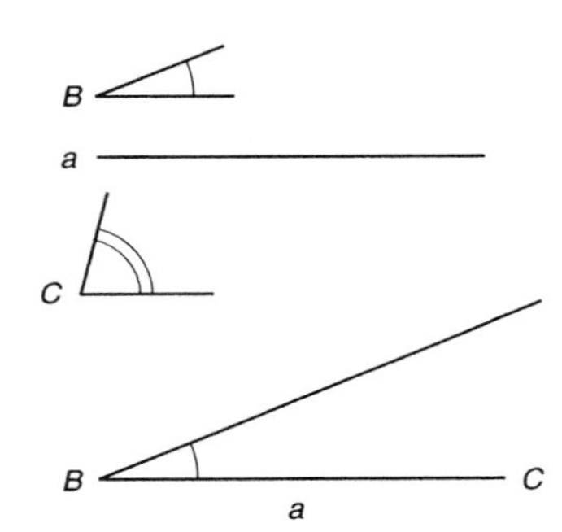

Figure 9: Parts for and the beginning of the ASA construction.

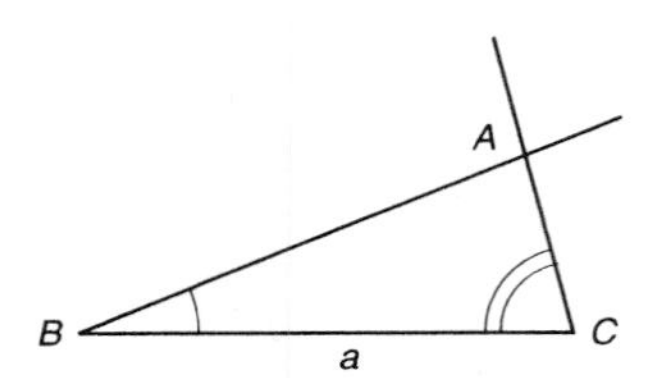

Figure 10: The completed ASA construction.

Any Two Angles Yield the Third. In the Angle-Side-Angle construction above we saw that the third angle was determined by the two given angles. Two angles of a triangle always determine the third.

(6.1.4) The Sum of the Angles of a Triangle. The angles of a triangle sum to a straight angle (180°), which is two right angles (90° each).

Proof: Given triangle ABC, construct a line through A parallel to BC (Figure 11). Then the angle labeled "1" is congruent to angle B (They are called "alternate interior angles"). The angle labeled "2" is congruent to angle C, for the same reason. Angle 1 plus angle BAC plus angle 2 form a straight angle. Therefore angle B plus angle A plus angle C sum to a straight angle.

Example 1: Let $\angle A = 32°$ and $\angle C = 16°$. Determine $\angle B$.

$\angle A + \angle B + \angle C = 180°$.

Therefore, $\angle B = 180° - 32° - 16° = 132°$.

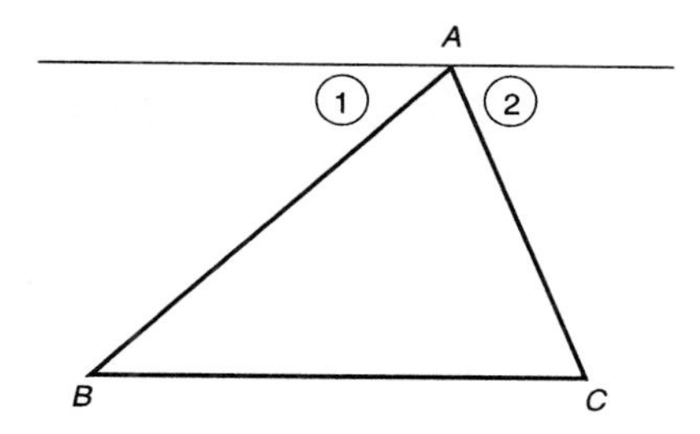

Figure 11: The angles of a triangle sum to a straight angle.

This is one of the key measurement theorems in geometry and trigonometry. The others are the Pythagorean Theorem (later in this section) and the Laws of Sines and Cosines from trigonometry (in Section 6.3).

Given two angles and a side, the side is either included (as in ASA, 6.1.3, above) or not included. If the side is not included, we use the previous result to obtain the third angle. Then all three angles will be known and the known side will be included between two of them. Then we can use the above ASA construction.

(6.1.5) Angle-Angle-Side (AAS). Two angles and a not-included side determine a triangle. The construction uses the Angle-Side-Angle construction (6.1.3) after determining the third angle (problem B13).

Combining ASA and AAS, we see that a triangle is determined by a side and any two angles.

(6.1.6) Angle-Side-Side (ASS). Two sides and an angle which is not included *almost* determine a triangle. This is the tricky case. There may be *two* triangles which have the given angle and two sides. In geometry this is easy to see, but, in trigonometry, when your calculator reports a single number (instead of *two* relevant numbers), you may forget that the other number is the one you really want. Inspect this geometric construction to see how two triangles can result, instead of just one.

Begin with an angle, B, and two sides, a and b (Figure 12). One side is opposite the angle, so the angle is not an "included" angle (it is not between the given sides as in SAS).

1) Lay out a line anywhere and reproduce length a on it. Call the points at the ends B and C.

2) Reproduce angle B at point B and create the ray extending from B (Figure 13).

3) With center at C, draw the circle of radius b.

4) Label with A and A' the (possibly two) points where the circle intersects the ray (Figure 13).

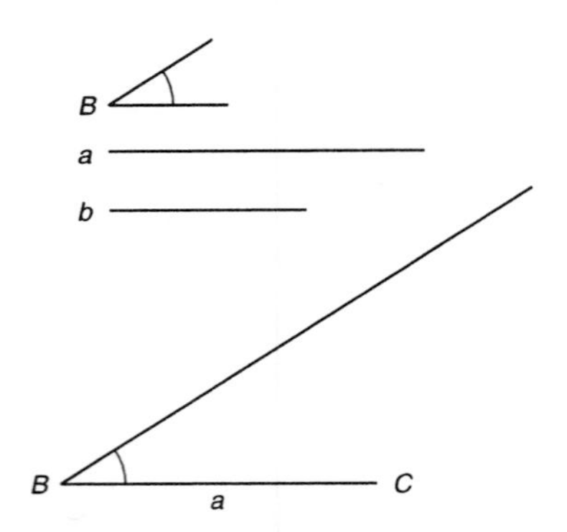

Figure 12: Parts for, and part of, the ASS construction.

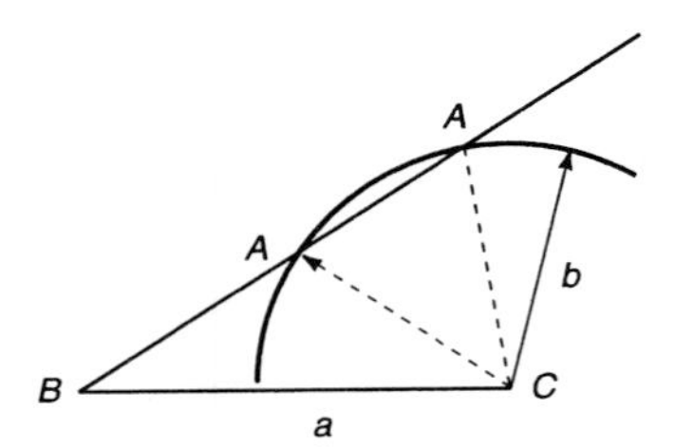

Figure 13: The completed ASS construction, with two triangles.

This construction uses two sides and an angle that is not included. I use the abbreviation "ASS" for this case is used instead of "SSA" (which would be the same case) because it reflects the order in the construction above. The angle is laid out on the left, with a side to its right, and then the remaining side to the right of that: Angle-Side-Side.

Note that, in ASS, if the side opposite the given angle (side b in Figure 13) is long enough, two triangles fit the conditions, one with an acute ($< 90°$) angle at A and the other with an obtuse ($> 90°$) angle at A'. It can be shown that the sum of angle A and angle A' is 180° (Problem B11). We will see that both angles have the same sine function value. For example, if angle A were 70°, then angle A' would be 110°, and sin 70° = sin 110°. This will be important when we use the so-called "Law of Sines" to try to determine angle A from the value of its sine. There will be two possible answers, just like there are two possible triangles in Figure 13.

<u>Calculator Exercise 1</u>: Evaluate sin 70°. Evaluate sin 110°. Then take inverse sine of each.

For this exercise make sure your calculator is in degree mode (as opposed to radian mode). Evaluate sin 110° = .9397. Sin 70° is exactly the same. Now, without clearing the display of your calculator, apply $\sin^{-1}$ to the image. ("$\sin^{-1}$" is called "inverse sine" or "arcsine", which is also written "arcsin".)

$\sin^{-1}(.9397) = 70°$.

The calculator responds with an acute angle that has the given sine value. But there is also an obtuse angle (110°) with the proper sine value -- it's just that your calculator will not report it. If you want an obtuse angle, you have to think of that yourself. The angle can be obtained from the angle displayed by subtracting it from 180°.

Watch for this ambiguous case when we get to the Law of Sines.

<u>(6.1.7) Angle-Angle-Angle (AAA)</u>. Three angles do *not* determine a triangle, but they do determine its shape. Triangles with the same shape (that is, with the same angles), but not necessarily the same size, are said to be <u>similar</u> (Figure 14, mirror images are also similar). Similar triangles have proportional sides. For example (Figure 15), if angles A, B, and C are congruent to angles D, E, and F, respectively, and side d is twice as long as side a, then side e will be twice as long as b and side f will be twice as long as side c. This proportionality is fundamental to trigonometry.

Figure 14: Similar triangles.

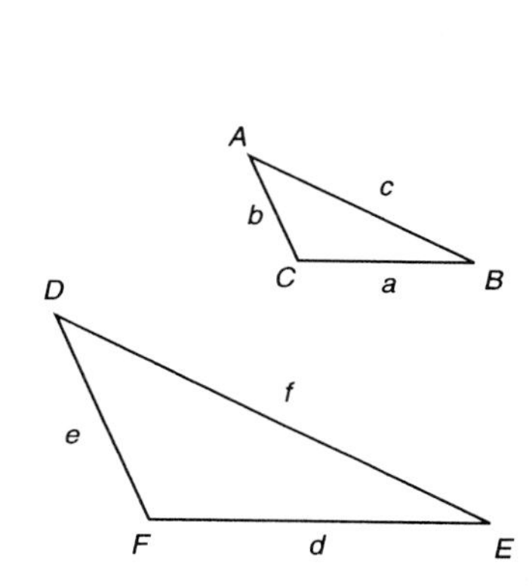

Figure 15: Two similar triangles. Their sides are proportional.

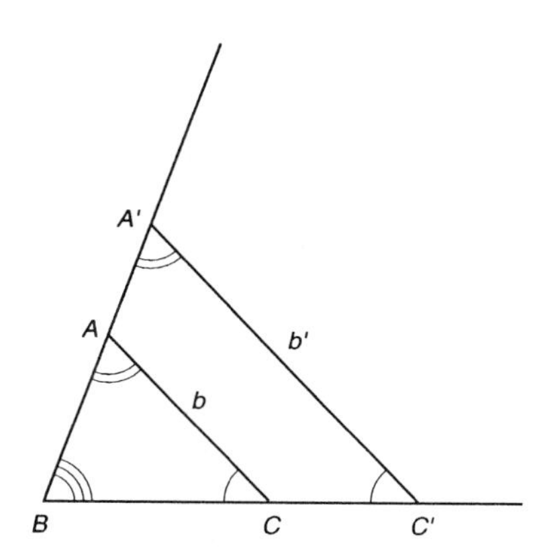

Figure 16: Similar triangles.

Figure 16 illustrates proportionality another way. Imagine angle B and its rays to be fixed in place, but let the endpoints of side b, A and C, move along the rays so side b remains parallel to the illustrated side, but changes length. Then b and a and c change the same way. In Figure 16, b' is 1.67 times as long as b. Therefore a' is 1.67 times as long as a, and c' is 1.67 times as long as c. All sides change in proportion.

So, to describe a triangle, we need at least one side, to fix the size. Three angles alone are not enough.

<u>**The Pythagorean Theorem**</u>. The famous measurement theorem of geometry is the Pythagorean Theorem.

<u>(6.1.8) The Pythagorean Theorem</u>. Let sides a, b, and c be opposite angles A, B, and C, respectively.

$$a^2 + b^2 = c^2 \text{ if and only if angle } C \text{ is a right angle.}$$

Do *not* think of the Pythagorean Theorem as just "$a^2 + b^2 = c^2$." In most of trigonometry that abbreviation would be false! The connection to a *right* triangle is essential! The Law of Cosines (Theorem 6.3.4) in trigonometry is specifically designed to modify this result to so that non-right triangles can be measured too.

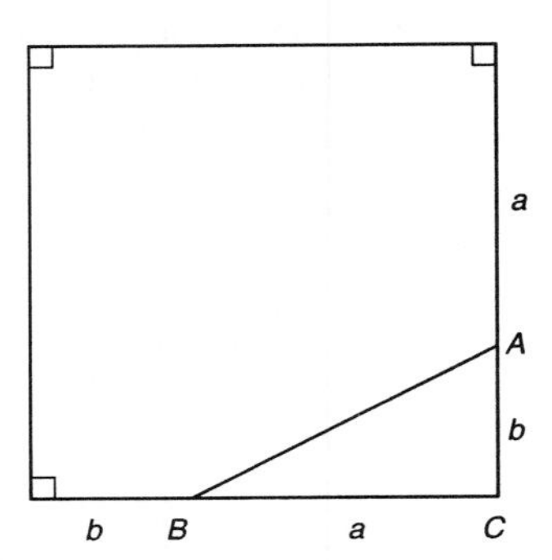

Figure 17: A right triangle and a square formed from its extended sides.

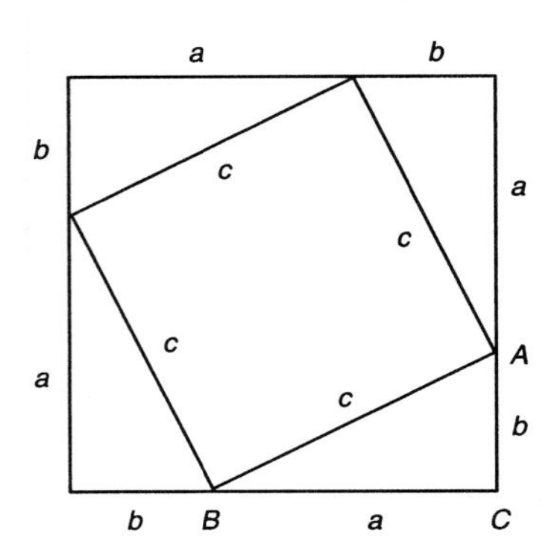

Figure 18: The right triangle with the square subdivided into a square and four triangles.

Proof: Consider the right triangle ABC with right angle at C as in Figure 17. Extend CA and CB by lengths a and b, respectively, and form the surrounding square. Then area of the square can be computed two ways. On the one hand, its sides are "$a + b$" so its area is "$(a + b)^2$."

On the other hand, the same square can be regarded as the sum of the square on the hypotenuse and four triangles, each congruent to the original triangle (Figure 18). Its area must be the same either way.

The area in Figure 18 is $c^2 + 4(\frac{1}{2}ab)$. Therefore,

$$(a + b)^2 = c^2 + 4(\tfrac{1}{2}ab),$$

$$a^2 + 2ab + b^2 = c^2 + 2ab \quad \text{[multiplying each expression out]},$$

$$a^2 + b^2 = c^2 \quad \text{[subtracting } 2ab\text{]}.$$

This proves the "if" half of Theorem 6.1.8. The "only if" half follows, after a clever construction, from SSS (problem B12).

The Pythagorean Theorem can be used to determine the hypotenuse from two legs, or a leg given the hypotenuse and the other leg.

Example 2: Suppose angle C is a right angle and $a = 65$ and $b = 23$. Find c, the hypotenuse.

$c^2 = a^2 + b^2 = 65^2 + 23^2 = 4754$. So $c = \sqrt{4754} = 68.95$.

Example 3: Suppose angle C is a right angle, $c = 92$ and $b = 37$. Find a.

$a^2 + b^2 = c^2$, so $a^2 = c^2 - b^2 = 92^2 - 37^2 = 8040$. Therefore $a = \sqrt{8040}$ $= 89.67$.

Example 4: $a = 24$, $b = 25$, and $c = 7$. Is it a right triangle?

A quick glance shows that c is the shortest side, so c^2 cannot be the sum of a^2 and b^2. But that is *not relevant*. The letters have been used differently. Often we use "c" for the hypotenuse of a right triangle, but that is only when "C" is the right angle. There are many cases when we need to use "a" or some other letter in the position of "c" in the theorem. After all, theorems are stated with dummy variables and their results apply to other letters. Of course, in a right triangle the hypotenuse is the longest side. So, if this were a right triangle, "b" would play the role of "c" in the theorem. So we need to know if $a^2 + c^2 = b^2$.

$a^2 + c^2 = 24^2 + 7^2 = 625$, which is $25^2 = b^2$. Therefore, it is a right triangle, with right angle at B.

Example 5: Let $a = 40$, $b = 50$, and $\angle C = 78°$. Find c.

Do *not* assume "$a^2 + b^2 = c^2$." That is a dangerous abbreviation of the Pythagorean Theorem which holds *only for right triangles*, and this is not a right triangle. We will need a different theorem to do this problem. The so-called Law of Cosines (coming up in Section 6.3) will handle it easily.

Example 6: A rope is tied tight straight between two trees 10 feet apart. If it is pushed sideways at the middle, it stretches one inch. How much will it be deflected from its previous line?

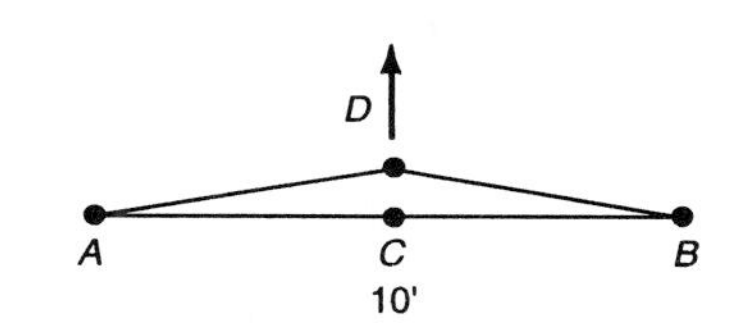

Figure 19: A 10-foot rope stretched one inch sideways. (View from above)

One inch in ten feet doesn't seem like much. Let's determine the deflection as described by the distance CD in Figure 19. Since it is pushed sideways "at the middle," let C be midway between the trees at A and B, so AC is 5 feet. Let D be the midpoint of the stretched rope. AD = 5 feet ½ inch. We can determine CD from the Pythagorean Theorem, since triangle ACD is a right triangle.

To avoid mixing feet and inches, use inches.

$AD^2 = AC^2 + CD^2$. $60.5^2 = 60^2 + CD^2$. $CD^2 = \sqrt{(60.5^2 - 60^2)} = 7.76$ (inches). So the sideways deflection is 7.76 inches (all from a one inch stretch!)

<u>**Recognizing the Cases**</u>. A triangle has six parts. Sometimes three parts determine the other three.

Example 7: Let $a = 12$, $b = 7$, and $c = 16$. Is a triangle determined?

This fits SSS (side-side-side). We can construct the triangle using the method in 6.1.1.

Example 8: Let $a = 18$, $b = 4$, and $\angle C = 134°$. Is a triangle determined?
Angle C is between sides a and b, so this is a case of SAS (Aide-Angle-Side). SAS always determines a triangle, as shown in the construction 6.1.2 (problem A7).

Example 9: Let $\angle A = 39°$, $\angle C = 55°$, and $b = 25$. Is a triangle determined?
Yes, b is the side between angles A and C, so this fits ASA (angle-side-angle), which determines a triangle. The only way ASA would not determine a triangle is if the two angles add up to too much. The sum of the angles in a triangle is 180°. The first two must leave room for the third, so they must sum to less than 180°. If that condition is met, the construction in 6.1.3 shows there will be a triangle (problem A8).

Example 10: Let $\angle A = 120°$, $\angle B = 20°$, and side $a = 12$. Is a triangle determined?
This time the given side is not between the given angles, so it does not yet fit ASA. But the third angle, $\angle C$, is easy to determine. Because the angles sum to 180°, it must be 40°. Now the given side *is* between two known angles. So it fits ASA with angles 20° and 40° on either side. A triangle is determined (Figure 20).

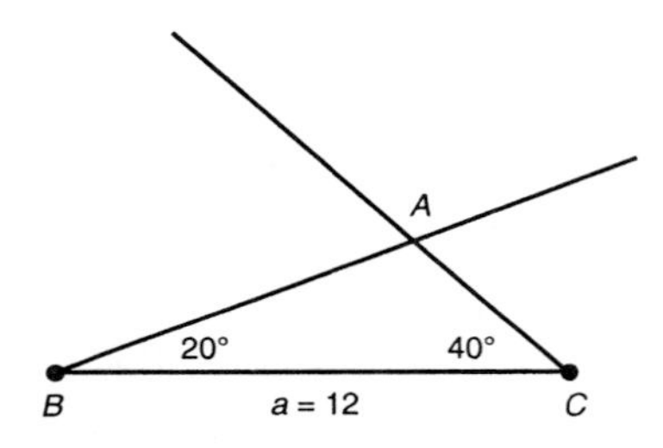

Figure 20: $\angle A = 120°$, $\angle B = 20°$, and $a = 12$. Determine $\angle C$ and use ASA.

Example 11: Let $b = 5$, $a = 10$, and $\angle B = 25°$. Is a triangle determined? Which case is it?
Angle B next to side a and opposite side b.
So we do not know the included angle. This case is ASS, which could be called SSA. There will be two possible triangles, although that is not obvious from the given measures. But we can see it from the picture (Figure 21).
In trigonometry, both the Law of Sines and the Law of Cosines will allow us to solve for the other sides and angles. The fact that the equations they create have two solutions shows that the triangle is not determined. There will be two triangles. (I hope you are getting curious about the Laws of Sines and Cosines, but you will have to wait).

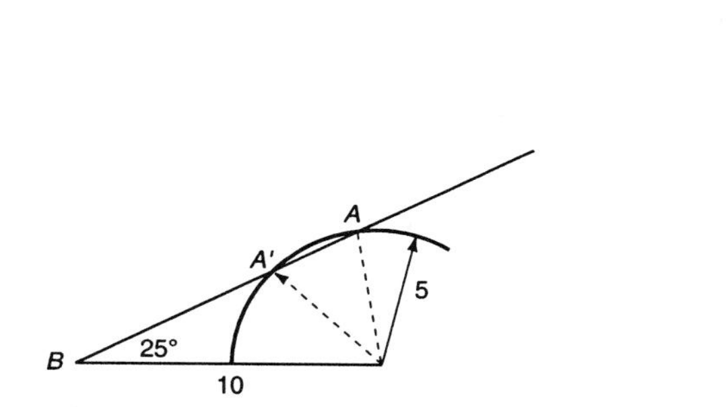

Figure 21: $b = 5$, $a = 10$, and angle $B = 25°$.

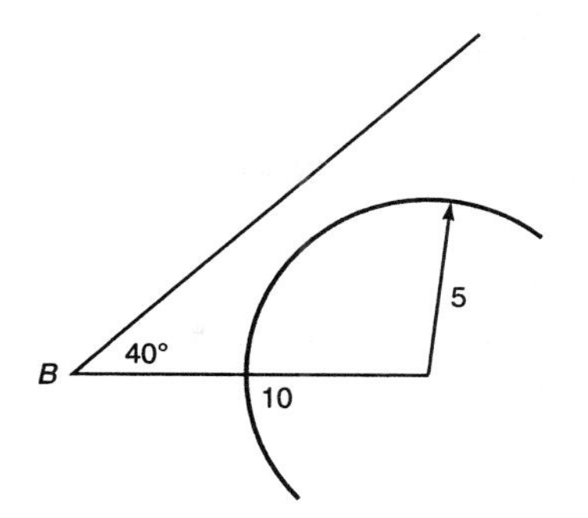

Figure 22: $b = 5$, $a = 10$, and angle $B = 40°$. b is not long enough to reach the ray for side c. There is no such triangle.

Example 12: Let $b = 5$ and $a = 10$ as in the previous example, but let angle $B = 40°$.

Again, this is ASS. Not only does ASS often give two triangles instead of one, it may not give any triangle. There may not be any triangle that fits because the second side may not be long enough to reach the ray formed by the angle. This happens in Figure 22.

In trigonometry, when we compute the measures of the parts, this behavior will be evident from both the Law of Sines and the Law of Cosines. The equations they create will have no solutions, which means there is no such triangle.

Figure 21 illustrates 2 solutions and Figure 22 illustrates no solutions. There is a case in between. If the radius were a bit shorter in Figure 21 or a bit longer in Figure 22, the arc could be made tangent to the ray so they intersected just once -- yielding exactly one solution (problem B14).

Example 13: Let $b = 12$, $a = 10$, and $\angle B = 25°$. Is a triangle determined?

Again, this is ASS. There are similarities with Example 12, which yielded two triangles. But in this example side b is much longer. In fact, it's too long to intersect side a twice on the same side of point B (Figure 23). So one triangle is determined (with vertex A), because the other candidate (with vertex A') would not really have an angle of 25° at B, rather an angle of

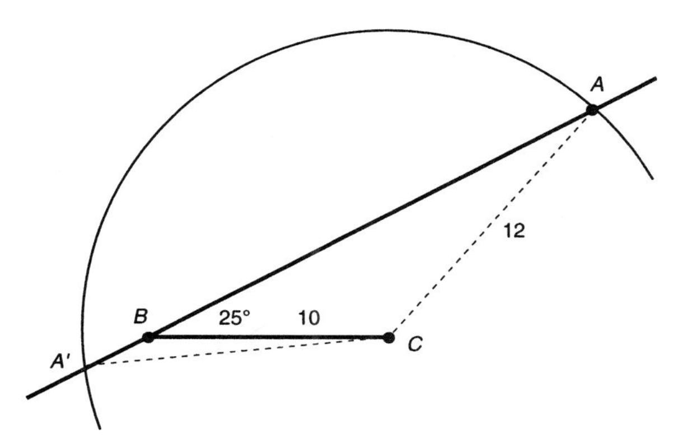

Figure 23: ASS. $b = 5$, $c = 12$, and angle $B = 25°$.

$180° - 25° = 155°$.

In trigonometry, when we compute the measures of the parts using the Law of Cosines, this case will be evident in that there will be two solutions for a, but one will be negative and therefore not really a legal side. Also, in the Law of Sines this case will be evident. Angle A will be a solution, but the angle at A' will not be a solution because it plus $\angle B$ will be more than 180°, which is illegal by Theorem 6.1.4.

Angle-Side-Side is by far the trickiest case. Beware.

Conclusion. Triangles have six parts. To solve a triangle means to find them from given parts. Triangles are determined by SSS, SAS, and ASA. ASS is the ambiguous case because there may be two solutions (or one, or none). Even in cases which supposedly determine a triangle (SSS, ASA) there may be no solutions if the parts do not fit certain conditions.

Terms: part (of a triangle), SSS, SAS, ASA, ASS, AAA, included angle, included side, Pythagorean Theorem.

****

Exercises for Section 6.1, "Geometry for Trigonometry":

A1.* a) What does it mean to "solve a triangle"?
b) In the context of solving a triangle, how many parts does it have?

^ ^ ^ ^

Identify the cases (SSS, ASA, etc.) if the following components are given:

A2. a) b, c, and $\angle A$. b) a, b, and c. c) $\angle B$, $\angle C$, and a.
A3. a) $\angle A$, $\angle B$, and $\angle C$ b) c, b, and $\angle A$ c) $\angle C$, a, and c.
A4. a) $\angle A$, $\angle B$, and b b) a, b, and $\angle C$. c) c, a, and $\angle C$.
A5. a) $\angle A$, $\angle C$, and b. b) $\angle A$, $\angle B$, and a. c) $\angle A$, $\angle B$, $\angle C$.
A6. a) $\angle C$, c, a. b) $\angle A$, b, c. d) $\angle A$, b, $\angle C$.

A7. Sketch the triangle in Example 9. A8. Sketch the triangle in Example 10.

^ ^ ^ ^ ^ ^ ^ ^ ^

B1.* a) What are the abbreviations of the cases that determine a triangle?
b) Which case is ambiguous?
c) Which case determines similar triangles?

B2.* Suppose two angles and a side which is not the included side are given.
a) Which case is most relevant? b) How do these parts fit that case, in spite of the fact that the side is not the included side?

B3.* a) What basic geometrical construction operations can you do with a straightedge?
b) What basic geometrical construction operations can you do with a compass?

B4. Outline a proof that the angles of any triangle sum to 180°.

B5. Outline a proof that "$a^2 + b^2 = c^2$" for a right triangle. Do the algebra.

B6. It is possible for ASS to yield exactly one triangle. How?

^ ^ ^ ^ Identify the cases (SSS, SAS, etc.) if the following components are given:

B7. a) d, $\angle C$, and e b) $\angle E$, $\angle F$, and d. c) $\angle D$, f, and d.

B8. a) $\angle D$, $\angle E$, and e b) d, e, and $\angle F$ c) d, f, and $\angle F$.

B9. a) a, b, $\angle A$. b) $\angle B$, $\angle C$, d. c) a, c, $\angle B$.

B10. Describe the geometric sequence of steps for reproducing a given angle in a new location, using only a compass and straightedge. Which geometric case is most relevant? [In trigonometry, which is really the current subject, it is common to use a protractor to measure and reproduce angles.]

B11. Show, in Figure 13, that angle BAC and angle $BA'C$ sum to 180 degrees.

B12. Show the "only if" of Theorem 6.1.8. Assume $a^2 + b^2 = c^2$ and show that C must be a right angle. Use SSS (6.1.1) after constructing a triangle with sides a and b and a right angle between them.

B13. The AAS construction is the most difficult. Given two angles and the side opposite one of them, how can you construct a triangle? [Do not gloss over the non-trivial steps.]

B14. In ASS there may be only one solution because the opposite side is just long enough to reach the ray extending from the angle, but not long enough to intersect the ray twice (as in Figure 13). Use the labeling of Figure 13. Use a trigonometric function to describe just how long side b would have to be in that case when there is only one solution.

Section 6.2. Trigonometric Functions

Trigonometric functions are important in two distinct contexts: triangles and circular motion. The context of circular motion includes all kinds of periodic behavior such as vibrations and waves of all kinds, including sound, light, and electrical waves.

This chapter is on applications of trigonometric functions to triangles. Nevertheless, their "circular" interpretation in the context of rotations is even more important in calculus. Therefore, we emphasize the "circular" definitions of trigonometric functions.

Many students have learned the trigonometric functions only in a right-triangle context. This context is very limiting, even if we only discuss triangles. Many triangles are not right triangles, and many have angles larger than 90 degrees that can not fit in a right triangle. Therefore, even for triangle-trigonometry, you need to understand the trigonometric functions of angles between 90 and 180 degrees. The "circular function" definitions below are the key.

Learn to interpret trigonometric functions in two distinct ways: 1) in right triangles where angles are limited to between 0 and 90 degrees, and 2) on the unit circle where angles are unlimited.

The sine function takes an angle and yields a number. You may express the angle in either of two ways: in degrees or in radians. For triangle trigonometry we prefer degrees; for calculus and circular trigonometry we prefer radians. In this chapter we will work with degrees.

(6.2.1) There are 360 degrees in a full circle, 180 degrees in a straight angle, and 90 degrees in a right angle (Figure 1).

Angles between $0°$ and $90°$ are said to be <u>acute</u>. Angles between $90°$ and $180°$ are said to be <u>obtuse</u>. Right triangles never contain obtuse angles, but many triangles do. Learn to recognize the approximate sizes of angles. There are only three common acute angles ($30°$, $45°$, and $60°$), so memorize their appearance and use them to help you estimate the measures of other angles (Figure 2).

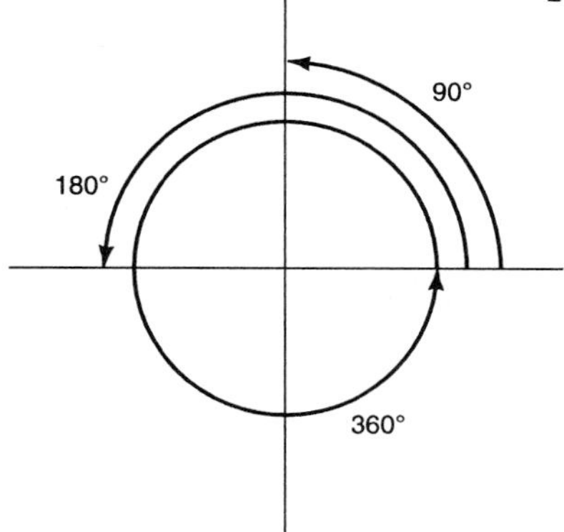

Figure 1: 360, 180, and 90 degree angles.

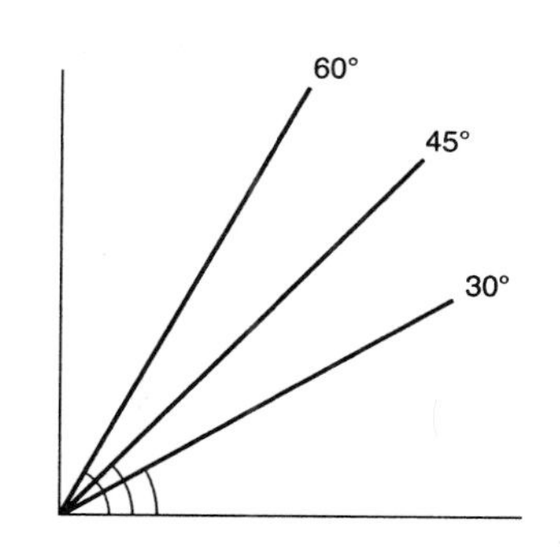

Figure 2: 30, 45, and 60 degree angles.

Consider a right triangle with the right angle at the lower right corner labeled C (Figure 3). In a *right* triangle the side opposite the right angle is called the hypotenuse and the other two sides are sometimes called the "legs". Since C is the right angle, the hypotenuse is c. Let θ (theta) be the angle at the lower left (also called $\angle B$). So "a" is the side (leg) adjacent to angle θ and "b" is the side (leg) opposite θ. This labeling is traditional.

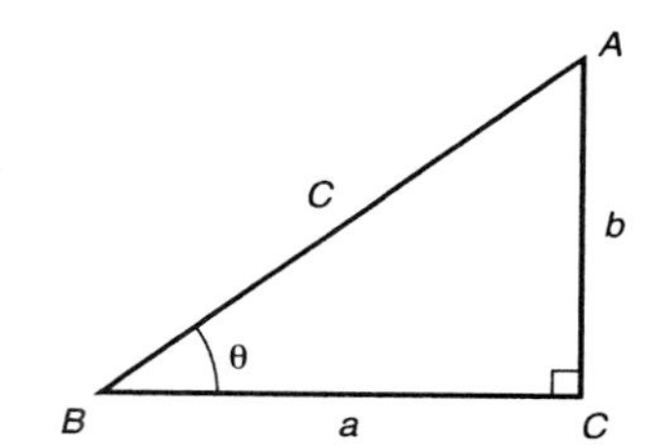

Figure 3: A right triangle labeled with letters.

The three basic trigonometric functions, sine, cosine, and tangent, can be interpreted in this limited context where angles are between 0 and 90 degrees. However, their *use* extends far beyond their *interpretation*. Be aware that they have a second, more general, definition coming up in 6.2.3 that applies to all angles.

For angle θ, $0° < \theta < 90°$,

$$\sin\theta = \frac{\textit{opposite}}{\textit{hypotenuse}} = \frac{b}{c}$$

$$(6.2.2) \qquad \cos\theta = \frac{\textit{adjacent}}{\textit{hypotenuse}} = \frac{a}{c} \qquad [\textit{in a right triangle}].$$

$$\tan\theta = \frac{\textit{opposite}}{\textit{adjacent}} = \frac{b}{a}$$

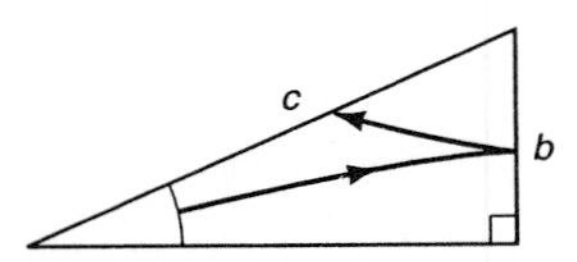

Figure 4: Sin θ is opposite over hypotenuse.

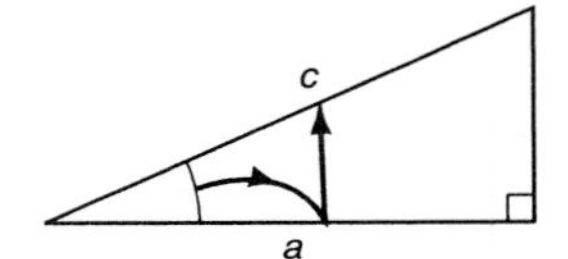

Figure 5: cos θ is adjacent over hypotenuse.

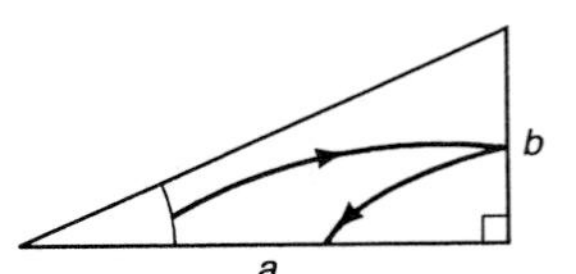

Figure 6: tan θ is opposite over adjacent.

These ratios depend upon the angle, θ, but not on the size of the triangle (Figure 7). Also, these ratios are pure numbers, without units. For example, if the two sides b and c are measured in inches, then their quotient, b/c, has "inches/inches" which cancel.

Everyone memorizes these three *right-triangle* interpretations:

sine is opposite over hypotenuse.
cosine is adjacent over hypotenuse.
tangent is opposite over adjacent.

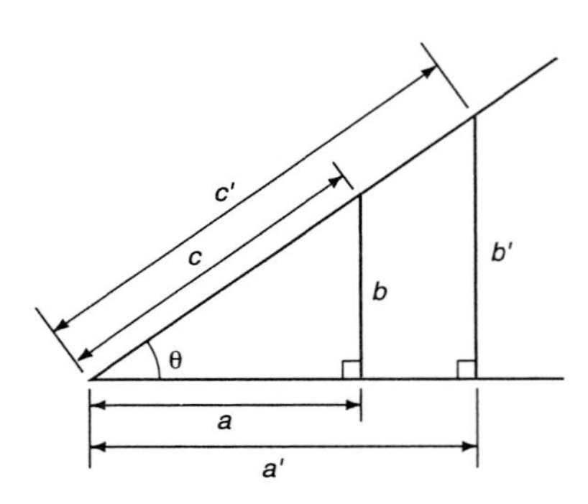

Figure 7: Similar triangles have proportional sides.

There is an easy way to extend these basic concepts to angles that do not fit in right triangles, including obtuse angles and angles greater than 180°. The idea is to put the angle at the center of circle with radius 1, instead of in a triangle.

Draw the unit circle (radius 1, $x^2 + y^2 = 1$) centered at the origin. Let the angle θ be at the origin, with one ray (the "initial side") along the positive x-axis and the other ray (the "terminal side") extending out to cut the circle at point P which depends upon θ ($P(\theta)$, Figure 8). The location of P can be described two ways: by the angle θ, or by its x and y coordinates. Trig functions are, by definition, relationships between these three numbers, θ, x, and y (Figure 9).

(6.2.3) $\sin \theta = y$, the vertical coordinate
$\cos \theta = x$, the horizontal coordinate
$\tan \theta = y/x = (\sin \theta)/(\cos \theta)$, (for $x \neq 0$).

For any angle θ, $P(\theta) = (x, y) = (\cos \theta, \sin \theta)$.
$\sin^2\theta + \cos^2\theta = 1$, a "Pythagorean" identity

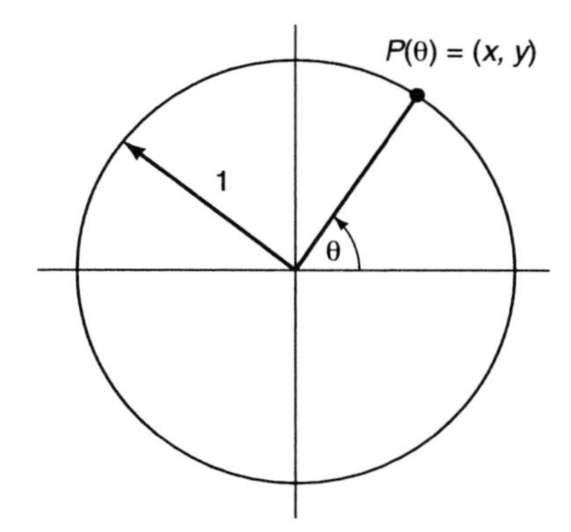

Figure 8: A unit circle, angle θ, and $P(\theta) = (x, y)$.

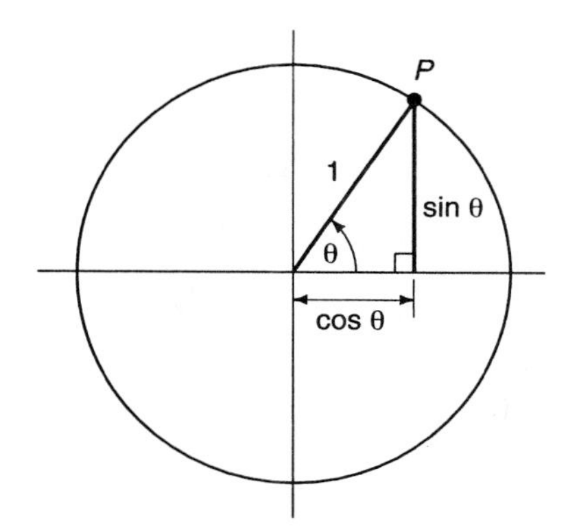

Figure 9: A unit circle, θ, cos θ, and sin θ. **Memorize this figure.**

sin θ is the vertical coordinate of *P*.
cos θ is the horizontal coordinate of *P*.
tan θ is (sin θ)/(cos θ).

Inspection of Figure 9 shows that these definitions duplicate the right-triangle interpretations for angles between 0 and 90 degrees. However, these definitions apply to all angles, including those angles greater than 90 degrees and even negative angles. By the way, negative angles are possible; they are just oriented clockwise instead of the usual counterclockwise (Figure 12).

Example 1: Consider an angle of 120° (Figure 10). The x-value of P is negative, so cos 120° is negative. From the picture it appears to be about -1/2 (remember the radius is 1). With a good picture you can almost read the sine and cosine of an angle right from the picture. Just estimate the horizontal (cosine) and vertical (sine) coordinates of P. Sin 120° is the y-value, which appears to be a bit less than .9. Check these estimates using a calculator (problem A49).

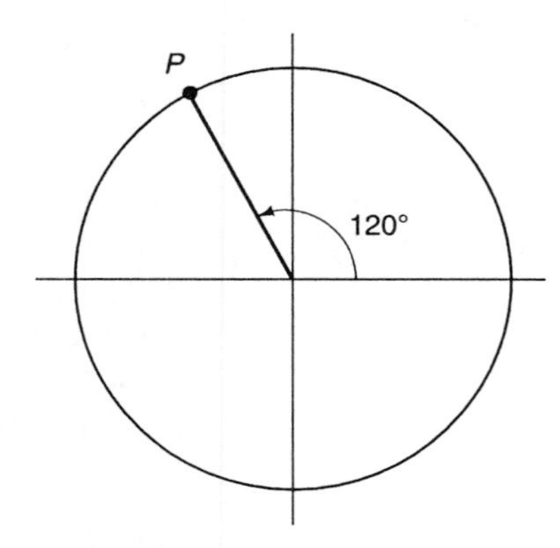

Figure 10: A unit circle and $\theta = 120°$

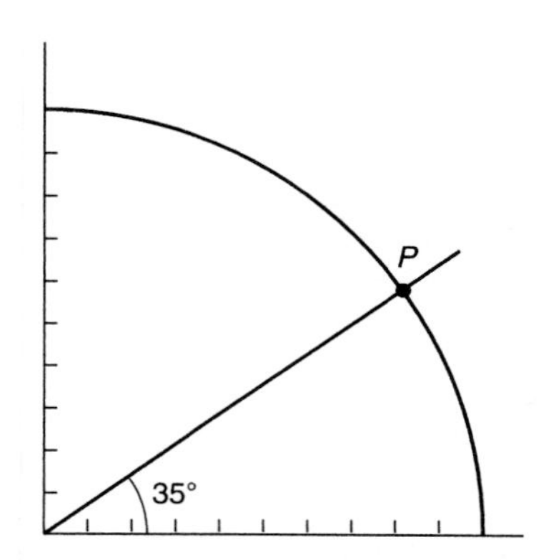

Figure 11: The first quadrant of a unit circle and $\theta = 35°$.

Example 2: Estimate sin 35° and cos 35° from Figure 11 which has the angle accurately drawn.

From Figure 11 it appears that the horizontal coordinate of *P* is about 8/10 of the radius. Estimate cos 35° by .8. The vertical coordinate appears to be about 6/10 of the radius. Estimate sin 35° by .6. Use your calculator to check the accuracy of these crude estimates (problem A50).

Example 3: Estimate sin 300° and cos 300° from a unit-circle picture.

A 300 degree angle is in the fourth quadrant. A full circle is 360°, so a 300° angle is 60° short of a full circle. Go backwards 60° from the positive *x*-axis to locate the ray (Figure 12). So the angle -60° creates the same *P* that 300° creates.

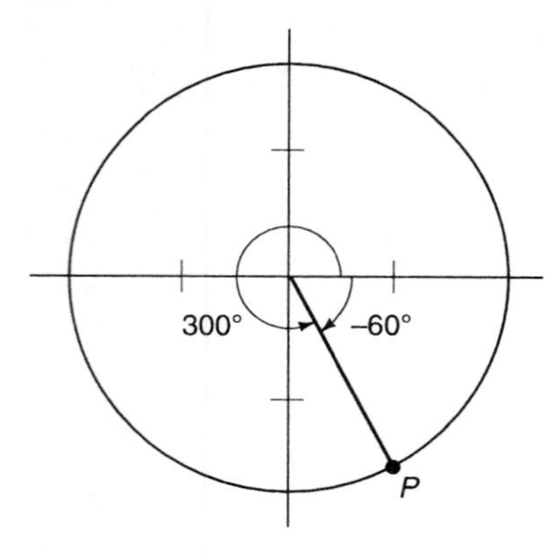

Figure 12: A unit circle, $\theta = 300°$, and $\theta = -60°$.

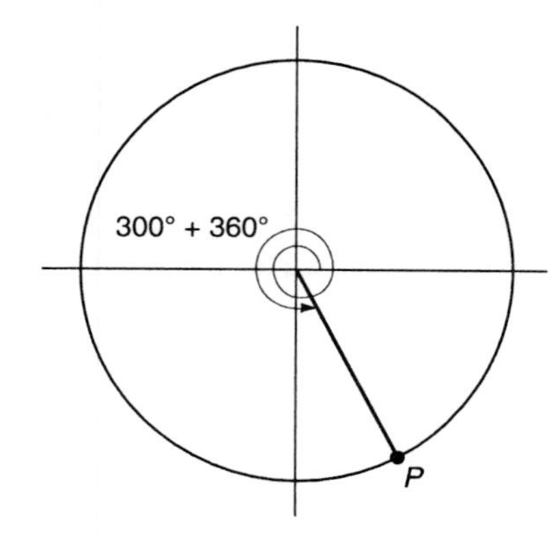

Figure 13: A unit circle, $\theta = 660° = 300° + 360°$.

From Figure 12, the *x*-value (cosine) is positive and the *y*-value (sine) is negative. The cosine appears to be near 1/2 and the sine near -.9. Use your calculator to check the accuracy of these estimates.

Two angles with the same *P* and therefore the same terminal side are said to be "coterminal". Angles are coterminal if and only if they differ by an integer multiple of 360 degrees. This concept is valuable in the context of rotation.

In the context of rotation, think of P as a point on the outside of a rotating shaft. The angle describes how P got to where it is. If the angle is 300°, it got there by rotating counterclockwise somewhat less than a full revolution. However, if the angle is -60°, P got there by rotating clockwise 60 degrees. The same terminal position can result from an infinite number of different rotations. Take any rotation that yields that position and add or subtract 360 degrees and the position remains unchanged (Figure 13). The x and y values of P remain unchanged; therefore the values of the trigonometric functions remain unchanged.

Sine. The graphs of sine and the other trigonometric functions result from this rotational, circular, interpretation. There are three numbers being related: the angle of P, the vertical coordinate of P, y, and the horizontal coordinate of P, x. Unfortunately, we must now switch letters so that the angle can be denoted by "x". Most graphics calculators use "x" for angles. So what was the "x" coordinate in the unit-circle pictures must be called something else. The horizontal coordinate of P will be called "cos x," and the vertical coordinate will be called "sin x."

Imagine the angle θ developing over time. That is, the terminal side to P is not simply fixed in place, but rotates into position. As it does so, the coordinates of P change. The graphs of sine and cosine exhibit this change as time changes on the horizontal axis (called the x-axis, but it often would be nice if it were the t, for time, axis).

Let's deal with sine first. Remember that sine is the vertical coordinate of P associated with an angle. Figure 14 shows both a unit circle and a traditional graph of sin x, side by side.

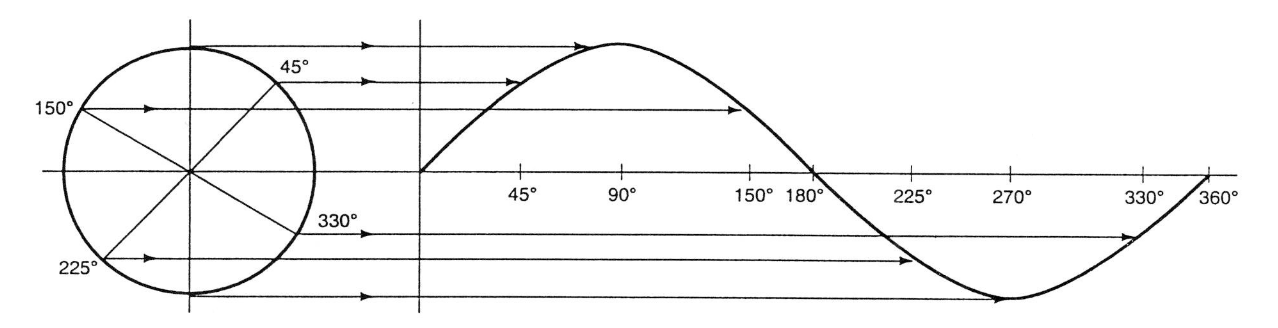

Figure 14: sin x is the vertical coordinate of P associated with angle x on the unit circle.

Imagine time (x) elapsing and P moving around the circle. The height of P ($\sin x$) goes up from 0 to 1 (at 90°) and then back down to 0 (at 180°) It continues down to -1 (at 270°) and then goes back up to 0 (at 360°). The graph of "$\sin x$" plots this up and down motion. This explains why $\sin x$ repeats every 360 degrees -- the same points are retraced.

Example 4: Use a unit-circle picture to estimate the solution for θ to $\sin \theta = .4$.

On a unit-circle picture, $\sin \theta = y$. So draw in the line $y = .4$ and see where it intersects the unit circle. It will intersect it twice, at P and P' (Figure 15), so there will be two solutions between 0° and 360° -- one in the first quadrant and one in the second. The first looks to be about 20°, the second about 160°. Check these estimates with a calculator (problem A51).

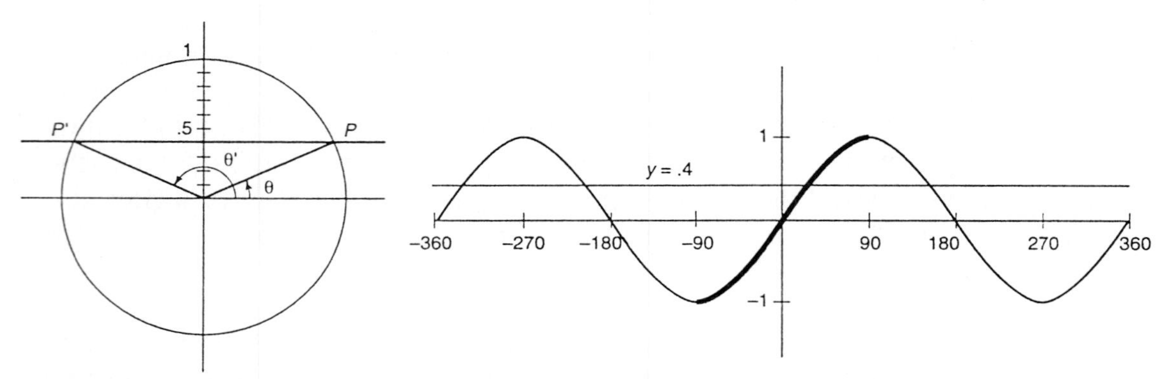

Figure 15: $\sin \theta = .4$. P and P' yield two solutions.

Figure 16: $y = \sin x$ and $y = .4$. $\sin^{-1}$ is defined on the emphasized region. $-360° \le x \le 360°$.

Figure 16 graphs both $y = \sin x$ and $y = .4$. The picture shows there are four solutions to "$\sin x = .4$" between -360° and 360°. The inverse sine function will find one of them -- the one in the emphasized region.

__Inverse Sine__. The inverse sine function is denoted by "$\sin^{-1}$" (read "inverse sine" or "arc sine"), which is sometimes written "arcsin" ("arc sine"). Any picture that shows sine also illustrates inverse sine by reading it backwards. A unit-circle picture such as Figure 15 illustrates both "$\sin \theta = y$" and "$\sin^{-1} y = \theta$" (read θ from the *right* half of the unit circle). Figure 16 uses different letters. It illustrates "$\sin x = y$" (where "x" is an angle) when it is read from x to y. However, when it is read from y to x *on the emphasized region*, it illustrates "$\sin^{-1} y = x$." Figure 17 illustrates both sine and inverse sine using an acute angle x: "$\sin x = c$" and "$x = \sin^{-1} c$."

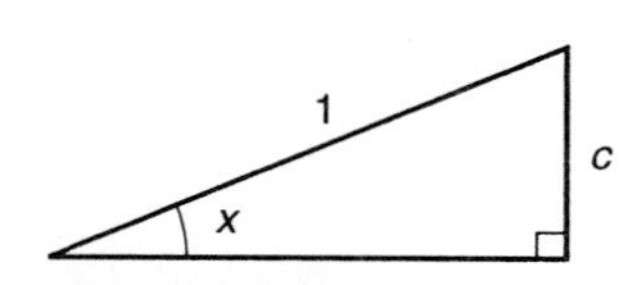

Figure 17: $\sin x = c$. $x = \sin^{-1} c$.

For acute angles inverse sine undoes what

sine does.

Example 5: Solve $\sin x = .8$.

In this problem x is an angle. $x = \sin^{-1}(.8) = 53.13°$. But this is not the only answer. The other important answer is $x = 180\text{-}53.13° = 126.87°$. Remember that, on the unit circle picture, $\sin x$ is the height, and heights occur twice (Figure 18). Also, all other angles coterminal with these two are also solutions. So $x = 53.13+360°$ also solves it, and so do $x = 53.13\text{-}360°$, $x = 126.87+360°$, and $x = 126.36\text{-}360°$. But these coterminal solutions are not as important as the two solutions between 0° and 180°. Next we discuss why the second solution is 180° minus the first solution.

By inspecting a unit-circle picture you can see the relationship among the numerous solutions for x to "$\sin x = c$" (Figure 18). One solution is $\sin^{-1}c$. **By definition**, $\sin^{-1}$ yields the solution on the **right** half of the unit circle expressed as an angle between -90° and 90°. That is the solution in the emphasized region of Figure 16. Every height occurs once on the right half on the unit circle. For the second solution you want to see where the same height occurs on the left half.

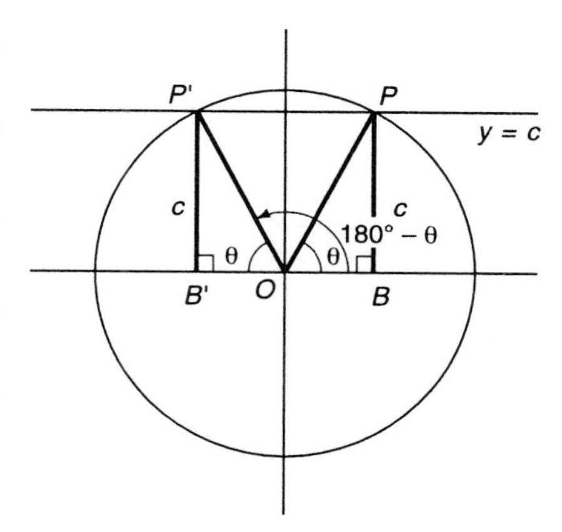

Figure 18: Two solutions to $\sin x = c$.

Let θ be a first-quadrant angle associated with P on the unit circle (Figure 18). Then $180° - \theta$ is a second quadrant angle. Associate it with P'. By ASA, the picture has two congruent triangles, triangle BOP and triangle $B'OP'$. Therefore BP and $B'P'$ are congruent, and θ and $180 - \theta$ have the same sine. This argument can be generalized to cover all angles. The result is:

(6.2.4) $$\sin \theta = \sin(180° - \theta).$$

So, if θ solves "$\sin x = c$" so does $180° - \theta$, because their sine values are the same. Combining this result with the fact that P remains the same when the angle is increased or decreased by 360°, or a multiple of 360°, you find

(6.2.5) $$\sin x = c \text{ iff } \begin{aligned} x &= \sin^{-1}c, \text{ or} \\ x &= 180° - \sin^{-1}c, \text{ or} \\ x &= \sin^{-1}c \pm 360n°, \text{ for some integer } n, \text{ or} \\ x &= 180° - \sin^{-1}c \pm 360n°, \text{ for some integer } n. \end{aligned}$$

The first two are the most important. They are angles which can fit in a triangle. The other solutions can occur in rotational problems, but not in triangle problems. If you want to be concise, you could state this with just the last two lines, since they include the first two when $n = 0$.

Example 4, revisited. Find x when $\sin x = .4$. The solutions can be seen using graphs of "$y = \sin x$" and "$y = .4$" as in Figure 16. The line intersects the curve many times; there are many solutions to "$\sin x = .4$." Look at the symmetry of the bumps. The rise of the graph from 0° to 90° is exactly the same shape as the rise of the graph read backwards from 180° back toward 90°. Therefore, for any solution at angle θ on the rise in the first quadrant there is a corresponding solution back toward 90° from 180°, at 180° - θ. That is another explanation of 6.2.4 and the first two lines of 6.2.5.

Example 6: In Figure 19, $\sin A = .913$. Find $\angle A$.

$\sin^{-1} .913 = 65.9°$. But angle A is obviously not 65.9°, it's obtuse. So you must want the other solution. $\angle A = 180 - 65.9° = 114.1°$.

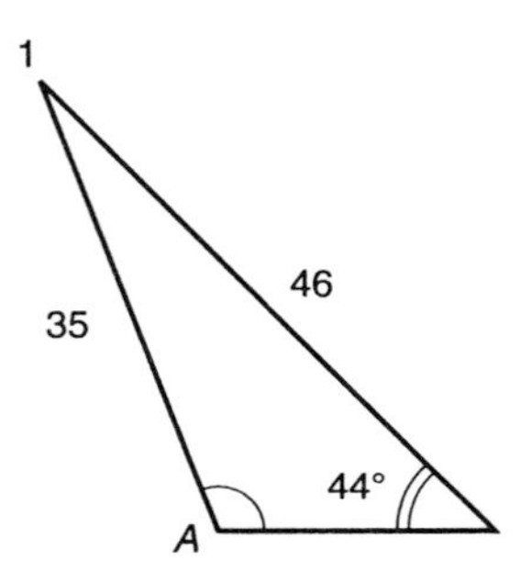

Figure 19: $\sin A = .913$.

There is a parallel here with solving equations like "$x^2 = 30$." You use a keystroke sequence to find one solution ($\sqrt{30} = 5.377$), but you know in your head that there is a second solution (with a negative sign). To solve "$\sin x = c$" you use a keystroke sequence to find one solution ($\sin^{-1} c$), but you know in your head there is second solution (180° minus the first one).

The range of the sine function (the set of possible y-values) is [-1, 1]. Because the inverse function reverses the order, this is the domain of the inverse sine function. If you try to evaluate "$\sin^{-1} 2$" you will not obtain a real number. Try it on your calculator and see.

Some people remember what the sine function is by memorizing the phrase "opposite over hypotenuse." What is a good way to remember what the inverse sine is? In the first quadrant,

"$\sin^{-1} x$" is "the angle whose sine is x."

For example, the sentence "$\sin^{-1}(1/2) = 30°$" could be thought of as "The angle whose sine is 1/2 is thirty degrees," although it is usually read "Inverse sine of 1/2 is thirty degrees," or "Arc sine of 1/2 is thirty degrees." This is the inverse sentence of "$\sin 30° = 1/2$." We would never write "$\sin^{-1} \theta$" because θ represents an angle and the inverse sine function has angles as *images*, not arguments.

(6.2.6) The domain of $\sin^{-1}$ is the interval [-1, 1].
The range of $\sin^{-1}$ is the interval [-90°, 90°].
$\sin(\sin^{-1}x) = x$, for all x in the domain.
$\sin^{-1}(\sin x) = x$ iff $-90° \leq x \leq 90°$.

The third line computes "the sine of the angle whose sine is x." Obviously, that is x. The fourth line is trickier, because sine is not one-to-one. Its inverse is read from the **right** half of the unit circle.

Example 7: $\sin^{-1}(\sin 42°) = 42°$. [42° is between -90° and 90°.]
$\sin^{-1}(\sin(-14°)) = -14°$. [-14° is between -90° and 90°.]
$\sin^{-1}(\sin 170°) = 10°$. By 6.2.4, 170° and 10° have the same sine, and 10° is the angle in the designated region from -90° to 90° (Figure 16).
$\sin^{-1}(\sin 340°) = -20°$. Note that 340° and -20° are coterminal, and therefore have the same sine value, and -20° is in the designated region from -90° to 90°.
$\sin(\sin^{-1}(.88)) = .88$. [This is not the tricky direction.]

Solving Right Triangles. Trigonometric functions can be used to solve right triangles.

Example 8: A triangle has a right angle at C, $\angle B = 23°$, and hypotenuse $c = 480$. Find b.

I would sketch the triangle (Figure 20). It helps me keep clear which side is which, and it helps me estimate the magnitude of the unknown side so I can check my answer.

Side b is opposite angle B. In a right triangle, sine is opposite over hypotenuse:

$$\sin B = b/c. \quad \sin 23° = b/480.$$
$$b = 480 \sin 23° = 187.6.$$

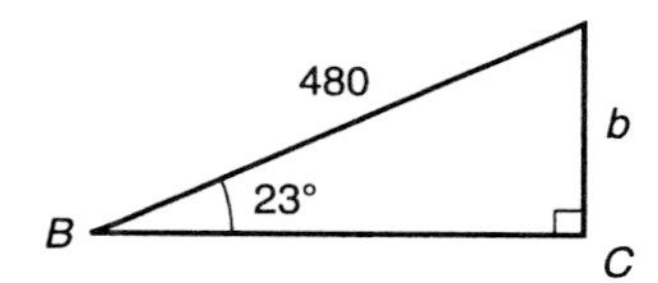

Figure 20: $\angle B = 23°$. $\angle C = 90°$. $c = 480$.

In this example we solved "sine is opposite over hypotenuse" for the opposite side given the hypotenuse. This procedure is so common that you may wish to memorize the result: In a *right* triangle,

(6.2.7) "opposite is hypotenuse times sine."

In the next example we use the opposite side to determine the hypotenuse.

Example 9: A triangle has a right angle at C, $\angle A = 69°$, and $a = 55$. Find the hypotenuse, c.

Sketch it. Side a is opposite angle A. In a right triangle, sine is opposite over hypotenuse: $\sin A = a/c$. $\sin 69° = 55/c$. $c = 55/(\sin 69°) = 58.9$.

This problem uses the letters differently than Formula 6.2.2. Remember, the letters are just dummy variables. We tend to use "$\angle C$" for the right angle and "a" for the adjacent side, but there is no guarantee that the labeling will be that way.

Example 10: $\angle B = 51°$, $\angle A = 90°$, and $b = .046$. Which side can we find with sine?

Side b is opposite angle B, and angle A is the right angle. So side a is the hypotenuse. We can find side a.

$$\sin 51° = .046/a. \quad a = .046/(\sin 51°) = .059.$$

Example 11: Consider the triangle in Figure 21. $c = 50$, $\angle B = 20°$, and $a = 50$. Find b.

Side b is opposite the given angle. Can we use "sine is opposite over hypotenuse" to find b?

No. This triangle does not have a hypotenuse. The term "hypotenuse" applies to the side opposite the right angle in a *right* triangle. This is not a right triangle and therefore has no hypotenuse. But it is a special type of triangle. Triangles with two equal sides are said to be <u>isosceles</u>. In this example, $a = c = 50$, so the triangle is isosceles. By drawing an auxiliary line we can subdivide it into two congruent right triangles. Bisect angle B. The ray will also be a perpendicular bisector of side b (Figure 22). Now we can use sine, because there is a right triangle. The angle is half of $\angle B$, and the opposite side is half of b. Now c is a hypotenuse. "sine is opposite over hypotenuse."

$\sin(\angle B/2) = (b/2)/c$.

$\sin 10° = (b/2)/50 = b/100$.

$b = 100 \sin 10° = 17.36$.

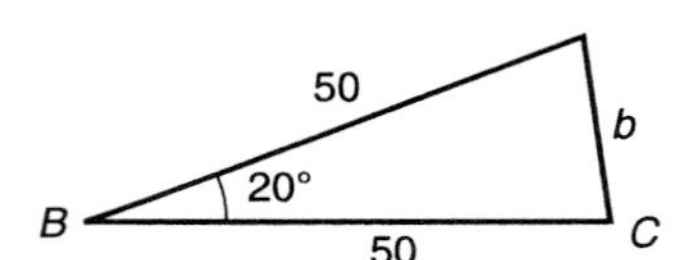

Figure 21: $c = 50$, $\angle B = 20°$, and $a = 50$.

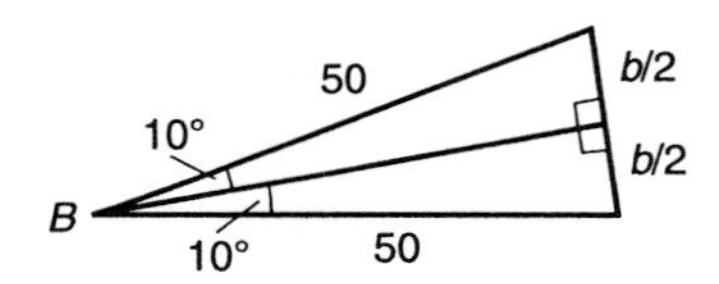

Figure 22: Bisecting the angle and opposite side of an isosceles triangle.

Example 12: Consider a circle with radius 5 (Figure 23). A chord AB makes a central angle AOB of 72°. How long is the chord (the line segment with ends on the circle)?

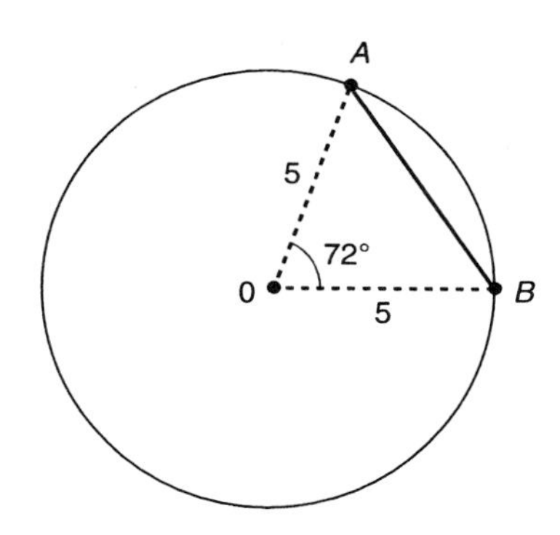

Figure 23: A circle of radius 5 and a chord with a 72° central angle.

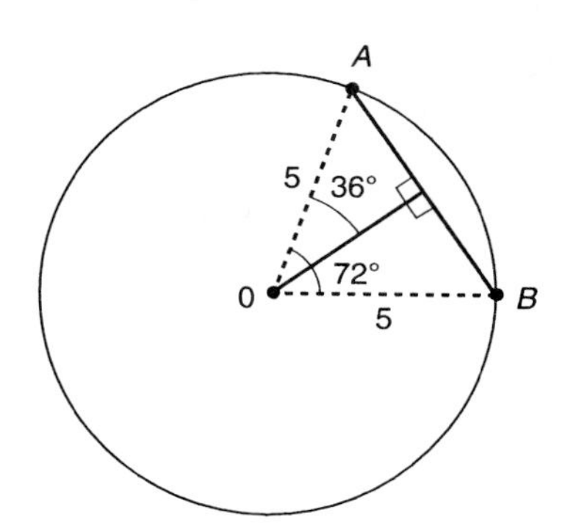

Figure 24: Figure 23, with the central angle bisected.

Chord AB is opposite angle AOB, but this type of "opposite" is not the "opposite" in the right-triangle interpretation of sine because triangle AOB is not a right triangle. But it is isosceles (the two radii are equal), so it can be split into two right triangles by bisecting the angle and the chord (Figure 24). Now there is a right triangle where half the chord is "opposite" half the angle.

$\sin(\frac{1}{2}\angle AOB) = \frac{1}{2}AB/OA.$

$\sin 36° = \frac{1}{2}AB/5.$

$AB = (5 \sin 36°)/\frac{1}{2} = 5.88.$

Example 13: $\angle A = 13°$, $a = 8.4$, and $b = 35.2$. Find c.

Side a is opposite angle A. Can we use "sine is opposite over hypotenuse"?

No. This thought is a *right* triangle thought, but this triangle is not a right triangle. We don't know $\angle C$. This is a case of Angle-Side-Side. In the next section we will use the so-called Law of Sines to find the two possibilities for c. Wait for the Law of Sines.

Cosine. Cosine, like sine, has both a right-triangle interpretation and a unit-circle definition. Angles greater than 90° are very important for the cosine function, but do not fit in right triangles, so it is important to learn the unit-circle definition.

The idea of point P rotating around the unit circle as the angle increases is important for cosine, as it was for sine. Only this time the horizontal coordinate is tracked. If we call the angle θ, the cosine of θ is x. Unfortunately, we often need to call the angle "x" (for instance, on calculators) and we can not use the same letter for the horizontal coordinate of P. Oh well. When the angle is "x", call the horizontal coordinate of P "$\cos x$," which is what it is, by definition (Figure 25).

At angle 0, the horizontal value is 1, so $\cos 0° = 1$. At angle 90°, the horizontal coordinate is 0, so cos 90° = 0. Similarly, cos 180° = -1, cos 270° = 0, and cos 360° repeats cos 0° = 1. The graph of cosine has exactly the same shape as the graph of sine, only the graph of sine is shifted left 90 degrees (Figure 26). We will be able to prove this later using a trigonometric identity.

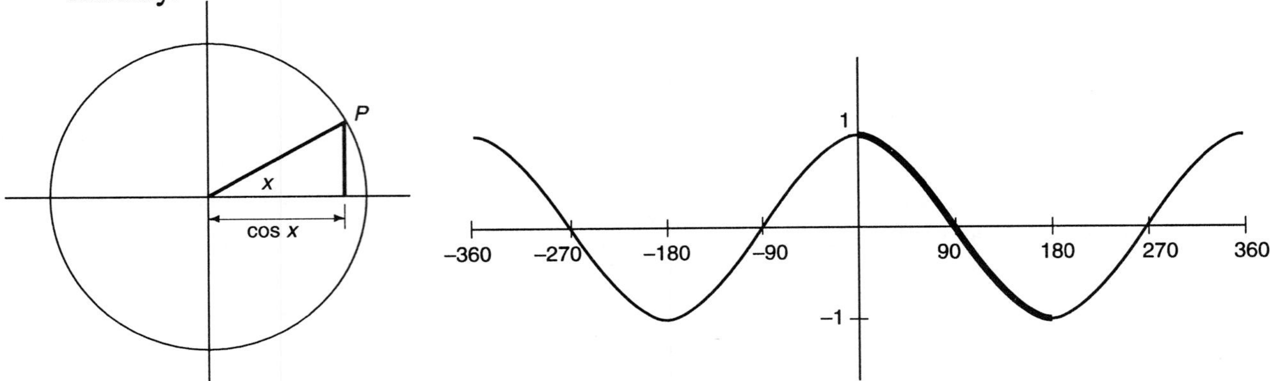

Figure 25: A unit circle, angle x, and cos x.

Figure 26. cos x. $-360° \leq x \leq 360°$. The region where $\cos^{-1}$ is defined is emphasized.

In a right triangle, cosine is adjacent over hypotenuse. Figure 27 is a right triangle picture that illustrates both "cos x = c" and "x = $\cos^{-1} c$" ("inverse cosine of c" or "arc cosine of c") for acute angles only.

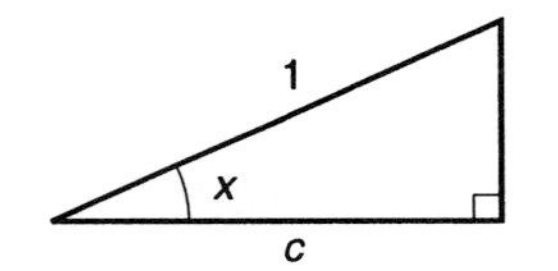

Figure 27. $\cos x = c$. $\cos^{-1} c = x$.

Example 14: $a = 15$, $\angle B = 39°$, and $\angle C = 90°$. Find c.

Of course, you should be sketching the figures in the examples before you read the solution. Since c is opposite the right angle, c is the hypotenuse. The leg next to angle B is a, so a is the adjacent side. It is appropriate to use cosine.

$$\cos 39° = 15/c. \quad c = 15/(\cos 39°) = 19.3.$$

Example 15: $\angle B = 90°$, $b = 93$, and $\angle A = 88.7°$. Find c.

This time "c" is not the hypotenuse. b is, because it is opposite the right angle. Side b is the hypotenuse and the unknown side c is adjacent to angle A. Use cosine.

$$\cos 88.7° = c/93. \quad c = 93 \cos 88.7° = 2.11.$$

In this example we solved "cosine is adjacent over hypotenuse" for the adjacent side given the hypotenuse. This procedure is so common that you may wish to memorize the result: In a *right* triangle,

(6.2.8) "adjacent is hypotenuse times cosine."

Inverse Cosine. Sometimes sides are used to find an angle. The inverse cosine function ($\cos^{-1}$, also written "arccos" and called "arc cosine") is used to solve problems where an unknown angle has a known cosine value.

Example 16: $\angle C = 90°$, $c = 10$, and $a = 3$. Find angle B.

The hypotenuse is c and a is adjacent to angle B.

$$\cos B = 3/10 = .3. \quad \angle B = \cos^{-1}.3 = 72.54°.$$

This is the complete answer. In triangles there is no second answer with inverse cosine as there is with inverse sine. Do not misunderstand. For any value of c, there is more than one solution to "$\cos x = c$," but only one is between 0° and 180° so it can fit into a triangle. The others must be greater than 180° degrees or negative.

(6.2.9) $\cos x = c$ iff $x = \cos^{-1}c$, or
$x = -\cos^{-1}c$, or
$x = \cos^{-1}c \pm 360n$, for some integer n, or
$x = -\cos^{-1}c \pm 360n$, for some integer n.

Turn to the unit circle to understand this complicated result. Remember that cosine is the horizontal coordinate, which must be between -1 and 1. For any horizontal coordinate between -1 and 1 there are *two* points on the circle (P and P', or Q and Q', Figure 28). **By definition**, the inverse cosine *function* selects only one -- the one between 0 and 180 degrees -- the one on the *top* half of the circle. The other one is easy to find using the symmetry of the picture. If P is on the top half of the unit circle at angle x and P' is directly below it on the unit circle, then P' can be regarded as being at angle $-x$. This yields the second line of 6.2.9. Lines 3 and 4 come from adding or subtracting multiples of 360 degrees to the two main solutions. Recall that adding or subtracting 360 degrees to an angle makes a coterminal angle, which therefore has the same trigonometric function values.

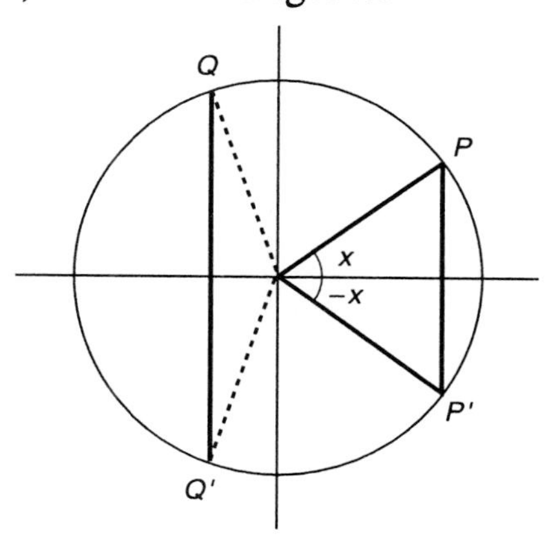

Figure 28: $\cos x = \cos(-x)$.

Example 17: Solve $\cos x = .2$.

By 6.2.9, $x = \cos^{-1}.2 = 78.46°$ is the only solution which could fit in a triangle However, there are more solutions which could be rotational angles, including -78.46° (from line 2), and 78.46+360° = 438.46° and -78.46+360° = 281.54°. To state them all we can write "$x = \pm 78.46° \pm 360n°$, for any integer n." The case $n = 0$ handles lines 1 and 2 of 6.2.9. Thus 6.2.9 could be abbreviated as

(6.2.9, again) $\cos x = c$ iff $x = \pm \cos^{-1} c \pm 360n°$, for some integer n,

where the "plus or minus" on the "$\pm \cos^{-1} c$" term consolidates lines 1 and 2 of the previous version.

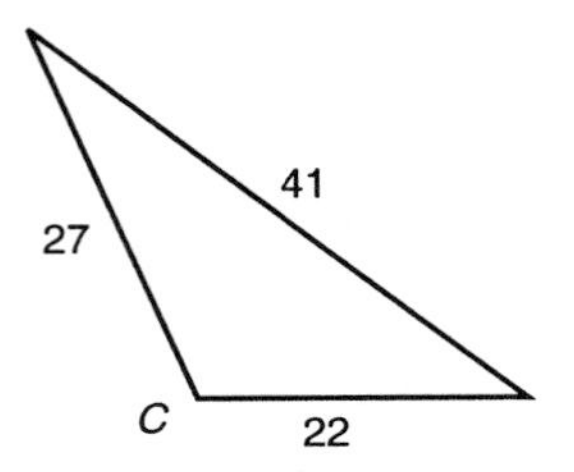

Figure 29: $\cos C = -.394$.

Example 18: In Figure 29, $\cos C = -.394$. Find $\angle C$.

In a triangle, $\cos C = -.394$ iff $\angle C = \cos^{-1}(-.394) = 113.2°$.

For the record,

(6.2.10) The domain of $\cos^{-1}$ is the interval [-1, 1].
The range of $\cos^{-1}$ is the interval [0°, 180°].
$\cos(\cos^{-1}(x)) = x$, for all x in the domain.
$\cos^{-1}(\cos x) = x$ iff $0° \leq x \leq 180°$.

The third line computes "the cosine of the angle whose cosine is x." Obviously, that is x. The fourth line is trickier because cosine is not one-to-one and its inverse is read from the **top** half of the unit circle (Figure 28).

<u>**Tangent**</u>. In a right triangle, "Tangent is opposite over adjacent." Trigonometric ratios do not depend upon the size of the triangle. Ratios are easy to visualize and interpret when the denominator is 1. So one way to visualize tangent for acute angles is to let the adjacent side be 1 in a right triangle. Figure 30 illustrates both "$\tan x = c$" and "$x = \tan^{-1} c$" ("inverse tangent" or "arc tangent") for acute angles only.

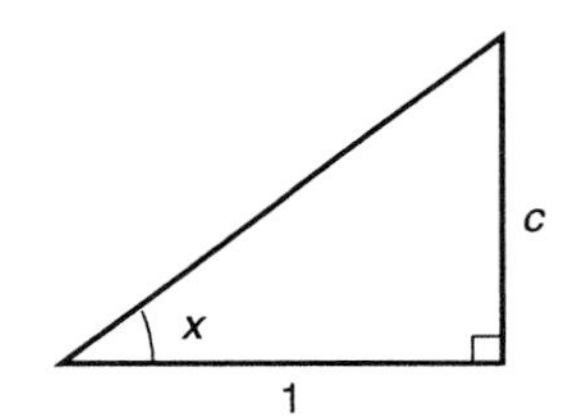

Figure 30: $\tan x = c$.
$x = \tan^{-1} c$.

Another way to visualize tangent is to augment the unit circle with a line perpendicular to the x-axis through the point (1, 0) as in Figure 31. Then extend the ray through $P(\theta)$ until it crosses the line. It crosses the line at height "$\tan \theta$."

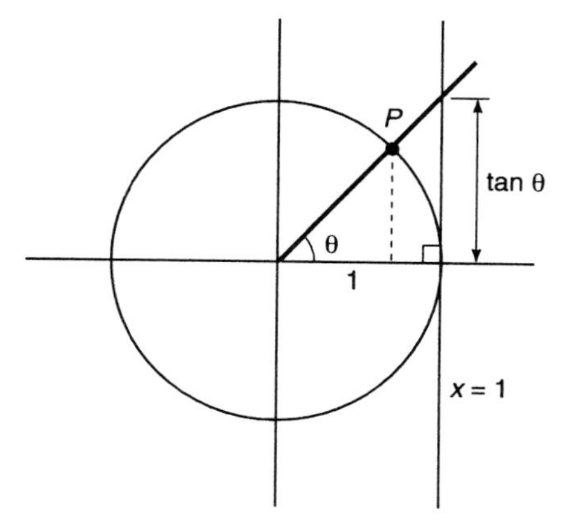

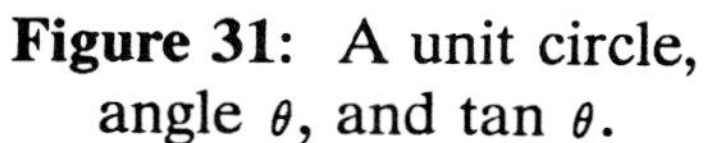
Figure 31: A unit circle, angle θ, and tan θ.

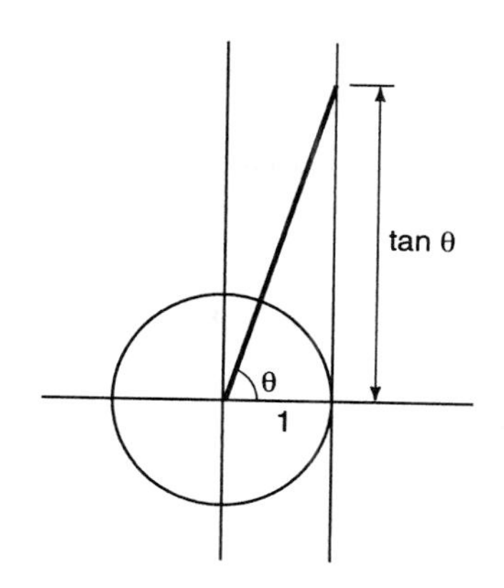

Figure 32: As θ approaches 90°, tan θ becomes large.

As angle θ increases to 90°, tan θ becomes arbitrarily large. Look at the picture. As angle θ increases to 90°, its ray cuts the line $x = 1$ further and further up (Figure 32). If θ is negative in the fourth quadrant, the ray cuts the line below the x-axis and y is negative. Tangent is negative in the fourth quadrant.

This "cut the line" interpretation is also valid for angles in the second and third quadrants. For any P on the unit circle, tan $\theta = y/x$. In the second quadrant y is positive and x is negative, so tangent is negative. Extend the ray backwards to cut the line $x = 1$ and the negative y-value there will be tan θ (Figure 33).

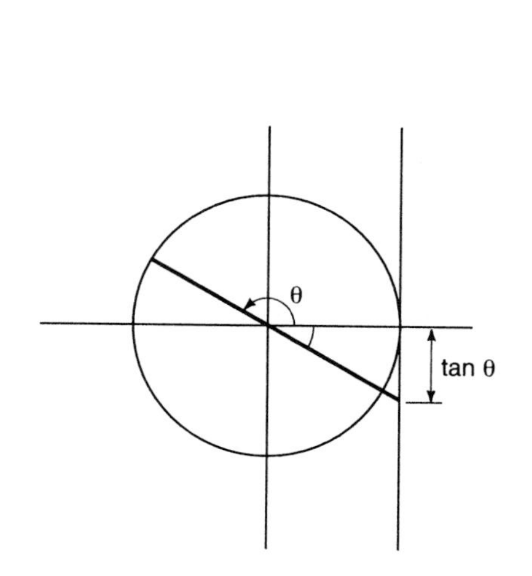

Figure 33: tan θ for θ in the second quadrant, and for θ in the fourth quadrant.

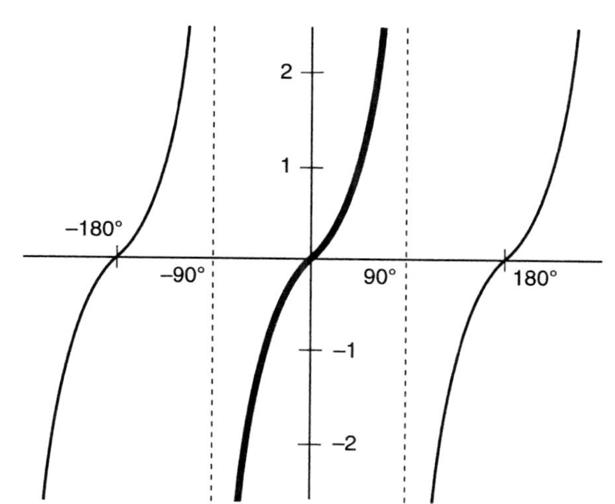

Figure 34: tan x. [-270°, 270°] by [-2.5, 2.5]. The region where $\tan^{-1}$ is defined is emphasized.

To plot the usual graph of "tan x," we need to use "x" for the angle, not for the horizontal coordinate. The usual graph is in Figure 34. Note that it repeats every 180 degrees.

Example 19: $\angle C = 90°$, $b = 50$, and $\angle A = 80°$. Find a.

Of course, you are making sketches of the triangles in these problems without pictures, aren't you?

From the point of view of A, a is the opposite side and b is the adjacent side. In a right triangle, "tangent is opposite over adjacent."

$$\tan 80° = a/50. \quad a = 50 \tan 80° = 283.6.$$

Example 20: $\angle B = 90°$, $a = 12$, and $\angle A = 60°$. Find c.

This time angle B is the right angle. From the point of view of angle A, a is the opposite side and c is the adjacent side.

$$\tan 60° = 12/c. \quad c = 12/(\tan 60°) = 6.93.$$

Example 21: $\angle C = 90°$, $a = 10$, and $b = 20$. Find angle B.

From the point of view of angle B, b is the opposite side and a is the adjacent side.

$$\tan B = 20/10 = 2. \quad \angle B = \tan^{-1} 2 = 63.43°.$$

Inverse Tangent. The inverse tangent is simpler than the inverse sine or inverse cosine. The values of tangent repeat every 180 degrees (sine and cosine repeat every 360 degrees). In Figure 34 the continuous curve between -90° and 90° is evident, and the fact that both 90° and -90° form vertical asymptotes is very noticeable. They are easy to explain: Since, on the unit circle, tangent = y/x, there will be a vertical asymptote when $x = 0$, which occurs at 90° and also at -90°.

(6.2.11)

$$\tan x = c \text{ iff } x = \tan^{-1} c, \text{ or}$$
$$x = \tan^{-1} c \pm 180n°, \text{ for some integer } n.$$

Line 1 is more important than line 2, however, line 2 actually includes line 1, since n could be 0.

Example 22: In a triangle, $\angle C = 90°$, $a = 14$, and $b = 3$. Find $\angle A$.

Side a is opposite angle A, and side b is adjacent to angle A. So,

$$\tan \angle A = a/b = 14/3. \quad \angle A = \tan^{-1}(14/3) = 77.9°.$$

Example 23: In a triangle, $\angle C = 80°$, $c = 10$, and $b = 8$. Find $\angle B$.

As always, b is opposite angle B and c is adjacent to angle B. But we cannot use "tangent is opposite over adjacent" here, because that only applies to right triangles. Wait for the Law of Sines.

Reference Angles. Acute angles are the most common, followed by obtuse angles. Other angles are less often needed, but they do come up. Any trig function of any angle can be given in terms of an acute angle called its "reference angle." Knowledge of how to do this is important for solving for angles, because the inverse trigonometric functions on your calculator do not return all solutions, but only one. You are expected to generate the others using reference angles.

Let θ be any angle. In a unit-circle picture of θ, the positive acute angle formed by the terminal side and the x-axis is called the reference angle for θ. Figure 35 shows four angles with the same reference angle, α (alpha). Figure 36 gives four angles with reference angle β (beta).

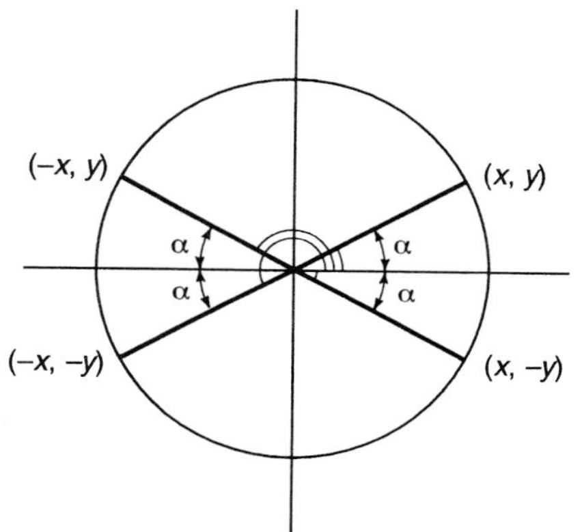

Figure 35: Four angles with reference angle α.

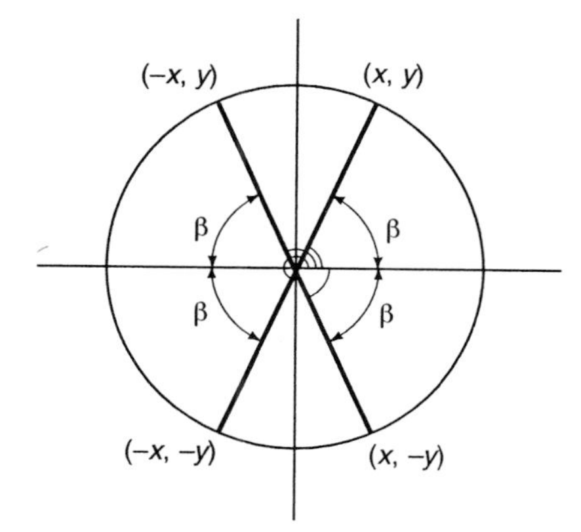

Figure 36: Four angles with reference angle β.

Note that the x-values (cosines) of the four angles all have the same absolute value, and their y-values (sines) all have the same absolute value too. So the sine or cosine of any one of them gives the sine or cosine of the others, if the right sign is attached. Tangents also all have the same absolute value. With any trig function value for the reference angle, all you need to find that trig function value for the original angle is knowledge of the sign in that quadrant. You are expected to know, or be able to figure out, the signs of the trig functions in the various quadrants (Table 6.2.14).

Example 24: $\angle D = 195°$. Give its reference angle, and give the three basic trig function values in terms of their values at the reference angle.

Angle D is 15 degrees into the third quadrant. It is 15 degrees past 180 degrees. Its reference angle is 15 degrees.

sin 195° = -sin 15° (sine is negative in the third quadrant, since $y < 0$).
cos 195° = -cos 15° (cosine is negative in the third quadrant, since $x < 0$).
tan 195° = tan 15° (tangent is positive in the third quadrant, since y and x are both negative, so y/x is positive).

Example 25: $\angle B = 100°$. Give its reference angle, and give the three basic trig function values in terms of their values at the reference angle.

Angle B is in the second quadrant, 80 degrees back from 180 degrees. The reference angle is 80°.

sin 100° = sin 80° (sine is positive in the second quadrant).
cos 100° = -cos 80° (cosine is negative in the second quadrant).
tan 100° = -tan 80° (tangent is negative in the second quadrant).

<u>**Basic Facts**</u>. There are only a few famous angles. Mathematicians memorize the figures that yield the sines, cosines, and tangents of the three acute angles that work out really nicely: 30°, 45°, and 60° (Figures 37 and 38. Note that the 30° right triangle has the 60° angle in it, so *two* pictures suffice for *three* key angles).

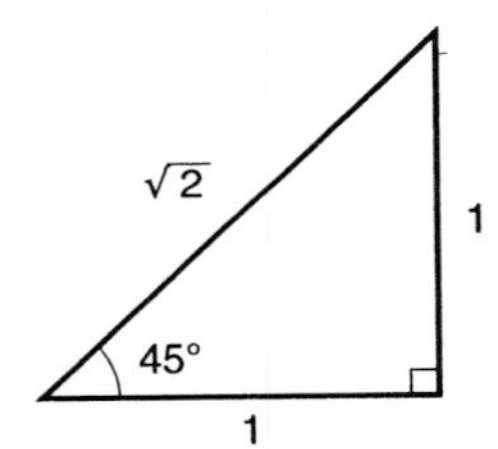

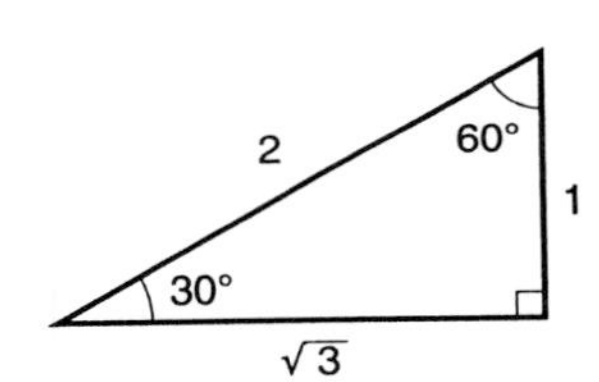

Figure 37: A 45° right triangle. **Figure 38**: A 30° - 60° right triangle.
Remember these two pictures.

Here are the values:

Table 6.2.12

θ	sin θ	cos θ	tan θ
0°	0	1	0
30°	1/2 = .5	$\sqrt{3}/2$ = .866	$1/\sqrt{3}$ = .577
45°	$\sqrt{2}/2 = 1/\sqrt{2}$ = .707	$\sqrt{2}/2 = 1/\sqrt{2}$ = .707	1
60°	$\sqrt{3}/2$ = .866	1/2 = .5	$\sqrt{3}$ = 1.732
90°	1	0	does not exist

Figure 37 exhibits a 45° right triangle. Both legs are 1. The hypotenuse follows from the Pythagorean Theorem. The values of the trig functions can be read from the picture.

Figure 38 exhibits a 30° right triangle with convenient side lengths -- the opposite is 1 and the hypotenuse is 2. sin 30° = 1/2 is the one memorable value from which you can reconstruct the picture. The values of cosine and tangent follow from the picture. Figure 38 also has a 60° angle. From the point of view of the 60° angle, the difference is only in the labeling of which side is "adjacent" and which "opposite." The adjacent side is 1 and the hypotenuse is 2: cos 60° = 1/2. Remember the values of the trig functions of these famous angles by computing them from these figures.

If you remember that, on a unit-circle figure, sine is the vertical coordinate and cosine is the horizontal coordinate, you should have no trouble remembering the signs of the trig functions in the various quadrants.

Table 6.2.13

[second quadrant]	$\sin\theta > 0$ $\cos\theta < 0$ $\tan\theta < 0$	$\sin\theta > 0$ $\cos\theta > 0$ $\tan\theta > 0$	[first quadrant]
[third quadrant]	$\sin\theta < 0$ $\cos\theta < 0$ $\tan\theta > 0$	$\sin\theta < 0$ $\cos\theta > 0$ $\tan\theta < 0$	[fourth quadrant]

A Pythagorean Identity. The unit circle has equation $x^2 + y^2 = 1$. Since $x = \cos\theta$ and $y = \sin\theta$ (6.2.3 and Figure 9), this immediately yields a famous trig identity:

(6.2.3, again) $$\sin^2\theta + \cos^2\theta = 1.$$

For clarity we write "$\sin^2\theta$" for "$(\sin\theta)^2$." We could confuse the sound of "$\sin(\theta^2)$" [sine (pause) theta squared] with "$(\sin\theta)^2$" [sine theta (pause) squared], but nothing else sounds like "$\sin^2\theta$" (sine squared theta).

Note that this is merely a disguised version of the Pythagorean Theorem. The hypotenuse is 1, so the square of the legs sum to the square of the hypotenuse.

Example 26: $\sin\theta = .6$. Find $\cos\theta$.

We could solve for θ using a calculator and then find cosine of it, however, we want to use the Pythagorean identity "$\sin^2\theta + \cos^2\theta = 1$."

$.6^2 + \cos^2\theta = 1$. $\cos^2\theta = 1 - .6^2 = .64$. So $\cos\theta = \pm.8$.

Which sign (plus or minus) to use depends upon which quadrant θ is in. For instance, if θ were a first quadrant angle, we would select +.8 because cosine is positive in the first quadrant. However, if θ were a second quadrant angle, cosine would be negative and we would select -.8.

Example 26, extended: $\sin \theta = x$, and θ is a first quadrant angle. Find $\cos \theta$.
$\sin^2\theta + \cos^2\theta = 1$. $x^2 + \cos^2\theta = 1$. $\cos^2\theta = 1 - x^2$.

$$\cos \theta = \sqrt{1 - x^2}\ .$$

The negative solution was discarded since we were told the angle is a first quadrant angle.

Example 26, another way: $\sin \theta = x$, and θ is a first quadrant angle. Find $\cos \theta$.

This problem can be done by drawing a picture and using the Pythagorean Theorem. After all, the trig identity "$\sin^2\theta + \cos^2\theta = 1$" is really just the Pythagorean Theorem with special notation.

Draw a right triangle and label it to illustrate "$\sin \theta = x$" (Figure 39). Now use the Pythagorean Theorem to determine the other leg. It will be $\sqrt{1 - x^2}$. Then read the cosine from the picture: "adjacent over hypotenuse" is

$$\cos \theta = \sqrt{1 - x^2}\,/1 = \sqrt{1 - x^2}, \text{ as before.}$$

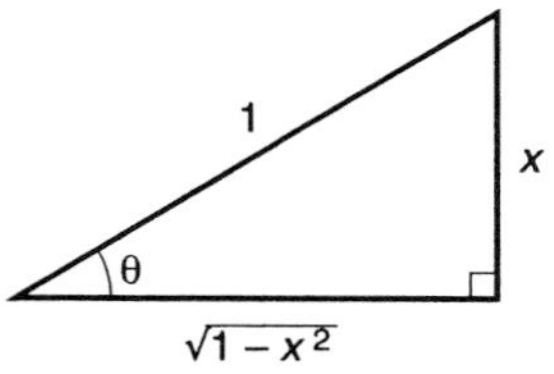

Figure 39: $\sin \theta = x$.
$\sin^{-1}x = \theta$.
$\cos \theta = \sqrt{(1 - x^2)}$.

<u>**Conclusion**</u>. The right-triangle interpretations of sine, cosine, and tangent are useful, but not sufficient for advanced mathematics. Learn the unit-circle interpretations. Remember that many triangles are not right triangles, so be careful to use the right-triangle interpretations only in right triangles.

Terms: acute, obtuse, right, hypotenuse, opposite, adjacent, sine, cosine, tangent, initial side, terminal side, coterminal, reference angle, inverse sine, inverse cosine, inverse tangent, isosceles.

****

Exercises for Section 6.2, "Trigonometric Functions":

A1.* State the unit-circle definition of
a) sine b) cosine c) tangent

A2.* State the right-triangle interpretation of
a) sine b) cosine c) tangent

A3.* In which quadrants are they positive?
a) sine b) cosine c) tangent

A4.* There are two interpretations of trigonometric functions in this section. Which is preferred for calculus?

A5.* Sketch the graph of sin x and label multiples of right angles.

A6.* Sketch the graph of cos x and label multiples of right angles.

A7.* In the unit-circle definition, we read $\sin^{-1}$ off the ________ half of the unit circle. We read $\cos^{-1}$ off the ________ half of the unit circle.

A8.* Define "coterminal."

^ ^ ^ ^ Give the exact value (not a decimal answer)

A9. a) tan 45° b) sin 30° c) cos 60°.
A10. a) sin 0° b) cos 0° c) tan 0°.
A11. a) sin 30° b) cos 30° c) tan 30°.
A12. a) sin 60° b) cos 60° c) tan 60°.
A13. a) sin 90° b) cos 90° c) tan 90°.
A14. a) sin 120° b) cos 135° c) cos 150°.

^ ^ ^ ^ Give the quadrant of the angle

A15. a) 170° b) -40° c) 57° d) -210° e) 350°.
A16. a) -12° b) 370° c) -100° d) 100° e) 210°.

^ ^ ^ ^ Give the reference angle of the angle

A17. a) 170° b) -40° c) 57° d) -210° e) 350°.
A18. a) -12° b) 370° c) -100° d) 100° e) 210°.
A19. a) 92° b) 23° c) 200° d) 178° e) 290°.
A20. a) -40° b) -100° c) 165° d) -190° e) 130°.

^ ^ ^ ^ The figure illustrates θ. Estimate a) sin θ. b) cos θ.

A21. **Figure 40** A22. **Figure 41** A23. **Figure 42** A24. **Figure 43.**

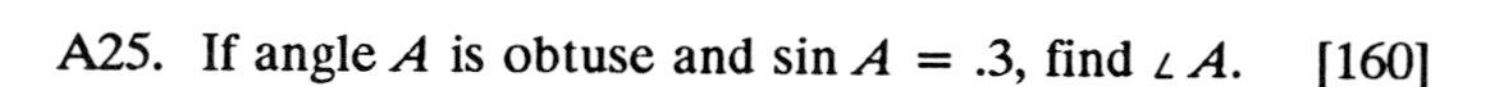

A25. If angle A is obtuse and sin A = .3, find $\angle A$. [160]

A26. If angle A is obtuse and sin A = .8, find $\angle A$. [130]

A27. If cos A = .3, find $\angle A$, $0° \le \angle A \le 180°$. [73]

A28. If cos A = -.37, find $\angle A$, $0° \le \angle A \le 180°$. [110]

^ ^ ^ ^ Right-triangle solutions. Let $\angle C = 90°$.

A29. $\angle A = 17°$ and $b = 12$. Find c. [13]

A30. $\angle B = 26°$ and $a = 45$. Find c. [50]

A31. $\angle B = 21°$ and $b = 5$. Find a. [13]

A32. $\angle A = 49°$ and $a = 90$. Find b. [78]

A33. $\angle B = 79°$ and $c = 100$. Find b. [98]

A34. $\angle A = 33°$ and $c = 100$. Find a. [54]

A35. $\angle A = 32°$ and $b = 50$. Find a. [31]

A36. $\angle B = 5°$ and $a = 100$. Find b. [8.7]

A37. $\angle B = 56°$ and $b = 2$. Find c. [2.4]

A38. $\angle A = 89°$ and $a = 100$. Find c. [100]

A39. $\angle A = 4°$ and $b = 100$. Find c. [100]

A40. $\angle B = 6°$ and $a = 10$. Find c. [10]

^ ^ ^ ^ Sketch a unit-circle picture to illustrate

A41. $\sin x = .8$

A42. $\cos x = .2$

A43. $\cos x = .4$

A44. $\sin x = .3$

A45. $\tan x = 2.$

A46. $\tan x = -.5$

A47. $\cos x = -.9$

A48. $\sin x = .2$

A49. Check the estimates in Example 1 using a calculator. How far off were they?
A50. Check the estimates in Example 2 using a calculator. How far off were they?
A51. Check the estimates in Example 4 using a calculator. How far off were they?

A52. How many degrees does the hour hand of a clock rotate in one hour?
A53. How many degrees does the minute hand of a clock rotate through in one minute?
A54. How many degrees does the hour hand of a clock rotate through in one minute?
A55. How many degrees of longitude does the earth rotate through in one hour?

^ ^ ^ ^ ^ ^ ^ ^ ^

B1.* a) Draw a unit-circle picture to illustrate "$\sin x = c$" for $0 < c < 1$.
b) Use it to explain why that equation has two solutions which could be angles in a triangle.
c) Draw and label a right-triangle picture to illustrate "$\sin x = c$."
d) Will your picture in part (c) work for second quadrant angles?

B2.* a) Draw a unit-circle picture to illustrate "$\cos x = c$" for $-1 < c < 1$.
b) Use it to explain why that equation has two solutions between 0 and 360 degrees, but only one could fit in a triangle.
c) Draw and label a right-triangle picture to illustrate "$\cos x = c$."
d) Will your picture in part (c) work for second quadrant angles?

B3.* a) Draw a unit-circle picture to illustrate "$\tan x = c$" for $0 < c$.
b) Draw and label a right-triangle picture to illustrate "$\tan x = c$."

B4.* Draw a picture to illustrate "$\sin \theta = \sin(180° - \theta)$."

B5.* If $\sin x = c$, x is not uniquely determined. a) Illustrate this with a unit circle picture.
b) Illustrate this with a graph of $\sin x$.

B6.* If $\cos x = c$, x is not uniquely determined. a) Illustrate this with a unit circle picture.
b) Illustrate this with a graph of $\cos x$.
c) If x were limited to being an angle in a triangle, would it be uniquely determined?

B7.* a) Illustrate "$\sin^2 x + \cos^2 x = 1$." b) What is the geometric name of this theorem?

B8.* Some people remember sine as "opposite over hypotenuse." What is "inverse sine"?

B9.* Draw and label a picture that gives the trig functions of 30° and 60°.

B10.* Draw and label a picture that gives the trig functions of 45°.

B11. a) Evaluate sin 160°. b) Evaluate $\sin^{-1}$(sin 160°).
c) Why isn't the answer to part b) "160°"? d) For which values of θ is $\sin^{-1}(\sin \theta) = \theta$?

B12. Sketch the graph of tan x and label multiples of right angles.

B13. Why do values of sine repeat every 360°?

B14. For which values of θ is $\cos^{-1}(\cos \theta) = \theta$?

B15. Use a unit-circle picture to help do this problem. Express the following in terms of the same trigonometric function of θ.
a) $\sin(-\theta)$. b) $\cos(-\theta)$ c) $\tan(-\theta)$.

B16. Use a unit-circle picture to help do this problem. Express the following in terms of the same trigonometric function of θ.
a) $\sin(180° + \theta)$ b) $\cos(180° + \theta)$ c) $\tan(180° + \theta)$.

B17. Use a unit-circle picture to help do this problem. Express the following in terms of the same trigonometric function of θ.
a) $\sin(180° - \theta)$ b) $\cos(180° - \theta)$ c) $\tan(180° - \theta)$

B18. In a unit circle, a) How long is a chord associated with central angle 40°? [.69]
b) What is the central angle of a chord of length .8? [47]

B19. In a unit circle, a) How long is a chord associated with central angle 110°? [1.6]
b) What is the central angle of a chord of length 1.5? [97]

B20. In a circle of radius 1000, a) How long is a chord associated with central angle 5°?
b) What is the central angle of a chord of length 10? [44, .57]

^ ^ ^ ^ Let θ be in the first quadrant. [See "Example 26, another way."]
B21. $\sin \theta = x$. Give $\cos \theta$. B22. $\sin \theta = x$. Give $\tan \theta$.
B23. $\cos \theta = x$. Give $\sin \theta$. B24. $\cos \theta = x$. Give $\tan \theta$.

B25. Solve for x in [0°, 360°).
a) $\cos(2x) = .5$ [...,150,...] b) $\tan(3x) = 1$ [...,75,...] c) $\sin(4x) = .5$.

B26. Solve for x in [0°, 360°).
a) $\sin(2x) = .5$ b) $\tan(x - 30°) = 1$ c) $\cos(4x) = .5$.

B27. A regular pentagon (5 sides) is inscribed in a unit circle. How long are its sides? [1.2]

B28. Hold out your arm and stick up your thumb. If your thumb is 2 feet from your eyes and the top part of your thumb is 1¼ inches high, what angle does it make with your eyes?

B29. Solve graphically: $\sin x < 1/2$ in [0°, 360°).

B30. The text draws a parallel between solving "$\sin A = c$" and solving "$x^2 = d$." What is the parallel?

B31. Draw a triangle. Draw an altitude to side c. Use it and the definition of sine to derive "$(\sin A)/a = (\sin B)/b$."

B32. The sun makes an angle of about ½° side-to-side. Approximately how long does it take for the sun to completely disappear after its bottom rim just touches the horizon?

B33. Suppose you want to solve for a side of a right triangle, given the angle at B and one side. How many distinct types of such problems are there? (For example, 1) find the adjacent side given the hypotenuse, 2) find the opposite side given the adjacent side, ...)

B34. a) Give the usual formula for the area of a triangle.
b) Draw a picture of a typical triangle with base b and sketch in an altitude.
c) Use the definition of sine to derive the formula for the area of a SAS triangle (6.2.5).

Section 6.3. Solving Triangles

At first it may seem that right triangles are the proper setting for trigonometric functions, but we will see that trigonometric functions are equally valuable in non-right triangles. In Section 6.1 we used geometry to describe when a triangle is determined and when it is not. This section gives the trigonometric theorems that correspond to those geometric results, regardless of whether the triangle is a right triangle. The Law of Sines and the Law of Cosines use trigonometric functions to determine the remaining angles and sides of a triangle given three parts.

Consider any triangle oriented above one horizontal side (called the "base," which can be any side, depending upon how the triangle is turned). Call the base "b" and the vertex opposite it "B". Both the Law of Sines and the Law of Cosines are easy to derive by drawing a perpendicular line from B to the base (Figures 1A and 1B).

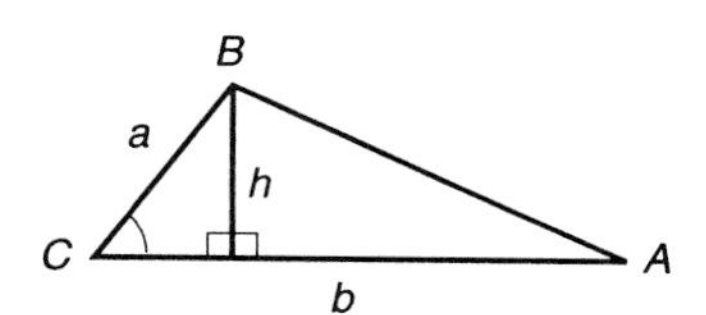

Figure 1A: Angle C is acute.

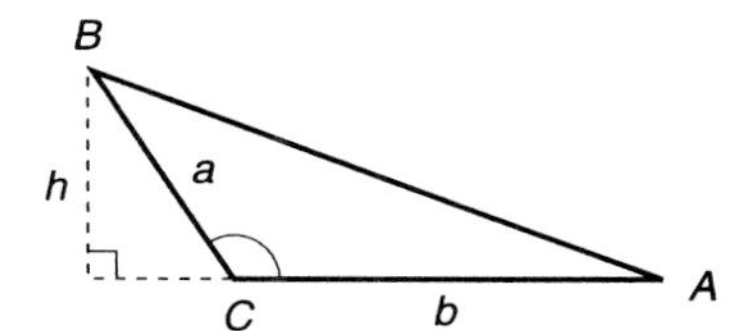

Figure 1B: Angle C is obtuse.

This line is sometimes called an "altitude" and its length is commonly known as the "height" and labeled "h". A simple formula for the area of a triangle in the Side-Angle-Side case follows almost immediately from the same picture and the usual formula for the area of a triangle.

<u>**Area**</u>. The area of a triangle is given by the well-known formula

(6.3.1) $$\text{area} = (1/2) \times \text{base} \times \text{height} = \tfrac{1}{2}bh,$$

where the height must be measured perpendicular to the base. This perpendicular creates a right triangle. Suppose a triangle is labeled as in Figure 1, with side b as the base and side a adjacent to a known angle, $\angle C$. The geometric case is Side-Angle-Side. Then the height ("altitude") h is opposite $\angle C$ in a right triangle with hypotenuse a. So $\sin C = h/a$, and

$h = a \sin C$.

Then the above area formula can be rewritten:

<u>(6.3.2) SAS Area Formula</u>: area $= (1/2)ab \sin C$.

Of course, the variables are dummy variables. The result applies to any example where the given angle is included between the given sides (Problems B3 and B10).

Example 1: Find the area of the triangle with angle $B = 71°$, $c = 4$, and $a = 9$.

The letters do not match 6.3.2, but that is irrelevant. The SAS Area Formula applies whenever two sides and the included angle are given, regardless of the notation for them.

$$\text{area} = (1/2)4(9)\sin 71° = 17.01.$$

The area formula 6.3.2 has a factor of "sin C". If C is a right angle, $\sin C = 1$ and the formula simplifies to "$\frac{1}{2}ab$", which is just "(1/2) × base × height, since a would be the height. Formula 6.3.2 extends the well-known area formula 6.3.1 to triangles described by two sides and the included angle (SAS).

<u>**The Law of Sines**</u>. The Law of Sines is a powerful result with a simple proof. Consider sides a and b and angles A and B in Figures 2A and 2B. How are these four quantities related?

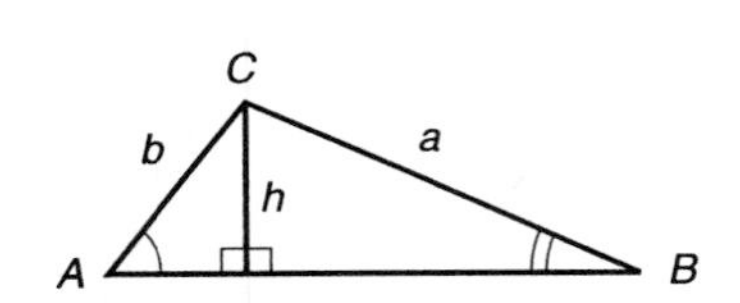

Figure 2A: Angle A is acute.

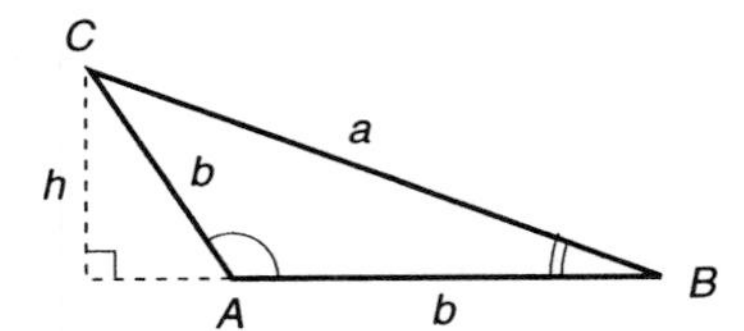

Figure 2B: Angle A is obtuse.

Draw in an altitude from C to side c. Now angles A and B are in *right* triangles and we can use the interpretation of sine twice:

$$\sin A = h/b. \quad \sin B = h/a.$$

Now isolate h in each:

$$b \sin A = h. \quad a \sin B = h.$$

Setting these expressions for h equal,

$$b \sin A = a \sin B.$$

Dividing through by ab, this yields

$$(\sin A)/a = (\sin B)/b.$$

If, instead, you divide through by $(\sin A)(\sin B)$, this yields

$$b/(\sin B) = a/(\sin A).$$

The two versions are equivalent.

Of course, this applies to c and $\sin C$ too (without further proof), because all the letters are just dummy variables. The key is that the four parts consist of two angles and the two sides opposite them. If you know three of the four, there will be only one unknown.

(6.3.3) The Law of Sines: In any triangle, if side a is opposite angle A, side b opposite angle B, and side c opposite angle C, then

A) $$\frac{a}{\sin A} = \frac{b}{\sin B} = \frac{c}{\sin C},$$ which is equivalent to

B) $$\frac{\sin A}{a} = \frac{\sin B}{b} = \frac{\sin C}{c}.$$

There is no significant difference between the two equivalent ways of stating the Law of Sines -- either will do. Part A is convenient for solving for unknown sides, because the unknown will be in the numerator. Part B is convenient for solving for unknown angles, because the angle will be in the numerator.

If you know two angles and one of the opposite sides, the first version makes it easy to solve for the unknown side.

Example 2: Let $\angle A = 70°$, $\angle B = 80°$, and side $b = 100$. Find a (Figure 3). Use the first version.

$$\frac{a}{\sin A} = \frac{b}{\sin B}. \qquad \frac{a}{\sin 70°} = \frac{100}{\sin 80°},$$

$$a = (\sin 70°)\left(\frac{100}{\sin 80°}\right),$$

$$a = 95.42.$$

Geometrically this case is described by "two angles and an opposite side" (AAS, 6.1.5). As Example 2 shows, it is an easy case trigonometrically.

If we know two sides and one of the opposite angles, the triangle is "almost" determined. This is the dangerous geometric Angle-Side-Side case (6.1.6) which may or may not determine a triangle.

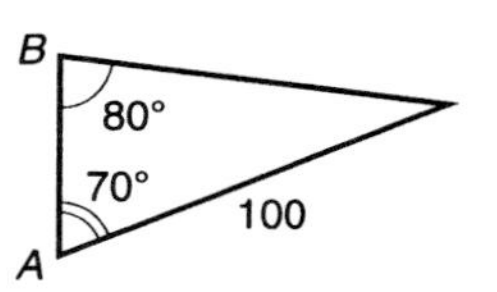

Figure 3: $\angle A = 70°$, $\angle B = 80°$, and $b = 100$.

Example 3: Suppose $\angle C = 39°$, $c = 16$, and $a = 20$. Find $\angle A$.

Do not thoughtlessly use the Law of Sines. Think about the geometric case. This example gives two sides and a *non-included* angle. Dangerous! Angle-Side-Side!

Use the Law of Sines, but beware the possibility of two, one, or no answers. To solve for an angle, I prefer the second version of 6.3.3 which has the angles on top. Use the c, C, a, and A equation.

$$\frac{\sin A}{a} = \frac{\sin C}{c}. \qquad \frac{\sin A}{20} = \frac{\sin 39°}{16}$$

$$\sin A = 20\left(\frac{\sin 39°}{16}\right) = .78665.$$

Here is the dangerous part. It is too easy to compute the inverse sine of this number ($\sin^{-1}.78665 = 51.87°$) and feel finished. But there is a second answer -- and it is often the one we want. Remember that "$\sin x = c$" has two solutions -- the one your calculator provides using the inverse sine function, and 180° minus that one (6.2.5). So $\angle A = 51.87°$ or $180° - 51.87° = 128.13°$.

Figure 4 shows the two triangles: CAB (acute A) and $CA'B$ (obtuse A').

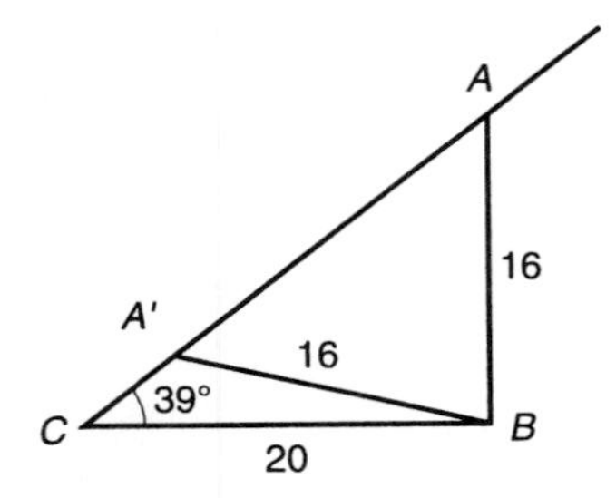

Figure 4: $\angle C = 39°$, $c = 16$, and $a = 20$.

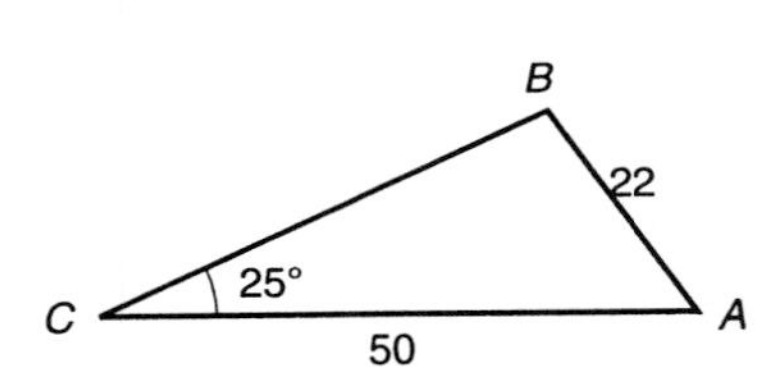

Figure 5: Find Angle A.

Example 4: Find angle A in Figure 5.

This is not given as a right triangle, so do not employ right-triangle thoughts. Note an angle and opposite side are given, which suggests using the Law of Sines, even though both angle A and its opposite side are unknown. Since b is given, work with angle B first. Then use the fact that the sum of the angles is 180° to find the remaining angle, $\angle A$.

$$\frac{\sin B}{b} = \frac{\sin C}{c}. \qquad \frac{\sin B}{50} = \frac{\sin 25°}{22},$$

$$\sin B = 50\left(\frac{\sin 25°}{22}\right) = .9605\ .$$

Here is the dangerous part again. Solving for $\angle B$,

$$\sin^{-1} .9605 = 73.84°,$$

and the second answer is $180° - 73.84° = 106.16°$. We accept the second answer because the picture shows angle B to be obtuse, not acute. Now

$$\angle A = 180° - \angle C - \angle B = 180° - 25° - 106.16° = 48.84°.$$

The range of the sine function is [-1, 1]. There are no solutions to "$\sin x = c$" if $c > 1$ or $c < -1$. In Example 4 we had to solve "$\sin B = .9605$," where .9605 is not much less than 1. Certain examples lead to unsolvable equations. They imply that there is no such triangle.

Example 5: Let $\angle B = 50°$, $b = 10$, and $a = 15$. Find $\angle A$.

By the Law of Sines,

$$\frac{\sin A}{15} = \frac{\sin 50°}{10} .$$

Solving, $\sin A = 15(\sin 50°)/10 = 1.15$. There is no such angle. There is no such triangle. The given parts are a case of ASS and this time the opposite side is not long enough to reach the ray of the angle (Figure 6).

Figure 6: $\angle B = 50°$, $b = 10$, and $a = 15$.

Example 6: Let $\angle B = 32°$, $c = 50$, and $b = 70$. Find $\angle C$.

Perhaps by now you have learned to identify the geometric case before blindly applying the trigonometric formula. Again, the given parts fit ASS. The Law of Sines is still appropriate, but we will not use it without thinking.

$$\frac{\sin C}{50} = \frac{\sin 32°}{70} . \qquad \sin C = 50\left(\frac{\sin 32°}{70}\right) = .3785 .$$

Now we are wary. We remember that there are two solutions to "$\sin C = .3785$." We will not be fooled.

$$\angle C = \sin^{-1} .3785 = 22.24°, \text{ or } 180° - 22.24° = 157.76°.$$

But, if we think there are two answers we will have been fooled after all. Look again. Think.

In trig it is good to draw a picture (Figure 7). In this example the side opposite the known angle is longer than the adjacent side. We saw this as a case of ASS which has only one solution (Example 6.1.19). The geometric construction has the arc of length b so long that it cuts the ray from angle B only once -- the second cut would be on the "backside," where the angle would not really be angle B. There is only one triangle that meets the given specifications. How can we see that in the Law of Sines?

The sum of the angles in a triangle is 180 degrees. Angle B is given as 32 degrees, so there is no room for a 157.76 degree angle -- the total would be too much. So 157.76° is an extraneous solution. The solution must be 22.24°.

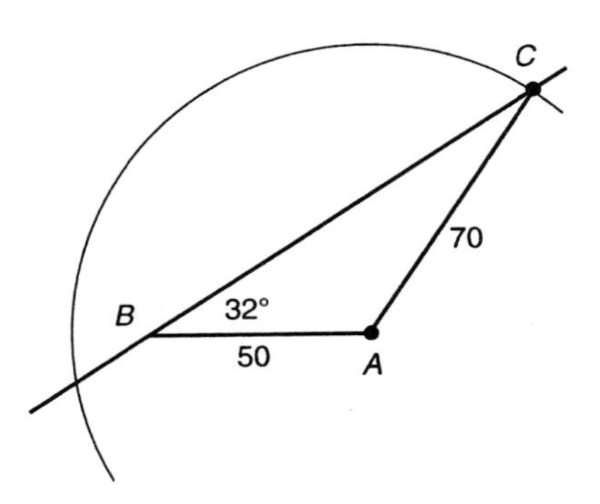

Figure 7: $\angle B = 32°$, $c = 50$, and $b = 70$.

Examples 3 through 6 show that the Law of Sines is tricky because Angle-Side-Side is a tricky geometric result. Example 2 shows that it handles "two angles and an opposite side" by using Angle-Side-Angle after determining the third angle. The other geometric cases are handled with the Law of Cosines.

The Law of Cosines. The Law of Cosines generalizes the Pythagorean Theorem to include non-right triangles. It is derived by using the Pythagorean Theorem twice.

(6.3.4) The Law of Cosines: A) $c^2 = a^2 + b^2 - 2ab \cos C$.

B) $$\cos C = -\frac{(c^2 - a^2 - b^2)}{2ab}.$$

Part A alone is usually called the Law of Cosines. It is written to emphasize solving for an unknown side opposite a known angle. Part B need not be learned separately -- it is what you would obtain if you solved for "$\cos C$" in Part A. It is an equivalent equation reorganized to emphasize solving for an unknown angle when all three sides are known (problem B6).

Note the similarity of Part A with the Pythagorean Theorem. The usual "$c^2 = a^2 + b^2$ " is there, but with a bit more on the end to compensate for angle C possibly not being a right angle. When angle C is a right angle, $\cos C = \cos 90° = 0$ and the extra term disappears. This theorem includes the Pythagorean Theorem!

If angle C is acute, $\cos C$ is positive and the term "$-2ab \cos C$" subtracts some amount from $a^2 + b^2$, so c is shorter than in a right triangle (Figure 8). If angle C is obtuse, $\cos C$ is negative, and then the term "$-2ab \cos C$" *adds* some amount to $a^2 + b^2$, so c is longer than in a right triangle (Figure 9).

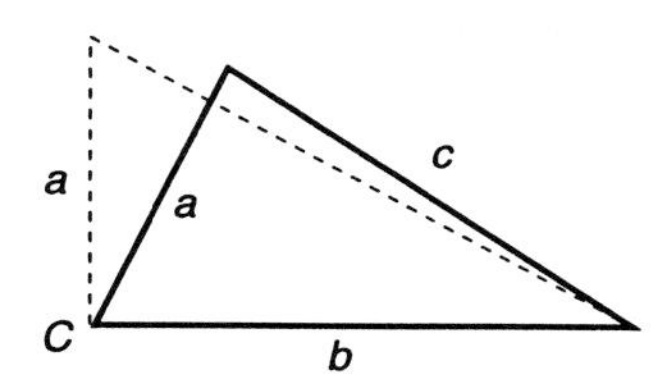

Figure 8: Angle C is acute. Side c is shorter than in a right triangle.

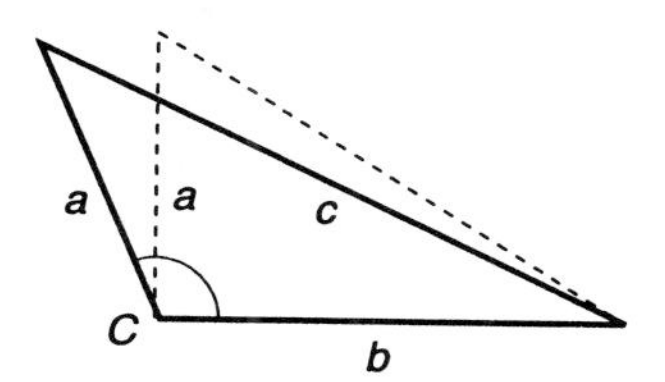

Figure 9: Angle C is obtuse. Side c is longer than in a right triangle.

Of course, the letters are dummy variables and the theorem can be used to solve for the third side in any case where two sides and the included angle are known, and to solve for any angle if all the sides are known. The letters used to express the theorem are arbitrary (Problem B4). The important part is that Part A is set up for the geometric case of Side-Angle-Side (SAS) and Part B is set up for the geometric case of Side-Side-Side (SSS).

Example 7: Let $a = 5$, $b = 3$, and $\angle C = 150°$. Find c.

The given angle C is between the given sides a and b, so the geometric case is SAS and the Law of Cosines applies.

$$c^2 = a^2 + b^2 - 2ab \cos C = 5^2 + 3^2 - 2(5)(3) \cos 150° = 59.98,$$

so c is the square root of 59.98. $c = 7.74$. (We do not need to worry about negative solutions, since sides of triangles are never negative.)

Example 8: Let $a = 9$, $b = 6$, and $c = 4$. Find $\angle C$ (Figure 10).

All three sides are given, so the geometric case is SSS and Part B of the Law of Cosines applies.

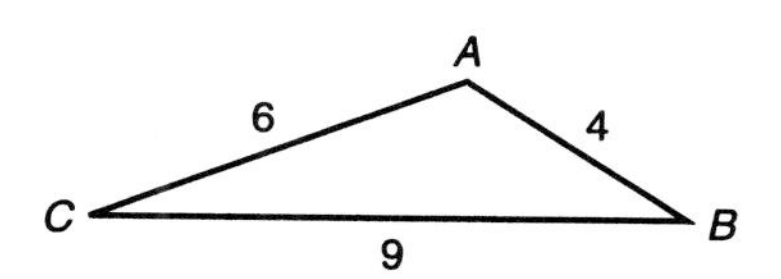

Figure 10: $a = 9$, $b = 6$, and $c = 4$.

$$\cos C = -\frac{(c^2 - a^2 - b^2)}{2ab}$$

$$= -\frac{(4^2 - 9^2 - 6^2)}{2(9)(6)} = .9352 .$$

This is solved for $\angle C$ by taking the inverse cosine.

$$\cos^{-1} .9352 = 20.74°.$$

Angle C is 20.74°.

There is no ambiguity with this as there is in taking the inverse sine in the Law of Sines. Taking the inverse cosine in the Law of Cosines always gives the right answer. This is because triangle angles (angles between 0 and 180 degrees) correspond to the top half of the unit circle, and there is only one point and one angle on the top half corresponding to "cos $\theta = x$," for any x between -1 and 1 (Figure 6.2.15. There is a second point on the bottom half, but it does not correspond to a triangle angle). However, in the Law of Sines there are two angles corresponding to "sin $\theta = y$" for y between 0 and 1, and both of those angles are between 0 and 180 degrees. That is why the Law of Sines yields an ambiguous answer for an unknown angle.

Part B of the Law of Cosines, as stated, emphasizes solving for an angle labeled C. If you want to solve for angle B, change the letters.

Example 8, continued: Let $a = 9$, $b = 6$, and $c = 4$. Find $\angle A$ (Figure 10, again).

The given parts are the same as in Example 8, but the question is different. This time we want to find angle A instead of angle C. Use the Law of Cosines, rewritten to emphasize angle A opposite side a (instead of angle C opposite side c):

$$\cos A = -\frac{(a^2 - b^2 - c^2)}{2bc}$$

$$= -\frac{(9^2 - 6^2 - 4^2)}{2(6)(4)} = -.6042 \, .$$

The cosine of obtuse angles is negative. $\cos^{-1}(-.6042) = 127.17°$. Angle A is 127.17°.

Triangles have six parts, which are often determined by three. When you use either the Law of Sines or the Law of Cosines to determine a fourth, there are usually several ways to determine the fifth and sixth parts.

More about Example 8: Suppose $a = 9$, $b = 6$, $c = 4$, and $\angle C = 20.74°$. Find angle A (Figure 11).

Usually a triangle is determined by three parts. Here, four are given. These are the three sides given in Example 8 and angle C which was determined in that example. Does knowing angle C help us find angle A?

In "Example 8, continued" we were able to find angle A by the Law of Cosines without using angle C. So we do not need angle C. But, knowing angle C gives us another option.

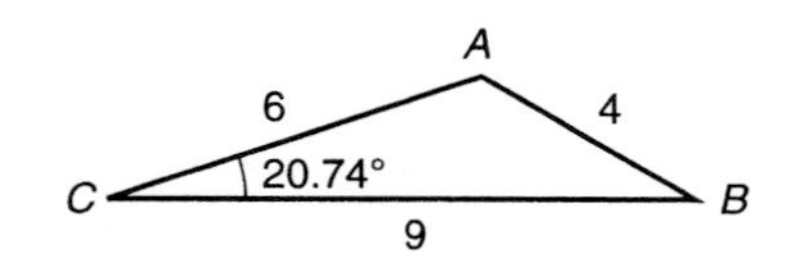

Figure 11: $a = 9$, $b = 6$, $c = 4$, and $\angle C = 20.74°$.

The other option is to use the Law of Sines to find angle A from the given side a, angle C, and side c. Then, once you get a second angle, the third follows easily from the sum of the angles in a triangle, Theorem 6.1.4.

The point is that, given four parts (instead of only three), there will be more than one way to find the rest.

Example 3, revisited: Suppose $\angle C = 39°$, $c = 16$, and $a = 20$. Find $\angle A$ (Figure 4, again, reproduced as Figure 12).

The Law of Cosines does not handle this problem directly. It has four variables, three of which are sides. It could handle "Find b" (the third side). Then we could find $\angle B$ and then $\angle A$, but it is not ready to "Find $\angle A$." So, change the problem to "Find b" (first).

From Part A,

$$c^2 = a^2 + b^2 - 2ab \cos C$$
$$16^2 = 20^2 + b^2 - 2(20)b \cos 39° \ .$$

This is a quadratic in b. To solve it, put it in "standard" form and treat "b" as the unknown "x".

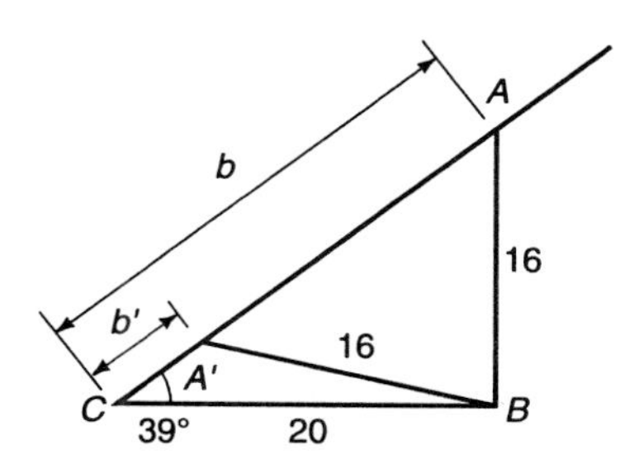

Figure 12: $\angle C = 39°$. $c = 16$, and $a = 20$.

$$b^2 - (40 \cos 39°)b + (20^2 - 16^2) = 0 \ .$$

Now you have a chance to use the Quadratic Formula by placeholder position, rather than by letter. The "b" you are solving for is not the "b" in "negative b plus or minus ..." formula.

$$\frac{40 \cos 39° \pm \sqrt{(-40 \cos 39°)^2 - 4(1)(20^2 - 16^2)}}{2(1)} = 25.42 \ \textit{or} \ 5.66 \ .$$

In the Law of Cosines the ambiguity of ASS appears in the two sides as answers to a quadratic equation. In the Law of Sines it appears in the two angles as answers to an equation for sine of the angle.

From the answer we know there are two triangles which satisfy the given conditions. How could we find $\angle A$ given one of the possibilities for side b? Now that we have four parts (a, b, c, and $\angle C$), there are several ways to obtain the rest.

1) Use b, c, $\angle C$, and the Law of Sines to obtain $\angle B$, and then use the sum of the angles is 180° to obtain $\angle A$.

2) Use a, b, c, and the Law of Cosines, Part B, with b and angle B emphasized in place of c and angle C.

Proof of the Law of Cosines. The proof of the Law of Cosines is not hard. It follows from using the Pythagorean Theorem twice on the same sort of picture we used for the SAS Area Formula and the Law of Sines Proof (Figures 13A and 13B).

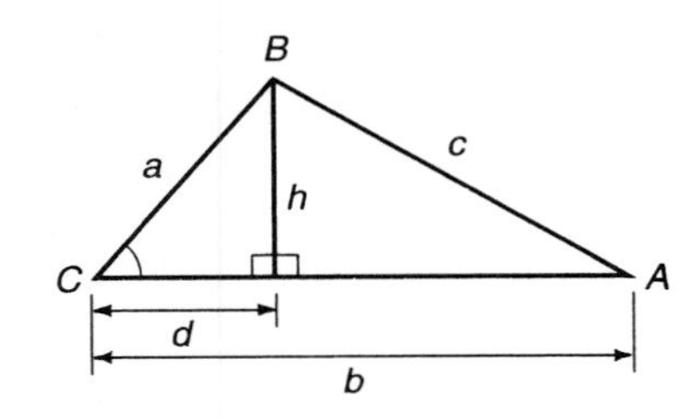

Figure 13A: Angle C is acute. **Figure 13B**: Angle C is obtuse.

Label h, the altitude, and d, the directed distance (positive in Figure 13A, negative in Figure 13B) from point C to the place where the altitude intersects the base, b.

Using the two right triangles,

(1) $a^2 = h^2 + d^2.$

(2) $c^2 = h^2 + (b - d)^2$ [d is negative in Figure 13B.]

$= h^2 + b^2 - 2bd + d^2.$

Now use (1) to replace $h^2 + d^2$ in (2):

(3) $c^2 = a^2 + b^2 - 2bd.$

Now note $\cos C = d/a$, so $d = a \cos C$. Substituting this into (3):

$$c^2 = a^2 + b^2 - 2ab \cos C,$$

which is the desired result (6.3.4).

<u>**The Geometric Cases**</u>. Reconsider the geometric cases to determine when the Law of Sines is appropriate and when the Law of Cosines is appropriate.

1) Side-Side-Side

The Law of Sines has two angles in it. We can solve for one, but not both. The Law of Sines won't work for SSS. For SSS use the Law of Cosines to obtain any desired angle.

2) Side-Angle-Side

The Law of Sines has two sides opposite two angles. In SAS neither of the given sides is opposite the given angle. The Law of Sines won't work, but the Law of Cosines is set up to handle this case and determine the third side.

3) Angle-Side-Angle or Angle-Angle-Side

In ASA the two angles determine the third. Then, and in AAS, the side is opposite a known angle, so the Law of Sines applies. The Law of Cosines has three variable sides -- knowing one is not enough -- it does not apply.

4) Angle-Side-Side

This case has a known side opposite a known angle. The Law of Sines will work, and so will the Law of Cosines. If the geometric construction is ambiguous, the equations resulting from these laws will be ambiguous too.

Conclusion. If a triangle is geometrically determined, it can be solved using the Laws of Sines and Cosines.

Key Terms: Law of Sines, Law of Cosines.

****

Exercises for Section 6.3, "Solving Triangles":

A1.* The Law of Cosines generalizes the _____________ Theorem.

A2.* a) The Law of Cosines is set up to handle two geometric cases that yield a unique answer. Which two?
b) There is another geometric case it can handle, although it may not give a unique solution. Which case?

A3.* a) When does the Law of Sines give an unambiguous answer?
b) When do you have to worry that the Law of Sines might give an ambiguous answer?

A4.* What must be given to make it easy to compute the area of a triangle? [There are two alternatives.]

^ ^ ^ ^

First identify the geometric case. Then solve.
A5. $\angle A = 12°$, $a = 20$, and $\angle B = 75°$. Find b. [93]
A6. $\angle C = 60°$, $c = 56$, and $\angle B = 72°$. Find b. [61]
A7. $\angle A = 40°$, $a = 90$, and $b = 100$. Find $\angle B$. [46,...]
A8. $\angle B = 18°$, $b = 20$, and $c = 30$. Find $\angle C$. [28,...]
A9. $\angle C = 55°$, $a = 100$, and $b = 200$. Find c. [160]
A10. $\angle A = 125°$, $c = 4$, and $b = 9$. Find a. [12]
A11. $a = 10$, $b = 12$, and $c = 20$. Find $\angle C$. [130]
A12. $a = 90$, $b = 80$, and $c = 40$. Find $\angle B$. [63]
A13. $a = 100$, $b = 80$, and $\angle A = 37°$. Find c. [150]
A14. $\angle C = 40°$, $c = 10$, and $b = 4$. Find $\angle B$. [15]
A15. $\angle A = 120°$, $a = 7$, and $b = 3$. Find $\angle B$. [22]
A16. $\angle C = 165°$, $b = 200$, and $c = 300$. Find $\angle B$. [9.9]
A17. $\angle A = 70°$, $\angle B = 105°$, and $c = 300$. Find b. [3300]
A18. $\angle B = 2°$, $\angle C = 5°$, and $a = 1000$. Find c. [720]

A19. Find the area: $\angle C = 67°$, $a = 90$, and $b = 150$. [6200]
A20. Find the area: $\angle B = 135°$, $a = 3$, and $c = 4$. [4.2]

A21. $a = 10$, $b = 20$, and the area of the triangle is 80. Find $\angle C$. [53,...]
A22. $b = 5$, $c = 3$, and the area of the triangle is 3.2. Find $\angle A$. [25,...]

^ ^ ^ ^ ^ ^ ^ ^

B1.* Essay: a) When is the Law of Sines applicable? b) When does it give unique results? c) Which geometric case is dangerous?

B2.* Essay: a) When is the Law of Cosines applicable? b) When does it give unique results? c) Which geometric case is dangerous?

B3.* Short essay: a) When is the SAS Area Theorem applicable? b) Rewrite it for the case when $\angle B$ is given.

B4.* Rewrite the Law of Cosines to emphasize solving for
a) Side b given sides a and c and angle B. b) angle B.

B5.* Rewrite the Law of Cosines to emphasize solving for
a) Side a given sides b and c and angle A. b) angle A.

B6.* Show the steps in converting the Law of Cosines, Part A, into Part B.

B7.* In each case, which theorem would be most appropriate for the first step in solving the triangle? a) SSS b) SAS c) ASA d) AAS e) SSA

B8.* Explain why solving for a triangle-angle can yield an ambiguous answer in the Law of Sines but not in the Law of Cosines.

B9. If $\angle B$ is a right angle and a and $\angle A$ are given, then b can be determined from the right-triangle interpretation of sine. Show how b also can be determined by the Law of Sines. [That is, the Law of Sines generalizes the right-triangle interpretation of sine.]

B10. Figure 1 illustrates Theorem 6.3.2 for an acute angle C. Do the same for an obtuse angle C. What changes in the proof?

B11. $\angle C = 21°$, $c = 5$, and $a = 6$. Given these parts, which part would be easiest to find next using (a) the Law of Sines and (b) the Law of Cosines?
c, d) Do those two computations.
e) How, in each computation, can you tell that this is a tricky problem?

B12. $\angle B = 25°$, $b = 15$, and $c = 9$. Given these parts, which part would be easiest to find next using (a) the Law of Sines and (b) the Law of Cosines?
c, d) Do those two computations.
e) How, in each, can you tell that this is a tricky problem?

B13. Find the area: $\angle A = 20°$, $\angle B = 85°$, and $c = 10$. [18]
B14. Find the area: $\angle B = 50°$, $\angle C = 110°$, and $b = 20$. [84]
B15. Find the area: $a = 9$, $b = 10$, $c = 15$. [44]
B16. Here is a problem [do not actually do it]: "$a = 12$, $b = 10$, and $\angle B = 30°$. Find the Area." a) What is tricky about this problem? b) If you did it, what would you do first? Next? How would you finish it?

B17. Suppose a, b, and the area of the triangle are given. a) Outline a plan for finding side c. b) Is c uniquely determined?

B18. Prove the Law of Sines (acute angle case, without looking).

B19. Prove the Side-Angle-Side Area Formula (acute angle case, without looking).

B20. a) $\angle C = 70°$, $c = 2$, and $b = 5$. Find $\angle B$.
b) Explain what happened trigonometrically and why it happened, geometrically.

B21. $\angle A = 55°$, $a = 17$, and $c = 20$. Find $\angle C$. [75]
B22. $\angle B = 126°$, $a = 4$, and $c = 6$. Find $\angle A$. [21]
B23. $\angle C = 119°$, $b = 10$, and $\angle A = 20°$. Find c. [13]
B24. $\angle B = 37°$, $\angle A = 109°$, and $c = 100$. Find a. [170]
B25. $\angle A = 22°$, $c = 5$, and $a = 4$. Find $\angle B$. [5.9,...]
B26. $\angle B = 59°$, $a = 10$, and $b = 9$. Find $\angle C$. [72,...]

B27. (Heron's Formula for the area of a triangle in the SSS case.) Let a, b, and c be the sides of a triangle (not necessarily a right triangle). Define $s = \frac{1}{2}(a + b + c)$ [the "semiperimeter"]. Then the area of the triangle is given by Heron's Formula:
$A = \sqrt{(s(s-a)(s-b)(s-c))}$. a) Use it to find the area of a 3-4-5 triangle.
b) Prove this formula using a perpendicular to side c and the Pythagorean Theorem twice on the right triangles formed. [It's quite long, but the mess simplifies very nicely in the end.]

B28. Prove the Law of Cosines (acute angle case, without looking).

B29. With his contact lenses, David, a basketball player, can focus his eyes precisely on an object (the hoop) 21 feet away. Suppose that his brain computes the distance using the angle between the lines of sight of his eyes and, when his muscles are responding well, he can then shoot three-point baskets pretty regularly. Without his lenses, David's two eyes cannot focus so well, and their angle may be a bit off. Suppose his eyes are 60 millimeters apart and his eyes focus at an angle 1/200 of a degree too wide. What happens to his shots when his brain and muscles are responding well to the erroneous input?

Section 6.4. Solving Figures

Trigonometry can be used to measure distances and angles indirectly in the real world. Some examples are so simple that they use only the right-triangle interpretations of trigonometric functions. Others are more complex and apply to geometric figures with more than three sides. The key to finding the sides and angles of geometric figures with four or more sides is to break them up into triangular regions and then use triangle-solving techniques.

Solving right triangles is relatively easy.

Example 1: Sarah wishes to determine the height of a building. From 200 feet away, the angle to the top from ground level is 35°. How tall is the building (Figure 1)?

The angle (35°) measured upward from level is called the "angle of elevation." In Figure 1 the unknown height is h and opposite the known angle. The known side (200 feet) is adjacent the known angle. Therefore the tangent function applies.

$\tan 35° = h/200$. $h = 200 \tan 35° = 140.0$ (feet).

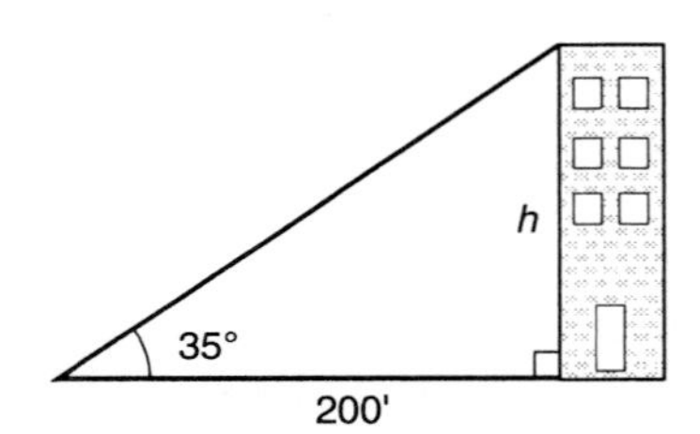

Figure 1: A building 200 feet away.

Occasionally angles are measured downward from level.

Example 2: A parked car is spotted at an angle of depression of 13.2° from a window at an elevation of 100 feet above the car (Figure 2). How far away is the car from the window?

The term "angle of depression" refers to the angle below level. The hypotenuse is the unknown distance, say d, and the opposite is known. This suggests using sine.

$\sin 13.2° = 100/d$. $d = 100/(\sin 13.2°) =$ 437.9 (feet).

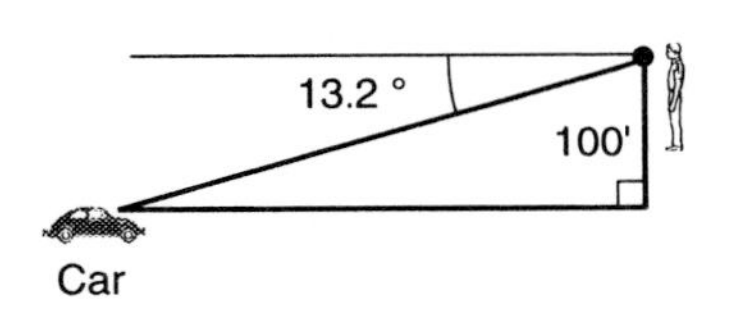

Figure 2: Angle of depression 13.2° from a height of 100 feet.

Solving Triangles. Geometric figures may have more than three sides, but they are still solved by the techniques developed for triangles. When an example is complicated, the idea is to break it into triangular components and solve those successively. Success depends upon being able to "see" how to build from known parts (usually three of them) to some unknown part, which then becomes known for determining the next. The upcoming examples illustrate the sequence of steps without actually computing the numbers.

In the next five examples, suppose we know the parts labeled 1, 2, and 3, and we wish to solve for the remaining three parts labeled 4, 5, and 6. The problem is to decide which of parts 4, 5, and 6 to solve for first, and which method to use to solve for it.

Example 3: Decide how to solve the triangle in Figure 3, given the first three parts.

The first thing to do is to decide on the geometric case. It is Angle-Side-Angle. Neither the Law of Sines nor the Law of Cosines will work. Use the sum of the angles in a triangle to determine angle 5 first. Then the Law of Sines will work to find either side 4 or side 6.

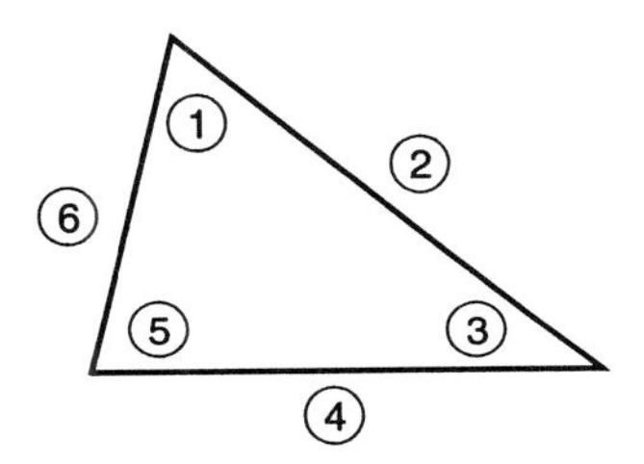

Figure 3: Parts 1, 2, and 3 are given.

Example 4: Decide how to solve the triangle in Figure 4, given the first three parts.

This is a case of Side-Angle-Side. The Law of Cosines makes it easiest to find side 5 next. Then either the Law of Cosines or the Law of Sines could be used to find the remaining angles.

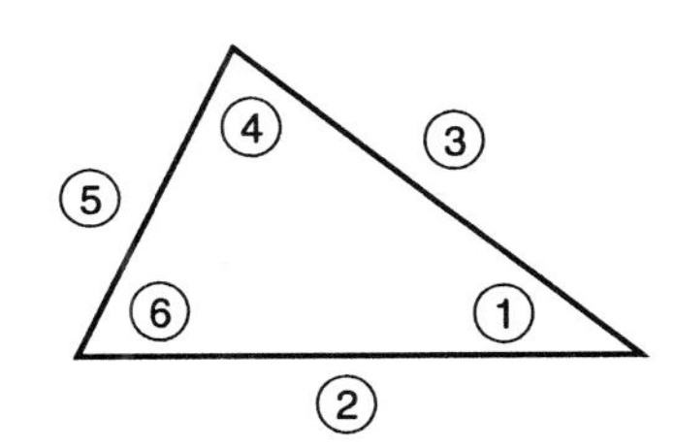

Figure 4: Parts 1, 2, and 3 are given.

Example 5: Decide how to solve the triangle in Figure 5, given the first three parts.

This is a case of Side-Side-Side. The Law of Cosines makes it easy to find any of the three unknown angles next. Then, it could be used again to find another angle. The third would follow easily from the sum of the angles in a triangle. The second angle could also be determined using the Law of Sines.

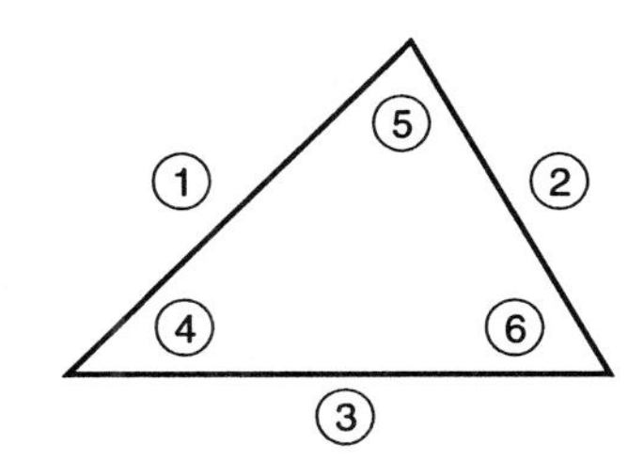

Figure 5: Parts 1, 2, and 3 are given.

Example 6: Decide how to solve the triangle in Figure 6, given the first three parts.

This is a case of Angle-Angle-Side. Whenever two angles are given, the third is easy to obtain using the sum of the angles in a triangle. So find angle 5 first. Then the Law of Sines will yield unambiguous results for the unknown sides.

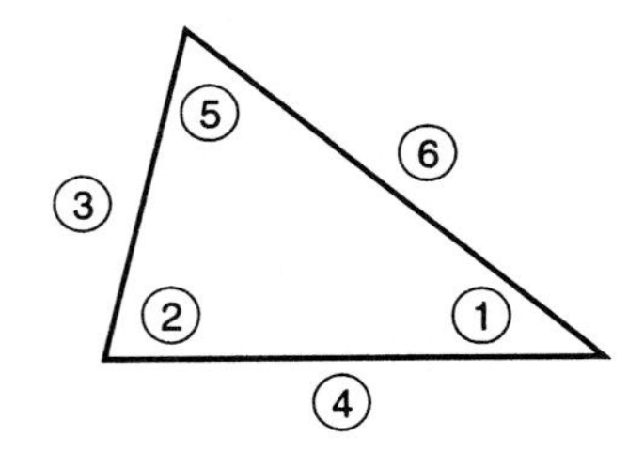

Figure 6: Parts 1, 2, and 3 are given.

Example 7: Decide how to solve the triangle in Figure 7, given the first three parts.

This is a case of Angle-Side-Side. Be careful. In this case you have two options. You may use the Law of Sines to find angle 5 opposite side 2. There may be two answers to the equation. From the picture, it looks like the obtuse angle is the answer we want.

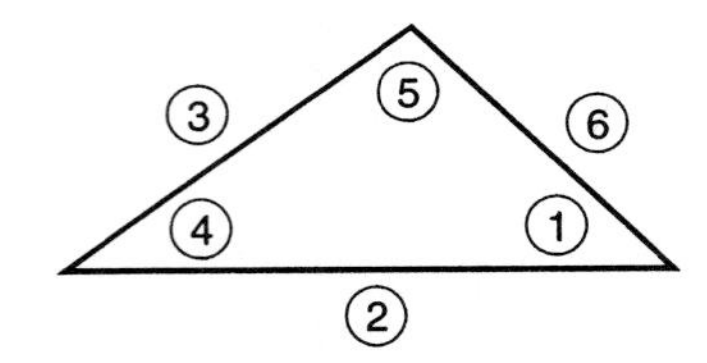

Figure 7: Parts 1, 2, and 3 are given.

Instead, you may use the Law of Cosines to find the third side, side 6. There may be two answers to the resulting quadratic equation because there are two triangles with parts 1, 2, and 3, although Figure 7 only pictures 1 of them. From the picture, it looks like side 3 could be the same length if angle 4 were larger, side 6 longer, and angle 5 acute instead of obtuse. Since side 6 could be longer, we must want the shorter of the two solutions for side 6.

If we have the original three parts and a solution for angle 5, the remaining angle follows from the sum of the angles in a triangle, and the remaining side follows easily from the Law of Sines.

If we have the original three parts and a solution for side 6, we may use either the Law of Cosines or the Law of Sines to find the remaining angles.

More Complex Figures. When geometric figures are not triangles, it is often necessary to subdivide them into triangles. More complex figures are solved with the same techniques we employ on triangles.

Example 8: Figure 8 illustrates three known sides and two known angles of a four-sided geometric figure. We can determine the fourth side and the two unknown angles by first subdividing the figure into two triangles as in Figure 9, and then solving each of the triangles.

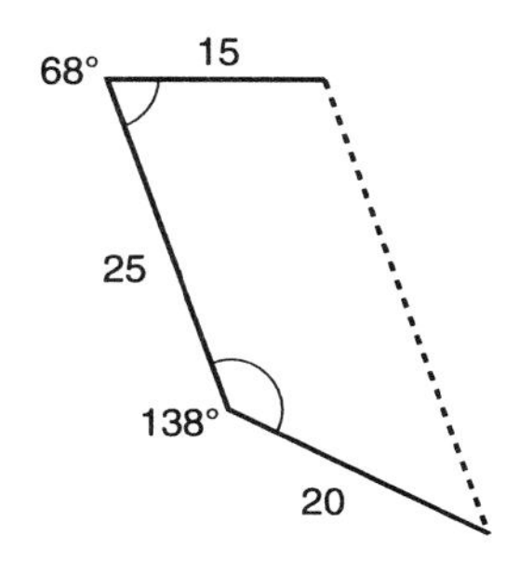

Figure 8: A four-sided figure.

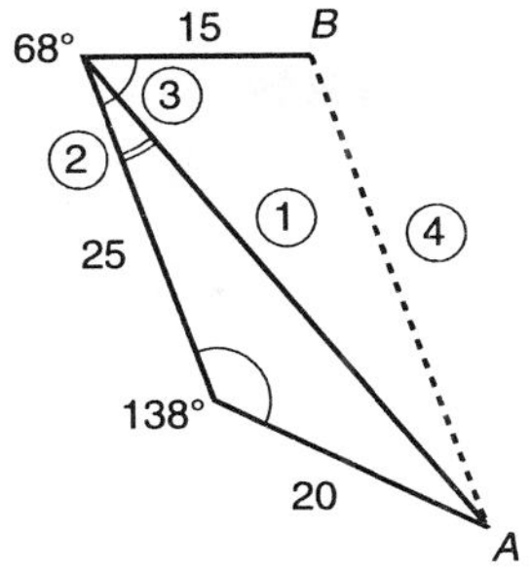

Figure 9: The figure subdivided into two triangles, and a solution sequence.

To find the fourth (dashed) side, we could first find the diagonal (labeled "(1)" in Figure 9) by SAS. Then we could find angle 2 using the Law of Cosines or the Law of Sines. Angle 3 follows easily by subtracting from 68°. Then side 4 follows from SAS and the Law of Cosines.

Calculator Exercise 1: Learn to use your calculator's memories to store and retrieve values. That way you will not need to enter them again when you want them for the next calculation. Furthermore, you will be less tempted to round off too much, because you will not be rounding off at all! You will be using all the decimal places the calculator computed.

Follow the calculations in Example 8. Side 1 is

$$\sqrt{20^2 + 25^2 - 2(20)(25)\cos 138}\ .$$

Result: 42.04931421.

Store this in a memory, say, in C. Now you can retrieve the value of C whenever you need it. You need it to compute angle 2.

Angle 2 follows from the Law of Cosines (or Sines). Here is the Law of Cosines approach:

$$\cos(\text{angle } 2) = -(20^2 - 25^2 - C^2)/(2(25)C).$$

Result: .9480034873.

This need not be stored, since we are going to use it only once -- immediately. Take the inverse cosine of that to find angle 2 and subtract from 68° to find angle 3. Result: 49.4422741 (degrees).

Store this in, say, T (for no good reason).

Now side 4 follows from SAS, given side 1 (stored as C), angle 3 (stored as T), and the other side of length 15.

Side 4 is computed by:

$$\sqrt{15^2 + C^2 - 2(15)C \cos T}\ .$$

Result: 34.24782406.

This has more decimal places than we need, so now, and not before, is a good time to "round off." Say, 34.25. People who round off to two decimal places at each step often find they no longer have two accurate decimal places by the time they get to the end. Use your calculator's memory rather than rounding off at each step.

Example 8, continued: Suppose we want to continue Example 8 and find angle B (Figure 9). How would you do it?

There are several ways to find angle B. We could use SSS and the Law of Cosines, now that we have sides 1 and 4 and the side of length 15. We would use the version with side 1 opposite angle B. Also, we could use the Law of Sines, now that we have side 4 opposite angle 3 and side 1 opposite angle B. Still another way would be to find the angle at A first and then subtract the two known angles from 180 degrees.

Trigonometry is useful for indirect measurement.

Example 9: Suppose you want to determine the elevation of the top of a mountain above a level plain. Of course, you can not measure from the top to the bottom, since the distance you want is perpendicular to the plain down to a point underneath the mountain. So here are the measurements you take (Figure 10).

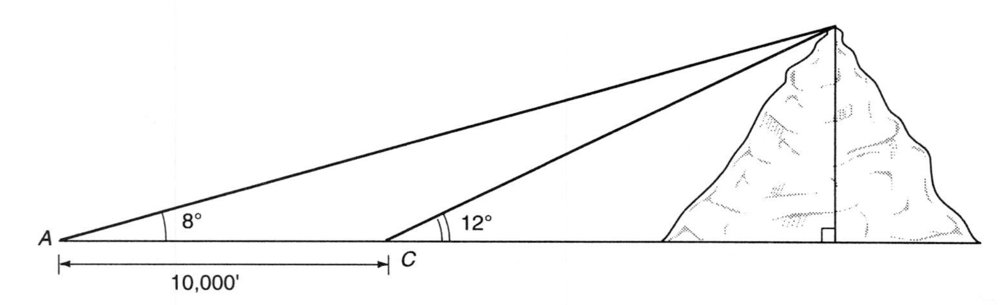

Figure 10: The elevation of a mountain. (The vertical scale has been exaggerated to make room for labeling small angles.)

You find a point on the plain where you can measure the angle of elevation to the top of the mountain. At point C the angle is 12°. Then you back off 10000 feet to point A and measure the angle of elevation from there. It is 8°. How high is the mountain above the plain?

There are several ways to do this problem. Before doing any calculations, look at the given parts and plan a sequence of steps that will find h. Labeling the appropriate parts on your picture will be helpful. Figure 11 has useful labels.

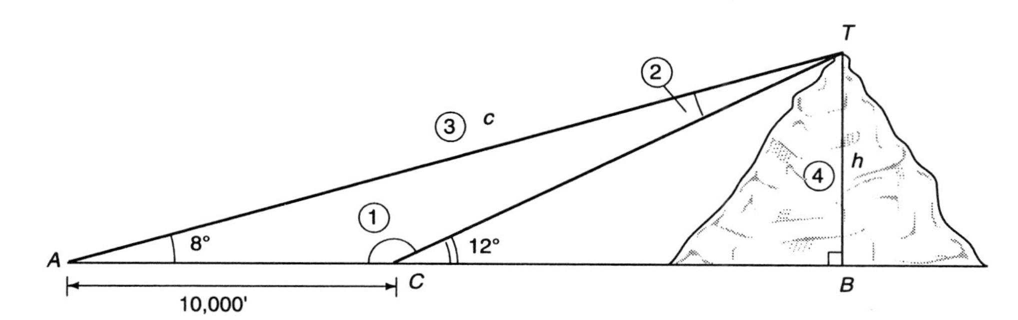

Figure 11: Figure 10, with the addition of useful labels.

In Figure 11 it is easy to find angle 1 and then angle 2. Then triangle ACT can be solved for side 3 (side c) using the Law of Sines. Then triangle ABT (a right triangle) can be solved for side h, given its hypotenuse c and angle A.

The above plan works. The plan is the key. The steps are then just "plug in" steps. Here they are for this example.

Angle 1 is 180° - 12° = 168°. Angle 2 is 180° - 168° - 8° = 4°. By the Law of Sines:

$$\frac{c}{\sin 168°} = \frac{10000}{\sin 4°}.$$

$$c = \sin 168°(10000)/(\sin 4°) = 29{,}805.36 \text{ (feet)}.$$

Store this in C, or, at least, do not erase it yet.

Now

$$\sin 8° = h/c. \quad h = c \sin 8° = 4148.1 \text{ (feet)}.$$

The mountain top is 4148 feet above the level plain. There are other strategies for doing this problem (Problems B4 and B5).

Example 10: Given the marked sides and angles in Figure 12, outline a plan to find x. Illustrate the sequence of steps and name the theorem you intend to apply at each step.

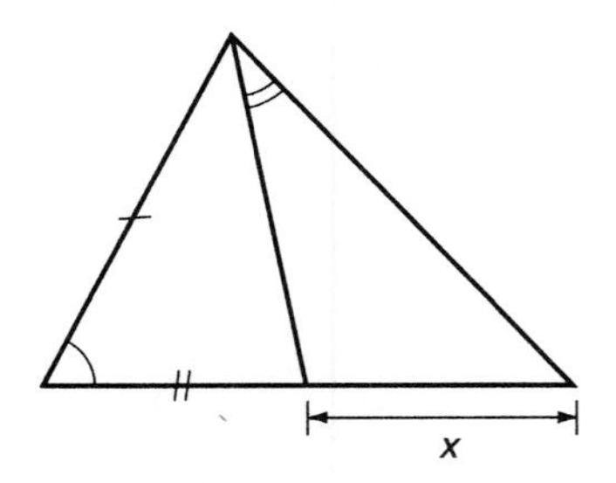

Figure 12: Find x, given the marked sides and angles.

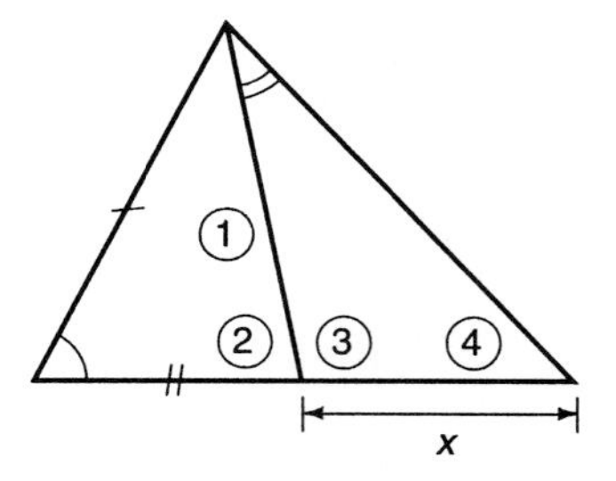

Figure 13: The sequence of steps.

Figure 13 illustrates the sequence of steps.
Step 1: Find side 1 using the Law of Cosines.
Step 2: Find angle 2 using the Law of Cosines or the Law of Sines.
Step 3: Find angle 3 (a straight angle is 180 degrees).
Step 4: Find angle 4 (angles of a triangle sum to 180 degrees).
Step 5: Find x using the Law of Sines.

To solve a complicated problem it is best to outline a plan before worrying about the details. The plan is the interesting part; the steps are just "plug in" calculations.

<u>Navigation</u>: Planes and ships can navigate by keeping track of directions and distances. In navigation, directions are related to North, South, East, and West. One way to give a direction is to state the <u>bearing</u> as either one of these four primary directions or as an acute angle away from North or South toward East or West. For example, "N 75° W" (Read "North seventy five degrees West") would mean 75° away from North toward West. That direction is mostly West. The bearing "S 20° E" would be mostly South, but 20° toward East.

Example 11: Starting from her dock in Key Largo in Florida, Pam navigates S 60° E for 10 miles. How far south of her dock is she? How far east?

Figure 14 illustrates her trip. In circular trigonometry it is common for the hypotenuse to be given, as in this example, and the sides to be determined from the angle. Then it is appropriate to use the following interpretations: In a *right* triangle,

(6.4.1) "opposite is hypotenuse times sine,"

and

(6.4.2) "adjacent is hypotenuse times cosine."

The hypotenuse is 10 and the southward component of her position is the adjacent side. Therefore she has traveled

$$10 \cos 60° = 5 \text{ (miles) south.}$$

The eastward component is the opposite side. She has traveled

$$10 \sin 60° = 8.66 \text{ (miles) east.}$$

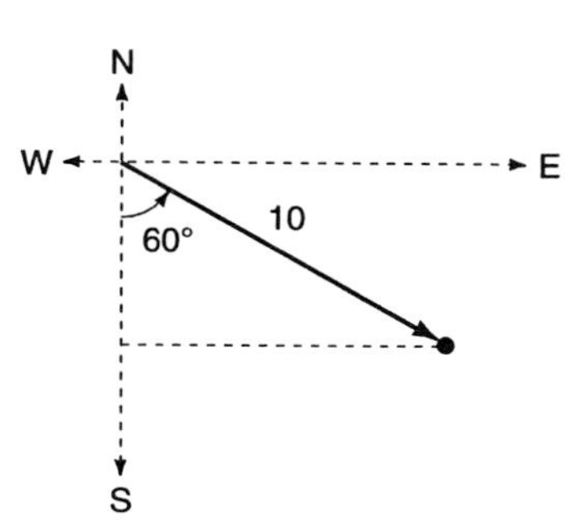

Figure 14: A 10-mile trip S 60° E.

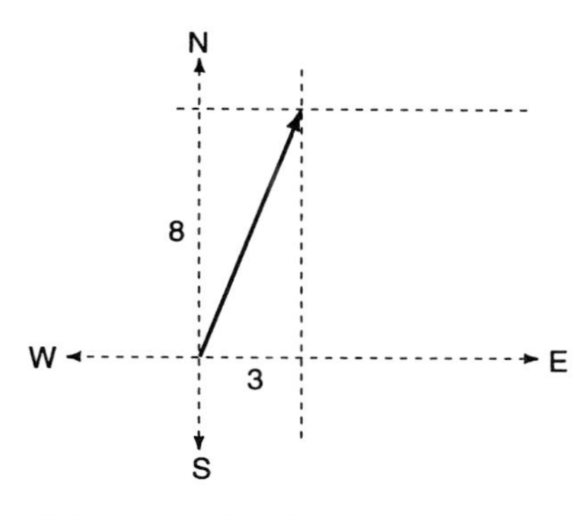

Figure 15: A trip 3 miles east and 8 miles north.

Given the opposite and adjacent sides, the angle is determined by its tangent.

Example 12: The next day Pam piloted straight to a position 3 miles east and 8 miles north of her dock (Figure 15). What bearing did she travel?

In this method of reporting directions we measure the angle away from North. Then 3 is the opposite side and 8 is the adjacent side.

$$\tan \theta = 3/8. \qquad \theta = \tan^{-1}(3/8) = 20.56°.$$

She traveled N 20.56° E.

The directional ideas in Examples 11 and 12 can be combined to handle more complicated problems.

Example 13: Tamara flew from base N 20° E for 30 miles and then S 55° W for another 22 miles (Figure 16). Where was she then? Which bearing should she fly to go directly back to her base?

There is more than one way to solve this problem. We will give two distinctly different approaches. The key is to plan a sequence of steps to obtain the answer. Pause here and think of a plan.

Both plans given below use the fact from geometry that, when a line cuts two parallel lines, the "alternate interior angles" are equal (Figure 17). We will treat all North-South lines as parallel, although that is not strictly true on the curved surface of the earth.

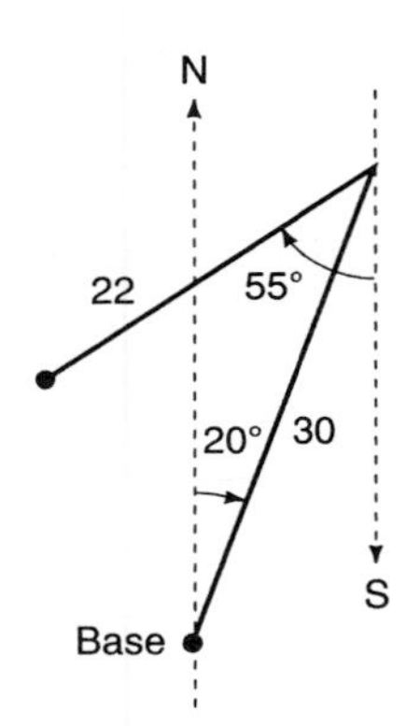

Figure 16: Two legs of a flight.

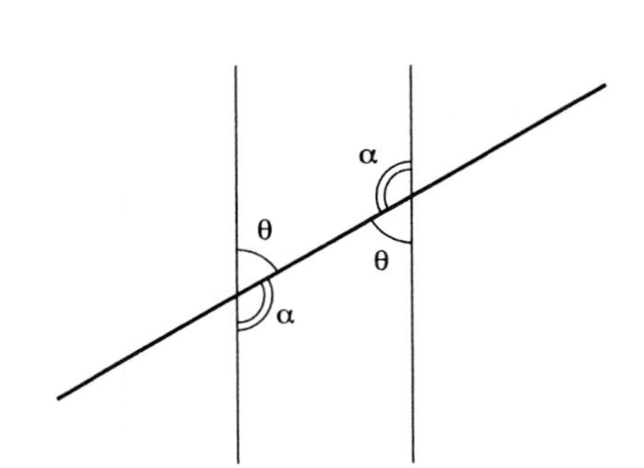

Figure 17: Alternate interior angles are equal.

Here is one plan. The leg back to base makes the third side of a triangle. From the given angles in Figure 16, we can determine the angle between the two given legs of the flight (Figure 18). Then, triangle ABC is determined by Side-Angle-Side and we can find the opposite side (b, which is the distance back to base) using the Law of Cosines. Then we can find angle A using the Law of Sines. From angle A, by subtracting 20°, we can find θ, which will give the bearing back to base: S θ° E.

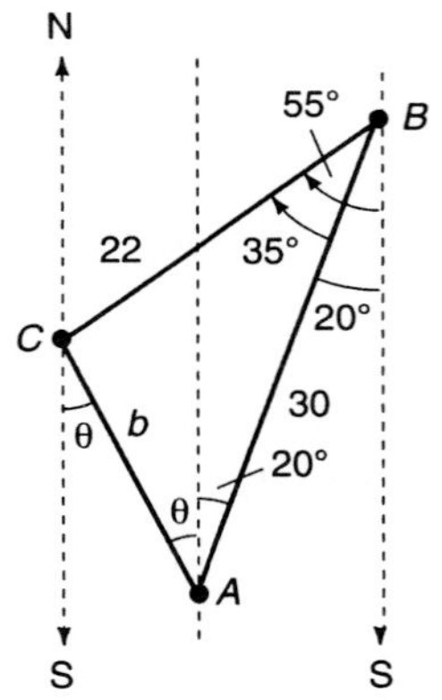

Figure 18: Figure 16 with auxiliary lines.

Here are the computations.

Angle B is 55°- 20° = 35°.

$$b^2 = 30^2 + 22^2 - 2(30)(22)\cos 35°.$$

$$b = 17.40.$$

She is 17.40 miles from base.
Now, by the Law of Sines,

$$\frac{\sin A}{22} = \frac{\sin 35°}{17.40}.$$

$$\sin A = 22(\sin 35°)/17.40 = .7252.$$

$$\angle A = 46.49°. \qquad \theta = 46.49° - 20° = 26.49°.$$

Therefore, she should fly S 26.49° E to return to base.

Example 13, again: Here is a second plan to find the distance and bearing back to base (Figure 16). As we saw in Example 11, for each leg of her flight we can extract the distance traveled in each primary direction from the bearing and distance of the leg. For this problem, we can do this for each leg and add the contributions to find the total distances in the primary directions.

Then use the idea of Example 12 to extract the bearing from the position.

The auxiliary lines and labels in Figure 19 are helpful.

Leg 1: north: $30 \cos 20°$ (a) east: $30 \sin 20°$ (b).

Leg 2: north: $-22 \cos 55°$ (c) east: $-22 \sin 55°$ (d)

[On Leg 2, the minus signs are because South is the negative of North and West is the negative of East.]

Total: north: $30 \cos 20° - 22 \cos 55° = 15.57$ (e) east: $30 \sin 20° - 22 \sin 55° = -7.76$ (f)

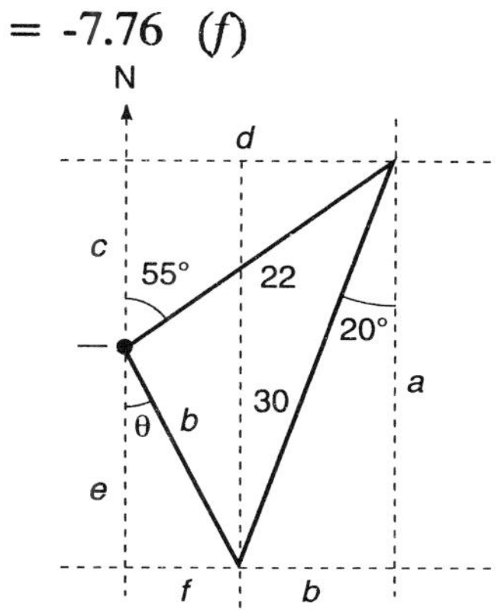

Figure 19: Figure 16 with auxiliary lines and labels.

So to return to base she will have to fly 15.57 miles south and 7.76 miles east. The bearing is S θ° E where θ is in the picture.

$$\tan \theta = 7.76/15.57 = .498.$$
$$\theta = \tan^{-1}.498 = 26.49°.$$

She should fly S 26.49° E.

The Pythagorean Theorem gives the distance: $\sqrt{(15.57^2 + 7.76^2)} = 17.40$ miles.

Both techniques work. They are about equally long. Take your pick.

Bearings from two known positions can be used to locate a third position.

Example 14: Sue lives 3 miles Northeast of Amy. When Sue spots a tornado at N 75° W, Amy sees it at N 20° W. At that instant, how far is the tornado from Amy?

This requires a picture (Figure 20). "Northeast" is half way between North and East, so the direction from Amy to Sue is 45° east of North. Plan how you could do this problem. Label the picture with angles and sides you could determine. Fill in auxiliary lines if useful. Your picture should develop from Figure 20 to a more complex picture with important parts labeled, and unimportant parts not labeled (Figure 21).

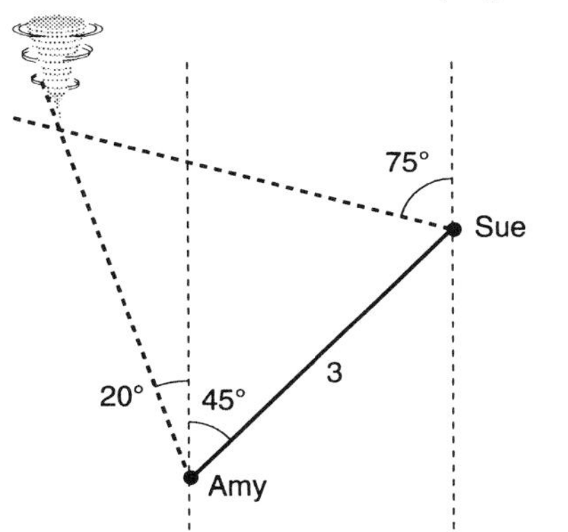

Figure 20: A tornado sighting.

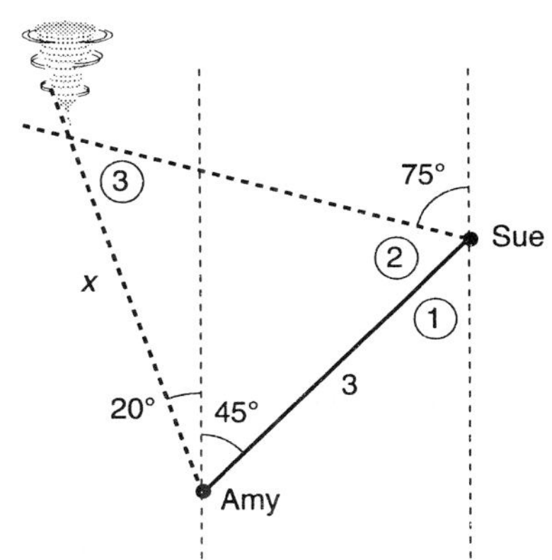

Figure 21: The order of solution.

By alternate interior angles, angle 1 in Figure 21 is also 45°. Then angle 2 can be found, and then angle 3. Finally the Law of Sines yields side x.

$$\text{angle } 2 = 180° - 75° - 45° = 60°.$$
$$\text{angle } 3 = 180° - 60° - 20° - 45° = 55°.$$

From the Law of Sines

$$\frac{x}{\sin 60°} = \frac{3}{\sin 55°}.$$

Solving for x, $x = 3.17$ miles. The tornado is 3.17 miles from Amy.

There is another way to state a bearing by giving the clockwise (not counterclockwise) angle in degrees away from North. So North is bearing 0, East is bearing 90, South is bearing 180, and West in bearing 270. So, when you watch an old World War II submarine movie and the skipper says "bearing two seven zero" he just means West, and "bearing three zero zero" means 30 degrees north of West, which would be "N 60° W" is the notation we just discussed.

In this section we have subdivided complex figures into simpler component figures. Here is a challenging problem about area that utilizes the idea of subdivision in a new way.

Example 15: Find angle B such that the area shaded in Figure 22 is its maximum for the given sides and right angles. That is, treat angle B as adjustable, and find how to adjust it to maximize the area.

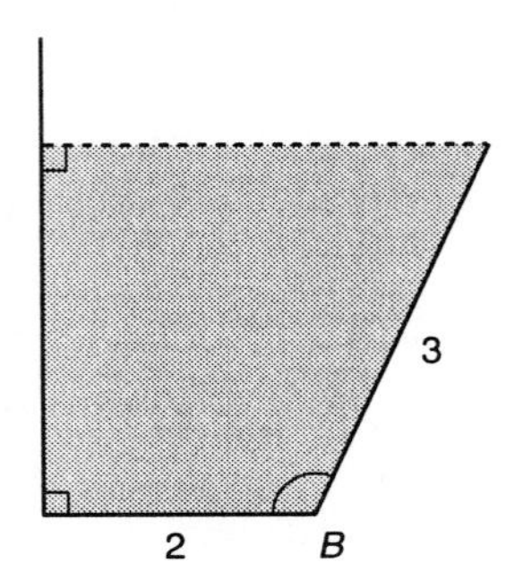

Figure 22: Find angle B to maximize the shaded area.

Many students cannot do this challenging problem without help. Pause here and think what you would do.

This is a "word" problem. Recall the standard word-problem advice. The first step is not to think about solving. It is to think about evaluating. The problem asks about the angle and area. How would you evaluate the area, if you knew the angle? When you can answer that, in algebraic notation, you will have a formula ("Build your own formula").

There is more than one way to do this problem and we will exhibit two distinctly different approaches. But both begin by following the same principle of solving word problems: Build your own formula.

You need the relationship between angle and area. Given the angle, what is the shaded area? Figure 23 includes a well-chosen auxiliary line and labels. Angle B is $90° + \theta$. There are other ways to subdivide the shaded region into triangles, but the goal is to find a formula for area, and this way the area can be written as the sum of areas of two simple regions, a rectangle and a right triangle.

Now In Figure 23,

$$\text{Area} = 2h + (1/2)xh.$$

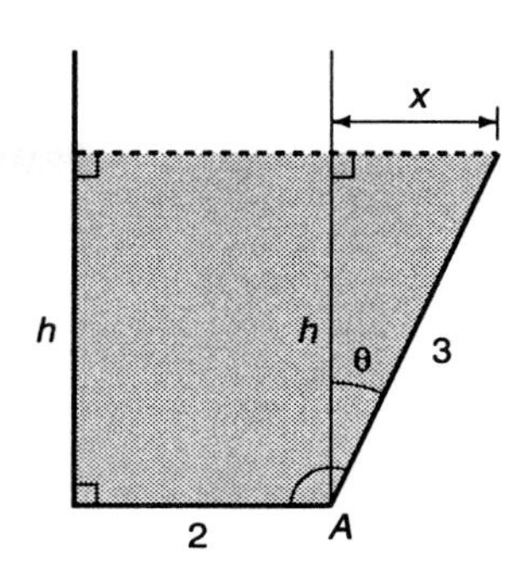

Figure 23: Figure 22 with some auxiliary lines and labels.

We can maximize this graphically as soon as it is expressed in terms of only one variable, x or h or θ. First we write x and h in terms of θ.

$$\cos\theta = h/3. \quad \sin\theta = x/3.$$

$$h = 3\cos\theta. \quad x = 3\sin\theta.$$

$$\text{Area} = 2(3\cos\theta) + (1/2)(3\sin\theta)(3\cos\theta).$$

$$\text{Area} = 6\cos\theta + 4.5(\sin\theta)(\cos\theta).$$

Now graph this and find the value of θ which maximizes it. Then angle B will be that answer plus 90 degrees. For the numerical value, do problem A4.

There is a second approach which finds θ last. It is possible to express the area in terms of x (or h) alone, solve for the value of x which maximizes the area, and then solve for the value of θ which produces that x. Of course, this approach yields the same answer. Again,

$$\text{Area} = 2h + (1/2)xh = (2 + x/2)h.$$

We want only one variable, so express h in terms of x (or vice versa). Use the Pythagorean Theorem.

$$x^2 + h^2 = 3^2. \quad h = \sqrt{9 - x^2}.$$

$$\text{Area} = (2 + x/2)\sqrt{9 - x^2}.$$

Now graph this to find the value of x that yields the maximum area. For the numerical value, do problem A5. Then solve for θ in

$$\sin\theta = x/3$$

to find the value of θ which yields the maximum. Again, angle B is $\theta + 90°$

Slope. The slope of a line is often denoted by "m", as in the famous slope-intercept equation "$y = mx + b$." The slope describes how steep a line is. Another way to describe the steepness of a line would be to give the angle it makes with the positive x-direction (angle θ in Figure 24). Since the slope is "the rise over the run," which, from the point of view of angle θ is "opposite over adjacent," a simple relationship follows:

If θ is the angle a line "$y = mx + b$" makes with the positive x-direction,

(6.4.3) $\quad \tan\theta = m$, the slope.

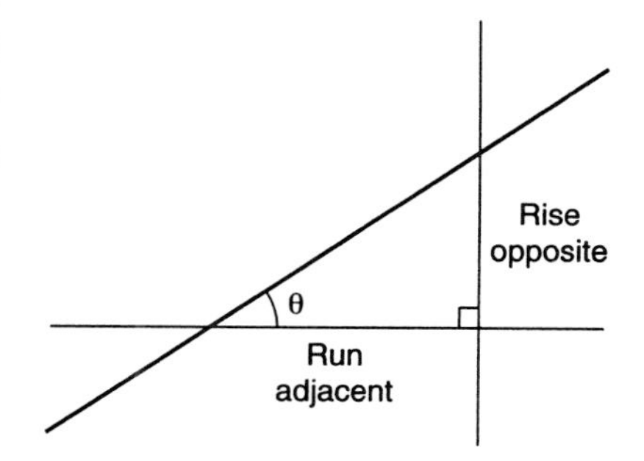

Figure 24: The slope and angle of a line.

Example 16: Find the angle that the line "$x + 2y = 6$" makes with the positive x-direction.

First we need the slope. $x + 2y = 6$ iff $2y = 6 - x$ iff $y = 3 - x/2$. The slope is -1/2. Therefore, $\tan \theta = -1/2$.

$\theta = \tan^{-1}(-1/2) = -26.56°$.

The line slopes downward at an angle of -26.56 degrees to the positive x-direction.

The slope of a line is the tangent of θ,
the angle it makes with the positive x-axis.

Conclusion: Trigonometry is the area of mathematics which assigns measures to angles and sides of geometric figures. Given three parts of a triangle, it is often possible to solve for the remaining three using trigonometry. Given any configuration of known sides and angles, it is appropriate to develop a plan for solving for the unknown sides and angles before worrying about the numbers. If the figure is more complex than a triangle, it may be helpful to subdivide it into triangular components.

Terms: bearing.

****

Exercises for Section 6.4, "Solving Figures":

A1.* Name the four most useful trigonometric measurement theorems (that are not simply definitions of trig functions).

A2.* Give the three-letter abbreviations for the geometric cases that a) determine a triangle, and b) "almost" determine a triangle.

A3. In Example 13, find $\angle C$.

A4. In Example 15, graph the area in terms of θ and find the value of θ which maximizes the area. [θ = ..., maximum area = 7.2]

A5. a) In Example 15, graph the area in terms of x and find the value of x which maximizes the area. b) Then find the associated angle B. [x = ..., maximum area = 7.2]

A6. Samantha wants to measure across the Grand Canyon from observation Deck A on her side to observation deck B on the other side. She picks a third point on her side of the canyon, point C, and measures the distance from A to C, and angles BCA and BAC. How will she figure the distance from A to B?

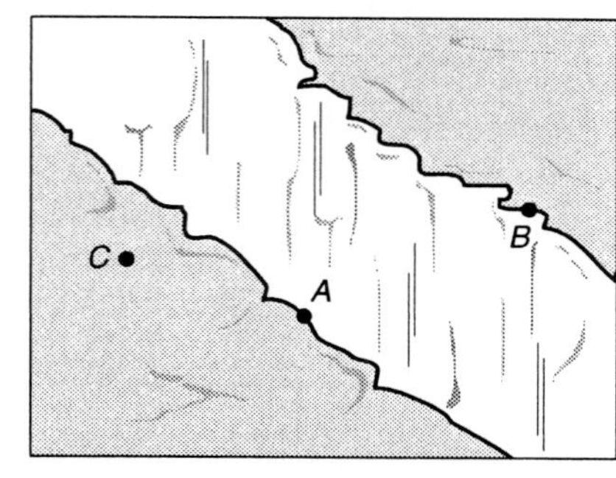

A7. Ruby steers N 82° W for 12 miles. How far north does she go? How far west? [1.7,...]

A8. Rufus pilots his plane S 32° E for 30 miles. How far south did he go? How far East?

A9. Rose ends up 6 miles south and 4 miles east of home. Which bearing would lead straight back home? [N 34° ...]

A10. Liz ends up 3 miles north and 4 miles east of home. Which bearing would lead straight back home? [...53° W]

A11. If you stand 100 feet away from a tower than is 1000 feet tall, what angle is it to the top? [84]

A12. The moon is 240,000 miles away and it is 2160 miles in diameter. What is the apparent angle the width of the moon makes, viewed from the earth? (Continued in A13.) [.52]

A13. a) The sun is 93,000,000 miles away and it is 865,000 miles in diameter. What is the apparent angle the width of the sun makes, viewed from the earth? b) Compare this to the answer in A12 and note what this has to do with solar eclipses. [.53]

A14. Find the angle that the line "$y = 5x + 7$" makes with the positive x-direction. [79]

A15. Find the angle that the line "$3y - 2x + 7 = 4$" makes with the positive x-direction. [34]

A16. Find the angle that the line "$y - 4 = x/3 + 1$" makes with the positive x-direction. [18]

^ ^ ^ ^ Given parts 1, 2, and 3 of the figure, which part would be easiest to find next, and which theorem would you use? If there is more than one part which can be easily found, mention them all.

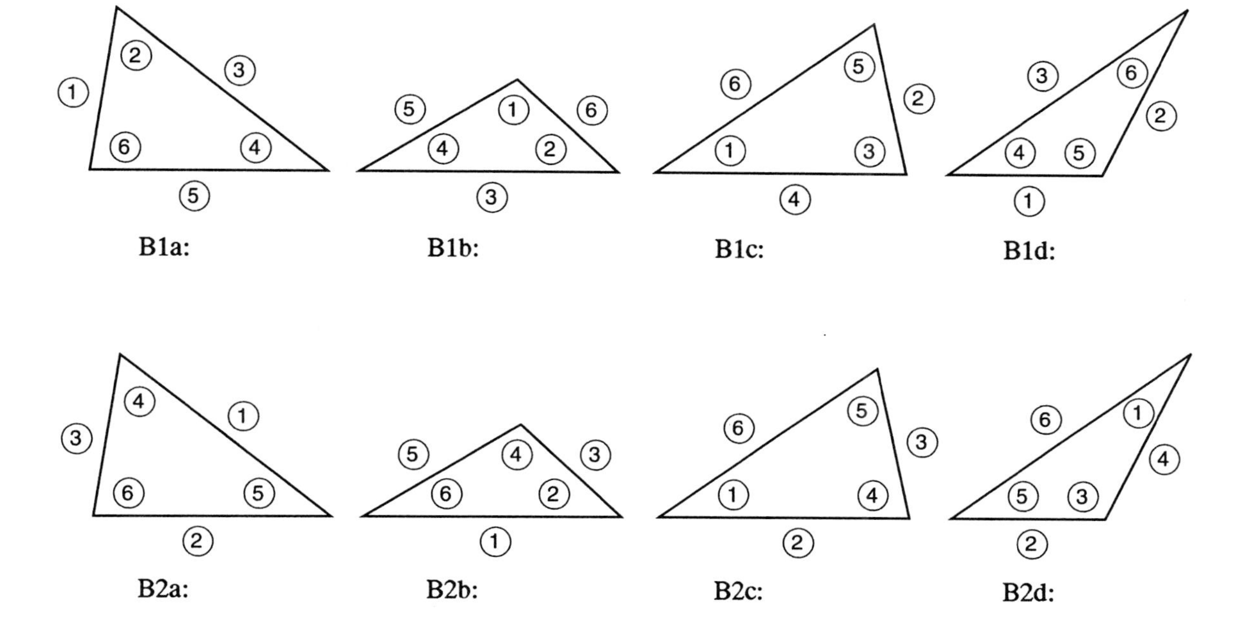

B1a: B1b: B1c: B1d:

B2a: B2b: B2c: B2d:

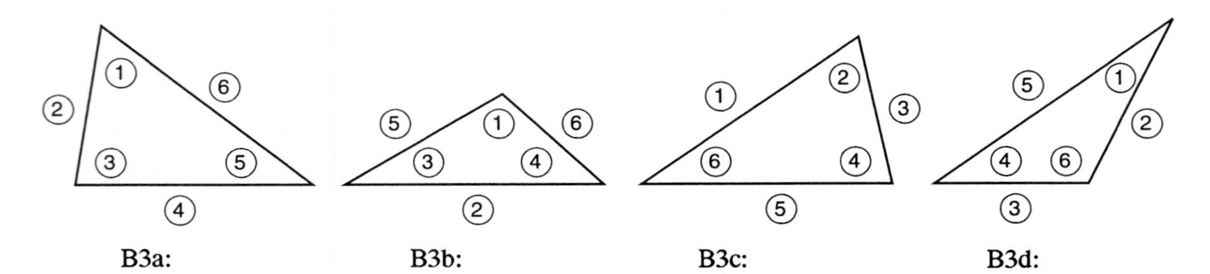

B3a: B3b: B3c: B3d:

B4. Here is a picture (Figure 25) of the type of problem in Example 9 (Figures 10 and 11). Outline a plan to find h by finding CT instead of AT. Label the steps and name the law you would use for each.

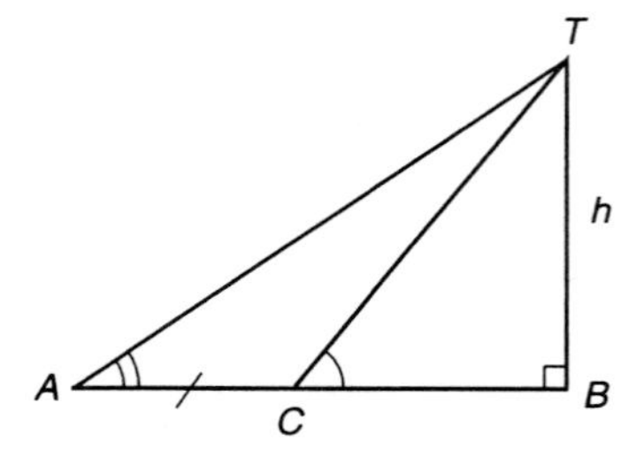

Figure 25.

B5. a) Set up two equations for finding h in Figure 25 (again) by using the tangent function twice after labeling the segment CB.
b) Solve for h with these equations using the data from Example 9 (Figure 10).

B6. Given the marked side and the three marked angles in Figure 26, outline a plan to find x. Illustrate the sequence of steps and name the theorem you intend to apply at each step.

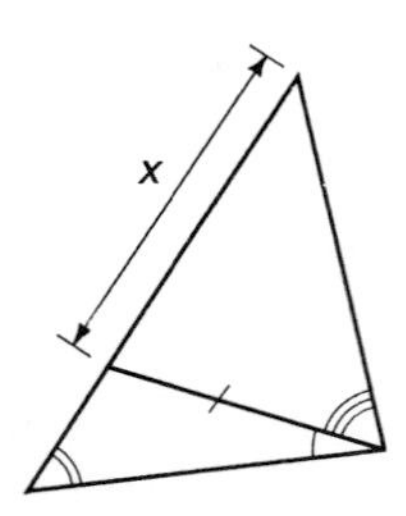

Figure 26:

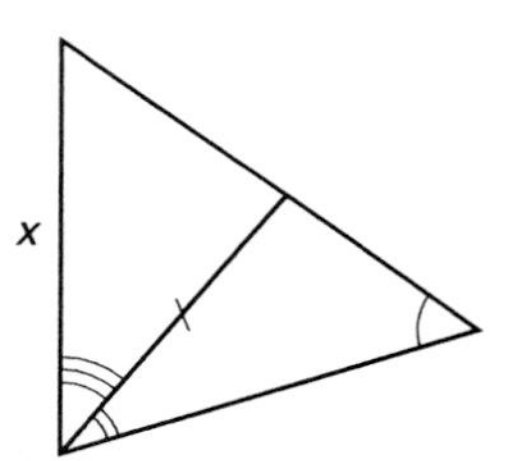

Figure 27:

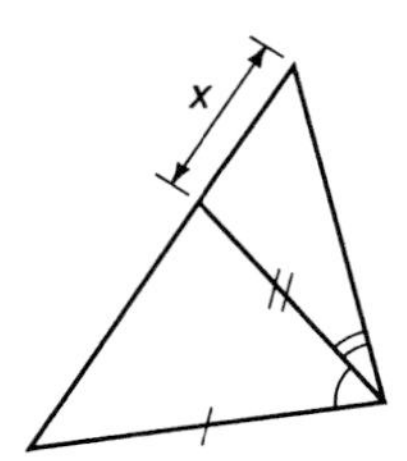

Figure 28:

B7. Given the marked side and the three marked angles in Figure 27, outline a plan to find x. Illustrate the sequence of steps and name the theorem you intend to apply at each step.

B8. Given the two marked sides and the two marked angles in Figure 28, outline a plan to find x. Illustrate the sequence of steps and name the theorem you intend to apply at each step.

B9. Given the two marked sides and the two marked angles in Figure 29, outline a plan to find x. Illustrate the sequence of steps and name the theorem you intend to apply at each step.

B10. Given the three marked sides and the two marked angles in Figure 30, outline a plan to find x. Illustrate the sequence of steps and name the theorem you intend to apply at each step.

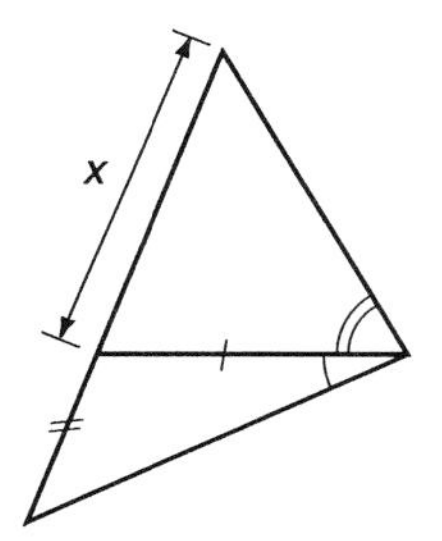

Figure 29:

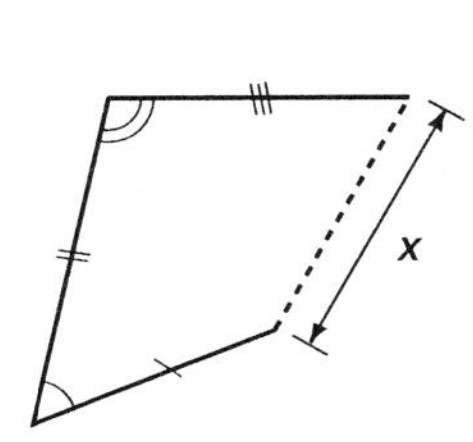

Figure 30:

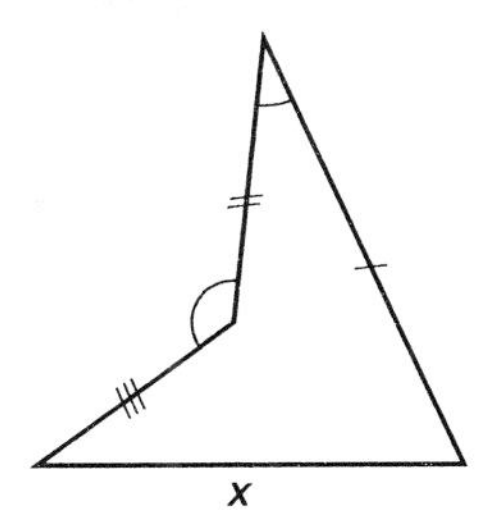

Figure 31:

B11. Given the three marked sides and the two marked angles in Figure 31, outline a plan to find x. Illustrate the sequence of steps and name the theorem you intend to apply at each step.

B12. Given the four marked angles and one marked side in Figure 32, outline a plan to find the distance from C to D.

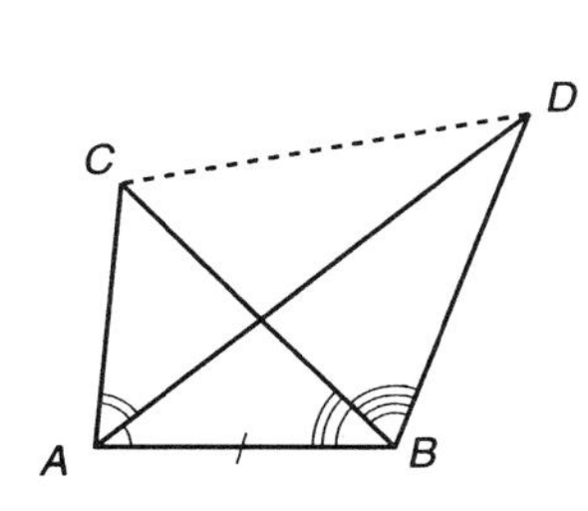

Figure 32:

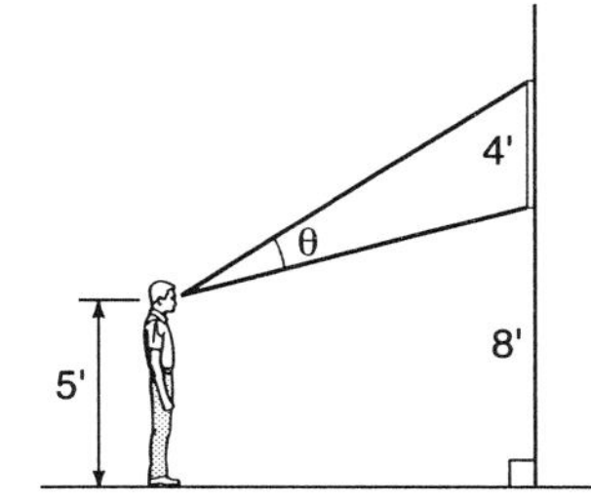

Figure 33:

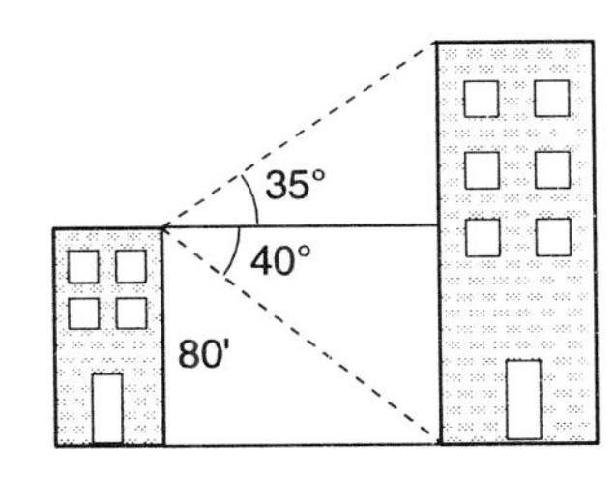

Figure 34:

B13. A picture that is 4 feet high is mounted on a wall so its bottom edge is 8 feet from the ground. A person whose eye level is 5 feet from the ground stands back and looks at it. From which distance will the angle subtended (the angle θ in Figure 33) be largest?
[x = ..., θ = 24]

B14. Compute the height of the taller building in Figure 34. From 80 feet up, the angle of elevation to the top is 35° and the angle of depression to the bottom is 40°. [150]

B15. A boat heads N 33° E for 2 miles and then N 75° E for 3 more miles. What bearing would have led straight from the initial point to the terminal point? [N 58°...]

B16. A boat heads from its dock S 30° E for 20 miles and then turns around to come back. If it returns 18 miles at a bearing of N 25° W, how far and in what direction is the dock?
[..., N 67° W]

B17. A World War II submarine is 1 mile due south of a ship steaming 20 miles per hour due east when it fires a torpedo which goes straight at 60 miles per hour. At what bearing should the torpedo be aimed? [19]

B18. A hunter shoots at ducks flying past at 40 miles per hour in a straight line that is 20 yards away at its closest. His shotgun pellets fly at 900 feet per second. If he shoots when the ducks are closest to him, how much should he lead the ducks? [3.9 feet]

B19. In Example 10, Step 2, you could use either the Law of Cosines or the Law of Sines. What reason might you have to prefer a) the Law of Cosines? b) the Law of Sines?

B20. Luann flies from base 25 miles S 70° E and then 20 miles S 35° E. How far from base is she and which bearing heads directly back to base? [..., N 68° W]

B21. Solve for θ in Figure 35.

B22. Outline a plan to find θ in Figure 36.

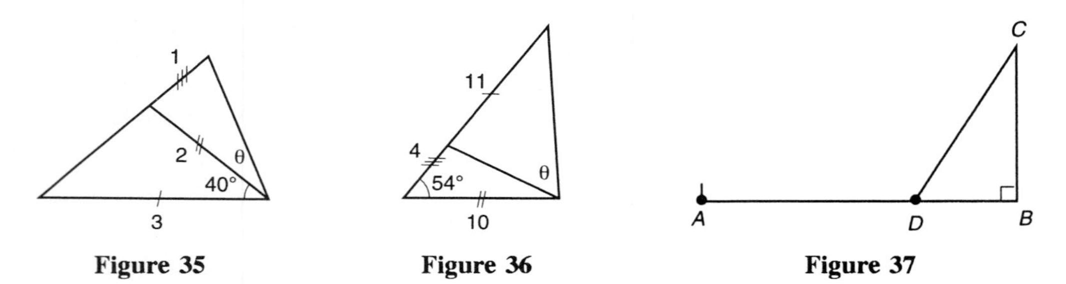

Figure 35 **Figure 36** **Figure 37**

B23. Consider an isosceles triangle with base between the two equal angles and angle θ opposite the base. The base can be computed in terms of the two equal sides and θ using the Law of Cosines or, alternatively, by bisecting angle θ to create two right triangles. a) Do it both ways and equate the two answers to find a trigonometric identity for $\sin(\theta/2)$. b) Rewrite it to give an identity for $\cos(2\alpha)$.

B24. Line segment AB is of length 2 (Figure 37). Line segment BC is perpendicular to AB and has length 1. If the sum of the lengths of AD and DC is 2.6, how long is AD?

B25. (Figure at right) The elevation of point A is 5100 feet. Point B is 600 feet from point A, at an angle of elevation of 28°. Point C is at an angle of elevation of 62° from point B, and at an angle of elevation of 49° from point A. What is the elevation of point C?

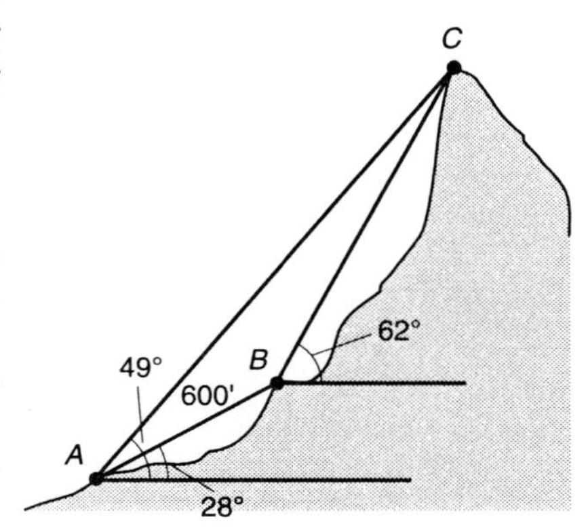

^ ^ ^ ^ Assume the orbits of Venus and Earth are circles about the sun. The radius of the orbit of Venus is 67,000,000 miles and of Earth is 93,000,000 miles.

B26. Viewed from Earth, what is the maximum possible angle between Venus and the Sun?

B27. Viewed from Earth, if the angle between Venus and the Sun is 26°, how far is Venus from Earth?

CHAPTER 7

Trigonometry for Calculus

Section 7.1. Arc Length and Radians

We have seen in Chapter 6 that trigonometry is the branch of mathematics which concerns the relationships of the sides and angles of triangles. This chapter treats trigonometry another way -- as the branch of mathematics which concerns circular motion and other types of periodic motion such as waves (including sound and light waves), the swing of a pendulum, and vibrations of all kinds.

Motion produces change, and calculus is the mathematical subject which studies change. So the aspect of trigonometry which regards trigonometric functions as "circular" functions in the context of motion is important in calculus. The functions we studied for triangle trigonometry are also used as circular functions and the unit-circle definitions of the trigonometric functions apply in both contexts. One minor difference is that in circular trigonometry and calculus it is common to measure angles in radians instead of in degrees. The subject of this section is the close connection between radians, arc length, and the trigonometric functions.

Arc Length. The definition of π (the Greek letter "pi", representing approximately 3.14159) is the ratio of the circumference to the diameter of a circle. That ratio is independent of the size of the circle. So the well-known formula for the circumference, "$C = \pi d$," is based on the definition of π (Figure 1). Since the diameter is twice the radius, this formula is equivalent to "$C = 2\pi r$," which is the form used most in trigonometry, since in trigonometry we describe a circle in terms of its radius, rather than its diameter.

The arc length of a circle is its circumference.

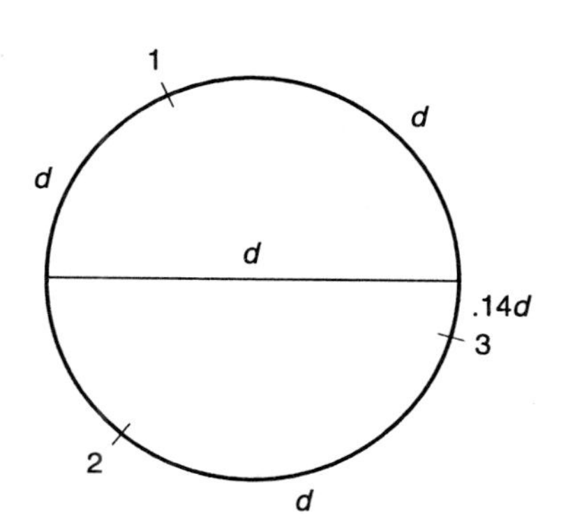

Figure 1: $C = \pi d$.
π is approximately 3.14.

Example 1: Find the circumference of a circle of radius 10.
Plug in to the formula "$C = 2\pi r$":
$C = 2\pi(10) = 62.8$. The arc length of the whole circle is 62.8.

Example 2: Find the arc length of a semicircle (half a circle) with radius 10.
Take the answer to Example 1 and multiply by 1/2 because it is only half a circle. $(1/2)62.8 = 31.4$.

An arc of a fraction of a circle can be described by the radius and the <u>central</u> angle, that is, the angle at the center. In Figure 2, the central angle θ and radius r are associated with the arc AB. The arc "subtends" the angle and the angle "subtends" the arc. <u>Subtend</u> means to "be opposite to" or "enclose." The circumference is the arc length of the whole circle, which has central angle 360°. Using degrees to measure the angle,

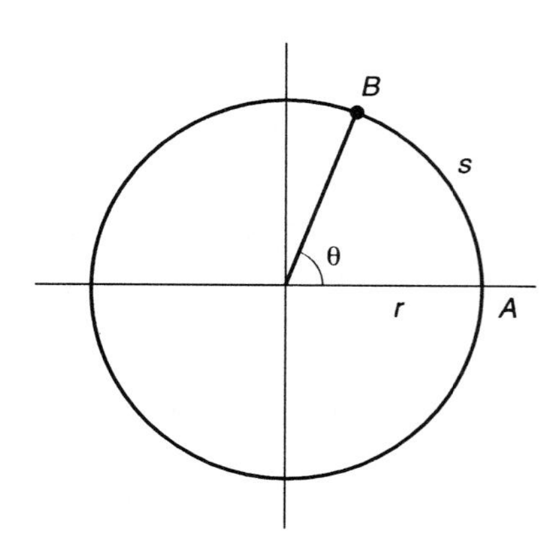

Figure 2: Central angle θ and arc AB of length s.

$$\frac{\textit{arc length}}{2\pi r} = \frac{\theta^\circ}{360^\circ}. \tag{7.1.1}$$

For some reason, arc length is usually denoted by "s." Solving this for arc length, s,

$$s = 2\pi r\theta/360 = \pi r\theta/180, \tag{7.1.2}$$
if θ is expressed in degrees.

You don't need to memorize 7.1.2. Just remember the argument which resulted in Formula 7.1.1 and derive 7.1.2 from it whenever you need it.

Figure 3 illustrates that the arc length is proportional to the radius. For a fixed angle, θ, when the radius doubles the arc length also doubles.

Example 3: The central angle of an arc of radius 10 is 72 degrees. How long is the arc?

Call the arc length s. $s/(2\pi r) = \theta/360$.
$s/(20\pi) = 72/360$.
$$s = (20\pi)72/360 = 12.57.$$
Of course, 7.1.2 is equivalent to 7.1.1, so it would yield the same answer.

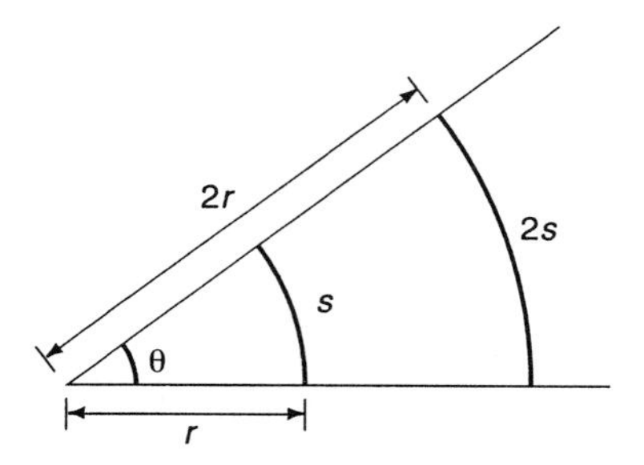

Figure 3: Arc length is proportional to the radius.

Our method of subdividing the central angle of a whole circle into 360 parts comes from the ancient Babylonians who lived in what is now Iraq. We even have a table of a trigonometric function written on a clay tablet which is dated to between 1900 and 1600 BC. The Babylonians used a number system with base 60 (instead of our base 10) and therefore tended to measure things using multiples of 60. A circle is easily divided into 6 parts by swinging a radius from any point on the circle as in Figure 4. The Babylonians divided each of these parts into 60, for a total of 6 times 60 = 360 parts in a circle. Therefore, the central angle of each of the 6 original parts is 60°, and the points in Figure 4 form equilateral triangles with three 60° angles each.

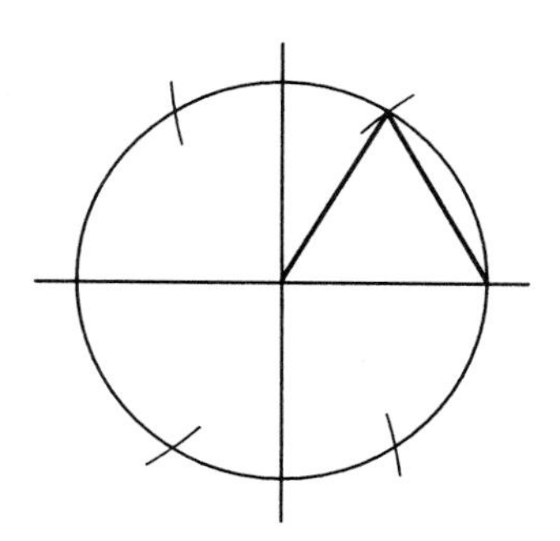

Figure 4: A circle divided into six parts.

For a circle with a given radius, the angle determines the arc length, and the arc length determines the angle.

Example 4: The length of an arc of a circle of radius 300 is 500. What is the central angle?

From 7.1.1 or 7.1.2, $s = 2\pi r\theta/360$. Plugging in, $500 = 2\pi(300)\theta/360$.

$$\theta = 500(360)/[2\pi(300)] = 95.5°.$$

Example 5: How big is a degree? For example, if you form an angle of 1° with two yardsticks by putting their left ends together, how far apart would their right ends be?

A yard is 36 inches. The circumference of a circle with radius 36 inches would be $2\pi(36) = 226.2$ inches. A degree is 1/360 of a circle, so divide the circumference into 360 parts. $226.2/360 = .628$ inches, which is about 5/8 inch. So 5/8 inch is the arc length of a 1° arc with radius 36 inches. Put the other ends 5/8 inches apart to create a 1° angle.

As the next example shows, the arc length is not exactly the straight-line distance between the ends, but it is very close if the angle is small, and 1° is a small angle (problem B8).

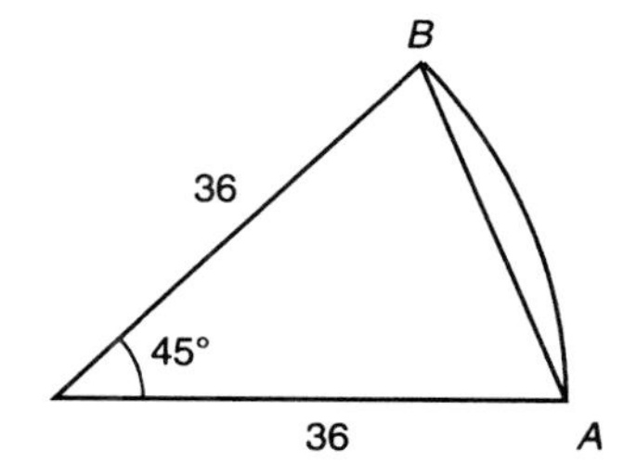

Figure 5: A 45° angle. Compare arc length to linear distance between points.

Example 6: Form a 45° angle with two yardsticks by putting their left ends together (Figure 5). Compare the distance between their right ends to the arc length for a circular 45° arc with radius 36 inches.

From the picture it is obvious that the

linear distance is less than the arc length. Of course, "A line is the shortest distance between two points." The arc length is 1/8 of the circumference: (1/8)226.2 = 28.27 inches. On the other hand, the linear distance AB is calculated by bisecting the central angle (as in Example 12, Section 6.2).
$AB = 2(36)\sin(45°/2) = 27.55$ (inches).

The straight-line distance of 27.55 inches is a little less than the arc length of 28.27 inches, but they are comparable. This tells us that, even when the angle is as great as 45°, the arc length is similar to the length of the corresponding chord. When the central angle is only 1°, as in the previous example, the arc length and the chord length are very similar (problem B8). However, the difference between the arc length and the straight-line distance increases dramatically for larger angles (problem B17).

A degree is a small unit of measure, but still smaller units are often necessary. The Babylonians divided degrees into 60 parts called minutes (or "minutes of arc") and minutes are, in turn, divided into 60 parts called seconds (or "seconds of arc"). The abbreviation for minutes is a prime symbol (′) and for seconds is a double prime symbol (′′). So 3° 47′ 13′′ would be three degrees, 47 minutes, and 13 seconds. Yes, our subdivision of time comes from the Babylonians too.

Example 7: How long is a minute of arc on the surface of the Earth?

To answer this we need the size of the Earth. The diameter of the Earth averages 7912.176 "statute" (usual) miles of 5280 feet each.

The circumference of the Earth is therefore about $7912.176\pi = 24{,}856.834$ miles. To find the length of a minute of arc on the surface this would be divided into 360 parts to find a degree and another 60 to find a minute.

$$24{,}856.834/[360(60)] = 1.150779 \text{ (statute miles)},$$

which, in feet, is

$$1.150779(5280) = 6076.115 \text{ feet}.$$

This length is called a "nautical" mile (as opposed to the usual "statue" mile used on land).

The speed of ships is measured in "knots." One knot is, by definition, one nautical mile per hour. (Speed is not measured in "knots per hour," just "knots.") So a ship that goes 40 knots goes about 40(1.15) = 46 (statute) miles per hour.

Example 8: What is the circumference of the Earth in kilometers?

The circumference of the Earth in miles (a little less than 25,000 miles) was given in Example 7. If we knew the conversion factor, we could convert miles to kilometers to solve the problem. But, there is an easier way.

The original definition of a kilometer was given in terms of the size of the Earth. It was to be 1/10,000 of the distance from the equator to the North pole. Therefore, even though the technical definition of a kilometer has changed since then (it's now atomic), one quarter of the way around the Earth is very nearly 10,000 kilometers. Therefore the circumference is about 40,000 kilometers. This must be easy for people on the metric system to memorize.

Radians. A radian is a unit of measure for angles -- like a degree, except larger. There are about 57.3 degrees in 1 radian. The word "radian" comes from the word "radius." When a central angle in a circle subtends an arc of length equal to the radius, the angle is 1 radian (Figure 6). By definition, the radian measure of a central angle is the ratio of the arc length to the radius.

The radian measure of a full circular angle is 2π (about 6.28) radians because the arc length of a full circle is its circumference, $2\pi r$, when its radius is r, so their ratio is $2\pi r/r = 2\pi$. A straight angle is π radians, and a right angle is $\pi/2$ radians. Figure 6 divides a circle into radians. Figure 7 illustrates some famous angles.

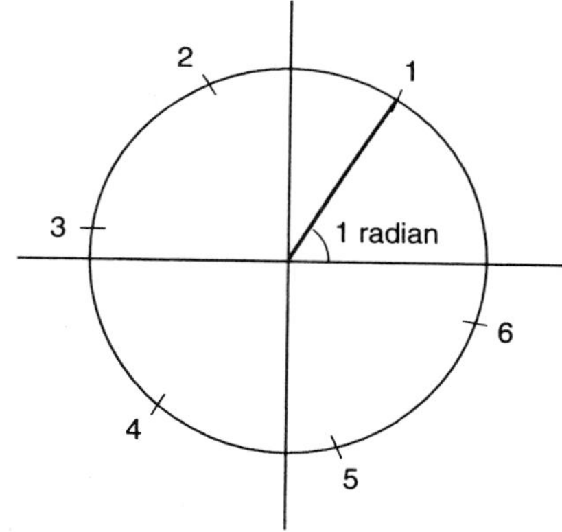

Figure 6: $2\pi = 6.28$ radians in a full circle.

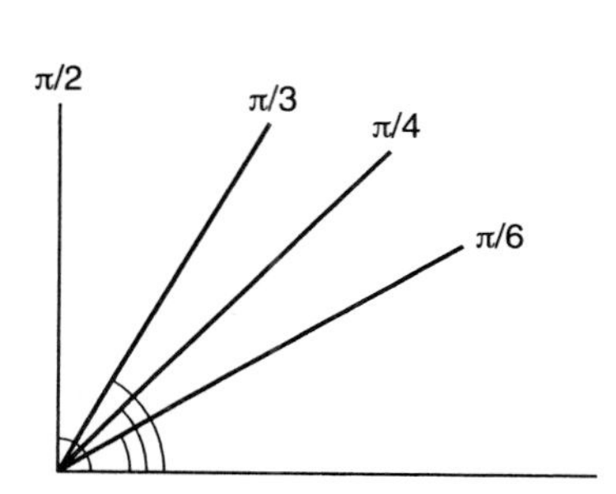

Figure 7: Angles $\pi/6$, $\pi/4$, $\pi/3$, and $\pi/2$ (in radians).

Table 7.1.3. Names of Famous Angles

degrees	360°	180°	90°	60°	45°	30°	0
radians	2π	π	$\pi/2$	$\pi/3$	$\pi/4$	$\pi/6$	0.

When these important angles are expressed in radians we almost always carry along the "π" and avoid decimal form. Using the form with π, you can immediately understand an angle as a part of a straight angle. For example, $\pi/4$ radians is one-quarter of a straight angle. In decimal form this would be 0.785 radians, which is much less illuminating. Keeping the "π" may seem awkward at first, but you will get used to it. After a while, the names of the famous angles may become as familiar in radians as they are in degrees.

Arc length and central angle are related in degrees by Formulas 7.1.1 and 7.1.2. Radian measure of angles is another way to relate arc length and angle without using degrees. It is the modern approach used in calculus.

It is conventional that the word "radians" need not be used or even abbreviated when giving the measure of an angle in radians. In particular, do *not* use "r" to abbreviate "radians" (we need "r" for the radius.) However, the word "degrees" or the symbol "°" for it must be used when giving the measure of an angle in degrees. So "sin 1" refers to the sine of 1 radian, whereas "sin 1°" refers to the sine of 1 degree. Angles expressed in radian measure frequently have "π" in them, such as "$\pi/2$," or "2π." Then the appearance of the number serves to remind you that the angle is measured in radians. If you are using a calculator to evaluate trigonometric functions, be sure to put it in the mode (degree or radian) which matches the measure of the angles you are using.

Here is the definition of <u>radian</u> written with symbols:

(7.1.4) $$\theta = s/r$$
$$s = r\theta$$

(where s is arc length and θ is expressed in radians).

These formulas are simpler than Formulas 7.1.1 and 7.1.2 which use degrees to express the same relationship. On a unit circle where the radius is 1, the arc length and the central angle have the same numerical value because $\theta = s/r = s$ when $r = 1$. If the radius is not 1, the numerical value of the arc length is the radian measure of the central angle multiplied by the radius.

Example 4, revisited: The length of an arc of a circle of radius 300 is 500. What is the central angle?

In radians, this is easy. The radian measure of a central angle is simply the arc length divided by the radius: 500/300 = 5/3 (radians).

Example 9: An arc of length 5 is part of a circle of radius 10. What is its central angle?

The arc length is half the radius, so the angle is 1/2 radian. According to Formula 7.1.4, the angle, in radians, is s/r.

$s/r = 5/10 = 1/2$. The angle is 1/2 radian.

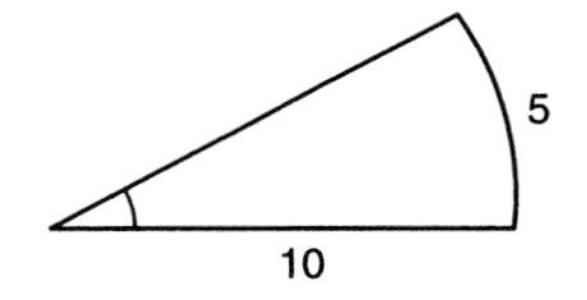

Figure 8: Radius 10. Arc length 5. Angle 1/2 radian.

It is possible to estimate the radian measure of an angle by eye. In this computation both s and r have the same units. If one is measured in centimeters, so is the other. Therefore, in the ratio, the units cancel and the quotient has no units. Radians do not have units the way measures of length, area, time, or speed do.

To think in radians, compare the arc length to the radius, regardless of the unit of length. If they are the same, the angle is one radian. If the arc length is less than the radius, the angle measure is correspondingly less.

Example 10: Estimate the radian measure of the angle in Figure 9.

With the vertex as center, draw in an arc which cuts both rays. Compare the length of the arc to the radius. In the picture it looks like the arc length is about 1/3 the radius. So an estimate of the measure of the angle is about 1/3 radian.

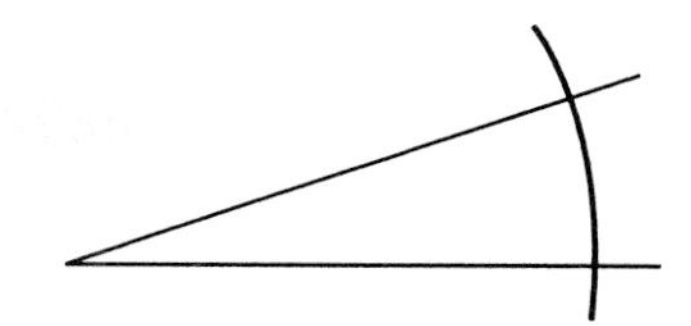

Figure 9: An angle. The arc length is 1/3 the radius. Therefore, the angle is 1/3 radian.

Of course, radians can be converted to degrees and degrees to radians. The result is

<u>(7.1.5) Degree-Radian Conversion</u>:

<u>A</u>) 360 degrees = 2π radians.

<u>B</u>) $$\frac{360 \textit{ degrees}}{2\pi \textit{ radians}} = 1 . \qquad \frac{2\pi \textit{ radians}}{360 \textit{ degrees}} = 1 .$$

This could be written with "180 degrees = π radians," which is the same thing with a factor of "2" removed.

Part B is merely Part A rewritten to emphasize how it is used to change units: Multiply the given amount by 1. Multiplying by 1 does not change an amount, but it changes the units used to express the amount.

Example 11: Change 5 degrees to radians.

$$5 \textit{ degrees} = 5 \textit{ degrees} \left(\frac{2\pi \textit{ radians}}{360 \textit{ degrees}}\right)$$
$$= \frac{5(2\pi)}{360} \textit{ radians} = \frac{\pi}{36} \textit{ radians} .$$

$\pi/36$ radians is 1/36 of a straight angle. If we convert this to decimal form it would be .087 radians, which is not very illuminating. Leave it as "$\pi/36$."

To convert from radians to degrees, multiply by the factor of 1 in 7.1.5 with "radians" in the denominator and "degrees" in the numerator.

Example 12: Convert 1 radian to degrees.

$$1 \textit{ radian} = 1 \textit{ radian} \left(\frac{360 \textit{ degrees}}{2\pi \textit{ radians}}\right)$$
$$= \left(\frac{360}{2\pi}\right) \textit{ degrees} = 57.3 \textit{ degrees} .$$

So one radian is about 57.3 degrees. Figure 6 illustrates the size of a radian.

Sectors. The part of a circular disk between the center and an arc is known as a sector. A sector is a region shaped like a piece of pie (Figure 10). The area of a sector can be given in terms of the area of the circle and the central angle. The argument is similar to the argument for the length of an arc. The area is proportional to the central angle, just as the arc length is proportional to the central angle.

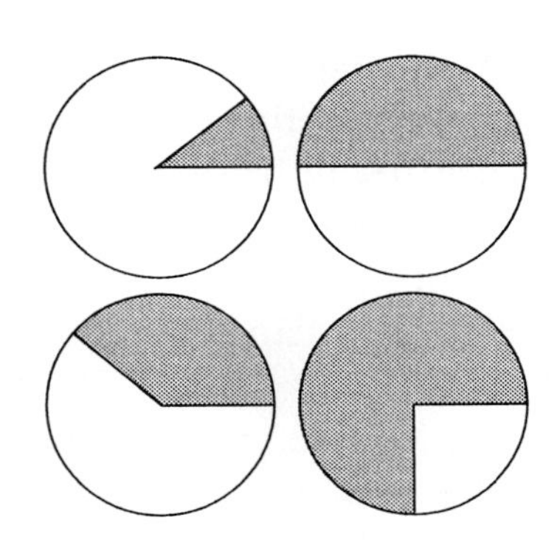

Figure 10: Shaded sectors.

$$\frac{\textit{area of a sector}}{\textit{area of a circle}} = \frac{\textit{angle of the sector}}{\textit{angle of the circle}}.$$

The area of a circle is given by the well-known formula $A = \pi r^2$. If a sector has radius r and central angle θ, then the area of the sector satisfies

$$\frac{\textit{area of a sector}}{\pi r^2} = \frac{\theta}{\textit{angle of a circle}},$$

where the "angle of the circle" depends upon the units used to express θ. It is either 360° or 2π radians. Solving for area,

$$\text{(7.1.6A)} \qquad \textit{area} = \left(\frac{\theta\ \textit{degrees}}{360\ \textit{degrees}}\right)\pi r^2 = \frac{\theta\pi r^2}{360}, \quad \text{if } \theta \text{ is in degrees,}$$

and

$$\text{(7.1.6B)} \qquad \textit{area} = \left(\frac{\theta\ \textit{radians}}{2\pi\ \textit{radians}}\right)\pi r^2 = \frac{\theta r^2}{2}, \quad \text{if } \theta \text{ is in radians.}$$

The π's cancel in the radian version. If θ were 360° or 2π (a full circle), then the formulas would yield πr^2, as they should.

Example 13: Find the area of a sector of radius 50 feet and central angle 75°.
$A/(\pi r^2) = \theta°/360°$, so $A/(\pi 50^2) = 75°/360°$ and $A = (75/360)\pi 50^2 = 1636$ (square feet).

Why Radians? School children learn about angles measured in degrees. Protractors are labeled in degrees. The famous angles (such as 90°, 30°, and 45°) are easy to express in degrees. Triangle angles are measured in degrees. So why bother to learn about radians? Why switch?

As we saw in Examples 9 and 10, radians are a natural method of measuring angles in circles based on the radius. It is only the occurrence of

"π" that in any way makes them awkward. But π is a simple angle -- angle π is a straight angle. Angle $\pi/2$ is therefore one half of a straight angle. Angle $\pi/4$ is one fourth of a straight angle. Simple.

If you are willing to carry along the written "π", angles in radian measure are natural. But that is not all. The trigonometric functions have some very nice properties when angle measure is expressed in radians. Here are some ideas we discussed in the context of applications of polynomials.

Trigonometric functions are defined on the unit circle. For example, $\sin x$ is the vertical coordinate of the point $P(x)$ associated with central angle x. (We have switched from angles called "θ" to angles called "x".) A glance at a unit-circle picture (Figure 11) shows that, if the angle is small, the value of sine and the arc length are very similar. This close relationship is visible if the angle is measured in terms of the arc length, that is, if the angle is given in radians.

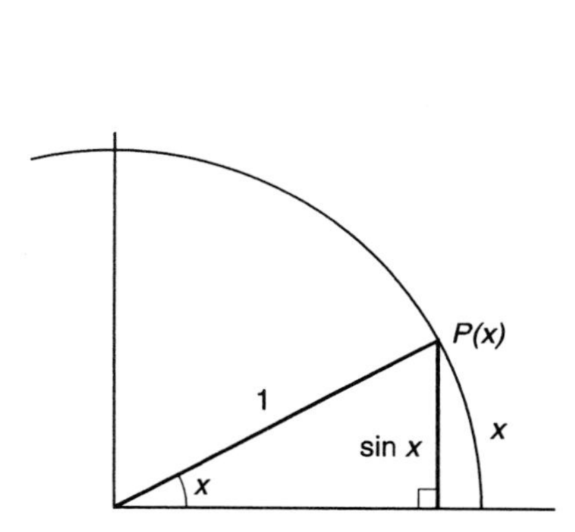

Figure 11: x and $\sin x$, where x is both an angle measured in radians and an arc length.

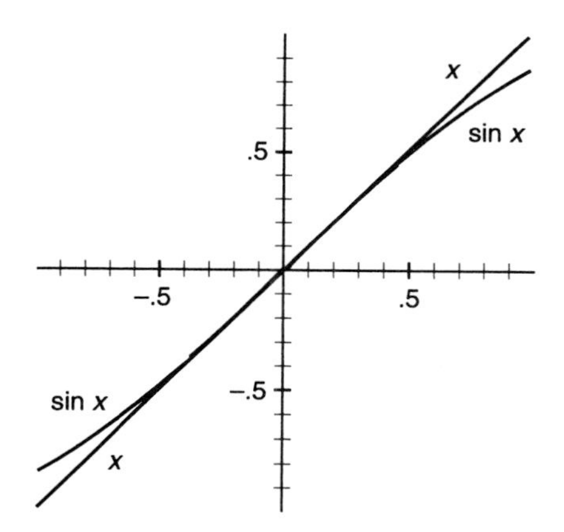

Figure 12: x and $\sin x$. (in radians) [-1, 1] by [-1, 1].

Calculator Exercise 1: Put your calculator in radian mode. Evaluate sine at several argument values, including small values, and compare the images to the arguments.

The results should not be a surprise. Figure 11 shows that, on a unit circle, the sine value will be slightly less than the arc length, which is also the measure of the angle. The table gives some numbers.

x	0	.01	.02	.1	.2	.3	.5	1
$\sin x$	0	.0099998	.0199987	.09983	.1987	.2955	.479	.841
$x - \sin x$	0	.0000002	.0000013	.00017	.0013	.0045	.021	.159

As the angle becomes larger, the difference becomes greater. For small values of x, x itself is a good approximation for $\sin x$. If you use *radians* to graph "x" and "sin x" on the same graph, they are very similar on a region near the origin (Figure 12). This simple relationship does not hold when you use degrees.

So one major value of radians is that radian measure yields simple polynomial approximations for the trigonometric functions (problem B9).

In calculus, measures of angles are often denoted by "x". This is not the same usage of "x" that we used to define cosine of an angle, where "x" was the horizontal coordinate. Do not be surprised to see angles and lengths labeled "x" (instead of "θ") where "x" is *not* the horizontal coordinate. Example 14 is done in terms of "x" and not "θ" because we like to express polynomials using variable "x".

The process of evaluating the expression "sin x" for some x is not trivial. Well, for you it may be simply the press of a button. But how does the calculator do it? You know ways to add, subtract, multiply, and divide by hand, but the process of evaluating "sin .2" or "sin 37°" is not at all like those methods. Before the advent of computers and calculators, people looked up the values of sine in tables that had been laboriously computed by hand. You can see symmetry and repetition in the graph of sin x (Figure 13).

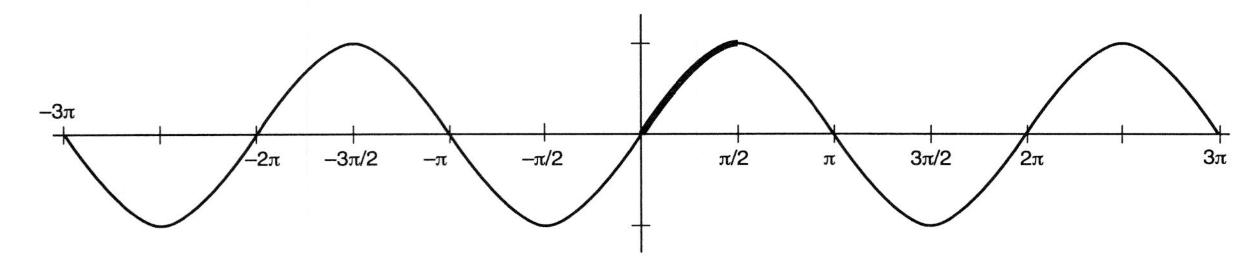

Figure 13: sin x. $-3\pi \le x \le 3\pi$.

If we could determine sin x for all x in the emphasized interval (from 0 to $\pi/2$, which is 0° to 90°), the images at all other values of x would follow (using trig identities such as those developed in the next section). Now, rather than reproduce the original computations in that interval, most calculators have an "approximating polynomial" for "sin x". Over that interval the polynomial is selected to give values extremely close to the tabled values. The polynomial is quite complex, but at least it can be evaluated using only the four arithmetic functions: addition, subtraction, multiplication, and division (For one such polynomial, see Example 20 in Section 4.1).

Calculator Exercise 2: The unit circle interpretation of tangent shows why tan x is also very similar to x for small values of x (Figure 14). Graph x and tan x in radians together and see.

x	.01	.1	.2	.5	1	1.2
$\tan x$	.010000033	.10033	.2027	.546	1.557	2.57
$\tan x - x$	.000000033	.00033	.0027	.046	.557	1.37

From the table it is evident that x is close to $\tan x$ for small values of x, but not for larger values of x. If you graph "x" and "$\tan x$" on the same graph, they are very similar on a region near the origin (Figure 15).

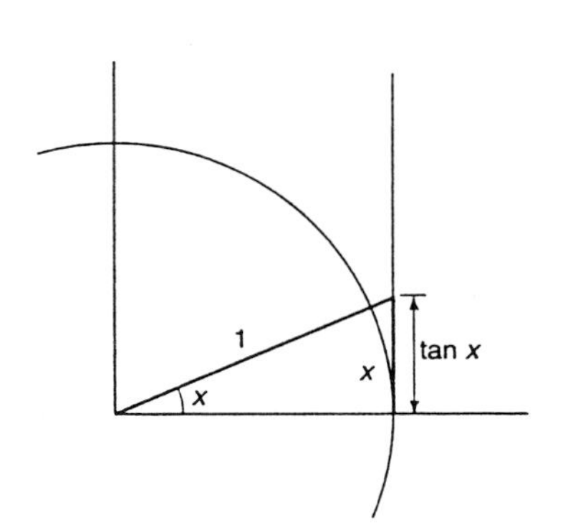

Figure 14: The unit circle, x, and $\tan x$.

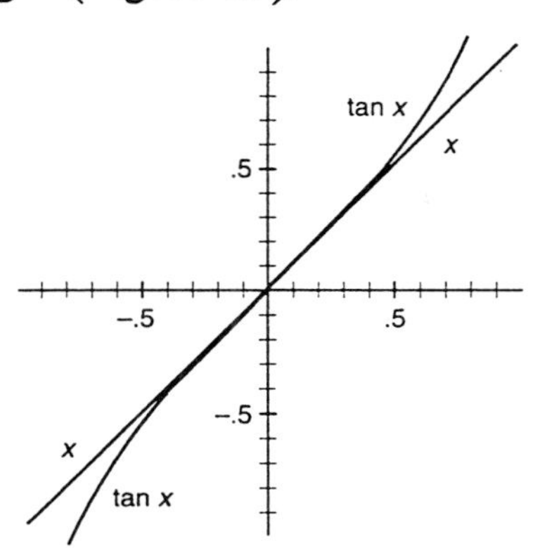

Figure 15: x and $\tan x$. (in radians) [-1, 1] by [-1, 1].

The previous two examples only hint at the advantages of measuring angles using radians instead of degrees. In calculus, you will discover that the properties of trigonometric functions are distinctly simpler when angles are measured in radians.

Conclusion. Radians are a natural way to measure angles in terms of arc length. Although the use of angles which are multiples or fractions of π may seem awkward at first, it turns out to be very convenient in calculus.

Terms: central angle, subtend, arc length, radian, sector.

****

Exercises for Section 7.1, "Arc Length and Radians":

A1.* Give the corresponding radian measure.
a) $90°$ b) $45°$ c) $30°$ d) $60°$.

A2.* Give the corresponding degree measure to these angles given in radians
a) $\pi/3$ b) $\pi/2$ c) $\pi/4$. d) $\pi/6$.

A3. a) Convert to radian measure $25°$. b) Convert to degree measure $\pi/8$ radians.

A4. a) Convert to radian measure 10°. b) Convert to degree measure $\pi/12$ radians.

A5. A pulley has a 4 inch diameter. How many radians does it turn when 3 feet of rope are pulled around it? [18]

A6. From 100 yards away, how wide must something be to subtend an angle of 1°? [1.7]

A7. a) Convert 5° 26′ 45″ to decimal degrees. b) Convert 0.276° to minutes and seconds of arc.

^ ^ ^ ^ Answer these in radians.
A8. How many radians does the hour hand of a clock sweep out in one hour?
A9. How many radians does the minute hand of a clock sweep out in one minute?
A10. How many radians does the sun move across the sky in one hour?

A11. How long is a second of arc on the surface of the earth? [100]

^ ^ ^ ^ ^ ^ ^ ^ ^

B1.* Define "radian." Give the basic equivalence between radians and degrees.

B2.* Give the proportion from which the formula for the area of a sector can be derived.

B3.* If x is expressed in radians, sin x is nearly x for small x. Draw a picture to illustrate why.

B4.* If x is expressed in radians, tan x is nearly x for small x. Draw a picture to illustrate why.

B5. For very small x expressed in radians, sin x is nearly x. a) Use this to estimate sin 1° by converting to radians. b) Compare your estimate to the real value of sin x obtained with your calculator. How far off is it?

B6. A bike wheel has a 26 inch diameter. How many radians does it go through in a one mile long bike ride?

B7. A bike is geared so that the pedal gear has 48 teeth and the wheel gear has 18 teeth. When the pedal turns through 1 radian, how many radians will the wheel gear turn through?

B8. In Examples 5 and 6 the linear distance between the ends of the yardsticks was not equal to the arc length, but it was similar. Example 5 found the arc length subtended by an angle of 1°. Find the linear distance and compare it to the arc length. Form the ratio of the two.

B9. A better approximation than "x" for sin x (for small values of x) is given by a cubic polynomial "$x - x^3/6$." Compare "sin x" to "$x - x^3/6$" graphically. Which is larger for $x > 0$? Discover when the ratio of the smaller to the larger is at least .99. [$x < 1.0$]

B10. An archaeologist finds a broken piece of a pottery cup and asks you to determine the cup's outside diameter. You take the measurements in the picture. What is the diameter? [3.0]

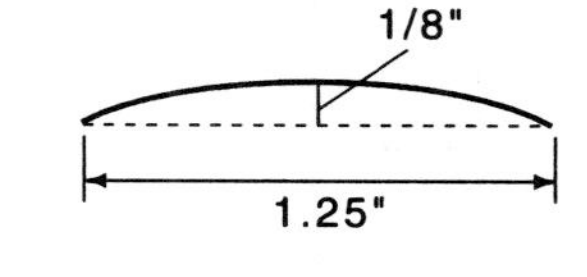

B11. Suppose you are standing on the seashore and you look out to the horizon. Presume that you can see out to where the tangent line to the earth would come to eye level (exaggerated in the picture). If your eyes are 6 feet above sea level, how far out can you see? (Assume the diameter of the earth is 7900 miles.)

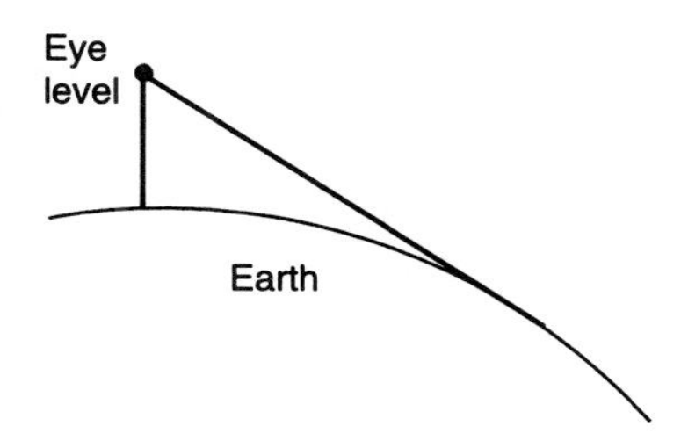

B12. Use graphs of "sin x" and "$x - x^3/6 + x^5/120$" (in radians) to find the greatest positive value of x such that they differ by at most .01.

B13. Use graphs of "cos x" and "$1 - x^2/2 + x^4/24$" (in radians) to find the greatest positive value of x such that they differ by at most .01.

B14. The sun is about ½° wide. How long does the sun take to move its own width across the sky?

B15. a) Redo problem B11 if eye level is h feet above sea level and letting the radius of the Earth be r. Then try to simplify your answer by ignoring any term that is much smaller than the other terms. Plug in for the actual value of r only after you have simplified your expression.
b) If your eye is 20 feet above sea level, as a boat approaches from over the horizon, when will the top of its 50-foot mast come into view (neglect refraction of light rays).

B16. A belt goes around two wheels, one with diameter of 20 inches and the other with diameter of 6 inches. Their centers are 22 inches apart. How long is the belt?

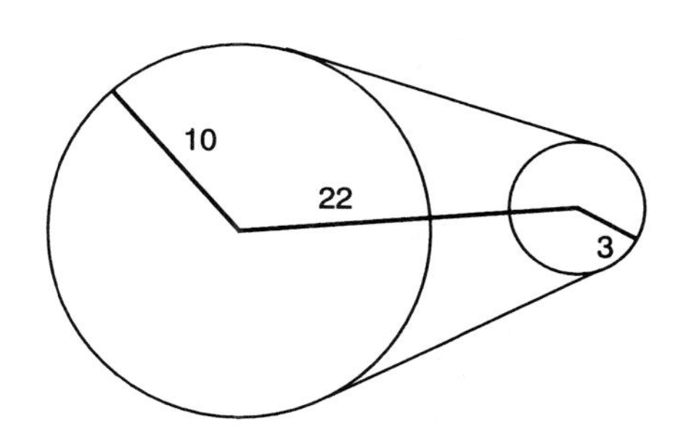

B17. Consider two points on the unit circle, A and B. Use evaluate-and-compare to determine when the arc length between them (measured counterclockwise) is twice the straight-line distance between them.

Section 7.2. Trigonometric Identities

Identities are about operations and order. They provide alternative sequences of operations with which expressions may be evaluated. For example, we can express cosine in terms of sine. Some identities relate trigonometric functions at different arguments. For example, one gives the cosine of half an angle in terms of the cosine of the angle. Why do you study trig identities?

Occasionally you will need to use a trig identity. For example, they are used numerous times in calculus. But, more importantly, you study them to understand trigonometry in a broad sense. Working with identities forces you to become familiar with the definitions and basic properties of the trigonometric functions. This familiarity is the true goal of this section.

This is a long section with a substantial number of identities. Its emphasis is entirely on identities you would actually see and use in calculus. There are a great many trig identities that do not have much use in calculus, and they are omitted here. But that still leaves quite a few to learn and understand. Good pictures help. It has been said, "A picture is worth a thousand words." We can extend that bit of wisdom to trigonometry and say, "A picture is worth several trig identities." Learn to draw and understand trig pictures and many of the results of this section will follow easily.

You can "see" many identities just by using the definitions of the functions on the unit circle. The equation of the unit circle is "$x^2 + y^2 = 1$." Let $P(\theta)$ be the point on the unit circle at angle θ counterclockwise from the positive x-axis (Figure 1). Then, by definition,

(7.2.1)
$$\sin \theta = y,$$
$$\cos \theta = x,$$

and
$$\tan \theta = \frac{y}{x} = \frac{\sin \theta}{\cos \theta}.$$

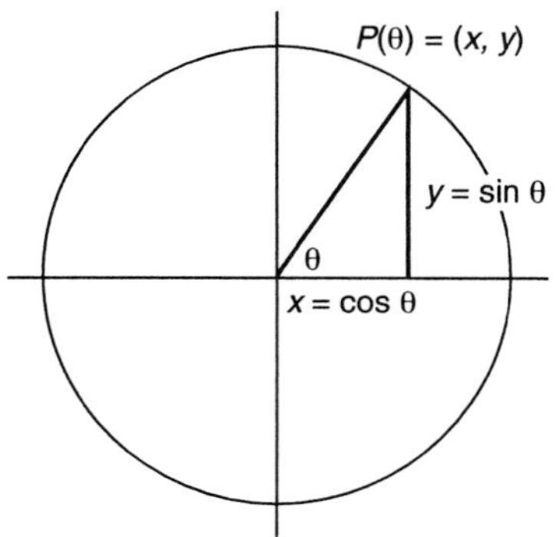

Figure 1: The unit circle, θ, $P(\theta)$, x, and y.

The equation of the unit circle immediately yields the famous

<u>Pythagorean identity</u>:
(7.2.2)
$$\sin^2 \theta + \cos^2 \theta = 1.$$

Example 1: Let θ be acute and such that $\cos\theta = 0.3$. Find $\sin\theta$. Find $\tan\theta$.

Of course, we could use a calculator to solve for θ and then evaluate sine of that angle. But this section is concentrating on theory, not the shortcuts that result from that theory. Use the Pythagorean Identity.

$\sin^2\theta + \cos^2\theta = 1$. $\sin^2\theta + .3^2 = 1$. $\sin^2\theta = 1 - .3^2 = .91$. $\sin\theta = \pm\sqrt{.91} = \pm.954$.

This is as far as the identity can go. One solution is extraneous. Since θ is given as acute, and we know that the sine of acute angles is positive, the only correct answer is: $\sin\theta = .954$.

Given the values of sine and cosine, the value of tangent follows easily. It is their quotient (by 7.2.1). $\tan\theta = .954/.3 = 3.18$.

Acute angles, that is, angles that fit in right triangles, are the easiest to deal with. The basic right-triangle interpretations such as "sine is opposite over hypotenuse" work for acute angles, although these interpretations don't work for other angles (Interpret them with a unit circle). The next goal of this section is to show how the trigonometric functions of angles which are *not* acute can be given in terms of the trigonometric functions of angles that *are* acute by using the so-called "reference" angles.

Angles, such as θ and $\theta + 360°$, which correspond to the same points on the unit circle are said to be coterminal (because they terminate in the same place). Their trigonometric functions are therefore the same. Let n denote an integer.

Coterminal Identities:

(7.2.3) $\sin(\theta \pm 360n°) = \sin\theta$, and $\cos(\theta \pm 360n°) = \cos\theta$.

This identity is commonly used for the first step in evaluating the sine or cosine of large arguments.

Of course, all these identities which mention degrees have perfect parallels when the angles are given in radians. For example, in 7.2.3, just replace "$360n°$" with "$2\pi n$".

Example 2: Find sin 870° by finding a smaller coterminal angle.

This problem gives you some idea of how calculators work. Here we will do explicitly a step that your calculator will do (invisibly) inside its guts. Convert the angle to another angle with the same trig function values by subtracting or adding some multiple of 360° that yields a convenient angle.

870° is not in the interval from 0° to 360°. By subtracting two times 360° we will obtain an angle that *is* in that interval.

$$\sin 870° = \sin(870 - 2(360)°) = \sin 150°.$$

Now we will need to find sin 150°. We will resume this example shortly.

<u>**Reference Angles**</u>. In a unit-circle picture of an angle, the positive acute angle formed by its terminal side and the x-axis is called its <u>reference angle</u>. Consider the four terminal sides at angle θ away from the x-axis as in Figure 2. Let the coordinates of $P(\theta)$ be (x, y). By geometry, the coordinates of the others are as labeled (Figure 3). Note that the x-values of all four points have the same absolute valves, and the y-values of all four points have the same absolute values. Those values are the cosines and sines of four angles between 0° and 360°. So we can see that each angle has the same cosine (x-value), except possibly for a change of sign, and each angle has the same sine (y-value), except possibly for a change of sign. Since tangent is sine over cosine, each angle has the same tangent, except possibly for a change of sign.

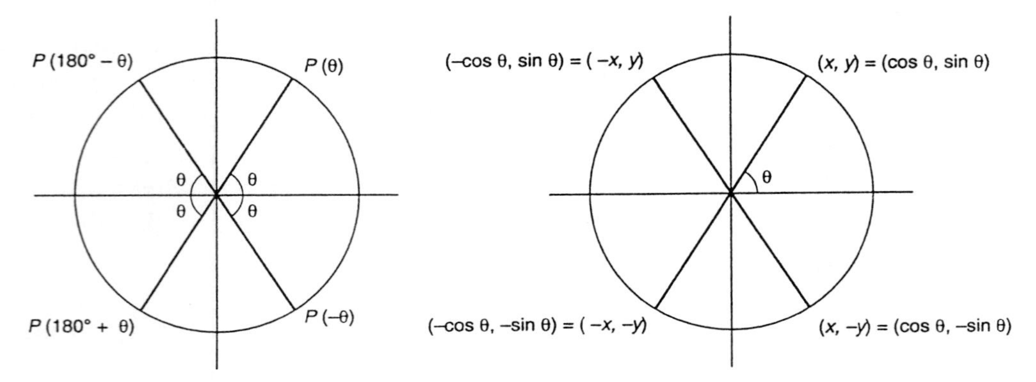

Figure 2: Four angles with reference angle θ.

Figure 3: The sines and cosines of four angles in terms of x and y.

Figure 3 is a picture worth at least nine (9!) trigonometric identities. From the coordinates of $P(\theta)$ and $P(-\theta)$ we find

<u>Negatives (Odd-Even Identities)</u>:

(7.2.4) $\cos(-\theta) = \cos\theta$. $\sin(-\theta) = -\sin\theta$. $\tan(-\theta) = -\tan\theta$.

The result for tangent follows because tangent is sine over cosine, so

$$\tan(-\theta) = \frac{\sin(-\theta)}{\cos(-\theta)} = \frac{-\sin\theta}{\cos\theta} = -\tan\theta .$$

Using the terminology developed for even and odd power monomials in Section 4.2, these facts say that cosine is an even function and sine and tangent are odd functions. The graph of cosine is symmetric about the y-axis Figure 5), and the graphs of sine and tangent are "point symmetric" about the origin (Figures 4 and 6).

The identities in 7.2.4 help find values of the trigonometric functions of angles expressed as negative angles in the fourth quadrant.

Example 3: Find sin(-30°).

Sin(-30°) = -sin 30°, and sin 30° is well-known. sin(-30°) = -1/2.

In Figure 3, from the coordinates of $P(\theta)$ and $P(180° - \theta)$ we find

Reference-Angle Identities:

(7.2.5) $$\sin(180° - \theta) = \sin \theta, \quad \cos(180° - \theta) = -\cos \theta, \quad \tan(180° - \theta) = -\tan \theta.$$

The result for tangent follows because tangent is sine over cosine.

This is particularly useful for angles between 90° and 180°.

Example 2, continued: Find sin 150°.

The reference angle of 150° is 30°. Because 150° is a second-quadrant angle between 90° and 180°, it is useful to treat 150° as 180°- 30°. Identity 7.2.5 tells us that sin 150° = sin 30°, which is well-known. Sin 150° = 1/2.

In Figure 3, from the coordinates of $P(\theta)$ and $P(\theta + 180°)$ we find

Reference-Angle Identities:

(7.2.6) $$\sin(\theta + 180°) = -\sin \theta. \quad \cos(\theta + 180°) = -\cos \theta. \quad \tan(\theta + 180°) = \tan \theta.$$

This is easy to see on a unit circle because $P(\theta)$ and $P(180° + \theta)$ are directly opposite one another, so both their horizontal and vertical values are negatives of one another.

Example 4: Express "tan 205°" in terms of the tangent of an acute angle.

Tan 205° = tan(205°-180°) = tan 25°.

Express "tan 160°" in terms of the tangent of an acute angle.

Tan 160° = tan(160°- 180°) = tan(-20°) = -tan 20°, by 7.2.4. The reference angle is 20° and the sign is changed because tangent is negative in the second quadrant.

Table 7.2.7. The Reference Angle for Angle θ

quadrant		reference angle
I	$0° \le \theta \le 90°$	θ
II	$90° < \theta \le 180°$	$180° - \theta$
III	$180° < \theta \le 270°$	$\theta - 180°$
IV	$270° < \theta \le 360°$	$360° - \theta$
	or $-90° \le \theta < 0°$	$-\theta$

There is no need to memorize this table because its contents are easily derived from Figure 3. Memorize Figure 3 instead.

<u>**Solving Trigonometric Equations**</u>. Identities 7.2.3, 4, 5, and 6 tell us there are many solutions to "$\sin x = c$," "$\cos x = c$," and "$\tan x = c$." Of course, the inverse function is designed to yield one solution. (It yields exactly one, because it is a function.) These identities yield the rest. In summary,

(7.2.8) $$\sin x = c \quad \text{iff} \quad x = \sin^{-1} c \pm 360n°, \text{ or} \quad x = 180° - \sin^{-1} c \pm 360n°,$$

where n is an integer that may be zero (Figure 4). This follows from 7.2.3 on coterminal angles which tells us that adding 360° to any solution yields another solution and from 7.2.5 which says that subtracting any solution from 180° yields another solution.

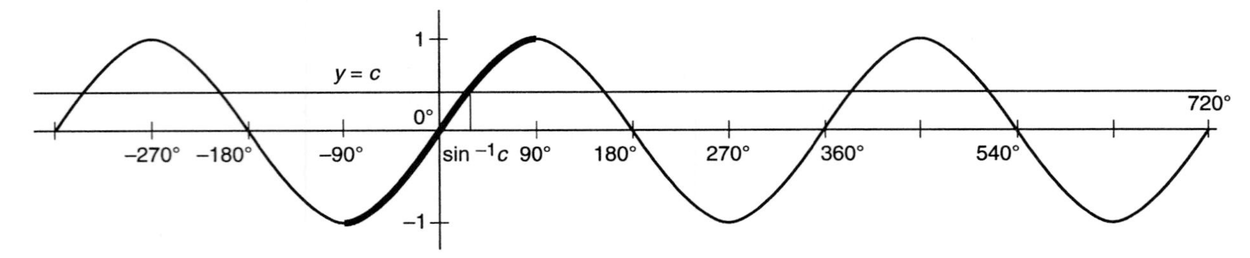

Figure 4: $y = \sin x$ and $y = c > 0$. $-360° < x < 720°$
The region where $\sin^{-1}$ is defined is emphasized.
Note how the solutions described in 7.2.8 are visible.

(7.2.9) $$\cos x = c \quad \text{iff} \quad x = \pm \cos^{-1} c \pm 360n°,$$

where n is an integer that may be zero (Figure 5). This follows from 7.2.3 on coterminal angles which tells us that adding 360° to any solution yields another solution and from 7.2.4 which says that the negative of any solution is another solution.

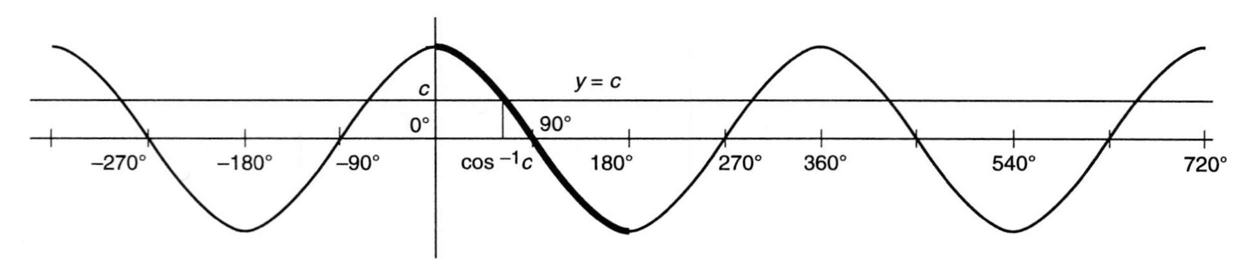

Figure 5: $y = \cos x$ and $y = c > 0$. $-360° < x < 720°$
The region where $\cos^{-1}$ is defined is emphasized.
Note how the solutions described in 7.2.9 are visible.

(7.2.10) $\tan x = c \text{ iff } x = \tan^{-1} c \pm 180n°$,

where n is an integer that may be zero (Figure 6). This follows from 7.2.6 which says that adding 180° to any solution yields another solution.

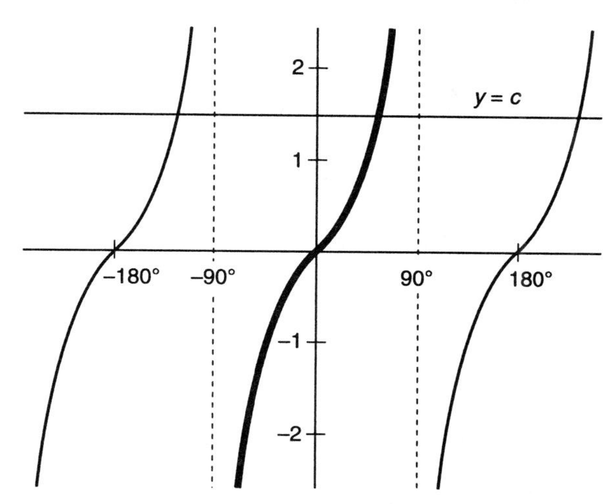

Figure 6: $y = \tan x$ and $y = c > 0$. $-270° < x < 270°$, $-2.5 < y < 2.5$.
The region where $\tan^{-1}$ is defined is emphasized.
Note how the solutions described in 7.2.10 are visible.

A picture is worth a thousand words. To remember these results, simply remember the unit-circle picture in Figure 3.

Example 5: Solve $\sin x = 1/2$.

We know one solution: $x = 30°$. Theorem 7.2.8 tells us how to generate the others. They include $30+360° = 390°$, $30\text{-}360° = -330°$, $180\text{-}30° = 150°$, and $150+360° = 510°$. We can write them all using "n" for an arbitrary integer: "$x = 30° \pm 360n°$ or $x = 150° \pm 360n°$."

Example 6: Solve "$\tan(5x) = 1$" for x between 0 and 180 degrees.

We use the inverse tangent function first.

$$5x = \tan^{-1} 1 \pm 180n° = 45° \pm 180n°.$$

So,

$$x = 9° \pm 36n°.$$

The solutions between 0 and 180 degrees are: 9°, 45°, 81°, 117°, and 153°.

These identities show that the value of any trigonometric function at any angle can be determined by its value at the acute reference angle.

<u>**Values at Acute Angles**</u>. The values of the trigonometric functions are well-known at 30°, 45°, and 60°. The next identities permit us to combine known values in various ways to determine unknown values. For example, the "sum

identities" will allow us to obtain sin 75° = sin(45°+ 30°) as a combination of the sines and cosines of 45° and 30° (Example 7).

The Sum and Difference Identities. Given the values of the trigonometric functions at α and β, what are the values of the trigonometric functions at $\alpha+\beta$ and $\alpha-\beta$? The results are next.

Sum Identities

(7.2.11) $$\sin(\alpha+\beta) = (\sin \alpha)(\cos \beta) + (\cos \alpha)(\sin \beta).$$

(7.2.12) $$\cos(\alpha+\beta) = (\cos \alpha)(\cos \beta) - (\sin \alpha)(\sin \beta).$$

(7.2.13) $$\tan(\alpha + \beta) = \frac{\tan \alpha + \tan \beta}{1 - (\tan \alpha)(\tan \beta)}.$$

Of course, since order matters, you do not expect $\sin(\alpha+\beta)$ to be $\sin \alpha + \sin \beta$. We will prove the sum identities below. The difference identities are proved immediately from them by treating "$\alpha-\beta$" as "$\alpha+(-\beta)$" and replacing, according to 7.2.4, $\cos(-\beta)$ with $\cos \beta$ and $\sin(-\beta)$ with $-\sin(\beta)$. The results are:

Difference Identities

(7.2.14) $$\sin(\alpha - \beta) = (\sin \alpha)(\cos \beta) - (\cos \alpha)(\sin \beta).$$

(7.2.15) $$\cos(\alpha - \beta) = (\cos \alpha)(\cos \beta) + (\sin \alpha)(\sin \beta).$$

(7.2.16) $$\tan(\alpha - \beta) = \frac{\tan \alpha - \tan \beta}{1 + (\tan \alpha)(\tan \beta)}.$$

The difference identities are very similar to the sum identities -- only the signs on the second terms are changed.

The sum and difference identities have numerous applications (problem B39). You will see them in calculus and you will see them a few sections later in this text where they apply to multiplying and dividing complex numbers. It is remarkable that trigonometry appears even in a very algebraic context such as solving quadratic and polynomial equations where complex numbers arise.

Theoretically, the sum identities can be used to find the sine of a sum of angles, given the sines and cosines of the component angles.

Example 7: Find sin 75° explicitly using a sum identity.

The values of sine and cosine are known explicitly at 45° and 30° (6.2.13).

$$\sin 75° = \sin(45°+30°) = (\sin 45°)(\cos 30°) + (\cos 45°)(\sin 30°)$$
$$= (\sqrt{2}/2)(\sqrt{3}/2) + (\sqrt{2}/2)(1/2) = (\sqrt{6} + \sqrt{2})/4.$$

Example 8: Find cos 15° explicitly using a difference identity.

$$\cos 15° = \cos(45° - 30°) = (\cos 45°)(\cos 30°) + (\sin 45°)(\sin 30°)$$
$$= (\sqrt{2}/2)(\sqrt{3}/2) + (\sqrt{2}/2)(1/2) = (\sqrt{6} + \sqrt{2})/4.$$

Proof. Not all identities have proofs as simple as the reference angle proofs. The following proof of the sum identities is long, but straightforward and illuminating. It determines $\sin(\alpha+\beta)$ and $\cos(\alpha+\beta)$ simultaneously. Following the details will help you learn trigonometry. The identity for tangent is proved in Example 14.

Consider the unit circle. Let angle α terminate at point A and $\alpha+\beta$ terminate at point P as in Figure 7. By definition, the coordinates of A are $(\cos \alpha, \sin \alpha)$ and the coordinates of P are $(\cos(\alpha+\beta), \sin(\alpha+\beta))$. The coordinates of P are just what we want to know for the sum identities.

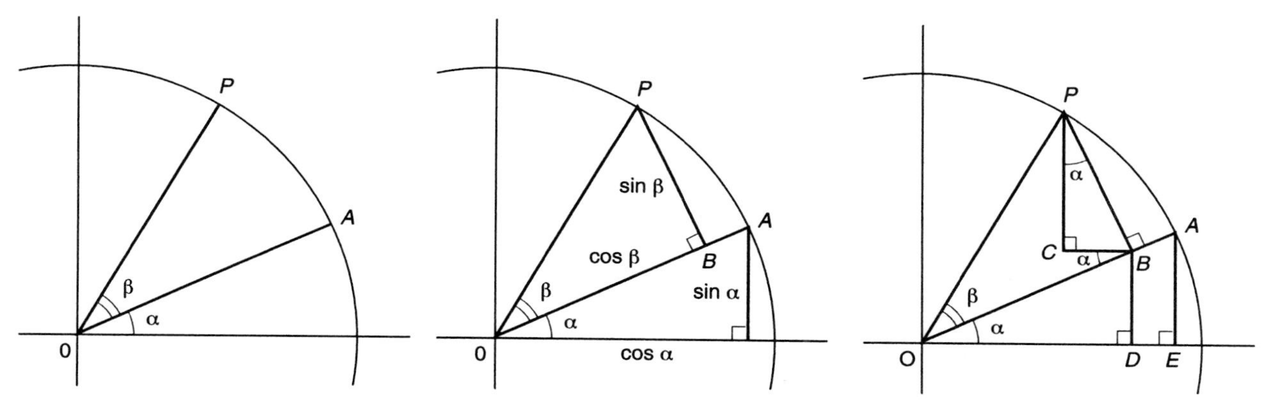

Figure 7: α and point A, β and point P corresponding to angle $\alpha+\beta$.

Figure 8: Figure 7 again, plus a few key lines and labels.

Figure 9: Figure 8 again, with more labels.

Now, as in Figure 8, drop a perpendicular from P to the line OA, intersecting the line at point B. Now OB is $\cos \beta$ and BP is $\sin \beta$. The problem is to express the vertical and horizontal location of P in terms of only the lengths labeled in Figure 8.

Figure 9 exhibits the needed auxiliary lines. We will locate B and then C on BC which is parallel to the x-axis. Then we will treat the vertical coordinate of P as $DB+CP$. And we will treat the horizontal coordinate of P as $OD - CB$.

Locate point B first. Consider triangle ODB which is similar to triangle OEA.

$$\frac{OB}{OA} = \frac{DB}{EA} = \frac{OD}{OE}.$$

$$\frac{\cos\beta}{1} = \frac{DB}{\sin\alpha} = \frac{OD}{\cos\alpha}.$$

Therefore,

$$DB = (\sin\alpha)(\cos\beta) \quad \text{and} \quad OD = (\cos\alpha)(\cos\beta).$$

Note that these are the first terms in the sum identities 7.2.11 and 12 for sine and cosine, respectively.

Now consider triangle *PCB*. Because *CB* is parallel to the *x*-axis and *PC* is parallel to the *y*-axis, the angles are α as labeled. Using "opposite is hypotenuse times sine" and "adjacent is hypotenuse times cosine," we obtain

$$CP = (\sin\beta)(\cos\alpha) \quad \text{and} \quad CB = (\sin\beta)(\sin\alpha).$$

These are the second terms in the sum identities for sine and cosine, respectively. Add *DB* and *CP* to obtain the vertical coordinate of *P*, $\sin(\alpha+\beta)$. This yields 7.2.11. Subtract *CB* from *OD* to obtain the horizontal coordinate of *P*, $\cos(\alpha+\beta)$. This yields 7.2.12.

This is not the only possible proof. There are alternative proofs that are more algebraic and make less use of a picture (problem B31). But I think this is the most illuminating derivation of the sum identities. From here we easily can derive the important "double angle" and "half angle" identities.

Double- and Half-Angle Identities. Given the trigonometric functions of θ, what are their values at 2θ? At $\theta/2$? Of course, since order matters, you do not expect to simply multiply or divide the values by 2. We begin with "double" angles.

In the sum identity, let α and β both be θ. Then $\alpha+\beta = 2\theta$ and the double angle formulas result from 7.2.11 and 12:

(7.2.17) Double-Angle Identities

$$\sin 2\theta = 2(\sin\theta)(\cos\theta).$$

$$\cos 2\theta = \cos^2\theta - \sin^2\theta$$

Also, $$\cos 2\theta = 1 - 2\sin^2\theta,$$

and $$\cos 2\theta = 2\cos^2\theta - 1.$$

$$\tan(2\theta) = \frac{2\tan\theta}{1-\tan^2\theta}.$$

The two alterative cosine lines follow from "$\sin^2\theta + \cos^2\theta = 1$," since $\cos^2\theta = 1 - \sin^2\theta$ and $-\sin^2\theta = \cos^2\theta - 1$. There is another method of proof in problem B29. The proof of the tangent identity is problem B17.

Two very important identities for calculus follow merely by rearranging the double-angle cosine formulas.

<u>(7.2.18) Squared-Function Identities:</u>

$$\sin^2\theta = (1 - \cos 2\theta)/2. \qquad \cos^2\theta = (1 + \cos 2\theta)/2.$$

You will see these in calculus when it come time to integrate "$\sin^2 x$" or "$\cos^2 x$."

These, in turn, yield the half-angle formulas by changing dummy variables. Replace "2θ" by "θ" and therefore replace "θ" by "$\theta/2$".

$$\sin^2(\theta/2) = (1 - \cos\theta)/2,$$
$$\cos^2(\theta/2) = (1 + \cos\theta)/2.$$

Now remove the squaring to obtain the half angle identities (problem B18).

<u>(7.2.19) Half-Angle Identities</u>

$$\sin\left(\frac{\theta}{2}\right) = \pm\sqrt{\frac{1 - \cos\theta}{2}}. \qquad \cos\left(\frac{\theta}{2}\right) = \pm\sqrt{\frac{1 + \cos\theta}{2}}.$$

Example 9: Find cos $22\frac{1}{2}°$ explicitly.

Treat $22\frac{1}{2}°$ as $45°/2$. We already know that $\cos 45° = \sqrt{2}/2$. From the half angle identity,

$$\cos 22\frac{1}{2}° = \cos\left(\frac{45°}{2}\right) = \sqrt{\frac{1 + \frac{\sqrt{2}}{2}}{2}} = \sqrt{\frac{2 + \sqrt{2}}{4}} = \frac{\sqrt{2 + \sqrt{2}}}{2}.$$

By determining more and more values we could fill out a trigonometric table for arguments between 0° and 90°. For example, with this value for $\cos 22\frac{1}{2}°$ we could find $\cos 11\frac{1}{4}°$ using the half-angle identity again, and we could find $\cos 52\frac{1}{2}°$ ($52\frac{1}{2} = 30 + 22\frac{1}{2}$) using a sum identity. To extend this process to fill out the entire trigonometric table would entail a great deal of work. Fortunately, that work has already been done by our predecessors long ago.

<u>**Other Trigonometric Functions**</u>. You may be wondering why I have not mentioned the trigonometric functions cosecant (csc), secant (sec), and cotangent (cot) until now. The reason is that they are simply reciprocals of sine, cosine, and tangent, respectively, so they provide little that is new and interesting. More often than not it is appropriate to simply rewrite them in terms of sine, cosine, or tangent and work from there. Glance at your calculator keyboard and note which trig functions it displays. You will see sine, cosine, and tangent. You will not see cosecant, secant, or cotangent. If you need to work with them, you will probably want to use the next result to

convert them to more familiar trigonometric functions instead. Their unit-circle interpretations are also given (Figure 1).

<u>(7.2.20) Reciprocal Identities</u>:

$$\csc\theta = 1/(\sin\theta) = 1/y.$$
$$\sec\theta = 1/(\cos\theta) = 1/x.$$
$$\cot\theta = 1/(\tan\theta) = x/y.$$

Example 10: Evaluate csc 30°, sec 80°, and cot 12°.

csc 30° = 1/(sin 30°) = 1/(1/2) = 2.
sec 80° = 1/(cos 80°) = 1/.174 [from a calculator] = 5.76.
cot 12° = 1/(tan 12°) = 1/.213 [from a calculator] = 4.70.

When sine takes on positive values less than 1, cosecant, its reciprocal, takes on positive values greater than 1. Because the range of sine is the interval [-1, 1], the range of cosecant is $(-\infty, -1]\cup[1, \infty)$, which is also the range of secant. There are no solutions to equations such as "csc x = 1/2" or "sec x = -2/3." The values 1/2 and -2/3 are not in their ranges.

To solve equations with these trig functions, convert the equation to one with the familiar trig functions.

Example 11: Solve "csc x = 2," "sec($3x$) = 4," and "cot(x^2) = 1/2."

In each case we can use reciprocal properties.

csc x = 2 iff 1/(csc x) = 1/2 iff sin x = 1/2. Now solve this.

sec($2x$) = 4 iff 1/(sec($2x$)) = 1/4 iff cos($2x$) = 1/4. Now solve this.

cot(x^2) = 1/2 iff 1/(cot(x^2)) = 1/(1/2) = 2 iff tan(x^2) = 2. Now solve this. Every text graphs cosecant and secant, although they are of little use (Figures 10 and 11).

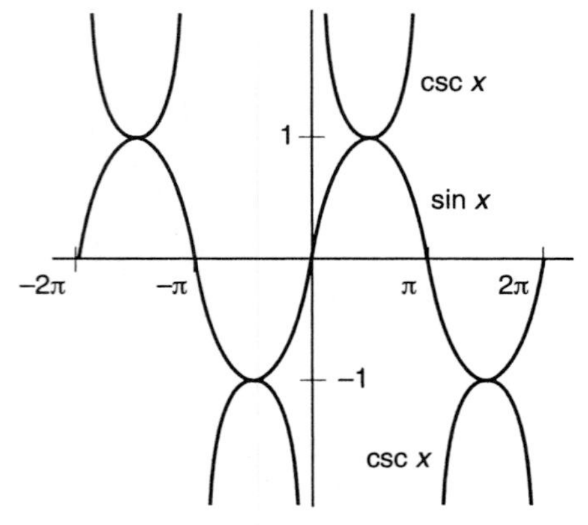

Figure 10: sin x and its reciprocal, csc x.

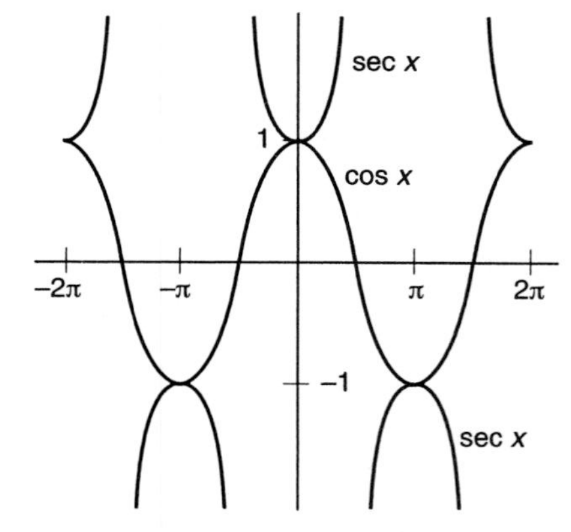

Figure 11: cos x and its reciprocal, sec x.

You may have wondered why three trig functions have the names of three others with "co" as a prefix: sine and cosine, tangent and cotangent, secant and cosecant. The "co" refers to complementary angles. Angles are said to be complementary if and only if they sum to a right angle. In a right triangle the two non-right angles are complementary. Consider Figure 12 with complementary angles θ and α. The sine of θ is "opposite over hypotenuse" from the point of view of θ. But from the point of view of angle α that is "adjacent over hypotenuse." Thus the sine of θ is the cosine of α. So "cosine" is an abbreviation of the archaic term "complemental sine."

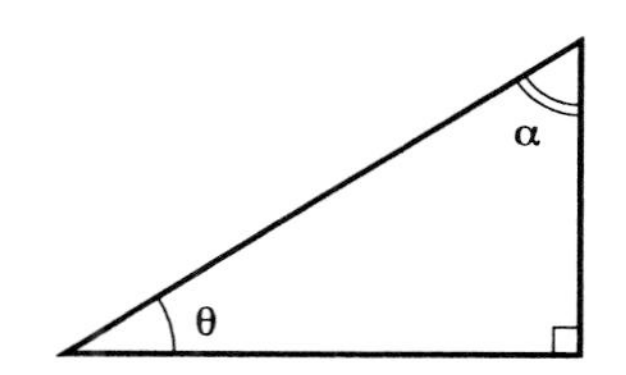

Figure 12: Complementary angles θ and α.

Similarly, the tangent of any angle is the cotangent of the complementary angle, and the secant of any angle is the cosecant of the complementary angle (problems B19 and 20).

(7.2.21) Cofunction Identities.

$$\sin(90° - \theta) = \cos\theta, \quad \cos(90° - \theta) = \sin\theta,$$
$$\text{and} \quad \tan(90° - \theta) = \cot\theta = 1/(\tan\theta).$$

This result still holds even if angle θ will not fit in a right triangle. A proof uses the difference identities (or symmetry on the unit circle):

$$\begin{aligned}\sin(90° - \theta) &= (\sin 90°)(\cos\theta) - (\cos 90°)(\sin\theta)\\ &= 1(\cos\theta) - 0(\sin\theta)\\ &= \cos\theta.\end{aligned}$$

The cosine result follows similarly (problem B15). The tangent result follows because tangent is sine over cosine.

Secant. Secant is the reciprocal of cosine, by 7.2.20. Therefore it satisfies, "secant is hypotenuse over adjacent." There is one important usage of secant. In calculus it is used to rewrite "$x^2 + 1$" and "$x^2 - 1$" as perfect squares so that their square roots can be written without the square root sign. Of course, the square root sign does not really go away, it is just concealed in a new notation which is often convenient. Here is how.

Return to the unit-circle definitions where "$\sin\theta = y$," "$\cos\theta = x$," and $x^2 + y^2 = 1$. Dividing through by x^2,

$$\frac{x^2 + y^2}{x^2} = \frac{1}{x^2}$$

$$\frac{x^2}{x^2} + \frac{y^2}{x^2} = \frac{1}{x^2}$$

$$1 + (\frac{y}{x})^2 = (\frac{1}{x})^2$$

Tangent is y over x (7.2.1), and secant is $1/x$ (7.2.20), so, switching sides, we obtain more "Pythagorean identities" to go with 7.1.2:

(7.2.22) Pythagorean Identities:

$$\sec^2 \theta = \tan^2 \theta + 1.$$
$$\tan^2 \theta = \sec^2 \theta - 1.$$

Example 12: Use 7.2.21 to rewrite " $\sqrt{x^2 + 1}$ " using "trigonometric substitution" so that it is expressed without a square root.

Looking at 7.2.21, it appears that, if "$\tan \theta$" plays the role of "x", then "$x^2 + 1$" will be "$\tan^2 \theta + 1$" which is "$\sec^2 \theta$." This has a nice square root without a square root sign. So, let $\tan \theta = x$. Then

$$\sqrt{x^2 + 1} = \pm \sec \theta .$$

This technique is useful in calculus for integrating expressions with square roots.

The tangent-secant Pythagorean identities (7.2.21) are easy to interpret in a right triangle. Let the adjacent side be 1, so that $\tan \theta = x$ is the opposite side (Figure 13). Any size acute angle will do for the picture; θ can be small or large, it doesn't matter. Then, by the Pythagorean Theorem, the hypotenuse is $\sqrt{x^2 + 1}$. Then "cosine is adjacent over hypotenuse," so secant, which is the reciprocal of cosine, is "hypotenuse over adjacent." The adjacent is 1, so $\sec \theta = \sqrt{x^2 + 1}$.

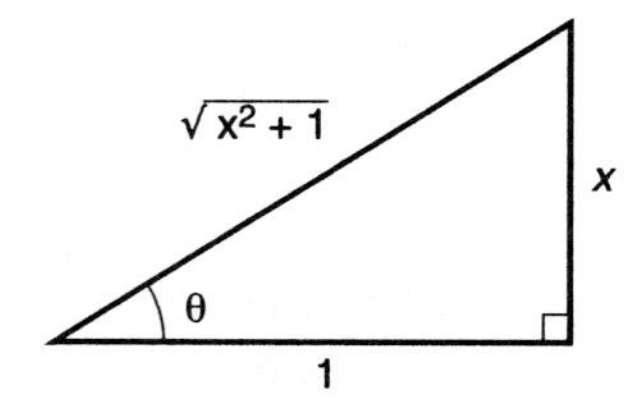

Figure 13: $\tan \theta = x$.

An effective way to deal with trigonometric expressions with these less-familiar functions is to rewrite the expression in terms of sine and cosine.

Example 13: Rewrite "$(\sec \theta)(\cot \theta)$" in terms of sine and cosine.

$$(\sec\theta)(\cot\theta) = \left(\frac{1}{\cos\theta}\right)\left(\frac{1}{\tan\theta}\right) = \left(\frac{1}{\cos\theta}\right)\left(\frac{1}{\left(\frac{\sin\theta}{\cos\theta}\right)}\right)$$

$$= \left(\frac{1}{\cos\theta}\right)\left(\frac{\cos\theta}{\sin\theta}\right) = \frac{1}{\sin\theta}.$$

If we want to convert the final expression to "csc θ", we may.

Example 14: Prove the sum identity for tangent, 7.2.13,

$$\tan(\alpha + \beta) = \frac{\tan\alpha + \tan\beta}{1 - (\tan\alpha)(\tan\beta)},$$

given the sum identities for sine and cosine, 7.2.11 and 12.

Recall that "tangent is sine over cosine" (7.2.1). Proof:

$$\tan(\alpha + \beta) = \frac{\sin(\alpha + \beta)}{\cos(\alpha + \beta)}$$

$$= \frac{(\sin\alpha)(\cos\beta) + (\cos\alpha)(\sin\beta)}{(\cos\alpha)(\cos\beta) - (\sin\alpha)(\sin\beta)}.$$

Now, divide through both the top and bottom by the first term in the denominator, so that term becomes 1 (as in the finished identity).

$$= \frac{\frac{(\sin\alpha)(\cos\beta)}{(\cos\alpha)(\cos\beta)} + \frac{(\cos\alpha)(\sin\beta)}{(\cos\alpha)(\cos\beta)}}{1 - \frac{(\sin\alpha)(\sin\beta)}{(\cos\alpha)(\cos\beta)}}$$

Some terms cancel and the sum identity follows immediately by "tangent is sine over cosine."

The difference identity for tangent follows by treating "α-β" as "α+(-β)" and using 7.2.4 (problem B16). The half-angle formula follows using "tangent is sine over cosine" (Problem B18).

<u>**A Picture is Worth Several Trig Identities**</u>. Identities relate trig functions. For example, the Pythagorean identity "$\sin^2 x + \cos^2 x = 1$" allows us to express sine in terms of cosine, or vice versa, as noted in Example 1. In calculus, we may also want secant in terms of tangent, or cosine in terms of tangent, or any other combination of two trigonometric functions. Rather than memorize all the possible trig identities, they can be derived on demand with the following approach in which "A picture is worth several trig identities."

All the trigonometric identities that relate two functions at the same argument can be derived from pictures in which one side has length 1.

Example 15: $\sin \theta = x$. Find $\cos \theta$ in terms of x.

We could substitute "x" in for "$\sin \theta$" in "$\sin^2 \theta + \cos^2 \theta = 1$," and solve for "$\cos \theta$", but that would miss the point. The point of this problem is to see how to do problems like this even if you do not recall the relevant trig identity.

Sketch a right triangle and label it to satisfy "$\sin \theta = x$" (Figure 14), letting the hypotenuse be 1. The size of the picture, and the size of angle θ do not matter, sine "θ" and "x" do not have particular values. Then use the Pythagorean Theorem to find the remaining side. The adjacent side will be $\sqrt{1 - x^2}$, as labeled. Now the absolute value of any trig function can be read from the picture. Cosine is adjacent over hypotenuse.

$$\cos \theta = \sqrt{1 - x^2} / 1 = \sqrt{1 - x^2}.$$

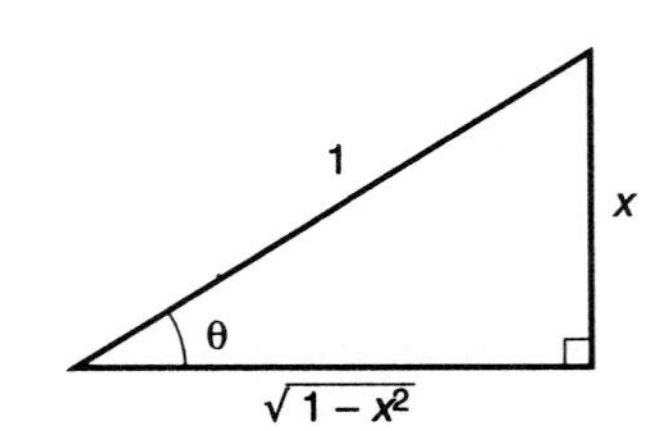

Figure 14: $\sin \theta = x$.

We should be careful to note that in quadrants other than the first quadrant the sign of the answer might have to be changed. So, for example, if θ were in the second quadrant where cosine is negative we would want to change the sign.

The reason for the change of the sign goes back to the idea of reference angles and that fact that only angles between 0 and 90 degrees fit in right triangles. A right-triangle picture can give us the absolute value of the answer, but it cannot give us the sign. If we had done this problem with the Pythagorean identity 7.2.2, we would have had to solve an equation with "$\cos^2 \theta$" in it and it would have been clear that the sign of "$\cos \theta$" is not determined until you know the quadrant of θ.

Example 16: Express tangent in terms of cosine.

This asks for a trig identity which is easy to derive. The method is straightforward and based on labeling a good picture. Denote $\cos \theta$ by some letter, say x. Illustrate this relationship for an acute angle (Figure 15). Then use the Pythagorean Theorem to find the other side. Read $\tan \theta$ from the picture in terms of x. Substitute "$\cos \theta$" in for "x" to obtain the desired relationship.

In the picture the opposite side is $\sqrt{1 - x^2}$. Thus

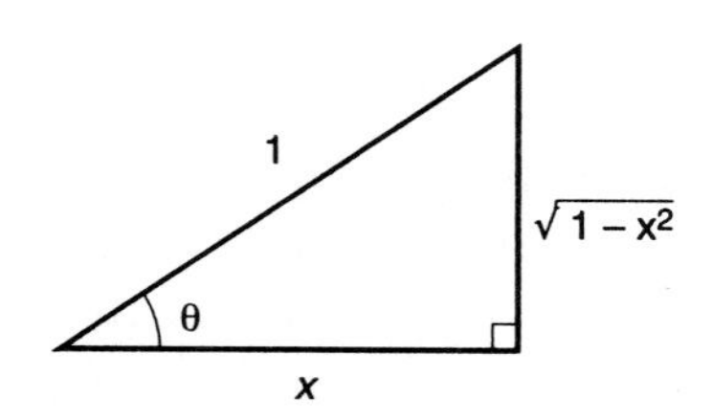

Figure 15: $\cos \theta = x$.

$$\tan \theta = \frac{\sqrt{1 - x^2}}{x} = \frac{\sqrt{1 - \cos^2 \theta}}{\cos \theta}.$$

This holds for acute angles -- the most important case. If the angle is not acute, this can be interpreted as a picture of the reference angle and the sign of the right side may need to be changed, depending upon the quadrant of the angle.

Of course, this is also the result of using both "tangent is sine over cosine" and "sine squared plus cosine squared equals 1." But the point is that the result follows from a good picture and the definitions without using these trig identities.

What do you need to know to do problems like Examples 15 and 16? First, you need to know the definitions of the trigonometric functions. Then you need to know how to illustrate the relationship between an angle and the value of a given trig function. Then you need to know the Pythagorean Theorem. That's all (except for remembering that in some quadrants the sign may need to be changed).

Sometimes these problems are disguised as inverse problems.

Example 17: Find $\tan(\sin^{-1}x)$.

Name "$\sin^{-1}x$" something. Since it is an angle, call it θ. Illustrate that relationship in a picture (Figure 16). Now, as before, read the value of the tangent from the picture.

$$\tan(\sin^{-1}x) = \frac{\pm x}{\sqrt{1-x^2}}.$$

The most important case is in the first quadrant where the sign is plus, but the sign would be minus in the second quadrant.

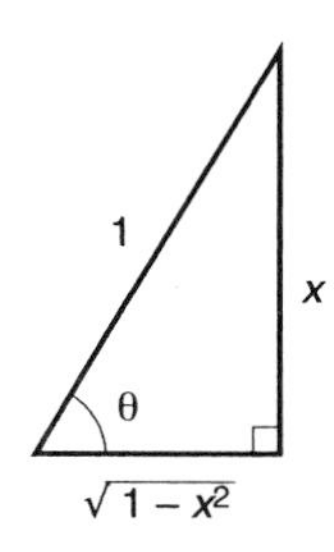

Figure 16: $\sin\theta = x$.

Example 18: Suppose $\sec\theta = x/2$. Find $\tan\theta$ in terms of $x > 0$.

We could use Identity 7.2.21, "$\tan^2\theta = \sec^2\theta - 1$" and solve for "$\tan\theta$," but the purpose of this example is to show you how a picture can yield the same result (if you know the definitions of tangent and secant).

First draw a right triangle to illustrate the given relationship: "$\sec\theta = x/2$." Since "secant is the reciprocal of cosine" (7.2.20) and "cosine is adjacent over hypotenuse," therefore "secant is hypotenuse over adjacent." So label the picture with angle θ, hypotenuse "x", and adjacent "2" (Figure 17). Then, use the

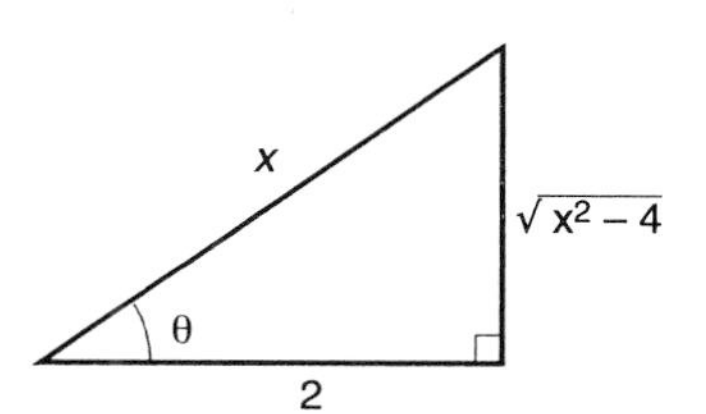

Figure 17: $\sec\theta = x/2$.

Pythagorean Theorem to find the remaining side. Then you can read the value of any trigonometric function of θ from the picture. Because "tangent is opposite over adjacent,"

$$\tan \theta = \frac{\sqrt{x^2 - 4}}{2}.$$

Conclusion. A picture is worth several trig identities. Learn Figures 1, 3, and 4 through 6. Learn to use figures like the ones in Examples 15 through 18 to derive simple trig identities. Some identities are too complicated to be easily derived from a picture. These you have to learn, or, at least, be aware of. They include the sum and difference identities (7.2.11-16), the double-angle and half-angle identities (7.2.17 and 19), and the identities for squares of sine and cosine (7.2.18). But primary and most important is simply knowing the definitions of the trigonometric functions (7.2.1 and 20) and how to illustrate them on the unit circle and in a right triangle.

Terms: coterminal, reference angle, secant, cosecant, cotangent, complementary angles.

****

Exercises for Section 7.2, "Identities":

A1.* In the unit-circle definition, express these in terms of x and y.
a) $\sin \theta$ b) $\cos \theta$ c) $\tan \theta$.

A2.* These are the reciprocals of which functions?
a) secant b) cosecant c) cotangent.

A3.* a) State the sine-cosine Pythagorean identity.
b) State the secant-tangent Pythagorean identity.

A4.* a) Define "coterminal." b) What is interesting about coterminal angles?

A5.* True or False?
a) $\sin 2x = 2 \sin x$. b) $\cos 2x = 2 \cos x$.
c) $\sin(\alpha+\beta) = \sin \alpha + \sin \beta$. d) $\cos(\alpha+\beta) = \cos \alpha + \cos \beta$.
e) What conclusion can you draw from parts a) through d)? [Two words will do.]

A6.* True or False?
a) $\sin x = \sin(-x)$ b) $\cos x = \cos(-x)$ c) $\tan x = \tan(-x)$.

A7.* Write in terms of $\sin x$, $\cos x$, or $\tan x$, whichever is most convenient.
a) $\sec x$. b) $\csc x$. c) $\cot x$.

A8.* What is the purpose of identities?

^ ^ ^ ^ Find the coterminal angle between 0 and 360°.

A9. a) -210° b) 1000°. A10. a) 250° b) -350°.

^ ^ ^ ^ Find the reference angle.

A11. a) 40° b) 170° c) 200° d) -50°.
A12. a) 100° b) 300° c) 19° d) 260°.

^ ^ ^ ^ Give the complementary angle:

A13. a) 23° b) 80°. A14. a) 30° b) 59°.

A15. θ is in the second quadrant and cos θ = -.6. Algebraically find sin θ and tan θ.

A16. θ is in the first quadrant and tan θ = 2. Algebraically find sin θ and cos θ.

A17. Is the function even, odd, or neither?
a) sine b) cosine c) tangent.

^ ^ ^ ^ Draw a picture to illustrate and find

A18. a) $\sin(\cos^{-1}.4)$ b) $\tan(\sin^{-1}.9)$. A19. a) $\cos(\sin^{-1}.7)$ b) $\sin(\tan^{-1}3)$.

^ ^ ^ ^ Find an angle in the first quadrant with the same trig function values, except possibly for the sign.

A20. a) 100° b) 345° c) 250° d) 175°.
A21. a) 220° b) 178° c) -92° d) 110°.

A22. Solve for two solutions, and give at least two decimal places. $\sin(x - \pi/6) = -1/3$.

A23. Find x in the given quadrant (use degrees).
a) cos x = -.4, III b) sin x = .9, II. c) tan x = .2, III.

A24. Find x in the given quadrant (use degrees).
a) cos x = .8, IV b) sin x = -.6, IV d) tan x = -5, II.

A25. Evaluate a) sec 40°. [1.3] b) csc 20°. [2.9] c) cot 80°. [.18]
A26. Evaluate a) sec 87°. [19] b) csc 5°. [11] c) cot 5°. [11]

A27. Solve for x in the first quadrant (use degrees).
a) sec x = 3. [71] b) csc x = 4. [14] c) cot x = 1/2. [63]

A28. Solve for x in the first quadrant (use degrees).
a) sec x = 1.5. [48] b) csc x = 2.5. [24] c) cot x = 10. [5.7]

^ ^ ^ ^ ^ ^ ^ ^ ^

B1.* Draw a right triangle to illustrate these and fill in the remaining side.
a) sin $\theta = x$. b) cos $\theta = x$. c) tan $\theta = x$.

B2.* Draw a right triangle to illustrate these and fill in the remaining side.
a) sec $\theta = x$. b) csc $\theta = x$. c) cot $\theta = x$.

B3.* State the simplest trig identity for
a) tan x b) sec x c) csc x d) cot x.

B4.* a) Define "reference angle." b) What is interesting about reference angles?

B5. Is the function even, odd, or neither?
a) secant b) cosecant c) cotangent.

^ ^ ^ ^ Assume the angles are first quadrant angles. Find
B6. $\tan(\sin^{-1}x)$ B7. $\cos(\tan^{-1}x)$
B8. $\sin(\cos^{-1}(2x))$ B9. $\sec(\tan^{-1}(x/4))$

^ ^ ^ ^ Solve for x in $[0°, 360°)$:
B10. $\csc x = 2$. [...,150] B11. $\sec(2x) = 4$. [38,...] B12. $\cot(x/2) = 1/2$. [130]

B13. Solve algebraically "$\cos^2 x = \sin x + .4$" for (at least) two solutions. [25,...]

B14. Use a difference identity to find sin 15° algebraically.

B15. Use the difference identity for cosine to prove the result for $\cos(90° - \theta)$ in 7.2.21.

B16. Use 7.2.13 and 7.2.4 to derive the difference identity for tangent, 7.2.16.

B17. a) Use 7.2.13 to derive the double-angle identity for tangent, 7.2.17.
b) Discuss what happens when $\theta = 45°$.

B18. Use "tangent is sine over cosine" to derive the <u>half-angle identity for tangent</u>:
$\tan(\theta/2) = (\sin\theta)/(1 + \cos\theta)$.

B19. State, using a variable, the trig identity, "The tangent of any angle is the cotangent of the complementary angle.

B20. State, using a variable, the trig identity, "The secant of any angle is the cosecant of the complementary angle."

B21. Find $\sin^{-1}(\sin x)$ for $0 \le x < 2\pi$. Be careful to cover all cases.

B22. Find $\cos^{-1}(\cos x)$ for $0 \le x < 2\pi$. Be careful to cover all cases.

B23. If the coordinates of $P(\theta)$ are (a, b), what are the coordinates of $P(90° - \theta)$?

^ ^ ^ ^ Use the figure to the right for problems B24-28. Solve algebraically for $\theta > 0$ (use degrees) with the help of trigonometric identities if the following relationships hold.

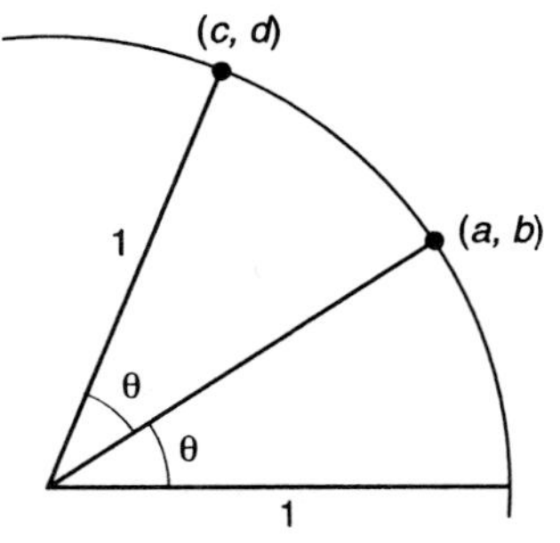

B24. $d = 5b/3$. [34]
B25. $d/c = 3b/a$. [30]
B26. $a = 2c$. [33]
B27. $a = c + .2$. [22]
B28. Solve graphically for θ if $b + .2 = c$. [25]

B29. To prove 7.2.15, "$\cos(2\theta) = 1 - 2\sin^2\theta$", consider an isosceles triangle with angle 2θ and adjacent sides equal to 1. Compute the length of the opposite side two ways and equate the results. One way is to use the Law of Cosines. The other is to bisect the angle and use the definition of sine.

B30. Set up equations for θ in the picture. Solve them algebraically, using the double-angle identity for tangent (Problem B17). [18]

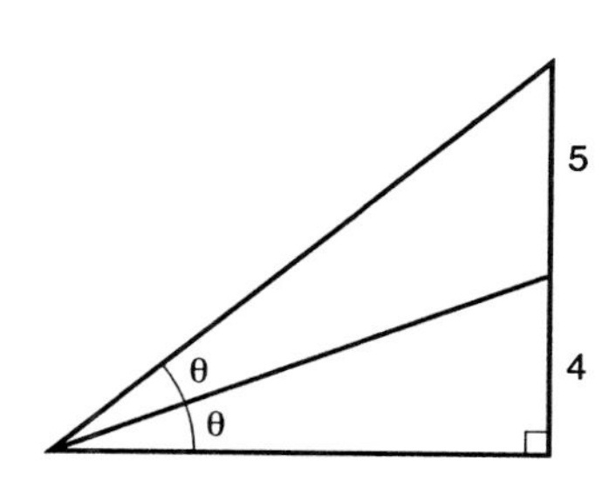

B31. (For a proof of 7.2.15, which leads to a proof of all the sine and cosine sum and difference identities.) On a unit circle locate $P(\alpha)$ and $P(\beta)$, where $\alpha > \beta$. Find the straight-line distance between them. Locate $P(\alpha - \beta)$ and $P(0)$ and find the straight-line distance between them. Set those two distances equal and simplify to obtain 7.2.15.

^ ^ ^ ^ Use trigonometric substitution to rewrite the expression as a trigonometric function of θ without a square root symbol.

B32. $\sqrt{x^2 + 4}$. B33. $\sqrt{x^2 - 9}$. B34. $\sqrt{1 - x^2}$. B35. $\sqrt{5 - x^2}$. B36. $\sqrt{4 - 3x^2}$.

B37. Create a Pythagorean identity for cosecant.

B38. Consider the graph of x^2 to be a reflecting surface. Let a ray of light come in from above, parallel to the y-axis and reflect off the curve (see the figure). Where does the ray intersect the y-axis? To do this, use the fact from physics that "the angle of incidence is equal to the angle of reflection" and the fact from calculus that the slope of the curve x^2 at any point (x, x^2) is $2x$. You might first try the particular case of a ray parallel to the y-axis which reflects off the curve at (1, 1).

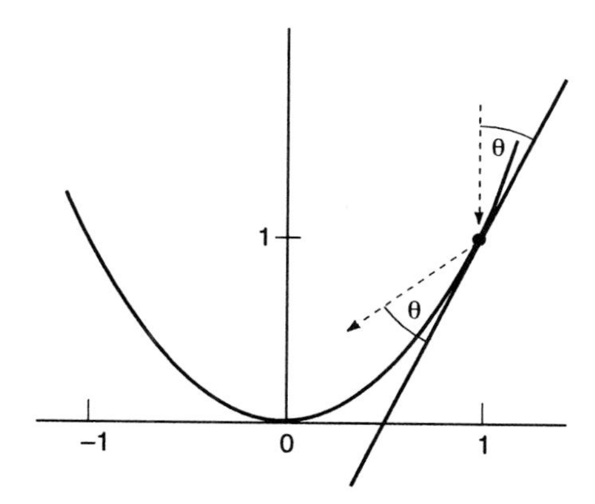

B39. The area of the parallelogram in the figure is given by "$A = ad - bc$." Prove this result using the SAS Area Formula (6.3.2) and the difference of angles identity, 7.2.14.

^ ^ ^ ^ Prove the following trig identities. You may use lower-numbered ones to prove higher-numbered ones. B40-43 are "Sum-to-Product" identities. B44-47 are "Product-to-Sum" identities.

B40. $\sin a + \sin b = 2\cos[(a-b)/2]\sin[(a+b)/2]$.

B41. $\sin a - \sin b = 2\cos[(a+b)/2]\sin[(a-b)/2]$.

B42. $\cos a + \cos b = 2\cos[(a+b)/2]\cos[(a-b)/2]$.

B43. $\cos a - \cos b = 2\sin[(a+b)/2]\sin[(b-a)/2]$.

B44. $(\sin a)(\sin b) = (1/2)[\cos(a-b) - \cos(a+b)]$.

B45. $(\sin a)(\cos b) = (1/2)[\sin(a+b) + \sin(a-b)]$.

B46. $(\cos a)(\sin b) = (1/2)[\sin(a+b) - \sin(a-b)]$.

B47. $(\cos a)(\cos b) = (1/2)[\cos(a+b) + \cos(a-b)]$.

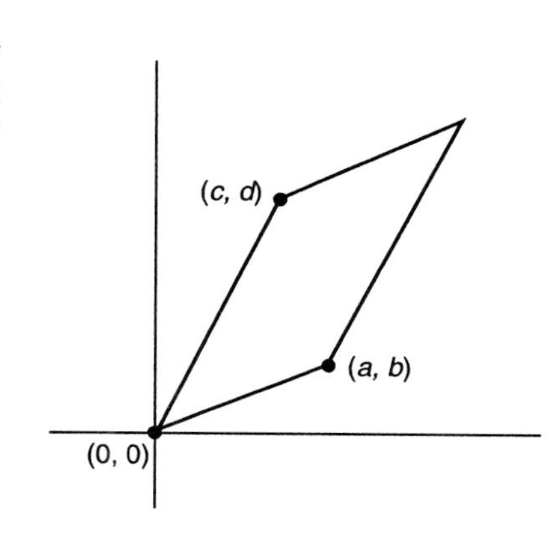

Section 7.3. Waves

The shape of the graph of "sin t" (Figure 1) is called a "sine wave." (In this section we will use the letter "t" for time, instead of "x".) By definition, a sine wave describes one type of periodic motion -- the vertical motion of a point rotating uniformly about a circle. Circular motion has two components as time elapses: the point goes both up-and-down and side-to-side. The sine wave isolates the up-and-down part and ignores the side-to-side part. Think of the horizontal axis in Figure 1 as time and the vertical axis as the vertical position of a rotating or oscillating point.

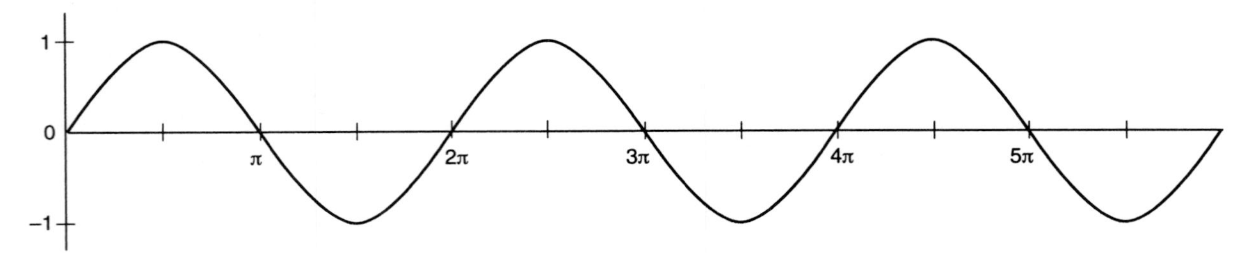

Figure 1: sin t. $0 \le t \le 6\pi$.

Modifications of sine waves are well-suited to describing other types of periodic behavior -- behavior that repeats after some fixed amount of time. A function f is periodic if and only if there is a positive number, p, such that $f(t + p) = f(t)$ for all t. For example, $f(t)$ could describe the horizontal position of a swinging pendulum of a clock at time t. If the pendulum swings back and forth once a second, then $f(t)$ repeats every second and p could be 1. Of course, its position will also repeat every two seconds, so p could be 2, and its position repeats every 10 seconds, so p could be 10. But the repetition every 1 second is the most important. If f is a periodic function, there will be numerous possible values of p that satisfy "$f(t) = f(t + p)$, for all t." We keep track of the smallest such p.

(7.3.1) The period of f is the least positive p that satisfies "$f(t) = f(t + p)$, for all t."

A function can be periodic without being exactly a sine wave. For example, a regular heartbeat can produce a repeating up and down wave that is not as simple as a sine wave.

Example 1: Sound waves are sine waves. That is, when you hear a single note at constant volume, the pressure at your eardrum goes up and down according to a sine wave.

Pluck a string on a guitar and watch it vibrate at the middle. The position of the string changes very rapidly back and forth. The position is described by a sine wave with a short period.

Electricity coming out of a wall plug in the U.S. has a frequency of 60 cycles per second (in England it is 50 cycles per second). That is, it repeats 60 times a second. It can be described by a sine wave with a period of 1/60 second.

Other electrical phenomena, such as radio and television waves, *x*-rays, gamma rays, light waves (as in lasers that read music off of compact disks), microwaves (for cooking) are described by sine waves. Radio waves repeat from thousands to millions of times a second. Visible light repeats on the order of about 10^{15} times a second. X-rays repeat on the order of 10^{20} times per second.

Even if a periodic motion is not exactly described by a sine wave, sine waves are often involved in combination with other functions.

Example 2: The position of a piston in an engine running at constant speed is periodic (Figure 2, position y). The piston is connected to a shaft which is going around at a constant angular speed, so the connection (P) moves up and down in a sine wave. That is, the vertical component of P moves in a sine wave. But the piston itself does not move the same vertical distance that P does because it is connected to the rotating point with a piston rod which is not constantly vertical (problem B9). So, in Figure 2, y is closely related to a sine wave but it is not exactly a sine wave.

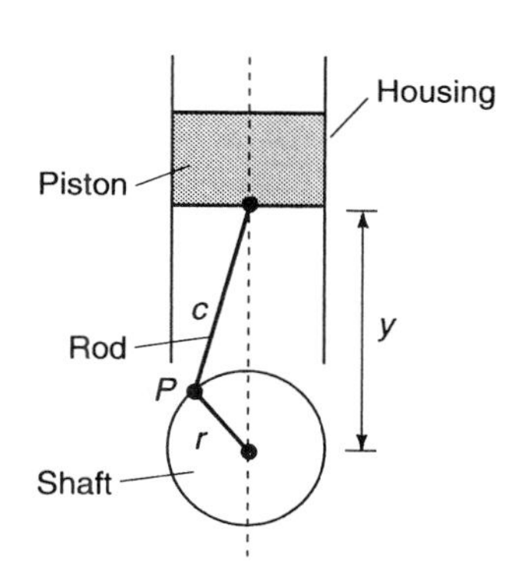

Figure 2: A piston connected to a rotating shaft.

Sine waves are important. Any one of the applications in Examples 1 and 2 could be studied for a long time (in another course). We will just introduce some of the key terms and indicate some fascinating properties of sine waves.

Again, imagine the argument of a sine wave to be time. When $P(t)$ goes around the circle once, one revolution (or cycle) has been completed. The amount of time one revolution takes is the period. The frequency of a particular sine wave is the number of periods (cycles, revolutions) per unit of time, for example, "60 cycles per second" or "60 revolutions per second."

Example 3: The period of "sin t" is 2π, because its values repeat after 2π. Its frequency is $1/(2\pi)$, because that is the number of periods (revolutions) that occur in one unit of time.

Sometimes sine waves are written with the "2π" included so the period and frequency will be simpler numbers. The period of "$\sin(2\pi t)$" is 1. As t goes from 0 to 1, the function goes though a full cycle (2π). Its frequency is 1/1 = 1.

The period of "$\sin(5(2\pi t))$" is 1/5. Scale changes were discussed in Section 2.3 on composition of functions. Multiplying by 5 before applying sine makes the graph 1/5 as wide and the period 1/5 as much. Therefore the frequency is 5 times as much. For instance, as t goes from 0 to 1, the argument of sine goes from 0 to 10π, 5 periods. Its frequency is 5 and its period is 1/5.

(7.3.2) Period p corresponds to frequency $1/p$.
"$\sin(k(2\pi t))$" has period $1/k$ and frequency k.
"$\sin(kt)$" has period $2\pi/k$ and frequency $k/(2\pi)$.

If the frequency is described in "cycles per second," the period is described in "seconds (per cycle)."

Example 4: The musical note concert A has frequency 440 cycles per second. Give a sine wave with its frequency. What is its period?

The frequency of "$\sin(2\pi(440t))$" is 440 (cycles per second). Its period is 1/440 second.

So many types of real-world waves have such short periods that we often describe sine waves in terms of frequency rather than period so the numbers are big rather than tiny.

Example 5: The frequency of a local FM radio station is 93.7 million cycles per second. A "cycle per second" in the context of electricity is also known as a "hertz" after an important scientist, Heinrich Hertz, who worked in the late 1800's. (Think of something important and you might have something named after you, too.) The prefix "mega" is used for million, so this would be called "93.7 megahertz."

The distance that a wave travels in one period is called a <u>wavelength</u>. Electromagnetic waves (including radio and light waves) travel at the speed of light, about 300,000 kilometers per second (186,000 miles per second). That's fast.

Example 5, continued: What is the wavelength of a 93.7 megahertz radio wave?

The wavelength is distance per cycle. Divide "distance per second" by "cycles per second" to obtain "distance per cycle."

$$\frac{300{,}000 \textit{ kilometers per second}}{93{,}700{,}000 \textit{ cycles per second}} = .0032 \textit{ kilometers per cycle}$$
$$= 3.2 \textit{ meters per cycle} .$$

The wavelength is 3.2 meters.

Example 6: Visible light has frequency on the order of 10^{15} cycles per second. What is the wavelength?

$$\frac{300{,}000 \textit{ kilometers per second}}{10^{15} \textit{ cycles per second}} = 3 \times 10^{-10} \textit{ kilometers per cycle}$$
$$= 3 \times 10^{-7} \textit{ meters per cycle} .$$

This is on the order of 1/10,000 inch. A short wavelength!

Example 7: What is the wavelength of the musical note concert A?

For this we need to know the frequency of concert A (440 cycles per second) and the speed of sound. The speed of sound is about 1100 feet per second in air at 0° Celsius at sea level. (It's slower at warmer temperatures and at higher altitudes.)

$$\frac{1100 \textit{ feet per second}}{440 \textit{ cycles per second}} = 2.5 \textit{ feet per cycle} .$$

The wavelength is about 2½ feet.

The previous facts about the speed of sound and the speed of light can be combined to give a quick way to tell how far away lightning is. If lightning is, say 5000 feet away, the light flash will arrive is almost no time, but the sound will travel only about 1100 feet per second. The sound will arrive about 5000/1100 = 4.5 seconds later than the light. This type of computation can be converted into a simple formula. Since a mile is 5280 feet, sound takes close to 5 seconds to travel a mile, 10 seconds to travel two miles, etc. So the method to tell how far away lightning is: Count the time in seconds from the times you see the flash until you hear the thunder. Divide by 5 to determine how far away the lightning was in miles.

Example 8: Rose sees a lightning bolt and hears the thunder 7 seconds later. How far away was the bolt?

7/5 = 1 and 2/5. It was about one and two fifths miles away.

Describing Sine Waves. Sine waves are described by three features: amplitude, period, and phase shift. The frequency determines the period, and so does the wavelength, so they can be given in place of the period. These features produce scale and location changes of the sort discussed in Section 2.3 on composition of functions.

Let the basic function of the following discussion be the fundamental sine wave (in radians), "sin t," with period 2π. It has amplitude (maximum value) 1. Let "A" (greater than 0) denote a constant ("A" for "amplitude"). "A sin t" differs from "sin t" only in the vertical scale. Its maximum value is A -- its amplitude is A (Figure 3).

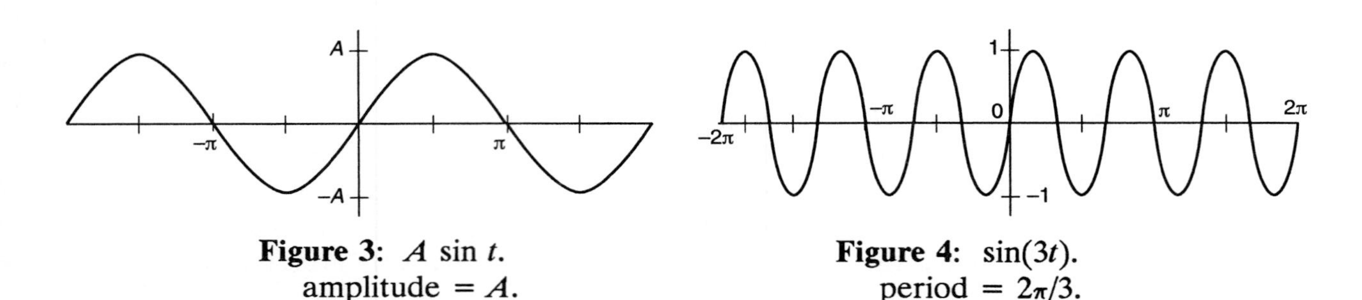

Figure 3: A sin t.
amplitude = A.
$-2\pi \leq t \leq 2\pi$.

Figure 4: sin($3t$).
period = $2\pi/3$.
$-2\pi \leq t \leq 2\pi$.

Let "B" denote another constant ("B" for no good reason). "sin(Bt)" differs from "sin t" only in horizontal scale. Since sine repeats every 2π, sin(Bt) repeats every $2\pi/B$. If B is greater than 1, compared to "sin t," the period of "sin Bt" is shorter and the frequency is greater (Figure 4).

The graph of "sin t" has image zero when $t = 0$. Think of the graph as starting at the origin. Because the graph of "sin(t - c)" has argument zero when $t = c$, it is said to have phase shift c. The graph is the same shape as the graph of "sin t", but shifted *right* c units (Figure 5). When a sine wave has phase shift c, it begins at c the way the usual sine graph begins at 0.

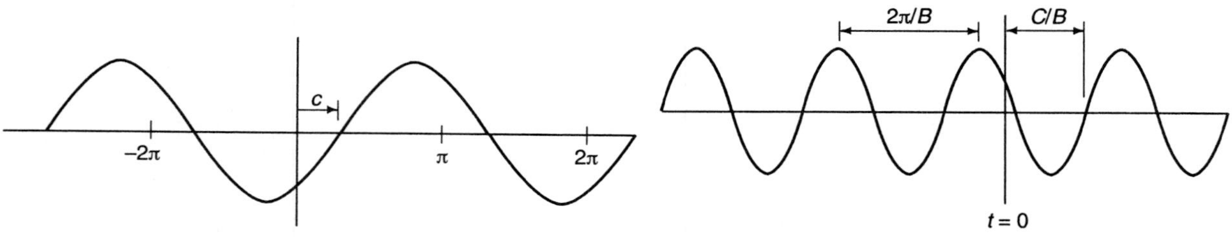

Figure 5: sin(t - c).

Figure 6: sin(Bt - C).

The function "sin(Bt - C)" combines a period change and a phase shift (Figure 6). The period is easy to see -- it has period $2\pi/B$. However, the phase shift is not simply C because the horizontal scale change created by the "B" occurs first. When is the argument of sine zero? It is zero when Bt - C

$= 0$, that is, when $t = C/B$. The phase shift is C/B. The graph begins at C/B with an upward bump the way the usual sine curve begins at 0.

These three types of scale changes and location shifts can be expressed in one expression. Let $B > 0$. Then

(7.3.3) $$y = A \sin(Bt - C)$$

is a sine wave with amplitude $|A|$ (absolute value, in case A is negative), period $2\pi/B$, and phase shift C/B.

Any curve with this equation is called a "sinusoid," or a "sinusoidal" curve. Any motion satisfying this equation is said to be "simple harmonic motion."

Example 9: Give the amplitude, period, and phase shift of

$$y = 500 \sin(20t - 100).$$

The amplitude is 500, the period is $2\pi/20 = \pi/10$, and the phase shift is $100/20 = 5$.

As you have seen, sine and cosine waves have the same shape and the same period, 2π (360°, 7.2.3). Actually, you don't hear the term "cosine waves" much because they are just shifted sine waves.

(7.3.4) $$\cos \theta = \sin(\theta + 90°) \quad \text{[in degrees]}$$
$$= \sin(\theta + \pi/2) \quad \text{[in radians]}.$$

This tells us that the graph of cosine is the graph of sine shifted *left* $\pi/2$. The identity follows from trigonometric identities in the previous section. From 7.2.19, $\cos \theta = \sin(90° - \theta)$, which is $\sin(180° - (90° - \theta))$ [by 7.2.5], which is $\sin(\theta + 90°)$.

Example 10: When a spring is stretched or compressed and let go it vibrates. Describe the motion.

This is one case where it is convenient to express the function using cosine because $\cos 0 = 1$. We can let the stretch (position, positive or negative) at time t, measured from the resting position, be $y(t)$. Then the initial stretch at time 0 is $y(0)$. Use the model "$y(t) = A \cos(Bt)$," where B determines the frequency of the vibration.

Suppose a spring is stretched 4 inches and let go and it vibrates up and down 5 times a second. Give the formula for the position of its end.

Let the direction the spring is stretched be the positive y-direction. So $y(0) = 4$ (inches). All that remains is to determine B, that is, to adjust the frequency. The problem says that five cycles occur in 1 second. By 7.3.2 (which works for cosines exactly as for sines), using $k = 5$, the formula is $y(t) = 4 \cos(5(2\pi t))$.

<u>**Adding Sine Waves**</u>. Some of the most fascinating features of sine waves are their additive properties. We will use sound waves to illustrate them. For this subsection you will want to get out your graphing calculator and play around with these ideas.

Most sound you hear consists of a mixture of frequencies, but to begin discussion, consider a single frequency. We will consider combining two sound waves of the same frequency, but with possibly different amplitudes (volumes) and different phase shifts. To make things easy, we will not bother to use a realistic scale; we will use "sin t" as the basic sinusoid.

Any sinusoid of the form "A sin(t - c)" has the same frequency as "sin t." What do you hear when you hear two sources of sound producing the same note? What happens when you add two such curves together?

Of course, if two people hit two tuning forks and both tuning forks produce middle C, you will still hear middle C, but probably louder. That is, you expect the amplitude to change, but not the frequency. Mathematically, this real-life result corresponds to examples such as the following.

Example 11: Use radians to graph "sin t + sin(t - 1)".

Figure 7 graphs three graphs. You recognize "sin t." The graph of "sin(t - 1)" is similar, but shifted right 1 unit (1 radian is about 1/6 period). The graph of their sum appears to have the same period, but be higher (which corresponds to louder). It does not cross the t-axis at the same place as either, so it has a different phase shift.

In this example, the amplitude of the sum is greater than 1, but not as great as 2, which would be the sum of the amplitudes. The amplitude is about 1.755.

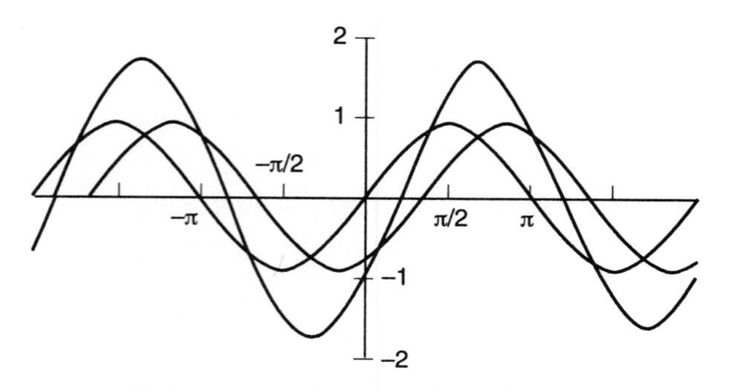

Figure 7: sin t, sin(t - 1) and their sum (bold). [-2π, 2π].

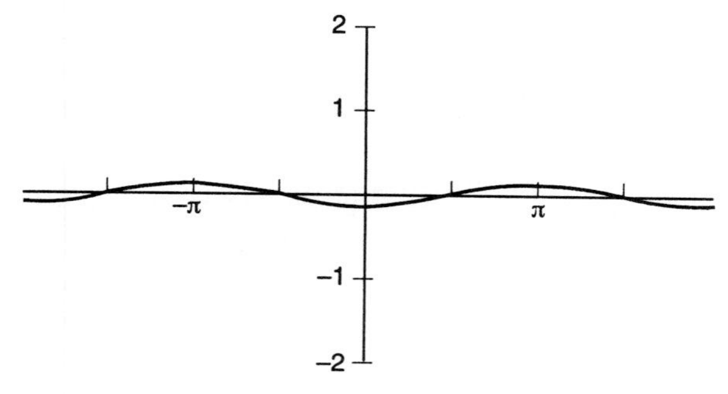

Figure 8: sin t + sin(t - 3) [-2π, 2π].

Example 12: Use radians to graph "sin t + sin(t - 3)."

Figure 8 shows that the sum is *less* than either. That is because the components are almost completely "out of phase" (Figure 9). The components, rather than working together to produce a larger sum, work against one another and tend to cancel each other. The amplitude of the sum is about .141, far less than the sum of the amplitudes and even less than the amplitude of either component.

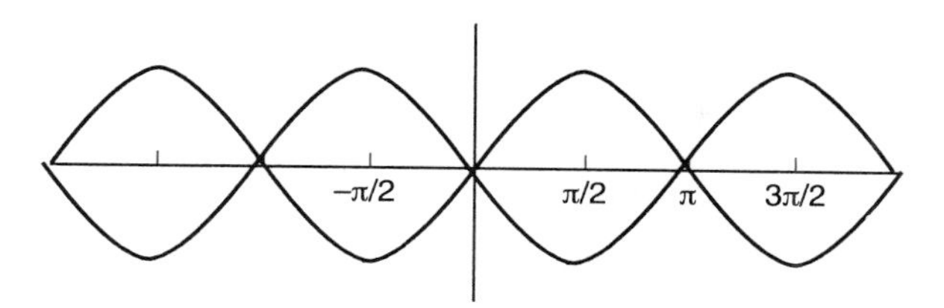

Figure 9: sin t, sin(t - 3).
[-2π, 2π]. They are out of phase.

Example 12 illustrates some very "high-tech" ideas. If an out-of-phase wave of the same frequency can be added to a wave, it can cancel the original wave. This "phase-shift" idea will eventually have applications in the car and in the home. There will be "silence-generating machines." The machine will analyze incoming sound for amplitude, frequency, and phase shift, and in an instant generate canceling waves from a speaker. If you are in the right position (say, the car driver's seat), you will hear very little. The machine will have turned sound into silence. You can already buy headphones that use this principle to *cancel* (not just muffle) the stressful monotonous drone of jet-engines during air travel.

Example 13: Graph "sin t + 2 sin(t - 4)."

Even if the components do not have the same amplitudes, the sum is again sinusoidal, as long as the frequencies of the components are the same. Figure 10 shows that, again, the sum has the same frequency as the components. The amplitude is about 1.55.

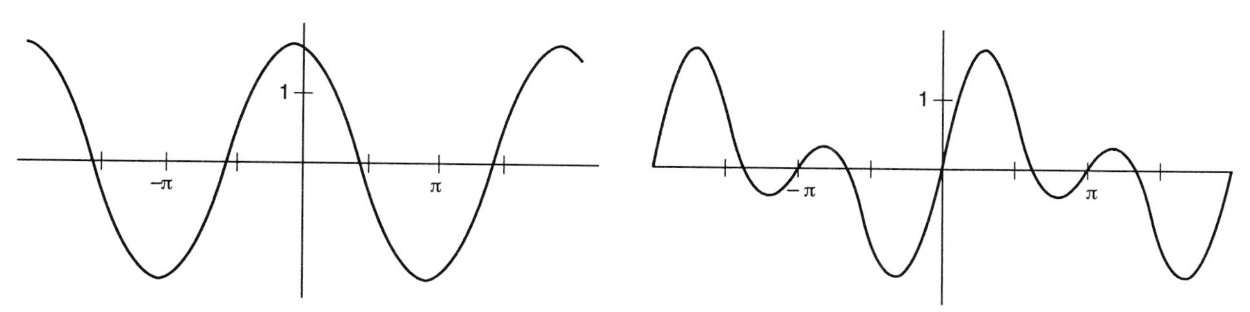

Figure 10: sin t + 2 sin(t - 4).
[-2π, 2π].

Figure 11: sin t + sin(2t).
[-2π, 2π].

The previous examples concerned waves that are not "in phase." That is, the components of the sum had different phase shifts. If two sines waves are in phase, the amplitude of the sum is the sum of the amplitudes. For example, "$\sin t + 2 \sin t$" is clearly just "$3 \sin t$," which has amplitude 3, which is the sum of the amplitudes.

When waves with different frequencies are added together, the result can be complicated and no longer simply sinusoidal.

Example 14: Graph "$\sin t + \sin(2t)$."

The graph is more complicated and not simply sinusoidal (Figure 11). It is a mathematical challenge to figure out how to take such a curve and decompose it into a sum of sinusoidal curves. Imagine you are a sound engineer helping design sound recording and analyzing machines. For example, you may wish to filter out "noise" of certain frequencies. But the music already consists of numerous frequencies at various amplitudes and all phase shifts. Sound hitting a microphone does not come labeled by its component frequencies. It's just sound. How can your machine, which "hears" it all at once, find the components at the frequencies it wants to filter out?

In the ocean there are many sources of sound. How can sonar (say, in a submarine) be analyzed to isolate the components of interest from the "background noise." Tough question. If you want to know, you will have to take more math!

Things are mathematically much simpler if all waves are of the same frequency. Normal light, say, from a flashlight, consists of many waves of many frequencies added together. Even if they were all of the same frequency, they would not add very efficiently because they would still be of different phase shifts. If fact, some components would cancel others, as in Example 12 (Figure 8). The reason laser light is so powerful is that lasers emit light waves *of the same frequency* that are *in phase*, so they add together efficiently and do not cancel each other.

Here is a theorem about the phase shift required to cancel a wave.

(7.3.5) $$\sin t + \sin(t \pm \pi) = 0, \text{ for all } t.$$

Example 15: What wave, when added to "$2 \sin(t - 1)$" will cancel it?

We need of wave of the same amplitude, but out of phase.

The wave "$2 \sin(t - 1 - \pi)$" will do it. I'd show a graph, but there wouldn't be much to see. It would be zero.

Example 16: What wave, when added to "$\sin(12t - 4)$" will cancel it?

Let "$12t - 4$" here play the role of "t" in 7.3.5. So the answer is simply

$$\sin(12t - 4 - \pi).$$

The original wave had a phase shift of 4/12 (= 1/3). The canceling wave has a phase shift of $(4 + \pi)/12$.

Waves that reinforce or cancel one another play an important role in physics in the study of sound and light (problems B11-14).

Conclusion: Sines waves are described by their amplitude, period (or frequency or wavelength), and phase shift. If waves of the same frequency are out of phase, they can cancel one another.

Terms: sine wave, periodic, period, cycle, revolution, frequency, wavelength, amplitude, sinusoid, phase shift, in phase, out of phase.

****

Exercises for Section 7.3, "Waves":

A1. Use the speed of 1100 feet per second to compute how many miles per hour sound travels.

A2. How far away is lightning if you hear the thunder
a) 15 seconds after you see the flash? b) 2 seconds after you see the flash?

A3. Some microwaves have a frequency of 5×10^{10} cycles per second. What is the corresponding wavelength?

A4. Some x-rays have a frequency of 10^{18} cycles per second. What is the corresponding wavelength?

A5. Identify the amplitude, period, and phase shift.
a) $2 \sin(3t - 4)$ b) $5 \sin(t/6 - 7)$.

A6. Identify the amplitude, period, and phase shift.
a) $8 \sin(t/9 + 10)$ b) $11 \sin(12t - 13)$.

A7. When a typical car engine is idling, approximately what is its frequency?

A8. What is the frequency of household electricity?

^^^^^^^^^

B1.* Give the relationships between period, frequency, and wavelength.

B2.* Explain how *adding* a wave to a given wave can cancel the given wave. (Subtracting would be easy to understand.)

B3. Find a wave, which when added to the given wave, would cancel it.
a) $3 \sin(t - 2)$ b) $2 \sin(1000t)$

B4. Find a wave, which when added to the given curve, would cancel it.
a) $\sin t + 2\sin(t + 4)$ b) $\sin 2t + 2\sin t$.

B5. a) Convert c cycles per second to radians per second. "Radians per second" serves as other units for measuring angular speed. b) If a wheel spins around a fixed shaft at d radians per second, what is the linear speed of a point on its circumference?

B6. True or false? Explain. If $f(x)$ has maximum value M an $g(x)$ has maximum value m, then $h(x) = f(x) + g(x)$ has maximum value $M + m$.

B7. Graph (in radians) "$\sin x + \sin(x - 2)$." From the graph, approximate its amplitude and phase shift. [..., 1.0]

B8. Graph (in radians) "$\sin x + 2\sin(x + 1)$." From the graph, approximate its amplitude and phase shift. [2.7,...]

B9. In Figure 2, determine y in terms of time if the shaft is rotating at 50 cycles per second, $r = 2$, and $c = 6$. Begin time when P is at the top.

B10. a) Use the identity in exercise 7.2.B40 to prove this result about the combination of two sine waves of the same frequency and amplitude but different phase shifts.
$\sin(Bt + k) + \sin(Bt + j) = 2\cos((k\text{-}j)/2)\sin(Bt+(k+j)/2)$.
b) What is the amplitude of the sum? How must the phase shifts be related to reach the maximum possible amplitude?

B11. When two sound waves are very similar, but not identical, in frequency, they will reinforce each other at times and nearly cancel each other at other times. For example, if one note is played at 440 cycles per second and a another note is played at 438 cycles per second, there will be an audible increase and decrease in amplitude twice a second known as a "beat." This can be illustrated with a very wide graph of the sum of two sine waves. However, your calculator does not have enough columns of pixels to display such graphs. Near $x = 0$ the graph of "$\sin x + \sin(1.01x)$" displays reinforcement (graph it and see). Here is the problem: Find the approximate location of the smallest positive x-interval (of width, say 4π) where this graph displays nearly complete canceling.

B12. (See B11). Suppose 2 sine waves of amplitude 1 begin in phase, one at 440 cycles per second and the other at 438 cycles per second. a) Give the expression for their sum. b) Find a window on your graphics calculator that clearly displays both non-canceling and the near-canceling which occurs periodically. c) At which value of t does the first such instance of that near-canceling occur? d) When is the next instance of that near-canceling?

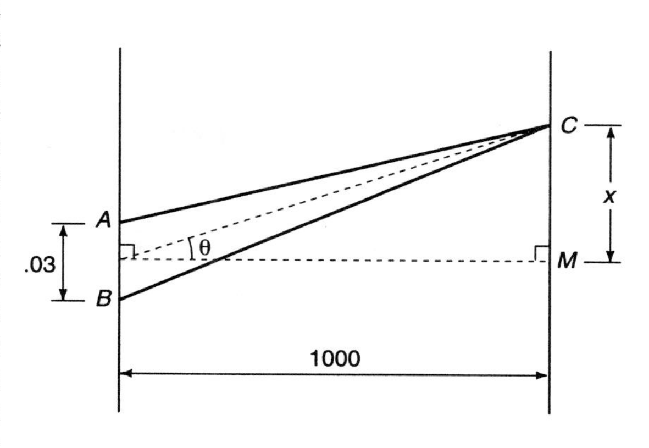

B13. Imagine a wave coming from the left and passing through two slits, A and B, in phase (See the figure, which is not to scale). From each slit the waves spread out to reach the points on the line CM, which is parallel to the line containing the slits. Since the distance from A to C is not the same as the distance from B to C (except when C is at M, right in the middle), the waves which were in phase when they passed through the slits may not be in phase when they reach C. When the waves are light waves, the difference in distance causes a pattern of alternating light and dark bands along the line CM because of the reinforcing or canceling of the two waves. Suppose the slits are .03 mm (millimeter) apart, and the line through C and M is 1000 mm from the slits, and the wavelength of the light passing through the slits in phase is 0.00051mm. Find the smallest distance, x, from M to C such that the waves are completely out of phase when they reach C. Hint: If you treat AC and BC as parallel, and x as small relative to L, the approximation will be good enough and substantial simplifications will occur. Remember, the picture is not to scale.

B14. (See B13.) If the distance between the slits is d, the wavelength is w, and the distance to the line is L, find a simple formula for the distance, x, from M to C such that the waves are completely out of phase when they reach C. If you treat AC and BC as parallel, and x as small relative to L, the approximation will be good enough and substantial simplifications will occur. Remember, the picture is not to scale.

B15. Here is a trigonometric identity for adding waves of the same frequency:
(7.3.6) $c \sin(ax) + d \cos(ax) = A \sin(ax + b)$, where
$A = \sqrt{c^2 + d^2}$ and $\cos b = c/A$ and $\sin b = d/A$.
a) Find an expression equivalent to $\sqrt{3} \sin x + \cos x$ that involves only the sine fucntion.
b) Graph the original expression and the result of part (a) and see if the amplitude and phase shift you found are right.

CHAPTER 8

Other Topics

Section 8.1. Complex Numbers

Complex numbers arise in the context of quadratic equations such as "$x^2 = -4$" and "$x^2 + 2x + 2 = 0$." Since $x^2 \geq 0$ for all real-valued x, the equation "$x^2 = -4$" cannot have any real-valued solutions. The usual way to solve "$x^2 = -4$" would be to take the square root, but negative numbers such as -4 do not have real-valued square roots. Similarly, the usual way to solve a quadratic equation such as "$x^2 + 2x + 2 = 0$" is to use the Quadratic Formula which yields

$$\frac{-2 \pm \sqrt{2^2 - 4(1)(2)}}{2(1)} = \frac{-2 \pm \sqrt{-4}}{2} .$$

Again, the square root of a negative number appears.

Since the square roots of negative numbers are not "real" numbers, one reasonable approach to solving these two equations is to say they have no solutions. Mathematicians accepted this as certainly true and completely obvious for many centuries.

But, it is equally obvious that you cannot take 7 stones away from a pile of 5 stones. Unfortunately, "obvious" facts have a habit of obstructing the development of mathematics. But, as you know, when negative numbers were finally invented, they turned out to have many useful but unsuspected applications (although not to taking 7 stones from a pile of 5 stones, which cannot be done). For example, now we are all familiar with negative numbers in the context of money, checking accounts, and credit-card balances.

Another, bold, approach to solving these quadratic equations is to create solutions by the simple device of inventing a solution to the equation "$x^2 = -1$." Call that invented solution "i" (pronounced as it looks, "eye"), so $i^2 = -1$. Assume that the usual properties of combinations of arithmetic operations apply to this new type of number. Then $(2i)^2 = 2^2 i^2 = 4i^2 = 4(-1) = -4$. So $2i$ will be a solution to the equation "$x^2 = -4$." Similarly, $-2i$ will be a second solution.

Reconsider the "solution" to the equation "$x^2 + 2x + 2 = 0$" provided by the Quadratic Formula. Using our new-found square root of negative four and the usual arithmetic operations on this new type of number,

$$\frac{-2 \pm \sqrt{-4}}{2} = \frac{-2 \pm 2i}{2} = -1 \pm i \,.$$

With this abstract approach all quadratic equations that have no real-valued solutions will have two "complex-valued" solutions.

You may be wondering if this approach yields anything important, or if it is just a way to create a useless solution to a useless equation. Well, you can probably guess the answer. The reason math books have section on complex numbers is, of course, that in some subjects they turn out to be extremely useful. For example, they are an essential fixture of electrical engineering (where our "*i*" is called "*j*", because they need to use the symbol "*i*" for something else). They are critical to the study of laser physics. In fact, in any subject where electro-magnetic waves occur, the theory is likely to use complex numbers. And, in mathematics, complex numbers have remarkable connections to trigonometry and rotations of two-dimensional figures. Like the applications of negative numbers, the applications of complex numbers have appeared in unanticipated places. For example, they are now used in the generation of fractals. Fractals are fantastically complicated images that arise from remarkably simple computer instructions (Figure 1, problem B29). Because of their promise for encoding a large amount of visual information in a few lines of computer code (and because they can be entrancingly beautiful), they are currently receiving much attention.

We cannot develop the subjects of electrical engineering, laser physics, or fractals here. But, every subject has to start somewhere. Complex numbers start with *i*.

Figure 1: A Fractal image.

<u>Definition 8.1.1 (Complex Numbers)</u>: There is a solution to "$x^2 = -1$" called i. The <u>complex</u> <u>numbers</u> consist of all numbers of the form "$a + bi$" where "a" and "b" are real numbers.

The use of the letters "a" and "b" to represent real numbers in this form is traditional. If we wish to denote a complex number by a single letter, we will use "z" or "w", never "a" or "b". So, in traditional notation, $z = a + bi$.

Complex numbers can be added, subtracted, multiplied, and divided according the usual properties of these operations on real numbers.

Example 1: Some complex numbers in "$a + bi$ form" are

$1 + i$ [$a = 1$ and $b = 1$],

$7 - 5.67i$ [$a = 7$ and $b = -5.67$],

$-1/2 - i/4$ [$a = -1/2$ and $b = -1/4$.],

Other complex numbers in "$a + bi$ form" are

i [$a = 0$ and $b = 1$],

3 [$a = 3$ and $b = 0$], and

0 [$a = 0$ and $b = 0$].

The complex numbers include the real numbers. All real numbers are complex numbers, but not all complex numbers are real numbers.

The solution to "$x^2 = -1$," i, is often said to be an imaginary number. With a good imagination, you can make it "real" to you. Just work with it enough and, like other abstractions, it will take on reality as its usefulness becomes apparent. Because $i^2 = -1$, i is often called "the square root of negative one."

Numbers of the form "bi" where "b" is real are said to be pure imaginary numbers. In the complex number "$a + bi$," "a" is the real part and "bi" is the imaginary part.

Two complex numbers "$a + bi$" and "$c + di$" are said to be equal if and only if $a = c$ and $b = d$, that is, if their real parts are the same and their imaginary parts are the same.

Real numbers can be graphed on a one-dimensional number line. Complex numbers cannot. They are two-dimensional and require two dimensions for graphing. When dealing with real numbers, two dimensions are enough to display two real numbers, so we can graph functions -- one dimension is used for the argument and the other for the image. However, complex numbers need two dimensions just to graph one number. It is difficult to graph complex-valued functions of complex numbers (such as $f(z) = z^2$) because four dimensions are involved (two for the argument and two for the image) -- too many for a standard picture. We will just graph the complex numbers themselves (not complex functions).

To graph a complex number let the horizontal axis be the real axis and the vertical axis be the imaginary axis. Use a square scale. Plot $z = a + bi$ in the usual position of the ordered pair (a, b) (Figure 2). When the plane is used to plot complex numbers in this manner it is called the complex plane. Figure 3 plots a few complex numbers.

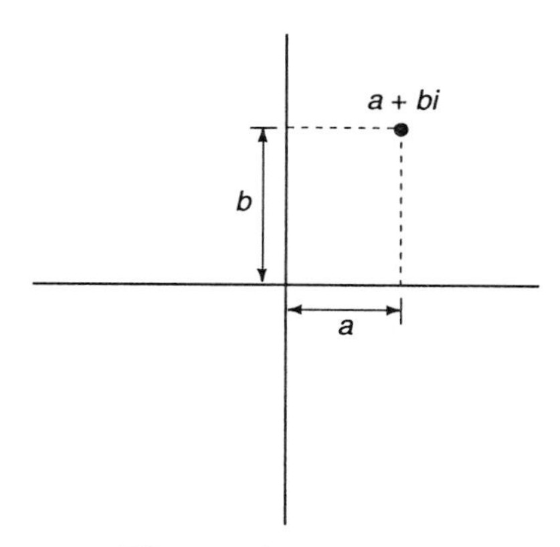

Figure 2: $a + bi$
[-5, 5] by [-5, 5].

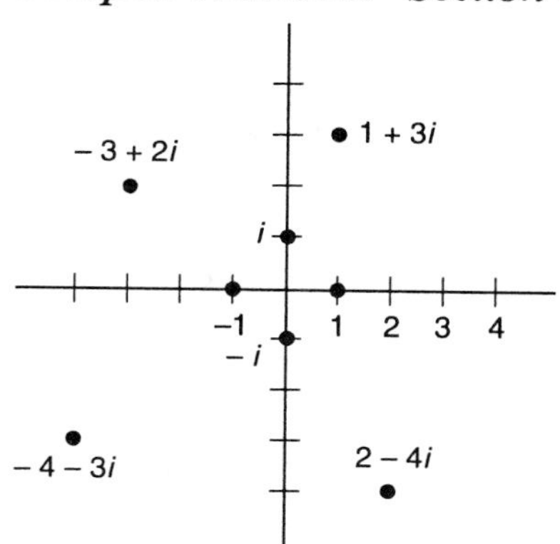

Figure 3: A few complex numbers.
[-5, 5] by [-5, 5]

The idea that complex numbers consist of two essentially distinct parts is critical to working with them. For example, to add or subtract two complex numbers, add or subtract their real and imaginary parts separately. There is no trick to adding or subtracting complex numbers. Simply consolidate like terms.

Example 2: $(2 + i) + (1 + 3i) = 2 + 1 + i + 3i = 3 + 4i$ (Figure 4).
$(2 + i) - (1 + 3i) = 2 - 1 + i - 3i = 1 - 2i$.

The abstract form of this simple idea is next.

Property 8.1.2 (Addition and Subtraction):

$$(a + bi) + (c + di) = (a + c) + (b + d)i.$$
$$(a + bi) - (c + di) = (a - c) + (b - d)i.$$

Graphically, adding has two common interpretations. To add $z + w$, plot z and then imagine the location of z to be a new origin (Figure 4). Then plot w relative to the new origin. That will be the position of $z + w$ relative to the original origin. This is much like adding numbers on the real number line, where, to add 2 + 3 you may find position "2" and treat it as a new origin. Then the position "3" relative to that new origin is position "5" relative to the original origin.

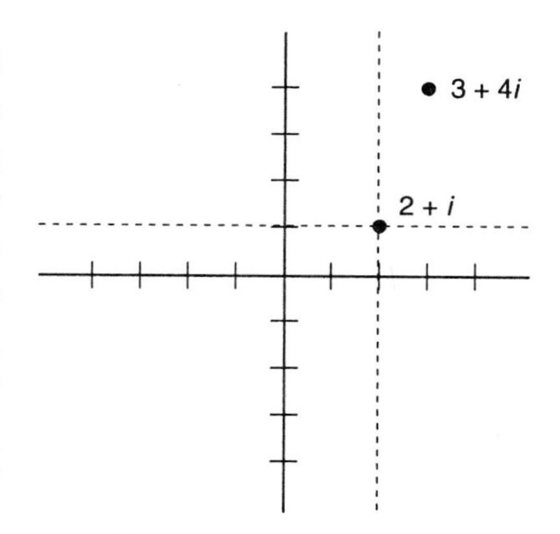

Figure 4: Adding complex numbers.
$(2+i) + (1+3i) = 3+4i$.
[-5, 5] by [-5, 5].

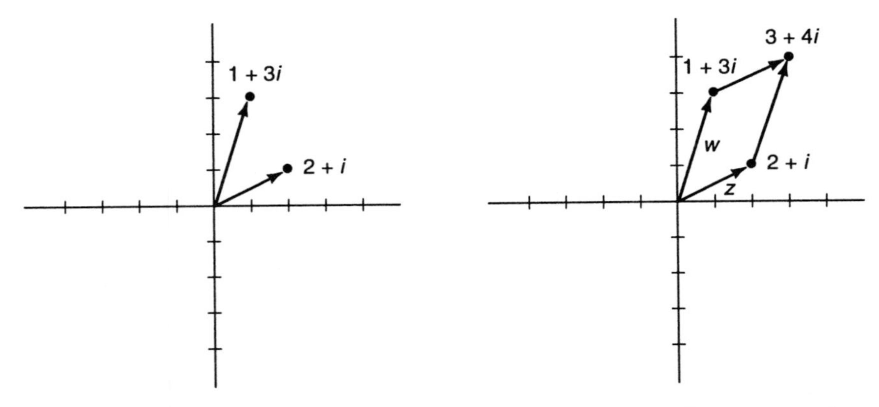

Figure 5: $1 + 3i$ and $2 + i$. [-5, 5] by [-5, 5].

Figure 6: The sum of $1 + 3i$ and $2 + i$. [-5, 5] by [-5, 5]

This addition may be visualized another way. Plot both z and w and draw arrows (vectors) from the origin to their locations (Figure 5). Then slide the vector for w so its tail is at z (Figure 6). Then the head of the vector w will be at $z + w$. This is the so-called "parallelogram" approach because the construction can create a parallelogram (Figure 6). By the Commutative Property, $z + w = w + z$, and the drawing for these two expressions creates opposite sides of a parallelogram.

Graphically, the function $f(z) = z + w$ ("add w") produces a translation (that is, a location shift).
Points are translated the distance and direction of w from the origin.

Multiplication. The key to multiplication of complex numbers is that the Distributive Property and the Extended Distributive Property still hold. Therefore, all you have to do is "multiply out" the expressions, consolidate like terms, and remember that $i^2 = -1$.

Example 3: $(3 + 2i)(7 + 4i) = 21 + 12i + 14i + 8i^2$ [using FOIL, 3.3.2]
[Recall that $i^2 = -1$, so $8i^2 = -8$]
$= 21 - 8 + (12 + 14)i$
$= 13 + 26i.$

Example 4: $(1 - i)(3 + 2i) = 3 + 2i - 3i - 2i^2$
$= 3 + 2 + (2 - 3)i$
$= 5 - i.$

This process can be expressed abstractly.

$$\begin{array}{r} a + bi \\ \underline{c + di} \\ adi + bdi^2 \\ \underline{ac + bci \quad\quad} \\ ac + adi + bci + bdi^2 \end{array} = (ac - bd) + (ad + bc)i.$$

<u>Property 8.1.3</u>: $(a + bi)(c + di) = (ac - bd) + (ad + bc)i.$

Rather than memorize this identity with all its dummy variables, simply use the Extended Distributive Property (FOIL) to multiply complex numbers.

<u>Property 8.1.4</u>: If $p \geq 0$, the solution to $z^2 = -p$ is $z = \pm i\sqrt{p}$.

By the multiplication property 8.1.3,

$$(i\sqrt{p})^2 = (i\sqrt{p})(i\sqrt{p}) = i^2(\sqrt{p})^2 = -p.$$

So $i\sqrt{p}$ is a solution to "$x^2 = -p$." $-i\sqrt{p}$ is another.

Square roots of negative numbers may appear in the Quadratic Formula.

Example 5: Solve $x^2 + 4x + 8 = 0$.

$$\begin{aligned} x^2 + 4x + 8 = 0 \quad &\textit{iff} \quad x = \frac{-4 \pm \sqrt{4^2 - 4(1)(8)}}{2(1)} \\ &\textit{iff} \quad x = \frac{-4 \pm \sqrt{-16}}{2} \\ &\textit{iff} \quad x = \frac{-4 + i\sqrt{16}}{2} \qquad [\textit{by } 8.1.4] \\ &\textit{iff} \quad x = \frac{-4 \pm 4i}{2} \\ &\textit{iff} \quad x = -2 + 2i\ . \end{aligned}$$

The answer is in "$a + bi$" form. The last line follows by straightforward division. As expected, multiplication or division of a complex number by a real number, c, simply multiplies or divides both parts. This is stated abstractly next.

<u>8.1.5 (Corollary to 8.1.3)</u>: $c(a + bi) = ca + cbi$

$$(a + bi)/c = a/c + (b/c)i, \text{ if } c \neq 0.$$

Example 6: $3(1 + i) = 3 + 3i$ (Figure 7).

$(-2 + 4i)/4 = -1/2 + i$, where $a = -1/2$ and $b = 1$ (Figure 7).

Figure 7 illustrates that multiplying a complex number by a real number changes its distance from the origin, but not its direction from the origin.

Multiplying by $c > 1$ expands the point away from the origin by a factor of c. Multiplying by c between 0 and 1 contracts the point toward the origin.

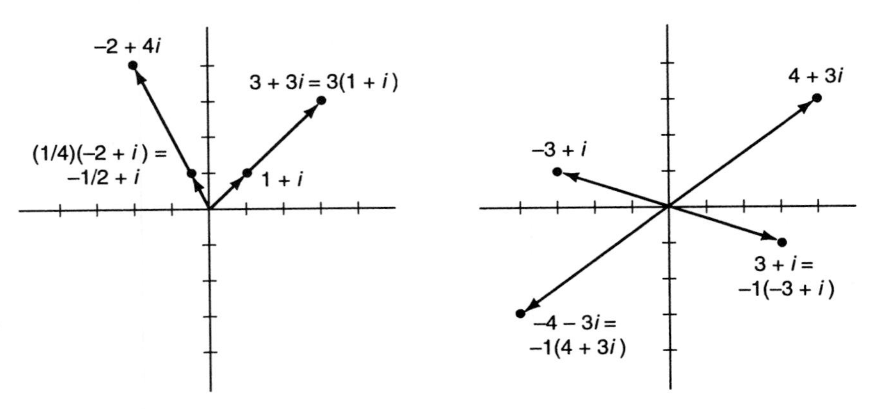

Figure 7: Multiplying by a real number. [-5, 5] by [-5, 5].

Figure 8: Multiplying by -1. [-5, 5] by [-5, 5].

Multiplying by -1 reverses the direction of the point from the origin.

Example 7: $-1(4 + 3i) = -4 - 3i$ (Figure 8).
$-1(3 - i) = -3 + i$ (Figure 8)

Subtraction can be regarded as addition of the negative.

$$z - w = z + (-1)w = z + (-w).$$

That is, we can distribute the minus sign like usual. Graphically, we may interpret $z - w$, using the parallelogram idea, as $z + (-w)$, where $-w$ has the length of w but the opposite direction.

Example 8: $(1 + 2i) - (3 + i) = -2 + i$ (Figure 9).
$(1 + 2i) - (-2 - 2i) = 3 + 4i$ (Figure 10).

For the first example, Figure 9 plots $1+2i$, $3+i$, and $-(3+i)$. Then the parallelogram process is used to subtract $3+i$ by *adding* $-(3+i)$.

For the second example, to subtract $-2-2i$, plot it and then $-(-2-2i)$. Then *add* $-(-2-2i)$.

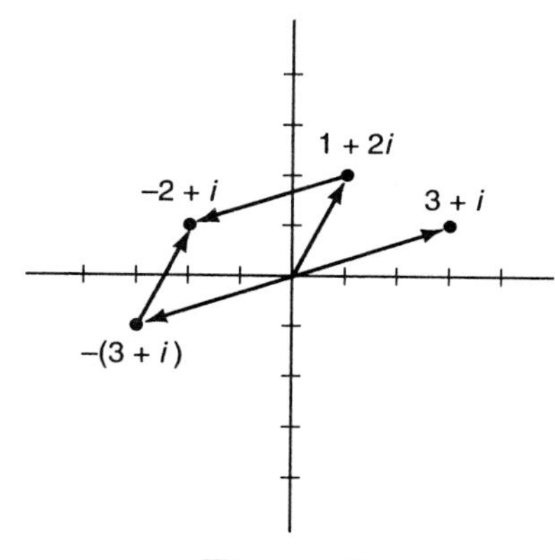

Figure 9:
$(1+2i) - (3+i) = -2+i$
[-5, 5] by [-5, 5].

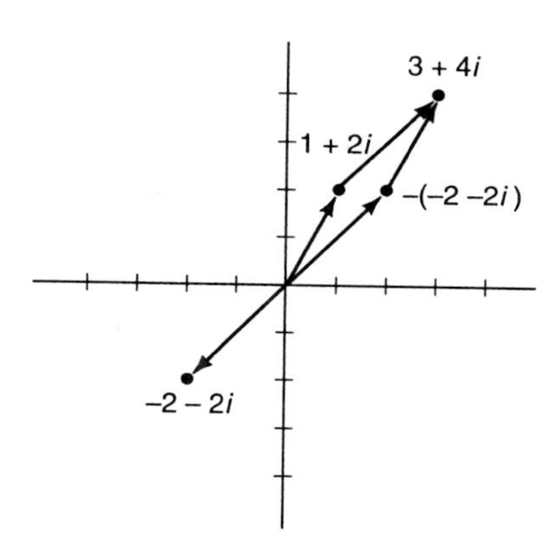

Figure 10:
$(1+2i) - (-2-2i) = 3+4i$.
[-5, 5] by [-5, 5].

So far we have seen that addition, subtraction, and multiplication by a real number are straightforward to interpret in the complex plane. The unexpected properties lie ahead.

Multiplication and Rotation. In the complex plane multiplication has a truly remarkable interpretation which is not at all evident from the algebraic form of multiplication expressed in Property 8.1.3. We will show that multiplication by a complex number corresponds to a *rotation* and a *contraction* or *expansion* toward or away from the origin. Before we formalize this thought, consider a simple example.

Example 9: Consider multiplication by i. Multiply several complex numbers by i and inspect their graphs to see what the graphical effect of multiplying by i is.

Any complex numbers will do. For instance, consider 2 [P], $4i$ [Q], and $-3 - i$ [R] (Figure 11).

$2(i) = 2i$ [P']
$(4i)i = 4i^2 = -4$ [Q']
$(-3 - i)i = -3i - i^2 = 1 - i$ [R']

Every time the product is the same distance away from the origin as the original number, but rotated $\pi/2$ (90°) counterclockwise. **Multiplying any number by i rotates the number by $\pi/2$ (90°).**

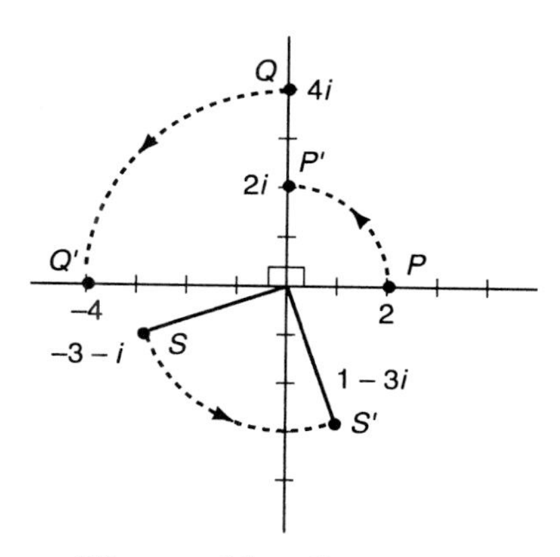

Figure 11: Complex numbers multiplied by i.

This is related to why $i^2 = -1$. Think of i^2 as i times i. Rotating i $\pi/2$ yields -1. $i^2 = -1$.

<u>**Trigonometric Form**</u>. To see why multiplication and rotation are related, we use an alternative method to describe the position of a complex number in the complex plane. The idea is to describe the point by its distance from the origin and its angle with the positive real axis (Figure 12). Because angles and distances are used, complex numbers described this way are said to be in "trigonometric form" which is also known as "polar form" (the pole is the origin).

The distance from $z = a + bi$ to the origin is called its <u>modulus</u> (or, its <u>absolute value</u>), which is often denoted "$|z|$" or "r" (for "radius"). By the Pythagorean Theorem,

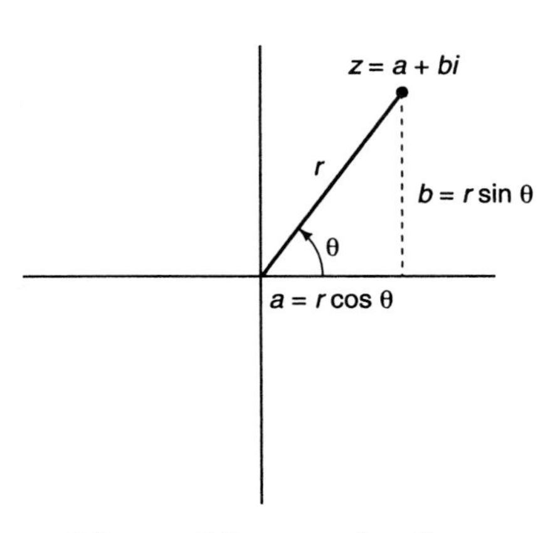

Figure 12: r and θ for "trigonometric form" of complex numbers. [Memorize this picture. All of 8.1.7 follows from it.]

(8.1.6) $$|z| = \sqrt{a^2 + b^2} = r.$$

$z = 0$ if and only if $r = 0$.

Example 10: $|3 + 4i| = \sqrt{3^2 + 4^2} = \sqrt{25} = 5$.

$|-1 - 3i| = \sqrt{(-1)^2 + (-3)^2} = \sqrt{10}$.
$|2.34i| = 2.34.$
$|-5.67| = 5.67.$

The angle the line from the origin through z makes with the positive real axis is called the <u>argument</u> of z ("arg z"). Being an angle, the argument of z is often denoted by "θ" (Figure 13). All the results in the next list follow immediately from the picture. You do not have to memorize all these facts -- just remember the picture. "A picture is worth several complex-number facts."

(8.1.7) Let $z = a + bi$. Then
arg $z = \theta$ [which is simply the usual notation].
$\tan \theta = b/a$, if $a \neq 0$. θ is not defined if $z = 0$.
$\cos \theta = a/r$ and $\sin \theta = b/r$, if $z \neq 0$ (that is, if $r \neq 0$).
$a = r \cos \theta$ and $b = r \sin \theta$, if $z \neq 0$.
If $a > 0$, θ is in the first or fourth quadrant.
If $a < 0$, θ is in the second or third quadrant.
If $a = 0$ and $b > 0$, $\theta = \pi/2$.

If $a = 0$ and $b < 0$, $\theta = -\pi/2$.

The second last line mentions "$\pi/2$" rather than "90°" because trigonometric form is usually expressed using radian measure (which has advantages over degree measure when applied to advanced material). Take this opportunity to become a bit more comfortable with radian measure. Actually, θ is not uniquely determined, since adding 2π (360°) to any argument yields another coterminal angle. But, whenever convenient, we prefer to use θ between $-\pi$ and π, or, sometimes, between 0 and 2π.

Example 11: $\arg(1 + i) = \pi/4$.

To see this, think of a picture of the location of $1 + i$ in the complex plane (Figure 13). From the picture (or 8.1.7), $\tan\theta = b/a = 1/1 = 1$. So, $\theta = \tan^{-1}(1) = \pi/4$.

$\arg(3) = 0$, because positive real numbers make angle 0 with the positive x-axis. Or, from 8.1.7, $\tan\theta = b/a = 0/3 = 0$, so $\theta = \tan^{-1}0 = 0$.

$\arg(-2) = \pi$ (Figure 13), because negative numbers make angle π with the positive x-axis.

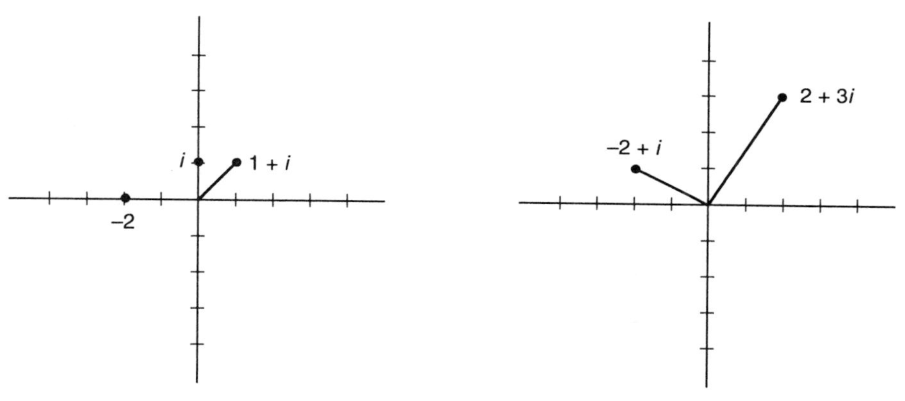

Figure 13: $1 + i$, -2, i. [-5, 5] by [-5, 5].

Figure 14: $2 + 3i$, $-2 + i$. [-5, 5] by [-5, 5].

In the last example ($\arg(-2)$) the equation for θ from 8.1.7 ($\tan\theta = 0$) has two important solutions: 0 and π. We want $\theta = \pi$ rather than $\theta = 0$, because $a < 0$. By inspection, $z = -2$ is at angle π from the positive real axis. Real numbers have argument 0 or π.

$\arg(i) = \pi/2$ [second last line of 8.1.7, Figure 13].

$\arg(2 + 3i) = \tan^{-1}(3/2)$ (= .983 radians = 56.31°, Figure 14).

We expressed this answer as "$\tan^{-1}(3/2)$" because it is not a famous angle and the given decimal form for it is not illuminating.

$\arg(-2 + i) = \tan^{-1}(-1/2) + \pi = -.464 + \pi = 2.68$ (153.43°, Figure 14). Fortunately, we do not often do examples that are this tricky. For $z = -2 + i$, $\tan\theta = 1/(-2) = -1/2$. But θ is not simply $\tan^{-1}(-1/2)$ because θ is in the second quadrant and $\tan^{-1}(-1/2)$ is in the fourth quadrant (expressed as a negative angle). So adding π gives the second quadrant angle with the same tangent value (recall identity 7.2.10 that says tangent repeats every π radians).

Complex Numbers in Trigonometric Form. Using "$\cos\theta = a/r$" and "$\sin\theta = b/r$" from Figure 12 (or 8.1.7), we can rewrite "$a + bi$."

(8.1.8) $$a + bi = r\cos\theta + (r\sin\theta)i = r(\cos\theta + i\sin\theta).$$

"$r(\cos\theta + i\sin\theta)$" is the trigonometric form (or "polar form") of a complex number $z = a + bi$, where $r = |z|$ as in 8.1.6 and θ is an argument of z as in 8.1.7.

We write "$i\sin\theta$" rather than "$(\sin\theta)i$" to avoid the extra parentheses.

The factor "$\cos\theta + i\sin\theta$" appears a lot in this subject. Since only the "θ" varies in the "$\cos\theta + i\sin\theta$" factor, it is often abbreviated to emphasize the θ. Some authors write "$\cos\theta + i\sin\theta$" as "cis θ," where the three letters c, i, and s abbreviate the phrase "cosine plus i sine." However, this abbreviation is not common in higher mathematics.

Exponential Form. Actually, there is no real need to abbreviate the very common expression "$\cos\theta + i\sin\theta$" at all, because this trigonometric form is both correct and illuminating. Nevertheless, it is often abbreviated because it is a bit long. It turns out, in calculus, that the exponential function with base e that we studied in Chapter 5 yields a shorter and more convenient equivalent expression. We cannot prove it here, but the key identity is

(8.1.9) $$e^{i\theta} = \cos\theta + i\sin\theta.$$

This is a complex number on the unit circle in the complex plane, at angle θ from the positive real axis (Figure 15). Since $\cos\pi = -1$ and $\sin\pi = 0$, the famous equation which relates e, i and π follows:

(8.1.10) $$e^{i\pi} = -1.$$

The short exponential form follows immediately from 8.1.8 and 8.1.9:

(8.1.11) $$r e^{i\theta} = r(\cos\theta + i\sin\theta).$$

"$e^{i\theta}$" conveniently abbreviates "$\cos\theta + i\sin\theta$." "Exponential form" and "trigonometric form" are two equivalent notations. Everything expressed in the remainder of this section using "exponential form" is nothing more or less than "trigonometric form" rewritten using 8.1.11.

This point is r times as far from the origin and in the same direction as $e^{i\theta}$ on the unit circle (Figure 15). Formula 8.1.11 gives two notations for the same information.

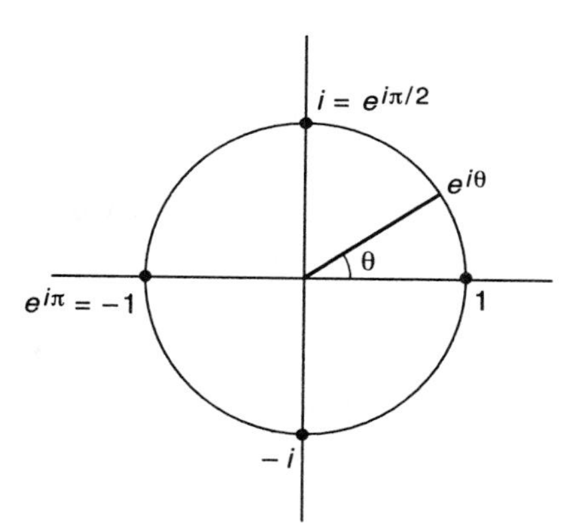

Figure 15: $\cos\theta + i\sin\theta = e^{i\theta}$, on the unit circle in the complex plane.

Example 12: $3 = 3(\cos 0 + i \sin 0) = 3\,e^{0i}$. The real number 3 is at angle 0 from the positive real axis (Figure 16).

$-4 = 4(\cos\pi + i\sin\pi) = 4\,e^{i\pi}$.
The real number -4 is at angle π from the positive real axis (Figure 16).

$2i = 2[\cos(\pi/2) + i\sin(\pi/2)] = 2\,e^{i\pi/2}$.
The imaginary number $2i$ is at angle $\pi/2$ from the positive real axis (Figure 16).

$1 + i = \sqrt{2}[\cos(\pi/4) + i\sin(\pi/4)] = \sqrt{2}\,e^{i\pi/4}$, because $r = \sqrt{(1^2 + 1^2)} = \sqrt{2}$ and $\tan\theta = 1/1$. Therefore $\theta = \pi/4$, since θ is in the first quadrant (Figure 16).

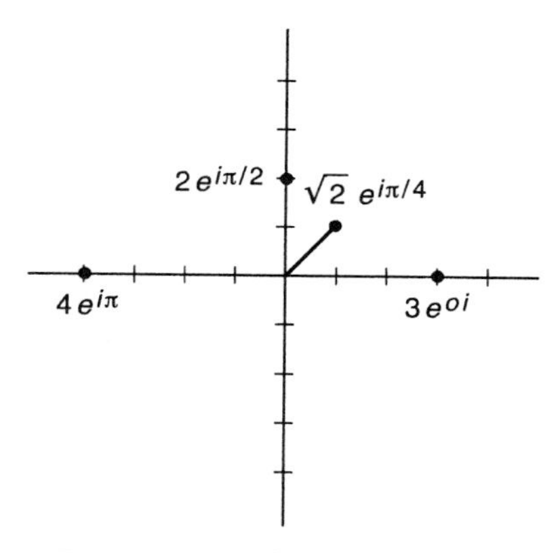

Figure 16. Some points expressed in exponential form. [-5, 5] by [-5, 5].

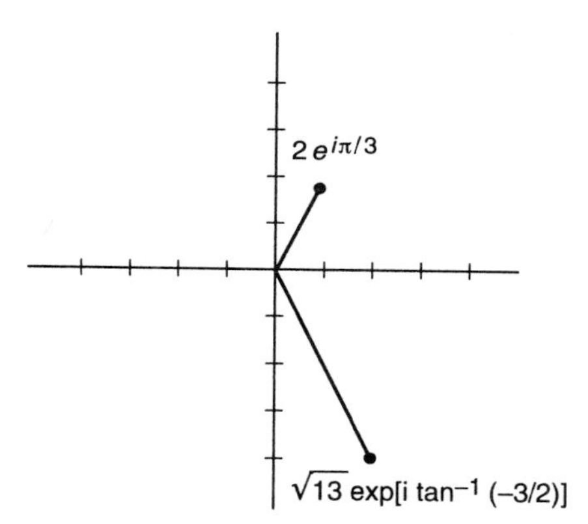

Figure 17. Some points expressed in exponential form. [-5, 5] by [-5, 5].

$1 + i\sqrt{3} = 2[\cos(\pi/3) + i\sin(\pi/3)] = 2\,e^{i\pi/3}$, because $r = \sqrt{(1^2 + \sqrt{3}^2)} = 2$ and $\tan\theta = \sqrt{3}/1$. Therefore $\theta = \pi/3$, since θ is in the first quadrant (Figure 17).

$2 - 3i = \sqrt{13}\exp[i\tan^{-1}(-3/2)] = \sqrt{13}\,e^{-.98i}$, because $r = \sqrt{(2^2 + 3^2)} = \sqrt{13}$ and $\tan\theta = -3/2$. Therefore, $\theta = \tan^{-1}(-3/2)$, (which is about -.98) since θ is in the fourth quadrant (Figure 17). The notation "$\exp(i\theta)$" instead of "$e^{i\theta}$" in this example allows us to see the tiny exponent more clearly.

Multiplication. Multiplying complex numbers in trigonometric or exponential form yields a remarkable and elegant result:

To multiply complex numbers in trigonometric or exponential form, multiply their moduli and *add* their arguments.

Remember that something similar holds for real numbers: $e^x e^y = e^{x+y}$, which says to multiply exponentials we may add their arguments (5.2.7 or 5.1.1). With complex numbers, addition of arguments is addition of angles. With complex numbers, the angle of a product is the *sum* of the angles of the numbers. The next theorem states this in both trigonometric and exponential notation. It looks long, but it is easily interpreted.

Theorem 8.1.12:
$r_1(\cos \theta_1 + i \sin \theta_1)r_2(\cos \theta_2 + i \sin \theta_2) = r_1 r_2[\cos(\theta_1 + \theta_2) + i \sin(\theta_1 + \theta_2)]$.

$$[r_1 \exp(i\theta_1)][r_2 \exp(i\theta_2)] = r_1 r_2 \exp[i(\theta_1 + \theta_2)].$$

Formula 8.1.12 has two lines which say the same thing in two different notations. The second is shorter and shows that exponential notation is convenient. The theorem explains why multiplying by a complex number produces a rotation. Adding angles is equivalent to rotating about the origin.

Example 9, revisited: Multiplying by i produces a rotation of $\pi/2$ (90°) as we saw in Example 9 (Figure 11). The modulus of i is 1 and its argument is $\pi/2$ (90°). So, $i = 1 \exp(i\pi/2)$, according to 8.1.11. Multiplying any complex number by i leaves the modulus unchanged (since $r = 1$) and adds $\pi/2$ to the argument, so the product is the original number rotated by $\pi/2$ (problem B19).

Proof of 8.1.12: This proof relies on regular multiplication of complex numbers in "$a + bi$" form (8.1.3), and then uses the sum-of-angles trigonometric identities for sine and cosine (7.2.11 and 12). Note that "trigonometric form" is virtually "$a + bi$" form, but with the "a" and "b" written using trigonometric functions. The first line of the proof expresses the equivalence of the two forms given in 8.1.11.

$$[r_1 \exp(i\theta_1)][r_2 \exp(i\theta_2)] = r_1(\cos \theta_1 + i \sin \theta_1)\, r_2(\cos \theta_2 + i \sin \theta_2),$$

Now, factor out the r_1 and the r_2, and multiply out the rest using FOIL (8.1.3).

$$= r_1 r_2 \{(\cos \theta_1)(\cos \theta_2) + i(\cos \theta_1)(\sin \theta_2) + i(\sin \theta_1)(\cos \theta_2) + i^2(\sin \theta_1)(\sin \theta_2)\}.$$

Now, two terms of the sum are real and two are imaginary. Recall $i^2 = -1$.

Grouping like terms, this

$$= r_1 r_2 \{(\cos \theta_1)(\cos \theta_2) - (\sin \theta_1)(\sin \theta_2) + i[(\cos \theta_1)(\sin \theta_2) + (\sin \theta_1)(\cos \theta_2)]\}.$$

Look up the two "sum identities" for cosine and sine (7.2.12 and 7.2.11). Note how they fit perfectly. Therefore, this is

$$= r_1 r_2 \{\cos(\theta_1 + \theta_2) + i \sin(\theta_1 + \theta_2)\}$$

$$= r_1 r_2 \exp[i(\theta_1 + \theta_2)].$$

The proof is complete.

Aside. It is truly remarkable that "imaginary" numbers should have such a close connection to trigonometry, which began as the study of triangles. Who would have thought that trigonometry and the square root of negative one would be related by such as basic concept as multiplication? Mathematicians have discovered over the ages that abstract concepts that are suggested by mathematics (such as negative numbers and the square root of negative one) often turn out later to be useful and practical in unexpected real-world contexts. Mathematicians use this argument to support "pure" research -- because they know that, no matter how "pure" research may seem at first, there is always a chance that important practical applications will follow over time.

Example 13: Multiply several numbers by the complex number $z = \sqrt{2} + i\sqrt{2}$ and note the graphical effect.

For instance, multiply 3, $1 + 2i$, and $-2 - i$, by that number (P, Q, and R in Figure 18).

Visually, this is easy. The absolute value of the factor z is 2, so the products will be twice as far from the origin. The angle (argument) of z is $\pi/4$ (45°), so the new points will be rotated $\pi/4$. Done. These effects are clearly visible in Figure 18.

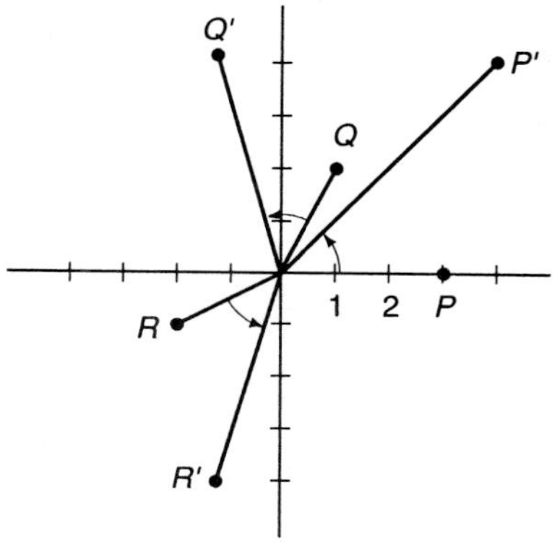

Figure 18: Multiplication by $\sqrt{2} + i\sqrt{2}$. [-5, 5] by [-5, 5].

To compute the numerical values of the products in rectangular form takes some work. Here are the three results:

$Pz = 3(\sqrt{2} + i\sqrt{2}) = 3\sqrt{2} + (3\sqrt{2})i$ [= P', Figure 18].

$Qz = (1 + 2i)(\sqrt{2} + i\sqrt{2}) = -\sqrt{2} + (3\sqrt{2})i$ [= Q', Figure 18].

$Rz = (-2 - i)(\sqrt{2} + i\sqrt{2}) = -\sqrt{2} - (3\sqrt{2})i$ [= R', Figure 18].

Let z be any complex number. Graphically,
multiplying a number by z produces a
rotation around the origin
and an expansion away from the origin or a contraction toward the origin.
The rotation is through the angle arg(z).
The expansion or contraction is by a factor of $|z|$.

<u>Complex Conjugates</u>. When solving quadratic equations, complex numbers arise in "complex conjugate" pairs.

<u>Definition (8.1.13)</u>: The <u>complex conjugate</u> of $a + bi$ is $a - bi$. The complex conjugate of z is often denoted by $\bar{z}$ ("z bar") (Figure 19).

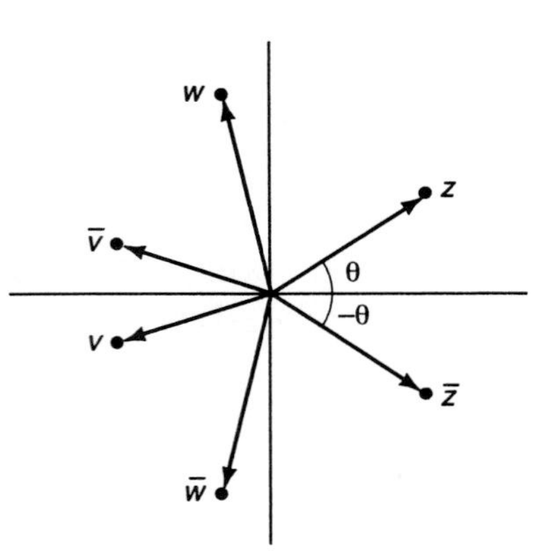

Figure 19: Complex conjugates.

In the Quadratic Formula, the square root term has a "plus or minus" on it:

$$\frac{-b \pm \sqrt{b^2 - 4ac}}{2a} = \frac{-b}{2a} \pm \frac{\sqrt{b^2 - 4ac}}{2a}.$$

If the interior of the square root is negative, the plus sign yields one complex number and the minus sign yields its complex conjugate.

Graphically, the complex conjugate of z is symmetric with z about the real (horizontal) axis. Their real parts are the same and their complex (vertical) parts are opposites (Figure 19).

Example 14: The complex conjugate of "$1 + 2i$" is "$1 - 2i$". Their product is $(1 + 2i)(1 - 2i) = 1 + 4 = 5$, a real number. The cross product term drops out. Also, by 8.1.6, $|1 + 2i| = \sqrt{5}$. So, $z\bar{z} = |z|^2$.

The product of a complex number and its conjugate is always a real number, and it is always the square of the modulus (absolute value).

(8.1.14) $\quad z\bar{z} = (a + bi)(a - bi) = a^2 + b^2 = |z|^2$, a real number.

Multiplying it out shows the product is a real number. There is also an illuminating trigonometric explanation. A complex number and its complex conjugate have the same modulus, but opposite angles (Figure 19). So, from 8.1.12, multiplying them together yields the product of the moduli (that is the $|z|^2$ part) and the sum of the angles, which is 0 ($\theta + -\theta = 0$). Angle 0 implies the product is a real number.

Division. So far we have discussed addition, subtraction, and multiplication. Dividing complex numbers by *real* numbers is like multiplication by real numbers; it is done term-by-term, as expected (8.1.5). However, division by a non-real complex number is not quite so easy. In trigonometric form, division is, as expected, the inverse of multiplication, but there is a trick to obtaining the quotient in "$a + bi$" form.

In trigonometric or exponential form, to divide complex numbers, divide the moduli and *subtract* the argument of the denominator from the argument of the numerator (problem B25). That is, assuming $r_2 \neq 0$,

$$\text{(8.1.15)} \qquad \frac{r_1 \exp(i\theta_1)}{r_2 \exp(i\theta_2)} = \left(\frac{r_1}{r_2}\right) \exp[i(\theta_1 - \theta_2)] .$$

Example 15: $$\frac{6 \exp(i\frac{\pi}{3})}{2 \exp(i\frac{\pi}{4})} = 3 \exp\left[i\left(\frac{\pi}{3} - \frac{\pi}{4}\right)\right] = 3 \exp\left(i\frac{\pi}{12}\right)$$

(Figure 20).

Division of complex numbers in $a + bi$ form is tricky when the object is to express the result in $a + bi$ form. The key is to multiply the top and bottom by the complex conjugate.

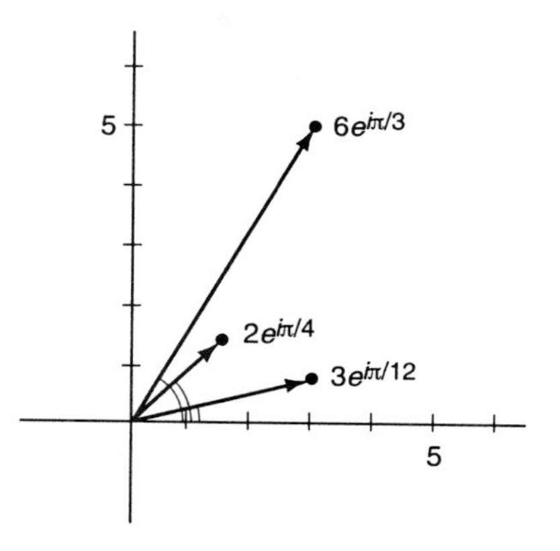

Figure 20: $6 \exp(i\pi/3)$, $2 \exp(i\pi/4)$, and their quotient.

Example 16: Write $\dfrac{2 + i}{3 - 2i}$ in $a + bi$ form.

$$\begin{aligned} \frac{2 + i}{3 - 2i} &= \left(\frac{2 + i}{3 - 2i}\right)(1) \\ &= \left(\frac{2 + i}{3 - 2i}\right)\left(\frac{3 + 2i}{3 + 2i}\right) \\ &= \frac{(2 + i)(3 + 2i)}{9 + 4} . \end{aligned}$$

Now the division problem has been converted to a straightforward multiplication problem. Use 8.1.3 and 8.1.5 (problem A25).

The point of using the complex conjugate is to make the denominator a real number, so that the remaining division will be by a real number. To finish off, multiply out the top using 8.1.3. In general, the process can be described by the next theorem.

Theorem 8.1.16:

$$\frac{a + bi}{c + di} = \left(\frac{a + bi}{c + di}\right)\left(\frac{c - di}{c - di}\right) = \frac{(a + bi)(c - di)}{c^2 + d^2} = \frac{ac + bd + (bc - ad)i}{c^2 + d^2}.$$

This is close enough to "$a + bi$" form. The pattern is too complicated to remember in this abstract form. Instead, simply remember to multiply both the top and bottom by the complex conjugate of the bottom.

Example 17: Evaluate, in $a + bi$ form, $\frac{1 + i}{1 - i}$.

$$\frac{1 + i}{1 - i} = \left(\frac{1 + i}{1 - i}\right)\left(\frac{1 + i}{1 + i}\right) = \frac{0 + 2i}{1^2 + 1^2} = i.$$

Problem B20 asks for a trigonometric, visual, explanation of why this quotient reduces to i.

Conclusion. Complex numbers are expressed in three common forms, "$a + bi$" form, trigonometric (polar) form, and exponential form. Trigonometric and exponential forms are especially valuable for multiplication and division, but are not convenient for addition or subtraction. Multiplication of complex numbers is closely related to rotation in two dimensions.

Terms: complex number, $a + bi$ form, real part, imaginary part, complex plane, modulus, argument, trigonometric form, exponential form, complex conjugate.

****

Exercises for Section 8.1, "Complex Numbers":

A1.* Given two complex numbers in "$a + bi$" form, how can you obtain their sum?

^ ^ ^ ^Sketch the location in the complex plane of

A2. a) $P = 2 + i$ b) $Q = -1 + 2i$ c) $R = 3i$

A3. a) $P = 2 - i$ b) $Q = -2i$ c) $R = -1 - i$

A4. a) $P = \cos \pi + i \sin \pi$. b) $Q = \cos(\pi/3) + i \sin(\pi/3)$

A5. a) $P = \cos(-\pi/2) + i \sin(-\pi/2)$ b) $Q = \cos(3\pi/4) + i \sin(3\pi/4)$

A6. a) $P = e^{i\pi/6}$ b) $Q = e^{i\pi}$.
A7. a) $P = e^{i\pi/2}$ b) $Q = e^{i\pi/4}$.
A8. a) $P = 3e^{i\pi/3}$ b) $Q = 2e^{i5\pi/4}$.
A9. a) $P = 3[\cos(\pi/6) + i\sin(\pi/6)]$. b) $Q = 2[\cos(3\pi/4) + i\sin(3\pi/4)]$
A10. a) $P = 4[\cos(5\pi/6) + i\sin(5\pi/6)]$. b) $Q = 3[\cos(-\pi/4) + i\sin(-\pi/4)]$

^ ^ ^ ^ Simplify:
A11. a) $4 + i + 5 + 7i$ b) $4 + i - (5 + 7i)$
A12. a) $(4 + i)(3 + 2i)$. b) $(4 + i)/(3 - 2i)$.
A13. a) $(2 + 3i)(5 - i)$. b) $(2 + 3i)/(5 + i)$.
A14. a) $(1 + 3i)i$ b) $(1 + 2i)/i$.

^ ^ ^ ^ Write the given complex number in exponential form.
A15. a) i b) -1 c) $2 + 2i$ d) $1 + 4i$ e) $-1 + i\sqrt{3}$.
A16. a) $-i$ b) 1 c) $1 + i\sqrt{3}$ d) $-1 + i$. e) $2 + 3i$.

A17. Write the complex numbers is A15 in trigonometric form.
A18. Write the complex numbers is A16 in trigonometric form.

^ ^ ^ ^ Convert these to "$a + bi$" form.
A19. a) $3\exp(i\pi/6)$ b) $4\exp(i3\pi/4)$. A20. a) $5\exp(\pi i/4)$. b) $2\exp(i2\pi/3)$.

^ ^ ^ ^ Simplify:
A21. $[2\exp(i\pi/4)][6\exp(i3\pi/4)]$ A22. $[5\exp(i\pi/12)][2\exp(i\pi/6)]$

A23. Divide: $\dfrac{20\exp(3\pi i/4)}{4\exp(\pi i/3)}$ A24. Divide: $\dfrac{3\exp(\pi i/2)}{60\exp(\pi i/6)}$

A25. Simplify the expression from Example 16: $(2 + i)(3 + 2i)/(9 + 4)$.

A26. Draw and label a picture of the complex plane to illustrate "$e^{i\pi} = -1$."

A27. What is the argument (angle) of negative real numbers in the complex plane? Explain how multiplying a real number by a negative real number fits the idea of rotation in 8.1.12.

^ ^ ^ ^ ^ ^ ^ ^

B1.* Sketch a picture that illustrates how to convert back and forth between "$a + bi$" form and trigonometric or exponential form.

B2.* a) State the two most important facts for converting a number in "$a + bi$" form to a number in trigonometric or exponential form.
b) State the two most important facts for converting a number in trigonometric or exponential form to a number in "$a + bi$" form.

B3.* Given two complex numbers in "$a + bi$" form, how can you obtain their product?

B4. Exponential and trigonometric forms are very similar. Are there any essential differences?

^ ^ ^ ^ The pictures locate P, Q, and R in the complex plane. Find the location of the resulting point when each is multiplied or divided by the indicated complex number.

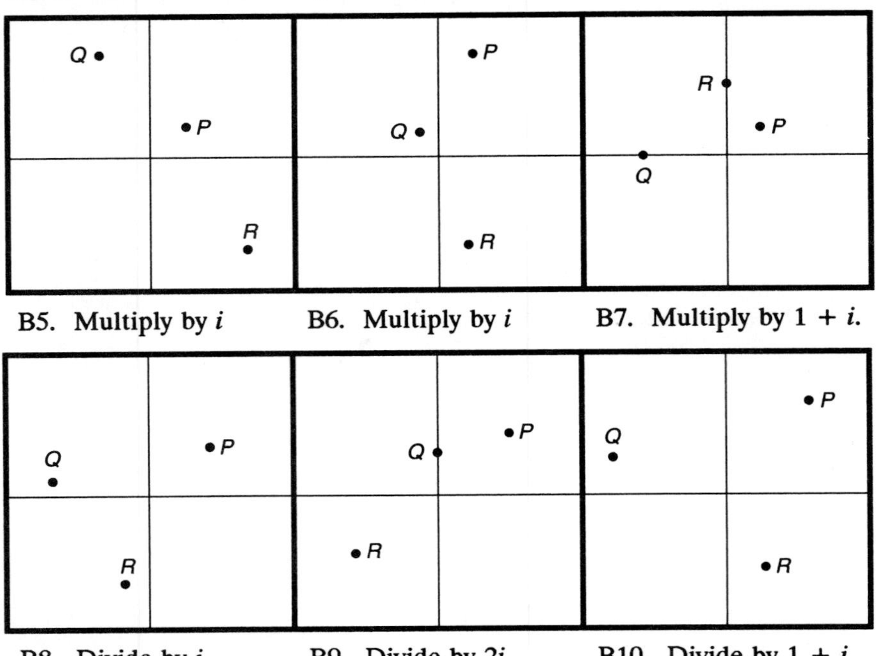

B5. Multiply by i B6. Multiply by i B7. Multiply by $1 + i$.

B8. Divide by i B9. Divide by $2i$ B10. Divide by $1 + i$

B11.* a) Thinking only of real numbers, what is the graphical effect on the number line of the function, "multiply by c"?
b) In the complex plane, what is the graphical effect of the function "multiply by w"?

B12. Using the points in problem B5, illustrate the result of "add i" to each. Assume the window is [-5, 5] by [-5, 5].

B13. Using the points in problem B7, illustrate the result of "add $1 - i$" to each. Assume the window is [-5, 5] by [-5, 5].

B14.* What is the graphical effect of the function "add w"?

B15.* Given two complex numbers in "$a + bi$" form, how can you obtain their quotient?

B16.* Given two complex numbers in exponential form, how can you obtain their product?

B17.* Give the "trigonometric form" explanation of why $i^2 = -1$.

B18. The modulus function has properties on the complex numbers that the absolute value function has on the real numbers. So it can reasonably be called the "absolute value" function on the complex numbers. Prove a) The modulus of a real number is its absolute value. b) $|z| \geq 0$.

B19. Use abstract notation and 8.1.12 to show that multiplying z by i produces a rotation of $\pi/2$, as illustrated in Example 9 (Figure 11) and as argued below 8.1.12.

B20. Explain, trigonometrically and visually, why the quotient in Example 17 is simply i.

B21. The Quadratic Formula gives two solutions to a quadratic equation. Express their product.

[The remaining problems are challenging.]
^ ^ ^ ^ ^ ^ ^ ^ The pictures below illustrate P and Q on a square scale, where $Q = wP$. Estimate the solution for the unknown complex number w. (You do not need to know the scale!)

B22. B23. B24.

B25. Prove 8.1.15 on division of complex numbers in trigonometric form.

B26. Find $\sqrt{i}$ in "$a + bi$" form.

B27. [Part (b) is a famous theorem about complex numbers known as DeMoivre's Theorem.] a) Use trigonometric form to express z^2 in terms of the modulus and argument of z. b) Use exponential or trigonometric form to express z^n in terms of the modulus and argument of z. c) Recall that $1 = e^{2k\pi i}$, for any integer k. Use 3 different values of k to obtain 3 distinct solutions to "$z^3 = 1$." d) Plot them in the complex plane.

B28. a) Use the ideas in B27 to find n "n^{th} roots of unity" by solving "$z^n = 1$."
b) If all n n^{th} roots are plotted in the complex plane a pattern results. What pattern?

B29. Fractals. Fractal images are increasingly common because fast computers are able to do the immense amount of computation required to create the pictures. Here is the idea behind the particular fractal in Figure 1. It uses repeated composition with the same simple function, $f(z) = z^2 - 1$. Consider all possible complex numbers, z, one by one. Pick one, apply f to it, then apply f again to the image, and then again and again. So the initial point is repeatedly moved in the complex plane. Now consider the question, "Does it move further and further from the origin, (move off to "infinity") or does its orbit (sequence of images) stay in a bounded region near the origin?" Theoretically, we could do this for every point in the plane and keep track of the points that move off to infinity. There will be a region of such points, and the "Julia set" is the boundary of that region, which is highly irregular for most functions, f, even when f appears to be simple. For example, Figure 1 graphs the Julia set of the very simple f given by $f(z) = z^2 - 1$. Here is the problem: Find all the points in the complex plane ("fixed points") that have the same image as argument when this function is applied.

Section 8.2. Polar Coordinates

Polar coordinates are a way to locate points in the plane. Rectangular coordinates use two numbers to locate positions in the plane, one for the horizontal position and one for the vertical position. Polar coordinates also use two numbers, one for the distance from the origin and the other for the angle with the positive x-axis.

Fix a point for the origin, O (Figure 1). The origin is the pole of "polar coordinates." Fix a ray (half line, OA, like a positive x-axis) from the pole and label it with a scale. To locate the point P in this "polar coordinate system," give the ordered pair (r, θ), where r is a directed (positive or negative) distance and θ is an angle measured counterclockwise as illustrated in Figure 1.

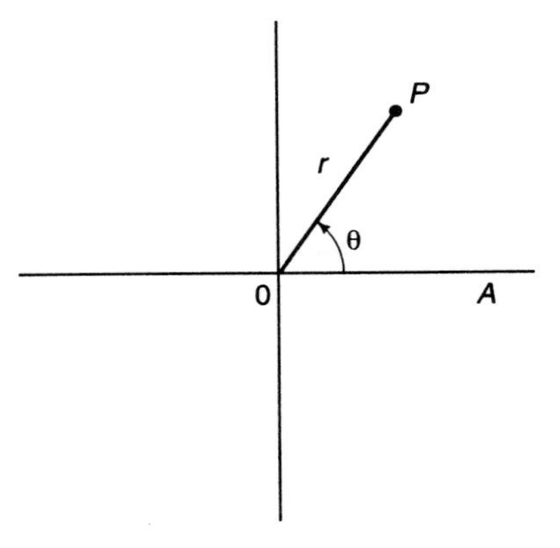

Figure 1: Polar coordinates. P is the point (r, θ).

It is possible to associate many different angles with any point P. Figure 2 illustrates $(3, \pi/6)$. Figure 3 labels the same point with a negative angle. The angles are coterminal (terminate in the same place).

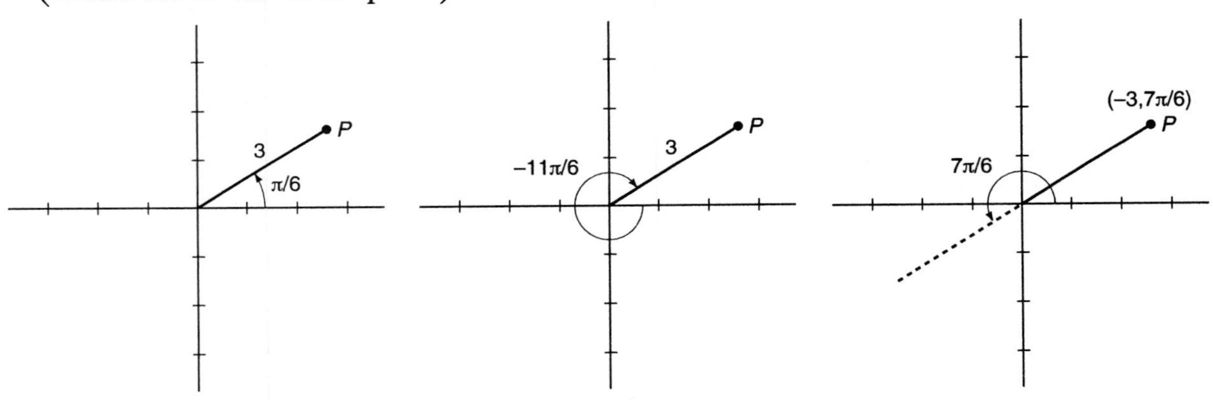

Figure 2: $P = (3, \pi/6)$ [-4, 4] by [-4, 4].

Figure 3: $P = (3, -11\pi/6)$ [-4, 4] by [-4, 4].

Figure 4: $(-3, 7\pi/6)$ [-4, 4] by [-4, 4].

Usually we think of "r" as representing distance from the origin, but it is actually "directed" distance because r can be negative. Figure 4 illustrates the same point as Figures 2 and 3, but with the angle in the opposite direction and with r negative. Figures 2, 3 and 4 show that one major difference between rectangular coordinates and polar coordinates is that points do not have a unique representation in polar coordinates.

In rectangular coordinates the grid is determined by setting x equal to a constant (for the vertical grid lines) and y equal to a constant (for the horizontal grid lines). In polar coordinates the grid is determined by setting r equal to a constant r_0 (to obtain circles with radius r_0 centered at the origin) and setting θ equal to a constant θ_0 (to obtain lines through the origin at angle θ_0 with the positive x-axis, Figure 5).

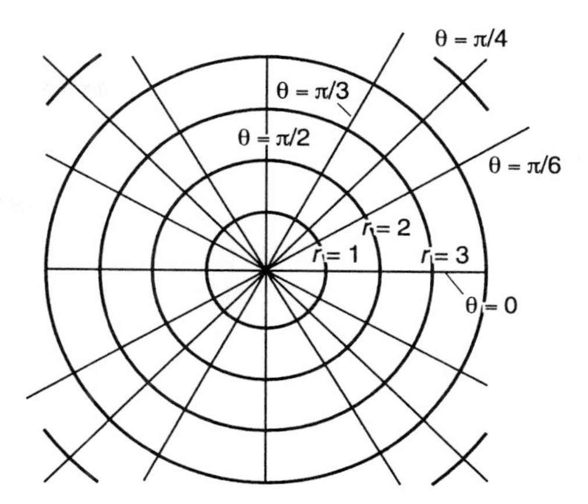

Figure 5: The polar coordinate grid.

Example 1: Graph "$r = 2$" in polar coordinates.

Since θ is not mentioned, θ can be anything. Figure 6 graphs all points 2 units from the origin.

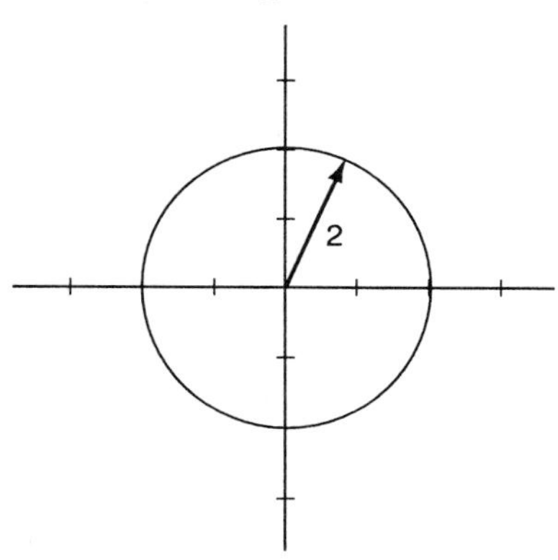

Figure 6: $r = 2$.
[-4, 4] by [-4, 4].

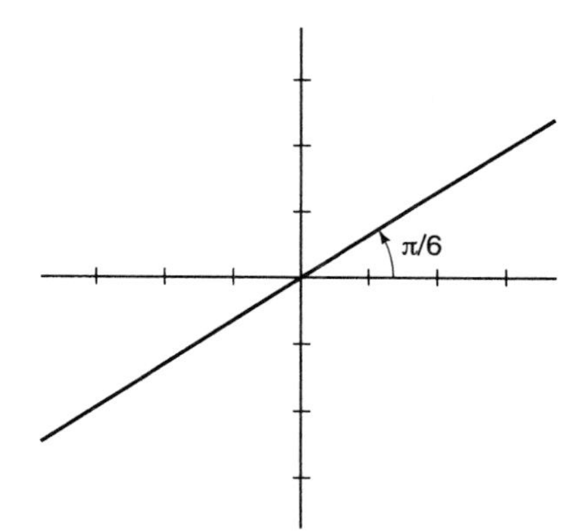

Figure 7: $\theta = \pi/6$.
[-4, 4] by [-4, 4].

Example 2; Graph "$\theta = \pi/6$" in polar coordinates.

Since r is not mentioned, r can be anything, including negative values. Figure 7 graphs all points at angle $\pi/6$ with the positive x-axis.

<u>Conversion Between Coordinate Systems</u>. The relationship between polar and rectangular representations follows easily from Figure 8. Conversion from polar to rectangular coordinates yields a unique rectangular coordinate representation:

(8.2.1) $$x = r \cos \theta \quad \text{and} \quad y = r \sin \theta.$$

Converting from rectangular to polar coordinates does not yield a unique polar coordinate representation. Given the rectangular representation (x, y), r and θ satisfy

(8.2.2) $r^2 = x^2 + y^2$ and $\tan\theta = y/x$.

These equations yield two solutions for r (positive or negative) and an infinite number of solutions for θ, including two in $[0, 2\pi)$. This means we must make some choices. Usually, but not always, we chose the positive value of r. Then we choose θ to match the quadrant of (x, y).

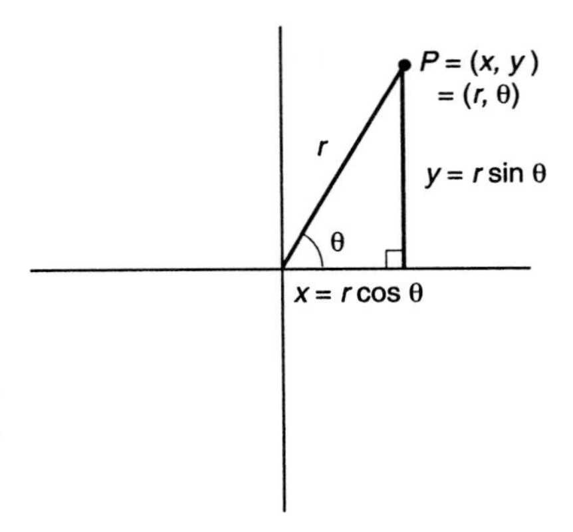

Figure 8: Polar and rectangular representations of *P*.

Example 3: Give the polar-coordinate representation of the rectangular-coordinate point (1, 1) (Figure 9).

From 8.2.2,

$r^2 = 1^2 + 1^2 = 2$ and $\tan\theta = 1/1 = 1$.

Now, make some choices. Our preference for positive r and first quadrant angles leads us to prefer $r = \sqrt{2}$ and $\theta = \tan^{-1} 1 = \pi/4$. So one polar representation of the point is $(\sqrt{2}, \pi/4)$.

Example 4: Give the polar-coordinate representation of the rectangular-coordinate point (-1, -1) (Figure 9).

$r^2 = (-1)^2 + (-1)^2 = 2$ and $\tan\theta = (-1)/(-1) = 1$. These are the *same equations* for r and θ as in Example 3, but the point is different. Clearly we must choose a different solution because the point is in the third quadrant.

Again, choose $r = \sqrt{2}$. Then we must choose a third quadrant angle θ with $\tan\theta = 1$. Tangent repeats every π radians, so $\tan(\pi/4+\pi)$ is also equal to 1. A polar-coordinate representation is $(\sqrt{2}, \pi/4+\pi) = (\sqrt{2}, 5\pi/4)$. Another is $(-\sqrt{2}, \pi/4)$.

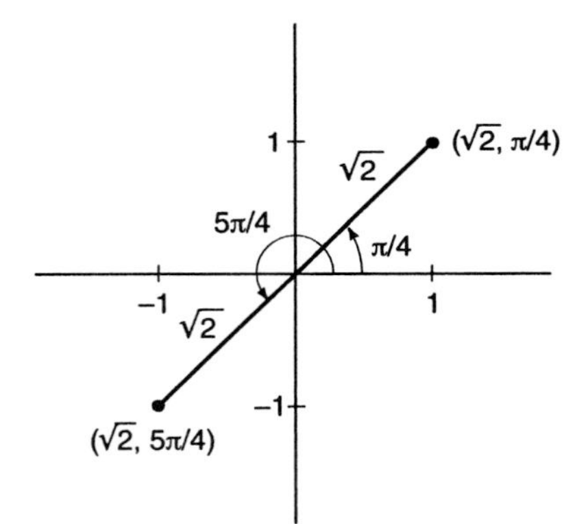

Figure 9: (1, 1), (-1, -1) and their polar-coordinate representations. [-2, 2] by [-2, 2]

To convert from rectangular coordinates, if we choose $r > 0$, a polar-coordinate representation of (x, y) is

(8.2.3)

$$(\sqrt{x^2 + y^2}\,,\ \tan^{-1}(\frac{y}{x})) \text{ if } (x, y) \text{ is in the first or fourth quadrant, and}$$

$$(\sqrt{x^2 + y^2}\,,\ \tan^{-1}(\frac{y}{x}) + \pi) \text{ if } (x, y) \text{ is in the second or third quadrant.}$$

Since angles that differ by 2π, or any multiple of 2π, are coterminal, any θ may be replaced by $\theta \pm 2n\pi$. That is, in polar coordinates,

(8.2.4) (r, θ) and $(r, \theta + 2n\pi)$ represent the same point.

Instead of memorizing these formulas, learn Figure 8, which contains the information of all the formulas in this section.

Also, since angles that differ by π are opposite, the same point is obtained by also using the opposite directed distance.

(8.2.5) (r, θ) and $(-r, \theta + \pi)$ represent the same point.

Example 5: Give several alternative polar-coordinate representations of $(5, 2\pi/3)$.

One alternative is to add 2π to the angle: $(5, 2\pi/3) = (5, 2\pi/3+2\pi) = (5, 8\pi/3)$. Another alternative is to subtract 2π: $(5, 2\pi/3) = (5, 2\pi/3 - 2\pi) = (5, -4\pi/3)$. A third alternative is to use a negative r as in 8.2.5: $(5, 2\pi/3) = (-5, 2\pi/3 + \pi) = (-5, 5\pi/3)$.

Conversion from polar-coordinates to rectangular coordinates is not tricky.

Example 6: Give the rectangular-coordinate representation of the polar-coordinate point $(2, \pi/6)$ (Figure 2).

From 8.2.1, $x = r \cos \theta = 2 \cos(\pi/6) = 2(\sqrt{3}/2) = \sqrt{3}$. $y = r \sin \theta = 2(1/2) = 1$. The unique rectangular-coordinate representation is $(\sqrt{3}, 1)$.

Well-known Polar-Coordinate Graphs. Most graphs are given in rectangular coordinates, but there are occasionally reasons to use polar coordinates. Section 8.1 on complex numbers has an application to multiplying complex numbers. Polar-coordinate graphs are most appropriate in applications where the distance to a particular point is key. For example, when studying the effect of the gravitational pull of the sun on the planets, the distance of the planets to the sun is important. By locating the origin at the sun, that distance becomes r and the position in orbit around the sun is described by (r, θ). We will study this polar representation of orbits in Section 9.3 on conic sections.

To express polar-coordinate functions, r is usually treated as a function of θ (rather than θ as a function of r). Here are some graphs that are easy to express in polar coordinates.

Example 7: Graph $r = \cos\theta$ in polar coordinates.

The graph is the circle in Figure 10. We can show it is a circle by converting to rectangular coordinates.

Multiply by r to obtain

$$r^2 = r\cos\theta.$$

By 8.2.2 and 8.2.1, this is

$$x^2 + y^2 = x.$$
$$x^2 - x + y^2 = 0.$$
$$x^2 - x + 1/4 + y^2 = 1/4.$$
$$(x - 1/2)^2 + y^2 = (1/2)^2.$$

This is the standard form (3.1.13) of a circle centered at (1/2, 0) with radius 1/2.

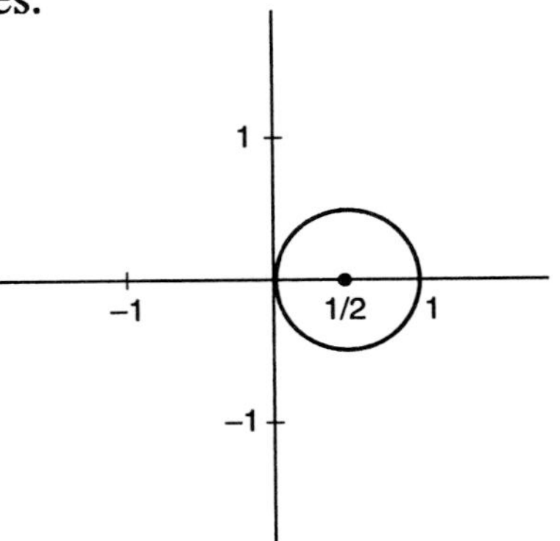

Figure 10: $r = \cos\theta$. [-2, 2] by [-2, 2]. $0 \le \theta < 2\pi$ or $0 \le \theta < \pi$.

Since $\cos\theta$ repeats every 2π, there is no need to use a domain larger than $[0, 2\pi)$ (one complete revolution). Actually, this particular graph is complete after only half a revolution. The first quadrant angles yield the top half. In the second quadrant cosine is negative, so $r = \cos\theta$ is negative. Therefore, rather than yielding points in the second quadrant, second quadrant angles yield points in the opposite quadrant, the fourth quadrant. These form the bottom half of the circle. Then, when θ is a third quadrant angle, $\cos\theta$ is still negative and the points in the first quadrant are retraced. Finally, when θ is a fourth quadrant angle, $\cos\theta$ is positive and the fourth quadrant points are retraced.

Calculator Exercise 1: Watch the graph of $r = \cos\theta$ develop on a graphics calculator using domain $[0, 2\pi)$. Which points are created first? Which are created last?

If your calculator does not have a "polar" coordinate graphing mode, you may use "parametric" mode instead. To graph "$r = f(\theta)$" parametrically, simply let $x = f(\theta)\cos\theta$ and $y = f(\theta)\sin\theta$.

For example, to graph "$r = \cos\theta$" parametrically, let $x = (\cos\theta)^2$ [this is $r\cos\theta$] and $y = (\cos\theta)(\sin\theta)$ [this is $r\sin\theta$].

Example 8: Graph $r = 1 + \sin\theta$.

See Figure 11. This shape is called a "cardioid" from the Greek "kardia" meaning "heart." It is not convenient to express in rectangular coordinates.

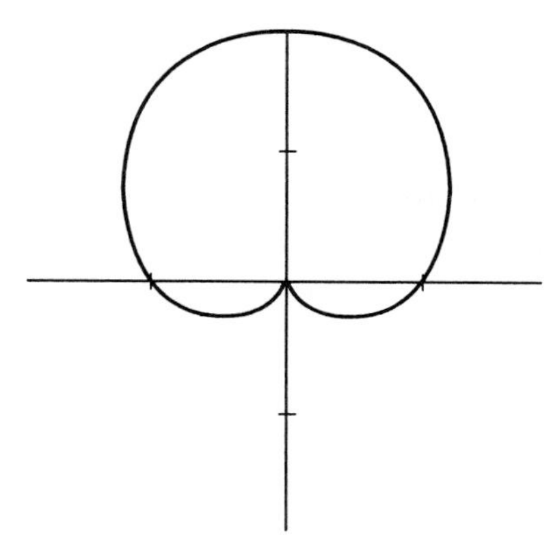

Figure 11: $r = 1 + \sin\theta$.
[-2, 2] by [-2, 2].
$0 \leq \theta < 2\pi$.

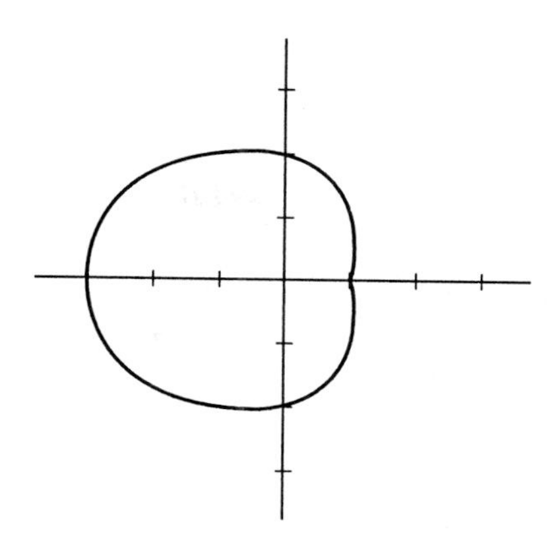

Figure 12: $r = 2 - \cos\theta$.
[-4, 4] by [-4, 4].
$0 \leq \theta < 2\pi$.

Example 9: Graph $r = 2 - \cos\theta$.

In contrast to Example 8 ($r = 1 + \sin\theta$), here r is never zero. This shape is called a "limaçon," as are other shapes with similar equations (Problems B17-20. Some even have two loops, one inside other (problems B18 and B19).

Example 10: The simple equation "$r = \theta$" forms the "Spiral of Archimedes." In it the distance from the origin equals the angle, so as the angle increases the curve spirals out (Figure 13) (problem B42).

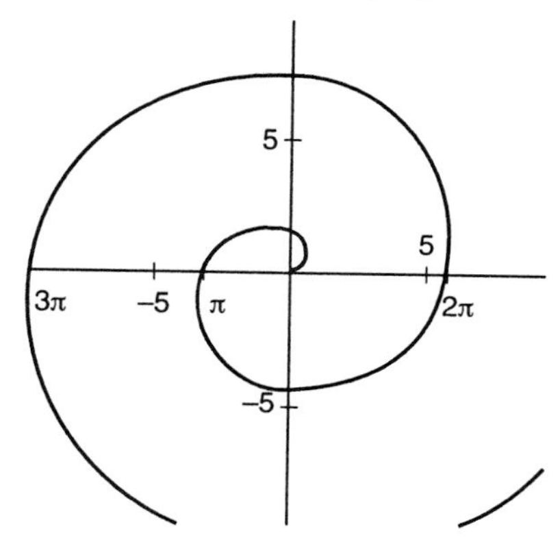

Figure 13: A Spiral of Archimedes. $r = \theta$.
[-10, 10] by [-10, 10].
$0 \leq \theta$. Points for $\theta > 12$ are off the window.

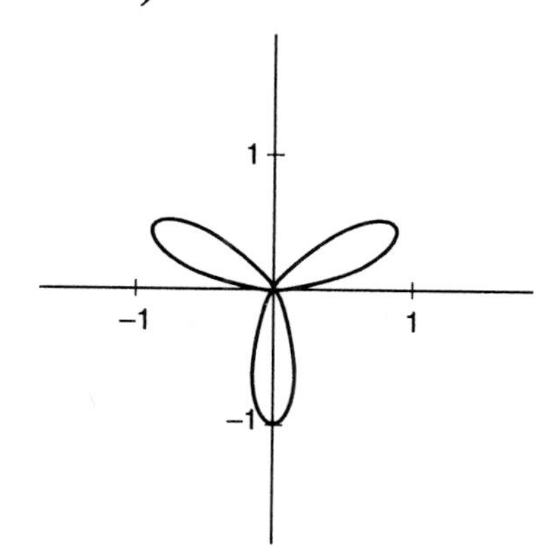

Figure 14: $r = \sin(3\theta)$.
[-2, 2] by [-2, 2].
$0 \leq \theta < 2\pi$.

Example 11: Various leaf shapes can be described by "$r = \sin(n\theta)$" for various integer values of n. Figure 14 illustrates "$r = \sin(3\theta)$."

Calculator Exercise 2: Watch the graph of "$r = \sin(3\theta)$" develop on a graphics calculator. Which leaf develops second? (problem A33).

Symmetry. Symmetry about the x- or y-axis is often visible in polar-coordinate graphs (Figures 10, 11, and 12). "cos θ" is symmetric about $\theta = 0$, which is the x-axis, so equations expressed in terms of "cos θ" have graphs that are symmetric about the x-axis (Figures 10 and 12, see also problem B2). "sin θ" is symmetric about $\theta = \pi/2$, which is the y-axis, so equations expressed in terms of "sin θ" have graphs that are symmetric about the y-axis (Figure 11, see also problem B3).

Example 12: Figure 15 exhibits the graph of $r = 1 + \sin(2\theta)$. It uses the sine function, but its graph is not symmetric about the y-axis. Why not?

The function "sin θ" is symmetric about $\pi/2$. But the argument of sine in the expression "$1 + \sin(2\theta)$" is not θ, it is 2θ. So symmetry occurs about θ such that $2\theta = \pi/2$, that is, about $\theta = \pi/4$ (45°, the line $y = x$). In Figure 15 you can see the symmetry about the diagonal line $\theta = \pi/4$.

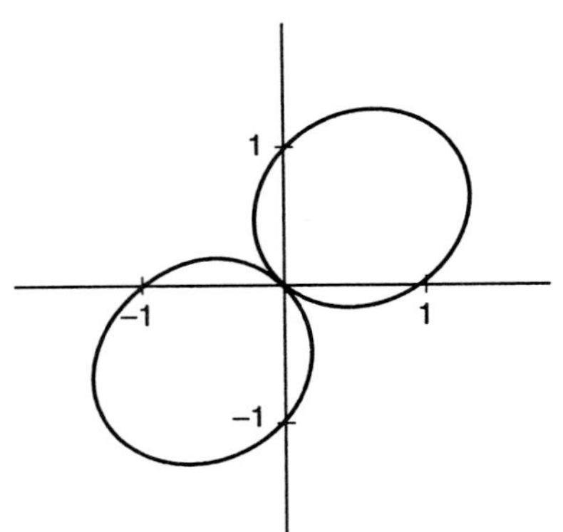

Figure 15: $r = 1 + \sin(2\theta)$.
[-2, 2] by [-2, 2].
$0 \le \theta < 2\pi$.

Another type of symmetry in Figure 15 is point symmetry about the origin -- for every point there is another equidistant on the opposite side of the origin. That is, when (r, θ) is on the graph, so it $(r, \theta + \pi)$. This always happens when r is a function of sine or cosine of 2θ, as in Figure 15 (Problem B4).

There is another simple cause of point symmetry. The two points (r, θ) and $(-r, \theta)$ are point symmetric about the origin. Therefore, for any $f(\theta)$, the graph of "$r^2 = f(\theta)$" will be point symmetric about the origin because the negative of any r that solves it will also solve it (problems A37, A38, and B27).

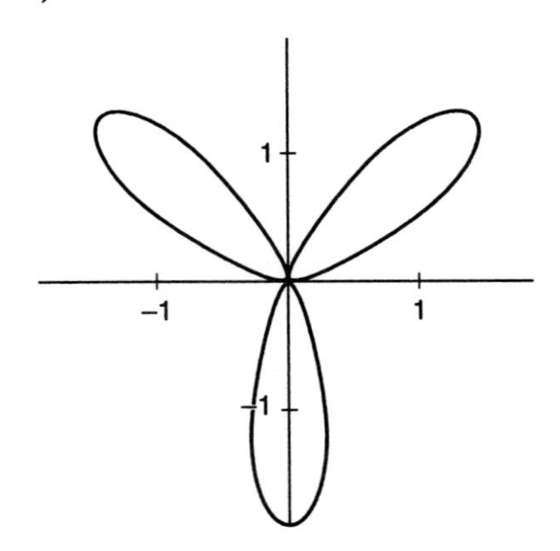

Figure 16: $r = 2\sin(3\theta)$.
[-2, 2] by [-2, 2].
[Compare with Figure 14.]

Scale: The graphs of "$r = f(\theta)$" and "$r = cf(\theta)$" differ by the scale factor c. The second graph is c times as large, expanded *radially*.

Example 11, modified: Figure 14 gives the graph of "$r = \sin(3\theta)$" from Example 11. Figure 16 shows the graph of "$r = 2\sin(3\theta)$"

would be twice the size, expanded away from the origin.

Conclusion. The relationship between rectangular and polar coordinates is illustrated in Figure 8: $x = r \cos \theta, y = r \sin \theta, r^2 = x^2 + y^2$, and $\tan \theta = y/x$. Polar-coordinate representations of points are not unique.

****

Exercises for Section 8.2, "Polar Coordinates":

^ ^ ^ ^The point is given in rectangular coordinates. Convert it to polar coordinates with $r \geq 0$ and $0 \leq \theta < 2\pi$.

A1. (2, 2) A2. (-2, 0) A3. (0, 4) A4. (0, -3)
A5. (3, 0) A6. (-2, 2) A7. $(1, \sqrt{3})$ A8. $(\sqrt{3}, 1)$

^ ^ ^ ^The point is given in polar coordinates, using radians. Convert it to rectangular coordinates.

A9. $(3, \pi/2)$ A10. $(4, -\pi/2)$ A11. (2, 0) A12. $(5, \pi)$
A13. $(4, \pi/4)$ A14. $(6, \pi/3)$ A15. $(2, \pi/6)$ A16. $(-2, 5\pi/6)$

^ ^ ^ ^The point is given in polar coordinates. Give three alternative polar-coordinate representations of the point.

A17. $(2, \pi)$ A18. $(0, \pi/2)$ A19. (5, 0) A20. (-4, 0)

^ ^ ^ ^Sketch the graph.

A21. $r = 5$. A22. $r = 4$ A23. $\theta = \pi/12$ A24. $\theta = -\pi/4$

^ ^ ^ ^Use the cited figure to sketch the graph

A25. $r = 3 \cos \theta$ (Figure 10) A26. $r = 2 + 2 \sin \theta$ (Figure 11)
A27. $r = 1 - (\cos \theta)/2$ (Figure 12) A28. $r = 5 \sin(3\theta)$ (Figure 14)

^ ^ ^ ^Determine the length of one leaf.

A29. $r = 4 \sin(3\theta)$ A30. $r = 7 \cos(2\theta)$

A31. Compare the graphs of "$r = f(\theta)$" and "$r = 5f(\theta)$."

A32. Watch the graph of $r = \sin(2\theta)$ develop on a graphics calculator using domain $[0, 2\pi)$. a) The points that appear in the fourth quadrant are associated with angles in which quadrant? b) How can that be?

A33. Do Calculator Exercise 2: Watch the graph of $r = \sin(3\theta)$ develop on a graphics calculator using domain $[0, 2\pi)$. a) Which leaf develops second? b) Which values of θ yield points in the second leaf? c) The points that appear in the third quadrant are associated with angles in which quadrant? d) How can that be?

A34. "$r = 1 + \sin \theta$" can be regraded as "$r = f(\sin \theta)$." Give f.

A35. "$r = 2 - \cos \theta$" can be regarded as "$r = f(\cos \theta)$." Give f.

A36. "$r = \sin \theta$" can be regarded as "$r = f(\sin \theta)$." Give f.

A37. a) Graph "$r^2 = \theta$, $0 \le \theta \le 2\pi$." b) What symmetry does it have?

A38. a) Graph "$r^2 = \cos\theta$." b) Give all the symmetries it has.

^ ^ ^ ^ ^ ^ ^ ^ ^

B1.* Sketch a figure that exhibits how to convert between rectangular and polar coordinates.

B2. a) Draw a picture to illustrate why (r, θ) and $(r, -\theta)$ are symmetric about the x-axis. b) Draw a unit-circle rectangular-coordinate picture to illustrate why $\cos(-\theta) = \cos\theta$. c) If $r = f(\cos\theta)$ for some f (as in Figures 10 and 12), and (r, θ) is on the graph, what other point is automatically on the graph? Why? d) Sketch a figure of your choice (your choice of f) to illustrate the resulting symmetry about the x-axis.

B3. a) Draw a picture to illustrate why $(r, \pi/2 + \theta)$ and $(r, \pi/2 - \theta)$ are symmetric about the y-axis. b) Draw a unit-circle rectangular-coordinate picture to illustrate why $\sin(\pi/2 + \theta) = \sin(\pi/2 - \theta)$. c) If $r = f(\sin\theta)$ for some f (as in Figure 11), and $(r, \pi/2 - \theta)$ is on the graph, what other point is automatically on the graph? Why? d) Sketch a figure of your choice (your choice of f) to illustrate the resulting symmetry about the y-axis.

B4. a) Draw a picture to illustrate why (r, θ) and $(r, \theta+\pi)$ are point symmetric about the origin. b) If $r = f(\sin(2\theta))$ for some f and (r, θ) is on the graph, why is $(r, \theta+\pi)$ automatically on the graph? [Hint: "$\sin(\alpha+2\pi) = \sin\alpha$" is the relevant trig identity.] c) Sketch a figure of your choice (your choice of f) to illustrate the resulting point symmetry about the origin.

^ ^ ^ ^ The figure locates $P = (r, \theta)$. Sketch the figure and on it locate

B5. a) $A = (-r, \theta)$. b) $B = (r, -\theta)$. c) $C = (r, \theta+\pi)$ d) $D = (r, \theta+\pi/2)$

B6. a) $A = (2r, \theta)$. b) $B = (r, \theta - \pi/2)$ c) $C = (r, -\theta)$ d) $D = (-r, \theta)$

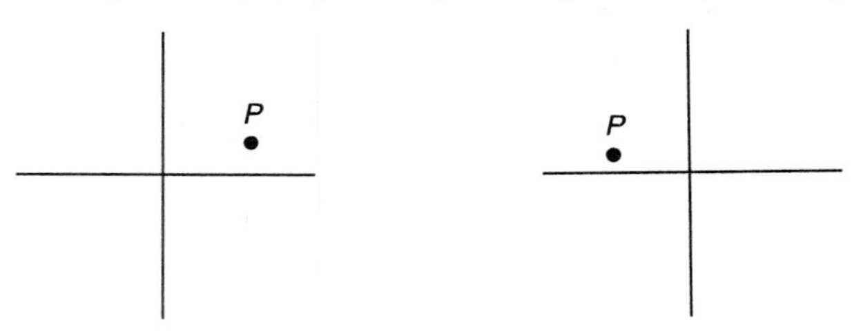

Figure for B5: **Figure for B6:**

B7. Express "$x = c$" in polar coordinates. B8. Express "$y = k$" in polar coordinates.

^ ^ ^ ^ Sketch the graph of

B9. $r = \sin\theta$ B10. $r = 1 + \cos\theta$ B11. $r = 2 - \sin\theta$ B12. $r = \cos(3\theta)$

^ ^ ^ ^ (Following B9-B12) Each of the graphs in B9-B12 is the same shape as a graph in an example in this section, but rotated about the origin. For each give the figure number of the similar graph and give the rotation from the text graph to the new graph.

B13. Problem B9. B14. Problem B10. B15. Problem B11. B16. Problem B12.

^ ^ ^ ^ Sketch the graph.

B17. $r = 3 - 2 \sin \theta$.

B18. $r = 2 - 3 \sin \theta$.

B19. $r = 2 - 3 \cos \theta$.

B20. $r = 2 + \cos \theta$.

B21. $r = 3 \sin(2\theta)$.

B22. $r = 4 \cos(2\theta)$.

B23. $r = 4 \sin(4\theta)$.

B24. $r = 5 \cos(4\theta)$.

B25. $r^2 = 5 \cos(2\theta)$

B26. $r = 7 \cos(3\theta)$.

B27. a) Draw a picture to illustrate why (r, θ) and $(-r, \theta)$ are point symmetric about the origin. b) If $r^2 = f(\theta)$ and (r, θ) is on the graph, what other point is automatically on the graph? Why? c) Sketch a figure of your choice to illustrate the resulting point symmetry about the origin.

B28. The graph of "$r = f(\sin(2\theta))$" is symmetric about $\theta = \pi/4$ for any f. a) Why?
b) Give an f (your choice) and sketch the graph, including the line of symmetry.

B29. The graph of "$r = f(\cos(2\theta))$" is symmetric about the x-axis for any f. a) Why?
b) Give an f (your choice) and sketch the graph, including the line of symmetry.

B30. Suppose we wish to graph "$r = f(\sin(3\theta))$" for some f. There will be a line of symmetry of the graph in the first quadrant. a) Use trig to show which line it is.
b) Sketch "$r = \sin(3\theta)$" and note the line of symmetry on the graph (To extend this result, see also Problem B41).

B31. Suppose we wish to graph "$r = f(\cos(3\theta))$" for some f. There will be a line of symmetry of the graph in the first quadrant (not just $\theta = 0$). a) Use trig to show which line it is. b) Sketch "$r = \cos(3\theta)$" and note the line of symmetry on the graph (To extend this result, see also Problem B40).

B32. Look at graphs of $r = \sin(2\theta)$, $r = \sin(3\theta)$, and $r = \sin(4\theta)$.
a) Guess how many leaves the graph of $r = \sin(n\theta)$ has for integer values of n.
b) Look at a graph of $r = \sin(5\theta)$ to check your guess. Was your guess right for $n = 5$?

B33. Let L be the line $x = -2$. Let P be a point to the right of line L. a) Find a polar-coordinate equation for the set of all points P such that the distance from P to the line L is equal to the distance from P to the origin. b) Solve for r. c) What shape is the curve?

B34. Let L be the line $x = -3$. Let P be a point to the right of line L. Find a polar-coordinate equation for the set of all points P such that the distance from P to the line L is twice the distance from P to the origin. Solve for r.

B35. Let L be the line $x = -3$. Let P be a point to the right of line L. Find a polar-coordinate equation for the set of all points P such that the distance from P to the line L is half the distance from P to the origin. Solve for r.

B36. Recall that the rectangular-coordinate graphs of "$\sin x$" and "$\cos x$" have the same shape. Either can be expressed as the other shifted left or right. Now consider the graph of any equation "$r = f(\cos \theta)$" for some f (for instance, Figure 10, $r = \cos \theta$ and Figure 12, $r = 2 - \cos \theta$). a) What does the graph of "$r = f(\sin \theta)$" [cosine is replaced by sine] look like compared to the old one? State this result clearly. b) State the relevant trig identity.
c) Explain why the trig identity justifies your result in part (a).

B37. Consider any graph with equation "$r = f(\sin\theta)$" for some f (for instance, Figure 11, $r = 1 + \sin\theta$). a) What does the graph of "$r = f(\cos\theta)$" [sine is replaced by cosine] look like compared to the old one? State this result clearly. b) State the relevant trig identity. c) Explain why the trig identity justifies your result in part (a).

B38. Suppose a graph is symmetric about both the x- and y-axes. Prove that it must be point symmetric about the origin.

B39. Suppose the unit square with sides connecting (0, 0), (1, 0), (1, 1), and (0, 1) [rectangular coordinates] were expressed in polar coordinates by "$r = f(\theta)$." Would the graph of "$r = 2f(\theta)$" be the graph of a square, or would the shape change? Indicate why or why not.

B40. Suppose we wish to graph "$r = f(\cos(3\theta))$." For any f there will be lines of symmetry of the graph because of the symmetry of cosine. a) Find all those the lines of symmetry. b) Sketch "$r = \cos(3\theta)$" and note the lines of symmetry on the graph.

B41. Suppose we wish to graph "$r = f(\sin(3\theta))$." For any f there will be lines of symmetry of the graph because of the symmetry of sine. a) Find all those lines of symmetry. b) Sketch "$r = \sin(3\theta)$" and note the lines of symmetry on the graph.

B42. One of the most famous problems of geometry is to find a procedure that will trisect any angle (that is, divide it into three equal angles). It has been proven that it cannot be done with "Greek" rules, that is, using only a compass and straightedge. However, given a Spiral of Archimedes, any angle can be trisected. How? [By the way, the Spiral of Archimedes cannot be constructed with a compass and straightedge.]

B43. Suppose we wish to graph "$r = f(\cos(n\theta))$." For any f, there will be lines of symmetry of the graph because of the symmetry of cosine. Find all those lines of symmetry.

Section 8.3. Parametric Equations

Parametric equations are useful for describing motion. Motion produces change, and calculus is the mathematical subject which studies change. So, parametric equations are useful in calculus.

The usual type of graph is described in "functional" form, that is, y is given as a function of x by an equation of the form "$y = f(x)$." If this type of equation is used to describe a path of a moving object, two severe limitations leap to mind. One is that, because f is a function, the path can have only one y-value for each x-value, so the functional equation cannot describe a non-functional path such as in Figure 1.

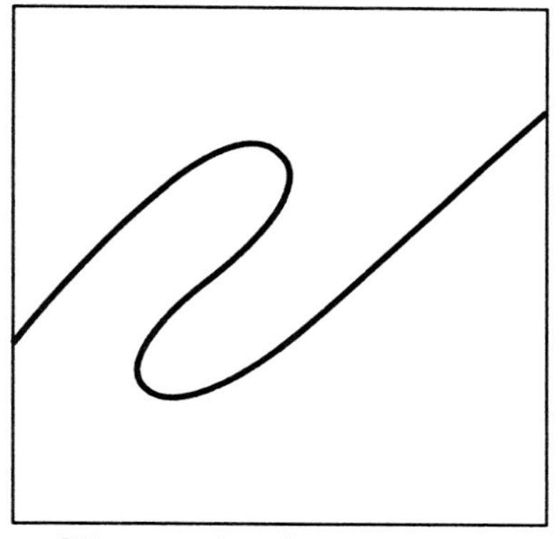

Figure 1: A path that cannot be described functionally by "$y = f(x)$."

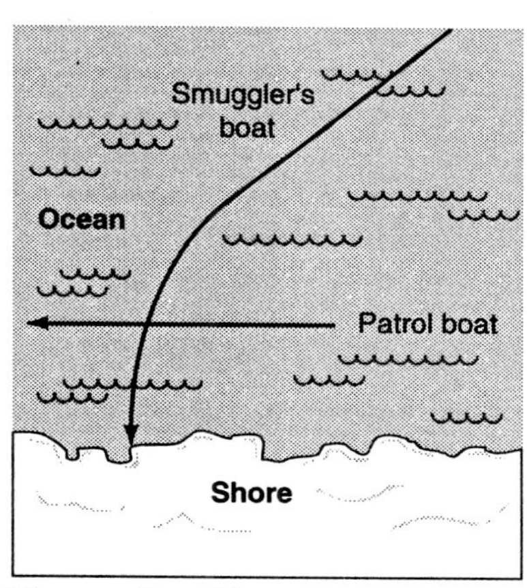

Figure 2: The routes of two boats. Do the boats meet?

A second limitation of a functional description of a path is that it does not contain information about *when* the object was at any point on the path. Figure 2 shows two paths, the route of a smuggler's boat from offshore to the coast and the route of a patrol boat. Will the patrol boat intercept the smuggler's boat?

We cannot tell. Figure 2 does not have information on *when* the boats will be at the various locations described by the paths.

Parametric equations use a parameter (often "t" for time) and two equations of the form "$x = f(t)$ and $y = g(t)$" to describe the (x, y) points on the path in terms of the parameter,

so that location and time are related. This contrasts with a functional description of the path in which the y-value is described in terms of the x-

value alone.

Parametric Equations of Circles. There is more than one way to describe a circle. A circle centered at the origin with radius r has "standard form"

$$x^2 + y^2 = r^2,$$

according to 3.1.12. The so-called "functional form" requires y to be given in terms of x. Solving for y, the circle is given by

$$y = \pm\sqrt{r^2 - x^2} \,.$$

This is the form commonly entered into graphics calculators. Another approach is to describe x and y separately in terms of some parameter, say, "θ" or "t". Figure 3 reminds us that in polar coordinates $x = r \cos \theta$ and $y = r \sin \theta$.

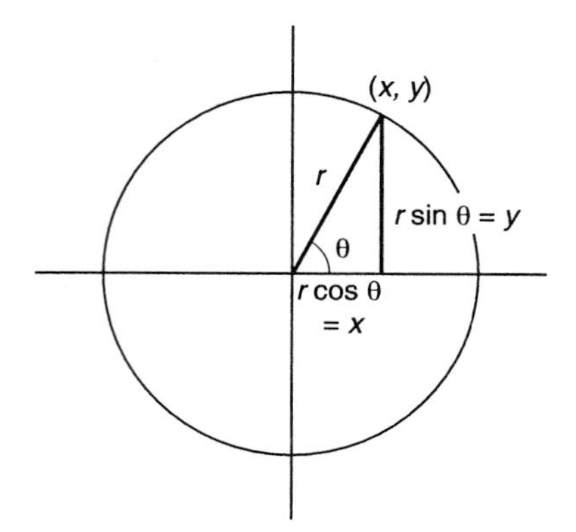

Figure 3: $x = r \cos \theta$ and $y = r \sin \theta$.

Example 1: Suppose a wheel rotates counterclockwise 1 radian per second. Describe the position of a point 5 centimeters from the center at time t, if its initial position is at the rightmost point on the circle.

The speed of rotation and orientation are such that $\theta = t$ in the usual polar coordinate system. The radius is 5, so parametric equations for x and y are given by

$$x = 5 \cos t \quad \text{and} \quad y = 5 \sin t.$$

Figure 4 plots the path with the addition of labels relating position to time.

Calculator Exercise 1: Learn how to use "Parametric" mode on your calculator. For example, plot the graph in Figure 4 and watch it develop as t increases (problem A1, B24).

Parametric equations give x and y in terms of t. It may be possible to convert parametric form to functional form by eliminating the t.

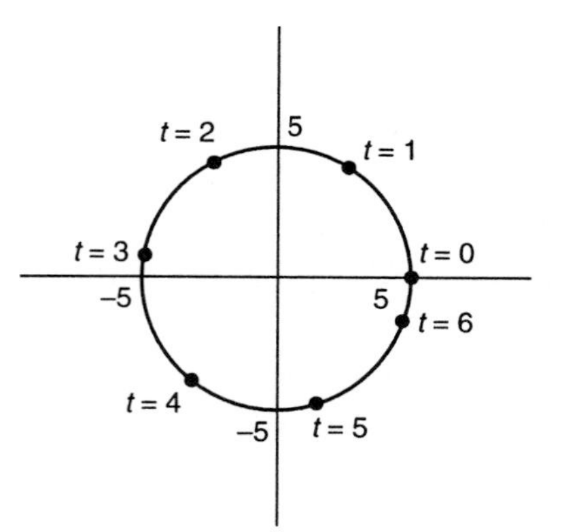

Figure 4: A circle of radius 5.
$x = 5 \cos t$. $y = 5 \sin t$.
[-10, 10] by [-10, 10].

Example 1, continued: Let $x = 5 \cos t$ and $y = 5 \sin t$ be a parametric description of a path. If we wish to describe the path without reference to time, we can take the two equations and eliminate t.

Squaring x and y and adding the result,

$$x^2 + y^2 = (5 \cos t)^2 + (5 \sin t)^2 = 25(\cos^2 t + \sin^2 t) = 25.$$

Of course, "$x^2 + y^2 = 25$" is the equation of the circle with radius 5 centered at the origin.

(8.3.1) $$x = r \cos t \text{ and } y = r \sin t, \quad 0 \le t < 2\pi,$$

are parametric equations of a circle centered at the origin with radius r.

Example 1, continued further: If we want a different speed than one revolution in 2π units of time, we can replace "t" with some function of t. For example, let

$$x = 5 \cos(2\pi t) \text{ and}$$
$$y = 5 \sin(2\pi t).$$

Now, as t changes from 0 to 1 the arguments of sine and cosine change from 0 to 2π, so one revolution occurs in 1 unit of time, which is quite a bit faster than in Example 1. The relationship between x and y is unchanged, but their relationship to time is changed (Figure 5). The increase in speed is easy to see as the picture develops on a graphics calculator (try it).

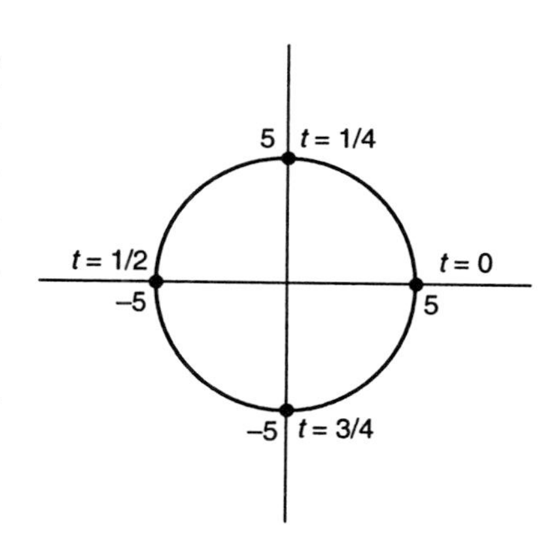

Figure 5: $x = 5 \cos(2\pi t)$ and $y = 5 \sin(2\pi t)$. [-10, 10] by [-10, 10].

More about Example 1: The previous parametric descriptions are of uniform motion about the circle. If we want to describe the same path, but with some sort of accelerating motion, we can replace "t" with some function of t, for example, "t^2".

Let

$$x = 5 \cos(t^2) \text{ and}$$
$$y = 5 \sin(t^2).$$

Again, the relationship between x and y is not changed. But the angle (angle t^2, playing the role of θ in polar coordinates) is increasing more and more rapidly (Figure 6).

time t	angle t^2	change in angle during the previous 1/2 second
0	0	
1/2	1/4	1/4
1	1	3/4
3/2	9/4	5/4
2	4	7/4

The rotation is speeding up.

In this example the "t" of Example 1 has been replaced by "t^2" in both equations. If "t" is replaced *in both equations* by the same function of t, the graph will still be a circle. The relationships of x to time and y to time would change, but the relationship of x to y would not (problem B4).

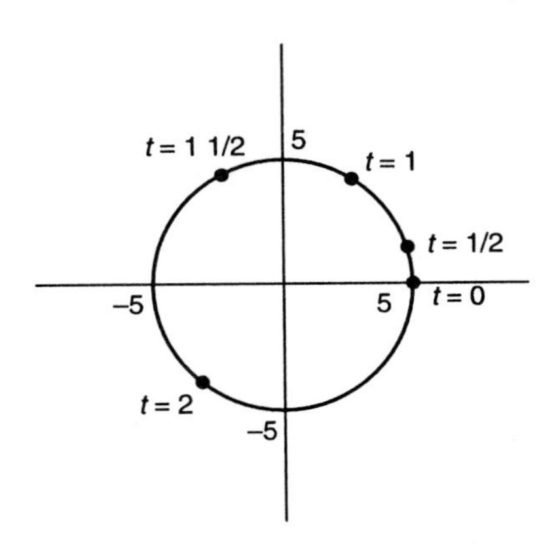

Figure 6: $x = 5\cos(t^2)$ and $y = 5\sin(t^2)$. [-10, 10] by [-10, 10].

For any function f such that $[0, 2\pi)$ is a subset of its range,

(8.3.2)
$$x = r\cos(f(t)), \text{ and}$$
$$y = r\sin(f(t)),$$

are parametric equations of a circle centered at the origin with radius r (problem B5).

Example 2: Suppose a rod 15 centimeters long is pinned to a rotating wheel at a radius of 5 centimeters (point A, Figure 7). The piston (point P) at the other end of the rod slides vertically up and down. Describe the path of point P.

The position of P depends upon the angle θ in the figure. The motion is vertical, $x = 0$, regardless of θ. The vertical component of P can be determined in more than one way. We could use the Law of Cosines. Or we could treat the y-value as OC plus CP. By trigonometry, OC is $5\sin\theta$ and CA is $5\cos\theta$. Now, CP can be determined with the aid of the Pythagorean theorem: $CP^2 + CA^2 = 15^2$.

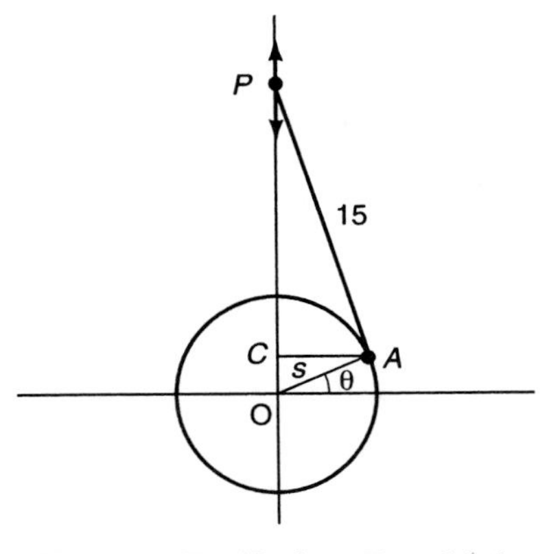

Figure 7. Point P, which moves vertically, is connected by a rod of length 15 to a point, A, that rotates with radius 5.

$$CP = \sqrt{15^2 - (5\cos\theta)^2}$$

Therefore, adding OC and CP,

$$y = 5\sin\theta + \sqrt{15^2 - (5\cos\theta)^2}\ .$$

Now, suppose the wheel rotates 20 times a second. 20 revolutions is $20(2\pi)$ radians. So $\theta = 20(2\pi)t$. Parametric equations of the motion of P are given by

$$x = 0 \quad \textit{and} \quad y = 5\sin(40\pi t) + \sqrt{15^2 - (5\cos(40\pi t))^2}\ .$$

Figure 8 plots the vertical component of the motion as a function of time (problem A2).

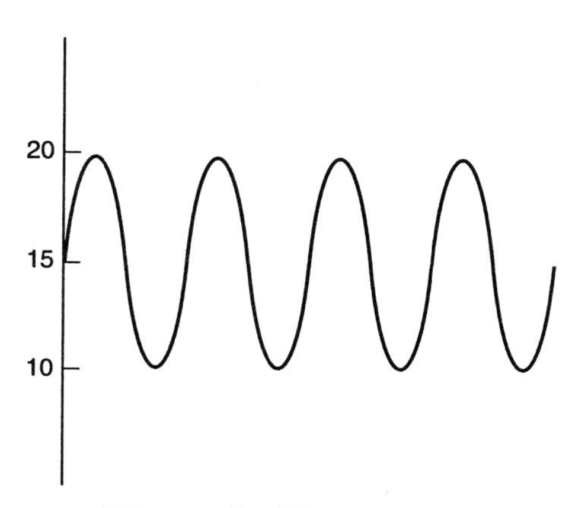

Figure 8: The vertical position of P in Example 2 as a function of time. [0, .2] by [5, 25].

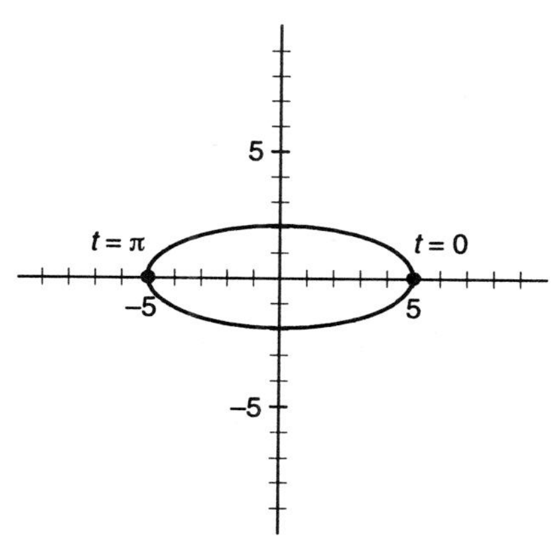

Figure 9: The ellipse described parametrically by $x = 5\cos t$ and $y = 2\sin t$. [-10, 10] by [-10, 10].

Example 3: Let $x = 5\cos t$ and $y = 2\sin t$. Graph the curve and find y in terms of x.

The graph is the ellipse in Figure 9. We can see that it is an ellipse by eliminating t. Divide by the constant and square.

$(x/5)^2 = \cos^2 t$. $(y/2)^2 = \sin^2 t$.

Adding, since $\sin^2 t + \cos^2 t = 1$, we obtain

$$\frac{x^2}{5^2} + \frac{y^2}{2^2} = 1$$

which is the standard form of the equation of an ellipse (3.2.10) (problem B7).

<u>**Projectiles**</u>. The positions of projectiles are often represented parametrically. The vertical and horizontal components of the motion of ball or artillery shell can be treated independently. If air resistance is neglected, we can obtain simple answers. (Air resistance is an important factor, so neglecting it makes the answers wrong. Nevertheless, it is a good "first approximation" and the real effect of air resistance is complicated and cannot be derived mathematically until after calculus.)

Example 4: Assume that, due to the force of gravity, the vertical coordinate of a projectile is given by

$$y = y(t) = -16t^2 + 100t + 5.$$

This is the usual formula (from Example 3.2.132) with distance measured in feet, time measured in seconds, initial upward speed 100 feet per second, and initial vertical coordinate 5 feet above ground level at $y = 0$.

If the initial speed in the x-direction is 800 feet per second, without air resistance the object would continue the same horizontal speed, so

$$x = x(t) = 800t,$$

from "distance equals rate times time," assuming the initial horizontal coordinate is zero.

We can use a calculator to plot the graph parametrically, but the standard window is not appropriate. Figure 10 shows the trajectory (path) in a suitable window.

When and where does the projectile hit the ground? Describe the path in functional form.

The projectile hits the ground when $y = 0$. So solve $-16t^2 + 100t + 5 = 0$ for t. Then plug that solution into "$x = 800t$" to find the corresponding x value (problem A3).

To describe the path in functional form, eliminate t. From the equation for x, $t = x/800$. Then t can be replaced by $x/800$ in the equation for y to give y in terms of x.

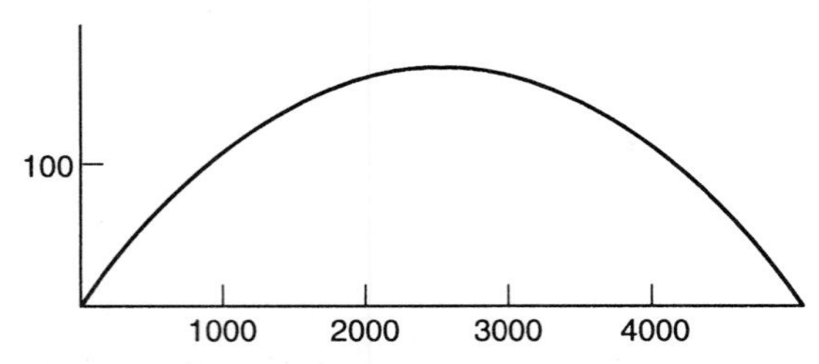

Figure 10: The path of a projectile. $x = 800t$. $y = -16t^2 + 100t + 5$. [0, 5000] by [0, 200].

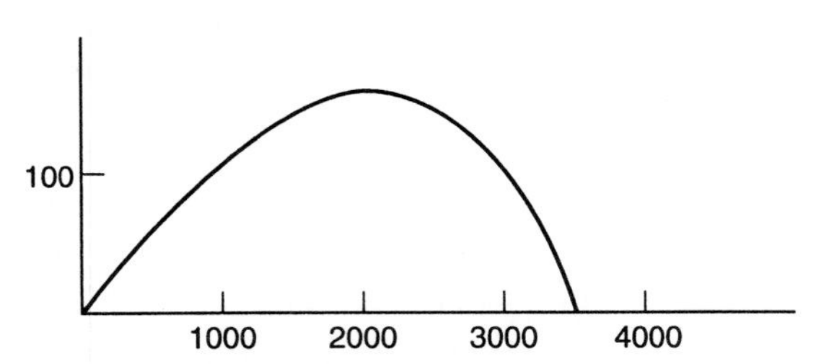

Figure 11: The path of a projectile. $x = 800t - 40t^2$. $y = -16t^2 + 100t + 5$. [0, 5000] by [0, 200].

Example 5: Suppose a projectile moves with the vertical motion described in Example 4,

$$y = -16t^2 + 100t + 5,$$

but the horizontal speed slows down as time passes so $x(t)$ is less than $800t$. For example, let $x = 800t - 40t^2$, $0 \leq t \leq 10$.

Plot the trajectory (problem A4).

Plot it parametrically (Figure 11). In this example it not so easy to eliminate t. But there is no real need to find or use a functional equation that gives y in terms of x. One of the advantages of parametric equations is that some curves are far easier to describe in parametric equations.

Parametric Equations of Lines. The point-slope formula for lines follows from similar triangles as discussed in Section 3.1 (Figure 3.1.3). Parametric equations for lines follow similarly. Figure 12 illustrates two similar triangles.

Parametric equations for lines give x and y in terms of a particular point on the line (labeled "(x_1, y_1)" and "B") and two direction numbers (labeled "a" and "b") which are the change in x-value and change in y-value to any other particular point (labeled "A") on the line.

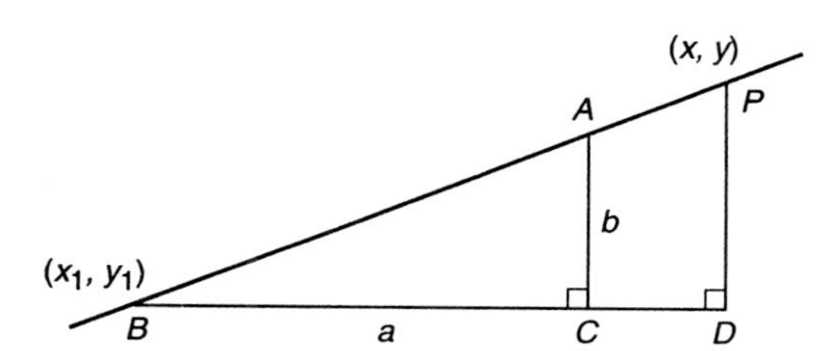

Figure 12: Similar triangles.

Let $P = (x, y)$ represent a general point on the line, and form the similar triangles in the picture. Now DB is some multiple of a. Call that multiple t, so $DB = at$, where "t" is the parameter (t looks to be about 1.3 in Figure 12). Then, by proportionality of sides of similar triangles, PD is bt. Therefore the coordinates of P satisfy $x = x_1+DB = x_1+at$ and $y = y_1+PD = y_1+bt$.

Therefore,

$$x = x_1 + at \text{ and } y = y_1 + bt \tag{8.3.3}$$

are parametric equations of the line through the point (x_1, y_1), with direction numbers a and b and with parameter t.

When $t = 0$, the point is A. When $t = 1$, the point is B. If $a = 0$, x does not change and the line is vertical. If $b = 0$, y does not change and the line is horizontal. The slope of the line is b/a, if $a \neq 0$.

Example 6: Find parametric equations of a line through (1, 5) and (3, 2).

Sketch a figure (Figure 13). x changes 2 units when y changes -3 units, so let the direction numbers a and b be 2 and -3. Parametric equations are:

$x = 1 + 2t$ and $y = 5 - 3t$.

Example 6, continued. Suppose an object undergoing uniform linear motion is at (1, 5) at time 0 and at (3, 2) at time 5. Find parametric equations of its path.

The points are the same as before, but we need to change the time scale.

The equations from Example 6 yield (1, 5) when $t = 0$ as we want, but they yield (3, 2) at time $t = 1$, not at time $t = 5$. So, simply change the scale by a factor of 5.

$$x = 1 + 2(t/5) \text{ and } y = 5 - 3(t/5).$$

Regrouping, the equations are

$$x = 1 + (2/5)t \text{ and } y = 5 - (3/5)t.$$

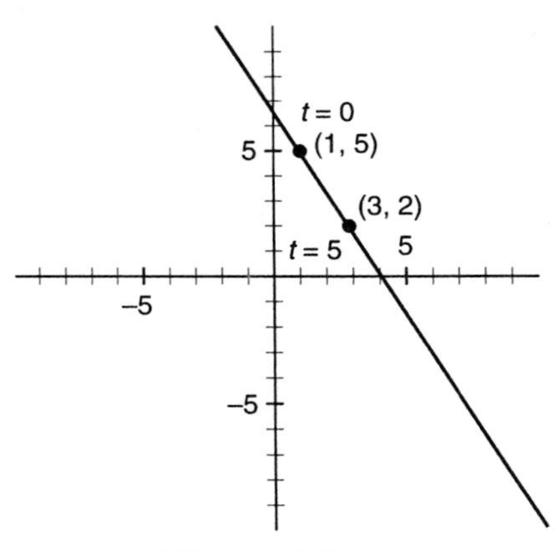

Figure 13:
$x = 1 + 2t$ and $y = 5 - 3t$.
[-10, 10] by [-10, 10].

By similar triangles as in Figure 12, we see that if a and b are direction numbers of a line, so are ka and kb for any $k \neq 0$. Here the direction numbers of the same line are 1/5 the direction numbers used in Example 6.

Formula 8.3.3 describes uniform motion along a line. If we want to describe non-uniform motion, simply replace "t" by some function of t (problem B3).

Inverses. Inverses of relations given parametrically are particularly easy to state -- simply interchange the expressions for x and y.

Example 7: The equations $x = t^3 - 5t$ and $y = 6 \cos(t - 4)$ yield the relation graphed in Figure 14 (solid line). Therefore the inverse relation is given by

$$x = 6 \cos(t - 4) \text{ and } y = t^3 - 5t$$

(Figure 14, dashed line). Of course, as we saw in Section 2.3 on inverses, the points of the inverse relation are the mirror image through the diagonal line $y = x$ of the points of the relation.

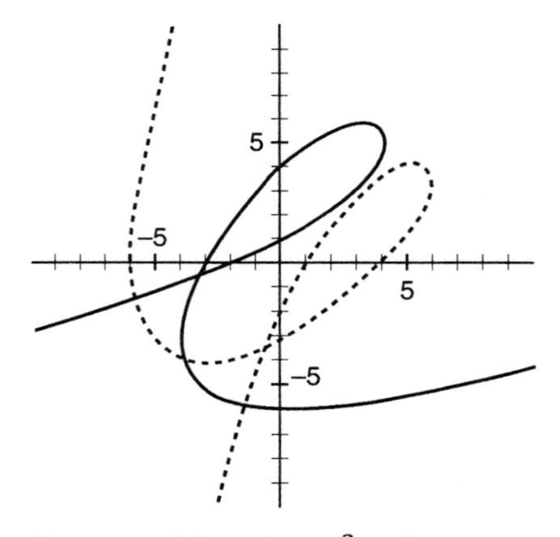

Figure 14: $x = t^3 - 5t$ and $y = 6 \cos(t - 4)$ [solid], and its inverse [dashed] [-10, 10] by [-10, 10].

Functional form can always be converted to parametric form.

(8.3.4) The points given in functional form by
$y = f(x)$ for x in some domain
are the same points given in parametric form by
$x = t$ and $y = f(t)$ for t in the same domain.

Using this idea, we can express and graph the inverse of any function given in functional form, even if its inverse cannot be expressed functionally.

Example 8: Let $y = x^3 - 5x + 3$ (Figure 15, solid line). Because some values of y (for example, $y = 2$) correspond to more than one value of x, the inverse is not a function and cannot be expressed functionally. Nevertheless, we can plot the inverse on a calculator by thinking of the function parametrically as $x = t$ and $y = t^3 - 5t + 3$. Then its inverse can be plotted as $x = t^3 - 5t + 3$ and $y = t$ (Figure 15, dashed line).

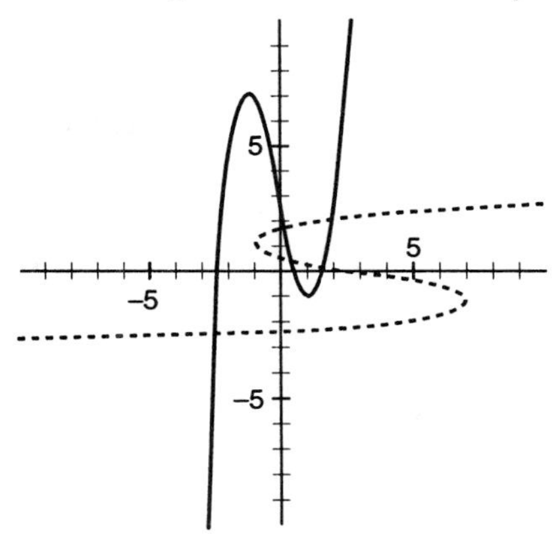

Figure 15: $y = x^3 - 5x + 3$ and its inverse [dashed]. [-10, 10] by [-10, 10].

<u>**Conclusion**</u>: Parametric equations have some advantages over equations in functional form. They can express non-functional relationships and they can relate positions to time. Also, some relationships that could be expressed functionally are easier to express parametrically.

Terms: parametric equations, direction numbers.

****

Exercises for Section 8.3, "Parametric Equations":

A1. Do Calculator Exercise 1. That is, use your calculator to plot the parametric equations in Example 1 (Figure 4), "Example 1, continued further" (Figure 5), and "More about Example 1" (Figure 6). The final graphs are the same. Comment on the differences in how the plots *develop* as your calculator plots them. [If your calculator plots the graphs too rapidly to see the difference in time of development, see problem B26 for a way to slow it down.]

A2. a) In Figure 7, what is the minimum possible y value of point P? b) At what value of θ does it occur? c) What is the maximum y value? d) At what value of θ does it occur?

A3. In Example 4, when and where does the projectile hit the ground? Describe the path in functional form.

A4. In Example 5, where does the projectile hit the ground?

^ ^ ^ ^ Identify the *type* (no details) of shape described by the parametric equations:

A5. $x = 2 \cos t$ and $y = 2 \sin t$.

A6. $x = 20 \cos t$ and $y = 20 \sin t$.

A7. $x = 6 \cos(3t)$ and $y = 6 \sin(3t)$.

A8. $x = 7 \cos(e^t)$ and $y = 7 \sin(e^t)$.

A9. $x = 5 - 2t$ and $y = 4 + 9t$.

A10. $x = 4 + t$ and $y = 12 + 5t$.

A11. $x = t^2$ and $y = 3t^2$.

A12. $x = e^t$ and $y = 5 - e^t$.

A13. $x = t$ and $y = t^2$.

A14. $x = 3t$ and $y = t^2 - 4$.

A15. $x = t^2$ and $y = 4t$.

A16. $x = t^2 + t$ and $y = t - 6$.

A17. $x = 2 \cos t$ and $y = 3 \sin t$. A18. $x = 7 \cos t$ and $y = 4 \sin t$.
A19. $x = \cos t$ and $y = 3 \cos t$. [Be careful.]
A20. $x = 2 \sin t$ and $y = 5 \sin t$. [Be careful.]
A21. $x = 5 \sin t$ and $y = 5 \cos t$. A22. $x = 4 \sin t$ and $y = 6 \cos t$.

A23. Watch the two graphs develop on a graphics calculator to answer this question: What is the difference between the development of the graph of "$x = 5 \sin t$ and $y = 5 \cos t$" and the graph of "$x = 5 \cos t$ and $y = 5 \sin t$" (from Example 1)?

A24. Watch the two graphs develop on a graphics calculator to answer this question: What is the difference between the development of the graph of "$x = \sin t$ and $y = \sin t$" and the graph of "$x = \cos t$ and $y = \cos t$."

A25. Watch the two graphs develop on a graphics calculator to answer this question: What is the difference between the development of the graph "$x = 1 + 2t$ and $y = 5 - 3t$" for all t (from Example 6) and the graph of "$x = 3 - 2t$ and $y = 2 + 3t$" for all t [where the other point is treated as (x_1, y_1) and the negatives of the direction numbers are used]?

A26. Watch the two graphs develop on a graphics calculator to answer this question: What is the difference between the graph of "$x = 5 \cos t$ and $y = 2 \sin t$" (from Example 3) and the graph of "$x = 5 \cos(t - \pi)$ and $y = 2 \sin(t - \pi)$".

^ ^ ^ ^ Find parametric equations of
A27. A line through (-1, 6) and (2, 4).
A28. A line through (5, 10) and (1, 2).
A29 A circle with radius 6 centered at the origin.
A30. A circle with radius 2 centered at the origin.
A31. An ellipse centered at the origin with $a = 2$ and $b = 8$.
A32. An ellipse centered at the origin with $a = 10$ and $b = 3$.

^ ^ ^ ^ Give the slope of the line.
A33. $x = 5 + 3t$ and $y = -7 - 2t$. A34. $x = -2 + 6t$ and $y = 4t$.
A35. $x = 12 - 3t^2$ and $y = 20 + t^2$. A36. $x = -1 + 2e^t$ and $y = 9 + e^t$.

^ ^ ^ ^ Convert the parametric equations to functional form.
A37. $x = 4 + 3t$ and $y = 2 - t$. A38. $x = 5 - 6t$ and $y = 2 + 3t$.
A39. $x = 7 \sin t$ and $y = 7 \cos t$. A40. $x = 5 \cos t$ and $y = 3 \sin t$.

^ ^ ^ ^ ^ ^ ^ ^

B1.* Give two advantages of parametric equations over equations for y in terms of x ("functional" form).

B2.* a) Relate the "m" of "$y = mx + b$" to the "a" and "b" which are direction numbers in 8.3.3 of the parametric equations of that line. b) Does a given line have a unique a and b? Explain.

B3. Formula 8.3.2 is more general than Formula 8.3.1, because the motion in 8.3.2 need not be uniform. Generalize Formula 8.3.3 the same way, that is, generalize the formula for parametric equations of a line to the case of non-uniform motion.

B4. Prove that equations 8.3.2 yield a circle.

B5. Equations 8.3.2 are introduced with the phrase "For any function f such that $[0, 2\pi)$ is a subset of its range." a) Why is the range of f relevant? b) Give a (very) simple example of an f and its domain such that the equations 8.3.2 do not yield a complete circle.

B6. Give parametric equations of a circle centered at (h, k) with radius r.

B7. Inspect Example 3 and then give parametric equations for *general* ellipses centered at the origin with horizontal and vertical semi-axes of lengths a and b (corresponding to Formula 3.2.10). [Continued in B8.]

B8. [After B7]. Give parametric equations for general ellipses centered at (h, k) with horizontal and vertical semi-axes of lengths a and b (corresponding to Formula 3.2.11).

^ ^ ^ ^ Find parametric equations of a circle centered at the origin with radius 5 such that
B9. $t = 0$ corresponds to the leftmost point on the circle.
B10. $t = 0$ corresponds to the topmost point on the circle and 1 revolution occurs when $t = 1$.

^ ^ ^ ^ Find parametric equations of uniform motion along a line such that
B11. $t = 0$ corresponds to the origin and $t = 10$ corresponds to $(1, 5)$.
B12. $t = 0$ corresponds to $(0, 3)$ and $t = 2$ corresponds to $(2, 1)$.

^ ^ ^ ^ Convert the parametric equations to functional form and identify the type of shape of the graph.
B13. $x = 3 \sin t$ and $y = 7 \sin t$ [consider the domain.]
B14. $x = 5 \sin(t^2)$ and $y = 10 \sin(t^2)$ [consider the domain.]
B15. $x = 50t$ and $y = -9.8t^2 + 40t$.
B16. $x = t - 5$ and $y = t^2 + 4t$.
B17. $x = e^t$ and $y = 2e^t$ [consider the domain.]
B18. $x = t^2$ and $y = 5t^2$ [consider the domain.]

B19. Suppose a projectile moves with vertical component given by $y = -16t^2 + 200t + 50$ and horizontal coordinate given by $x = 2500t - 300t^{1.5}$. If ground level is $y = 0$, where does it hit the ground?

^ ^ ^ ^ Find the equations for graphing on a calculator the inverse relation of
B20. $f(x) = x^3 - 4x - 2$.
B21. $f(x) = \cos x$, all x.
B22. $f(x) = 5 \sin x$, all x.
B23. $f(x) = x^4 - 5x^2$.

B24. This section discussed parametric equations of lines in the two-dimensional plane. The idea of lines in three-dimensional space is precisely similar. We need three direction numbers (instead of two) and a point that the line goes through. The equations are

$$x = x_1 + at, y = y_1 + bt, \text{ and } z = z_1 + ct.$$

Find parametric equations of a line through $(1, 2, 3)$ and $(4, 6, 8)$.

B25. Give parametric equations for any graph that somewhat resembles Figure 1.

B26. Learn how to adjust the speed at which graphs develop on your graphics calculator. On some models, in "parametric" mode, the window includes the domain of t and a "Tstep" entry which can be used to adjust how much t is advanced between evaluations. If the advance is smaller, more t-values are used and the graph develops more slowly. Here is the problem: Graph $x = 8\cos(t^2)$ and $y = 8\sin(t^2)$ for $0 \le t \le 2\pi$. a) What is the shape graphed? b) How many times around is the shape traced? c) On many calculators, the resulting shape is a bit (or a lot) fuzzy. Why? d) What does this have to do with the amount t is advanced between calculations?

B27. Find parametric equations of the path of a moon revolving around a planet which is revolving about a sun. To make the graph fit on your calculator screen, assume the distance from the planet to the sun is 8 times the distance from the planet to the moon, and assume the moon goes around the planet 12 times during one revolution of the planet around the sun.

CHAPTER 9

Conic Sections

Section 9.1. Conic Sections: Parabolas

Curves called conic sections have been extensively studied for over two thousand years. They include parabolas, circles, ellipses, and hyperbolas (Figures 1A,B, and C).

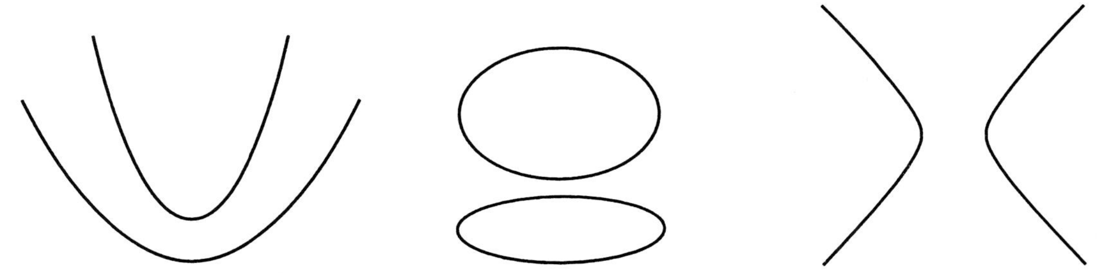

Figure 1A: Two parabolas. **Figure 1B:** Two ellipses **Figure 1C:** One hyperbola.

By definition, a conic section is a curve formed as the intersection of a cone and a plane (Figures 2A,B,C).

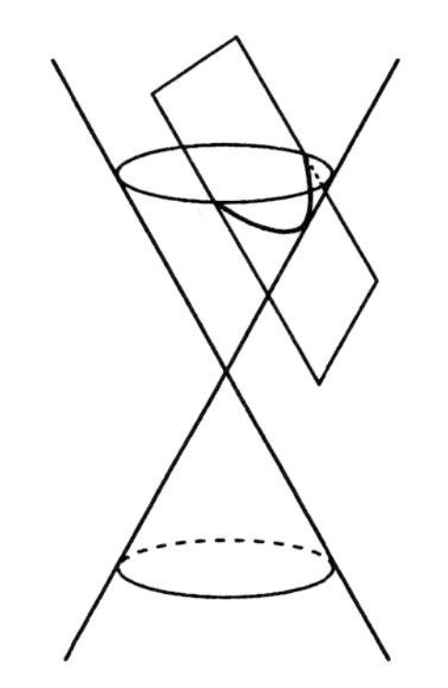

Figure 2A: A parabola as a cross section of a cone.

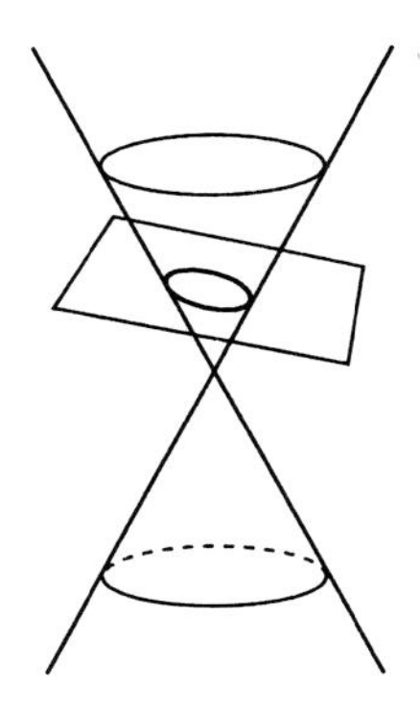

Figure 2B: An ellipse as a cross section of a cone.

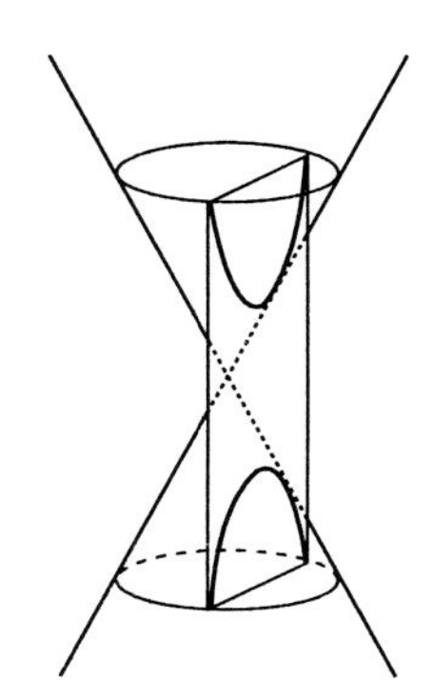

Figure 2C: A hyperbola as a section of a cone.

Different types of curves are formed depending upon the angle at which the cone is cut. Parabolas result when the plane is parallel to the edge of the cone (Figure 2A). Ellipses result when the plane is tilted to cut only one part of the cone (Figure 2B). Hyperbolas result when the plane is tilted to cut both parts of the cone (Figure 2C).

Conic sections are not merely mathematical curiosities; they are important shapes of nature. The Polish astronomer Kepler (1571-1630) discovered that planets move in ellipses about the sun. Man-made satellites and some comets also move in elliptical orbits, and some comets move in orbits that form one branch of a hyperbola. Disregarding friction, projectiles move in parabolic paths. To explain why objects move in these orbits takes calculus and Newton's law of gravity.

Other applications concern vision and light. In the real world circles are extremely common. Viewed from an angle, a circle appears elliptical -- photographs and paintings illustrate circles as ellipses. The tip of a shadow of a fixed object (for example, the tip of a shadow of a pole) describes one branch of a hyperbola as the day passes (problem B16).

Parabolic, hyperbolic, and elliptical shapes all have remarkable reflective properties. Satellite dishes, microwave relay towers, telescopes, solar energy collectors and parabolic microphones utilize parabolic shapes.

By 200 BC the Greeks had, as a purely intellectual exercise, discovered numerous fascinating properties of conic sections, including the reflective properties. Ancient results about conic sections are strong support for the argument that pure research may lead to important discoveries, even if no application is in mind when the research is undertaken. Centuries passed before the important practical applications of conic sections were devised.

After straight lines, the next most important curves are these conic sections. There are alternative ways to define these curves. We will not use the ancient "conic section" approach. Instead, we will define these curves using two-dimensional geometric definitions well-suited to algebraic geometry.

Parabolas. We begin with a description of parabolas in geometric terms and then translate that definition into algebraic notation.

Definition 9.1.1: Let L be a line in the plane and F be a point in the plane not on that line. The set of points that are equidistant from L and F is called a parabola (Figure 3).

The line L is called the directrix ("Dih REK tricks") of the parabola and the point F is called the focus ("FOH kus") of the parabola.

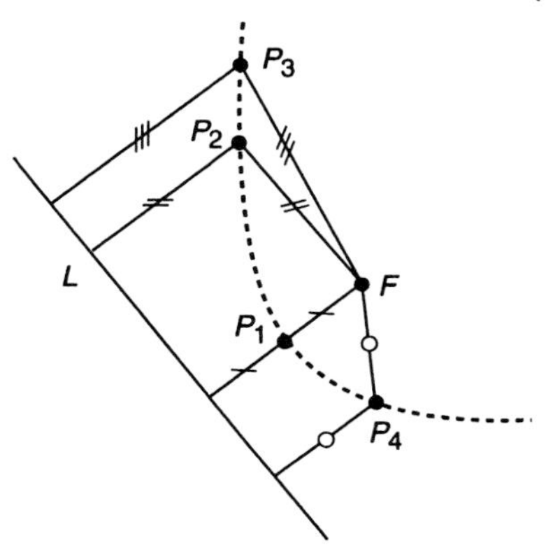

Figure 3: A line L, a point F, and a few points equidistant from both.

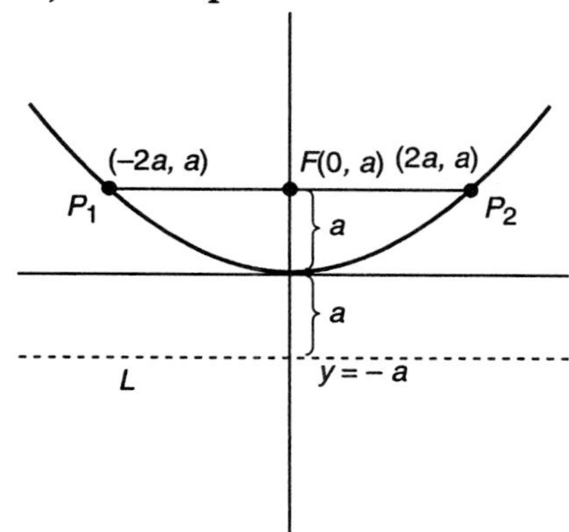

Figure 4: The axis-system pictured vertically.

If we want to express that geometric definition algebraically, we may. For that we need an axis system. It is most convenient to locate the axis system with the y-axis through the focus and perpendicular to the directrix, and the origin halfway (equidistant) between the focus and directrix. Figure 4 pictures the axis-system vertically as usual.

There are only two other points on the parabola other than the origin that are easy to find by inspection. The points $(-2a, a)$ [P_1 in Figure 4] and $(2a, a)$ [P_2] are also on the parabola because they are the same distance from the directrix and the focus. They are $2a$ units above the line (measured perpendicular to the line) and $2a$ units directly to the side of the point. The line segment through the focus connecting P_1 to P_2 is called the *latus rectum* (that is a Latin term) and serves as a description of how wide the parabola is. The latus rectum is $4a$ units long -- exactly twice the distance from focus to directrix.

The other points on the parabola are not so simple to discover. However, using algebra, we may treat the geometric description of a parabola as a word problem, build our own formulas, and then set up the equation.

Algebraic Description of a Parabola. To describe the geometric definition algebraically we need to express the two distances mentioned. Then we will set the distances equal to each other.

Let a point be denoted by (x, y). How far is it from the focus $(0, a)$?

By the distance formula 3.1.11, the distance from (x, y) to $(0, a)$ is (Figure 5)

$$\sqrt{(x - 0)^2 + (y - a)^2} = \sqrt{x^2 + y^2 - 2ay + a^2}.$$

Points (x, y) on the parabola will be in the top half of the plane (Figure 5), so the distance from (x, y) to the directrix $y = -a$ is simply $y + a$.

Now, set these two distances equal and simplify.

$$y + a = \sqrt{x^2 + y^2 - 2ay + a^2}.$$

Squaring both sides,

$$(y + a)^2 = x^2 + y^2 - 2ay + a^2.$$

$$y^2 + 2ay + a^2 = x^2 + y^2 - 2ay + a^2.$$

Simplifying,

(9.1.2A) $$4ay = x^2.$$

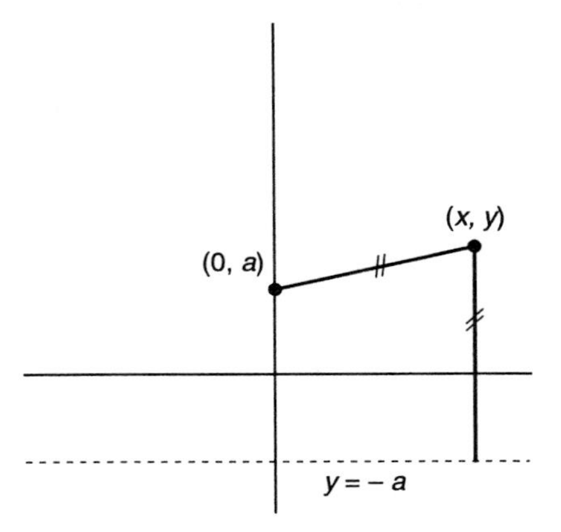

Figure 5: Focus $(0, a)$, directrix $y = -a$, and a point (x, y) on the parabola.

Dividing by $4a$, we can isolate y.

Vertical Parabola Centered at the Origin. An equation of a parabola with focus $(0, a)$ and directrix $y = -a$ is

(9.1.2B) $$y = \frac{x^2}{4a}.$$

The parabola goes through the origin and the focus and directrix are $|2a|$ units apart.

If a is positive, the parabola "opens upward," as parabolas usually do in pictures, for example, Figures 1A and 4. If a is negative, the focus is below the directrix and the parabola "opens downward" (Figure 9).

Parabola

$y = x^2/(4a)$

focus: $(0, a)$
directrix: $y = -a$
vertex: $(0, 0)$

distance from focus to directrix: $2a$

Example 1: The most well-known algebraic description of a parabola is "$y = x^2$." Locate its focus and directrix.

In 9.1.2B, "a" is a parameter which locates the focus and directrix. We are given that the parabola is "$y = x^2$." In 9.1.2B this is "$y = x^2/(4a)$." So,

$$y = x^2 = x^2/(4a).$$

Solving for a, $4a = 1$ and $a = 1/4$. Therefore, the focus is $(0, 1/4)$ and the directrix is $y = -1/4$.

Example 2: Find an equation of the parabola with focus (0, 1) and directrix $y = -1$.

Half way between the line and the focus is the vertex. Here, that point is the origin, so equation 9.1.2 fits. Since the distance between the focus and the origin is $2a$, which is here 2 units, $a = 1$. Plugging into Equation 9.1.2B, the equation is

$$y = x^2/4.$$

<u>**Location Shifts**</u>. If a parabola is defined geometrically on a "blank page" *before* the axis system is laid down, we can choose our axis system to fit the parabola as we did in Figure 4. Then the algebra is relatively easy. The resulting formula is 9.1.2B, which has only one parameter, a.

If the directrix and focus are given on a pre-existing axis system, the algebra is more complicated. Nevertheless, any parabola can be expressed algebraically in terms of location shifts, scale changes, or rotations of this basic parabola. Location shifts and scale changes are dealt with here exactly as in Section 2.3 on composition of functions and the part of Section 3.2 on circles and ellipses.

First consider parabolas where the axis of symmetry is still vertical, but the location is different. It is easy to algebraically shift the vertex to a new location (Figure 6). For example, to locate the vertex at (h, k) ["h" and "k" are the traditional letters used to locate the vertex], we can substitute "$x - h$" for "x" and "$y - k$" for "y" in 9.1.2. The result is immediate.

$$y - k = \frac{(x - h)^2}{4a} .$$

<u>Vertical Parabolas</u>. The equation of the standard geometric form of a parabola with vertical axis and vertex at (h, k) is

$$(9.1.3) \qquad y = \frac{(x - h)^2}{4a} + k .$$

The directrix is $y = k - a$ and the focus is $(h, k + a)$ (Figure 6).

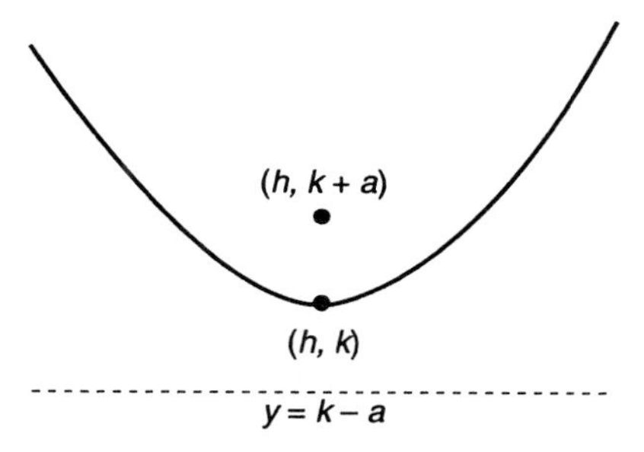

Figure 6: A vertical parabola that "opens upward" with vertex (h, k).

Example 3: Find the algebraic representation of a parabola with focus at (5, 7) and directrix $y = 1$.

Draw a picture (Figure 7).

Since the directrix is horizontal, the axis of symmetry will be vertical and the standard geometric form in 9.1.3 fits. The distance from the directrix $y = 1$ to the focus (5, 7) is 6. This distance is $2a$, so $a = 3$. The vertex is half way in between the focus and directrix, so the vertex is (5, 4). Therefore, the equation is

$$y = (x - 5)^2/12 + 4$$

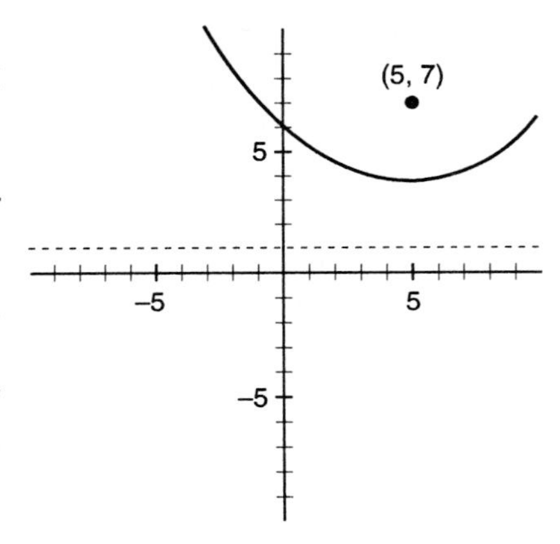

Figure 7: A parabola with focus (5, 7) and directrix $y = 1$. [-10, 10] by [-10, 10].

Example 4: Find the focus and directrix of the parabola with equation "$y = x^2/8 + x$."

Graph it (Figure 8). You should be able to approximate the answers just by looking at the graph.

This equation can be manipulated into the form in 9.1.3, which exhibits the three parameters h, k, and a. Complete the square (as in Section 3.2) to obtain

$$\begin{aligned} y = x^2/8 + x &= (1/8)(x^2 + 8x) \\ &= (1/8)(x^2 + 8x + 16 - 16) \\ &= (1/8)(x + 4)^2 - 2 \end{aligned}$$

The algebraic form which exhibits all three parameters would be

$$y = \frac{(x - -4)^2}{4(2)} + -2 .$$

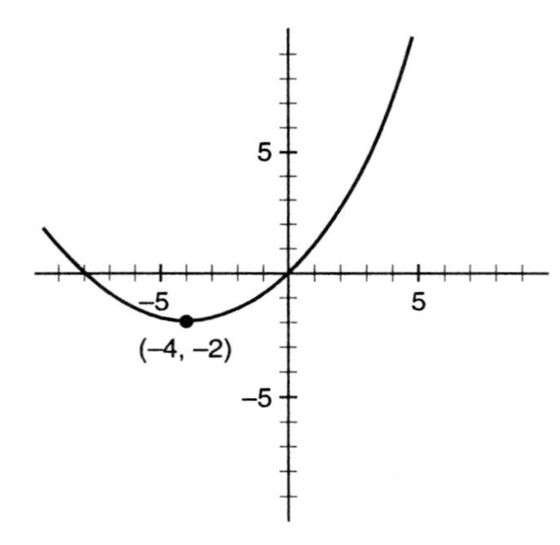

Figure 8: $y = x^2/8 + x$. [-10, 10] by [-10, 10].

So $h = -4$, $k = -2$, and $a = 2$. The focus is 2 units above the vertex and the directrix is horizontal two units below the vertex. Vertex: (-4, -2). Focus: (-4, 0). Directrix: $y = -2 - 2 = -4$.

A vertical parabola is upside down ("opens downward") if and only if a is negative.

Example 5: A parabola has vertex (4, 5) and focus (4, 0) (Figure 9). Find its equation.

The axis of symmetry is vertical, so 9.1.3 fits. The directed distance from the vertex to the focus is -5 (negative, because the focus is *below* the vertex) so $a = -5$. The center (4, 5) exhibits h and k. The equation is

$$y = -(x - 4)^2/20 + 5.$$

<u>Horizontal Parabolas</u>. The difference between a horizontal graph and a vertical graph is the difference between "x" and "y". Switch the letters and the role of the axes switches, so what was vertical becomes horizontal, and vice versa.

An equation of the standard geometric form of a parabola with horizontal axis and vertex at (h, k) is (switching "x" and "y" in 9.1.3):

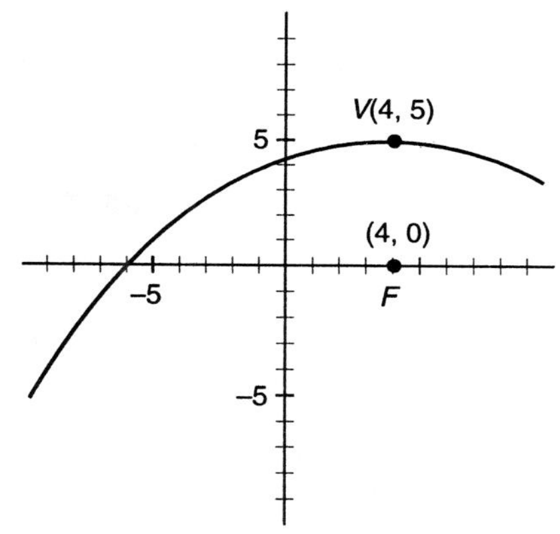

Figure 9: A parabola with vertex (4, 5) and focus (4, 0). [-10, 10] by [-10, 10].

(9.1.4) $$x = \frac{(y - k)^2}{4a} + h \; .$$

The directrix is $x = h - a$ and the focus is $(h + a, k)$ (Figure 10).

Solving 9.1.4 for y,

(9.1.5) $$y = \pm\sqrt{4a(x - h)} + k \; .$$

This is the functional form (with two functions) suitable for graphing horizontal parabolas with a graphics calculator.

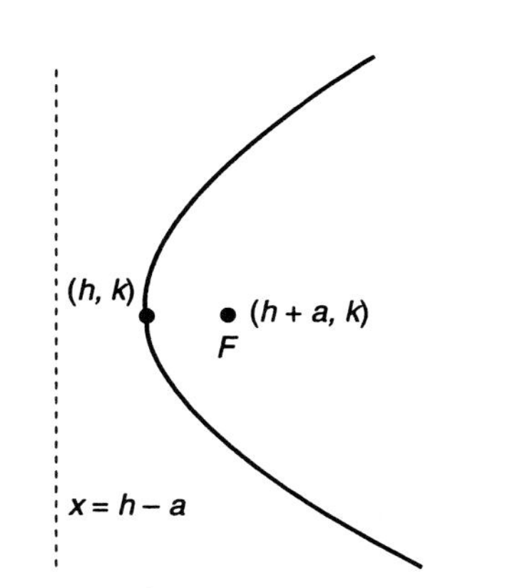

Figure 10: A horizontal parabola. Vertex: (h, k). Focus $(h + a, k)$ Directrix: $x = h - a$

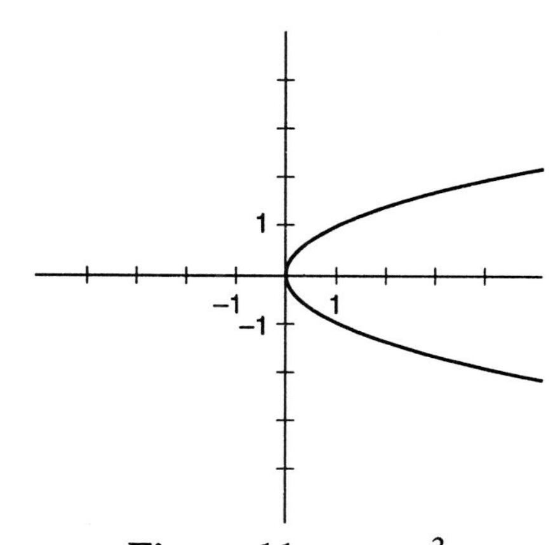

Figure 11: $x = y^2$. [-5, 5] by [-5, 5].

Example 6: The most well-known horizontal parabola is given by $x = y^2$ (Figure 11). We have seen the top half graphed many times as $y = \sqrt{x}$. Locate its focus and directrix.

This is just Example 1 with the letters switched. In Example 1 the focus of "$y = x^2$" was (0, 1/4). Here the roles are reversed; the focus is (1/4, 0). In

Example 1 the directrix of "$y = x^2$" was "$y = -1/4$." Here the roles are reversed. The directrix of "$x = y^2$" is "$x = -1/4$."

If you want to use 9.1.4 or 9.1.5 to do Example 6, "$x = y^2$," from scratch, you may. We are given $x = y^2$, which, by 9.1.4, is also $x = (y - k)^2/(4a) + h$. Comparing the two expressions, we see $k = 0$, $a = 1/4$, and $h = 0$.

Reflections. Parabolas have an important reflective property. Narrow-beam lights (such as some flash lights and search light beams) have a mirrored reflective surface behind the light source (bulb filament) so that light reflects straight forward (Figure 12). The shape that does this is a parabola. If the inner surface of a parabola is mirrored, light from the focus is reflected parallel to the axis of symmetry, regardless of where it hits the mirror.

Solar energy collectors, satellite dishes, and parabolic microphones also use this property, but in reverse. Parallel light rays, electromagnetic waves, or sound waves coming in toward the parabolic dish are reflected toward a single point, the focus (Figure 12).

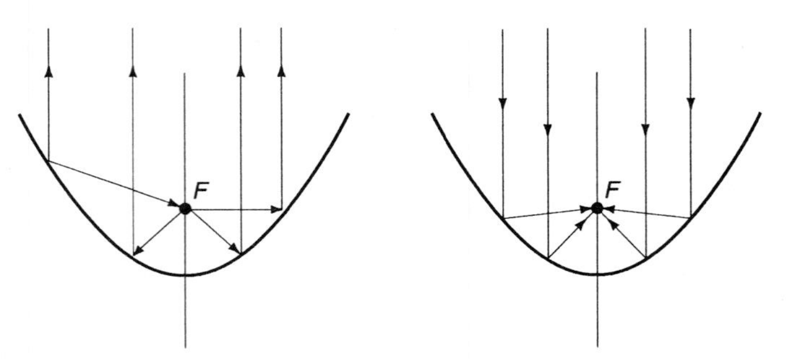

Figure 12: Reflections from a parabola.

This property can be discussed algebraically using a principle expressed by Fermat in 1650:

(9.1.6) "Light traveling from one point to another will follow a path such that, compared to other nearby paths, the time required will be a minimum or a maximum, or will remain unchanged."

Most examples require the time to be a minimum (Problem B14). In physics, this principle can be used to explain why light bends ("refracts") as it passes through the boundary between air and water, because light travels more slowly through water (Problem B13). Here we will use it to illustrate that light from the focus of a parabola reflects from the inner side of the parabola parallel to the axis of symmetry.

Example 7: This example would work with any parabola, but, for convenience, consider the famous parabola $y = x^2$ from Example 1. Imagine the inner surface to be reflective and the focus to be a light source (like the filament of a light). From Example 1, the focus is $F = (0, 1/4)$. Let A be a point above the parabola (Figure 13). Suppose a light ray from the focus reflects off the parabola and passes through point A. Where does that ray reflect off of the parabola?

By the Fermat Principle, we need to find the point P on the parabola such that the total travel time from F to P to A is minimum. This is the same as finding the point P on the parabola such that the total distance from F to P to A is minimum, since light travels at a constant speed (Figure 13).

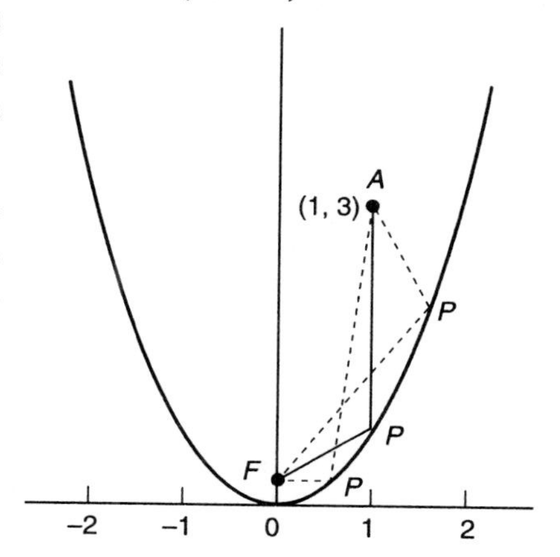

Figure 13: Light from the focus reflects off the parabola at P and goes through point A. Where is point P? [-2.5, 2.5] by [0, 5].

For this we must build a formula for the total distance, depending upon P. Then we can minimize the distance, perhaps graphically,

To simplify matters let's use a particular point A, say (1, 3). We need the distances from F to P and from P to A. Suppose we denote P by (x, y). Using the distance formula 3.1.11,

$$d(F, P) = \sqrt{(x - 0)^2 + (y - 1/4)^2} .$$

$$d(P, A) = \sqrt{(x - 1)^2 + (y - 3)^2} .$$

These are valid for any point (x, y), but the problem has a constraint (recall Section 3.5 on constraints). For a given x, we cannot use just any y; y must be on the parabola. For any x, y must be x^2. Substituting x^2 for y and adding,

$$\textit{total distance} = \sqrt{(x - 0)^2 + (x^2 - 1/4)^2} + \sqrt{(x - 1)^2 + (x^2 - 3)^2} .$$

Graph this distance (Figure 14).

When the total distance is a minimum, $x = 1$, which is the "1" from the point $A = (1, 3)$. That is, all reflected light rays are reflected parallel to the axis of symmetry.

In general (and we can prove this using calculus), regardless of the location of A, the point P on the curve that minimizes the total reflected distance from the focus always has the same x-value as A itself.

Conclusion. Conic sections are important. In addition to being cross sections of a cone cut by a plane, they can be described geometrically in terms of distance. Also, they have relatively simple algebraic equations with parameters that exhibit the center and scale if the axis system is laid out parallel to the axis of symmetry.

Terms: conic section, parabola, focus, directrix.

Figure 14: The distance traveled from F to P to A, in terms of the x-value of P. [-2, 2] by [0, 5].

****

Exercises for Section 9.1, "Conic Sections: Parabolas":

^^^^Rewrite the equation in "standard geometric form." Identify the vertex, axis of symmetry, focus, and directrix of each parabola. Then roughly sketch it.

A1. $y = x^2$. [F: (0, 1/4)]

A2. $y = 2x^2$. [dir: y = -1/4]

A3. $y = (x - 1)^2 + 2$. [F: (1, 9/4)]

A4. $y = (x + 3)^2 + 1$. [V: (-3, 1)]

A5. $y = x^2 + 4x$. [dir: y = -17/4]

A6. $y = x^2 - 6x$. [axis: x = 3]

A7. $y = 2 - x^2$. [V: (0, 2)]

A8. $y = -(x/3)^2$. [F: (0, -9/4)]

A9. $y = 2x^2 - 8x - 5$. [axis: x = 2]

A10. $y = 4x^2 + 12x + 1$. [V: (-3/2, -8)]

A11. $x = y^2$. [dir: x = -1/4]

A12. $x = y^2/8$. [F: (0, 2)]

A13. $x = (y - 2)^2 + 3$. [V: (3, 2)]

A14. $x = (y - 7)^2 + 4$. [F: (17/4, 7)]

A15. $x^2 + 2x - y = 5$. [dir: y = -25/4]

A16. $2x + y^2 = 6y$. [V: (9/2, 3)]

^^^^Find the equation of the parabola with the given properties.

A17. Vertex (0, 0), focus (0, 1).

A18. Vertex (0, 0), focus (0, -2).

A19. Vertex (0, 0), focus (1, 0).

A20. Vertex (0, 0), focus (-3, 0).

A21. Vertex (3, 4), focus (3, 10).

A22. Vertex (0, 2), focus (0, -4).

A23. Vertex (0, 0), directrix y = -2.

A24. Vertex (0, 0), directrix x = -1.

A25. Focus (1, 2), directrix y = 0.

A26. Focus (1, 2), directrix x = 0.

A27. Focus (2, 0), directrix y = -10.

A28. Focus (0, 1), directrix x = -2.

A29. Vertex (0, 0), axis y = 0, through (4, 1).

A30. Vertex (0, 0), axis x = 0, through (2, 10).

^^^^^^^^

B1.* a) Define *parabola*. b) Illustrate the definition with a sketch.

B2.* Sketch $y = (x - h)^2/(4a) + k$, labeling the vertex, axis of symmetry, focus, and directrix.

B3. A solar reflector is parabolic, 10 feet wide and 2 feet deep. Where is its focus?

B4. A parabolic microphone is 2 feet wide and 4 inches deep. Where is its focus?

B5. A parabolic searchlight is 1 foot wide and 6 inches deep. Where is its focus?

B6. A parabolic solar reflector is 6 inches deep and has the focus 2 feet in front of the vertex. How wide is it?

B7. A parabolic solar reflector is 3 inches deep and has the focus 18 inches in front of the vertex. How wide is it?

B8. Suppose sound travels about 1100 feet per second. Location A is 500 feet from location B. Suppose a burst of sound is recorded at location A 1/10 second before it is recorded at location B. Draw a sketch, locate an axis system on the sketch, and find the algebraic equation satisfied by the possible locations of the origin of the burst of sound. (Do not bother to simplify it.)

B9. Suppose sound travels about 1100 feet per second. Location A is 2200 feet from location B. Suppose a burst of sound is generated at location A, reflects off of point P, and is recorded at location B 3 seconds later. Draw a sketch, locate an axis system on the sketch, and find the algebraic equation satisfied by the possible locations of point P. (Do not bother to simplify it.)

B10. Let $y = x^2/4$ and P be a point on that curve. Let F be the point (0, 1), and B be the point (2, 5). a) Find the expression for the sum of the distances from B to P and from P to F. b) Use a graph to find the x-value of the point on the curve that minimizes that expression.

B11. Let $y = x^2/100$ and P be a point on that curve. Let F be the point (0, 25), and B be the point (5, 20). a) Find up the expression for the sum of the distances from B to P and from P to F. b) Use a graph to find the x-value of the point on the curve that minimizes that expression.

B12. Draw a picture to illustrate why the distance from (x, y) to the line $y = -a$ is $|y + a|$.

B13. A person with eye level 6 feet above the water sees an object that is actually 3 feet below the water and 10 feet away horizontally (see the figure). Light from the object reaching the person's eye will not travel in a straight line, because light travels more slowly in water. It takes 1.33 times as long for light to travel 1 foot under water as in air. a) Find the path such that light from the object to the person's eye travels the minimum time (according to Fermat's Principle, 9.1.6). Describe the path in terms of the horizontal distance from the eye to where the light leaves the water. b) The object appears to be not as deep as it really is. Why?

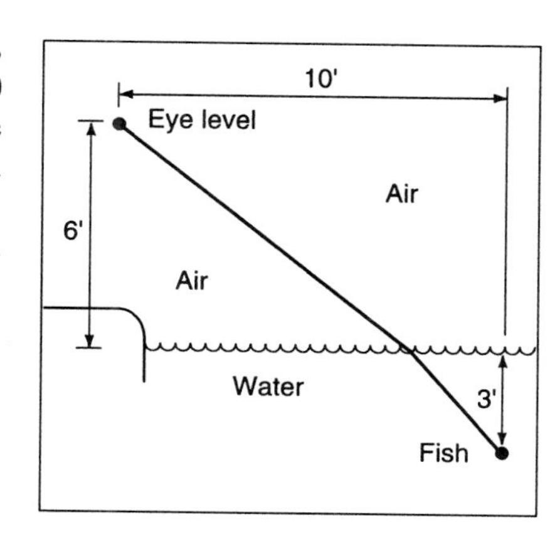

B14. Think of an example where a reflection follows the Fermat Principle and yields maximum time (instead of minimum time).

B15. Think of an example where a reflection follows the Fermat Principle and yields constant time (instead of minimum time).

B16. Use the idea of a "conic section" to explain why the shadow of the tip of a pointed post traces out one branch of a hyperbola on the ground as the day passes. [Hint: Where is the cone? What makes the section?]

Section 9.2. Ellipses and Hyperbolas

Ellipses and hyperbolas are conic sections that can be described with two-dimensional geometric definitions using distances from two fixed points. If the *sum* of the distances from any point on the curve to the two fixed points is a constant, the curve is an ellipse. If the *difference* is a constant, the curve is a hyperbola.

Definition 9.2.1. An ellipse is the set of all points in the plane such that the sum of the distances from two fixed points is a constant. Each of the fixed points is called a focus (pronounced "FOH kus"; the plural is "foci," pronounced "FOH sigh").

Figure 1 illustrates two fixed points, F_1 and F_2, and a curve of points P such that the sum of the lengths of the line segments from F_1 to P to F_2 is a constant. Using this definition, an ellipse can be drawn using a loop of string around two pins (serving as the foci). Use the point of a pencil to pull the string taut in any direction, forming a triangle with the point as the third vertex (Figure 2). That point will be on the ellipse, which will be drawn as the point moves through all positions at the extreme end of the triangular loop. I have drawn many ellipses this way and it works very well.

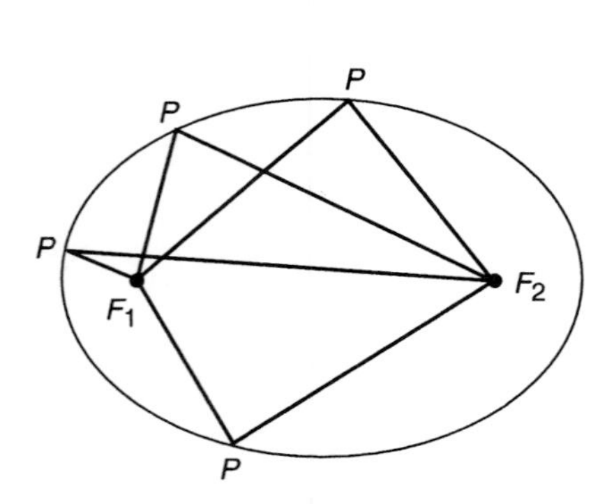

Figure 1: An ellipse. P varies, but the total distance from F_1 to P to F_2 is constant.

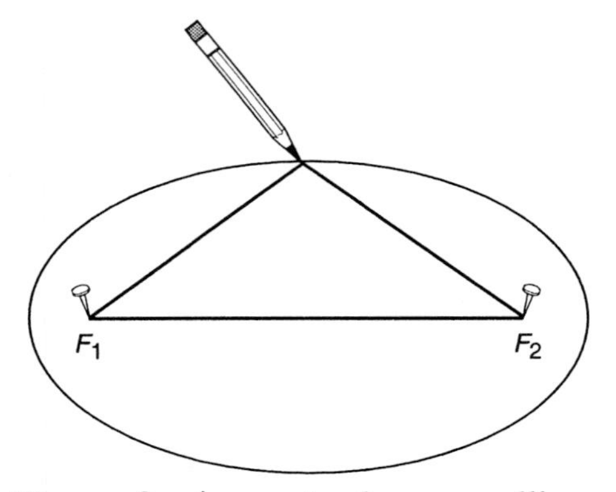

Figure 2: A way to draw an ellipse.

A circle is a special case of an ellipse when the two foci are the same point (the center). Then the distance from the center to any point on the circle and back to the center is a constant (twice the radius).

Applications. There are many applications of this definition based on distance. Some arise in the use of location-finding technology such as sonar and radar.

Example 1: Imagine we create a burst of sound at location F_1 and record when the sound arrives at location F_2 after it reflects off of a surface at an unknown location P. The sound travels from F_1 to P to F_2 (Figure 3). By keeping track of when the burst is generated at F_1 and when it is received at F_2 we can tell how long the sound travels. The total distance from F_1 to P to F_2 is determined by the time the sound travels (using "distance equals rate times time," because we know the speed of sound). Knowing only the *total* distance and the locations of F_1 and F_2, the position of P is not determined precisely, but it must be somewhere on a particular ellipse, according to the definition of *ellipse* in 9.2.1 (Figure 3).

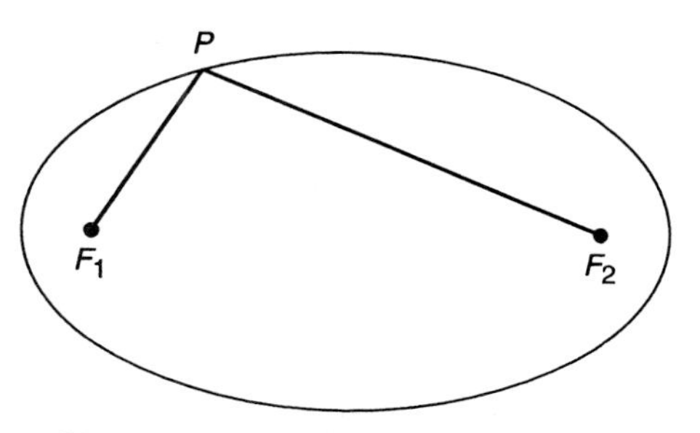

Figure 3: When the total distance from F_1 to P to F_2 is known, point P is on an ellipse with foci at F_1 and F_2.

Example 1, continued: So far, the only listening point is F_2. Now suppose we also listen at F_1 for the echo from P. The distance from F_1 to P and back to F_1 is determined by the travel time of the sound. So the distance from F_1 to P will be known. Then P must be on a circle with known radius centered at point F_1 (the circle in Figure 4). So, if we listen at both F_1 and F_2, then, P is on both the circle and the ellipse. Therefore the location of P almost determined; it is one of two points where the circle and ellipse intersect (Figure 4).

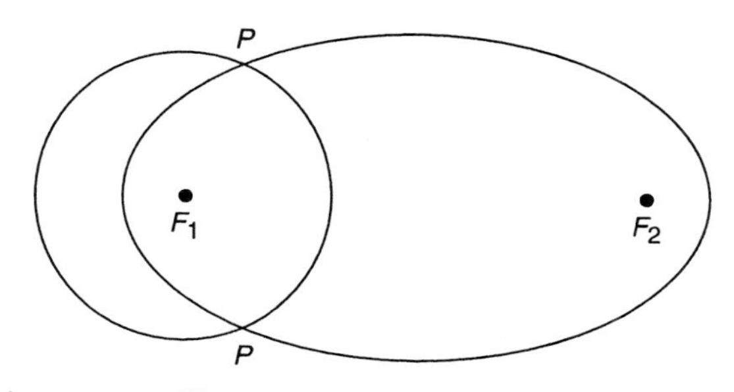

Figure 4: Point P is on an ellipse with foci at F_1 and F_2 and the total distance F_1 to P to F_2 is known. P is also on a circle with known radius centered at F_1. Therefore the location of P almost determined; it is one of two points.

When the burst is produced by the listener as in Example 1, the location-finding method is said to be "active." When the object at P itself generates the burst, there is a similar "passive" method of determining location. The object can be located on a hyperbola.

Definition 9.2.2. A hyperbola (high PER boh lah) is the set of all points in the plane such that the (absolute value of the) difference of the distances from two fixed points is a constant (Figure 5). Each of the fixed points is called a focus of the hyperbola.

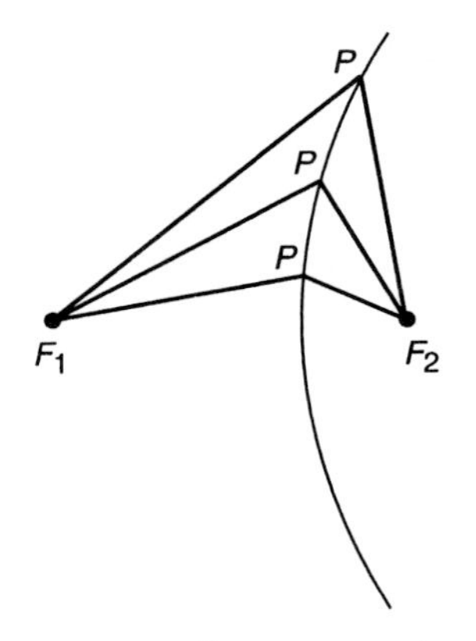

Figure 5: One branch of a hyperbola. P varies, but the difference between PF_1 and PF_2 is constant.

A hyperbola has two branches. Each consists of points closer to the focus inside that branch.

We continue our discussion of finding locations.

Example 2: Again, suppose two listening locations, F_1 and F_2, are fixed. Imagine a burst of sound originating at unknown location P at an unknown time (Figure 5). Can we use the arrival times of the sound at F_1 and F_2 to locate the point of origin?

If the sound arrives at both locations at the same instant, then the source P is equidistant from each. This is a special case resulting in a line of possible locations (Figure 6). So a line is a "degenerate" special case of a hyperbola. Usually we use the term "hyperbola" to apply to the other cases where the difference in distances is not zero.

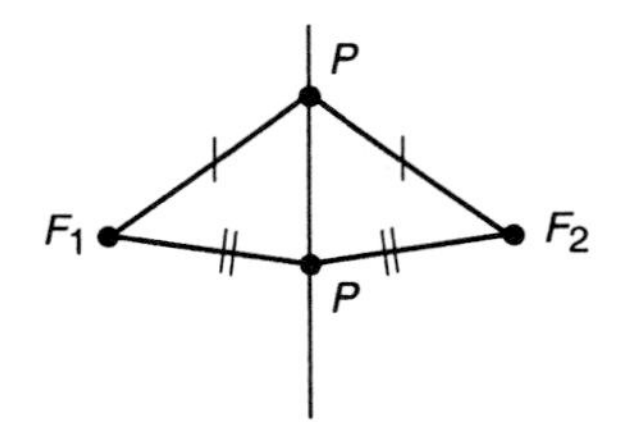

Figure 6: A line of points equidistant from F_1 and F_2.

Suppose the sound arrives at F_2 before it arrives at F_1, so the source is closer to F_2 (Figure 5). The difference in arrival times can be used to compute the difference between the distances from P to F_1 and from P to F_2. According to the definition of a hyperbola, that constant difference locates P somewhere on a particular hyperbola with F_1 and F_2 as foci, and on the branch closer to F_2.

This result is like knowing that an unknown point is on a particular ellipse -- it is not nearly enough to locate the point. In the previous example, when the sound was generated at the listening point, we needed a second listening point to almost determine the location of the reflecting object. Now, when the sound is generated at an unknown point source, we need a third listening point to almost determine the location of the source. Each pair (three points make three pairs of points) will locate the point on one branch of a hyperbola. The three hyperbolas are likely to intersect in only one point. Figure 7 illustrates branches of two hyperbolas on which the point P must lie.

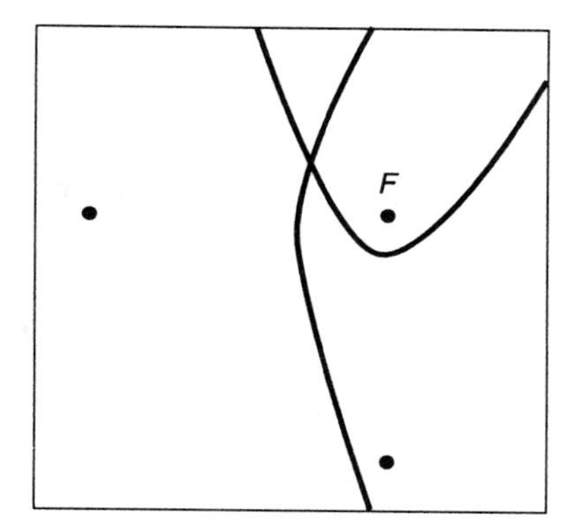

Figure 7: A point at the intersection of branches of two hyperbolas with focus F in common. The other 2 foci are marked.

This location-finding method has applications to locating submarines (using Sonar) and airplanes using radar waves (instead of sound waves). A method which is similar in spirit to this example is used by the "Global Positioning System" to precisely determine locations on earth. In that case transmissions from several satellites are timed.

Algebraic Descriptions of Ellipses and Hyperbolas. The geometric definitions of *ellipse* and *hyperbola* mention distances from two foci. Therefore, to describe them algebraically, we need to express both distances. For an ellipse, their *sum* is set equal to a constant. For a hyperbola, their *difference* is set equal to a constant.

It is convenient to center the ellipse or hyperbola at (0, 0) and have the x-axis down the middle by locating the foci at $(-c, 0)$ and $(c, 0)$, for some $c \geq 0$.

Suppose we call a point in the plane (x, y). How far is it from the focus $(c, 0)$ (Figure 8)? The formula is

$$\sqrt{(x - c)^2 + y^2}\ .$$

How far is it from $(-c, 0)$? The formula is

$$\sqrt{(x + c)^2 + y^2}\ .$$

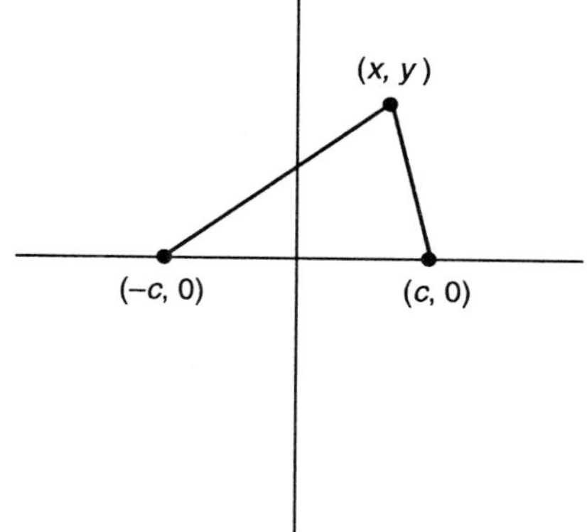

Figure 8: (x, y) and the two foci, $(c, 0)$ and $(-c, 0)$.

It is traditional to let the constant that is the sum or difference of these two distances be denoted by $2a$. For an ellipse, $2a$, the sum of the distances, must be greater than $2c$, which is the straight-line distance between the two foci. So, for an ellipse, $a > c$. For a branch of a hyperbola, the difference must be less than the distances between the two foci, so $2a$ must be less than $2c$ and $a < c$. The number $2a$ may be zero as a special case (a line, Figure 6). The equations are

(9.2.3, ellipse) $$\sqrt{(x - c)^2 + y^2} + \sqrt{(x + c)^2 + y^2} = 2a \ .$$

(9.2.4, hyperbola) $$\sqrt{(x - c)^2 + y^2} - \sqrt{(x + c)^2 + y^2} = \pm 2a \ .$$

This hyperbola equation with the minus sign determines the branch closer to $(c, 0)$ than to $(-c, 0)$. With the plus sign it determines the branch closer to $(-c, 0)$.

Neither of these equations is pleasing. Fortunately, they can be simplified by squaring twice. The procedure for eliminating two square roots from an equation was outlined in Example 9 of Section 4.3. You may wish to remind yourself of the general method in that example before you try to follow these messy details.

Details. We simplify the case of the ellipse here.

$$\sqrt{(x - c)^2 + y^2} + \sqrt{(x + c)^2 + y^2} = 2a$$
$$\textit{iff} \quad \sqrt{(x + c)^2 + y^2} = 2a - \sqrt{(x - c)^2 + y^2}$$

Squaring both sides

$$(x + c)^2 + y^2 = 4a^2 - 4a\sqrt{(x - c)^2 + y^2} + (x - c)^2 + y^2$$
$$x^2 + 2cx + c^2 + y^2 = 4a^2 - 4a\sqrt{x^2 - 2cx + c^2 + y^2} + x^2 - 2cx + c^2 + y^2$$

Several terms cancel.

$$4cx - 4a^2 = -4a\sqrt{x^2 - 2cx + c^2 + y^2}$$
$$cx - a^2 = -a\sqrt{x^2 - 2cx + c^2 + y^2}$$
$$(cx - a^2)^2 = a^2(x^2 - 2cx + c^2 + y^2)$$
$$c^2x^2 - 2a^2cx + a^4 = a^2x^2 - 2a^2cx + a^2c^2 + a^2y^2$$

Now, two terms cancel. Putting all the x's on the right,

$$a^2(a^2 - c^2) = (a^2 - c^2)x^2 + a^2y^2.$$

Now, because $a > c \geq 0$, $a^2 - c^2 > 0$. Thus we can let $b^2 = a^2 - c^2$.

$$a^2b^2 = b^2x^2 + a^2y^2.$$

Dividing through by a^2b^2, we obtain

$$\frac{x^2}{a^2} + \frac{y^2}{b^2} = 1 \ .$$

This is the "standard form" of an ellipse (3.2.10) that we first encountered in Section 3.2 on quadratics. There we recognized this equation as the equation of a circle with the scale changed. The graph of the unit circle is expanded by a factor of a in the x-direction and by a factor of b in the y-direction.

Ellipses Centered at the Origin. The "standard form" of ellipses centered at the origin is

$$\frac{x^2}{a^2} + \frac{y^2}{b^2} = 1 . \qquad (9.2.5)$$

The graph goes through $(a, 0)$, $(-a, 0)$, $(0, b)$, and $(0, -b)$. These four points make it easy to graph (Figures 9 and 10). The foci are c units from the origin, where $a^2 = b^2 + c^2$ if the major axis is horizontal and $b^2 = a^2 + c^2$ if the major axis is vertical.

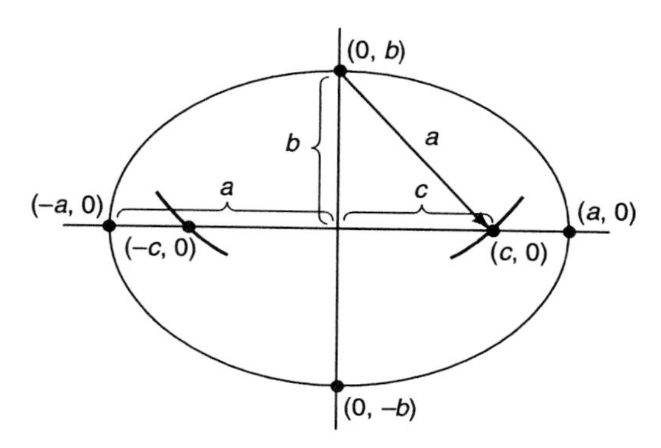

Figure 9: An ellipse,with $a > b$.
$x^2/a^2 + y^2/b^2 = 1$.
Foci at $(-c, 0)$ and $(c, 0)$.

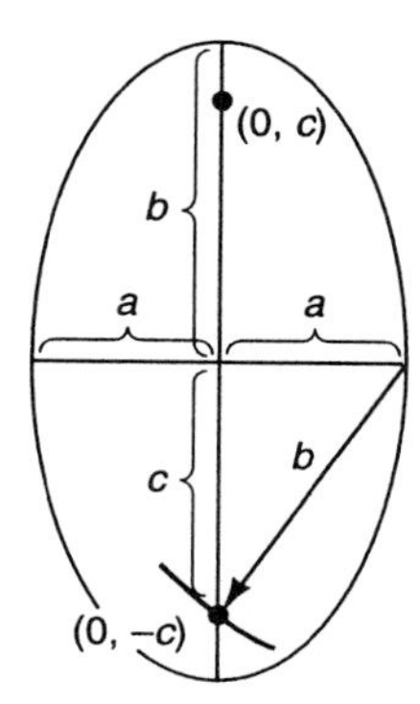

Figure 10: An ellipse with $b > a$.
$x^2/a^2 + y^2/b^2 = 1$.
Foci at $(0, c)$ and $(0, -c)$.

Ellipses are longest along the "major axis" through the foci. When $a > b$ as in Figure 9, the major axis is the horizontal line segment of length $2a$ from $(-a, 0)$ to $(a, 0)$. When $b > a$ as in Figure 10, the ellipse is taller than it is wide and the major axis is the vertical line segment of length $2b$ from $(0, -b)$ to $(0, b)$. Therefore, the greater of a and b is the length of the "semi-major" axis ("semi" means "half"). The distance from the foci to the center is denoted "c".

Through the center and perpendicular to the major axis is the "minor" axis. The smaller

---

Some texts always use "a" for the length of the semi-major (longer) axis, regardless of whether "a" is associated with "x" or "y". This text always associates "a" with the x-direction and "b" with the y-direction.

If $a > b$ (Figure 9), $c^2 = a^2 - b^2$.
The length is horizontal.
The foci are on the horizontal axis of symmetry at $(-c, 0)$ and $(c, 0)$.

If $b > a$ (Figure 10), $c^2 = b^2 - a^2$.
The length is vertical.
The foci are on the vertical axis of symmetry at $(0, -c)$ and $(0, c)$.

In both cases, $c^2 = \text{larger}^2 - \text{smaller}^2$.

---

of a and b is the length of the "semi-minor axis" (half of the minor axis).

Given the equation in standard form, ellipses are easy to sketch (Figures 9 and 10).

Example 3: Sketch the ellipse and locate its foci given the equation

$$x^2/25 + y^2/9 = 1.$$

Because the form 9.2.5 fits, it is centered at the origin. $a = 5$ and $b = 3$, so it goes through (-5, 0) and (5, 0) on the major axis, which is the x-axis. The ellipse goes through (0, 3) and (0, -3) on the minor axis, which is the y-axis (Figure 11). The foci are on the major axis at $(-c, 0)$ and $(c, 0)$, where $a^2 - b^2 = c^2$, so $c = 4$. The foci are at (-4, 0) and (4, 0).

When $a > b$, as in this example, a is both the distance from the center to the horizontal ends and the distance from $(0, b)$ to the foci. Given the graph, but not the foci, the foci can be easily determined by striking an arc of length a from the top of the ellipse and noting the two points where it crosses the horizontal axis (Figure 9).

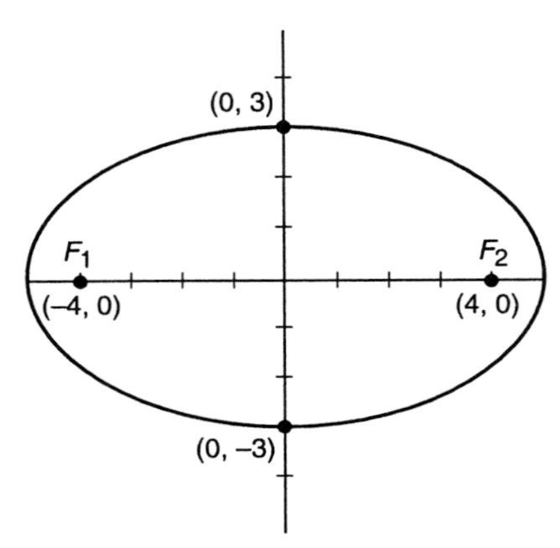

Figure 11: $x^2/25 + y^2/9 = 1$. [-5, 5] by [-5, 5].

Calculator Exercise 1: Learn how to graph ellipses on your calculator by solving for y in 9.2.5.

$$y = \pm b \sqrt{1 - \frac{x^2}{a^2}} \, . \tag{9.2.6}$$

This is the form to enter in to your calculator. You will need to graph two graphs, one for the upper half and one for the lower half. Use this approach to graph $x^2/25 + y^2/9 = 1$ on your calculator. Does your graph resemble Figure 11?

Example 4: Identify and sketch the ellipse with equation $3x^2 + 5y^2 = 40$.

The keys to "standard form" are that the right side is 1 and that the squared terms are alone on top with coefficient 1. Any multiple must be rewritten to be on the bottom.

$3x^2 + 5y^2 = 40$
$3x^2/40 + 5y^2/40 = 1$

$$\frac{x^2}{\left(\frac{40}{3}\right)} + \frac{y^2}{8} = 1 .$$

$$\frac{x^2}{\left(\sqrt{\frac{40}{3}}\right)^2} + \frac{y^2}{(\sqrt{8})^2} = 1 .$$

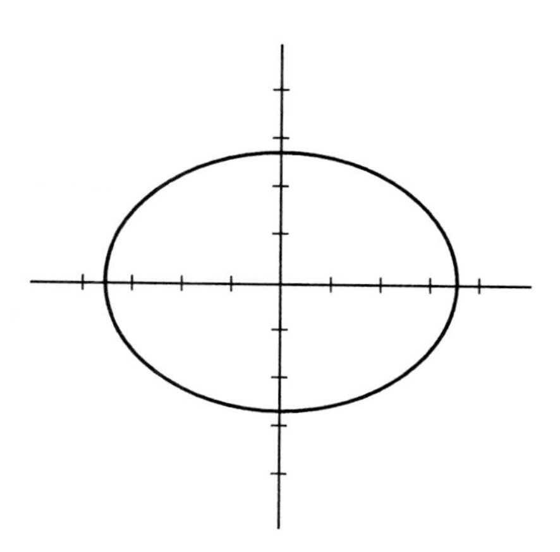

Figure 12: $3x^2 + 5y^2 = 40$. $a = 3.65$ and $b = 2.82$. [-5, 5] by [-5, 5].

This fits the form of an ellipse centered at the origin. From 9.2.5, $a = \sqrt{40/3} = 3.65$ and $b = \sqrt{8} = 2.82$. If we do not worry too much about the precise shape, the graph is easy to sketch using just the four points on the graph that the values of a and b provide (Figure 12).

Example 5: Identify the foci of the ellipse with equation

$$\frac{x^2}{25} + \frac{y^2}{169} = 1 .$$

Here the major (longer) axis is the y-axis and therefore the foci lie on the y-axis at $(0, c)$ and $(0, -c)$, where, again, c^2 is determined as the larger denominator minus the smaller denominator. $c^2 = 169 - 25 = 144$, so $c = 12$. The foci are (0, 12) and (0, -12) (Figure 13). Be sure not to draw the ellipse pointed at the end. Even long and thin ellipses are really quite rounded at the ends.

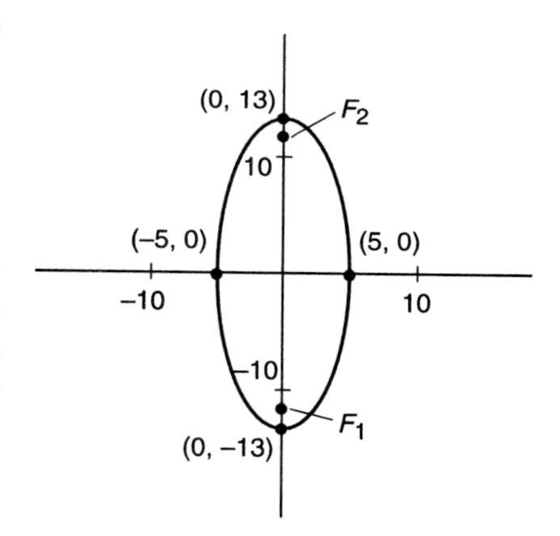

Figure 13: An ellipse with vertical major axis. $x^2/25 + y^2/169 = 1$. [-20, 20] by [-20, 20].

<u>Location Shifts</u>. To center an ellipse at (h, k), let "$x - h$" and "$y - k$" play the roles of "x" and "y" in 9.2.5 where the graph was centered at the origin.

The Standard Form of an Ellipse:

(9.2.7) $$\frac{(x - h)^2}{a^2} + \frac{(y - k)^2}{b^2} = 1 .$$

This ellipse is centered at (h, k) and goes through the points $(h - a, k)$, $(h + a, k)$, $(h, k + b)$ and $(h, k - b)$ (Figure 14).

If $a > b$, the major axis lies along $y = k$ (in the x-direction), and the foci are c units from the center on the major axis at $(h - c, k)$ and $(h + c, k)$, where $c^2 = a^2 - b^2$.

If $b > a$, the major axis lies along $x = h$ (in the y-direction), and the foci are c units from the center on the major axis at $(h, k - c)$ and $(h, k + c)$, where $c^2 = b^2 - a^2$.

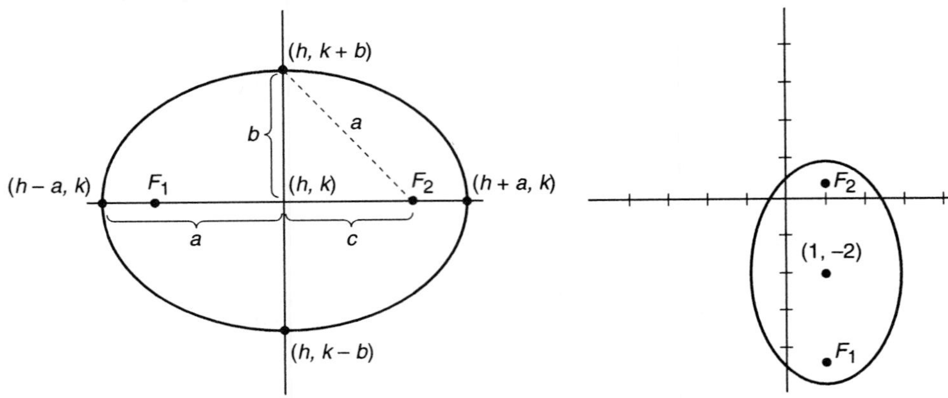

Figure 14: An ellipse with center (h, k), semi-major axis a, and semi-minor axis b.

Figure 15: The ellipse with center (1, -2), $b = 3$ and $a = 2$. [-5, 5] by [-5, 5].

Example 6: Identify the ellipse with equation $9x^2 - 18x + 4y^2 + 16y = 11$.

Aim for the standard form of 9.2.7. If the coefficients on x^2 and y^2 were the same, it would be a circle and we would complete the square to find the center. Since the coefficients are different, but the same sign, it is an ellipse. Complete the square to find the center.

$9x^2 - 18x + 4y^2 + 16y = 11.$
$9(x^2 - 2x) + 4(y^2 + 4y) = 11.$
$9(x^2 - 2x + 1 - 1) + 4(y^2 + 4y + 4 - 4) = 11.$
$9(x - 1)^2 - 9 + 4(y + 2)^2 - 16 = 11.$
$9(x - 1)^2 + 4(y + 2)^2 = 36.$

$$\frac{(x - 1)^2}{2^2} + \frac{(y + 2)^2}{3^2} = 1 .$$

The center is (1, -2) (Figure 15). Because $3 > 2$ and 3 is associated with y, the major axis is vertical. $c^2 = 3^2 - 2^2 = 5$, so $c = \sqrt{5}$. The foci are $\sqrt{5}$ from the center.

<u>The Equation of a Hyperbola</u>. The "standard form" of a hyperbola centered at the origin with foci at ($-c$, 0) and (c, 0) follows by simplifying Equation 9.2.4. We omit the details, which are very similar to the details included for the ellipse (problem B17).

<u>Hyperbolas Centered at the Origin with Foci on the x-axis</u>. The "standard form" of a hyperbola centered at the origin and foci at (c, 0) and ($-c$, 0) is

$$\frac{x^2}{a^2} - \frac{y^2}{b^2} = 1 . \tag{9.2.8}$$

where $a^2 + b^2 = c^2$. The lines $y = \pm(b/a)x$ are asymptotes (Figures 16 and 17).

This equation is very similar to the equation for an ellipse -- only the plus sign has become a minus sign.

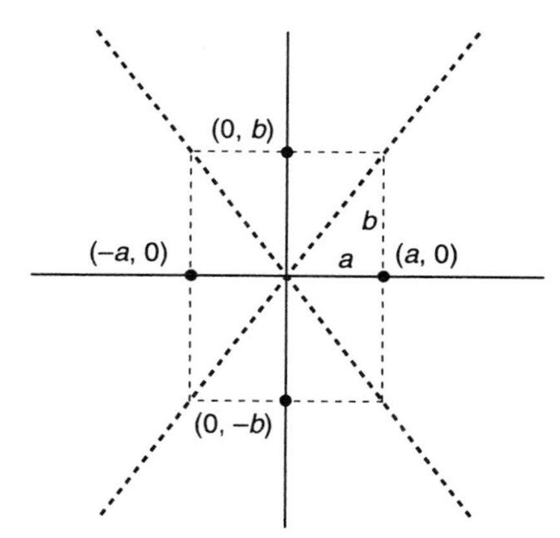

Figure 16: Sketching a hyperbola.
Step 1: Find (a, 0) and ($-a$, 0).
Step 2: Find (0, b) and (0, $-b$) and sketch in the rectangle, as shown.
Step 3: Lightly draw the asymptotes through the corners of the box and the origin.

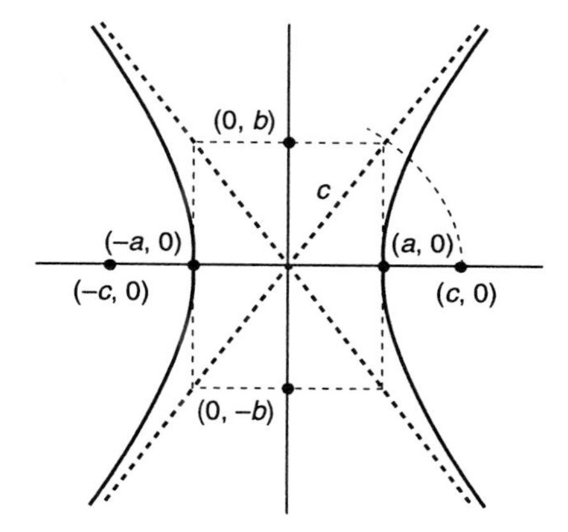

Figure 17: Sketching a hyperbola.
Step 4: Sketch the curve through (a, 0) sweeping up and down toward the asymptotes.
Step 5: Fill in the other half, symmetrically.

The graph goes through two easy points: ($-a$, 0) and (a, 0). These points on the curve are between the foci (the reverse of the situation for an ellipse, where the foci are between the points on the curve). There is no y-value for $x = 0$ because substituting in $x = 0$ yields the equation $-y^2/b^2 = 1$, which cannot be satisfied (because a negative number is never 1). Therefore, the two branches of the hyperbola do not meet -- there is a gap between them near $x = 0$.

Solving for y and simplifying

(9.2.9) $$y = \pm \frac{b}{a}\sqrt{x^2 - a^2}$$

This is the functional form suitable for graphing with a graphics calculator.

As x becomes large, $\sqrt{x^2 - a^2}$ approaches x, so, for large x, y is approximately $\pm(b/a)x$. The lines through the origin with equations $y = \pm(b/a)x$ are asymptotes with slopes $\pm b/a$. The curve sweeps up and down from the axis toward the asymptotes. To sketch the graph of Equation 9.2.8 it is essential to locate $(\pm a, 0)$ and the two asymptotes. An easy way to plot the asymptotes is to make the rectangle in Figure 16 and draw the diagonal line through the corners and the origin.

To locate the focus at $(c, 0)$, swing an arc centered at the origin through the upper right corner of the box and mark where the arc cuts the x-axis (Figure 17). This works because c is both the distance from the center to the foci and the distance from the center to the corner of the box (because $c^2 = a^2 + b^2$).

Calculator Exercise 2: Formula 9.2.9 gives the functional form which permits graphing a hyperbola with a graphics calculator. Identify a and b in the next example and use 9.2.9 to graph the equation on your calculator.

Example 7: Sketch the graph of the equation $x^2/25 - y^2/4 = 1$.

Here $a = 5$ and $b = 2$. The graph goes through (-5, 0) and (5, 0). The asymptotes have slope 2/5 and -2/5 (Figure 18).

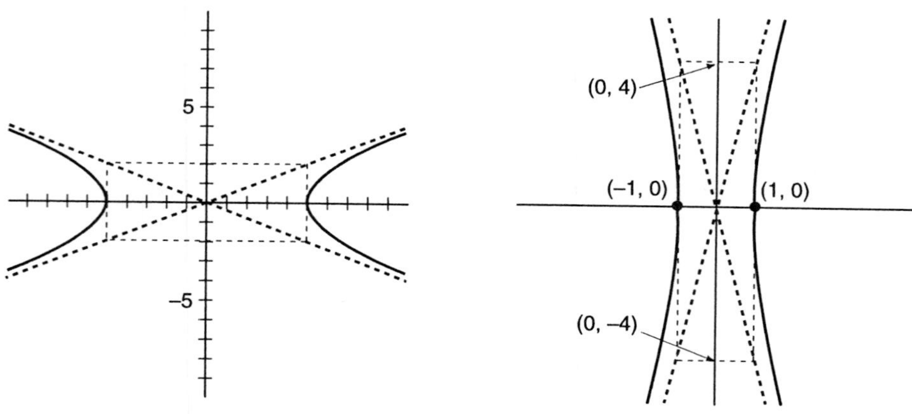

Figure 18: $x^2/25 - y^2/4 = 1$. [-5, 5] by [-5, 5].

Figure 19: $x^2 - y^2/16 = 1$. [-5, 5] by [-5, 5].

Example 8: Sketch the graph of the equation $x^2 - y^2/16 = 1$.

Here $a = 1$ and $b = 4$. The graph still opens around the x-axis because the "x^2" term is positive, as in Equation 9.2.8. The graph goes through (-1, 0) and (1, 0), and the asymptotes have slope 4 and -4 (Figure 19).

Equation 9.2.8 is *not* symmetric in x and y. One of the squared terms has a plus sign and the other a minus sign. The foci are on the axis corresponding to the plus sign. The hyperbola "opens" around the central axis, which is the horizontal axis in Figures 16 and 17. If the foci are vertically aligned, the roles of "x" and "y" are switched and the parabola opens around the vertical axis.

Example 9: Graph $y^2 - x^2 = 1$.

The graph opens around the y-axis (Figure 20).

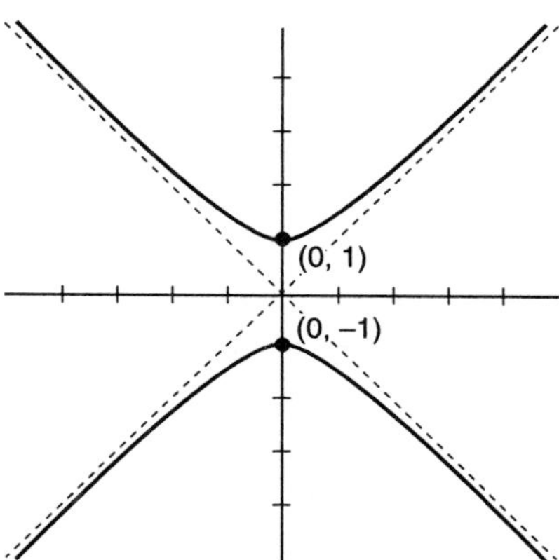

Figure 20: $y^2 - x^2 = 1$ opens around a vertical axis, corresponding to the positive y^2 term. [-5, 5] by [-5, 5].

<u>Scale Changes</u>. In Section 2.4 (Composition and Decomposition) we discussed scale changes in general, and in Section 3.1 (Lines, Distance, and Circles) we discussed circles and how, with scale changes, circles become ellipses. The "standard forms" given above can be interpreted as scale changes of just two basic graphs, the unit circle "$x^2 + y^2 = 1$" (Figure 21) and the basic hyperbola "$x^2 - y^2 = 1$" (Figure 22).

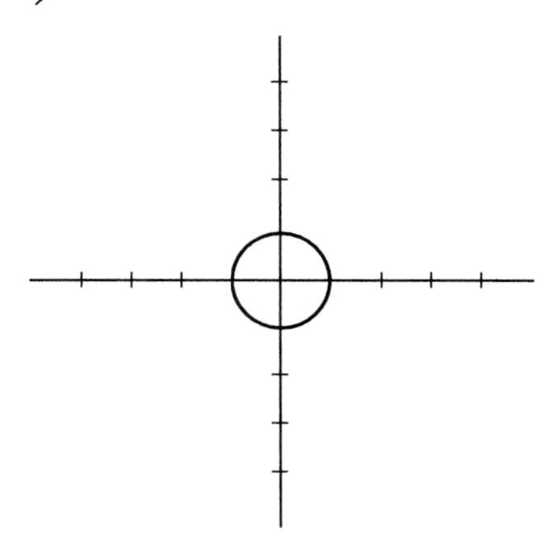

Figure 21: $x^2 + y^2 = 1$. [-5, 5] by [-5, 5].

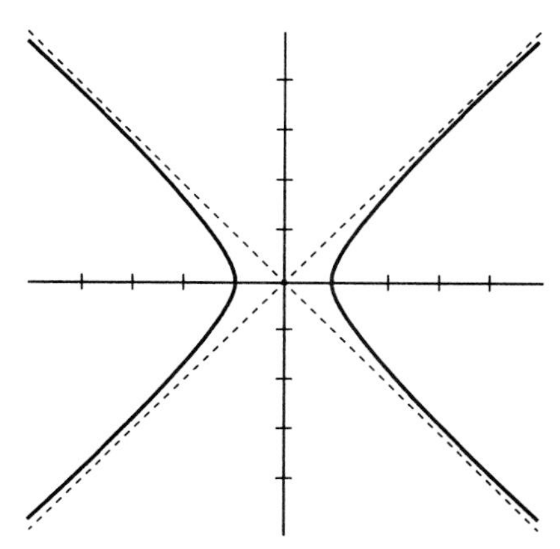

Figure 22: $x^2 - y^2 = 1$. [-5, 5] by [-5, 5].

If, in an equation, "x" is replaced by "x/a," the new graph becomes a times as wide. This is because x must be a times as great for "x/a" to assume the same value and yield the same image as the old "x". Similarly, replacing "y" by "y/b" yields a graph b times as tall. Can you see this by comparing Figures 9 through 13 to Figure 21? Can you see this by comparing Figures 18 and 19 to Figure 22?

__Location Changes__. To shift the graph so the center is (h, k), simply substitute "$x - h$" for "x" and "$y - k$" for "k" in the form 9.2.8.

Hyperbola with a Horizontal Axis: The "standard form" of a hyperbola with a horizontal axis is

$$\frac{(x - h)^2}{a^2} - \frac{(y - k)^2}{b^2} = 1 . \quad (9.2.10)$$

The center is (h, k), the axis is $y = k$, the foci are $(h - c, k)$ and $(h + c, k)$, where $c^2 = a^2 + b^2$. The graph goes through $(h + a, k)$ and $(h - a, k)$. The asymptotes are lines through the center with slopes $\pm b/a$.

Example 10: Identify and sketch the hyperbola with equation "$x^2 + 4x - 9y^2 = 32$."

Aim for standard form. First complete the square.

$x^2 + 4x - 9y^2 = 32$ is equivalent to
$x^2 + 4x + 4 - 9y^2 = 36$,
$(x + 2)^2 - 9y^2 = 36$,

$$\frac{(x + 2)^2}{6^2} - \frac{y^2}{2^2} = 1 .$$

The center is (-2, 0). $a = 6$ and $b = 2$. The axis is horizontal, since the "x^2" term has the positive sign. Figure 23 illustrates the graph, including the asymptotes.

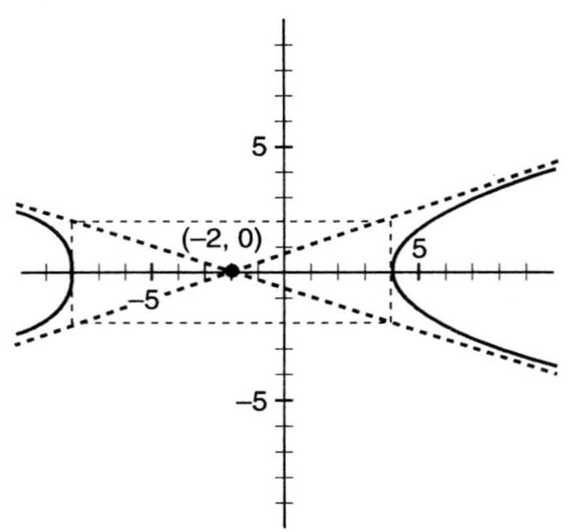

Figure 23: The graph of $x^2 + 4x - 9y^2 = 32$. [-10, 10] by [-10, 10].

Example 11: Identify and sketch the hyperbola with equation

"$5x^2 - 2y^2 - 16y = 12$."

The orientation of this hyperbola is not easy to guess. Aim for standard form. Complete the square.

$5x^2 - 2y^2 - 16y = 12$
$5x^2 - 2(y^2 - 8y) = 12$.
$5x^2 - 2(y^2 - 8y + 16 - 16) = 12$.
$5x^2 - 2(y^2 - 4)^2 = 12 - 32 = -20$.

Note the negative sign on "-20." To convert to "standard form" the right side must be (positive) 1. The signs must be changed.

$$2(y - 4)^2 - 5x^2 = 20.$$

$$\frac{(y-4)^2}{10} - \frac{x^2}{4} = 1$$

$$\frac{(y-4)^2}{(\sqrt{10}\,)^2} - \frac{x^2}{2^2} = 1.$$

The center is (0, 4) and it opens around a vertical axis (Figure 24).

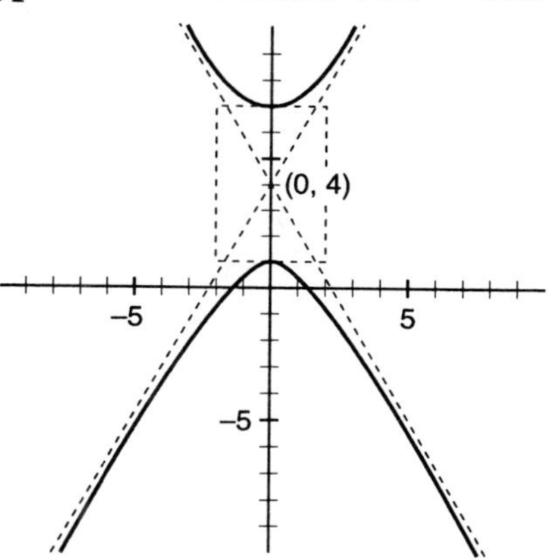

Figure 24:
$5x^2 - 2y^2 - 16y = 12.$
[-10, 10] by [-10, 10].

The equation we just obtained in Example 11 is very similar to the standard form for a horizontal hyperbola -- but the signs are changed on the squared terms. That switches the roles of the axes and makes the hyperbola vertical.

Hyperbola with a Vertical Axis: The "standard form" of a hyperbola with a vertical axis is

$$\frac{(y-k)^2}{b^2} - \frac{(x-h)^2}{a^2} = 1\,. \tag{9.2.11}$$

The center is (h, k), the axis is $x = h$, the foci are $(h, k + c)$ and $(h, k - c)$, where $c^2 = a^2 + b^2$. The graph goes through $(h, k + c)$ and $(h, k - c)$. The asymptotes are lines through the center with slopes $\pm b/a$.

Hyperbolas and ellipses have important reflective properties which are discussed next.

Reflective Properties. All ellipses have a remarkable reflective property. Every ray generated at one focus will reflect through the other focus, regardless of its original direction (Figure 25).

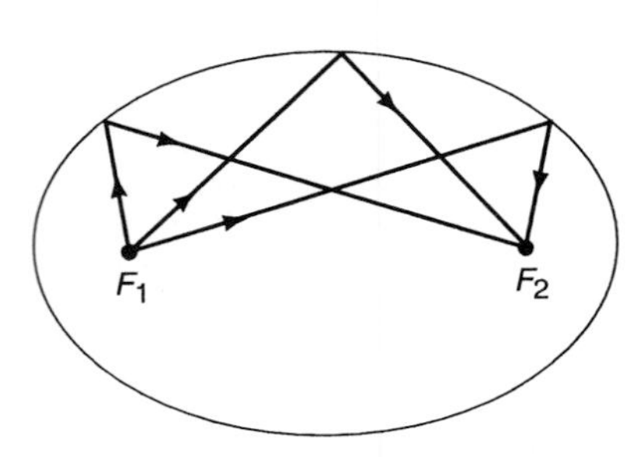

Figure 25: Rays generated at one focus of an ellipse reflect through the other focus.

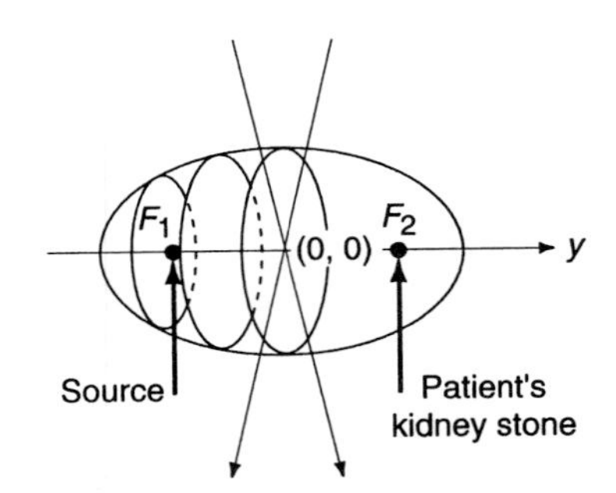

Figure 26: An ellipsoidal lithotripter. Shock waves reflect through the second focus at the same time.

Example 12 : A lithotripter is a medical machine for destroying painful kidney stones without surgery. It has a reflecting ellipsoidal surface (Figure 26). An ellipsoid is a 3 dimensional surface formed by rotating an ellipse about its major axis. The reflecting surface is not a complete ellipsoid. It encloses only one focus and leaves room for the other focus to be precisely positioned at the kidney stone inside the patient. Then a strong shock wave is generated at the enclosed focus. Waves reflecting off the ellipsoid in all directions pass through the other focus (the kidney stone) at the same time. The focused shock wave can pulverize the stone and then its fragments can pass through the system without surgery.

Hyperbolas also have a remarkable and useful reflective property. Consider one branch with the convex side reflective. All light rays aimed at the focus behind the branch will be reflected through the other focus, regardless of the original angle of the rays (Figure 27).

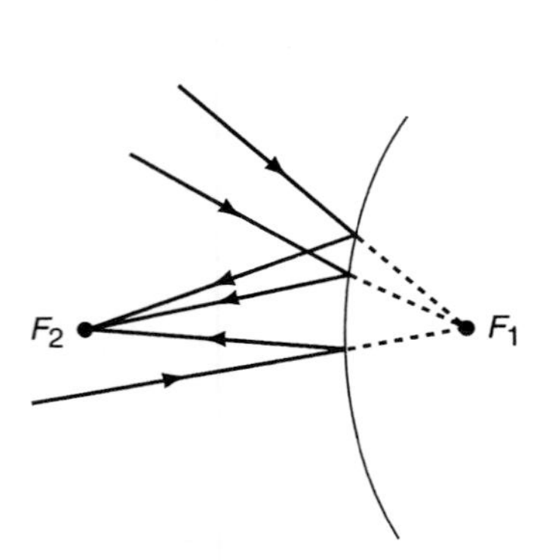

Figure 27: Rays aimed at a focus of a hyperbola reflect through the other focus.

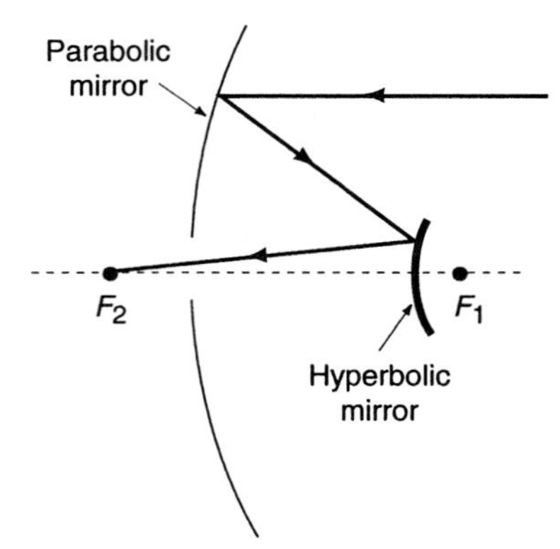

Figure 28: Parabolic and hyperbolic mirrors combined.

Example 13: A reflecting telescope with a parabolic mirror at one end will focus parallel rays at a point in front of the mirror (Figure 9.1.15). But you can't put your eye there; your head would block the incoming rays. Instead, put a convex hyperbolic mirror just in front of the focus so that the focus of the parabola is also the focus of the hyperbola (Figure 24). Then the rays reflected from the parabolic mirror will be aimed at a focus of the hyperbolic mirror. By the reflective property of hyperbolas, they will reflect through the other focus. Arrange it so that the second focus of the hyperbola is behind the parabolic mirror. Then, only a small hole in the big parabolic mirror is needed to focus the light behind the big mirror, where an eye or camera can be located.

Distinguishing the Types of Conic Sections. The algebraic equations for parabolas, circles, ellipses, and hyperbolas are somewhat similar. How can we tell which is which?

In the equation of a parabola, only one of "x" or "y" is squared (see 9.1.5 and 9.1.6). If both "x" and "y" are squared, expect a circle, ellipse, or hyperbola. (There are some special cases where the graph is a line, a point, or nothing at all.) How are circles, ellipses, and hyperbolas distinguished?

Consider an equation of the form

(9.2.12) $$ax^2 + bx + cy^2 + dy + k = 0.$$

Assuming $a \neq 0$ and $c \neq 0$ (that is, both squared terms exist)

1) If a and c are the same sign, expect an ellipse (problem B16).
2) If a and c are equal, expect a circle, which is a special ellipse.
3) If a and c are opposite signs, expect a hyperbola.

One of the squared terms may be missing.

4) If $a = 0$ or $c = 0$, but not both, expect a parabola.

To determine the equation in "standard form," the procedure is to first complete the square on x and also on y to locate the center. Then aim for one of the standard forms.

Example 14: $3x^2 + 12x + 3y^2 = 50$ is a circle. The coefficients on x^2 and y^2 are equal.

$5x^2 + 20x + y^2 = 50$ is an ellipse because the coefficients on x^2 and y^2 are not equal, but they have the same sign.

$5(x + 2)^2 - y^2 = 50$ is a hyperbola. The coefficients on x^2 and y^2 have opposite signs. Because the coefficient on x^2 is the same sign as the constant, the hyperbola opens around the x-axis.

$x + 2 + y^2 = 50$ is a parabola. There is no x^2 term.

Conclusion. Equations with terms in both x^2 and y^2 can determine circles, ellipses, or hyperbolas. These shapes have important properties. They are easily graphed after they are written in "standard form." Completing the square is an important step in attaining standard form.

Terms: ellipse, hyperbola, focus, semi-major axis, semi-minor axis, standard form.

****

Exercises for Section 9.2, "Ellipses and Hyperbolas":

^ ^ ^ ^Determine the type of conic section. Rewrite the equation in "standard form," determine a, b, and c, identify the center, and roughly sketch it.

A1. $x^2 + y^2/4 = 1$. [$c = \sqrt{3}$]
A2. $x^2/9 + y^2 = 1$. [$a = 3$]
A3. $x^2/4 - y^2 = 1$. [$a = 2$]
A4. $x^2 - y^2/9 = 1$. [$b = 3$]
A5. $y^2 - x^2/4 = 1$. [$c = \sqrt{5}$]
A6. $y^2/16 - x^2 = 1$. [$a = 1$]
A7. $4x^2 + 9y^2 = 36$. [$b = 2$]
A8. $4x^2 + y^2 = 25$. [$c = 4.3$]
A9. $4x^2 - 9y^2 = 36$. [$c = 3.6$]
A10. $4x^2 - y^2 = 25$. [$c = 5.6$]
A11. $x^2 - 4y^2 = 1$. [$c = .87$]
A12. $9x^2 - y^2 = 1$. [$c = .94$]
A13. $100x^2 + 25y^2 = 4$. [$c = .35$]
A14. $4x^2 - 9y^2 = 1$. [$b = 1/3$]
A15. $(x - 3)^2 + 4(y + 2)^2 = 4$. [$c = 2.2$]
A16. $9(x + 2)^2 - y^2 = 9$. [$c = 3.2$]
A17. $4x^2 - 8x + 9y^2 + 36y + 4 = 0$. [$c = 3.6$]
A18. $4x^2 + 8x + y^2 - 6y + 9 = 0$. [$c = 2.2$]

^ ^ ^ ^Find the equation of the ellipse with the following properties.

A19. Foci (-2, 0) and (2, 0), semi-major axis 3.
A20. Foci (0, 1) and (0, -1), semi-major axis 2.
A21. Center (1, 3), semi-major axis vertical and length 4, semi-minor axis 2.
A22. Center (3, 4), semi-major axis horizontal and length 5, semi-minor axis 1.

^ ^ ^ ^Find the equation of the hyperbola with the following properties.

A23. Foci at (-2, 0) and (2, 0), through (1, 0).
A24. Foci at (0, 1) and (0, -1), through (0, ½).
A25. Center at (1, 5), through (1, 9), with focus at (1, 10).
A26. Center at (-2, 3), through (0, 3), with focus at (1, 3).

^ ^ ^ ^Identify the type of conic section. Do as little work as possible.

A27. $x^2 - 5y^2 = 1$.
A28. $x^2 + 5y = 2$.
A29. $x^2 + y^2 + 10y = 50$.
A30. $x^2 + 6x + 3y^2 = 100$.
A31. $x + 5y^2 = 27$.
A32. $y^2 + 6y = x^2 + 10x$.
A33. $3x^2 + 9x + 3y^2 - 8y = 200$.
A34. $3x^2 + 9x + 4y^2 - 8y = 200$.
A35. $3x^2 + 9x - 4y^2 - 8y = 200$.
A36. $3x^2 + 9x - 8y = 200$.

^ ^ ^ ^ ^ ^ ^ ^ ^

B1.* a) Sketch an ellipse centered at the origin and label a, b, and c on it. Include a right triangle that relates a, b, and c as its sides and label the hypotenuse.

B2.* a) Define *ellipse*. b) Illustrate the definition with a sketch.

B3.* a) Sketch a hyperbola centered at the origin and label a, b, and c on it. Include a right triangle that relates a, b, and c as its sides and label the hypotenuse.

B4.* a) Define *hyperbola*. b) Illustrate the definition with a sketch.

B5. Suppose $x^2/c^2 - y^2/d^2 = 1$. What are the slopes of the asymptotes?

B6. Suppose $y^2/c^2 - x^2/d^2 = 1$. What are the slopes of the asymptotes?

B7. Let F be (3, 0) and L be the y-axis. Let P be a point such that the distance from F to P is 1/2 the distance from P to L. a) Find, by inspection, two points on the x-axis that satisfy the restriction on P. b) Find, by inspection, two points on the line $x = 3$ that satisfy the restriction on P. c) Use rectangular coordinates to set up an equation for all such points. d) Simplify the equation into standard form and identify the conic section.

B8. Let F be (3, 0) and L be the y-axis. Let P be a point such that the distance from F to P is twice the distance from P to L. a) Find, by inspection, two points on the x-axis that satisfies the restriction on P. b) Find, by inspection, two points on line $x = 3$ that satisfy the restriction on P. c) Use rectangular coordinates to set up an equation for all such points. d) Simplify the equation into standard form and identify the conic section.

^ ^ ^ ^ [For B9-B15] The eccentricity of an ellipse or hyperbola is

$$e = \frac{\textit{the distance between foci}}{\textit{the distance between vertices}},$$

where the vertices are the points on the major axis.

B9. For an ellipse, if $a > b$, $e = c/a$. Suppose the eccentricity of an ellipse is 1/2 and $a > b$. a) What is the ratio of a to b? b) If $a/b = 2$, what is the eccentricity?

B10. [See the definition of "eccentricity" above B9.] For a hyperbola that fits form 9.2.10, if the eccentricity is 3, what is the ratio of a to b? b) If $a/b = 3$, what is the eccentricity?

B11. For a hyperbola with horizontal axis, give a formula for the slopes of the asymptotes if the eccentricity is $e > 1$.

B12. For ellipses, give a formula for the ratio of length to width if the eccentricity is e.

B13. If the ratio of length to width of an ellipse is r, give a formula for the eccentricity.

B14. If the slopes of the asymptotes of a horizontal hyperbola are $\pm m$, give a formula for its eccentricity.

B15. Which hyperbola has greater eccentricity, the one in Figure 18 or the one in Figure 19?

B16. In 9.2.12 "special cases" can occur that do not produce the "expected" result.
a) Case 1 may produce a point or nothing. How?
b) Case 3 may produce a pair of lines. How?

B17. Inspect the details required to convert 9.3.3 into standard form (page 544). To convert 9.3.4 to standard form most steps would be very similar. At which step is the really significant difference? Redo that step to show you understand the difference.

B18. Let F be the origin and L be the line $x = -h$. Let P be a point such that the distance from F to P is e (a parameter, the "eccentricity," greater than 0) times the distance from P to L. Use rectangular coordinates to set up an equation for all such points. [Do not bother to simplify. Simplifying is quite hard.]

B19. In Example 1 the time of the burst was recorded at F_1 and compared to the time its reflection was received at F_2 to determine the total distance the sound traveled, and therefore, to determine the ellipse (Figure 3). There is a way to determine the total distance by recording time only at F_2, since the sound will be heard twice at F_2, one directly and once reflected. a) How can we do this, even if we fail to record when we generated the burst at F_1? b) For example, if sound travels at 1100 feet per second, F_1 is 2200 feet from F_2, and we record the sound twice at F_2, the second time 1/2 second after the first, what do we know that allows us to find the equation for all possible reflecting points?

B20. The graph of $y = 1/x$ is a hyperbola. a) What is its axis of symmetry?
b) Let $u = x - y$ and $v = x + y$. Rewrite "$y = 1/x$" in terms of u and v. c) Graph $u = 0$ and $v = 0$, and $y = 1/x$ together. [Note how, on the axis system created in Part (c), the graph looks like a standard hyperbola opening up around an axis.]

B21. [See Figure 4] If F_1 is the both center of a circle and the focus of an (non-circular) ellipse, can the circle and ellipse intersect in four locations, as some circles and ellipses can? Argue why it can or can not.

B22. (Short and interesting, but hard) A hyperbola can be determined as a cross-section of a cone (Figure 9.1.2C). Explain, using this fact, why the tip of the shadow of a pointed object moves in a branch of a hyperbola as the sun goes across the sky. For example, the tip of the shadow of the point sticking up from a flat sundial moves in a hyperbolic path.

Section 9.3. Polar Equations of Conic Sections

Parabolas, ellipses, and hyperbolas can be described using distances from a focus and a directrix. When the focus is located at the origin, polar coordinates use r for the distance from the focus. This makes conic sections easy to describe in polar coordinates.

There are several equivalent ways to define conic sections. Here is one that uses a focus and a directrix to simultaneously describe parabolas, ellipses, and hyperbolas.

Let L be a line (a directrix) and F be a point (a focus) not on that line. Let e, a parameter that is called the eccentricity of the conic section, be an arbitrary constant greater than 0. Consider the set of points P such that the distance from F to P is e times the distance from P to L.

$$\frac{\textit{distance from F to P}}{\textit{distance from P to L}} = e \,. \tag{9.3.1}$$

If $e = 1$, the curve is a parabola. If $e < 1$, the curve is an ellipse. If $e > 1$, the curve is a hyperbola.

If $e = 1$, we already know that $FP = PL$ defines a parabola (Figure 1, Definition 9.1.3). However, this way to define ellipses and hyperbolas is new. It yields simple polar-coordinate equations.

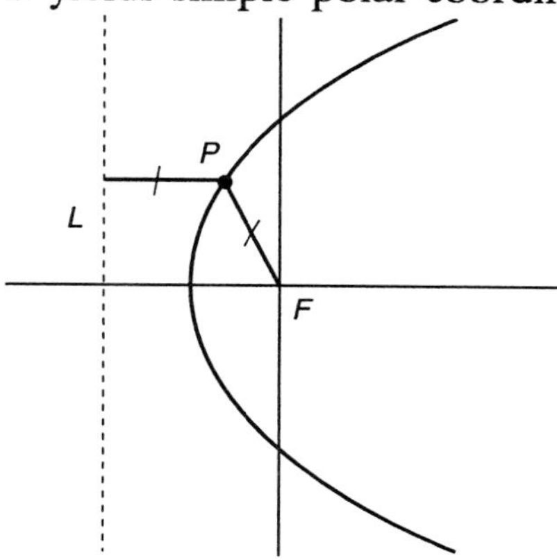

Figure 1: A parabola. $e = 1$.

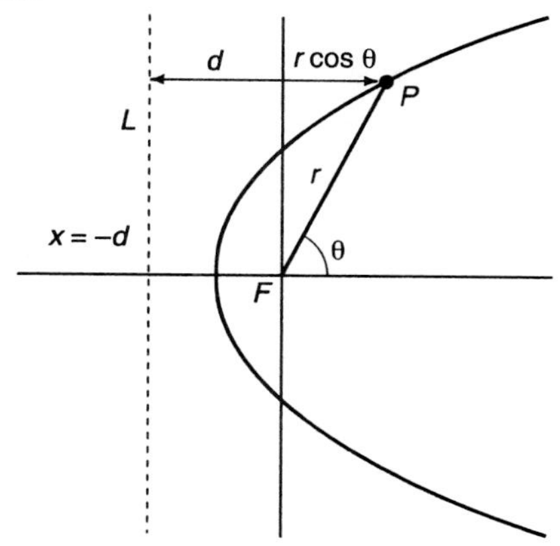

Figure 2: The distances from F to P and from P to L.

Orient the axis system so that F is at the origin and the directrix is $x = -d$ (Figure 2). Consider P to the right of the directrix as in Figure 2. Now express this: "The distance from F to P is e times the distance from P to L."

In polar coordinates, $P = (r, \theta)$ and the distance from F to P is simply r. The distance from P to L is $r \cos \theta + d$ (because $x = r \cos \theta$, Figure 2). Therefore,

$$r = e(r \cos \theta + d).$$

Solving for r in terms of θ,

$$r - er \cos \theta = ed.$$
$$r(1 - e \cos \theta) = ed.$$

(9.3.2)
$$r = \frac{ed}{1 - e \cos \theta}$$

This is the polar form of the equation of a conic section with focus (0, 0), directrix $x = -d$, and eccentricity $e > 0$. If $e < 1$, it is an ellipse. If $e = 1$, it is a parabola. If $e > 1$, it is a hyperbola.

Example 1: Let a focus be at the origin, the directrix be $x = -3$, and the eccentricity be 1/2. Find the polar equation of the conic section and sketch it.

The polar-coordinate equation follows directly from 9.3.2 (Figure 3):

$$r = \frac{(1/2)3}{1 - (1/2)\cos \theta}$$

Four points are easy to find because cosine is so simple at the major angles 0, $\pi/2$, π, and $3\pi/2$. When $\theta = 0$, $\cos \theta = 1$ and the polar-coordinate point (3, 0) is determined (it is also represented by (3, 0) in rectangular coordinates). When $\theta = \pi/2$, $\cos \theta = 0$, and the polar-coordinate point $(3/2, \pi/2)$ results, which is on the positive y-axis. When $\theta = \pi$, $\cos \theta = -1$ and the point $(1, \pi)$ results (which is (-1, 0) in rectangular coordinates). Finally, $\theta = 3\pi/2$, yields the point $(3/2, 3\pi/2)$ on the negative y-axis (which is (0, -3/2) in rectangular coordinates).

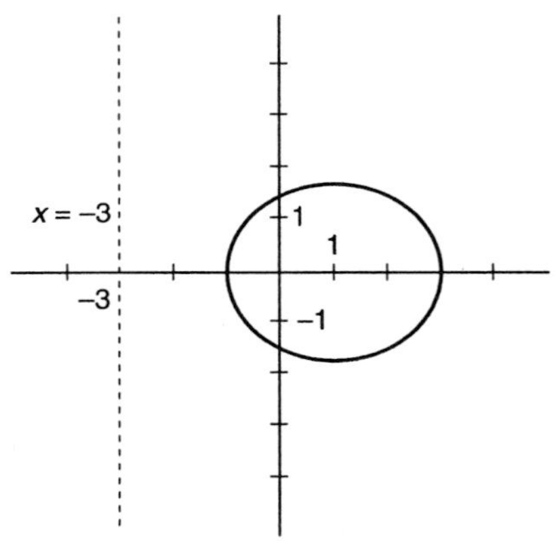

Figure 3: $r = (3/2)/[1 - (1/2)\cos \theta]$. [-5,5] by [-5,5]. $0 \leq \theta < 2\pi$. $e = 1/2$.

Note how these four points fit the fact that the eccentricity is 1/2. In rectangular coordinates, the point (3, 0) is 3 units from the origin and 6 units from the line $x = -3$, for a ratio of 1/2. On the negative x-axis, (-1, 0) is 1 unit from the origin and 2 units from the directrix, yielding the desired ratio 1/2. Points on the y-axis are all 3 units from the directrix. Therefore, to make the ratio 1/2, the points on the-axis must be 3/2 units from the origin: (0, 3/2) and (0, -3/2) in rectangular coordinates.

Example 2: Identify the type of conic section and give its eccentricity and directrix (Figure 4).

$$r = \frac{12}{3 - 2\cos\theta}.$$

Manipulate this into the form of 9.3.2. First get the "1" in the denominator.

$$r = \frac{12}{3 - 2\cos\theta} \quad \textit{iff}$$

$$r = \frac{4}{1 - (2/3)\cos\theta}.$$

So $e = 2/3 < 1$, so it is an ellipse. Now, since the numerator is ed, $(2/3)d = 4$, so $d = 6$. The directrix is $x = -6$.

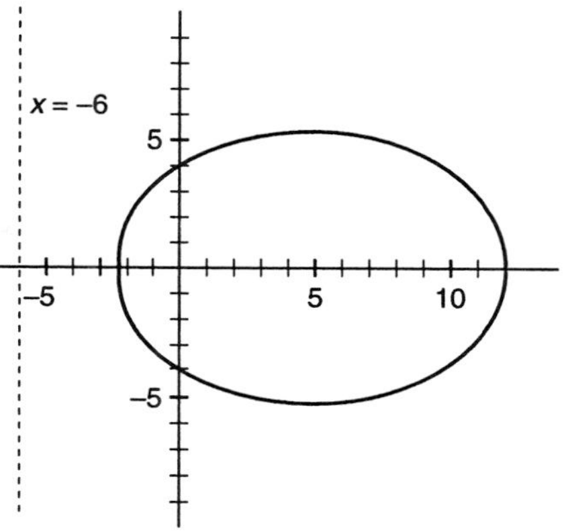

Figure 4: $r = 12/(3 - 2\cos\theta)$. [-7,13] by [-10,10]. $0 \le \theta < 2\pi$. $e = 2/3$.

Example 3: Identify the type of conic section and give its eccentricity and directrix (Figure 5).

$$r = \frac{5}{2 - 4\cos\theta}.$$

Again, get the "1" in the denominator.

$$r = \frac{5/2}{1 - 2\cos\theta}.$$

So $e = 2 > 1$ and the conic section is a hyperbola. $5/2 = ed = 2d$, so $d = 5/4$. The directrix is $x = -5/4$.

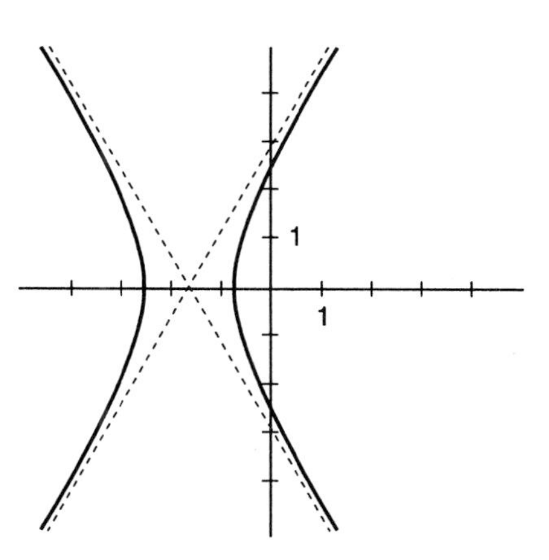

Figure 5: $r = 5/(2 - 4\cos\theta)$. [-5,5] by [-5,5]. $0 \le \theta < 2\pi$. $e = 2$.

Discovery 1: Graph the equation in Example 3 (Figure 5) with a graphics calculator in "Polar" mode. Which quarter of the hyperbola appears first? Second? Explain why, mathematically (problem B1). [If your calculator does not have a "Polar" mode, see problem B27.]

Figures 6 and 7 illustrate the effect of eccentricity on the shapes of conic sections.

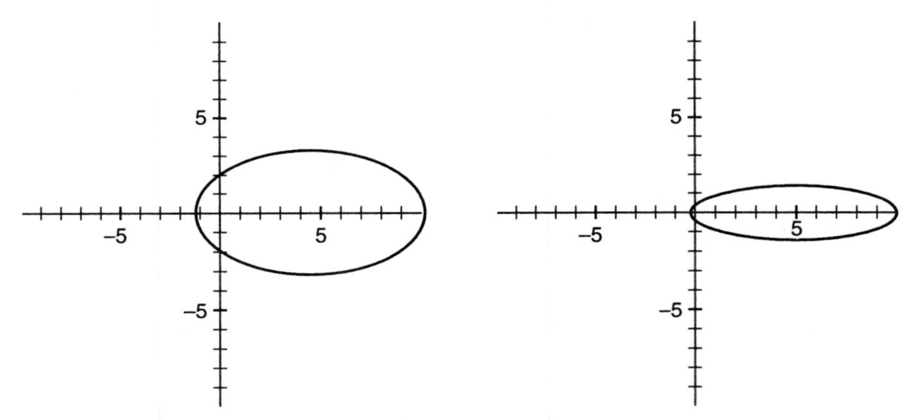

Figure 6: Ellipses with $e = .8$ and $.95$, respectively.

$$r = \frac{2}{1 - .8\cos\theta}. \qquad r = \frac{.5}{1 - .95\cos\theta}.$$

$[-10, 10]$ by $[-10, 10]$. $0 \le \theta < 2\pi$.

Discovery 2: Figure 6 displays the shapes associated with 2 different values of e. To get both ellipses approximately the same size in the same window, the numerators had to be different. Why? Use a graphics calculator to compare the graphs of

$$r = \frac{2}{1 - .8\cos\theta} \text{ and } r = \frac{2}{1 - .95\cos\theta}.$$

Why is the second so much wider than the first (problem B3)?

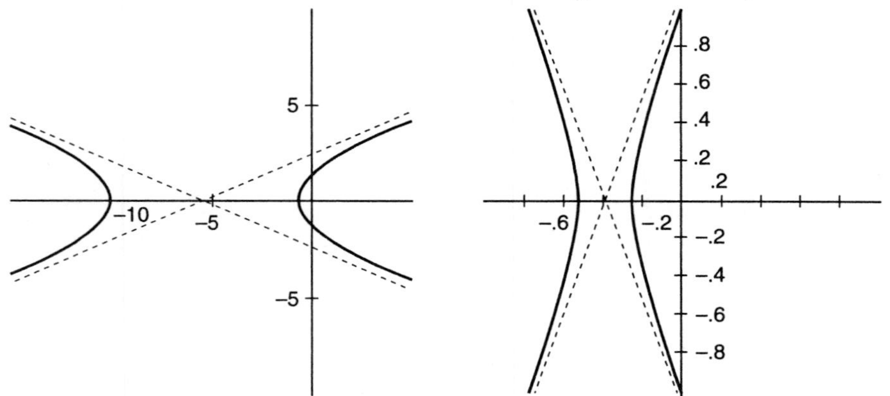

Figure 7: Hyperbolas with $e = 1.1$ and 3, respectively.

$$r = \frac{1}{1 - 1.1\cos\theta}. \qquad r = \frac{1}{1 - 3\cos\theta}.$$

$[-15, 5]$ by $[-10, 10]$. $0 \le \theta < 2\pi$. $[-1, 1]$ by $[-1, 1]$.

Other Orientations. The family of conic sections described by 9.3.2 had a vertical directrix $d > 0$ units to the *left* of the origin. Similar equations can be derived if the directrix is vertical and to the right of the origin, or if it is horizontal and above or below the origin.

Let the eccentricity be $e > 0$ and the directrix be $d > 0$ units from the focus at the origin.

If the directrix is vertical at $x = d$, an equation is

(9.3.3) $$r = \frac{ed}{1 + e\cos\theta}.$$

If the directrix is horizontal at $y = -d$, an equation is

(9.3.4) $$r = \frac{ed}{1 - e\sin\theta}.$$

Finally, if the directrix is horizontal at $y = d$, an equation is

(9.3.5) $$r = \frac{ed}{1 + e\sin\theta}.$$

These equations are given for positive values of d, but they also work for negative values of d (problems B7, B8, and B24).

A vertical directrix leads to symmetry about the x-axis. In this case r is written as a function of cosine, which is symmetric about $\theta = 0$, which is the x-axis. A horizontal directrix leads to symmetry about the y-axis. In that case, r is written as a function of sine, which is symmetric about $\theta = \pi/2$, which is the y-axis.

Example 4: Find the eccentricity and directrix of conic section with equation

$$r = \frac{5}{4 + 5\sin\theta}.$$

Aim for the "1" in the denominator.

$$r = \frac{5/4}{1 + (5/4)\sin\theta}$$

$$r = \frac{(5/4)1}{1 + (5/4)\sin\theta}.$$

Therefore, from 9.3.5, $e = 5/4 > 1$, the conic section is a hyperbola, and the directrix is $y = 1$ (Figure 8).

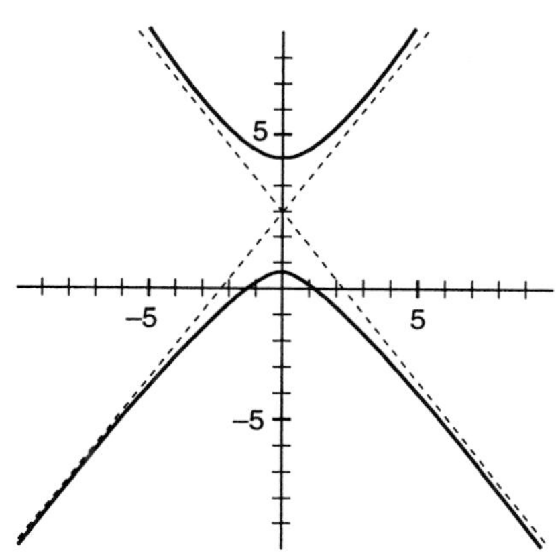

Figure 8: $r = 5/(4+5\sin\theta)$. [-10, 10] by [-10, 10]. $0 \le \theta < 2\pi$. $e = 5/4 = 1.2$.

The Major Axis. The semi-major axis played an important part in rectangular-coordinate equations of conics. Here we discover the length, a, of the semi-major axis in terms of e and d from equation 9.3.2.

Example 5: Find the length of the semi-major axis when the polar equation is

$$r = \frac{4}{1 - (1/3)\cos\theta}.$$

This equation fits 9.3.2, so the directrix is vertical and the major axis of length $2a$ is horizontal (9.2.7, Figure 9). Therefore $2a$ is the distance between the points on the x-axis, that is, the distance from the point where $\theta = 0$ to where $\theta = \pi$. Add the corresponding values of r:

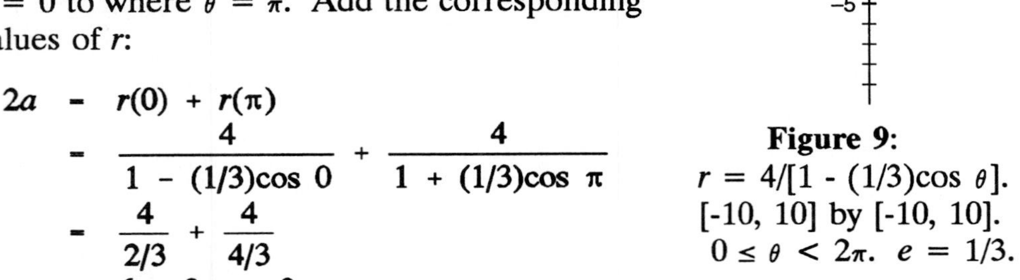

$$\begin{aligned} 2a &= r(0) + r(\pi) \\ &= \frac{4}{1 - (1/3)\cos 0} + \frac{4}{1 + (1/3)\cos\pi} \\ &= \frac{4}{2/3} + \frac{4}{4/3} \\ &= 6 + 3 = 9. \end{aligned}$$

Therefore $a = 9/2$.

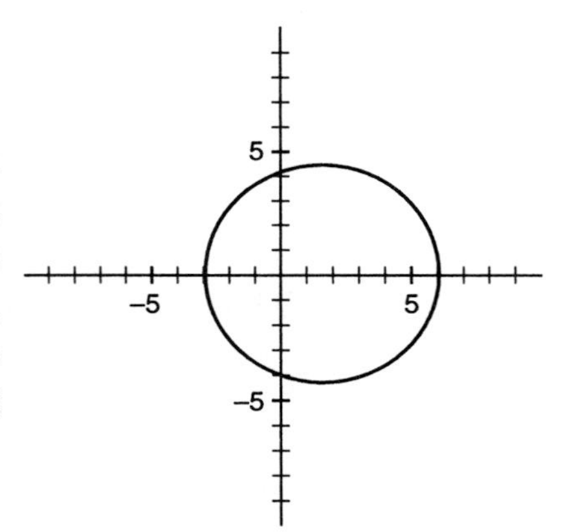

Figure 9:
$r = 4/[1 - (1/3)\cos\theta]$.
[-10, 10] by [-10, 10].
$0 \le \theta < 2\pi$. $e = 1/3$.

For general ellipses in the format 9.3.2,

$$\begin{aligned} 2a = r(0) + r(\pi) &= \frac{ed}{1 - e\cos 0} + \frac{ed}{1 - e\cos\pi} \\ &= \frac{ed}{1 - e} + \frac{ed}{1 + e} \\ &= ed\left(\frac{1 + e + 1 - e}{1 - e^2}\right) \\ &= \frac{2ed}{1 - e^2}. \end{aligned}$$

This generalizes to ellipses and hyperbolas in the forms 9.3.2-5 (Problem B4). If the directrix is $|d|$ units from the origin and the eccentricity is e, then

(9.3.6) $$a = \frac{e|d|}{|1 - e^2|}$$

where a is the length of the semi-major axis. Also,

(9.3.7) $$e|d| = a|1 - e^2|.$$

Example 6: Pluto moves about the sun in an ellipse with eccentricity 0.2485 and $a = 39.5$ astronomical units. (The distance from the Earth to the Sun defines 1 astronomical unit.) Give a polar equation of the orbit (Figure 10).

The parameters in polar form are e and d, and d only appears in the product ed. Use 9.3.7.

$$r = \frac{39.5(1 - .2485^2)}{1 - .2485 \cos \theta} = \frac{37.06}{1 - .2485 \cos \theta}.$$

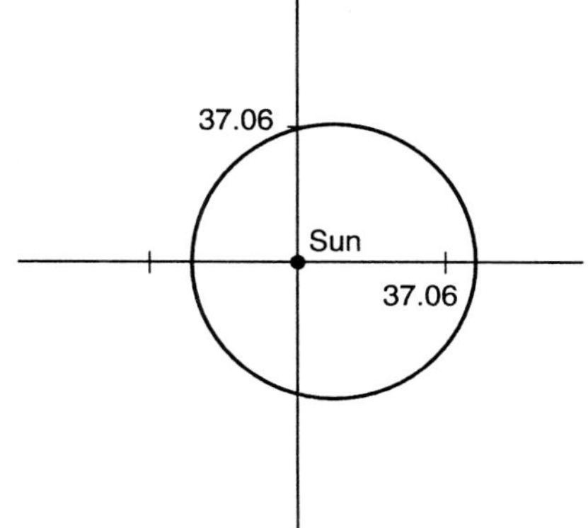

Figure 10: The orbit of Pluto. $e = .2485$.

The rectangular-coordinate equations for ellipses and hyperbolas in Section 9.2 emphasize the center, in contrast to the polar equations in this section which emphasize a focus. So the polar equations in this section are not simply polar versions of the rectangular-coordinate equations -- the location and parameters are different.

Example 5, revisited: Find the equation in standard rectangular-coordinate form if, as in Example 5 and Figure 9, $r = \dfrac{4}{1 - (1/3)\cos \theta}$.

To find rectangular-coordinate form we determine a, b, and the center (h, k) of form 9.2.7.

In Example 5 we found $a = 9/2$. Formula 9.3.6 would yield the same result in one step: $4/(1 - (1/3)^2) = 9/2$. The plan is to find the center next, which yields c, and then to find b from a and c.

The center is half way between the two endpoints of the major axis (Figure 9). The left endpoint has $x = -3$ and the right endpoint has $x = 6$, so the center has $x = 3/2$ -- it is $(3/2, 0)$ (which is expressed the same way in both coordinate systems).

This tells us that $c = 3/2$, because "c" denotes the distance from the center to a focus. Now we can find b using a and c: $c^2 = a^2 - b^2$ (from 9.2.7). Plugging in,

$$(3/2)^2 = (9/2)^2 - b^2.$$
$$b^2 = (9/2)^2 - (3/2)^2.$$
$$b = \sqrt{18} = 4.24.$$

Therefore, the rectangular-coordinate equation is

$$\frac{(x - 3/2)^2}{(9/2)^2} + \frac{y^2}{(\sqrt{18})^2} = 1 .$$

Calculator Exercise 1: Solve for y and graph this on your calculator using rectangular coordinates. Compare your graph to Figure 9.

Eccentricity in Rectangular Coordinates. Eccentricity is a parameter of standard polar form but not of standard rectangular-coordinate form. In Section 9.2 where rectangular form was used, ellipses and hyperbolas were defined with two foci and the *center* was at the origin. In this section, they are defined using a directrix and one focus, where the *focus* is at the origin. Of course, it is possible to relate the parameters with which the two forms are expressed (problem B32). For example, the eccentricity e of polar form has a simple rectangular-coordinate expression using the notation of Section 9.2.

$$e = \frac{\textit{distance between foci}}{\textit{length of major axis}} .$$

(9.3.8) If the major axis is horizontal, $e = c/a$.
If the major axis is vertical, $e = c/b$.

Recall that c is the distance from the center to a focus, and the length of the semi-major axis is a or b (9.2.5 and 9.2.7).

Example 7: Find the rectangular-coordinate equation of a hyperbola centered at the origin, with horizontal major axis, $a = 4$ and $e = 1.25$.

The parameters of rectangular-coordinates form are a and b (9.2.5). We can use the given a and e to find c from 9.3.8, and then a and c to get b from 9.2.5 (Figure 11).

$e = c/a$ [9.3.8]

$1.25 = c/4$.

$c = 5$.

$a^2 + b^2 = c^2$ [9.2.7]

$4^2 + b^2 = 5^2$.

$b = 3$.

The equation is, from 9.2.7,

$$\frac{x^2}{4^2} - \frac{y^2}{3^2} = 1 .$$

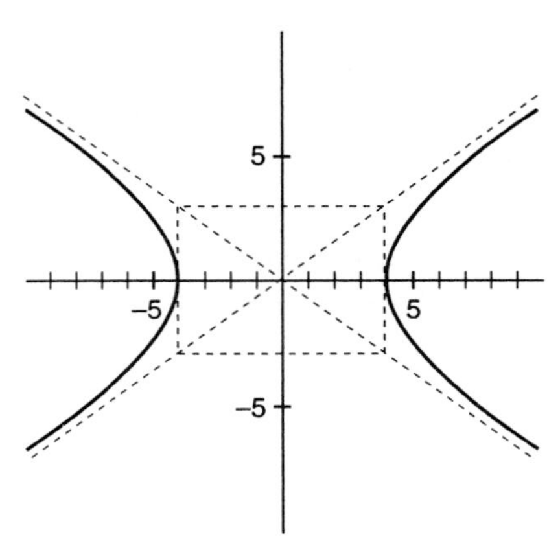

Figure 11: $x^2/4^2 - y^2/3^2 = 1$.
[-10,10] by [-10,10].
$e = 1.25$.

Example 8: Find the rectangular-coordinate equation of an ellipse with $e = .7$, one focus at the origin, the other focus on the positive x-axis, and semi-major axis 4 (Figure 12).

The center is c units from the focus, so the center is at $(c, 0)$ where $e = c/a$ from 9.3.8. Solving "$.7 = c/4$," we find $c = 2.8$.

Also, $b^2 = a^2 - c^2$ for a horizontal ellipse, so $b^2 = 4^2 - 2.8^2 = 8.16$, and $b = \sqrt{8.16} = 2.857$. Plugging in to standard rectangular form (9.2.6):

$$\frac{(x - 2.8)^2}{4^2} + \frac{y^2}{2.857^2} = 1 .$$

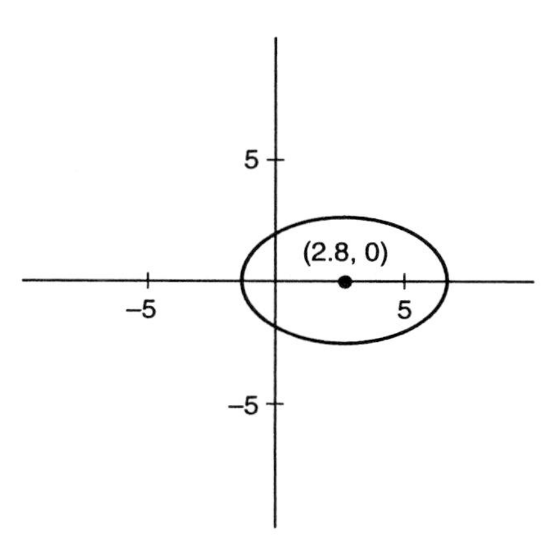

Figure 12: $e = .7$, $a = 4$, and one focus is at the origin. [-10, 10] by [-10, 10].

Circles. Ellipses with very small eccentricities are almost circles. Circles have eccentricity zero. But putting zero in for e in our polar forms yields "$r = 0$", an equation that describes the origin which is a single point, not a circle. So, how does the polar equation of a circle compare to the equation of an ellipse (9.3.2) with very small eccentricity?

The polar equation of a circle centered at the origin is "$r = k$," where k is the radius. θ does not appear. Reconsider equation 9.3.2:

$$r = \frac{ed}{1 - e \cos \theta}$$

To represent "$r = k$" the "$\cos \theta$" term would have to disappear, so e would have to be zero. But "ed" would have to be the radius k, which cannot hold if $e = 0$. So 9.3.2 does not hold for circles. But, we can imagine how, with smaller and smaller e, the effect of the "$e \cos \theta$" would become negligible. But the "ed" term on the top would also disappear unless we simultaneously adjust d (the location of the directrix) so that ed remains constant. So, to get ellipses that are similar in size to a given circle with fixed radius k, we need to increase d as we decrease e so that ed approaches k, the radius we want. That forces d to go to "infinity" as e goes down to zero, so ellipses that closely resemble circles have directrixes far from the origin (problems B21-23).

Conclusion. Polar coordinates make conic sections easy to describe if they have a focus at the origin. The eccentricity, e, is an important parameter.

Terms: parabola, ellipse, hyperbola, eccentricity, focus, directrix.

****

Exercises for Section 9.3, "Polar Equations of Conic Sections":

^ ^ ^ ^ Identify the type of conic section, its eccentricity, and sketch its graph.

A1. $r = 1/(1 - \cos\theta)$.
A2. $r = 4/(2 - 2\cos\theta)$.
A3. $r = 1/(1 - 2\cos\theta)$.
A4. $r = 6/(3 - 2\cos\theta)$.
A5. $r = 1/(1 - .5\cos\theta)$.
A6. $r = 5/(3 - 4\cos\theta)$.

^ ^ ^ ^ Identify the type of conic section, its eccentricity, and its directrix.

A7. $r = 5/(1 + \cos\theta)$. [dir: $x = 5$]
A8. $r = 1/(5 + 5\sin\theta)$. [dir: $y = 1/5$]
A9. $r = 7/(3 - 2\sin\theta)$. [$e = 2/3$]
A10. $r = 50/(7 + 2\sin\theta)$. [dir: $y = -25$]
A11. $r = 11/(9 + 10\sin\theta)$. [$e = 10/9$]
A12. $r = 3/(2 + 3\cos\theta)$. [dir: $x = 1$]

^ ^ ^ ^ Find a polar equation of the conic satisfying the given conditions. All have a focus at the origin (pole).

A13. Parabola, directrix $y = -2$.
A14. Parabola, directrix $x = -10$.
A15. $e = .95$, directrix $x = 3$.
A16. $e = .1$, directrix $y = -2$.
A17. Directrix $y = 4$, $e = 1.5$.
A18. Directrix $x = 10$, $e = 1.01$.

Table of eccentricities and distances (in astronomical units) of the planets from the sun.

Mercury $e = .206$, $a = .387$.
Venus $e = .0068$, $a = .723$.
Earth $e = .0167$, $a = 1$.
Mars $e = .093$, $a = 1.52$.
Jupiter $e = .048$, $a = 5.20$.
Saturn $e = .056$. $a = 9.54$.
Uranus $e = .047$, $a = 19.2$.
Neptune $e = .0085$, $a = 30.07$.
Pluto $e = .2485$, $a = 39.5$.

Find a polar equation of the orbit of

A19. Mercury
A20. Mars
A21. Venus.
A22. Jupiter

A23. Find the polar-coordinate equation of the ellipse in Example 8.

^ ^ ^ ^ ^ ^ ^ ^ ^

B1. Do Discovery 1: Graph the equation in Example 3 (Figure 5) with a graphics calculator in polar mode. Which quarter of the hyperbola appears first? Second? Explain why, mathematically.

B2. Find the eccentricity of the famous hyperbola: $x^2 - y^2 = 1$.

B3. Do Discovery 2: Figure 6 displays the shapes associated with 2 different values of e. To get both pictures on about the same scale, the numerators had to be different. Why? Use a graphics calculator to compare the graphs of

$$r = \frac{2}{1 - .8\cos\theta} \quad \text{and} \quad r = \frac{2}{1 - .95\cos\theta}.$$

Why, mathematically, is the second so much wider than the first?

B4. For an ellipse in the form 9.3.2, $2a = r(0) + r(\pi)$, as in Example 5. a) Why isn't it $r(0) - r(\pi)$? b) Express $2a$ in terms of r for a hyperbola in the form 9.3.2.

B5.* [About 9.3.2 and 9.3.3] "$r = f(\cos \theta)$" denotes a composite function, for any function f. Because of the "$\cos \theta$" part, it always has a certain symmetry. a) What symmetry does the graph have? b) Why?

B6.* [About 9.3.4 and 9.3.3] "$r = f(\sin \theta)$" denotes a composite function, for any function f. Because of the "$\sin \theta$" part, it always has a certain symmetry. a) What symmetry does the graph have? b) Why?

B7. [Compare 9.3.2 and 9.3.3] a) Give the equation if the directrix is $x = 3$ and $e = 2$, using 9.3.3. b) Graph it. c) Suppose we treat $x = 3$ as $x = -(-3)$ and use 9.3.2. Give the equation. d) Graph it. e) How do the two graphs differ? f) Explain the answer to part (e) mathematically.

B8. [Compare 9.3.4 and 9.3.5] a) Give the equation if the directrix is $y = 5$ and $e = 3$ using 9.3.5. b) Graph it. c) Suppose we treat $y = 5$ as $y = -(-5)$ and use 9.3.4. Give the equation. d) Graph it. e) How do the two graphs differ? f) Explain the answer to part (e) mathematically.

B9. Equations 9.3.2 and 9.3.4 look just alike except the first "cos" where the second has "sin". State a relevant trig identity of the form "$\sin \theta = \cos(\theta - c)$" and use it to explain why the second graph is just the first graph rotated through 90°.

^ ^ ^ ^The following equations yield hyperbolas on $[0, 2\pi)$. Which angles yield the left branch of the hyperbola?

B10. $r = 1/(1 - 2 \cos \theta)$.

B11. $r = 5/(1 - 3 \cos \theta)$.

B12. For a hyperbola with horizontal axis, give the slopes of the asymptotes if the eccentricity is 2. [$m = \pm 1.7$]

B13. For a hyperbola with horizontal axis, show the slopes of the asymptotes are given by

$$m = \pm\sqrt{e^2 - 1}\ .$$

B14. Let $r = 6/(1 - 2\cos \theta)$. a) Use 9.3.6 to find a. b) Find b and then b/a. c) For which angle θ is the denominator zero? d) What happens, graphically, as θ approaches that value? e) Find tangent of that angle. f) comment on why the answers to (b) and (e) are related.

B15. If the slopes of the asymptotes of a horizontal hyperbola are $\pm 1/4$, give its eccentricity.

B16. If the slopes of the asymptotes of a horizontal hyperbola are $\pm m$, show its eccentricity is $e = \sqrt{(m^2 + 1)}$.

B17. For an ellipse, give the ratio of width to length if the eccentricity is 1/2. [.87]

B18. For an ellipse with eccentricity e, show the ratio of width to length is $\sqrt{(1 - e^2)}$.

B19. If the ratio of width to length of an ellipse is 1/3, give the eccentricity. [.94]

B20. If the ratio of width to length of an ellipse is r, show the eccentricity is $\sqrt{(1 - r^2)}$.

B21. If $e = .1$ and the ellipse closely resembles a circle of radius 5, where is the directrix?

B22. If $e = .01$ and the ellipse closely resembles a circle of radius 3, where is the directrix?

B23. Suppose two ellipses are very similar to a circle with radius 1. One has $e = .02$ and the other has $e = .01$. Compare the locations of their directrixes. [Use the from 9.3.2, or work with 9.3.6. Rough approximations are encouraged.]

B24. Derive 9.3.4, paralleling the steps from which 9.3.2 was derived.

^ ^ ^ ^ Sketch the graph, locate the center, and determine the slopes of the asymptotes of:

B25. $r = \dfrac{1}{1 - 1.01 \cos \theta}$ [$m = \pm.14$] B26. $r = \dfrac{1}{1 + 1.2 \sin \theta}$ [$m = \pm1.5$]

^ ^ ^ ^ Sketch the graph, locate the center, and determine the semi-major axis and the semi-minor axis of

B27. $r = \dfrac{2}{1 - .7 \cos \theta}$ [C: (2.4, 0)] B28. $r = \dfrac{5}{1 + .7 \sin \theta}$ [C: (0, -6.9)]

^ ^ ^ ^ Find the rectangular-coordinate equation of a conic section with the given characteristics.

B29. $e = 1/2$, one focus is at the origin, the other focus is on the positive x-axis, and the semi-major axis is 3.

B30. $e = 3$, one focus is at the origin, the other focus is on the negative x-axis, and the semi-major axis is 2.

B31. Suppose you wish to graph $r = f(\theta)$ using a graphics calculator, but it does not have "Polar" mode. You may use "Parametric" mode, in which x and y are entered separately in terms of θ. Because $x = r \cos \theta$ and $y = r \sin \theta$, to graph "$r = f(\theta)$" you may enter:

$$x = f(\theta) \cos \theta. \quad y = f(\theta) \sin \theta.$$

To graph in parametric mode the graph in Example 3, what must you enter in for "x" and what for "y"?

B32. The definition of eccentricity is given in 9.3.1 where ellipses and hyperbolas are defined in terms of a directrix and one focus (instead of the two foci used in Section 9.2). a) Sketch an ellipse with a focus at the origin and vertical directrix at $x = -d$. Label the rectangular coordinate parameters on the sketch. b) Find d in terms of e, a, b, and c.
c) Find e in terms of a, b, and c.

B33. a) Use the definition of e from 9.3.1 and *rectangular* coordinates to set up a rectangular-coordinate equation for a parabola. b) Simplify it.

B34. a) Use the definition of e from 9.3.1, directrix $x = -d$, and *rectangular* coordinates to set up a rectangular-coordinate equation for an ellipse with eccentricity $e < 1$. b) Solve for y. c) Let $e = 1/2$ and $d = 3$ in part (b) and graph the equation.

B35. a) Use the definition of e from 9.3.1, directrix $x = -d$, and *rectangular* coordinates to set up a rectangular-coordinate equation for a hyperbola with eccentricity $e > 1$. b) Solve for y. c) Let $e = 2$ and $d = 3$ in part (b) and graph the equation.

INDEX

List of Trigonometric Identities

(7.2.1) $$\tan\theta = \frac{\sin\theta}{\cos\theta}.$$

Pythagorean Identity:
(7.2.2) $\sin^2\theta + \cos^2\theta = 1.$

Coterminal Identities:
(7.2.3) $\sin(\theta \pm 360n°) = \sin\theta$, and $\cos(\theta \pm 360n°) = \cos\theta$.

Negatives (Odd-Even Identities):
(7.2.4) $\cos(-\theta) = \cos\theta$. $\sin(-\theta) = -\sin\theta$. $\tan(-\theta) = -\tan\theta$.

Reference Angle Identities:
(7.2.5) $\sin(180° - \theta) = \sin\theta$, $\cos(180° - \theta) = -\cos\theta$,
$\tan(180° - \theta) = -\tan\theta$.
(7.2.6) $\sin(\theta + 180°) = -\sin\theta$. $\cos(\theta + 180°) = -\cos\theta$.
$\tan(\theta + 180°) = \tan\theta$.

Solving Trigonometric Equations:
(7.2.8) $\sin x = c$ iff $x = \sin^{-1}c \pm 360n°$, or
$x = 180° - \sin^{-1}c \pm 360n°$.
(7.2.9) $\cos x = c$ iff $x = \pm\cos^{-1}c \pm 360n°$.
(7.2.10) $\tan x = c$ iff $x = \tan^{-1}c \pm 180n°$.

Sum Identities:
(7.2.11) $\sin(\alpha+\beta) = (\sin\alpha)(\cos\beta) + (\cos\alpha)(\sin\beta)$.
(7.2.12) $\cos(\alpha+\beta) = (\cos\alpha)(\cos\beta) - (\sin\alpha)(\sin\beta)$.

(7.2.13) $$\tan(\alpha+\beta) = \frac{\tan\alpha + \tan\beta}{1 - (\tan\alpha)(\tan\beta)}.$$

Difference Identities:
(7.2.14) $\sin(\alpha - \beta) = (\sin\alpha)(\cos\beta) - (\cos\alpha)(\sin\beta)$.
(7.2.15) $\cos(\alpha - \beta) = (\cos\alpha)(\cos\beta) + (\sin\alpha)(\sin\beta)$.

(7.2.16) $$\tan(\alpha-\beta) = \frac{\tan\alpha - \tan\beta}{1 + (\tan\alpha)(\tan\beta)}.$$

Double-Angle Identities:
(7.2.17) $\sin 2\theta = 2(\sin\theta)(\cos\theta)$.
$\cos 2\theta = \cos^2\theta - \sin^2\theta$
Also, $\cos 2\theta = 1 - 2\sin^2\theta$,
and $\cos 2\theta = 2\cos^2\theta - 1$.
$\tan 2\theta = (2\tan\theta)/[1 - \tan^2\theta]$.

Squared-Function Identities:
(7.2.18) $\sin^2\theta = (1 - \cos 2\theta)/2$. $\cos^2\theta = (1 + \cos 2\theta)/2$.